Lecture Notes in Computer Science 12836

More information about this subseries at http://www.springer.com/series/7409

De-Shuang Huang · Kang-Hyun Jo ·
Jianqiang Li · Valeriya Gribova ·
Vitoantonio Bevilacqua (Eds.)

Intelligent Computing Theories and Application

17th International Conference, ICIC 2021
Shenzhen, China, August 12–15, 2021
Proceedings, Part I

 Springer

Editors
De-Shuang Huang
Tongji University
Shanghai, China

Kang-Hyun Jo
University of Ulsan
Ulsan, Korea (Republic of)

Jianqiang Li
Shenzhen University
Shenzhen, China

Valeriya Gribova
Far Eastern Branch of the Russian Academy
of Sciences
Vladivostok, Russia

Vitoantonio Bevilacqua
Polytechnic University of Bari
Bari, Italy

ISSN 0302-9743   ISSN 1611-3349 (electronic)
Lecture Notes in Computer Science
ISBN 978-3-030-84521-6   ISBN 978-3-030-84522-3 (eBook)
https://doi.org/10.1007/978-3-030-84522-3

LNCS Sublibrary: SL3 – Information Systems and Applications, incl. Internet/Web, and HCI

This Springer imprint is published by the registered company Springer Nature Switzerland AG
The registered company address is: Gewerbestrasse 11, 6330 Cham, Switzerland

Preface

The International Conference on Intelligent Computing (ICIC) was started to provide an annual forum dedicated to the emerging and challenging topics in artificial intelligence, machine learning, pattern recognition, bioinformatics, and computational biology. It aims to bring together researchers and practitioners from both academia and industry to share ideas, problems, and solutions related to the multifaceted aspects of intelligent computing.

ICIC 2021, held in Shenzhen, China, during August 12–15, 2021, constituted the 17th International Conference on Intelligent Computing. It built upon the success of ICIC 2020 (Bari, Italy), ICIC 2019 (Nanchang, China), ICIC 2018 (Wuhan, China), ICIC 2017 (Liverpool, UK), ICIC 2016 (Lanzhou, China), ICIC 2015 (Fuzhou, China), ICIC 2014 (Taiyuan, China), ICIC 2013 (Nanning, China), ICIC 2012 (Huangshan, China), ICIC 2011 (Zhengzhou, China), ICIC 2010 (Changsha, China), ICIC 2009 (Ulsan, South Korea), ICIC 2008 (Shanghai, China), ICIC 2007 (Qingdao, China), ICIC 2006 (Kunming, China), and ICIC 2005 (Hefei, China).

This year, the conference concentrated mainly on the theories and methodologies as well as the emerging applications of intelligent computing. Its aim was to unify the picture of contemporary intelligent computing techniques as an integral concept that highlights the trends in advanced computational intelligence and bridges theoretical research with applications. Therefore, the theme for this conference was "Advanced Intelligent Computing Technology and Applications". Papers that focused on this theme were solicited, addressing theories, methodologies, and applications in science and technology.

ICIC 2021 received 458 submissions from authors in 21 countries and regions. All papers went through a rigorous peer-review procedure and each paper received at least three review reports. Based on the review reports, the Program Committee finally selected 192 high-quality papers for presentation at ICIC 2021, which are included in three volumes of proceedings published by Springer: two volumes of *Lecture Notes in Computer Science* (LNCS) and one volume of *Lecture Notes in Artificial Intelligence* (LNAI).

This volume of LNCS includes 71 papers.

The organizers of ICIC 2021, including Tongji University and Shenzhen University, China, made an enormous effort to ensure the success of the conference. We hereby would like to thank all the ICIC 2021 organizers, the members of the Program Committee, and the referees for their collective effort in reviewing and soliciting the papers. We would like to thank Ronan Nugent, executive editor from Springer, for his frank and helpful advice and guidance throughout as well as his continuous support in publishing the proceedings. In particular, we would like to thank all the authors for contributing their papers. Without the high-quality submissions from the authors, the success of the conference would not have been possible. Finally, we are especially grateful to the International Neural Network Society and the National Science Foundation of China for their sponsorship.

August 2021De-Shuang Huang
Kang-Hyun Jo
Jianqiang Li
Valeriya Gribova
Vitoantonio Bevilacqua

Organization

General Co-chairs

De-Shuang Huang   Tongji University, China
Zhong Ming   Shenzhen University, China

Program Committee Co-chairs

Kang-Hyun Jo   University of Ulsan, South Korea
Jianqiang Li   Shenzhen University, China
Valeriya Gribova   Far Eastern Branch of Russian Academy of Sciences,
                             Russia

Organizing Committee Co-chairs

Qiuzhen Lin   Shenzhen University, China
Cheng Wen Luo   Shenzhen University, China

Organizing Committee Members

Lijia Ma   Shenzhen University, China
Jie Chen   Shenzhen University, China
Jia Wang   Shenzhen University, China
Changkun Jiang   Shenzhen University, China
Junkai Ji   Shenzhen University, China
Zun Liu   Shenzhen University, China

Award Committee Co-chairs

Ling Wang   Tsinghua University, China
Abir Hussain   Liverpool John Moores University, UK

Tutorial Co-chairs

Kyungsook Han   Inha University, South Korea
Prashan Premaratne   University of Wollongong, Australia

Publication Co-chairs

Vitoantonio Bevilacqua   Polytechnic of Bari, Italy
Phalguni Gupta   Indian Institute of Technology Kanpur, India

Special Session Co-chairs

Michal Choras   University of Science and Technology in Bydgoszcz,
                                      Poland
Hong-Hee Lee   University of Ulsan, South Korea

Special Issue Co-chairs

M. Michael Gromiha   Indian Institute of Technology Madras, India
Laurent Heutte   Université de Rouen, France
Hee-Jun Kang   University of Ulsan, South Korea

International Liaison Co-chair

Prashan Premaratne   University of Wollongong, Australia

Workshop Co-chairs

Yoshinori Kuno   Saitama University, Japan
Jair Cervantes Canales   Autonomous University of Mexico State, Mexico

Publicity Co-chairs

Chun-Hou Zheng   Anhui University, China
Dhiya Al-Jumeily   Liverpool John Moores University, UK

Exhibition Contact Co-chairs

Qiuzhen Lin   Shenzhen University, China

Program Committee

Mohd Helmy Abd Wahab   Universiti Tun Hussein Onn Malaysia, Malaysia
Nicola Altini   Polytechnic University of Bari, Italy
Waqas Bangyal   University of Gujrat, Pakistan
Wenzheng Bao   Xuzhou University of Technology, China
Antonio Brunetti   Polytechnic University of Bari, Italy
Domenico Buongiorno   Politecnico di Bari, Italy
Hongmin Cai   South China University of Technology, China
Nicholas Caporusso   Northern Kentucky University, USA
Jair Cervantes   Autonomous University of Mexico State, Mexico
Chin-Chih Chang   Chung Hua University, Taiwan, China
Zhanheng Chen   Shenzhen University, China
Wen-Sheng Chen   Shenzhen University, China
Xiyuan Chen   Southeast University, China

Wei Chen	Chengdu University of Traditional Chinese Medicine, China
Michal Choras	University of Science and Technology in Bydgoszcz, Poland
Angelo Ciaramella	Università di Napoli, Italy
Guojun Dai	Hangzhou Dianzi University, China
Weihong Deng	Beijing University of Posts and Telecommunications, China
YanRui Ding	Jiangnan University, China
Pu-Feng Du	Tianjing University, China
Jianbo Fan	Ningbo University of Technology, China
Zhiqiang Geng	Beijing University of Chemical Technology, China
Lejun Gong	Nanjing University of Posts and Telecommunications, China
Dunwei Gong	China University of Mining and Technology, China
Wenyin Gong	China University of Geosciences, China
Valeriya Gribova	Far Eastern Branch of Russian Academy of Sciences, Russia
Michael Gromiha	Indian Institute of Technology Madras, India
Zhi-Hong Guan	Huazhong University of Science and Technology, China
Ping Guo	Beijing Normal University, China
Fei Guo	Tianjin University, China
Phalguni Gupta	Indian Institute of Technology Kanpur, India
Kyungsook Han	Inha University, South Korea
Fei Han	Jiangsu University, China
Laurent Heutte	Université de Rouen Normandie, France
Jian Huang	University of Electronic Science and Technology of China, China
Chenxi Huang	Xiamen University, China
Abir Hussain	Liverpool John Moores University, UK
Qinghua Jiang	Harbin Institute of Technology, China
Kanghyun Jo	University of Ulsan, South Korea
Dah-Jing Jwo	National Taiwan Ocean University, Taiwan, China
Seeja K R	Indira Gandhi Delhi Technical University for Women, India
Weiwei Kong	Xi'an University of Posts and Telecommunications, China
Yoshinori Kuno	Saitama University, Japan
Takashi Kuremoto	Nippon Institute of Technology, Japan
Hong-Hee Lee	University of Ulsan, South Korea
Zhen Lei	Institute of Automation, CAS, China
Chunquan Li	Harbin Medical University, China
Bo Li	Wuhan University of Science and Technology, China
Xiangtao Li	Jilin University, China

Hao Lin	University of Electronic Science and Technology of China, China
Juan Liu	Wuhan University, China
Chunmei Liu	Howard University, USA
Bingqiang Liu	Shandong University, China
Bo Liu	Academy of Mathematics and Systems Science, CAS, China
Bin Liu	Beijing Institute of Technology, China
Zhi-Ping Liu	Shandong University, China
Xiwei Liu	Tongji University, China
Haibin Liu	Beijing University of Technology, China
Jin-Xing Liu	Qufu Normal University, China
Jungang Lou	Huzhou University, China
Xinguo Lu	Hunan University, China
Xiaoke Ma	Xidian University, China
Yue Ming	Beijing University of Posts and Telecommunications, China
Liqiang Nie	Shandong University, China
Ben Niu	Shenzhen University, China
Marzio Pennisi	University of Eastern Piedmont Amedeo Avogadro, Italy
Surya Prakash	IIT Indore, India
Prashan Premaratne	University of Wollongong, Australia
Bin Qian	Kunming University of Science and Technology, China
Daowen Qiu	Sun Yat-sen University, China
Mine Sarac	Stanford University, USA
Xuequn Shang	Northwestern Polytechnical University, China
Evi Sjukur	Monash University, Australia
Jiangning Song	Monash University, Australia
Chao Song	Harbin Medical University, China
Antonino Staiano	Parthenope University of Naples, Italy
Fabio Stroppa	Stanford University, USA
Zhan-Li Sun	Anhui University, China
Xu-Qing Tang	Jiangnan University, China
Binhua Tang	Hohai University, China
Joaquin Torres-Sospedra	UBIK Geospatial Solutions S.L., Spain
Shikui Tu	Shanghai Jiao Tong University, China
Jian Wang	China University of Petroleum, China
Ling Wang	Tsinghua University, China
Ruiping Wang	Institute of Computing Technology, CAS, China
Xuesong Wang	China University of Mining and Technology, China
Rui Wang	National University of Defense Technology, China
Xiao-Feng Wang	Hefei University, China
Shitong Wang	Jiangnan University, China
Bing Wang	Anhui University of Technology, China
Jing-Yan Wang	New York University Abu Dhabi, Abu Dhabi

Pufeng Du
Atif Mehmood
Jonggeun Kim
Eun Kyeong Kim
Hansoo Lee
Yiqiao Cai
Wuritu Yang
Weitao Sun
Guihua Tao
Jinzhong Zhang
Wenjie Yi
Lingyun Huang
Chao Chen
Jiangping He
Wei Wang
Jin Ma
Liang Xu
Vitoantonio Bevilacqua
Huan Liu
Lei Deng
Di Liu
Zhongrui Zhang
Qinhu Zhang
Yanyun Qu
Jinxing Liu
Shravan Sukumar
Long Gao
Yifei Wu
Tianhua Jiang
Lixiang Hong
Tingzhong Tian
Yijie Ding
Junwei Wang
Zhe Yan
Rui Song
S. A. K. Bangyal
Giansalvo Cirrincione
Xiancui Xiao
X. Zheng
Vincenzo Randazzo
Huijuan Zhu
Dongyuan Li
Jingbo Xia
Boya Ji
Manilo Monaco
Xiaohua Yu

Zuguo Yu
Jun Yuan
Punam Kumari
Bowei Zhao
X. J. Chen
Takashi Kurmeoto
Pallavi Pandey
Yan Zhou
Mascot Wang
Chenhui Qiu
Haizhou Wu
Lulu Zuo
Juan Wang
Rafal Kozik
Wenyan Gu
Shiyin Tan
Yaping Fang
Alexander Moopenn
Xiuxiu Ren
Aniello Castiglione
Qiong Wu
Junyi Chen
Meineng Wang
Xiaorui Su
Jianping Yu
Lizhi Liu
Junwei Luo
Yuanyuan Wang
Xiaolei Zhu
Jiafan Zhu
Yongle Li
Xiaoyin Xu
Shiwei Sun
Hongxuan Hua
Shiping Zhang
Xiangtian Yu
Angelo Riccio
Yuanpeng Xiong
Jing Xu
Chienyuan Lai
Guo-Feng Fan
Zheng Chen
Renzhi Cao
Ronggen Yang
Zhongming Zhao
Yongna Yuan

Chuanxing Liu
Panpan Song
Joao Sousa
Wenying He
Ming Chen
Puneet Gupta
Ziqi Zhang
Davide Nardone
Liangxu Liu
Huijian Han
Qingjun Zhu
Hongluan Zhao
Rey-Sern Lin
Hung-Chi Su
Conghua Xie
Caitong Yue
Li Yan
Tuozhong Yao
Xuzhao Chai
Zhenhu Liang
Yu Lu
Jing Sun
Hua Tang
Liang Cheng
Puneet Rawat
Kulandaisamy A.
Jun Zhang
Egidio Falotico
Peng Chen
Cheng Wang
Jing Li
He Chen
Giacomo Donato Cascarano
Shaohua Wan
Cheng Chen
Jie Li
Ruxin Zhao
Jiazhou Chen
Guoliang Xu
Congxu Zhu
Deng Li
Piyush Joshi
Syed Sadaf Ali
Kuan Li
Teng Wan
Hao Liu

Yexian Zhang
Xu Qiao
Lingchong Zhong
Wenyan Wang
Xiaoyu Ji
Weifeng Guo
Yuchen Jiang
Van-Dung Hoang
Yuanyuan Huang
Zaixing Sun
Honglin Zhang
Yu-Jie He
Rong Hu
Youjie Yao
Naikang Yu
Giulia Russo
Dian Liu
Cheng Liang
Iyyakutti Iyappan Ganapathi
Mingon Kang
Xuefeng Cui
Hao Dai
Geethan Mendiz
Brendan Halloran
Yue Li
Qianqian Shi
Zhiqiang Tian
Ce Li
Yang Yang
Jun Wang
Ke Yan
Hang Wei
Yuyan Han
Hisato Fukuda
Yaning Yang
Lixiang Xu
Yuanke Zhou
Shihui Ying
Wenqiang Fan
Zhao Li
Zhe Zhang
Xiaoying Guo
Zhuoqun Xia
Na Geng
Xin Ding
Balachandran Manavalan

Lianrong Pu
Di Wang
Fangping Wan
Renmeng Liu
Jiancheng Zhong
Yinan Guo
Lujie Fang
Ying Zhang
Yinghao Cao
Xhize Wu
Chao Wu
Ambuj Srivastava
Prabakaran R.
Xingquan Zuo
Jiabin Huang
Jingwen Yang
Qianying Liu
Tongchi Zhou
Xinyan Liang
Xiaopeng Jin
Yumeng Liu
Junliang Shang
Shanghan Li
Jianhua Zhang
Wei Zhang
Han-Jing Jiang
Kunikazu Kobayashi
Shenglin Mu
Jing Liang
Jialing Li
Zhe Sun
Wentao Fan
Wei Lan
Josue Espejel Cabrera
José Sergio Ruiz Castilla
Rencai Zhou
Moli Huang
Yong Zhang
Joaquín Torres-Sospedra
Xingjian Chen
Saifur Rahaman
Olutomilayo Petinrin
Xiaoming Liu
Lei Wang
Xin Xu
Najme Zehra

Zhenqing Ye
Zijing Wang
Lida Zhu
Xionghui Zhou
Jia-Xiang Wang
Gongxin Peng
Junbo Liang
Linjing Liu
Xiangeng Wang
Y. M. Nie
Sheng Ding
Laksono Kurnianggoro
Minxia Cheng
Meiyi Li
Qizhi Zhu
Pengchao Li
Ming Xiao
Guangdi Liu
Jing Meng
Kang Xu
Cong Feng
Arturo Yee
Kazunori Onoguchi
Hotaka Takizawa
Suhang Gu
Zhang Yu
Bin Qin
Yang Gu
Zhibin Jiang
Chuanyan Wu
Wahyono Wahyono
Kaushik Deb
Alexander Filonenko
Van-Thanh Hoang
Ning Guo
Deng Chao
Jian Liu
Sen Zhang
Nagarajan Raju
Kumar Yugandhar
Anoosha Paruchuri
Lei Che
Yujia Xi
Ma Haiying
Huanqiang Zeng
Hong-Bo Zhang

Yewang Chen
Sama Ukyo
Akash Tayal
Ru Yang
Junning Gao
Jianqing Zhu
Haizhou Liu
Nobutaka Shimada
Yuan Xu
Shuo Jiang
Minghua Zhao
Jiulong Zhang
Shui-Hua Wang
Sandesh Gupta
Nadia Siddiqui
Syeda Shira Moin
Ruidong Li
Mauro Castelli
Ivanoe De Falco
Antonio Della Cioppa
Kamlesh Tiwari
Luca Tiseni
Ruizhi Fan
Grigorios Skaltsas
Mario Selvaggio
Xiang Yu
Huajuan Huang
Vasily Aristarkhov
Zhonghao Liu
Lichuan Pan
Zhongying Zhao
Atsushi Yamashita
Ying Xu
Wei Peng
Haodi Feng
Jin Zhao
Shunheng Zhou
Changlong Gu
Xiangwen Wang
Zhe Liu
Pi-Jing Wei
Haozhen Situ
Xiangtao Chen
Hui Tang
Akio Nakamura
Antony Lam

Weilin Deng
Xu Zhou
Shuyuan Wang
Rabia Shakir
Haotian Xu
Zekang Bian
Shuguang Ge
Hong Peng
Thar Baker
Siguo Wang
Jianqing Chen
Chunhui Wang
Xiaoshu Zhu
Yongchun Zuo
Hyunsoo Kim
Areesha Anjum
Shaojin Geng
He Yongqiang
Mario Camana
Long Chen
Jialin Lyu
Zhenyang Li
Tian Rui
Duygun Erol Barkana
Huiyu Zhou
Yichuan Wang
Eray A. Baran
Jiakai Ding
Dehua Zhang
Insoo Koo
Yudong Zhang
Zafaryab Haider
Vladimir Shakhov
Daniele Leonardis
Byungkyu Park
Elena Battini
Radzi Ambar
Noraziah Chepa
Liang Liang
Ling-Yun Dai
Xiongtao Zhang
Sobia Pervaiz Iqbal
Fang Yang
Si Liu
Natsa Kleanthous
Zhen Shen

Chunyan Fan
Jie Zhao
Yuchen Zhang
Jianwei Yang
Wenrui Zhao
Di Wu
Chao Wang
Fuyi Li
Guangsheng Wu
Yuchong Gong
Weitai Yang
Yanan Wang
Bo Chen
Binbin Pan
Chunhou Zheng
Bowen Song
Guojing Wu
Weiping Liu
Laura Jalili
Xing Chen
Xiujuan Lei
Marek Pawlicki
Hao Zhu
Wang Zhanjun
Mohamed Alloghani
Yu Hu
Baohua Wang
Hanfu Wang
Hongle Xie
Guangming Wang
Fuchun Liu
Farid Garcia-Lamont
Hengyue Shi
Po Yang
Wen Zheng Ma
Jianxun Mi
Michele Scarpiniti
Yasushi Mae
Haoran Mo
Gaoyuan Liang
Pengfei Cui
Yoshinori Kobayashi
Kongtao Chen
Feng Feng
Wenli Yan
Zhibo Wang

Ying Qiao
Qiyue Lu
Dong Li
Heqi Wang
Tony Hao
Chenglong Wei
My Ha Le
Yu Chen
Naida Fetic
Bing Sun
Zhenzhong Chu
Meijing Li
Wentao Chen
Mingpeng Zheng
Zhihao Tang
Li Keng Liang
Alberto Mazzoni
Liang Chen
Meng-Meng Yin
Yannan Bin
Wasiq Khan
Yong Wu
Juanjuan Shi
Shiting Sun
Xujing Yao
Wenming Wu
Na Zhang
Anteneh Birga
Yipeng Lv
Qiuye Wang
Adrian Trueba
Ao Liu
Bifang He
Jun Pang
Jie Ding
Shixuan Guan
Boheng Cao
Bingxiang Xu
Lin Zhang
Mengya Liu
Xueping Lv
Hee-Jun Kang
Yuanyuan Zhang
Jin Zhang
Lin Chen
Runshan Xie

Zichang Tan
Fengcui Qian
Xianming Li
Jing Wang
Yuexin Zhang
Fan Wang
Yanyu Li
Qi Pan
Jiaxin Chen
Yuhan Hao
Xiaokang Wang
Jiekai Tang
Wen Jiang
Nan Li
Zhengwen Li
Yuanyuan Yang
Wenbo Chen
Wenchong Luo
Jiang Xue
Xuanying Zhang
Lianlian Zhong
Liu Xiaolin
Difei Liu
Bowen Zhao
Bowen Xue
Churong Zhang
Xing Xing Zhang
Yang Guo
Lu Yang
Jinbao Teng
Yupei Zhang
Keyu Zhong
Mingming Jiang
Chen Yong
Haidong Shao
Weizhong Lin
Leyi Wei
Ravi Kant Kumar
Jogendra Garain
Teressa Longjam
Zhaochun Xu
Zhirui Liao
Qifeng Wu
Nanxuan Zhou
Song Gu
Bin Li

Xiang Li
Yuanpeng Zhang
Dewu Ding
Jiaxuan Liu
Zhenyu Tang
Zhize Wu
Zhihao Huang
Yu Feng
Chen Zhang
Min Liu
Baiying Lei
Jiaming Liu
Xiaochuan Jing
Francesco Berloco
Shaofei Zang
Shenghua Feng
Xiaoqing Gu
Jing Xue
Junqing Zhu
Wenqiang Ji
Muhamad Dwisnanto Putro
Li-Hua Wen
Zhiwen Qiang
Chenchen Liu
Juntao Liu
Yang Miao
Yan Chen
Xiangyu Wang
Cristina Juárez
Ziheng Rong
Jing Lu
Lisbeth Rodriguez Mazahua
Rui Yan
Yuhang Zhou
Huiming Song
Li Ding
Alma Delia Cuevas
Zixiao Pan
Yuchae Jung
Chunfeng Mi
Guixin Zhao
Yuqian Pu
Hongpeng Ynag
Yan Pan
Rinku Datta Rakshit
Ming-Feng Ge

Mingliang Xue

Fahai Zhong

Shan Li

Qingwen Wu

Tao Li

Liwen Xie

Daiwei Li

Yuzhen Han

Fengqiang Li

Chenggang Lai

Shuai Liu

Cuiling Huang

Wenqiang Gu

Haitao Du

Bingbo Cui

Yang Lei

Xiaohan Sun

Inas Kadhim

Jing Feng

Xin Juan

Hongguo Zhao

Masoomeh Mirrashid

Jialiang Li

Yaping Hu

Xiangzhen Kong

Mixiao Hou

Zhen Cui

Na Yu

Meiyu Duan

Baoping Yuan

Umarani Jayaraman

Guanghui Li

Lihong Peng

Fabio Bellavia

Giosue' Lo Bosco

Zhen Chen

Jiajie Xiao

Chunyan Liu

Yue Zhao

Yuwen Tao

Nuo Yu

Liguang Huang

Duy-Linh Nguyen

Kai Shang

Wu Hao

Jiatong Li

Enda Jiang

Yichen Sun

Yanyuan Qin

Chengwei Ai

Kang Li

Jhony Heriberto Giraldo Zuluaga

Waqas Haider Bangyal

Tingting Dan

Haiyan Wang

Dandan Lu

Bin Zhang

Cuco Cristanno

Antonio Junior Spoleto

Zhenghao Shi

Ya Wang

Shuyi Zhang

Xiaoqing Li

Yajun Zou

Chuanlei Zhang

Berardino Prencipe

Feng Liu

Yongsheng Dong

Rong Fei

Zhen Wang

Jun Sang

Jun Wu

Xiaowen Chen

Hong Wang

Daniele Malitesta

Fenqiang Zhao

Xinghuo Ye

Hongyi Zhang

Xuexin Yu

Xujun Duan

Xing-Ming Zhao

Jiayan Han

Weizhong Lu

Frederic Comby

Taemoon Seo

Sergio Cannata

Yong-Wan Kwon

Heng Chen

Min Chen

Qing Lei

Francesco Fontanella

Rahul Kumar

Alessandra Scotto di Freca

Nicole Cilia

Annunziata Paviglianiti

Jacopo Ferretti

Pietro Barbiero

Seong-Jae Kim

Jing Yang

Dan Yang

Dongxue Peng

Wenting Cui

Wenhao Chi

Ruobing Liang

Feixiang Zhou

Jijia Kang

Huawei Huang

Peng Li

Yunfeng Zhao

Xiaoyan Hu

Li Guo

Lei Du

Xia-An Bi

Xiuquan Du

Ping Zhu

Young-Seob Jeong

Han-Gyu Kim

Dongkun Lee

Jonghwan Hyeon

Chae-Gyun Lim

Dingna Duan

Shiqiang Ma

Mingliang Dou

Jansen Woo

Shanshan Hu

Hai-Tao Li

Francescomaria Marino

Jiayi Ji

Jun Peng

Shirley Meng

Lucia Ballerini

Haifeng Hu

Jingyu Hou

Contents – Part I

Image and Signal Processing

Information Security

Neural Networks

Pattern Recognition

Virtual Reality and Human-Computer Interaction

Contents – Part II

Intelligent Computing in Computer Vision

Intelligent Control and Automation

Intelligent Modeling Technologies for Smart Cities

Knowledge Discovery and Data Mining

Machine Learning

Theoretical Computational Intelligence and Applications

Contents – Part III

Complex Diseases Informatics

Gene Regulation Modeling and Analysis

Intelligent Computing in Computational Biology

Protein Structure and Function Prediction

Evolutionary Computation and Learning

Multi-objective Optimization-Based Task Offloading and Power Control for Mobile Edge Computing

Yidan Chen, Xueyi Wang, Lianbo Ma[✉], and Ping Zhou

Northeastern University, Shenyang 110169, China

Abstract. With the rapid development of technologies such as IoT and 5G, the Internet industry has undergone significant changes and posed new challenges to the traditional computing model. Mobile edge computing (MEC) is considered as an effective solution to the challenge. However, as mobile devices generally have limited computing power and battery storage to cope with the increasing latency-sensitive and computationally intensive tasks, how to allocate computation and communication resources in the edge cloud to improve the quality of service (QoS) for mobile users is a challenging research. In this paper, we first formulate the optimization model of multi-user offloading and transmitted power control, aiming to reduce the energy consumption, task delay and price cost of mobile devices in a multi-channel wireless interference environment. Then we propose a multi-objective optimization algorithm based on theoretical analysis of the model and design a multi-objective offloading strategy based on NSGA-II. Our approach is able to obtain best trade-off among energy consumption, task delay and price. Experimental results on test instances shows the effectiveness and efficiency of our proposed algorithm.

Keywords: Mobile edge computing · Task offloading · Multi-objective optimization · Genetic algorithm · Hybrid coding · NSGA - II

1 Introduction

In recent years, cloud computing and wireless communication technologies have made rapid progress. With the continuous growth of devices connected to wireless networks, a large amount of data has been generated along the network edge [1]. At the same time, with the evolution of innovative applications such as face recognition [2], intelligent driving [3] and virtual reality [4], users have also put forward higher demands for service quality. The majority of these applications are high energy computation, computation-intensive and latency-sensitive [5], which are hard to accomplish on mobile devices with limited computing power and battery storage. This greatly limits the development of emerging industries.

Mobile cloud computing (MCC) systems have been proposed to address the computational power issue so that mobile devices can make the most of the strong computing power of cloud to reduce the local computational pressure, resulting in lower latency of

© Springer Nature Switzerland AG 2021
D.-S. Huang et al. (Eds.): ICIC 2021, LNCS 12836, pp. 3–17, 2021.
https://doi.org/10.1007/978-3-030-84522-3_1

task execution [6]. However, geographic location and network condition can also cause high latency during data transmission, and those latency-sensitive tasks such as work-flow applications are likely to be impossible to complete. To further reduce latency and ensure these tasks can be executed smoothly, fog computing [7–9] has been proposed as a new research area. Cisco defines that fog computing as complementing cloud comput-ing can migrate tasks to execute in the network edge. Edge computing is similar to fog computing, their terms are usually used interchangeably. MEC has received a high level of attention in recent years, and it reduces task latency and local energy consumption of mobile devices by building edge servers between the cloud and edge devices [10–12].

However, there are still many challenges in designing effective MEC systems [13]. To maximize the efficacy of the MEC system, we need to consider which user tasks should be offloaded to the edge and also decide on the optimal transmission power for offloading. In addition, the energy consumption, time delays and possible additional costs of task execution in the edge cloud also need to be considered in the system design. Most of the existing MEC methods focus on the study of a single objective and rarely consider multi-objective optimization. Swarm intelligence algorithms such as ant colony algorithm [14], bee colony algorithm [15] bring new ideas to solve multi-objective optimization problems.

Therefore, we detailed study the above problems and propose an efficient computa-tion offloading scheme in MEC, while the corresponding optimal transmission power is given. Our main contributions include:

- We jointly consider two problems of computation offloading and transmitted power control. The former of these is an integer programming and the latter is linear programming.
- We ensure that the energy consumption, time delay and cost in MEC are kept at the lowest level, which greatly enhances the user experience and satisfies the user psychology.
- We propose a multi-objective evolutionary optimization algorithm based on NSGA-II to solve this joint optimization problem. Extensive simulations and comparative experiments have shown that our algorithm can provide a superior offloading strategy and transmitted power for mobile users.

The rest of this paper is organized as follows. In Sect. 2 we summarize the rele-vant work. Section 3 describes the system model. Then, we propose the multi-objective computation offloading and transmitted power algorithm in Sect. 4. Section 5 gives simulation study to verify the performance of our proposed algorithm. Finally, Sect. 6 outlines the conclusion and describes our future work.

2   Related Work

In recent years, researchers have proposed multiple offloading schemes. Some of these studies focus on minimizing energy consumption. [16] proposed a joint radio resource allocation and offloading decision optimization to minimize the energy consumption. Xu et al. [17] considered data compression while solving the computation offloading and resource allocation problems to optimize the energy consumption. Yu et al. [18] proposed a novel mobile-aware partial offloading algorithm to dynamically calculate the amount of offloaded data using short-term mobile user's movement prediction, which minimizes the energy consumption while satisfying the delay constraint. Zhang et al. [19] schedule mobile services based on the particle swarm optimization algorithm (PSO) to minimize the total energy consumption of mobile devices.

Similarly, certain studies have worked on minimizing task latency. The delay minimization problem in a multi-user system with communication resources and computation resources allocation was studied by Ren et al. [20]. [21] given extra consideration to the problem of user mobility to minimize the task delay in MEC.

Some studies also consider energy consumption and task latency at the same time, but these studies tend to combine the two sub-problems into a single problem. [22] exploited the finite property of sensors to minimize processing delays and energy consumption of tasks in IoT. Since the wireless communication situation can have an impact on the data transmission process, Liu et al. [23] considered transmission power control while designing the offloading decision. Similar to the above, [24] proposed solutions in ultra-dense networks [25].

With the rapid development of multi-objective evolutionary optimization algorithms [26], these are often used to solve practical application problems. A multi-objective computation offloading approach was designed in [27] for workflow applications. In [28], an evolutionary algorithm is designed to find the best trade-offs between energy consumption and task processing delay. The optimization of joint energy consumption, processing delay, and price cost are further discussed in [29]. However, none of them consider the impact of transmission power on system performance.

3 System Model

In this section, we first introduce the MEC model as illustrated in Fig. 1. We assume that system consists of N mobile devices and one edge could server located at the edge of the network such as a radio base station, where each mobile device has a computation-intensive task waiting to be processed. These tasks can be performed locally or offloaded to the adjacent edge cloud. Next, the communication model, the computation model and the cost model are detailed described.

3.1 Network Model

Let $U = \{u_1, u_2, \cdots, u_N\}$ be the set of mobile devices. $T_n = (s_n, c_n, t_n)$ denotes the task of user, where s_n represents the size of the task T_n, i.e., the input data and associated processing code, c_n represents computation amount of the task T_n (e.g., the CPU cycles needed in total to accomplish the task), and t_n represents the maximum acceptable delay in completing the task. The actual execution time of the task must not exceed the maximum limit given. We describe the edge cloud as b, which has better computing power and storage space than mobile devices.

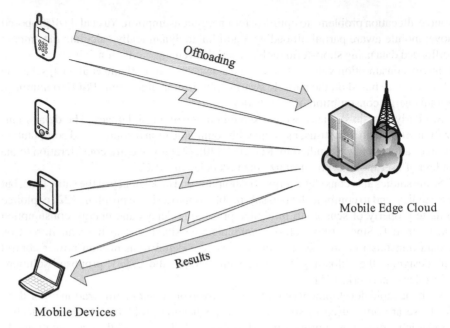

Fig. 1. Illustration of the multi-user MEC system in a multi-channel wireless environment.

3.2 Communication Model

Similar to many earlier studies such as [23] and [24], we divide the channel bandwidth B into M orthogonal sub-channels, denoted by $SC = \{1, 2, \ldots, M\}$. Mobile devices can select any sub-channel to offload their task to the edge cloud, or choose to execute it locally. Let $x_n = \{0, 1\}$ stand for the offloading decision variable of the user u_n, $x_n = 0$ denotes that the user u_n selects to solve the task locally, and $x_n = 1$ shows the task should be offloaded through selected sub-channel $sc_n \in SC$. Transmission power of the mobile user u_n during offloading is denoted by y_n. Let h_n stands for the channel gain between the user u_n and the edge cloud.

When $x_n = 1$, the user uplink data rate can be expressed as follows:

$$r_n^{ul} = B\log_2(1 + \frac{y_n h_n}{\sigma^2 + \sum_{i \in U \setminus \{n\}: sc_i = sc_n} y_i h_i}) \tag{1}$$

where σ^2 denotes the background noise power. When mobile user u_n choose to compute the task locally, the transmission power should be denoted by $y_n = 0$, there is no need to allocate the transmission power. From the above formula, we can see that as the transmitted power y_n increases, the uplink data rate r_{ul}^n increases accordingly. However, if the y_n is too large, it will cause interference to other user tasks transmitted through the same sub-channel.

Then the transmitted delay of mobile user u_n during offloading can be given by:

$$t_n^t = \frac{s_n}{r_n^{ul}} \tag{2}$$

Likewise, we can express the transmitted energy consumption of user u_n by:

$$\varepsilon_n^t = \frac{y_n s_n}{r_n^{ul}} \tag{3}$$

Because the size of the result of user tasks calculation is typically smaller than the offloaded data [23], the transmitted delay of the processing results is ignored.

3.3 Computation Model

The user can decide whether to perform its task locally or in the edge based on the local hardware and current network environment.

Local Computing. Let f_n^L denotes the computation capability of user u_n, its size depends on the local computation capability. The local execution time of computation task T_n can be represented as:

$$t_n^{exe,L} = \frac{c_n}{f_n^L} \tag{4}$$

In addition to execution latency, the energy consumption in local computing is also a very important measurement in MEC systems. According to [26], the energy consumption for a CPU cycle can be expressed as $\kappa(f_n^L)^2$, where κ is a coefficient related to the hardware architecture. So, the energy consumption during local execution of computation task T_n can be represented as:

$$\varepsilon_n^{exe,L} = \kappa(f_n^L)^2 c_n \tag{5}$$

in order to simplify the notation, we write $\kappa(f_n^L)^2$ as ρ_n, so the above equation can be simplified as $\varepsilon_n^{exe,L} = \rho_n c_n$.

From (4) and (5), we can discover there is the higher CPU computing capability f_n^L of this task T_n, the shorter execution delay of the task T_n, but the higher the local energy consumption. Given the limited energy stored locally on mobile devices, excessive energy consumption may make the user experience worse. It is therefore necessary to offload computation-intensive tasks to the edge cloud for execution.

Mobile Edge Computing. For mobile edge computing, we describe the allocated computing resource from b to user u_n as f_n^C. Then the edge cloud execution delay can be given by:

$$t_n^{exe,C} = \frac{c_n}{f_n^C} \tag{6}$$

3.4 Cost Model

If the mobile user u_n offload its task to b for execution, it must pay to the edge cloud providers, with the cost denoted by r_n, and if not, it will not have to pay this price. As shown below:

$$r_n = \begin{cases} 0 & x_n = 0 \\ d & x_n = 1 \end{cases} \tag{7}$$

4 Multi-objective Computation Offloading Algorithm Based on NSGA-II in MEC (MOPCA)

In this section, we demonstrate a scheme which considers computation resource and communication resource allocation of the edge cloud b. Firstly, we formulate a multi-objective optimization problem with constraints, then give the basic steps of the computation offloading method.

4.1 Objective Functions

Aiming at the limited wireless communication resources in the system, the limited battery storage and the limited computing resources in mobile devices. We mainly study the following three problems:

1. *Task Placement Problem.* We need to determine whether the mobile users' tasks will be executed locally or offloaded to the edge cloud. This is a binary decision problem.
2. *Resource Allocation Problem.* When the mobile user decides to offload, we have to determine how the edge cloud allocates communication resources and computation resources.
3. *Transmission Power Control Problem.* When the mobile user decides to offload, we have to determine how the mobile device determines the appropriate transmission power.

The objective of joint optimization problem is to minimize the energy consumption and task delay of mobile devices, as well as the total cost with meeting these constraints. It can be formulated by:

$$minimize \quad \mathrm{E} = \sum_{i=1}^{N} [(1 - x_i)\varepsilon_i^{exe,L} + x_i\varepsilon_i^t] \tag{8a}$$

$$minimize \quad \mathrm{T} = \sum_{i=1}^{N} [(1 - x_i)t_i^{exe,L} + x_i t_i^t + x_i t_i^{exe,C}] \tag{8b}$$

$$minimize \quad C = \sum_{i=1}^{N} [x_i r_i] \tag{8c}$$

$$subject\ to \quad x_i \in \{0, 1\}, \forall i = 1, \dots, N. \tag{8d}$$

$$0 \le y_i \le p_{\max}, \forall i = 1, \dots, N. \tag{8e}$$

$$sc_i \in \{1, 2, \cdots, M\}, \forall i = 1, \dots, N. \tag{8f}$$

$$y_i = 0, \forall x_i = 0. \tag{8g}$$

$$D_i \leq t_i, \forall i = 1, \ldots, N. \tag{8h}$$

where $p_{\max}$ stand for the maximum transmission power, and $D_n = (1 - x_n)t_n^{exe,L} + x_n t_n^t + x_n t_n^{exe,C}$ denotes the task delay spent on the actual execution of task T_n. We let E define the energy consumption of mobile devices, let T define the computation delay of task, and use C define the total cost of all mobile users. It's worth noting that each task's time delay for completion must not exceed its given upper limit.

The joint optimization problem is a multi-objective hybrid coding problem, and is difficult to solve directly. In what follow, we utilize NSGA-II [30] to solve this problem. Multi-objective optimization algorithms have advantages in solving large-scale high-dimensional problems [31]. Compared with other multi-objective genetic algorithms, NSGA-II has the advantages of fast operation and good convergence of the solution set.

4.2 The Main Steps of MOPCA

We design the MOPCA method in Algorithm 1. The populations are first generated randomly and calculated the fitness functions. Then, fast non-dominated sorting is used to stratify the individuals, and then the crowding distance of each individual is calculated to better distinguish the individuals' strengths and weaknesses. Finally, we generate new populations by binary tournament selection, crossover and mutation operations, after several iterations, until the termination condition is satisfied, which are described below:

Encoding. In this paper, genes in genetic algorithm stand for the offloading strategy for computational tasks, and multiple genes make up a complete individual, representing a solution to the optimization problem. In order to simultaneously consider the sub-channel selection during task offloading, for encoding we combine the offloading decision and channel selection into a single integer encoding, denoted by $x_n = \{0\} \cup SC$ in the range of $\{0, 1, \cdots, M\}$, and the corresponding power variable y_n is a floating point variable in $[0, p_{\max}]$, a hybrid coding is used for the above two variables. The hybrid encoded gene string contains two types of variables, which is encoded as shown in Fig. 2. The former is coded in integer and the latter is coded in floating point, as shown by $V = [x_n, y_n]$. A set of candidate solutions constitutes (individuals) a population.

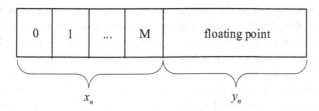

Fig. 2. Gene string by hybrid coding.

Fitness Functions and Constraints. The Fitness functions are the indicators used to determine the degree of merit of individuals in a group and is indicated by the formulas (8a), (8b) and (8c). The three equations represent the energy consumption, the total latency, and the total cost of mobile devices, respectively. Constraints are given by (8d), (8e), (8f), (8g) and (8h). This means that the actual completion delay of a task cannot exceed the upper limit of the delay for that task, while the transmission power cannot exceed the maximum value for that mobile device.

Initialization. In NSGA-II, an initial population P_0 with population size N_{pop} is first generated randomly according to the designed genetic coding, and parameters such as the number of iterations, crossover probability and mutation probability are determined. To speed up convergence, tasks in mobile devices with low local processing power are executed at the edge as much as possible during initialization in our approach (Line 1).

Fast Non-dominated Sorting. Fast non-dominance sorting refers to the stratification of populations based on the level of non-inferior solutions of individuals in the population. The specific process is traversing the entire population, with the number of dominated individuals per individual and the set of solutions dominated by that individual calculated until the population rank is fully divided (Line 5).

Crowding Distance Calculation. In order to be able to sort individuals in the same layer, the crowding distance of each individual needs to be calculated. The crowding distance can be found by calculating the sum of the distance differences between two individuals adjacent to it on each sub-objective function, as given by:

$$P[i]_{dis\,tan\,ce} = P[i]_E + P[i]_T + P[i]_C = \sum_{f=1}^{F} (|E[i+1] - E[i-1]|)$$

$$+ \sum_{f=1}^{F} (|T[i+1] - T[i-1]|) + \sum_{f=1}^{F} (|C[i+1] - C[i-1]|) \tag{9}$$

where $E[i+1]$, $T[i+1]$ and $C[i+1]$ represent the offloading strategy s_{i+1} to the functions (8a), (8b) and (8c), respectively.

Elitist Retention Strategy. To prevent the Pareto optimal individuals of the current population from being lost in the next generation, resulting in the genetic algorithm not converging to the global optimal solution, an elitist retention strategy is used. First, the parent population C_i and the offspring population D_i are synthesized into a new population R_i (Line 4). Then the whole layer is put into the parent population C_{i+1} in the order from lowest to highest according to the Pareto rank until the individuals in a certain layer cannot all be put in (Line 6–9), subsequently, they are arranged in order from largest according to the crowding distance of the individuals and put into the parent population C_{i+1} in turn until C_{i+1} is filled (Line 10–12).

Selection, Crossover and Mutation. Using binary tournaments as a method of selecting operations. The detailed steps are to first select half of the N_{pop} individuals at random, and then select the best adapted of them into the next generation population based on the fitness value of each individual. Then Crossover and mutation operations are performed to generate a new population Q_0. In order to reduce unnecessary computations, for the new population generated in each generation, the individuals that do not satisfy the constraint (8d), (8e), (8f), (8g) and (8h) are eliminated first, and then the remaining individuals are executed with sorting, crossover and mutation operations (Line 13).

Algorithm 1 MOPCA

Input:

N, M ; //number of mobile users and sub-channels

$T_i = (s_i, c_i, t_i)$ $i = (1, 2, ..., n)$; //task vector from mobile user u_i

N_{pop}, N_{gen} ; //population size and number of iterations

Output:

(x_i, y_i) $i = (1, 2, ..., N)$; //offloading strategies and transmission powers of mobile users

E, T, C ;

1. //initial the population $P_0, Q_0 = makingNewPop(P_0)$, $t = 0$

2. $G = 1$

3. While $G \leq N_{gen}$ do

4. $R_t = P_t + Q_t$

5. $F = fastNondo\min atedSort(R_t)$

6.   While $|P_{t+1}| + |F_i| \leq N_{pop}$ do

7. $P_{t+1} = P_{t+1} \cup F_i$

8. $i = i + 1$

9.   end while

10. $crowdingDis\tan ce(F_i)$ by (9)

11. $Sort(F_i)$

12. $P_{t+1} = P_{t+1} \cup F_i [1 : (N - |P_{t+1}|)]$

13. $Q_0 = makingNewPop(P_0)$

14. $t = t + 1$

15. $G = G + 1$

16. end while

17. return (x_{final}, y_{final}) $, E, T, C$

---

5 Simulation Study

5.1 Experiment Setup

Similar with the work in [23], we suppose that there is a wireless small cell and mobile devices connected to the edge cloud are randomly scattered within 50 m, and model the channel gains as: $h_n = L_n^{-\beta}$, where β is the path loss factor. Because of the disparities in computing power among different mobile devices, f_n^L is limited to the range of [0.5, 1.0]

Table 1. Experimental parameters.

Parameters	Value
Number of mobile users, N	30
Number of sub-channels, M	5
Transmission bandwidth of mobile users, B	10 MHz
Background noise, σ^2	-106 dBm
Path loss factor, β	3
Maximum of transmission power, p_{max}	0.1 W
Cost of edge cloud for each task, d	2
Data size for the offloading, s_n	5 MB
Number of CPU cycles for a task, c_n	1 Gigacycles

GHz. The computation resources f_n^C that the edge server can allocate to mobile users is 10 GHz. The other experimental parameters are given in Table 1.

The MOPCA is compared with three computation offloading strategies as follow:

1. Random Offloading to Edge Scheme (ROS): Mobile users decide to perform their tasks locally or offload to the edge cloud by randomly, and the sub-channels used in the offloading process are also generated by randomization. We use this scenario to evaluate the effect of offloading decisions in MOPCA.
2. All Offloading to Edge Scheme (AOES): All mobile users decide to offload their tasks to the edge cloud with the same transmission power, and the sub-channels used for offloading are divided equally. We use this scenario to situation where computing resources are abundant in the edge cloud, and mobile users have limited local processing power.

Fixed Transmission Power Scheme (FTPS): All mobile users use the fixed transmission power to offload with the same offloading scheme as MOPCA. We use the situation to verify the transmission power control effect of our proposed algorithm.

5.2 Performance Evaluation

We first consider how to balance the three objects of energy consumption, task latency, and total cost. In Fig. 3(a) to Fig. 3(c), we evaluate the impact of device amount on the offloading performance with ROS, AOES, FTPS and MOPCA under the condition that the number of mobile users is varied from 10 to 50 where the data size of task is fixed to 5 MB. Figure 3(a), Fig. 3(b) and Fig. 3(c) show the comparison of the average energy consumption, the average task delay and the total cost of different number of mobile users respectively, we can find that MOPCA shows a clear superiority to the other three algorithms. All four algorithms can maintain a better performance when there are few users. With the number of users increasing, the number of sub-channels used for communication remains constant, resulting in communication congestion and mutual interference in ROS and AOES, making the energy consumption and task latency grow rapidly in the system. And because of the impact of offloading strategy in MOPCA, optimal performance is always maintained even if the number of users keeps increasing. In FTPS, we used the same offloading strategy with MOPCA but fixed transmission power, the effect is not as good as MOPCA. The reason is that in MOPCA we dynamically adjust the transmission power according to the network environment and user scale. In Fig. 3(c), we can see that in AOES all users offload tasks to the edge cloud so the cost is the highest, ROS randomly decides whether offloading so the cost is the second highest, and MOPCA and FTPS have equal costs because they use the same offloading strategy that makes the users spend optimal.

In Fig. 4(a) to Fig. 4(c), we evaluate the impact of data size on the offloading performance with ROS, AOES, FTPS and MOPCA under the condition that the data size of tasks is varied from 1 MB to 9 MB where the number of mobile users is fixed to 30. Figure 4(a), Fig. 4(b) and Fig. 4(c) show the comparison of the average energy consumption, the average task delay and the total cost of different size of offloading data respectively. From Fig. 4(a) and Fig. 4(b), we can find when the data size of tasks is varied from 1 MB to 5 MB, energy consumption and task delay are increasing but from 5 MB to 9 MB, these two indicators began to gradually decline. Because an excessive amount of tasks can lead to higher energy and time consumption by the transfer process, so our proposed algorithm dynamically adjusts the policy to allow more tasks to be executed locally to ensure optimal performance. In Fig. 4(c) the total cost also appears to be decreasing. However, in AOES the total cost has been kept constant because the total number of mobile users remains the same.

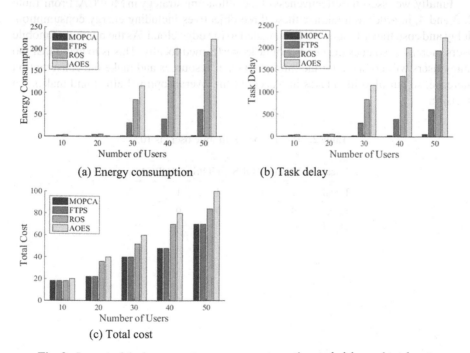

(a) Energy consumption

(b) Task delay

(c) Total cost

Fig. 3. Impact of device amount on energy consumption, task delay and total cost.

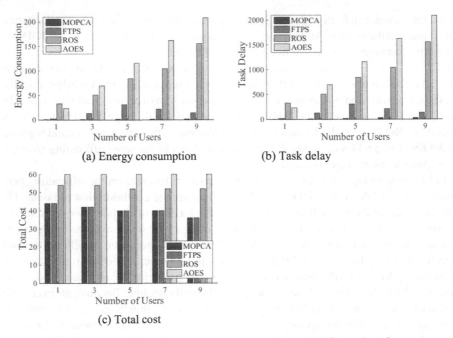

(a) Energy consumption   (b) Task delay

(c) Total cost

Fig. 4. Impact of data size on energy consumption, task delay and total cost.

Finally, we assess the effectiveness of the offloading strategy in MOPCA. From Table 2, 3 and 4, in order to minimize these three objectives including energy consumption, delay and cost, the tasks are mostly offloaded to the edge cloud. As the number of mobile users increases, so does the number of tasks performed locally. This is to avoid that too many users will compete for the limited channel resources and make the consumption increase, so our algorithm takes into account the overall optimal effect and makes the decision.

Table 2. The number of mobile users is 10.

Location	ROS	AOES	FTPS	MOPCA
Local	1	0	1	1
Edge cloud	9	10	9	9

Table 3. The number of mobile users is 30.

Location	ROS	AOES	FTPS	MOPCA
Local	4	0	10	10
Edge cloud	26	30	20	20

Table 4. The number of mobile users is 50.

Location	ROS	AOES	FTPS	MOPCA
Local	8	0	15	15
Edge cloud	42	50	35	35

6   Conclusion

In this paper, we investigate the multi-objective computation offloading problem for multiple users in MEC and jointly consider the effect of transmission power. In order to solve this joint problem, we propose a scheme MOPCA that figures out the optimal offloading strategy and transmission power under constraints. Extensive simulation studies show our proposed algorithm is more excellent than other schemes. In future work, we will improve our algorithm to adapt the ultra-dense network with multiple mobile devices and multiple edge cloud servers.

Acknowledgement. This work was supported in part by National Natural Science Foundation of China under Grant No. 61773103, and Intelligent Manufacturing Standardization and Test Verification Project "Time Sensitive Network (TSN) and Object Linking and Embedding Unified Architecture for Industrial Control OPC UA Fusion Key Technology Standard Research and Test Verification" project. Ministry of Industry and Information Technology of the People's Republic of China.

References

1. Hao, F., Min, G., Lin, M., Luo, C., Yang, L.T.: MobiFuzzyTrust: an efficient fuzzy trust inference mechanism in mobile social networks. IEEE Trans. Parallel Distrib. Syst. **25**, 2944–2955 (2014)
2. Chen, A., Xing, H., Wang, F.: A facial expression recognition method using deep convolutional neural networks based on edge computing. IEEE Access **8**, 49741–49751 (2020)
3. Tan, L.T., Hu, R.Q.: Mobility-aware edge caching and computing in vehicle networks: a deep reinforcement learning. IEEE Trans. Veh. Technol. **67**, 10190–10203 (2018)
4. Wang, J., Feng, Z., George, S., Iyengar, R., Pillai, P., Satyanarayanan, M.: Towards scalable edge-native applications. In: Proceedings of the 4th ACM/IEEE Symposium on Edge Computing, Arlington Virginia. pp. 152–165. ACM (2019)

5. Zheng, J., Cai, Y., Wu, Y., Shen, X.: Dynamic computation offloading for mobile cloud computing: a stochastic game-theoretic approach. IEEE Trans. Mobile Comput. **18**, 771–786 (2019)
6. Satyanarayanan, M., Bahl, P., Caceres, R., Davies, N.: The case for VM-based cloudlets in mobile computing. IEEE Pervasive Comput. **8**, 14–23 (2009)
7. Zhanikeev, M.: A cloud visitation platform to facilitate cloud federation and fog computing. Computer **48**, 80–83 (2015)
8. Chiang, M., Zhang, T.: Fog and IoT: an overview of research opportunities. IEEE Internet Things J. **3**, 854–864 (2016)
9. Sarkar, S., Chatterjee, S., Misra, S.: Assessment of the suitability of fog computing in the context of internet of things. IEEE Trans. Cloud Comput. **6**, 46–59 (2018)
10. Wang, C., Liang, C., Yu, F.R., Chen, Q., Tang, L.: Computation offloading and resource allocation in wireless cellular networks with mobile edge computing. IEEE Trans. Wireless Commun. **16**, 4924–4938 (2017)
11. Xia, W., Shen, L.: Joint resource allocation using evolutionary algorithms in heterogeneous mobile cloud computing networks. China Commun. **15**, 189–204 (2018)
12. Wei, F., Chen, S., Zou, W.: A greedy algorithm for task offloading in mobile edge computing system. China Commun. **15**, 149–157 (2018)
13. Mao, Y., You, C., Zhang, J., Huang, K., Letaief, K.B.: A survey on mobile edge computing: the communication perspective. arXiv:1701.01090 [cs, math]. (2017)
14. Zhao, Y., Lv, J.: A heuristic transferring strategy for heterogeneous-cached ICN. IEEE Access. **8**, 82421–82431 (2020)
15. Ma, L., Hu, K., Zhu, Y., Chen, H.: Cooperative artificial bee colony algorithm for multi-objective RFID network planning. J. Netw. Comput. Appl. **42**, 143–162 (2014)
16. Khalili, A., Zarandi, S., Rasti, M.: Joint resource allocation and offloading decision in mobile edge computing. IEEE Commun. Lett. **23**, 684–687 (2019)
17. Xu, D., Li, Q., Zhu, H.: Energy-saving computation offloading by joint data compression and resource allocation for mobile-edge computing. IEEE Commun. Lett. **23**, 704–707 (2019)
18. Yu, F., Chen, H., Xu, J.: DMPO: Dynamic mobility-aware partial offloading in mobile edge computing. Futur. Gener. Comput. Syst. **89**, 722–735 (2018)
19. Zhang, J., et al.: Hybrid computation offloading for smart home automation in mobile cloud computing. Pers. Ubiquit. Comput. **22**(1), 121–134 (2017). https://doi.org/10.1007/s00779-017-1095-0
20. Ren, J., Yu, G., Cai, Y., He, Y.: Latency optimization for resource allocation in mobile-edge computation offloading. IEEE Trans. Wireless Commun. **17**, 5506–5519 (2018)
21. Wang, Z., Zhao, Z., Min, G., Huang, X., Ni, Q., Wang, R.: User mobility aware task assignment for mobile edge computing. Futur. Gener. Comput. Syst. **85**, 1–8 (2018)
22. Ma, X., Lin, C., Zhang, H., Liu, J.: Energy-aware computation offloading of IoT sensors in cloudlet-based mobile edge computing. Sensors. **18**, 1945 (2018)
23. Liu, J., Li, P., Liu, J., Lai, J.: Joint offloading and transmission power control for mobile edge computing. IEEE Access. **7**, 81640–81651 (2019)
24. Chen, M., Hao, Y.: Task offloading for mobile edge computing in software defined ultra-dense network. IEEE J. Select. Areas Commun. **36**, 587–597 (2018)
25. Tong, L., Li, Y., Gao, W.: A hierarchical edge cloud architecture for mobile computing. In: IEEE INFOCOM 2016 - The 35th Annual IEEE International Conference on Computer Communications, San Francisco, CA, USA. pp. 1–9. IEEE (2016)
26. Ma, L., Cheng, S., Shi, Y.: Enhancing learning efficiency of brain storm optimization via orthogonal learning design. IEEE Trans. Syst. Man Cybern. Syst. 1–20 (2020). https://doi.org/10.1109/TSMC.2020.2963943
27. Xu, X., et al.: Multiobjective computation offloading for workflow management in cloudlet-based mobile cloud using NSGA-II. Comput. Intell. **35**, 476–495 (2019)

28. Bozorgchenani, A., Mashhadi, F., Tarchi, D., Salinas Monroy, S.: Multi-objective computation sharing in energy and delay constrained mobile edge computing environments. IEEE Trans. Mobile Comput. 1 (2020)

29. Peng, K., Zhu, M., Zhang, Y., Liu, L., Leung, V.C.M., Zheng, L.F.: A multi-objective computation offloading method for workflow applications in mobile edge computing. In: 2019 International Conference on Internet of Things (iThings) and IEEE Green Computing and Communications (GreenCom) and IEEE Cyber, Physical and Social Computing (CPSCom) and IEEE Smart Data (SmartData), Atlanta, GA, USA. pp. 135–141. IEEE (2019)

30. Deb, K., Pratap, A., Agarwal, S., Meyarivan, T.: A fast and elitist multiobjective genetic algorithm: NSGA-II. IEEE Trans. Evol. Comput. **6**, 182–197 (2002)

31. Ma, L., et al.: A novel many-objective evolutionary algorithm based on transfer matrix with Kriging model. Inf. Sci. **509**, 437–456 (2020)

An Evolutionary Neuron Model with Dendritic Computation for Classification and Prediction

Cheng Tang[1], Zhenyu Song[2], Yajiao Tang[3], Huimei Tang[4], Yuxi Wang[5], and Junkai Ji[4(✉)]

[1] Faculty of Engineering, University of Toyama, Toyama 930-8555, Japan
d2072006@ems.u-toyama.ac.jp
[2] College of Computer Science and Technology, Taizhou University, Taizhou 225300, China
songzhenyu@tzu.edu.cn
[3] College of Economics, Central South University of Forestry and Technology, Changsha 410004, China
[4] College of Computer Science and Software Engineering, Shenzhen University, Shenzhen 518060, China
{tanghm,jijunkai}@szu.edu.cn
[5] Taizhou People's Hospital, Taizhou 225300, China

Abstract. Advances in the understanding of dendrites promote the development of dendritic computation. For decades, the researchers are committed to proposing an appropriate neural model, which may feedback the research on neurons. This paper aims to employ an effective metaheuristic optimization algorithm as the learning algorithms to train the dendritic neuron model (DNM). The powerful ability of the backpropagation (BP) algorithm to train artificial neural networks led us to employ it as a learning algorithm for a conventional DNM, but this also inevitably causes the DNM to suffer from the drawbacks of the algorithm. Therefore, a metaheuristic optimization algorithm, named the firefly algorithm (FA) is adopted to train the DNM (FADNM). Experiments on twelve datasets involving classification and prediction are performed to evaluate the performance. The experimental results and corresponding statistical analysis show that the learning algorithm plays a decisive role in the performance of the DNM. It is worth emphasizing that the FADNM incorporates an invaluable neural pruning scheme to eliminate superfluous synapses and dendrites, simplifying its structure and forming a unique morphology. This simplified morphology can be implemented in hardware through logic circuits, which approximately has no effect on the accuracy of the original model. The hardwareization enables the FADNM to efficiently process high-speed data streams for large-scale data, which leads us to believe that it might be a promising technology to deal with big data.

Keywords: Dendritic neuron model · Firefly algorithm · Classification · Prediction · Hardwareization

1 Introduction

In the past 100 years, a vast number of biological studies have been conducted to determine the structure and function of the brain. In the brain, the cell bodies and connections

© Springer Nature Switzerland AG 2021
D.-S. Huang et al. (Eds.): ICIC 2021, LNCS 12836, pp. 18–36, 2021.
https://doi.org/10.1007/978-3-030-84522-3_2

are integrated into a variety of networks, and there are more than 10^4 neurons per cubic millimeter of the brain. The axon, the dendrite, and the soma/cell body are the essential components of neurons [12]. The axon is responsible for signal transmission among neurons through contact with dendrites at sites called synapses. The dendrite can be considered the input element of the neuron, which collects the signals from all the synapses formed with it. The soma/cell body plays the role of the central processing unit; if the membrane potential at the soma reaches a critical threshold, the cell is fired [3]. In 1943, McCulloch and Pitts pioneered the structural neuron model, which has long been regarded as the primary computational unit for research on artificial neural networks [25]. Due to the lack of the nonlinear mechanisms of synapses and dendrites, the McCulloch-Pitts model is considered to be oversimplified [24]. The inability of the single-layer McCulloch-Pitts model to implement certain fundamental calculation operators further exposes its limitations [14]. Due to advances in neuroscience techniques, great progress has been achieved in understanding neural computation. Numerous studies have suggested that the interaction among dendrites plays a crucial role in neural computing [30]. These experimental results have paved the way for more accurate neuron modeling. Mel et al. focused on the local mechanisms and decomposed the dendrite into numerous tiny computational units. The basic nonlinearity in the dendrites was modeled as a sigmoid function. Each dendrite integrated its input and performed a sigmoidal nonlinear operation on it. Thus, a two-layer neural network model was designed that gave each dendrite the ability to perform local computations [28]. In addition, a multilayer network based on neurons with active dendrites was proposed, in which the response of the spike patterns was selectively amplified [47]. In these models, the relevant synaptic signals have to be located in their corresponding positions in the dendrites [21]. It has long been assumed that synaptic interactions on dendrites can only be simulated by logical operations [13], but it has since been proven that the interactions of excitatory and inhibitory signals at the synapses constitute the basis of dendritic computation [9]. The model was verified by experiments in biological modalities such as the auditory and visual systems [41]. However, it does not include an effective pruning scheme to remove unnecessary synapses and dendrites. The morphological structure of the model is the same for any task; thus, the model cannot simulate the plasticity of the highly dynamic dendritic topology and is therefore obviously inconsistent with the characteristics of biological neurons [32].

To settle this issue, a neuron model in which the synapses and dendrites interact was proposed in our previous studies [39, 40]. Todo et al. used the backpropagation (BP) algorithm to enhance the model's ability to solve complex problems and then applied it to a two-dimensional multidirectional detection problem [43]. Subsequently, a dendritic neuron model (DNM) with dendritic nonlinearity was confirmed to effectively address the XOR problem [19], breast cancer classification [33], and liver disorder classification [18]. Zhou et al. applied the DNM to time series prediction problems [52, 53]. In addition, the DNM was verified to have outstanding performance in forecasting house prices [49], photovoltaic power [51], and the Istanbul stock and Taiwan futures exchange indexes [17] and in performing modeling and prediction for tourism [5, 50]. The approximation ability of the DNM was also studied and analyzed in [15, 37, 42]. It has been proven that the DNM can produce a wealth of nonlinear dendritic responses, therefore resulting in a strong nonlinear neural calculation function, which is highly consistent with many neu-robiological experimental results. Moreover, the model can learn and produce a unique dendritic structure for a specific task, solving the lack of neural plasticity from earlier dendritic models [44]. Thus, the DNM may lead to a new brand perspective for better explaining and understanding the principles underlying the calculations of neurons. Ji et al. demonstrated that the DNM can be further developed into an approximate logical neuron model, which has serious implications in improving the computational efficiency of the DNM [16]. In the improved DNM, the final dendritic morphology can be pruned for classification problems, and the specific, concise generated structure can be physically realized with logic circuits. The authors of [38] reported that logic circuits can mimic the excellent accuracy of the DNM in solving credit classification problems. To further improve the performance of the DNM, Qian et al. introduced the maximum relevance minimum redundancy method into the DNM and proposed a mutual information-based DNM, which was proven to achieve better performance on five classification datasets [29]. It must be mentioned that with the development of 5th-generation wireless sys-tems, big data will further penetrate into our lives and its quantity and description will increasingly generate from more fields [2, 27]. Therefore, the analysis of big data is a vast challenge in many fields, such as knowledge discovery and data mining [22, 23]. Fortunately, the logic operations have a huge advantage over floating-point operations in saving computing resources [4, 7], which makes the DNM more precious in various machine learning approaches. Thus, it is imperative and meaningful to investigate the DNM, as it will be vital for high-speed data stream processing in the era of big data.

Based on these prior studies, we observe that the performance of the DNM is depen-dent on the learning algorithm and is drastically limited by the BP algorithm. Due to its sensitivity to the initial values, the BP algorithm and its variants easily fall into local minima. Moreover, the transfer function must be differentiable, which leads to more computational resource consumption and overfitting problems [46]. In this study, a metaheuristic optimization algorithm, named the firefly algorithm (FA), is introduced into the DNM. The performance of the FA in training DNM is compared with the other three optimization algorithms and the conventional DNM. The experimental results on six classification datasets and six prediction datasets suggest that the FA provided excel-lent improvements for the DNM. In addition, the DNM trained by the FA (FADNM) is further compared with some competitive machine learning approaches. For the clas-sification problems, the multilayer perceptron (MLP), decision tree, the support vector machine (SVM) trained by the linear kernel function (SVM-l), SVM trained by the radial basis function (SVM-r), and SVM trained by the polynomial kernel function (SVM-p) are employed as the competitors. For the prediction problems, the MLP, Long short-term memory (LSTM), SVM-l, SVM-r, and SVM-p are adopted for comparisons. The main contributions of this study are summarized as follows: first, considering that the BP as the learning algorithm severely limits the performance of the conventional DNM, the FA is employed to train the FADNM for the first time. Second, the experimental results on six classification datasets and six prediction datasets determine that the FADNM is a successful improvement to the DNM, and the powerful search capabilities provide the FA with a better performance in training the DNM. Third, compared with other machine learning approaches, the FADNM is not only an excellent classifier but also has a prominent performance in time series prediction. Last but not least, the successful hardwareization of the FADNM on the classification problems without affecting the accuracy may lead to a promising technology for handling big data.

2   Dendritic Neuron Model

As shown in Fig. 1, the DNM mainly contains synapses, dendrites, a membrane, and a soma/cell body. The information in the DNM is transmitted directly from the synaptic layer to the dendritic layer to the membranous layer, and the soma/cell body layer receives the input from the membranous layer and produces the final result [20].

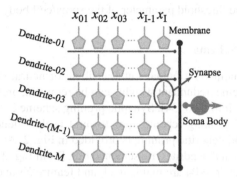

Fig. 1.   Structure of the DNM.

2.1   Model Description

The synaptic function $S_{i,j}$ of input X_i to the mth dendritic layer is given as follows:

$$S_{i,m} = \frac{1}{1 + e^{-k\left(w_{i,m}x_i - q_{i,m}\right)}},\tag{1}$$

where $i \in [1, 2, ..., I]$ and I denotes the number of features. $m \in [1, 2, ..., M]$, where M represents the number of dendritic layers. k indicates the user-defined distance parameter. $w_{i,m}$ and $q_{i,m}$ represent the weight and bias parameters, respectively. The corresponding threshold $\theta_{i,m}$ can be expressed as follows:

$$\theta_{i,m} = \frac{q_{i,m}}{w_{i,m}}.\tag{2}$$

We categorize the connection states according to the connection strength, i.e., direct connection ($0 < q_{i,m} < w_{i,m}$), inverse connection ($w_{i,m} < q_{i,m} < 0$), constant 1 connection ($q_{i,m} < 0 < w_{i,m}$ and $q_{i,m} < w_{i,m} < 0$), and constant 0 connection ($0 < w_{i,m} < q_{i,m}$ and $w_{i,m} < 0 < q_{i,m}$). We define the output D_m of the mth dendritic layer as follows [10]:

$$D_m = \prod_{i=1}^{I} S_{i,m}.\tag{3}$$

The membranous layer can be interpreted as an integration process; it receives all dendritic signals and sublinearly sums them. The function can be mathematically described as follows:

$$V = \sum_{m=1}^{M} D_m \tag{4}$$

We employ a sigmoid function as the activation function, and the final output of the model can be expressed as follows:

$$O = \frac{1}{1 + e^{-k(V - \theta_{soma})}}, \tag{5}$$

where θ_{soma} denotes the threshold parameter of the soma/cell body.

2.2 Neural Pruning Scheme

The neural pruning scheme is a fundamental attribute of neural networks. By implementing a pruning scheme, redundant weights can be removed, and the structure of the neural network can be simplified [34]. Synaptic pruning scheme can delete unnecessary synapses and dendritic pruning scheme aims to omit useless dendrites. A simplified example of synaptic and dendritic pruning is provided in Fig. 2. As shown in Fig. 2 (a), the redundant synapses and dendrites are screened out. From Fig. 2 (b), we can observe that dendrite-01 and dendrite-04 are maintained, and feature X_3 and X_4 can be deleted. Therefore, the neural pruning scheme can perform feature selection to some extent. According to the neural pruning scheme, a tidal neural morphology can be obtained.

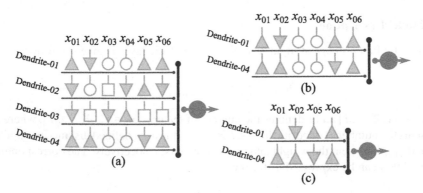

Fig. 2. An example of synaptic and dendritic pruning.

2.3 Logic Circuit Implementation

According to the neural pruning scheme, a unique structure with only direct connections and inverse connections can be obtained. In addition, the simplified structure can be replaced by comparators and logic AND, OR, and NOT gates. Each layer can be simulated by different logic components, as exemplified in Fig. 3. The direct connection can be replaced by a comparator, while an inverse connection can be simulated by a comparator and a NOT gate. Multiple synapses are connected to each dendrite, and the dendritic layer can be simulated by an AND gate. For the membranous layer, the dendrites connected to the membrane are connected through an OR gate. Finally, the result is transmitted to the soma/cell body, which is equivalent to a nonlinear mapping function, and it is simulated by a wire. Therefore, the structure can be implemented by a particular circuit. Compared with traditional classifiers, the logic circuit classifier has an unparalleled advantage when processing high-speed data streams because it involves no floating-point calculations.

3 Learning Algorithm

Taking into account the limitations of the BP algorithm, various other algorithms have been introduced for incorporation into the DNM. The firefly algorithm (FA) is employed as the learning algorithm, which is described in detail in this section. In the learning algorithm, we use the mean squared error (*MSE*) as the fitness function, which can be expressed as follows:

$$Fitness = \frac{1}{J} \sum\nolimits_{j=1}^{J} (T_j - O_j)^2, \tag{6}$$

where J is the number of samples. T_j denotes the target output of the jth sample, and O_j represents the corresponding actual output.

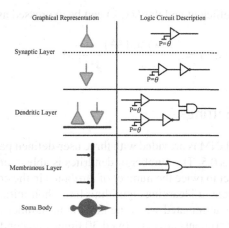

Fig. 3. Logic gate simulation of each layer.

3.1 Firefly Algorithm

The firefly algorithm (FA) is a nature-inspired metaheuristic algorithm, which mimics the flashing light of fireflies [48], and has been successfully utilized to solve numerous optimization problems [45]. In this algorithm, *Fitness* is used to describe the brightness I of each firefly, and the FA first initializes N fireflies randomly, where the ith ($i \in [1, 2, ..., N]$) firefly X_i consists of D elements, namely, $[x_{i,1}, x_{i,2}..., x_{i,D}]$. $x_{i,j}$ is related to the corresponding weight and bias parameters, and it can be obtained from the following equation:

$$x_{i,j} = L_j + rand(0, 1)(U_j - L_j), \tag{7}$$

where U_j and L_j represent the upper and lower bounds, respectively, of $x_{i,j}$.

The attraction of each firefly to the surrounding fireflies is proportional to its own light intensity. The attractiveness β can be described as follows:

$$\beta(i) = \beta_0 e^{-\gamma r^2}, \tag{8}$$

where β_0 denotes the attractiveness when r is 0, and the γ represents the light absorption parameter. We can observe that the attractiveness is also affected by the distance between fireflies. The distance $r_{i,k}$ between fireflies X_i and X_k is determined by the following:

$$r_{i,j} = \|X_i^t - X_k^t\|. \tag{9}$$

Next, the FA updates firefly positions by attracting individuals to each other, which can be expressed as follows:

$$X_i^{t+1} = X_i^t + \beta_0 e^{-\gamma r^2}(X_k^t - X_i^t) + \alpha sign\left[rand(0, 1) - \frac{1}{2}\right] \otimes L(s, \lambda), \tag{10}$$

where α is a scaling coefficient and the $L(s, \lambda)$ can be expressed as follows:

$$L(s, \lambda) = \frac{\lambda \Gamma(\lambda) \sin\left(\frac{\pi\lambda}{2}\right)}{\pi} \frac{1}{s^{1+\lambda}}, 1 < \lambda < 3. \tag{11}$$

4 Experimental Settings

In the experiments, the DNM is provided with three user-defined parameters. We set k to 5, and θ_{soma} is defined as 0.5. The number of dendrites is able to affect the performance of the DNM, which is set to twice the number of attributes in the corresponding dataset. To ensure fair comparisons, all learning algorithms have their initial population sizes set to 50. The simulation is terminated when the maximum number of fitness evaluations reaches 10^5, and all experiments are performed 30 times independently.

In the first part of the experiments, all models are implemented for six classification datasets [1], which contains the Iris dataset, Crab dataset, Liver dataset, Glass dataset, Cancer dataset, and Wine dataset. In the second part of the experiments, we employ six chaotic maps, namely, the Logistic map, Piecewise linear chaotic map, Singer map, Sine map, Sinusoidal map and Tent map to assess the performance of the models for prediction problems [35].

For the classification problems, the confusion matrix (also called error matrix) is a specific table layout that can visualize the performance of the classifier [36]. Six performance evolution criteria are adopted in this paper, namely, the accuracy, sensitivity (true positive rate, TPR), specificity (also called Recall), F_{score} [31], Cohen's Kappa coefficient (K) [26], and the area under the receiver opera characteristic (ROC) curve (AUC) [8]. For the prediction problems, the MSE, mean absolute percentage error ($MAPE$), mean absolute error MAE), and correlation coefficient (R) are employed to measure the prediction performance.

Furthermore, in order to compare the performances of the DNMs with different learning algorithms statistically, the Friedman Aligned Ranks test statistic is adopted in this work [6]. Another nonparametric statistical analysis, which is termed the Wilcoxon signed-rank test, is adopted to detect the significant difference between the FADNM and other machine learning methods [11]. The significance level is set to 0.05. We also employ the Bonferroni-Dunn procedure to further verify the confidence level of the statistical results, which is represented as p_{Bonf}.

5 Performance Comparison

5.1 Comparison with Other Learning Algorithms

First, the FADNM is compared with the conventional DNM and the DNMs trained by other optimization algorithms. For the classification problems, the accuracy is adopted as the evaluation criterion, and we compare the MSE of DNMs on the prediction datasets. The results yielded by the DNMs using the BP, cuckoo search (CS) algorithm, flower pollination algorithm (FPA), gravitational search algorithm (GSA), and FA learning algorithms on the classification and prediction datasets are summarized in Table 1.

It is easy to observe that the DNM trained by the FA algorithm obtains the best results on all datasets, which indicates that the FA algorithm has a stronger optimization ability in training the DNM. The results of the Friedman Aligned Ranks test statistic show that the FADNM ranks first in both classification and prediction problems. The corresponding p values show that the DNM trained by the FA algorithm is significantly better than the DNMs trained by the BP, CS, FPA, and GSA algorithms on both classification and prediction datasets. According to the corresponding p_{Bonf} values, the statistical results are further confirmed. The experimental results suggest that the DNMs quickly converge to their optimal performances for all datasets early in the training phase. The results in the Iris dataset are presented in Fig. 4. This result suggests that the training process can be terminated with a smaller maximum number of fitness evaluations to reduce learning costs. In addition, it is obvious that the convergence speed of the DNM trained by the FA algorithm ranks first and that FADNM has the best performance for all datasets. Based on these experimental results, we can conclude that compared with the remaining learning algorithms, the FA algorithm is the most effective in solving both classification and prediction problems. Comparison with other machine learning methods.

Furthermore, in order to investigate the effectiveness of the FADNM, we also compare the FADNM with other machine learning methods. In the classification problem experiments, the MLP, DT, SVM-l, SVM-r, and SVM-p are employed for comparison.

Table 1. Comparison of DNMs with different learning algorithms.

Classification datasets (accuracy)

Model	BP	CS	FPA	GSA	FA
Iris	94.00 ± 3.15	94.09 ± 1.84	93.56 ± 2.50	81.11 ± 11.35	**94.67 ± 2.10**
Crab	84.57 ± 9.92	74.73 ± 5.13	73.23 ± 6.43	70.33 ± 9.29	**84.97 ± 5.08**
Liver	68.23 ± 2.86	63.66 ± 3.91	64.12 ± 5.58	62.16 ± 5.28	**68.55 ± 3.51**
Glass	92.21 ± 2.94	92.37 ± 2.59	92.27 ± 2.55	91.65 ± 2.63	**93.89 ± 2.18**
Cancer	95.40 ± 1.51	95.06 ± 1.25	95.58 ± 1.09	94.90 ± 1.45	**96.29 ± 0.77**
Wine	77.23 ± 15.66	90.67 ± 2.83	87.64 ± 3.52	88.28 ± 6.70	**92.96 ± 2.72**
Ranking	3.1667	2.8333	3.3333	4.6667	1
z	2.373464	2.008316	2.556039	4.016632	–
p	0.017622	0.04461	0.010587	0.000059	–
p_{Bonf}	0.070488	0.178439	0.042349	0.000236	–

Prediction datasets (MSE)

Model	BP	CS	FPA	GSA	FA
Chaos-01	8.12e−02 ± 1.76e−02	9.07e−02 ± 4.08e−02	9.14e−02 ± 1.58e−02	1.08e−01 ± 3.48e−02	**6.62e−02 ± 2.27e−02**
Chaos-02	1.00e−01 ± 3.68e−02	7.96e−02 ± 9.32e−03	8.31e−02 ± 8.98e−03	8.53e−02 ± 2.17e−02	**5.77e−02 ± 1.68e−02**
Chaos-03	5.95e−02 ± 1.04e−02	6.20e−02 ± 6.64e−03	7.25e−02 ± 1.72e−02	1.01e−01 ± 3.77e−02	**4.81e−02 ± 1.49e−02**
Chaos-04	7.58e−02 ± 2.48e−02	5.36e−02 ± 8.81e−03	5.29e−02 ± 4.38e−03	6.89e−02 ± 1.32e−02	**3.98e−02 ± 1.82e−02**
Chaos-05	8.10e−02 ± 2.51e−02	4.94e−02 ± 2.47e−03	4.75e−02 ± 2.78e−03	8.03e−02 ± 3.90e−02	**4.62e−02 ± 3.37e−03**
Chaos-06	6.95e−02 ± 8.82e−03	6.91e−02 ± 5.54e−03	7.01e−02 ± 3.05e−03	7.99e−02 ± 1.65e−02	**6.13e−02 ± 6.01e−03**
Ranking	3.6667	2.6667	3.1667	4.5	1
z	2.921187	1.825742	2.373464	3.834058	–
p	0.003487	0.067889	0.017622	0.000126	–
p_{Bonf}	0.013948	0.271557	0.070488	0.000504	–

Table 2 lists the results of the MLP, DT, SVM-l, SVM-r, SVM-p and FADNM on classification datasets. For the Iris dataset, the FADNM achieves the highest accuracy, and the corresponding p values obtained by the Wilcoxon signed-rank test suggest that it significantly outperforms the MLP, SVM-l, and SVM-r. The best F_{score}, K and AUC are also obtained by the FADNM. Although the SVM-l has the highest sensitivity, the FADNM is slightly inferior to it and the specificity of the FADNM is drastically better than that of the SVM-l. The FADNM has the best accuracy, sensitivity, F_{score}, K and AUC on the Crab dataset. Compared with the SVM-p, the FADNM achieves a comparable performance in terms of specificity. The FADNM achieves the best accuracy, sensitivity, F_{score} and AUC on the Cancer dataset, and the p values denote that it is obviously better than the MLP, DT, SVM-l, SVM-r, and SVM-p in terms of accuracy. For the Liver and Glass datasets, the FADNM obtains the best accuracy, specificity, F_{score}, K and AUC, and it is significantly superior to the MLP, DT, SVM-l, SVM-r, and SVM-p in terms of accuracy. Although the SVM-l achieves the best sensitivity, it does not have satisfactory performance on the specificity. The FADNM achieves the best accuracy, sensitivity, F_{score} and AUC on the Cancer dataset, and the p values denote that it is obviously better than the MLP, DT, SVM-l, SVM-r, and SVM-p in terms of accuracy. For the Wine dataset, the FADNM has the best accuracy, specificity, F_{score}, K and AUC. The statistical results determine that the FADNM significantly outperforms the MLP and DT. Although the SVMs have better sensitivity than the FADNM, their specificity values are drastically worse than that of the FADNM. Moreover, the ROC curves of MLP, DT, SVM-l, SVM-r, SVM-p, and FADNM in the Iris dataset are also provided in Fig. 4. From this figure, we can observe that the FADNM has larger AUCs than the MLP, DT, SVM-l, SVM-r, and SVM-p on all datasets. Based on these experimental results, it can be concluded that the FADNM is an effective classifier.

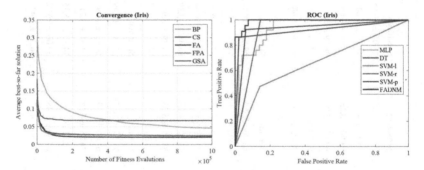

Fig. 4. ROCs of all methods on the classification datasets.

In the prediction problem experiments, the FADNM is compared with the MLP, LSTM, SVM-l, SVM-r, and SVM-p. The experimental results and corresponding statistical analysis are summarized in Table 3. For the Chaos-01 dataset, the FADNM achieves the best *MSE, MAPE, MAE* and *R*. The corresponding p values calculated by the Wilcoxon signed-rank test show that the FADNM is significantly better than the MLP, LSTM, SVM-l, SVM-r, and SVM-p in terms of the *MSE, MAPE, MAE* and *R*. The only exception that can be found that there is no significant difference between the FADNM and SVM-r in terms of the *R*. The FADNM obtains the best results of all indicators on the Chaos-02 and Chaos-05 datasets, and it is significantly superior to the MLP, LSTM, SVM-l, SVM-r, and SVM-p in terms of the *MSE, MAPE, MAE* and *R*. For the Chaos-03 dataset, the best *MSE, MAPE, MAE* and *R* are all achieved by the FADNM, and the FADNM obviously outperforms the MLP, LSTM, SVM-l, SVM-r, and SVM-p in terms

Table 2. Comparison of all methods for classification problems

Iris

Model	Accuracy	P	Sensitivity	Specificity	Fscore	K	AUC
MLP	76.53 ± 11.70	1.96e–06	93.53 ± 7.25	45.01 ± 30.65	84.350 ± 7.329	0.240 ± 0.056	0.8748 ± 0.098
DT	93.96 ± 3.70	2.49e–01	95.52 ± 4.27	91.28 ± 7.04	95.500 ± 2.663	0.326 ± 0.018	0.9340 ± 0.041
SVM-l	66.84 ± 4.52	8.84e–07	98.90 ± 3.25	3.54 ± 9.66	79.848 ± 3.066	0.185 ± 0.017	0.5122 ± 0.033
SVM-r	93.20 ± 3.35	2.54e–02	90.85 ± 5.01	98.19 ± 3.70	94.622 ± 2.754	0.325 ± 0.015	0.9452 ± 0.027
SVM-p	93.96 ± 2.88	1.17e–01	92.20 ± 4.69	97.74 ± 3.11	95.317 ± 2.265	0.328 ± 0.013	0.9497 ± 0.020
FADNM	94.67 ± 2.10	–	96.01 ± 2.07	92.16 ± 5.68	96.014 ± 1.557	0.329 ± 0.010	0.9871 ± 0.013

Crab

Model	Accuracy	P	Sensitivity	Specificity	Fscore	K	AUC
MLP	59.90 ± 11.62	1.23e–06	62.26 ± 22.40	59.82 ± 18.51	59.388 ± 15.851	0.214 ± 0.075	0.6912 ± 0.137
DT	79.93 ± 4.74	2.31e–04	81.20 ± 7.49	78.84 ± 8.42	79.915 ± 4.645	0.343 ± 0.031	0.8002 ± 0.047
SVM-l	76.17 ± 10.16	3.41e–04	78.06 ± 20.85	77.23 ± 26.04	75.908 ± 11.295	0.321 ± 0.063	0.7765 ± 0.090
SVM-r	77.30 ± 6.60	3.69e–06	77.58 ± 17.42	78.38 ± 10.34	76.701 ± 9.593	0.327 ± 0.041	0.7798 ± 0.058
SVM-p	82.33 ± 6.65	4.26e–02	79.22 ± 18.58	86.46 ± 13.59	81.075 ± 9.100	0.359 ± 0.042	0.8284 ± 0.061
FADNM	84.97 ± 5.08	–	86.29 ± 8.33	84.39 ± 8.95	84.776 ± 5.144	0.376 ± 0.033	0.9077 ± 0.053

Liver

Model	Accuracy	P	Sensitivity	Specificity	Fscore	K	AUC
MLP	54.16 ± 4.46	8.98e–07	77.50 ± 10.73	23.26 ± 11.49	65.856 ± 4.607	0.047 ± 0.077	0.5201 ± 0.075
DT	62.58 ± 4.45	1.02e–04	67.40 ± 7.69	56.12 ± 7.49	67.463 ± 4.793	0.245 ± 0.076	0.6176 ± 0.042
SVM-l	57.71 ± 2.67	9.94e–07	96.85 ± 4.67	4.31 ± 6.71	72.567 ± 1.892	0.057 ± 0.030	0.5058 ± 0.015
SVM-r	59.15 ± 4.35	9.06e–07	92.21 ± 8.76	14.87 ± 15.75	72.293 ± 2.526	0.110 ± 0.073	0.5354 ± 0.040
SVM-p	65.55 ± 4.23	1.12e–02	85.80 ± 9.48	39.00 ± 16.22	74.164 ± 2.866	0.267 ± 0.084	0.6240 ± 0.048
FADNM	68.55 ± 3.51	–	77.56 ± 4.87	56.33 ± 7.42	74.167 ± 3.161	0.342 ± 0.062	0.7209 ± 0.036

(continued)

Table 2. (*continued*)

Glass							
Model	Accuracy	p	Sensitivity	Specificity	Fscore	K	AUC
MLP	90.31 ± 2.42	1.26e−05	95.43 ± 2.90	73.27 ± 8.58	93.758 ± 1.707	0.396 ± 0.016	0.9387 ± 0.034
DT	91.34 ± 2.17	1.13e−04	95.10 ± 2.57	80.15 ± 9.12	94.291 ± 1.425	0.411 ± 0.018	0.8762 ± 0.041
SVM-l	91.18 ± 2.59	3.30e−04	97.39 ± 2.43	72.35 ± 9.99	94.346 ± 1.688	0.403 ± 0.020	0.8487 ± 0.047
SVM-r	91.93 ± 2.12	6.45e−04	96.61 ± 2.39	76.93 ± 6.90	94.783 ± 1.435	0.410 ± 0.017	0.8677 ± 0.034
SVM-p	92.74 ± 2.61	3.35e−02	95.82 ± 1.61	82.96 ± 10.04	95.312 ± 1.735	0.416 ± 0.019	0.8939 ± 0.049
FADNM	**93.89 ± 2.18**	–	96.76 ± 2.30	**85.20 ± 9.38**	**96.018 ± 1.433**	**0.428 ± 0.018**	**0.9663 ± 0.023**

Cancer							
Model	Accuracy	p	Sensitivity	Specificity	Fscore	K	AUC
MLP	93.46 ± 1.77	2.33e−06	86.40 ± 5.59	97.18 ± 1.23	90.027 ± 3.049	**0.259 ± 0.053**	0.9759 ± 0.017
DT	93.94 ± 1.58	1.28e−06	91.96 ± 4.28	95.01 ± 1.53	91.378 ± 2.371	0.216 ± 0.043	0.9949 ± 0.021
SVM-l	94.82 ± 1.01	1.90e−05	88.43 ± 2.31	**98.23 ± 0.78**	92.192 ± 1.474	0.235 ± 0.031	0.9333 ± 0.012
SVM-r	95.69 ± 1.07	9.57e−03	91.61 ± 2.53	97.84 ± 0.69	93.569 ± 1.544	0.214 ± 0.031	0.9473 ± 0.013
SVM-p	95.23 ± 0.99	1.75e−04	91.31 ± 2.87	97.27 ± 1.08	92.911 ± 1.544	0.219 ± 0.035	0.9429 ± 0.013
FADNM	**96.29 ± 0.77**	–	**95.94 ± 2.30**	96.48 ± 1.22	**94.693 ± 1.145**	0.177 ± 0.032	**0.9915 ± 0.003**

Wine							
Model	Accuracy	p	Sensitivity	Specificity	Fscore	K	AUC
MLP	89.51 ± 4.81	3.93e−03	92.59 ± 5.27	84.84 ± 7.59	91.498 ± 4.044	0.360 ± 0.027	0.9523 ± 0.034
DT	91.27 ± 2.50	4.94e−03	94.15 ± 3.43	86.95 ± 5.49	92.701 ± 2.293	0.371 ± 0.014	0.9055 ± 0.027
SVM-l	91.54 ± 7.15	2.86e−01	99.41 ± 1.50	80.62 ± 16.21	93.502 ± 5.331	0.370 ± 0.040	0.9002 ± 0.079
SVM-r	91.72 ± 8.13	6.68e−01	99.04 ± 1.90	81.94 ± 17.48	93.673 ± 5.827	0.371 ± 0.044	0.9049 ± 0.084
SVM-p	89.55 ± 8.92	5.94e−02	**99.60 ± 1.34**	76.29 ± 19.32	92.038 ± 6.455	0.359 ± 0.050	0.8794 ± 0.095
FADNM	**92.96 ± 2.72**	–	95.34 ± 3.38	**89.32 ± 6.11**	**94.253 ± 2.171**	**0.380 ± 0.016**	**0.9758 ± 0.014**

Table 3. Comparison of all methods for prediction problems

Chaos-01

Model	MSE	p	MAPE	P	MAE	P	R	P
MLP	7.91e-02 ± 2.14e-02	1.54e-02	2.13e-01 ± 4.50e-02	5.07e-05	2.39e-01 ± 3.73e-02	1.47e-04	6.14e-01 ± 1.60e-01	1.06e-02
LSTM	1.29e-01 ± 7.44e-02	3.33e-06	2.05e-01 ± 1.12e-01	9.01e-03	2.91e-01 ± 6.72e-02	2.48e-06	3.59e-01 ± 2.04e-01	4.44e-06
SVM-l	1.30e-01 ± 2.82e-17	9.13e-07	3.02e-01 ± 1.13e-16	9.13e-07	3.21e-01 ± 5.65e-17	9.13e-07	4.28e-02 ± 1.41e-17	9.13e-07
SVM-r	8.31e-02 ± 4.23e-17	6.05e-03	2.70e-01 ± 1.69e-16	9.13e-07	2.59e-01 ± 5.65e-17	9.13e-07	6.91e-01 ± 2.26e-16	1.72e-01
SVM-p	9.18e-02 ± 4.23e-17	1.12e-06	2.42e-01 ± 1.69e-16	1.01e-06	2.60e-01 ± 0.00e+00	9.13e-07	5.28e-01 ± 3.39e-16	1.24e-06
FADNM	**6.62e-02 ± 2.27e-02**	–	**1.58e-01 ± 3.56e-02**	–	**2.01e-01 ± 3.07e-02**	–	**7.29e-01 ± 1.88e-01**	–

Chaos-02

Model	MSE	p	MAPE	P	MAE	P	R	P
MLP	8.57e-02 ± 2.55e-02	9.85e-05	2.67e-01 ± 6.35e-02	1.47e-04	2.49e-01 ± 4.45e-02	1.49e-05	5.77e-01 ± 1.84e-01	3.03e-05
LSTM	7.55e-01 ± 3.09e+00	9.13e-07	3.82e-01 ± 7.76e-01	7.61e-03	4.60e-01 ± 5.73e-01	9.13e-07	1.98e-01 ± 1.43e-01	9.13e-07
SVM-l	1.33e-01 ± 2.82e-17	9.13e-07	4.16e-01 ± 5.65e-17	9.13e-07	3.25e-01 ± 0.00e+00	9.13e-07	3.25e-02 ± 0.00e+00	9.13e-07
SVM-r	8.80e-02 ± 5.65e-17	9.13e-07	2.56e-01 ± 1.69e-16	6.00e-05	2.69e-01 ± 1.13e-16	9.13e-07	7.91e-01 ± 3.39e-16	3.51e-02
SVM-p	1.12e-01 ± 7.06e-17	9.13e-07	3.62e-01 ± 1.69e-16	9.13e-07	2.97e-01 ± 0.00e+00	9.13e-07	4.00e-01 ± 0.00e+00	9.13e-07
FADNM	**5.77e-02 ± 1.68e-02**	–	**1.93e-01 ± 5.15e-02**	–	**1.93e-01 ± 1.80e-02**	–	**8.33e-01 ± 1.57e-01**	–

Chaos-03

Model	MSE	p	MAPE	P	MAE	p	R	P
MLP	7.83e-02 ± 2.44e-02	4.65e-05	1.72e-01 ± 4.03e-02	2.04e-06	2.35e-01 ± 4.72e-02	1.63e-05	5.56e-01 ± 1.81e-01	3.03e-05
LSTM	1.35e+00 ± 4.63e+00	9.13e-07	4.36e-01 ± 1.33e+00	7.64e-01	5.56e-01 ± 9.62e-01	9.13e-07	3.36e-01 ± 1.55e-01	9.13e-07
SVM-l	1.18e-01 ± 7.06e-17	9.13e-07	2.15e-01 ± 1.13e-16	9.13e-07	3.04e-01 ± 0.00e+00	9.13e-07	1.61e-01 ± 8.47e-17	9.13e-07
SVM-r	6.40e-02 ± 5.65e-17	9.13e-07	2.07e-01 ± 2.82e-17	9.13e-07	2.25e-01 ± 1.41e-16	9.13e-07	7.73e-01 ± 2.26e-16	1.88e-01
SVM-p	7.73e-02 ± 2.82e-17	9.13e-07	1.55e-01 ± 5.65e-17	9.13e-07	2.45e-01 ± 1.69e-16	9.13e-07	6.24e-01 ± 2.26e-16	9.13e-07
FADNM	**4.81e-02 ± 1.49e-02**	–	**1.02e-01 ± 1.12e-02**	–	**1.69e-01 ± 2.47e-02**	–	**7.93e-01 ± 1.18e-01**	–

(continued)

Table 3. (*continued*)

Chaos-04

Model	MSE	p	MAPE	P	MAE	p	R	P
MLP	5.89e−02 ± 1.46e−02	8.36e−05	1.87e−01 ± 3.94e−02	3.60e−05	2.00e−01 ± 2.87e−02	6.49e−06	5.51e−01 ± 1.62e−01	9.08e−05
LSTM	6.93e+00 ± 1.52e+01	1.24e−06	1.81e+00 ± 2.87e+00	2.77e−05	1.34e+00 ± 1.93e+00	1.12e−06	2.25e−01 ± 2.49e−01	1.37e−06
SVM-l	8.41e−02 ± 1.41e−17	9.13e−07	2.58e−01 ± 1.13e−16	9.13e−07	2.48e−01 ± 5.65e−17	9.13e−07	1.78e−01 ± 5.65e−17	9.13e−07
SVM-r	5.03e−02 ± 2.12e−17	7.61e−03	**1.21e−01 ± 2.82e−17**	9.50e−01	1.95e−01 ± 8.47e−17	2.48e−06	7.04e−01 ± 0.00e+00	1.99e−02
SVM-p	5.89e−02 ± 3.53e−17	5.07e−05	1.74e−01 ± 8.47e−17	2.96e−04	1.96e−01 ± 2.82e−17	2.48e−06	5.72e−01 ± 2.26e−16	7.70e−05
FADNM	**3.98e−02 ± 1.82e−02**	–	1.30e−01 ± 4.58e−02	–	**1.46e−01 ± 3.60e−02**	–	**7.62e−01 ± 1.93e−01**	–

Chaos-05

Model	MSE	p	MAPE	P	MAE	p	R	P
MLP	6.09e−02 ± 4.31e−03	9.13e−07	2.19e−01 ± 1.05e−02	9.13e−07	2.00e−01 ± 1.19e−02	9.13e−07	5.13e−01 ± 5.34e−02	9.13e−07
LSTM	1.53e−01 ± 3.28e−01	9.13e−07	2.27e−01 ± 1.42e−01	1.02e−03	2.63e−01 ± 8.08e−02	9.13e−07	2.77e−01 ± 1.21e−01	9.13e−07
SVM-l	6.29e−02 ± 2.82e−17	9.13e−07	2.17e−01 ± 1.13e−16	9.13e−07	2.10e−01 ± 1.41e−16	9.13e−07	4.88e−01 ± 2.26e−16	9.13e−07
SVM-r	5.92e−02 ± 2.82e−17	9.13e−07	2.11e−01 ± 8.47e−17	9.13e−07	2.00e−01 ± 0.00e+00	9.13e−07	5.51e−01 ± 3.39e−16	9.13e−07
SVM-p	6.87e−02 ± 2.82e−17	9.13e−07	2.81e−01 ± 5.65e−17	9.13e−07	2.22e−01 ± 8.47e−17	9.13e−07	4.38e−01 ± 1.13e−16	9.13e−07
FADNM	**4.62e−02 ± 3.37e−03**	–	**1.79e−01 ± 1.24e−02**	–	**1.51e−01 ± 8.45e−03**	–	**6.66e−01 ± 3.34e−02**	–

Chaos-06

Model	MSE	p	MAPE	P	MAE	p	R	P
MLP	8.00e−02 ± 3.77e−03	9.13e−07	1.68e−01 ± 8.87e−03	9.13e−07	2.35e−01 ± 8.09e−03	9.13e−07	2.05e−01 ± 6.42e−02	9.13e−07
LSTM	9.00e−02 ± 1.77e−02	9.13e−07	1.34e−01 ± 1.60e−02	1.00	2.38e−01 ± 2.03e−02	9.13e−07	2.08e−01 ± 7.72e−02	9.13e−07
SVM-l	7.96e−02 ± 1.41e−17	9.13e−07	1.63e−01 ± 1.65e−19	9.13e−07	2.21e−01 ± 1.13e−16	9.13e−07	2.11e−01 ± 1.41e−16	9.13e−07
SVM-r	7.22e−02 ± 4.23e−17	9.13e−07	**1.20e−01 ± 7.06e−17**	1.00	2.14e−01 ± 1.41e−16	9.13e−07	3.32e−01 ± 1.13e−16	9.13e−07
SVM-p	7.67e−02 ± 1.41e−17	9.13e−07	1.35e−01 ± 5.65e−17	1.00	2.29e−01 ± 8.47e−17	9.13e−07	2.31e−01 ± 8.47e−17	9.13e−07
FADNM	**6.13e−02 ± 6.01e−03**	–	1.49e−01 ± 5.36e−03	–	**1.89e−01 ± 1.44e−02**	–	**5.01e−01 ± 7.83e−02**	–

of all evaluation criteria. The exceptions can be observed that there is no significant difference between the FADNM and LSTM in terms of the *MAPE* or between the FADNM and SVM-r in terms of the *R*. The FADNM obtains the best *MSE*, *MAE*, and *R* on the Chaos-04 dataset, and it is obviously superior to the MLP, LSTM, SVM-l, SVM-r, and SVM-p in terms of the *MSE*, *MAE* and *R*. The FADNM significantly outperforms the MLP, LSTM, SVM-l, and SVM-p in terms of the *MAPE*. For the Chaos-06 dataset, the FADNM is significantly better than the MLP, LSTM, SVM-l, SVM-r, and SVM-p in terms of the *MSE*, *MAE*, and *R*.

5.2 Neural Pruning and Hardware Implementation

As mentioned above, for the classification problems, the neural morphology of the FADNM can be simplified according to the neural pruning scheme employed, resulting in a pithy dendritic structure. Furthermore, the unique, simple structure can be implemented into a hardware system with a logic circuit. Figure 5 depicts the neural structures after pruning for the Iris dataset. For the Iris dataset, only the dendrite-05 and dendrite-08 are maintained while the original structure of the FADNM has eight dendrites. Therefore, we can conclude that the pruning operation can drastically simplify the neural morphology of the FADNM. Furthermore, the simplified structures can be replaced by the logic circuit, which is also described in Fig. 5. It is worth noting that the logic circuit almost perfectly retains the classification accuracy of the FADNM. From Table 4, we can find that the hardware implementation has no obvious effect on the accuracy.

Table 4. Comparison of the FADNM and the logic circuit on classification datasets

Dataset	Iris	Crab	Liver	Glass	Cancer	Wine
Acc of the FADNM (%)	97.33	83.00	69.36	94.39	96.86	93.26
Acc of the logic circuit (%)	97.33	85.00	68.79	94.39	96.86	96.26

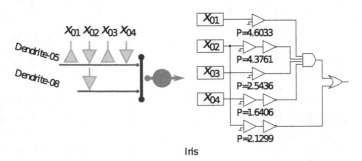

Fig. 5. Simplified structures of the FADNM for classification datasets.

6   Conclusions

To address the lack of neuronal plasticity in the models, the DNM with nonlinear synapses was designed in our previous research. While researching ways to improve the DNM, we found that the learning algorithm plays a key role, greatly affecting the performance of the DNM. Although the BP algorithm has long been regarded as a powerful learning algorithm for training artificial neural networks, its shortcomings still cannot be ignored. Therefore, the metaheuristic optimization algorithm is introduced into DNM as the learning algorithm to help it jump out of the local optima. In this study, we employed the FA as the learning algorithm to train the DNM. Five DNMs were ultimately evaluated for six classification datasets and six prediction datasets, and their effectiveness was verified. To objectively evaluate the performances of the learning algorithms, all DNMs were compared and ranked by statistical tests. The overall experimental results and corresponding statistical analysis demonstrated that the FA is the most effective learning algorithm for the DNM in solving both classification and prediction problems. The FADNM has high accuracy in solving classification problems and a low error in solving prediction problems, which proves its superior robustness to that of other algorithms. In addition, the efficiency of the FADNM was reaffirmed via its excellent convergence speed. For a more systematic assessment, the FADNM is further compared with other machine learning approaches. The experimental results and statistical analysis denote that the FADNM is not only an outstanding classifier but also has a superior performance in forecasting the time series. Moreover, according to the neural pruning scheme, the unnecessary synapses and superfluous dendrites can be discarded, and the structure of the FADNM can be drastically simplified. The resulting reconstructed structure can be replaced by the logic circuit, and the hardwareization has almost no impact on the accuracy. It may provide a feasible solution in the era of big data, where processing explosive information has become a daunting challenge. Furthermore, a large number of powerful machine learning methods are available, and their performance in training the DNM needs to be studied. More improvements and practical applications of the DNMs in more fields will be investigated in our future research.

Acknowledgment. This work was supported by a Project of the Guangdong Basic and Applied Basic Research Fund (No. 2019A1515111139), the Nature Science Foundation of the Jiangsu Higher Education Institutions of China (Grant No. 19KJB520015), the Talent Development Project of Taizhou University (No. TZXY2018QDJJ006), and the general item of Hunan philosophy and Social Science Foundation (20YBA260), namely Research on Financial Risk Management of Supply Chain in Hunan Free Trade Zone with Artificial Intelligence. The authors would like to thank the Otsuka Toshimi Scholarship Foundation for its support.

References

1. Asuncion, A., Newman, D.: Uci machine learning repository (2007)
2. Boccardi, F., Heath, R.W., Lozano, A., Marzetta, T.L., Popovski, P.: Five disruptive technology directions for 5g. IEEE Commun. Mag. **52**(2), 74–80 (2014)

3. y Cajal, S.R.: Histologie du système nerveux de l'homme & des vertébrés: Cervelet, cerveau moyen, rétine, couche optique, corps strié, écorce cérébrale générale & régionale, grand sympathique, vol. 2. A. Maloine (1911)
4. Chen, D.P.: High speed logic circuit simulator. US Patent 5,734,869 (1998)
5. Chen, W., Sun, J., Gao, S., Cheng, J.J., Wang, J., Todo, Y.: Using a single dendritic neuron to forecast tourist arrivals to Japan. IEICE Trans. Inf. Syst. **100**(1), 190–202 (2017)
6. Derrac, J., García, S., Molina, D., Herrera, F.: A practical tutorial on the use of nonparametric statistical tests as a methodology for comparing evolutionary and swarm intelligence algorithms. Swarm Evol. Comput. **1**(1), 3–18 (2011)
7. Dutta, S., Singh, D.: High-speed computation in arithmetic logic circuit. US Patent App. 10/005,551 (2003)
8. Flach, P.A., Hernández-Orallo, J., Ramirez, C.F.: A coherent interpretation of AUC as a measure of aggregated classification performance. In: ICML (2011)
9. Fortier, P.A., Bray, C.: Influence of asymmetric attenuation of single and paired dendritic inputs on summation of synaptic potentials and initiation of action potentials. Neuroscience **236**, 195–209 (2013)
10. Gabbiani, F., Krapp, H.G., Koch, C., Laurent, G.: Multiplicative computation in a visual neuron sensitive to looming. Nature **420**(6913), 320–324 (2002)
11. García, S., Molina, D., Lozano, M., Herrera, F.: A study on the use of non-parametric tests for analyzing the evolutionary algorithms behaviour: a case study on the CEC2005 special session on real parameter optimization. J. Heuristics **15**(6), 617 (2009)
12. Gerstner, W., Kistler, W.M., Naud, R., Paninski, L.: Neuronal dynamics: from single neurons to networks and models of cognition. Cambridge University Press (2014)
13. Gidon, A., et al.: Dendritic action potentials and computation in human layer 2/3 cortical neurons. Science **367**(6473), 83–87 (2020)
14. Haykin, S.: Neural Networks and Learning Machines, 3/E. Pearson Education India (2010)
15. He, J., Wu, J., Yuan, G., Todo, Y.: Dendritic branches of dnm help to improve approximation accuracy. In: 2019 6th International Conference on Systems and Informatics (ICSAI), pp. 533–541. IEEE (2019)
16. Ji, J., Gao, S., Cheng, J., Tang, Z., Todo, Y.: An approximate logic neuron model with a dendritic structure. Neurocomputing **173**, 1775–1783 (2016)
17. Jia, D., Zheng, S., Yang, L., Todo, Y., Gao, S.: A dendritic neuron model with nonlinearity validation on Istanbul stock and Taiwan futures exchange indexes prediction. In: 2018 5th IEEE International Conference on Cloud Computing and Intelligence Systems (CCIS), pp. 242–246. IEEE (2018)
18. Jiang, T., Gao, S., Wang, D., Ji, J., Todo, Y., Tang, Z.: A neuron model with synaptic nonlinearities in a dendritic tree for liver disorders. IEEJ Trans. Electr. Electron. Eng. **12**(1), 105–115 (2017)
19. Jiang, T., Wang, D., Ji, J., Todo, Y., Gao, S.: Single dendritic neuron with nonlinear computation capacity: a case study on XOR problem. In: 2015 IEEE International Conference on Progress in Informatics and Computing (PIC), pp. 20–24. IEEE (2015)
20. Koch, C.: Biophysics of computation: information processing in single neurons. Oxford University Press (2004)
21. Koch, C., Poggio, T., Torre, V.: Nonlinear interactions in a dendritic tree: localization, timing, and role in information processing. Proc. Natl. Acad. Sci. **80**(9), 2799–2802 (1983)
22. Lei, L., Zhong, Z., Zheng, K., Chen, J., Meng, H.: Challenges on wireless heterogeneous networks for mobile cloud computing. IEEE Wirel. Commun. **20**(3), 34–44 (2013)
23. Lomotey, R.K., Deters, R.: Towards knowledge discovery in big data. In: 2014 IEEE 8th International Symposium on Service Oriented System Engineering, pp. 181–191. IEEE (2014)
24. London, M., Häusser, M.: Dendritic computation. Annu. Rev. Neurosci. **28**, 503–532 (2005)

25. McCulloch, W.S., Pitts, W.: A logical calculus of the ideas immanent in nervous activity. Bull. Math. Biophys. **5**(4), 115–133 (1943)
26. McHugh, M.L.: Interrater reliability: the kappa statistic. Biochemia Med. **22**(3), 276–282 (2012)
27. Musolesi, M.: Big mobile data mining: good or evil? IEEE Internet Comput. **18**(1), 78–81 (2014)
28. Polsky, A., Mel, B.W., Schiller, J.: Computational subunits in thin dendrites of pyramidal cells. Nat. Neurosci. **7**(6), 621–627 (2004)
29. Qian, X., Wang, Y., Cao, S., Todo, Y., Gao, S.: MrDNM: a novel mutual information-based dendritic neuron model. Comput. Intell. Neurosci. **2019** (2019)
30. Reiff, D.F., Plett, J., Mank, M., Griesbeck, O., Borst, A.: Visualizing retinotopic half-wave rectified input to the motion detection circuitry of drosophila. Nat. Neurosci. **13**(8), 973–978 (2010)
31. Sasaki, Y., et al.: The truth of the f-measure. **2007** (2007)
32. Segev, I.: Sound grounds for computing dendrites. Nature **393**(6682), 207–208 (1998)
33. Sha, Z., Hu, L., Todo, Y., Ji, J., Gao, S., Tang, Z.: A breast cancer classifier using a neuron model with dendritic nonlinearity. IEICE Trans. Inf. Syst. **98**(7), 1365–1376 (2015)
34. Sietsma, J.: Neural net pruning-why and how. In: Proceedings of International Conference on Neural Networks, San Diego, CA, vol. 1, pp. 325–333 (1988)
35. Song, Z., Gao, S., Yu, Y., Sun, J., Todo, Y.: Multiple chaos embedded gravitational search algorithm. IEICE Trans. Inf. Syst. **100**(4), 888–900 (2017)
36. Stehman, S.V.: Selecting and interpreting measures of thematic classification accuracy. Remote Sens. Environ. **62**(1), 77–89 (1997)
37. Tang, C., Ji, J., Tang, Y., Gao, S., Tang, Z., Todo, Y.: A novel machine learning technique for computer-aided diagnosis. Eng. Appl. Artif. Intell. **92**, 103,627 (2020)
38. Tang, Y., Ji, J., Gao, S., Dai, H., Yu, Y., Todo, Y.: A pruning neural network model in credit classification analysis. Comput. Intell. Neurosci. **2018** (2018)
39. Tang, Z., Kuratu, M., Tamura, H., Ishizuka, O., Tanno, K.: A neuron model based on dendritic mechanism. IEICE **83**, 486–498 (2000)
40. Tang, Z., Tamura, H., Ishizuka, O., Tanno, K.: A neuron model with interaction among synapses. IEEJ Trans. Electron., Inf. Syst. **120**(7), 1012–1019 (2000)
41. Taylor, W.R., He, S., Levick, W.R., Vaney, D.I.: Dendritic computation of direction selectivity by retinal ganglion cells. Science **289**(5488), 2347–2350 (2000)
42. Teng, F., Todo, Y.: Dendritic neuron model and its capability of approximation. In: 2019 6th International Conference on Systems and Informatics (ICSAI), pp. 542–546. IEEE (2019)
43. Todo, Y., Tamura, H., Yamashita, K., Tang, Z.: Unsupervised learnable neuron model with nonlinear interaction on dendrites. Neural Netw. **60**, 96–103 (2014)
44. Todo, Y., Tang, Z., Todo, H., Ji, J., Yamashita, K.: Neurons with multiplicative interactions of nonlinear synapses. Int. J. Neural Syst. **29**(08), 1950,012 (2019)
45. Wang, H., et al.: Firefly algorithm with neighborhood attraction. Inf. Sci. **382**, 374–387 (2017)
46. Wang, X., Tang, Z., Tamura, H., Ishii, M., Sun, W.: An improved backpropagation algorithm to avoid the local minima problem. Neurocomputing **56**, 455–460 (2004)
47. Wang, Y., Liu, S.C.: Multilayer processing of spatiotemporal spike patterns in a neuron with active dendrites. Neural Comput. **22**(8), 2086–2112 (2010)
48. Yang, X.S.: Firefly algorithm, levy flights and global optimization. In: Bramer, M., Ellis, R., Petridis, M. (eds.) Research and Development in Intelligent Systems XXVI, pp. 209–218. Springer, London (2010). https://doi.org/10.1007/978-1-84882-983-1_15
49. Yu, Y., Song, S., Zhou, T., Yachi, H., Gao, S.: Forecasting house price index of China using dendritic neuron model. In: 2016 International Conference on Progress in Informatics and Computing (PIC), pp. 37–41. IEEE (2016)

50. Yu, Y., Wang, Y., Gao, S., Tang, Z.: Statistical modeling and prediction for tourism economy using dendritic neural network. Comput. Intell. Neurosci. **2017** (2017)
51. Zhao, K., Zhang, T., Lai, X., Dou, C., Yue, D.: A dendritic neuron based very short-term prediction model for photovoltaic power. In: 2018 Chinese Control and Decision Conference (CCDC), pp. 1106–1110. IEEE (2018)
52. Zhou, T., Chu, C., Song, S., Wang, Y., Gao, S.: A dendritic neuron model for exchange rate prediction. In: 2015 IEEE International Conference on Progress in Informatics and Computing (PIC), pp. 10–14. IEEE (2015)
53. Zhou, T., Gao, S., Wang, J., Chu, C., Todo, Y., Tang, Z.: Financial time series prediction using a dendritic neuron model. Knowl.-Based Syst. **105**, 214–224 (2016)

An Improved Genetic Algorithm for Distributed Job Shop Scheduling Problem

Sihan Wang[1], Xinyu Li[1(✉)], Liang Gao[1], and Lijian Wang[2]

[1] Huazhong University of Science and Technology, Wuhan 430074, People's Republic of China
sihanwang@hust.edu.cn, {lixinyu,gaoliang}@mail.hust.edu.cn
[2] AVIC China Aero-Polytechnology Establishment, Beijing, China

Abstract. Distributed job shop scheduling is an important branch in the field of intelligent manufacturing. But the literatures are still relatively limited since the issue is still in its infancy. Effective methods still need to be explored to obtain high-quality feasible schemes. This paper proposes an improved genetic algorithm (IGA) to solve distributed job shop scheduling problem (DJSP) considering minimize makespan. A crossover and a mutation strategies are designed for the problem-specific job assignment sub-problem in DJSP. The proposed crossover operator takes the similarity of individuals into consideration, thus the exploration and exploitation can be well balanced. The IGA was tested on TA benchmark set from the existing literature, and obtained 10 new best solutions. Extensive computational results and analyses indicate that the proposed algorithm is an efficient method for solving DJSP.

Keywords: Genetic algorithm · Genetic operator · Distributed job shop scheduling problem

1 Introduction

The Job Shop Scheduling problem (JSP) is a strongly Non-deterministic polynomial hard (NP-hard) problem [1]. In this problem, there are a set of jobs $J = \{J_1, ..., J_n\}$ that must be processed on a set $M = \{M_1, ..., M_m\}$ of machines. Each job J_i, $i = 1, ..., n$, consists of n_i operations that should be sequentially processed. Besides, each operation o_{ij} requires uninterrupted and exclusive use of its assigned machine for its whole processing time. The distributed job shop scheduling problem (DJSP) is an extension of JSP by allowing a job J_i to be processed on machines geographically distributed on f identical factories $F = \{F_1, ..., F_f\}$. The problem is to assign each job to a factory and to sequencing the operations on machines with the objective of minimizing one or more predefined performance criteria. As an extension of JSP, DJSP is also a NP-hard problem and has more complex solution space.

Since distributed manufacturing has become a common production model [2], DJSP gains more and more attention. But the literature is still relatively limited since the issue is still in its infancy. Excellent review of DE on the state-of-the-art research can be found in survey [3, 4]. From Ref [3], exact methods and approximate methods such as GA and

D.-S. Huang et al. (Eds.): ICIC 2021, LNCS 12836, pp. 37–47, 2021.
https://doi.org/10.1007/978-3-030-84522-3_3

Ant Colony Optimization are already used for solving DJSP. The previous literature tested the GA with a 6 jobs, 6 machines and 3 identical factories instance. The more effective method and experiment about large size benchmark instances is still waiting to be explore.

GA was proposed by John Holland at 1975 [5]. Since it has powerful global search-ing ability, GA has been applied to several scheduling problems [6–16]. Luo J et al. [12] made an investigation into minimizing total tardiness, total energy cost and disruption to the original schedule in the job shop with arrivals of urgent jobs. The particular work of Ref [13] adopted a modified genetic algorithm approach, it is proposed with operat-ing parameters which experiments show the advantages of the GA. Liang X et al. [15] combined GA with a simulated annealing algorithm (SA) component to overcome the common pre-mature convergence in solving JSP, thus improving the overall performance of GA and obtaining satisfactory results. In Ref [16], the authors propose a new hybrid genetic algorithm to solve JSP which contains a new selection criterion to tackle pre-mature convergence problem. Considering the research effort of GA and characteristics of DJSP, a novel improved GA, denoted as IGA, was proposed in this paper with two well-designed operators to solve DJSP.

The proposed algorithm is tested on the 10 benchmark instances provided by Taillard [17]. We compare IGA to the three algorithms in Ref [18], experiment results show that proposed algorithm can find high-quality solutions for all of the benchmark instances with less computational time. It means that the proposed IGA is very competitive to the state-of-the-art algorithms.

The remaining part of the paper is organized as follows. Section 2 gives details of the IGA algorithm. Computational results of benchmark instances are shown in Sect. 3 while Sect. 4 summarizes the concluding remarks.

2 IGA Algorithm

GA is a computational method based on the mechanics of biological genetic evolution. It contains the theory of evolution and biological processes such as selection, crossover, and mutation operators. For solving combinatorial optimization problems, GA is a good algorithmic framework. Based on GA, the IGA algorithm contains new crossover oper-ators and mutation operators for solving DJSP. The procedure of IGA is shown in Table 1.

The parameters include population size *Popsize*, the total number of generation *MaxGen*, reproduction probability p_r, crossover probability p_c, mutation probability p_m, degree of similarity p_s, and the run time of each instance *RunTime*. The following describes the encoding, decoding of the individual. The encoding method in Ref [19] is adopted. Because DJSP contains two sub-problems, the chromosome includes two strings carrying factory assignment information and operation sequence. For example in Fig. 1, there are three jobs and two factories. In this solution, J_1 and J_3 are assigned to factory 1, J_2 are assigned to factory 0. The sequence of operation in each factory following the operation sequence. In the decoding procedure, firstly generating job set $J_f = \{J_i\}, i \in \{1, ..., n\}$, then each factory is decoded to active schedule in turn [14].

Table 1. Procedure of IGA

Procedure of IGA

Step 1: Set the parameters of the proposed IGA;

Step 2: Initialization: initialize the population and set $Gen = 1$, Gen is the current generation;

Step 3: Evaluation: evaluate every individual in the population by the objective;

Step 4: Is the termination criteria satisfied? if yes, turn to step 7; else, turn to step 5;

Step 5: Generate the new population:

Step 5.1: Selection: generate the new population which $p_r \times Popsize$ individuals by elitist selection scheme and others by tournament selection;

Step 5.2: Crossover: set R as a random value, $R \in [0, 100000]$, if $R > p_c \times 100000$, copy parents to next generation; else, generate children by POX [14] or JBX [14] in same priority. During POX [14] and JBX [14], check the similarity of the factory-assignments of parents, if the same assignments higher than p_s of all elements, copy it from parents to children, else, generate the children's factory-assignments by factory assignment crossover operator;

Step 5.3: Mutation: generate the new population with the factory assignment mutation method and neighborhood mutation method with a p_m probability;

Step 6: Set $Gen = Gen + 1$ and go to Step 3;

Step 7: Output the best solution

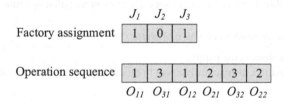

Fig. 1. Encode method

In GA, it is very important to employ good genetic operators which can effectively deal with the problem and efficiently generate excellent individuals in the population. The first step of IGA is selection operator. It is used to select the individuals according to the fitness. Elitist selection scheme and tournament selection scheme are both adopted in the selection procedure. In the elitist selection scheme, $p_r \times Popsize$ individuals are reproduced with good fitness in parents to the offspring. On the other hand, the tournament selection, a number of individuals will be selected randomly from the population and keep the best one to the offspring. The second step is crossover operator. Because factory assignment is one of sub-problem of DJSP and it impacts the fitness of each factory closely, a crossover operator has been designed for it, which embed in the crossover operators, POX and JBX. When the degree of similarity of parents lower than p_s, it will be used to generate the factory-assignments of children. The basic working procedure of factory-assignment crossover operator is described in Table 2 (two parents are denoted as P1 and P2; two offspring are denoted as C1 and C2):

Table 2. Procedure of factory assignment crossover operator

Procedure of factory assignment crossover operator
Step 1: select *length* consecutive elements in P1and save the positions as POS1, where *length* > 2;   **Step 2:**select *length* consecutive elements in P2 and save the positions as POS2;   **Step 3:** copy the elements which position in POS1 in sequence from P1 to C1, the remaining positions are complemented by the corresponding elements in P2;   **Step 4:** copy the elements which position in POS2 in sequence from P2 to C2, the remaining positions are complemented by the corresponding elements in P1

Precedence operation crossover (POX) and Job-based crossover (JBX) [14] are adopted for the operation sequence chromosome. At last but not least, two mutation operators have been adopted for the factory assignment and operation sequence separately. For the operation sequence, neighborhood mutation method [14] has been adopted. A random mutation operator has been designed for factory assignment. The procedure of factory assignment mutation is as follows (Table 3):

Table 3. Procedure of factory assignment mutation operator

Procedure of factory assignment mutation operator
Step 1: Select *r* positions in the parent;   **Step 2:** For each position, change the element to other factory in the factory set of the corresponding job

3 Computational Result

Our IGA algorithm is implemented in C++ and runs on an Intel Core i5-9400 processor with 2.90 GHz CPU and 16 GB RAM. As shown in Table 4, in our experiments, we set *Popsize*, *MaxGen*, p_r, p_c, p_m, p_s, and *RunTime* to 400, 200, 0.005, 0.9, 0.2, 0.7 and 10, respectively. These parameter values are determined by extensive preliminary experiments. The algorithm was tested using TA benchmark instances in Ref [16] and compared to the recent state-of-the-art three algorithms DAHACO, HACO and ACO in Ref [18]. The following is a partial TA instance. For example:

2 2
2 0 2 1 10 0 2 1 2 1 10 0 4
2 0 2 0 4 1 8 1 2 0 5 1 7
Frist line: there are two factories and 2 jobs.

Second line: for job 1 there are two available factories. For factory 0, there are 2 operations need to be processed, the first operation O_{11} should be processed on machine 1 for 10 s. And the second operation O_{12} should be processed on machine 0 for 2 s. As for factory 1, there are 2 operations need to be processed, the first operation O_{11} should be processed on machine 1 for 10 s. And the second operation O_{12} should be processed on machine 0 for 4 s. The third line has the same rule with the second line. Other instances in detail are shown in Table 7 at the end of the article.

Note that columns best(avg) and t(s) are the best(average) solution obtained and average computational time in seconds required by the proposed algorithm. The best-known solutions that can be obtained by each algorithm are indicated in bold.

Table 4. Parameters of IGA

Variable	Value
Popsize	400
MaxGen	200
p_r	0.005
p_c	0.9
p_m	0.2
p_s	0.7
RunTime	10

From Table 5, one observes that IGA outperforms DAHACO, HACO, ACO in terms of solution quality on TA benchmark with $f = 2$ and $f = 4$. For TA01–TA10 with $f = 2$, the minimize makespan of each instance proposed by DAHACO, HACO and ACO is over 1195, and the best solutions for TA01–10 obtained by IGA are less or more than 1000, lower than 1070. The optimal solution obtained by IGA better than DAHACO (the best of three comparison algorithms) in the range of [188, 385]. For example, the best solution of DAHACO for TA02 is 1213, IGA's solution for TA02 is 1025. The optimal solution of IGA is numerically reduced by 188 compared to the optimal solution of DAHACO. For TA07, the best solution of DAHACO is 1402, IGA's solution for TA07 is 1017. The optimal solution of IGA is numerically reduced by 385 compared to the optimal solution of DAHACO. For TA01TA10 with $f = 4$, only the minimize makespan of instance TA04 proposed by DAHACO is 980, others are over 1000, and the best solutions for TA01–10 obtained by IGA are less than 1000, more than 900. On behave of the objective function, IGA is competitive and efficiently compare with the recent state-of-the-art algorithms in the range of [62, 130]. For TA03, the best solution of DAHACO is 1015, IGA's solution for TA03 is 953. The optimal solution of IGA is numerically reduced by 62 compared to the optimal solution of DAHACO. For TA06, the best solution of DAHACO is 1054, IGA's solution for TA06 is 924. The optimal solution of IGA is numerically reduced by 130 compared to the optimal solution of DAHACO. The effectiveness of the proposed IGA is more clearly seen in the Fig. 2 and Fig. 3. For each of benchmark instances, IGA obtains better optimal results.

In terms of the computational time, IGA has a significant computational advantage, reducing the computational time spent compared to DAHACO. The computational time of HACO and ACO is not mentioned in the literature [18]. The computational time of IGA to search an optimal solution is about 9 s, it's almost ten percent of the cost of DAHACO. Table 6 shows the average relative percentage deviation (RPD) of the four algorithms grouped by f. It is used as performance measure in this study to show the stability of the algorithm. It can be calculated by follows:

$$RPD = \frac{Alg - Min}{Min} \times 100 \tag{1}$$

Where *Alg* is the makespan obtained by a given algorithm alternative on a given instance and *Min* is the lowest makespan obtained for the same instance. From Table 6, the RPD of IGA is much less than HACO and ACO, slightly higher in value than DAHACO. This phenomenon means that IGA is more stable than HACO and ACO, but slightly less stable than DAHACO.

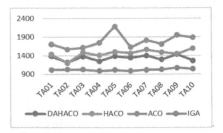

Fig. 2. Result of the Table 5 with $f = 2$

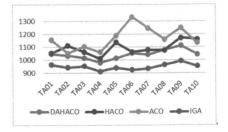

Fig. 3. Result of the Table 5 with $f = 4$

4   Discuss and Analysis

From the results of each experiment, the proposed IGA can obtain much better results for TA benchmark sets compare with other three state-of-the-art algorithms. Therefore, the above experimental results indicate that the proposed IGA can solve the DJSP effectively and obtain many good and useful results. The computational time of IGA has also been compared with other algorithms, IGA cost less computation time than other algorithm apparently. Therefore, the proposed IGA has both superior effectiveness and high efficiency for solving DJSP.

Table 5. The Comparison between IGA and other reference algorithms on TAdata

instance	size	2F						4F					
		DAHACO		HACO	ACO	IGA		DAHACO		HACO	ACO	IGA	
		Best	T(s)	Best	Best	Best(avg)	T(s)	Best	T(s)	Best	Best	Best(avg)	T(s)
TA01	15×15	1381	721	1433	1708	1016(1032)	9.15	1047	53	1053	1154	963(963)	9.78
TA02	15×15	1213	721	1195	1570	1025(1037.2)	9.51	1033	53	1112	1051	942(974)	9.75
TA03	15×15	1376	721	1484	1616	1025(1040.6)	9.475	1015	53	1062	1104	953(976)	9.71
TA04	15×15	1246	721	1406	1740	987(1006.6)	9.52	980	53	1011	1061	911(933)	9.72
TA05	15×15	1380	721	1505	2188	1008(1018)	9.42	1014	53	1137	1187	940(946)	9.61
TA06	15×15	1347	721	1467	1627	987(1024)	9.39	1054	53	1063	1333	924(958)	9.61
TA07	15×15	1402	721	1561	1811	1017(1036.7)	9.44	1042	53	1079	1246	935(969)	9.61
TA08	15×15	1295	721	1494	1713	1021(1051)	9.45	1077	53	1077	1159	963(992)	9.63
TA09	15×15	1434	721	1445	1947	1067(1074)	9.4	1112	53	1171	1248	992(1012)	9.6
TA10	20×15	1259	1893	1594	1886	1036(1044)	9.43	1044	175	1159	1133	953(961)	9.65

Table 6. Average RPD of the four algorithms grouped by f

F	DAHACO	HACO	ACO	IGA
2	**0,39**	13,12	57,08	1,77
4	**0,58**	6,39	34,58	2,45
Average	**0,48**	9,75	45,83	2,11

The reason are as follow. Firstly, GA as an evolutionary algorithm has powerful global searching ability. Secondly, in order to solve the DJSP effectively, effective genetic operators are employed and designed in this method. The population diversity was enhanced by the crossover and mutation operators worked on each solution. Therefore, the proposed approach has very good searching ability with little computational time.

Table 7. Instances in detail

TA01(*f*=2)

2 15

2 0 15 6 94 12 66 4 10 7 53 3 26 2 15 10 65 11 82 8 10 14 27 9 93 13 92 5 96 0 70 1

83 1 15 6 94 12 66 4 10 7 53 3 26 2 15 10 65 11 82 8 10 14 27 9 93 13 92 5 96 0 70 1 83

2 0 15 4 74 5 31 7 88 14 51 13 57 8 78 11 8 9 7 6 91 10 79 0 18 3 51 12 18 1 99 2 33

1 15 4 74 5 31 7 88 14 51 13 57 8 78 11 8 9 7 6 91 10 79 0 18 3 51 12 18 1 99 2 33

2 0 15 1 4 8 82 9 40 12 86 6 50 11 54 13 21 5 6 0 54 2 68 7 82 10 20 4 39 3 35 14 68

1 15 1 4 8 82 9 40 12 86 6 50 11 54 13 21 5 6 0 54 2 68 7 82 10 20 4 39 3 35 14 68

2 0 15 5 73 2 23 9 30 6 30 10 53 0 94 13 58 4 93 7 32 14 91 11 30 8 56 12 27 1 92 3

9 1 15 5 73 2 23 9 30 6 30 10 53 0 94 13 58 4 93 7 32 14 91 11 30 8 56 12 27 1 92 3 9

2 0 15 7 78 8 23 6 21 10 60 4 36 9 29 2 95 14 99 12 79 5 76 1 93 13 42 11 52 0 42 3

96 1 15 7 78 8 23 6 21 10 60 4 36 9 29 2 95 14 99 12 79 5 76 1 93 13 42 11 52 0 42 3 96

2 0 15 5 29 3 61 12 88 13 70 11 16 4 31 14 65 7 83 2 78 1 26 10 50 0 87 9 62 6 14 8

30 1 15 5 29 3 61 12 88 13 70 11 16 4 31 14 65 7 83 2 78 1 26 10 50 0 87 9 62 6 14 8 30

2 0 15 12 18 3 75 7 20 8 4 14 91 6 68 1 19 11 54 4 85 5 73 2 43 10 24 0 37 13 87 9 66

1 15 12 18 3 75 7 20 8 4 14 91 6 68 1 19 11 54 4 85 5 73 2 43 10 24 0 37 13 87 9 66

2 0 15 11 32 5 52 0 9 7 49 12 61 13 35 14 99 1 62 2 6 8 62 4 7 3 80 9 3 6 57 10 7 1 15

11 32 5 52 0 9 7 49 12 61 13 35 14 99 1 62 2 6 8 62 4 7 3 80 9 3 6 57 10 7

2 0 15 10 85 11 30 6 96 14 91 0 13 1 87 2 82 5 83 12 78 4 56 8 85 7 8 9 66 13 88 3 15

1 15 10 85 11 30 6 96 14 91 0 13 1 87 2 82 5 83 12 78 4 56 8 85 7 8 9 66 13 88 3 15

2 0 15 6 5 11 59 9 30 2 60 8 41 0 17 13 66 3 89 10 78 7 88 1 69 12 45 14 82 4 6 5 13

1 15 6 5 11 59 9 30 2 60 8 41 0 17 13 66 3 89 10 78 7 88 1 69 12 45 14 82 4 6 5 13

2 0 15 4 90 7 27 13 1 0 8 5 91 12 80 6 89 8 49 14 32 10 28 3 90 1 93 11 6 9 35 2 73

1 15 4 90 7 27 13 1 0 8 5 91 12 80 6 89 8 49 14 32 10 28 3 90 1 93 11 6 9 35 2 73

(*continued*)

Table 7. (*continued*)

2 0 15 2 47 14 43 0 75 12 8 6 51 10 3 7 84 5 34 8 28 9 60 13 69 1 45 3 67 11 58 4 87
1 15 2 47 14 43 0 75 12 8 6 51 10 3 7 84 5 34 8 28 9 60 13 69 1 45 3 67 11 58 4 87
2 0 15 5 65 8 62 10 97 2 20 3 31 6 33 9 33 0 77 13 50 4 80 1 48 11 90 12 75 7 96 14
44 1 15 5 65 8 62 10 97 2 20 3 31 6 33 9 33 0 77 13 50 4 80 1 48 11 90 12 75 7 96 14 44
2 0 15 8 28 14 21 4 51 13 75 5 17 6 89 9 59 1 56 12 63 7 18 11 17 10 30 3 16 2 7 0 35
1 15 8 28 14 21 4 51 13 75 5 17 6 89 9 59 1 56 12 63 7 18 11 17 10 30 3 16 2 7 0 35
2 0 15 10 57 8 16 12 42 6 34 4 37 1 26 13 68 14 73 11 5 0 8 7 12 3 87 2 83 9 20 5 97
1 15 10 57 8 16 12 42 6 34 4 37 1 26 13 68 14 73 11 5 0 8 7 12 3 87 2 83 9 20 5 97

5 Conclusion

In this paper, an IGA algorithm is presented to solve the benchmark problems designed for the distributed job shop scheduling problem. The IGA algorithm was able to produce new best known solutions of ten benchmark instances, taking less computation time. However, the stability of the algorithm still needs to be strengthened. For the future work, strategy of assigning jobs to suitable factories reasonably can be used to update the quality of optimal solution.

Acknowledgement. This work was supported by the National Key R&D Program of China under Grant No. 2018AAA0101704, National Natural Science Foundation for Distinguished Young Scholars of China (Grant No. 51825502) and the Program for HUST Academic Frontier Youth Team 2017QYTD04.

References

1. Garey, M.R., Johnson, D.S., Sethi, R.: The complexity of flow shop and job shop scheduling. Math. Oper. Res. **1**(2), 117–129 (1976)
2. Wang, L., Shen, W.: Process Planning and Scheduling for Distributed Manufacturing. Springer, London, VVI (2007). https://doi.org/10.1007/978-1-84628-752-7
3. Chaouch, I., Driss, O.B., Ghedira, K.: A survey of optimization techniques for distributed job shop scheduling problems in multi-factories. In: Silhavy, R., Senkerik, R., Kominkova Oplatkova, Z., Prokopova, Z., Silhavy, P. (eds.) CSOC 2017. AISC, vol. 574, pp. 369–378. Springer, Cham (2017). https://doi.org/10.1007/978-3-319-57264-2_38
4. Wang, L., Deng, J., Wang, S.Y.: Survey on optimization algorithms for the distributed shop scheduling. Control Decis. **31**(1), 1–11 (2016)

5. Holland, J.H.: Adaptation in Natural and Artificial Systems. The MIT Press Cambridge, London (1975)

6. Fang, Y., Xiao, X., Ge, J.: Cloud computing task scheduling algorithm based on improved genetic algorithm. In: 2019 IEEE 3rd Information Technology, Networking, Electronic and Automation Control Conference (ITNEC), Chengdu, China, pp. 852–856. IEEE (2019)

7. Chen, R., Yang, B., Li, S., et al.: A self-learning genetic algorithm based on reinforcement learning for flexible job-shop scheduling problem. Comput. Ind. Eng. **149**, 106778 (2020)

8. Ali, K.B., Telmoudi, A.J., Gattoufi, S.: Improved genetic algorithm approach based on new virtual crossover operators for dynamic job shop scheduling. IEEE Access **8**, 213318 (2020)

9. Kurniawan, B., Wen, S., Wei, W., et al.: Distributed-elite local search based on a genetic algorithm for bi-objective job-shop scheduling under time-of-use tariffs. Evol. Intell. **5**,1–15 (2020)

10. Ge, Y., Zhao, Z., Wang, A., et al.: An improved genetic algorithm based on neighborhood search for flexible job-shop scheduling problem. In: 2019 IEEE 10th International Conference on Mechanical and Intelligent Manufacturing Technologies (ICMIMT), Cape Town, South Africa, pp. 142–146. IEEE (2019)

11. Aquinaldo, S.L., Cucuk, N.R.: Yuniaristanto: Optimization in job shop scheduling problem using Genetic Algorithm (study case in furniture industry). IOP Conf. Ser.: Mater. Sci. Eng. **1072**(1), 12–19 (2021)

12. Luo, J., Baz, D.E., Xue, R., et al.: Solving the dynamic energy aware job shop scheduling problem with the heterogeneous parallel genetic algorithm. Future Gener. Comput. Syst. **108**(July), 119–134 (2020)

13. Kumar, P., Ghangas, G., Sharma, A., et al.: Minimising the makespan of job shop scheduling problem by using genetic algorithm (GA). Int. J. Prod. Eng. **6**, 27–39 (2020)

14. Li, X., Gao, L.: An effective hybrid genetic algorithm and tabu search for flexible job shop scheduling problem. Int. J. Prod. Econ. **174**(April), 93–110 (2016)

15. Liang, X., Du, Z.: Genetic algorithm with simulated annealing for resolving job shop scheduling problem. In: 2020 IEEE 8th International Conference on Computer Science and Network Technology (ICCSNT), Dalian, China, pp. 64–68. IEEE (2020)

16. Rafsanjani, M., Riyahi, M.: A new hybrid genetic algorithm for job shop scheduling problem. Int. J. Adv. Intell. Paradigms **16**, 157–171 (2020)

17. Taillard, E.D.: Benchmarks for basic scheduling problems. Eur. J. Oper. Res. **64**(2), 278–285 (1993)

18. Chaouch, I., Driss, O.B., Ghedira, K.: A novel dynamic assignment rule for the distributed job shop scheduling problem using a hybrid ant-based algorithm. Appl. Intell. **49**(5), 1903–1924 (2018). https://doi.org/10.1007/s10489-018-1343-7

19. Jia, H.Z., Nee, A., Fuh, J., et al.: A modified genetic algorithm for distributed scheduling problems. J. Intell. Manuf. **14**(3), 351–362 (2003)

An Improved Teaching-Learning-Based Optimization for Multitask Optimization Problems

Wei Guo, Feng Zou[✉], Debao Chen, Hui Liu, and Siyu Cao

Huaibei Normal University, Huaibei 235000, China

Abstract. In recent years, multitask optimization (MTO) plays an important role in real life, covering various fields such as engineering, finance, and agriculture. MTO can solve multiple optimization problems in the meantime and heighten the performance of solving each task. The teaching-learning-based optimization (TLBO) algorithm and its improved variants focus on solving single problem. In this article, a novel multitask teaching-learning-based optimization (MTTLBO) algorithm is proposed to handle multitask optimization (MTO) problems. Firstly, MTTLBO makes full use of knowledge transfer between different optimization problems to improve performance and efficiency on solving MTO problems. The concept of opposition-based learning (OBL) is introduced into the teaching stage which allow students to receive a wide range of knowledge and find individuals with great differences. MTTLBO is compared with some advanced evolutionary multitask algorithms. The experimental results state clearly that the MTTLBO algorithm owns excellent performance on nine sets of single-objective problems.

Keywords: Multitask optimization · Teaching-learning-based optimization · Opposition-based learning

1 Introduction

The goal of traditional evolution algorithms (EAs) is finding the optimal solution on a single optimization problem [1] and generate better individuals by evolutionary operators. Inspired by multitask learning in machine learning domain [2], evolutionary multitask (EMT) [3] or multifactorial optimization (MFO) [4] are proposed to resolve many optimization problems. The solution of each task was encoded in the same search space which makes full use of the commonness or complementarity between different tasks to transfer positive information between various tasks and promote the evolution of the population. For a certain task, there is some common useful knowledge on solving it, which may help solve another task.

With the increasing diversity and complexity of the optimization tasks, Gupta et al. [4] introduced the idea of multitask optimization (MTO) into evolutionary algorithm, and proposed multifactorial evolutionary algorithm (MFEA). Bali et al. [5] proposed a novel framework in MFEA (MFEA-II), which can adjust the degree of genetic transfer online to reduce the negative interaction between different optimization tasks and enhance the

© Springer Nature Switzerland AG 2021
D.-S. Huang et al. (Eds.): ICIC 2021, LNCS 12836, pp. 48–58, 2021.
https://doi.org/10.1007/978-3-030-84522-3_4

ability of problem solving. Zhou et al. [6] proposed a novel concept based on MFEA. The difference with MFEA is that a better crossover operator was selected flexibly for two parents with different skill factor. Li et al. [7] exploited an explicit multi-population evolutionary framework. The group is divided into multiple groups according to the skill factor and a group is responsible for a task. Then, a novel search engine was used to generate offspring. In addition, by calculating the success rate of information transmission in each generation, the value of random mating probability *rmp* can be flexibly adjusted. Wu et al. [8] applied EA to multitask optimization. The population was divided into multiple groups according to skill factor and some the worst individuals of a certain group was replaced by the best individuals in other groups. Then each group makes full use of information between different tasks to generate new offspring. Liang et al. [9] adopted two advanced strategies in MFEA. If two randomly selected parents with different skill factor, one of the parents is transformed into an individual closing to the other parent. Besides, the idea of OBL was introduced in MFEA. These strategies enhance the knowledge transfer efficiency and perform efficient exploration and exploitation. Gupta et al. [10] introduced the idea of multi-objective optimization (MOO) into the MFEA to deal with MOO problems. Bali et al. [11] established the probability model by integrating the data generated online during multi-task search in order to adjust the degree of genetic transfer online. Yao et al. [12] proposed an advanced strategy in MFEA. By decomposing MOO tasks into a series of single-objective optimization (SOO) tasks, a population could be made full use of optimizing all SOO tasks. Liang et al. [13] developed an advanced algorithm. For two randomly selected parents from different tasks, one of the parents was mapped into search space of the other task. Besides, the appropriate method to generate the offspring vector is based on the information in each generation.

TLBO has been widely used in many fields due to few parameters and high efficiency [14]. However, its research and application are not very mature and have some limitations. So, this paper developed a novel MFO with TLBO by making full use of their advantages to handle MTO problems. In additional, the opposition-based learning mechanism is introduced into the teaching stage to search for more promising positions and hence obtain the better solutions.

The rest of this paper is organized as follows. The concept of multitask optimization is described in Sect. 2. The basic process of TLBO is expounded clearly in Sect. 3. The basic process of MTTLBO is expounded clearly in Sect. 4. Section 5 carries out experiments and data analysis. Some conclusions are given in Sect. 6.

2 Multitask Optimization (MTO)

Multi-objective optimization (MOO) is commonly optimization method [15]. MOO problems only contain one task, which can be used to generate a series of pareto optimal solutions [16]. While the multitask optimization (MTO) problem contains multiple tasks which can be SOO problem or MOO problem. The optimal solutions of all tasks can be solved at the same time [17]. The definition of an MFO problem is provided as follows:

$$\{X_1, X_2, ..., X_k\} = \{\arg\min T_1(X_1), \arg\min T_2(X_2), ..., \arg\min T_k(X_k)\} \quad (1)$$

where X_i denotes an optimal solution of the i-th task $T_i (i = 1, 2, ..., k)$.

3 Teaching-Learning-Based Optimization

TLBO includes two basic processes: In the teaching stage, teachers guide students to improve their grades. In the learning stage, students communicate with each other to check for missing knowledge. The collection of all students is called a class. A certain subject of the student is regarded as a decision variable of the problem. The score of students corresponds to the fitness value. The student with the highest score corresponds to the teacher in the current class.

Teaching Stage: Teachers teach students according to the difference between the average value of themselves and students.

The average level X_{mean} of the class is given by

$$X_{mean} = \frac{1}{m}\left[\sum_{i=1}^{m} x_{i1}, \sum_{i=1}^{m} x_{i2}, ..., \sum_{i=1}^{m} x_{id}\right] \qquad (2)$$

where x_{ij} represents the the j-th dimensional position of the i-th student, and d represents the dimension.

The difference between Teacher and average level of the class is given by

$$difference = rand * (X_{teacher} - TF * X_{mean}) \qquad (3)$$

Then, the updating formula of the i-th student X_i in teaching stage is as follows:

$$X_{i\_new} = X_i + rand * (X_{teacher} - TF * X_{mean}) \qquad (4)$$

where $rand$ is a random value between 0 to 1 and teaching factor TF can be randomly 1 or 2 at the same probability.

Learning Stage: The updating equation of the i-th student X_i in learning stage can be given as follows:

$$\begin{cases} X_{i\_new} = X_i + rand \cdot (X_i - X_k) & if\ f(X_i) < f(X_k) \\ X_{i\_new} = X_i + rand \cdot (X_k - X_i) & otherwise \end{cases} \qquad (5)$$

where $X_{k \neq i}$ is randomly selected from the class, $f(X_i)$ and $f(X_k)$ is the fitness value of the i-th student and k-th student respectively.

After updating students in Teaching stage and Learning stage, their fitness values are recalculated. If the fitness value $f(X_{i\_new})$ is less than $f(X_i)$, then $X_i = X_{i\_new}$. Otherwise, this student will remain the same.

4 Multi-tasking Teaching-Learning Based Optimization

Edified by the multifactorial optimization, MTTLBO combine the main ideas of MFEA with the teaching process and learning process. In this paper, the skill factor τ indicates the task in which a certain student shows the best performance. Then, all students are divided into two groups P_τ and $P_{2/\tau}$ according to the skill factor. To clearly show the process of MTTLBO, its complete framework is given as shown in Fig. 1. The main parts of MTTLBO are introduced clearly as follows.

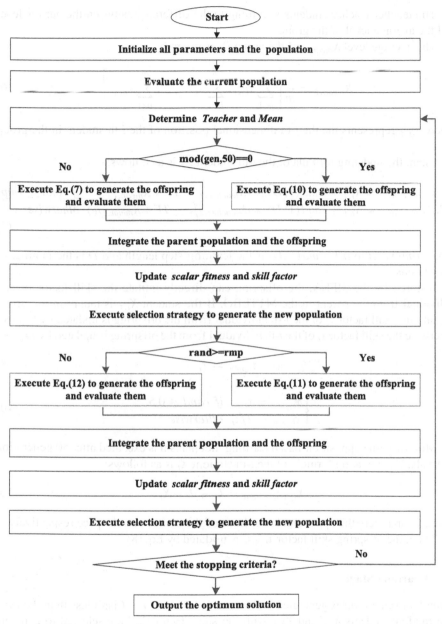

Fig. 1. The complete framework of MTTLBO algorithm

4.1 Teaching Stage

In the process of teaching, the learning strategy of the original teaching stage is executed, and then the opposition-based learning mechanism is executed when the algorithm is executed at intervals of 50 generations.

The teacher teaches students according to the difference between the her/his level and the average level of the group.

The average level $X_{mean,\tau}$ of the group P_τ is given by

$$X_{mean,\tau} = \frac{1}{m}\left[\sum_{i=1}^{m} x_{i1,\tau}, \sum_{i=1}^{m} x_{i2,\tau}, ..., \sum_{i=1}^{m} x_{id,\tau}\right] \qquad (6)$$

where $x_{ij,\tau}$ represents the the j-th dimensional position of the i-th student in the group P_τ.

Then, the updating formula of the i-th student X_i is as follows:

$$\begin{cases} X_{i\_new,\tau\_new} = X_{i,\tau} + rand(1,D).*(X_{teacher,\tau} - TF * X_{mean,\tau}) & if\ rand \geq rmp \\ X_{i\_new,\tau\_new} = X_{i,\tau} + rand(1,D).*(X_{teacher,2/\tau} - TF * X_{mean,2/\tau}) & otherwise \end{cases} \tag{7}$$

where rmp is set to 0.3, $rand(1,D)$ is the learning step length and D is the number of dimensions.

Inspired by the MFEA, the offspring can directly imitate the skill factor of any individual from the parent in the MTTLBO. If the student $X_{i,\tau}$ is taught in the group P_τ, and the skill factor τ_i of the i-th individual from the offspring is updated by Eq. (8), otherwise the skill factor τ_i of the i-th individual from the offspring is updated by Eq. (9).

$$\tau_{i\_new} = \tau_i \tag{8}$$

$$\begin{cases} \tau_{i\_new} = \tau_i & if\ rand \leq 0.5 \\ \tau_{i\_new} = 2/\tau_i & otherwise \end{cases} \tag{9}$$

Moreover, the opposition-based learning mechanism is executed after 50 generation intervals. The updating formula of the i-th student X_i is as follows:

$$X_{i\_new,\tau\_new} = U + L - X_{i,\tau} \tag{10}$$

where U and L are the upper and lower limit of the unified search space respectively.

Then, the offspring skill factor τ_{i_new} is updated by Eq. (8).

4.2 Learning Stage

A random value $rand$ is generated in the range of [0,1]. If $rand$ isn't less than the rmp, two random students $X_{i,\tau}$ and $X_{k,\tau}$ with the same factor skill are selected to generate offspring X_{i_new,τ_new} by Eq. (11).

$$\begin{cases} X_{i\_new,\tau\_new} = X_{i,\tau} + rand(1,D).*(X_{i,\tau} - X_{k,\tau})\ if\ \varphi_i > \varphi_k \\ X_{i\_new,\tau\_new} = X_{i,\tau} + rand(1,D).*(X_{k,\tau} - X_{i,\tau})\ otherwise \end{cases} \tag{11}$$

Where φ_i and φ_k is the scalar fitness of the i-th student and k-th student respectively. Then, the offspring skill factor τ_{i_new} is updated by Eq. (8).

If *rand* is less than the *rmp*, two random students $X_{i,\tau}$ and $X_{k,2/\tau}$ with the different factor skill are selected to generate offspring X_{i_new,τ_new} by Eq. (12).

$$\begin{cases} X_{i\_new,\tau\_new} = X_{i,\tau} + rand(1,D). * (X_{i,\tau} - X_{k,2/\tau}) \text{ if } \varphi_i > \varphi_k \\ X_{i\_new,\tau\_new} = X_{i,\tau} + rand(1,D). * (X_{k,2/\tau} - X_{i,\tau}) \text{ otherwise} \end{cases} \quad (12)$$

Then, the offspring skill factor τ_{i_new} is updated by Eq. (9).

4.3 Selection Mechanism

After the Teaching stage and Learning stage, parent population and the offspring are integrated into a new population M. For a certain task T_j, all individuals are sorted in ascending order according to factorial cost ψ_j^i to obtain factorial rank r_j^i of every individual. Generally, the factorial rank r_j^i can be defined as follows:

$$r_j^i = i, \; i = 1, 2, \ldots, 2N \quad (13)$$

where $2N$ symbolizes the number of the new population.

In multitask environments, the scalar fitness φ_i symbolizes the best factorial rank r_j^i of the i-th student among all the tasks. That is, the scalar fitness φ_i of every student can be defined as:

$$\varphi_i = \frac{1}{\min_{j \in \{1,\ldots,k\}} \left\{ r_j^i \right\}} \quad (14)$$

Finally, N individuals in the next generation are selected from new population M according to their scalar finesses.

5 Simulation Experiment

In order to test the effect of the MTTLBO algorithm, this experiment will use nine sets of SOMTO problems [1], and compare MTTLBO algorithm with SOTLBO, MFEA, MFDE and MFPSO algorithm. In order to make a fair comparison, the population size of all algorithms is set to100. The maximum number of evaluations is used as the stop criterion for terminating the run which is set to 100,000. Then the average results of 10 independent runs are used for comparison. Table 1 shows the parameter settings of all algorithms are as follows.

Table 1. The parameter settings of five algorithm

SOTLBO	MFEA	MFDE	MFPSO	MTTLBO
$TF = 1$ or 2	$RMP = 0.3$	$RMP = 0.3$	$RMP = 0.3$	$RMP = 0.3$
	$\eta_c = 2$	$F = 0.5$	$C_1 = C_2 = C_3 = 0.2$	$TF = 1$ or 2
	$\eta_m = 5$	$PCR = 0.9$	$\omega_{max} = 0.9, \omega_{min} = 0.4$	

Table 2 shows the average value and standard deviation of the objective function values of all algorithms independently run 10 times on the classic SOMTO test suite. Higher average target results are highlighted in bold.

Table 2. Comparison of results obtained by different algorithms on multi-tasking problems

Problem	Task	Metric	SOTLBO	MFEA	MFDE	MFPSO	MTTLBO
CI+HS	Griewank	Mean	3.85E−01	3.59E−01	1.55E−05	3.71E−01	**0.00E+00**
		Std	1.51E−01	5.97E−02	2.65E−05	7.79E−01	**0.00E+00**
	Rastrigin	Mean	3.21E+02	2.18E+02	3.58E−02	2.56E+02	**0.00E+00**
		Std	7.63E+01	4.52E+01	6.93E−02	5.36E+02	**0.00E+00**
CI+MS	Ackley	Mean	7.12E+00	4.65E+00	1.26E−02	2.66E+00	**2.20E−11**
		Std	8.36E−01	8.42E−01	3.86E−02	6.68E+00	**8.93E−12**
	Rastrigin	Mean	3.43E+02	2.18E+02	1.99E−01	2.87E+03	**0.00E+00**
		Std	8.59E+01	5.82E+01	6.30E−01	8.90E+03	**0.00E+00**
CI+LS	Ackley	Mean	2.12E+01	2.02E+01	2.12E+01	**7.21E+00**	1.10E+01
		Std	6.28E−02	1.05E−01	**3.43E−02**	9.63E+00	6.69E+00
	Schwefel	Mean	1.08E+04	**3.61E+03**	1.17E+04	4.82E+03	8.03E+03
		Std	6.79E+02	**3.81E+02**	1.24E+03	6.23E+03	5.53E+03
PI+HS	Rastrigin	Mean	2.94E+02	5.54E+02	8.15E+01	3.87E+02	**5.07E+01**
		Std	6.40E+01	1.32E+02	1.76E+01	1.94E+02	**1.51E+01**
	Sphere	Mean	1.97E+01	8.59E+00	**2.64E−05**	4.64E+03	1.81E+01
		Std	1.28E+01	2.46E+00	**2.49E−05**	4.05E+02	1.50E+01
PI+MS	Ackley	Mean	6.30E+00	3.67E+00	**1.54E−03**	3.67E+00	3.09E+00
		Std	1.51E+00	3.72E−01	**9.00E−04**	5.44E−01	8.14E−01
	Rosenbrock	Mean	9.40E+03	6.26E+02	**8.50E+01**	1.72E+02	5.99E+02
		Std	1.04E+04	1.22E+02	**2.02E+01**	2.08E+02	7.59E+02
PI+LS	Ackley	Mean	6.52E+00	2.00E+01	3.68E−01	1.52E−01	**2.50E−09**
		Std	9.86E−01	8.93E−02	6.15E−01	4.80E−01	**1.88E−09**
	Weierstrass	Mean	6.07E+00	2.11E+01	2.23E−01	1.95E−01	**8.21E−09**

(continued)

Table 2. (*continued*)

		Std	1.48E+00	2.92E+00	4.87E−01	5.85E−01	**1.09E−08**
NI+HS	Rosenbrock	Mean	8.27E+03	1.34E+03	7.87E+01	1.63E+02	**6.36E+01**
		Std	7.54E+03	1.25E+03	**3.15E+01**	2.54E+02	3.28E+01
	Rastrigin	Mean	3.50E+02	3.17E+02	2.14E+01	1.25E+02	**1.30E+01**
		Std	1.05E+02	1.25E+02	**1.63E+01**	1.97E+02	1.75E+01
NI+MS	Griewank	Mean	4.44E−01	4.34E−01	**6.65E−05**	7.30E−01	5.09E−01
		Std	1.77E−01	6.08E−02	**2.60E−05**	3.98E−01	2.83E−01
	Weierstrass	Mean	2.02E+01	2.69E+01	3.51E+00	1.27E+01	**5.46E−05**
		Std	2.42E+00	2.73E+00	1.35E+00	3.74E+00	**4.71E−05**
NI+LS	Rastrigin	Mean	3.16E+02	5.82E+02	9.26E+01	7.98E+02	**8.33E+01**
		Std	7.04E+01	1.10E+02	**1.94E+01**	4.65E+02	2.06E+01
	Schwefel	Mean	1.12E+04	**3.69E+03**	3.97E+03	1.05E+04	1.15E+04
		Std	4.23E+02	4.15E+02	1.09E+03	3.31E+03	**2.12E+02**

Compared with other EMT algorithms, MTTLBO shows excellent performance on 11 tasks out of 18 tasks on the classic single-objective MTO test. Compared with SOTLBO, MTTLBO exhibited superior performance on most of the MFO benchmarks except for Griewank function from benchmark NI+MS and Schwefel function from benchmark NI+LS. MFEA performed better than MTTLBO for Schwefel function from benchmark CI+LS, Sphere function from benchmark PI+HS, Griewank function from benchmark NI+MS and Schwefel function from benchmark NI+LS. MFDE performed better than MTTLBO for Sphere function from benchmark PI+HS, two tasks from benchmark PI+MS, Griewank function from benchmark NI+MS and Schwefel function from benchmark NI+LS. MFPSO performed better than MTTLBO for two tasks from benchmark PI+LS, Griewank function from benchmark NI+MS and Schwefel function from benchmark CI+LS, Rosenbrock function from benchmark PI+LS, and Schwefel function from benchmark NI+LS. Hence, the experimental data confirmed feasibility and superiority of the MTTLBO algorithm in solving MTO problems. In additional, the MTTLBO algorithm shows its advantages in the average target value of all tasks in the classic MTO test problem. Maybe that some better individuals be obtained by the opposition-based learning mechanism which can search for more promising positions when the algorithm fell into the local optima.

To observe and compare the searching efficiency of MTTLBO with those of other algorithms, the convergence curves on all MTO benchmarks are given in Fig. 2. For example, MTTLBO exhibited excellent performance over other EMT algorithms on the benchmarks CI+HS and PI+LS, and MTTLBO fell into local optimum on Schwefel from benchmark NI+LS.

56   W. Guo et al.

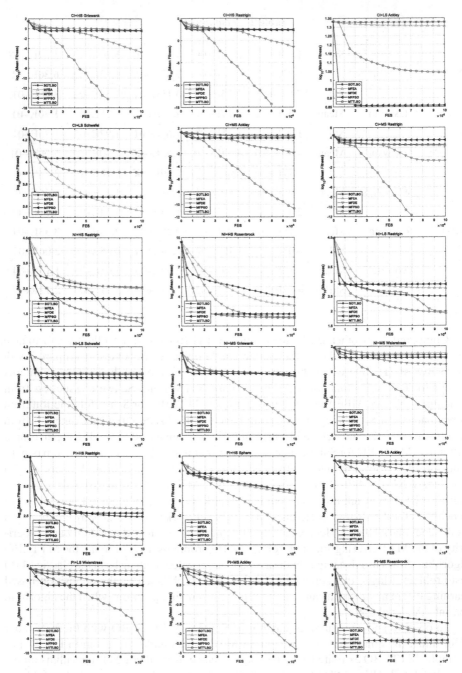

Fig. 2. The convergence curves on all MTO benchmarks obtained by different algorithms

6 Conclusions

In this paper, the MTO idea was introduced into the TLBO algorithm. By introducing opposition-based learning mechanism in the teaching stage, it can prevent the convergence from being too fast and falling into the local optimum. The experimental results show that, compared with the existing multi-tasking evolutionary algorithms, the MTTLBO algorithm achieves satisfactory results on SOO.

Acknowledgment. This work is partially supported by the National Natural Science Foundation of China (Grant No 61976101). This work is also partially supported by the University Natural Science Research Project of Anhui Province (Grant No. KJ2019A0593) and Graduate Education and Teaching Research Project of Huaibei Normal University (Grant No. 2019jyxm01).

References

1. Da, B., Ong, Y.S., Feng, L., et al.: Evolutionary multitasking for single-objective continuous optimization: benchmark problems, performance metric, and baseline results. arXiv:1706. 03470 (2017)
2. Chandra, R., Ong, Y.S., Goh, C.K.: Co-evolutionary multitask learning with predictive recurrence for multi-step chaotic time series prediction. Neurocomputing **243**(21), 21–34 (2017)
3. Ong, Y.S., Gupta, A.: Evolutionary multitasking: a computer science view of cognitive multitasking. Cogn. Comput. **8**(2), 125–142 (2016)
4. Gupta, A., Ong, Y.S., Feng, L.: Multifactorial evolution: toward evolutionary multitasking. Evol. Comput. IEEE **20**(3), 343–357 (2016)
5. Bali, K.K., Ong, Y.S., Gupta, A., Tan, P.S.: Multifactorial evolutionary algorithm with online transfer parameter estimation: MFEA-II. Evol. Comput. IEEE **24**(1), 69–83 (2020)
6. Zhou, L., Feng, L., Tan, K.C., et al.: Toward adaptive knowledge transfer in multifactorial evolutionary computation. Cybern. IEEE. https://doi.org/10.1109/TCYB.2020.2974100
7. Li, G.H., Lin, Q.C., Gao, W.F.: Multifactorial optimization via explicit multipopulation evolutionary framework. Inf. Sci. **512**, 1555–1570 (2020)
8. Wu, R.D., Tan, X.X.: Multitasking genetic algorithm (MTGA) for fuzzy system optimization. Fuzzy Syst. IEEE **28**(6), 1050–1061 (2020)
9. Liang, Z.P., Zhang, J., Feng, L., Zhu, Z.X.: A hybrid of genetic transform and hyper-rectangle search strategies for evolutionary multi-tasking. Expert Syst. Appl. **138**, 1–18 (2019)
10. Gupta, A., Ong, Y.S., Feng, L., Tan, K.C.: Multiobjective multifactorial optimization in evolutionary multitasking. Cybern. IEEE **47**(7), 1–14 (2016)
11. Bali, K.K., Gupta, A., Ong, Y.S., et al.: Cognizant multitasking in multiobjective multifactorial evolution: MO-MFEA-II. Cybern. IEEE **51**(4), 1784–1796 (2021)
12. Yao, S.S., Dong, Z.M., Wang, X.P., Ren, L.: A multiobjective multifactorial optimization algorithm based on decomposition and dynamic resource allocation strategy. Inf. Sci. **511**, 18–35 (2020)
13. Liang, Z.P., Dong, H., Liu, C., et al.: Evolutionary multitasking for multiobjective optimization with subspace alignment and adaptive differential evolution. Cybern. IEEE. https://doi.org/10.1109/TCYB.2020.2980888
14. Rao, R.V., Savsani, V.J., Balic, J.: Teaching-learning-based optimization algorithm for unconstrained and constrained real parameter optimization problems. Eng. Optim. **44**(12), 1447–1462 (2012)

15. Zhang, Q.F., Li, H.: MOEA/D: a multiobjective evolutionary algorithm based on decomposition. Evol. Comput. IEEE **11**(6), 712–731 (2007)
16. Deb, K., Pratap, A., Agarwal, S., Meyarivan, T.: A fast and elitist multiobjective genetic algorithm: NSGA-II. IEEE Trans. Evol. Comput. IEEE **6**(2), 182–197 (2002)
17. Yuan, Y., Ong, Y.S., Liang, F., et al.: Evolutionary multitasking for multiobjective continuous optimization: benchmark problems, performance metrics and baseline results. arXiv:1706.02776 (2017)

Emotion Recognition from Facial Expressions Using a Genetic Algorithm to Feature Extraction

Laura Jalili, Jair Cervantes[✉], Farid García-Lamont, and Adrián Trueba

Computer Sciences Department, UAEMEX (Autonomous University of Mexico State), 56259 Texcoco, Mexico
{lydominguezj,jcervantesc,fgarcial,atruebae}@uaemex.mx

Abstract. In this paper, we present a new method for emotion recognition from facial expressions. The proposed algorithm concentrates on only two specific areas (eyes and mouth), reducing features and descriptors and focusing only on these areas. The algorithm extracts characteristics from these two regions of the face and, in a subsequent process, eliminates the less significant characteristics or those that introduce noise into the classifier. The system allows obtaining a reduced set of features to improve the performance of the classification. In the experiments carried out, we obtain precisions of 99.56%. We evaluated the proposed algorithm on two benchmark datasets; we find that SVM consistently outperforms traditional machine learning techniques.

Keywords: Facial recognition · Vision system · Features selection

1 Introduction

The recognition of emotions is helpful for the identification of human behaviour. Understand human behaviour can help us to solve problems in several disciplines. Some of the applications are used in psychology, robotics, emotions in text, video games, intelligent environments, marketing [1], among many more. Last years, research in this area had increased, which has allowed the emergence of better techniques and the development of techniques to improve precision in systems to recognize emotions. Some research speaks of practical algorithms for improving precision, the aspects that influence are quality, and the extension of the data set, the preprocessing of the images, feature descriptors and the optimization of the parameters of the classifiers [2–4].

The most used techniques to improve precision are the feature selection stage and a fine selection of areas of interest. Several studies have concluded that feature selection intervenes in classification performance [5]. Feature selection techniques allow filtering the most discriminative characteristics and eliminating those with little discriminatory power or features that add noise to the classifier. The adequate feature selection stage has helped to increase the classification precision. Furthermore, the training and testing times can be reduced significantly. On the other hand, the identification of the face [6–8] and select areas of interest also reduces the processing and search for features in the entire image, concentrating only on small areas and reducing the processing significantly.

© Springer Nature Switzerland AG 2021
D.-S. Huang et al. (Eds.): ICIC 2021, LNCS 12836, pp. 59–71, 2021.
https://doi.org/10.1007/978-3-030-84522-3_5

In this paper, we identify the six basic emotions according to Ekman [9]. In the proposal, two areas of interest of the face are selected: the eyes and the mouth. From these areas, we extracted textural features, LBP and HoG. Finally, a genetic algorithm is used as a selector of characteristics. Both techniques allow to significantly reduce processing by considerably reducing the work area and features used. In the experiments carried out, we used two benchmark data sets. In the characteristic extraction stage, we use the HoG, LBP, and Haralick textural descriptors as information to identify emotions, and a genetic algorithm is used to identify the most discriminative features in the data set. The present work makes it possible to focus the search for features on only two areas of the face and reduce the number of features used to only those with high discriminative power. This work's objective is to reduce the region of interest and then reduce features by selecting the most discriminating and finding out the combination of features that can differentiate between types of facial emotions.

The work is structured in the following sections: Sect. 2 presents state of the art in the area of selection of characteristics for the recognition of emotions, analyzing the different selection algorithms, descriptors, and classifiers for the identification of emotions, also, to raise the different applications that these systems have today. In Sect. 3, we show the preliminaries. Section 4 presents our proposed methodology. In Sect. 5, we perform an analysis of the results. Finally, in Sect. 6, we present the conclusions and future work.

2   State of the Art

Identifying emotions is a research topic of great interest due to many implementations in different areas. Identifying the moods in people allows obtaining essential data, such as likes, approvals, rejections, and interest towards a specific situation or objects. These data are valuable because they can be raised in a large number of applications and investigations. The emotion classifier's accuracy depends on the quality of the data sets, the preprocessing, feature extraction techniques, and parameters selected in classifiers. Working with the most discriminative features allows us to reduce the identification time and increase precision. Most of the investigations seek to eliminate redundant features from their dataset by using algorithms. Some of these are genetic algorithms [10], ant colony system (ACS), Cuckoo search, sequential direct selection (SFS) [11]. PSO optimization, among others. As methods of selecting discriminative features. Some investigations perform combinations of metaheuristic search algorithms; for example, in [12], a non-dominated classification genetic algorithm II (NSGA-II) and a Cuckoo binary search for characteristics selection. The authors obtain subsets of the initial population using the Relief algorithm. In their results, the authors report an improvement in its times and an increase in its precision using SVM and treeBagger classification algorithms. However, many investigations are focused only on the genetic algorithm for the selection of the characteristics, in most of the investigations, they propose the extraction of characteristics using popular descriptors such as HoG, LBP, Transformed Fourier, Gabor, among others, and the classifiers most used in research are deep networks and SVMs. Research shows that the selection of characteristics with genetic algorithms decreases the error rate, improving identification [13].

Another essential aspect is the selection of the areas of interest of the face. The muscles of the face have different movements that together express emotions. The emotions are shown in the eyelids when manifesting an eye movement, or an expression of the eyes performs the movement in the muscles around the eyes, lifting the eyebrows, mouth movements, frowning cheek lift, and chin tension [14]. The expression of emotions is transmitted mainly on the face. However, the movement of the eyebrows and nose is showed in the eye area. The nose's movement is shown in the mouth area, and finally, the movement in the chin and cheeks is dependent on the movement applied in the mouth. However, several investigations identify emotions by taking the person's entire face to extract features. Other investigations are based on masks with crucial points on the entire face to identify the relationship of movements and interpret the sets of movements associated with an emotion. In this paper, we use specific areas of the face that we consider discriminatory to identify emotions. In our case, we consider the mouth and eyes as our areas of interest to identify emotions.

3   Proposed Method

Each of the steps of a vision system has a significant impact on its performance; that is, the success of a machine vision system depends on the inter-dependent processes. This paper tries to solve the following problems: extracting information from images from regions of interest, improving performance by reducing the features, and automatically recognizing facial emotions. In this section, we describe in detail the proposed methodology. Figure 1 shows a flow chart of the proposed methodology. The first step detects human faces in the image, in the second step detects the mouth and eyes in the face. The initial detection of the face allows the system to improve the detection of regions of interest. In the third step, the features of the mouth and eyes regions are extracted then a genetic algorithm is used to select the best features. This step allows to eliminate features with low discriminative power and features that introduce noise in the classifier. The genetic algorithm allows getting the best features with the best discriminative power and obtaining the best combination of features. Finally, the vector of selected features is used for the classification of facial emotions. In this paper, we compare the results with several classifiers. Each step is explained in detail in the following subsections.

3.1   Detection of Areas of Interest

Preprocessing is an essential step in face recognition that consists of several techniques such as illumination normalization, alignment, resized and cropped images to assure that the location of the face in a picture is the same in all pictures and reduce the time of training. In this paper, we work with two areas of interest for the identification of emotions. Based on practice and research, we determine that these two areas give the movements of the face. When we are moving the nose, immediately there are movements in the mouth. Similarly, the movement of the forehead or eyebrows is shown in the eye area. In addition, emotions are expressed directly through macro expressions or micro-expressions located in these areas of the eyes and mouth. That is why we decided to focus only on these two areas of interest. Figure 2 shows how the system works in two steps in the proposed method. First, the system works with Haar features to improve the performance of the Viola-Jones algorithm, which performs the identification of the

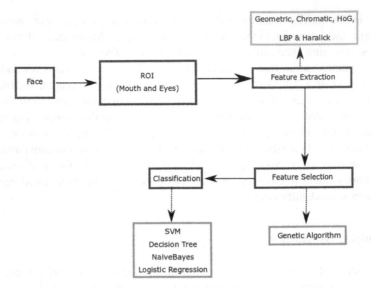

Fig. 1. Proposed Methodology

face. Then we used a cascade-type classifier trained with positive images of the eyes and mouths to automatically select the regions of interest. In this step, HoG and LBP type characteristics are extracted for the training of the cascade classifier. Once we identify the areas of interest, we cut and set aside working only in these areas.

Fig. 2. Selection of regions of interest

3.2 Features Extractors

For each area of interest, we extract Haralick, HoG, and LBP features. We show the arrangement of our feature vectors in Fig. 3, and we carry out a concatenation of our characteristic vectors of each area of interest.

Haralick Texture Features. The Haralick textural features are textural patterns such as roughness, the softness of the area, smooth or thick, wavy, or irregular. These characteristics allow us to identify an object within an image. These characteristics help us identify similar patterns in the object of interest, creating a relationship between shades in grayscale. The texture is obtained according to the consistent distribution of patterns and colours of an object. Haralick's method uses a co-occurrence matrix of gray levels using adjacent pixels. Texture analysis is performed by obtaining the frequency of the levels of pairs in the pixels with different distances and directions. Haralick obtains 14 texture descriptors.

LBP Features. The LBP descriptor is one of the most used techniques for pattern recognition due to its performance; even if there are somewhat complicated environments, LBP is invariant to lighting and translations. LBP works with monotonic changes in the scale of grays. This characteristic allows us to identify the same object or texture even though it has a darker or lighter tonality.

LBP descriptor works in a block size of 3×3. The center of the block is used as a threshold for the remaining pixels in the block and encoded the computed threshold value into a decimal value as follows:

$$LBP = \sum_{i=0}^{P-1} s(n_i - G_c)2^i \tag{1}$$

$$s(x) = \begin{cases} 1, \ if \ x > 0 \\ 0, \ otherwise \end{cases} \tag{2}$$

Where G_c represents the value of the center pixel in the block. P is the number of pixels in the block, n_i represents the ith neighboring pixel. Finally, a histogram of size 2^P is obtained from all the LBP code.

HoG Features. HoG descriptor is a technique for the extraction of shape features. Although it is also used for texture features, the algorithm works with the structure or shape of objects. HoG generally performs processes to obtain edges of objects. HoG gets the edges of the objects in the image. It calculates the magnitude and direction according to the intensity value of each pixel of a small spatial region referred as "cell". Finally, the algorithm obtains a histogram of the magnitude, depending on the direction angle of each pixel. These values allow us to distinguish different local changes around the pixels with the contrast change and the local shape.

Let L be an intensity (grayscale) function describing the image. The image is divided into cells of size n $\times$ n pixels, and the orientation $\theta_{x,y}$ of the gradient in each pixel is computed as follows:

$$\theta_{x,y} = tan^{-1} \frac{L(x, y+1) - L(x, y-1)}{L(x+1, y) - L(x-1, y)} \tag{3}$$

The orientations $\theta_i^j i = 1, \ldots, n^2$ in the cell j are quantized and accumulated into a m-bins histogram.

3.3 Feature Selection

The number of extracted features is enormous and, in some cases, can affect the classification accuracy by introducing noise into the classifier. We decided to apply a genetic

algorithm to select features to eliminate redundant characteristics with low discriminative power or not contribute to the classification precision. One of the most used feature
selection techniques is the genetic algorithm. This is a stochastic method for function
optimization inspired by the mechanics of natural genetics and species evolution. Genetic
algorithms operate on a population of individuals to produce better approximations in
each iteration.

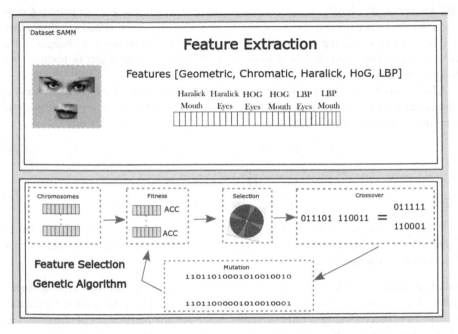

Fig. 3. Feature extraction and features selection of the proposed method.

For this stage, a basic genetic algorithm is implemented, using four different classification methods to obtain subsets of four features. In our problem, we use binary string
encoding, in which a chromosome represents the inclusion/exclusion of the set of features. The chromosome is represented as bit strings of 0's and 1's. In the chromosome, 1
represents that the feature is selected and 0 otherwise. The algorithm starts with a random
population of feature subsets (chromosomes). Each chromosome is evaluated, measuring its ability to predict a set of labels based on the accuracy. The algorithm replaces
the initial population with a new one that includes features from different chromosomes
that present a higher classification accuracy. This process is repeated until a number of
generations are reached, or accuracy is reached. At each generation, a new population
is created to improve the chromosome population by selecting individuals according to
their fitness level and recombining them together using operators of selection, mutation,
and crossover.

In our experiments, use binary encoding, elitism, size of population = 30, number of generations = 100, Crossover operator: 2-point crossover with probability =

0.7, mutation operator: Bit flip with probability $= 0.02$ and roulette wheel as selection operator.

In Fig. 2, we show the features extraction stage of two areas of interest selected and the final arrangement of our characteristic vectors. In this figure, we also mention our feature selection process with the genetic algorithm to optimize our precision and execution time.

3.4 Classification Techniques

In this section we describe the classification techniques used in our experiments.

Support Vector Machines (SVM). The SVM classifier has turned out to be one of the best classifiers used in many applications, and according to recent research, the SVM generally provides good precision for the recognition of emotions. The basic principle of an SVM is to create an optimal hyperplane over a set of data with linearly separable classes.

Logistic Regression. Logistic regression is a predictive algorithm based on the statistics and probability of the data. Logistic regression allows us to work with binary classes using a sigmoid activation function. The logistic regression estimates possible relationships between the dependent and independent variables of the data and defines the possible classification of the data according to the predisposition of the curve of an activation function. When the curve tends to be infinite positive, the classification is 1; otherwise, it is 0.

Random Forest. Decision trees are predictive algorithms, and this algorithm allows us to create a model in the form of a tree with the data. It is based on data impurity metrics and data probabilities preset in a class. The tree is made up of internal nodes and leaf nodes, where the internal nodes are based on the impurity metric; a decision is made, iteratively constructing bifurcations until reaching the leaf nodes where the complete path is associated with a class. Decision trees work by calculating the probability of each class, which is used to obtain a profit from each node. There are pre-labelled data for each of the existing classes to later calculate the impurity of each of the characteristics or variables of the data, being the nodes exhaustive for decision-making and new bifurcations. Decision trees are based on recursion. In each node, data are divided by eliminating characteristics that were already used as results of the nodes.

Naïve Bayes. Naive Bayes is a fast and easy to implement classification technique. The principal advantage of Naïve Bayes is that it only requires a small number of training data to estimate the parameters necessary for classification. On other hand, a naive Bayes classifier considers each of the features to contribute independently. It is a disadvantage because, in most real-life cases, the predictors are dependent, which impacts the classifier's performance. Despite their simplicity, naive Bayes classifiers have worked quite well in many complex real-world applications.

4 Experimental Results

In this section, the parameters selection technique is shown, also data normalization and experimental results obtained with the proposed method.

4.1 Data Sets

In the experiments, we use two data sets that are described as follows.

SAMM Database. The SAMM dataset was created at the University of Manchester is a dataset of actions and spontaneous micro-movements with demographic diversity and is based on the facial action coding system (FACS). This set contains 159 spontaneous micro-facial movements of 32 participants of 13 different ethnic groups, a total of 3634 images, and six emotions (Joy, Surprise, Anger, Disgust, Sadness, and Disgust) [15, 16].

SMIC Database. The SMIC dataset created by the University of Oulu is a spontaneous micro-expressions database, with 164 micro-expressions from 16 people. SMIC contains 1909 images and five emotions (Joy, Surprise, Disgust, Sadness, and Disgust) [17, 18, 19].

4.2 Feature Extraction and Selection

After of selection of the regions of interest, we extract features from the mouth and eyes images. Features are extracted one by one and concatenated into a single feature vector. The final feature vector T was stored in a m $\times$ 3824 size array containing m images with 3824 features. The features are obtained using LBP, textural and HoG features as described above. All the extracted features were normalized with mean zero and standard deviation equal to 1. In this stage, we use a GA to select the best features. For each classifier, we select the best features based on accuracy. In all the experiments, we use binary encoding, elitism, population with 30 individuals, 100 generations, 2-point crossover with probability = 0.7, bit flip mutation with probability = 0.02 and roulette wheel selection operator.

4.3 Parameter Selection

In all used classifiers, optimal parameters were obtained by cross-validation and grid search. Cross-validation is a model validation technique for assessing how the results of a statistical analysis will generalize to an independent data set. On the other hand, grid search exhaustively searches all parameter combinations obtaining the best. For SVM, the regularization parameter C which induced the best average test error was picked from the discrete values $10^{-4}, \ldots, 10^{4}$.

5   Results

In the experiments, all data sets were normalized, and cross-validation was used with k = 10. Table 1 shows the results obtained with 2 datasets. In the Table Acc represents the Accuracy, TP represents the true positives, FP represents the false positives, and Fm represents de f-measure metrics. For each classifier used, accuracies obtained with each individual set of characteristics are reported. The metric used to evaluate the classifier's performance was Accuracy, which is obtained from the classifier hits divided by the total of the data set.

Finally, we show the effectiveness of the proposed method using feature selection and with the entire dataset. Table 1, shows the results obtained with all the features and using only the features selected by the genetic algorithm. The first column describes the results obtained with all features (AF) and the reduced features with the GA (ReF). In the Table, SVM represents the results obtained with Support Vector Machines, in our experiments, we use the Sequential Minimal Optimization algorithm, RF represents the results obtained with Random Forrest, Bayes (Naïve Bayes), and LR represents the results with logistic regression.

In the Table, the maximal predictive accuracy to the SAMM dataset is obtained with SVM classifier 99.56% using the GA to reduce the features, and 98.72% with all the features. On the other hand, the maximal predictive accuracy to SMIC dataset is also obtained with SVM classifier 98.14% using the GA to reduce the features, and 97.68% with all the features.

From the experimental results, one can see that eliminating features in the dataset can significantly improve the performance of the classifiers and make the algorithm fast. Moreover, using individuals with all the features and elitism in the GA, guarantees that the GA improves the performance. The table shows that the results improve less than 1 percentage point when we select features. Only in the SMIC dataset using the Bayesian classifier, the improvement is 6% points.

Figures 4 and 5 show the confusion matrices obtained with the different classifiers (SVM, Bayes, random forest and regression logistic) for SAMM and SMIC datasets. Figure 4 shows that the best-predicted emotions with SVM are dislike and sadness with 100% and the worst predicted is surprise with 99.1%. However, each classifier has its difficulties in predicting some emotions. Figure 5 shows that best-predicted emotions are again obtained using SVM, and it is sadness with 100%, and the worst predicted is happiness with 96.9%. On the other hand, the worst results are obtained with Bayes classifier with the anger emotion as the best classified with 96.5% and sadness emotion as the worst predicted with 71.7%. Figure 5 shows that the worst predicted emotions are again obtained using the Bayes classifier with the SMIC dataset.

Experimental results presented in this section indicate that the proposed method can be employed for face recognition. Compared to the same method with all the features. The proposed method attained the highest classification accuracy using the GA on SAMM and SMIC datasets. In our experiments, we can observe that the use of elitism permits us to improve the accuracy. Due to the above, in the population generation, we generate an individual with all the features. GA starts with the individual with all features. In the worst case, at least the fitness of this individual will be preserved.

Table 1. Performance of the two datasets with different features.

	Subset	SAMM						SMIC					
		Acc	TP	FP	Recall	Fm	ROC	Acc	TP	FP	Recall	Fm	ROC
ReF	SVM	99.559	0.996	0.001	0.996	0.996	1.0	98.14	0.981	0.006	0.98	0.981	0.99
	Bayes	86.075	0.861	0.029	0.861	0.859	0.97	78.29	0.783	0.063	0.78	0.784	0.93
	RF	98.87	0.990	0.003	0.990	0.990	1.0	93.61	0.936	0.021	0.93	0.936	0.99
	LR	99.312	0.993	0.002	0.993	0.993	1.0	95.87	0.959	0.013	0.95	0.959	0.99
AF	SVM	98.721	0.987	0.003	0.987	0.986	0.99	97.68	0.977	0.007	0.97	0.977	0.99
	Bayes	85.828	0.858	0.029	0.858	0.856	0.95	72.76	0.728	0.073	0.73	0.729	0.89
	RF	98.514	0.985	0.004	0.985	0.985	1.0	93.11	0.931	0.023	0.93	0.931	0.99
	LR	98.348	0.983	0.005	0.983	0.981	0.99	95.32	0.953	0.014	0.95	0.953	0.99

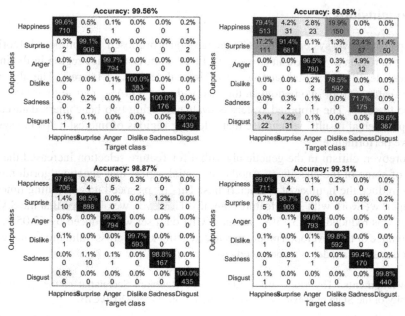

Fig. 4. Confusion matrices of the classifiers with SAMM dataset using reduced features with GA

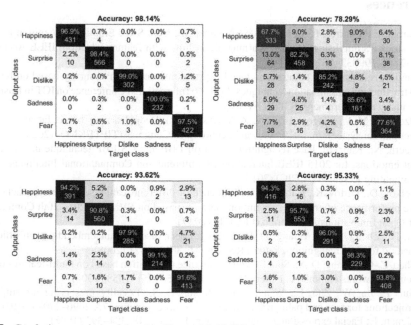

Fig. 5. Confusion matrices of the classifiers with SMIC dataset using reduced features with GA

6 Conclusions

In this paper, a novel system for emotion recognition is proposed. The proposed algorithm selects the region of interest in two stages. First, identify the face region in the image and then identify the mouth and eyes as principal regions of interest. The algorithm extract features from two regions in the face and uses a GA to reduce the search area and the features. The proposed system is evaluated using two benchmark datasets. In the experimental results, we can see that selecting regions of interest helps the system improve performance.

Moreover, elitism in the genetic algorithm for feature selection increased the precision of some of the classifiers. Another essential aspect that genetic algorithm gives us is to reduce the number of characteristics. This helps speed up computing times. In this paper, we work with 3824 features initially. However, we observed that the SAMM data set was reduced by 53.75%, and the SMIC data set was reduced by 52.05%. The performance of the developed model was tested using k fold cross-validation, and results revealed that the model is a good predictor.

Acknowledgments. This study was funded by the Research Secretariat of the Autonomous University of the State of Mexico with the research project 4996/2020/CIB.

References

1. Torres, E.P., Torres, E.A., Hernandez-Alvarez, M., Yoo, S.G.: Emotion recognition related to stock trading using machine learning algorithms with feature selection. IEEE Access **8**, 199719–199732 (2020)
2. Kauser, N., Sharma, J.: Automatic facial expression recognition: a survey based on feature extraction and classification techniques. In: 2016 International Conference on ICT in Business Industry & Government (ICTBIG) (2016)
3. Ozdemir, M.A., Elagoz, B., Soy, A.A., Akan, A.: Deep learning based facial emotion recognition system. In: 2020 Medical Technologies Congress (TIPTEKNO) (2020)
4. Neeru, N., Kaur, L.: Face recognition based on LBP and CS-LBP technique under different emotions. In: 2015 IEEE International Conference on Computational Intelligence and Computing Research (ICCIC) (2015)
5. Belkov, D., Purtov, K., Kublanov, V.: Influence of different feature selection approaches on the performance of emotion recognition methods based on SVM. In: 2017 20th Conference of Open Innovations Association (FRUCT) (2017)
6. Li, B., Zheng, C.-H., Huang, D.-S.: Locally linear discriminant embedding: an efficient method for face recognition. Pattern Recogn. **41**, 3813–3821 (2008)
7. Zhao, Z.-Q., Huang, D.-S., Sun, B.-Y.: Human face recognition based on multi-features using neural networks committee. Pattern Recogn. Lett. **25**, 1351–1358 (2004)
8. Gui, J., Jia, W., Zhu, L., Wang, S.-L., Huang, D.-S.: Locality preserving discriminant projections for face and palmprint recognition. Neurocomputing **73**, 2696–2707 (2010)
9. Ekman, P.: Facial expression and emotion. Am. Psychol. **48**, 384–392 (1993)
10. Cervantes, J., Li, X., Yu, W.: Imbalanced data classification via support vector machines and genetic algorithms. Connect. Sci. **26**, 335–348 (2014)
11. Liogiene, T., Tamulevicius, G.: SFS feature selection technique for multistage emotion recognition. In: 2015 IEEE 3rd Workshop on Advances in Information, Electronic and Electrical Engineering (AIEEE) (2015)

12. Yildirim, S., Kaya, Y., Kılıç, F.: A modified feature selection method based on metaheuristic algorithms for speech emotion recognition. Appl. Acoust. **173**, 107721 (2021)
13. Choudhary, D., Shukla, J.: Feature extraction and feature selection for emotion recognition using facial expression. In: 2020 IEEE Sixth International Conference on Multimedia Big Data (BigMM) (2020)
14. Yuki, M., Maddux, W.W., Masuda, T.: Are the windows to the soul the same in the east and west? Cultural differences in using the eyes and mouth as cues to recognize emotions in Japan and the United States. J. Exp. Soc. Psychol. **43**, 303–311 (2007)
15. Davison, A., Merghani, W., Yap, M.: Objective classes for micro-facial expression recognition. J. Imaging **4**, 119 (2018)
16. Yap, C.H., Kendrick, C., Yap, M.H.: SAMM long videos: a spontaneous facial micro- and macro-expressions dataset. In: 2020 15th IEEE International Conference on Automatic Face and Gesture Recognition (FG 2020) (2020)
17. Li, X., et al.: Towards reading hidden emotions: a comparative study of spontaneous micro-expression spotting and recognition methods. IEEE Trans. Affect. Comput. **9**, 563–577 (2018)
18. Li, X., Pfister, T., Huang, X., Zhao, G., Pietikainen, M.: A spontaneous micro-expression database: inducement, collection and baseline. In 2013 10th IEEE International Conference and Workshops on Automatic Face and Gesture Recognition (FG) (2013)
19. Pfister, T., Li, X., Zhao, G., Pietikainen, M.: Recognising spontaneous facial micro-expressions. In: 2011 International Conference on Computer Vision (2011)
20. Jalili, L.D., Morales, A., Cervantes, J., Ruiz-Castilla, J.S.: Improving the performance of leaves identification by features selection with genetic algorithms. In: Figueroa-García, J.C., López-Santana, E.R., Ferro-Escobar, R. (eds.) WEA 2016. CCIS, vol. 657, pp. 103–114. Springer, Cham (2016). https://doi.org/10.1007/978-3-319-50880-1_10

An Efficient Competitive Swarm Optimizer for Solving Large-Scale Multi-objective Optimization Problems

Yongfeng Li, Lingjie Li, Qiuzhen Lin[✉], and Zhong Ming

College of Computer Science and Software Engineering, Shenzhen University, Shenzhen 510860, China
{qiuzhlin,mingz}@szu.edu.cn

Abstract. Recently, the large-scale optimization problems have become a common research topic in the field of evolutionary computation. It is hard to find optimal solutions when solving large-scale multi-objective optimization problems (LSMOPs), due to the ineffectiveness of existing operators. In the other word, the search ability of most existing MOEAs on solving LSMOPs is still weak. To address this issue, an efficient competitive swarm optimizer with a strong exploration ability, denoted as E-CSO, is presented in this paper, which designs a novel three-particle-based particle updating strategy to improve the search efficiency. The experimental results validate the high efficiency and effectiveness of our proposed approach when solving various LSMOPs.

Keywords: Multi-objective optimization · Competitive swarm optimizer · Large-scale optimization · Particle swarm optimization

1 Introduction

Multi-objective optimization problems (MOPs) are commonly existed in real-world engineering applications [1, 2], which often involve two or more objectives conflicting with each other. There is no one single solution which can optimize all conflicting objectives. Hence, a set of trade-off optimal solutions called Pareto-optimal set (PS) is expected for each MOP and its mapping to the objective space is called Pareto-optimal front (PF). The main goal of solving MOPs is to find a set of solutions that can closely and evenly approximate its true PF. Over the last past several years, multi-objective evolutionary algorithms (MOEAs) have become very popular and shown some advantages for solving MOPs. Most MOEAs can be roughly categorized into three groups based on the used population update principle, including Pareto-based MOEAs [3, 4], decomposition-based MOEAs [5, 6] and indicator-based MOEAs [7, 8], which have been validated to be effective for solving MOPs with a few decision variables.

However, most existing MOEAs show an inefficient performance when solving some MOPs with a large number of decision variables (i.e., more than 100 decision variables), since the search space is exponentially related to the number of decision variables [9].

© Springer Nature Switzerland AG 2021
D.-S. Huang et al. (Eds.): ICIC 2021, LNCS 12836, pp. 72–85, 2021.
https://doi.org/10.1007/978-3-030-84522-3_6

That is, it is very difficult for most existing MOEAs to exploit the optimal solutions in such a huge search space, because they are more likely to encounter some problems that converge prematurely to a local optimum [10]. Recently, a number of studies have been presented to enhance the search ability of MOEAs for solving large-scale optimization problems (LSMOPs). These MOEAs can be roughly divided into the following two categories.

The first category of MOEAs for LSMOPs is based on decision variable grouping, which divides the decision variables into several groups and then optimizes each group of decision variable successively [11]. For example, Jia et al. [12] proposed a new contribution-based cooperative co-evolutionary method, called CBCCO, which can decompose and optimize nonseparable problems with overlapping subcomponents efficiently. In [13], a cooperative coevolution approach called BHCC was proposed, which tolerates decomposition errors and introduces cooperation between overall problem and sub-problems. In [14], an algorithm called HDG was proposed based on a new decomposition method called a deep grouping method, which not only considers the variable interaction but also takes the essentialness of the variable into account. In TRDG [15], when the interaction between two sets is detected, the variables in one of the sets are divided into three subsets based on the trichotomy method, and then the interaction between each subset and the other set is detected. ERDG [16] exploits the historical information on examining the interrelationship of the decision variables of a problem, so it can spend less computation.

The second category of MOEAs for LSMOPs is based on decision variable analysis, which divides the decision variables into different categories according to their contributions to the objective functions and then optimizes each type of decision variable by using different optimization strategies. For example, in [17], an algorithm called MOEA/DVA was proposed, which classifies the variables into position variables, distance variables and mixed variables and then divides the distance variables into a number of independent subcomponents. After that, MOEA/DVA optimizes each subcomponent of distance variables until population converges and then optimizes all the decision variables for better diversity, which can improve the solution quality on most difficult MOPs. In [18], an algorithm called LMEA was proposed based on a decision variable clustering method, which divides the decision variables into convergence-related variables and diversity-related variables, and then optimizes each decision variable by different strategies. In [19], an algorithm called LSMOEA/D incorporates the guidance of reference vectors into the control variable analysis and optimizes the decision variables using an adaptive strategy.

Nevertheless, most existing MOEAs for solving LSMOPs were proposed based on the division of decision variables, which may encounter some problems when solving some problems with complicated PFs or PSs. In addition, few MOEAs for solving LSMOPs optimize all the decision variables simultaneously, due to the inefficient search ability of existing search operators (i.e., simulated binary crossover [20] and differential evolution [21]). More recently, some competitive swarm optimizers (CSOs) have shown a promising performance for solving LSMOPs, because they have a strong exploration ability when compared to the traditional search operators. Inspired from that, we design an efficient competitive swarm optimizer in this paper, denoted as E-CSO, which proposes a novel three-particle-based particle updating strategy with a strong search ability. The main contributions of this paper are introduced as follows:

1) An effective particle updating strategy is proposed to enhance the search efficiency for solving LSMOPs in a huge search space, which involves three particles, including one winner particle (i.e., X_w) and two loser particles (i.e., X_{L_1} and X_{L_2}). Specifically, X_{L_2} is learning from X_w, while X_{L_1} is learning from X_w and X_{L_2} that has been updated by X_w. In this way, two-third of particles have been updated after one generation. That is, the convergence pressure is strengthened by using the proposed particle updating strategy.

2) A large-scale multi-objective CSO algorithm based on the proposed particle updating strategy is also presented in this paper, called E-CSO. The experimental results validate the effectiveness of the proposed E-CSO when compared to four state-of-the-art MOEAs on solving the large-scale multi-objective test problems LSMOP1-LSMOP9 [22].

The remaining parts of this paper are organized as follows. Section 2 introduces the related works which about competitive swarm optimizer (CSO). Section 3 introduces the details of the proposed E-CSO and an analysis on the proposed particle updating strategy. Section 4 provides the empirical results of the proposed E-CSO and four state-of-the-art MOEAs. Finally, the conclusions and future works are given in the Sect. 5.

2 Preliminary

2.1 Related Works of Competitive Swarm Optimizer

In so many popular heuristic algorithms, particle swarm optimization (PSO) is usually used for solving MOPs, because of the strong search ability of PSO and its easy implementation. Generally speaking, PSO is an intelligent optimization algorithm which simulates the behavior of bird flocking. Every particle in PSO stands for a solution, in which its velocity and position information are updated by using the particle update strategy. In these years, there are several PSO algorithms which are presented for solving MOPs [11, 23, 24] and have shown some promising performance.

However, most existing PSOs fail to solve the large-scale MOPs because of lacking of convergence pressure and population diversity. With decision variables increasing, although PSO has a large search space, it still can't solve the problem effectively. More recently, in order to enhance the ability of PSO solving problems that have a larger number of decision variables, a novel PSO variant, namely CSO [9], has been proposed for solving large-scale MOPs, which randomly selects two particles every time and the velocity and position of the worse particle, called loser particle, is updated by the better one, called the winner particle.

Recently, some CSO variants have been proposed, such as TPLSO [25], LLSO [26], SPLSO [27], LMOCSO [28] and so on. Specifically, particles in TPLSO will go through two process, which includes mass learning and elite learning. In mass learning, particles are divided into several groups. In each group, the particles compete with each other and the winner will be saved and enter into elite learning. In elite learning, particles in the swarm are sorted according to their fitness value. After that, each particle needs to learn from two particles which are better than itself. The above two processes form the entire evolution process of TPLSO. In LLSO, the particles are divided into four levels according to their fitness value. Afterwards, particles in each level learn from those particles which are in lower levels. In SPLSO, the entire process mainly includes Segment-Based Learning and Segment-Based Predominant Learning. In LMOCSO, the fitness value of particles in the swarm is first calculated. Then two particles are randomly selected every time. Afterwards, they compete with each other. At last, the worse particle will learn from the better particle.

Algorithm 1: The complete framework of the proposed E-CSO

Input: N (population size), Problem
Output: P (final population)
1    $R \leftarrow$ Uniformpoint (N, Problem.M);
2    $P \leftarrow$ Initpopulation(N);
3   **while** $fes < MaxFes$ **do**
4        $P' \leftarrow$ Updatingparticle(P); //**Algorithm 2**
5        $P \leftarrow$ EnvironmentalSelection($P \cup P'$, R);
6   **end while**
7   **return** P;

---

3 The Proposed E-CSO

3.1 The Complete Framework of E-CSO

The pseudo-code of the complete framework of the proposed E-CSO is given in Algorithm 1, which mainly includes initialization process, the process of updating particles and environmental selection process. The initialization procedure is activated in lines 1–2 of Algorithm 1. To begin with, uniform point function produces N uniform reference points according to the population size (N) and the number of objectives (M). Then, a population P is initialized in line 2. With the initialization completed, as shown in lines 3–6, E-CSO enters into a main loop of evolutionary process, which will be terminated until the fes is greater than or equal to $MaxFes$. In the loop of evolutionary process, the velocities and the positions for the particles in P are updated in line 4 by using the proposed efficient particle updating strategy. The details of the proposed particle updating strategy will be illustrated in Algorithm 2. Afterward, the environmental selection process is performed on the particle swarm, which selects the particles for next generation. Finally, all the particles in P will be outputted as the optimal solutions.

3.2 The Proposed Particle Updating Strategy in E-CSO

The pseudo-code of the proposed particle update process is given in Algorithm 2. Firstly, the fitness value of each particle in P is calculated in line 1, which adopts the shift-based density estimation (SDE) strategy [29] to construct the fitness value for each particle, formulated as below,

$$Fitness(p) = \min_{q \in P \setminus \{p\}} \sqrt{\sum_{i=1}^{M} (\max\{0, f_i(\vec{q}) - f_i(\vec{p})\})^2} \qquad (1)$$

where $f_i(\vec{p})$ denotes the ith objective value of p. Despite of a variety of density estimation techniques, they only measure the density of a solution by estimating its positional relationship with other solutions. However, SDE [29] not only considers the diversity of a solution in the population, but also considers the convergence of a solution in the population. In SDE, when it estimates the density of a solution i, if other solutions perform better than i for an objective, it will shift the solutions to the same position of i on this objective. Therefore, SDE not only reflects the diversity between solutions, but also reflects the convergence between solutions. Only the solution with both good diversity and convergence has a low crowding degree in SDE. The solution with either poor convergence or poor diversity has a high crowding degree in SDE. Besides, a solution with both poor convergence and poor diversity has a highest crowding degree in SDE. Then, in lines 3–13, the updating particle procedure steps into the main loop of updating particles until the i is equal to $|P|/3$. In the loop of updating particles, three particles are randomly selected from P in line 4. Afterward, in line 6, the three particles are sorted according to SDE. Then, X_w, X_{L2} and X_{L1} produced in lines 7–9. In lines 7–9 of Algorithm 2, SDE of the X_w is best among the three particles and SDE of the X_{L2} is in the middle. Therefore, SDE of the X_{L1} is the worst. With X_w, X_{L1}, X_{L2} produced, as shown in line 10, the X_{L2} and the X_{L1} update their velocities and positions using Eq. (2). At last, Offspring is returned as the final solutions. X_{L2} and X_{L1} update their positions and velocities using the following strategy:

$$V_{L_1}(t+1) = r_1 V_{L_1}(t) + r_2(X_w(t) - X_{L_1}(t)) + \psi r_3(X_{L_2}(t) - X_{L_1}(t))$$
$$V_{L_2}(t+1) = r_1 V_{L_2}(t) + r_2(X_w(t) - X_{L_2}(t))$$
$$X_{L_i}(t+1) = X_{L_i}(t) + V_{L_i}(t+1) + r_1(V_{L_i}(t+1) - V_{L_i}(t)) \qquad (2)$$

where r_1, r_2 and r_3 are uniformly randomly distributed values in [0,1]. ψ is a parameter within [0,1] that controls the influence of X_{L2}.

Algorithm 2: Updatingparticle(P)

Input: P
Output: Offspring
1   calculate the fitness (SDE) of each particle in P by (1);
2   Offspring←∅;
3   **for** each particle in P **do**
4        $\{X_w, X_{L_1}, X_{L_2}\}$←randomly select three particles from P;
5   P←$P \setminus \{X_w, X_{L_1}, X_{L_2}\}$;
6   sort the three particles according to SDE fitness value;
7   X_w ← the particle with the best fitness value in (1);
8   X_{L_1}←the particle with the worst fitness value in (1);
9   X_{L_2}←the particle with the middle fitness value in (1);
10   update X_{L_1} and X_{L_2} by using (2);
11   mutate X_{L_1}, X_{L_2} and X_w;
12   Offspring← $\{X_{L_1}, X_{L_2}$ and $X_w\}$ ∪ Offspring;
13   **end for**
14   **return** Offspring;

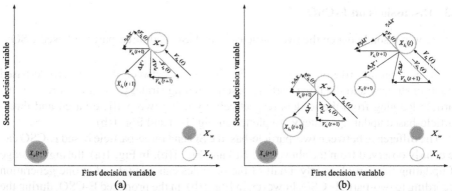

(a) The trajectory of X_{L_2} learning from X_w ; (b) The trajectory of X_{L_2} and X_{L_1} learning from X_w , X_w and X_{L_2} respectively.

Fig. 1. (a) The trajectory of X_{L_2} learning from X_w; (b) The trajectory of X_{L_2} and X_{L_1} learning from X_w, X_w and X_{L_2} respectively.

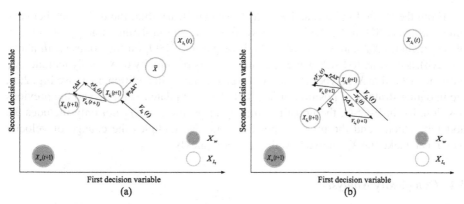

(a) the trajectory of X_{L_2} learning from X_w and $\bar{X}$; (b) the trajectory of X_{L_2} learning only from X_w .

Fig. 2. (a) The trajectory of X_{L_2} learning from X_w and $\overline{X}$; (b) the trajectory of X_{L_2} learning only from X_w

To conclude, E-CSO can accelerate the convergence speed and achieve a stable state faster, because E-CSO have more particles to be updated, which can move toward X_w faster and move toward right directions more quickly. In the meanwhile, the particles learn only from those particles which are better than themselves, which can guarantee that the particles will not move toward wrong directions. More discussions on the proposed E-CSO have been illustrated in Sect. 3.3 as below.

3.3 Discussions on E-CSO

(1) Comparison between the two-particle-based CSO and three-particle-based CSO:

To have an intuitive observation on the difference between the two-particle-based updating strategy and the three-particle-based updating strategy, the trajectories of one particle learning from the leaders respectively by using two-particle-based and three-particle-based updating strategy are plotted in Fig. 1(a) and Fig. 1(b)

The difference between two-particle-based CSO and three-particle based E-CSO can be clearly observed from the above Fig. 1(a) and Fig. 1(b). In Fig. 1(a), during the stage of updating the particles, only a half of the particles can be updated at one generation according to two-particle CSO, however, in Fig. 1(b), in the proposed E-CSO, during the same process, the positions of 2/3 particles can be updated. Evidently, more particles can be updated by using the proposed E-CSO. That is, E-CSO can accelerate the convergence speed. Thereby, more particles can get better to adapt themselves to the environment. Therefore, they can reach at a stable state in a faster speed and with less evaluations.

(2) Comparison between three particles with $\overline{X}$ and three particles without $\overline{X}$

From the above Fig. 2(a) and Fig. 2(b), we can clearly find the difference between three-particle CSO and E-CSO. In above Fig. 2(a), the updating strategy of velocity of X_{L_2} includes X_w and $\overline{X}$ according to three-particle CSO, on the contrary, in above Fig. 2(b), according to E-CSO, the updating strategy of velocity of X_{L_2} only includes X_w with the factor that $\overline{X}$ maybe slow its convergence speed. In addition, in above Fig. 2(a), the updating strategy of position of X_{L_2} includes the updated velocity and the previous position, but in E-CSO, the updating strategy of position of X_{L_2} not only includes the updated velocity and the previous position, but also includes the change of velocity which can make the X_{L_2} move toward X_w more quickly.

3.4 Complexity Analysis

As far as the computational complexity is concerned, from Algorithm 2, we can see that during one generation, it takes $O(NP^2)$ to calculate the fitness of each particle in P in line 1 of Algorithm 2, where NP is the swarm size. During the process of producing X_w, X_{L_2} and X_{L_1}, it takes $O(NP*D)$ in lines 2–9 of Algorithm 2, where D is the dimension of decision variables. In line 10 of Algorithm 2, $O(NP*D)$ is needed to update X_{L_2} and X_{L_1}. At last, it takes $O(NP*D)$ to mutate X_{L1}, X_{L2} and X_w in line 11. To sum up, the complexity of E-CSO is $O(NP^2 + NP*D)$.

As for the space complexity, from Algorithm 2, we can observe that at each generation, it takes $O(NP*D)$ space to calculate the fitness of each particle in P in line 1 of Algorithm 2. Afterwards, $O(NP*D)$ space is needed to produce X_w, X_{L_2} and X_{L_1} in lines 2–9. At last, it takes $O(NP*D)$ space to store the offspring. To sum up, the space complexity of E-CSO is $O(NP*D)$.

To conclusion, E-CSO remains efficient in time complexity and space complexity.

4 Experimental Studies

In this section, several experiments are conducted to investigate the performance of the proposed E-CSO in solving LSMOPs. First, the effectiveness of the proposed particle updating mechanism in E-CSO is empirically studied. Then, the experimental comparisons between the proposed E-CSO and several state-of-the-art MOEAs (i.e., MOEA/DVA [17], LMEA [18], WOF-NSGA-II [30] and LMOCSO [28] on solving LSMOPs are conducted.

4.1 Benchmark Problems and Performance Metrics

Regarding the adopted benchmark problems in our experiment, nine test problems (i.e., LSMOP1-LSMOP9 proposed in [22]) are used for experimental comparison. Among these problems, they have different characteristics. Specifically, LSMOP1-LSMOP4 are constructed by linear variable linkages in their PSs and a liner unit hyper-plane in their PFs, while LSMOP5-LSMOP8 show nonlinear variable linkages in their PSs and a concave unit hyper-sphere in their PFs. LSMOP9 has a nonlinear variable linkage in its PS and disconnected segments in its PF.

Regarding the number of objectives (m) for each test problem, we set $m = 2$ in our experiment. Regarding the number of decision variables (n), it varies from 100 to 500 (i.e., $n = \{100, 200, 500\}$). In addition, the number of subcomponents in each variable group n_k is set to 5.

Regarding the performance metric, the inverted generational distance (IGD) [31] is adopted to assess the performance of compared algorithms in our experiment. Please note that 10000 reference points are sampled on each Pareto front of LSMOP1-LSMOP9 for calculating IGD value, which is based on the methods suggested in [32]. Besides that, all experiments are performed on PlatEMO [33] and the mean and standard deviation of the IGD value are recorded after 20 independent runs by each compared algorithm. The Wilcoxon rank sum test with a significance level of 0.05 is adopted to perform statistical analysis on the experimental results, where "+", "−" and "~" respectively indicate that the result obtained by another compared algorithms is significantly better than, worse than and similar to that obtained by the proposed E-CSO.

4.2 The Compared Algorithms and Experimental Settings

To verify the effectiveness of the proposed E-CSO, four competitive algorithms for solving LSMOPs are included for comparison, including MOEA/DVA [17], LMEA [18], WOF-NSGA-II [30] and LMOCSO [28]. A brief introduction of each compared algorithms is given below:

1) MOEA/DVA [17]: MOEA/DVA introduces an interdependence variable analysis and a control variable analysis, which are used to decompose decision variables into a set of low-dimensional subcomponents and recognize the conflicts among objective function, respectively.
2) LMEA [18]: LMEA proposes a decision variable clustering method, which divides the decision variable into convergence-related variable and diversity-related variable.

3) WOF-NSGA-II [30]: WOF proposes a weight-based optimization framework that uses a problem transformation scheme to reduce the dimensionality of the search space.

4) LMOCSO [28]: LMOCSO proposes a two-stage-based particle updating strategy to update the position for each particle, which preupdates the position of each particle based on its previous velocity and then updates the position of each preupdated particle by learning from a leader particle.

Please note that some unique parameter settings for each compared algorithm are set as suggested in their corresponding references. In addition, some common parameter settings for all compared algorithms are uniformed, which are introduced as below:

1) Population size (N): The population size is set to 300 for all test problems with two objectives.

2) Termination criterion $(MaxFes)$: The maximum number of evaluations $(MaxFes)$ is adopted as the termination criterion for all compared algorithm, which is set to $MaxFes = 15000 \times n$ (n is the number of decision variables).

4.3   Comparisons Between E-CSO and Existing MOEAs

Table 1 provides a comparison of results in terms of the IGD values obtained by each compared algorithm for two-objective LSMOP1-LSMOP9 with 100, 200 and 500 decision variables. It is obvious that the proposed E-CSO showed better overall performance than the other four competitors. Specifically, E-CSO can obtain the best results on 11 out of the 27 cases, while MOEA/DVA, LMEA, WOF-NSGA-II, and LMOCSO only show the best results on 4, 0, 12 and 0 out of 27 cases. From the one-by-one comparisons in the last row of Table 1, E-CSO performed better than MOEA/DVA, LMEA, WOF-NSGA-II and LMOCSO in 18, 22, 15 and 10 out of 27 cases, respectively, and it was only respectively worse in 7, 5, 11 and 5 out of 27 cases. According to the experiments in Table 1, E-CSO performs well in LSMOP1 with 100 and 200 decision variables, LSMOP4, LSMOP5, LSMOP8 with 100 decision variables and LSMOP9 with 100 and 200 decision variables. E-CSO performs not well in LSMOP1 with 500 decision variables, LSMOP2, LSMOP3, LSMOP6, LSMOP7, LSMOP8 with 200 and 500 decision variables and LSMOP9 with 500 decision variables. Especially in LSMOP3, LSMOP6 and LSMOP7 with multimodal landscapes, E-CSO behaves badly mainly due to the lack of diversity. In other LSMOPs, E-CSO shows its superiority because of its fast convergence. Hence, the IGD results summarized in Table 1 has validated the effectiveness of the proposed E-CSO over other competitive MOEAs.

To visually support the above discussion, some final solution sets with the 10th best IGD values from all 20 runs are plotted in Figs. 3, 4, 5 for the nondominated solutions obtained by the five compared MOEAs on two-objective LSMOP9 with 100 decision variables, LSMOP1 with 200 decision variables and LSMOP5 with 500 decision variables based on IGD values. Some conclusions can be easily drawn from these figures. Regarding the different test problems with various decision variables, all the final solution sets obtained by the proposed E-CSO are distributed evenly in these representative problems with different characteristics, while the final solution sets obtained by other competitors show poor distributions. Hence, it is reasonable to conclude that the proposed E-CSO shows an obvious superiority over other compared algorithms in solving LSMOP test problems with different decision variables.

Table 1. IGD values of LMOCSO, WOF-NSGA-II, LMEA, MOEA/DVA, E-CSO on two-objective LSMOP1–LSMOP9, where the best result on each test instance is shown in a gray background

Problem	D	LMOCSO	WOF-NSGA-II	LMEA	MOEADVA	E-CSO
LSMOP1	100	3.0789e-3 (2.10e-4) -	4.0340e-1 (5.07e-2) -	1.4596e-2 (1.05e-3) -	4.7404e-3 (1.57e-4)-	2.1516e-3 (9.32e-5)
LSMOP1	200	4.7952e-3 (4.08e-4) -	4.2686e-1 (5.34e-2) -	1.4194e-1 (2.66e-1) -	5.0800e-3 (1.69e-4)-	3.3666e-3 (4.08e-5)
LSMOP1	500	1.4026e-2 (1.13e-3) -	4.1022e-1 (6.65e-2) -	5.2421e-2 (9.33e-2) -	6.4362e-3 (2.30e-4)+	1.0606e-2 (1.60e-3)
LSMOP2	100	4.8628e-2 (1.66e-2) =	2.5868e-2 (7.71e-3) +	1.2268e-1 (4.41e-2) -	1.8746e-1 (2.11e-3)-	3.9346e-2 (3.93e-3)
LSMOP2	200	4.1959e-2 (1.14e-2) =	1.7943e-2 (5.64e-4) +	1.0960e-1 (3.07e-2) -	1.3855e-1 (7.75e-4)-	3.7025e-2 (1.12e-2)
LSMOP2	500	3.1545e-2 (9.80e-4) +	1.1086e-2 (3.50e-4) +	6.0888e-2 (5.07e-2) -	6.2463e-2 (2.54e-4)-	3.3450e-2 (1.31e-3)
LSMOP3	100	6.9037e-1 (5.17e-2) =	6.2328e-1 (3.64e-2) +	1.8030e+0 (2.50e+0) -	7.8749e-1 (3.37e-2)-	7.0451e-1 (8.08e-3)
LSMOP3	200	7.8224e-1 (2.31e-1) -	6.5356e-1 (2.63e-3) +	3.0858e+0 (4.66e+0) -	8.3542e-1 (4.39e-2)-	7.0712e-1 (3.35e-5)
LSMOP3	500	8.4000e-1 (2.76e-1) =	6.6001e-1 (1.58e-2) +	4.0250e+0 (5.94e+0) -	9.1128e-1 (1.46e-1)=	9.8448e-1 (3.05e-1)
LSMOP4	100	2.7645e-2 (8.49e-4) =	5.0139e-2 (2.49e-3) -	1.4160e-1 (5.48e-2) -	4.4903e-2 (1.12e-2)-	2.6726e-2 (1.00e-3)
LSMOP4	200	1.6989e-2 (4.12e-4) =	3.7614e-2 (2.47e-3) -	8.7735e-2 (4.03e-2) -	4.2358e-2 (9.10e-3)-	1.6227e-2 (2.19e-4)
LSMOP4	500	1.1183e-2 (5.40e-4) =	2.5569e-2 (9.53e-4) -	4.3447e-2 (1.62e-2) -	2.6943e-2 (2.66e-3)-	1.0810e-2 (5.72e-4)
LSMOP5	100	3.9376e-3 (2.89e-4) -	1.3462e-1 (1.01e-1) -	3.7641e-1 (2.98e-1) -	1.0852e-2 (5.52e-4)-	3.2867e-3 (2.58e-4)
LSMOP5	200	5.3896e-3 (5.57e-4) -	7.5729e-2 (3.70e-2) -	4.1010e-1 (2.73e-1) -	1.2730e-2 (1.14e-3)-	3.7467e-3 (3.10e-4)
LSMOP5	500	9.4023e-3 (2.72e-3) -	3.9922e-2 (1.64e-2) -	4.6232e-1 (9.16e-2) -	1.2092e-1 (3.30e-1)-	5.7585e-3 (9.88e-4)
LSMOP6	100	7.5751e-1 (1.50e-2) +	4.4301e-1 (1.60e-1) +	7.6759e-1 (2.16e-1) +	8.6606e-1 (3.01e-2)+	1.0900e+0 (4.85e-2)
LSMOP6	200	7.6025e-1 (6.66e-3) +	4.8044e-1 (1.66e-1) +	7.7327e-1 (1.07e-1) +	6.6489e-1 (1.48e-2)+	9.0912e-1 (1.73e-2)
LSMOP6	500	7.5450e-1 (2.61e-3) +	3.2969e-1 (6.52e-2) +	5.8865e-1 (1.84e-1) +	5.1584e-1 (1.49e-2)+	8.1112e-1 (5.37e-3)
LSMOP7	100	1.2986e+0 (2.11e-1) =	9.3666e-1 (2.37e-1) =	1.2704e+0 (1.64e-1) -	6.0278e+0 (1.08e+0)-	1.1599e+0 (3.01e-1)
LSMOP7	200	1.6460e+0 (3.25e-1) =	9.3576e-1 (1.87e-1) +	1.3548e+0 (2.25e-1) +	4.4612e+0 (1.09e+0)-	1.8357e+0 (4.99e-1)
LSMOP7	500	1.7418e+0 (1.93e-1) =	9.3822e-1 (6.09e-2) +	1.2194e+0 (3.89e-1) +	2.9427e+0 (6.32e-1)-	1.7146e+0 (3.60e-1)
LSMOP8	100	4.7421e-2 (8.33e-3) -	1.2756e-1 (2.89e-2) -	9.0561e-2 (2.51e-2) -	4.5969e-2 (3.72e-3)-	3.9675e-2 (5.63e-3)
LSMOP8	200	5.1568e-2 (3.20e-3) =	6.1941e-2 (1.83e-2) -	6.5953e-2 (2.93e-3) -	4.5608e-2 (2.91e-3)+	4.8581e-2 (3.28e-3)
LSMOP8	500	2.6046e-2 (6.04e-4) =	4.2151e-2 (1.42e-2) -	4.1147e-2 (1.97e-3) -	2.2776e-2 (1.14e-3)+	2.5389e-2 (1.17e-3)
LSMOP9	100	1.5858e-2 (2.85e-3) -	8.1004e-1 (1.17e-16) -	7.1036e-2 (2.77e-1) -	1.1437e-1 (2.11e-2)-	1.0124e-2 (1.60e-3)
LSMOP9	200	9.3795e-2 (6.21e-2) =	8.1004e-1 (3.10e-16) -	6.8305e-1 (2.97e-1) -	8.1648e-2 (1.08e-2)=	6.3346e-2 (5.17e-2)
LSMOP9	500	1.0472e-1 (9.80e-3) +	8.0988e-1 (1.91e-4) -	5.4352e-1 (1.65e-1) -	4.6218e-2 (2.15e-2)+	1.2051e-1 (8.62e-3)
+/-/=		5/10/12	11/15/1	5/22/0	7/18/2	

'+', '-' and '=' indicate that the result is significantly better, significantly worse and statistically similar to that of E-CSO, respectively

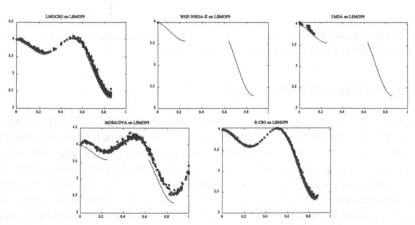

Fig. 3. The final non-dominated solutions obtained by the compared algorithms on two-objective LSMOP9 with 100 decision variables based on IGD values.

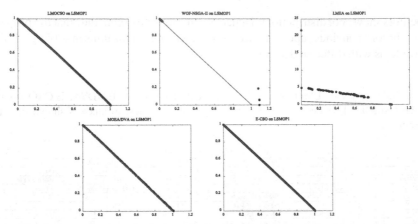

Fig. 4. The final non-dominated solutions obtained by the compared algorithms on two-objective LSMOP1 with 200 decision variables based on IGD values.

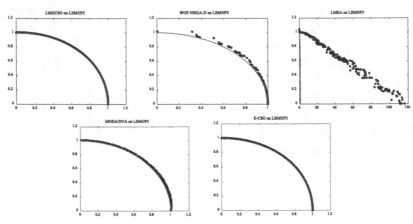

Fig. 5. The final non-dominated solutions obtained by the compared algorithms on two-objective LSMOP5 with 500 decision variables based on IGD values.

5　Conclusions and Future Work

In this paper, an efficient competitive swarm optimizer, namely E-CSO, is presented for solving LSMOPs, which uses a novel three-particle-based particle updating strategy to evolve the swarm. At the beginning of the proposed E-CSO, three randomly selected particles are compared with each other firstly, which is based on the fitness value. Then, these selected particles are marked as one winner particle (i.e., X_w showed the best results on the fitness value), two loser particles (i.e., X_{L_2} ranked the second based on the fitness value and X_{L_1} with the worst fitness value), respectively. After that, the loser particles are learning from the winner particle by using the proposed particle updating strategy. Specifically, X_{L_2} is learning from X_w, while X_{L_1} is learning from both X_w and X_{L_2} that has been updated by X_w already. In this way, there are total two-thirds of particles which were updated after one iteration. Hence, the convergence pressure in E-CSO is strengthened by using the proposed particle updating strategy, which is crucial for solving some LSMOPs with a huge search space. In the other word, the search ability in the proposed E-CSO is improved compared to other traditional CSOs. When compared to five competitive MOEAs (i.e., MOEA/DVA, LMEA, WOF-NSGA-II and LMOCSO), E-CSO showed an obvious superiority over other competitors when solving the LSMOP1-LSMOP9 test problems.

Regarding our future work, 1) the environmental selection in E-CSO will be studied in our future work. 2) The performance of E-CSO on solving some LSMOPs with more than 500 decision variables as well as some LSMOPs with some different characteristics will be further studied. Moreover, 3) the performance of competitive swarm optimizer that combines some transfer learning methods on tackling multi/many-tasking problems will also be considered in our future work.

Acknowledgements. This work was supported by the National Natural Science Foundation of China under Grants 61876110, 61836005, and 61672358, the Joint Funds of the National Natural Science Foundation of China under Key Program Grant U1713212, and Shenzhen Technology Plan under Grant JCYJ20190808164211203.

References

1. Hussain, A., Kim, H.: EV prioritization and power allocation during outages: a lexicographic method-based multi-objective optimization approach. IEEE Trans. Transp. Electrification (2021) https://doi.org/10.1109/TTE.2021.3063085
2. Zhou, J., Gao, L., Li, X.: Ensemble of dynamic resource allocation strategies for decomposition-based multi-objective optimization. IEEE Trans. Evol. Comput. (2021) https://doi.org/10.1109/TEVC.2021.3060899
3. Deb, K., Pratap, A., Agarwal, S., Meyarivan, T.: A fast and elitist multiobjective genetic algorithm: NSGA-II. IEEE Trans. Evol. Comput. 6(2), 182–197 (2002)
4. Bajaj, M., Sharma, N., Pushkarna, M., Malik, H., Alotaibi, M., Almutairi, A.: Optimal design of passive power filter using multi-objective Pareto-based firefly algorithm and analysis under background and load-side's nonlinearity. IEEE Access 9, 22724–22744 (2021)
5. Zhang, Q., Li, H.: MOEA/D. A multiobjective evolutionary algorithm based on decomposition. IEEE Trans. Evol. Comput. 11(6), 712–731 (2007)
6. Ma, X., Yu, Y., Li, X., Qi, Y., Zhu, Z.: A survey of weight vector adjustment methods for decomposition-based multiobjective evolutionary algorithms. IEEE Trans. Evol. Comput. 24(4), 634–649 (2020)
7. Liang, Z., Luo, T., Hu, K., Ma, X., Zhu, Z.: An indicator-based many-objective evolutionary algorithm with boundary protection. IEEE Trans. Cybern. (2020) https://doi.org/10.1109/TCYB.2019.2960302
8. Tian, Y., Cheng, R., Zhang, X., Cheng, F., Jin, Y.: An indicator-based multiobjective evolutionary algorithm with reference point adaptation for better versatility. IEEE Trans. Evol. Comput. 22(4), 609–622 (2018)
9. Cheng, R., Jin, Y.: A competitive swarm optimizer for large scale optimization. IEEE Trans. Cybern. 45(2), 191–204 (2015)
10. Oldewage, E., Engelbrecht, A., Cleghorn, C.: The merits of velocity clamping particle swarm optimisation in high dimensional spaces. In: IEEE Symposium Series on Computational Intelligence (SSCI), Honolulu, USA, pp. 1–8 (2017)

11. Li, X., Yao, X.: Cooperatively coevolving particle swarms for large scale optimization. IEEE Trans. Evol. Comput. **16**, 210–224 (2012)
12. Jia, Y., Mei, Y., Zhang, M.: Contribution-based cooperative co-evolution for nonseparable large-scale problems with overlapping subcomponents. IEEE Trans. Cybern. (2020) https://doi.org/10.1109/TCYB.2020.3025577
13. Ren, Z., Chen, A., Wang, M., Yang, Y., Liang, Y., Shang, K.: Bi-hierarchical cooperative coevolution for large scale global optimization. IEEE Access **8**, 41913–41928 (2020)
14. Liu, H., Wang, Y., Fan, N.: A hybrid deep grouping algorithm for large scale global optimization. IEEE Trans. Evol. Comput. **24**(6), 1112–1124 (2020)
15. Xu, H., Li, F., Shen, H.: A three-level recursive differential grouping method for large-scale continuous optimization. IEEE Access **8**, 141946–141957 (2020)
16. Yang, M., Zhou, A., Li, C., Yao, X.: An efficient recursive differential grouping for large-scale continuous problems. IEEE Trans. Evol. Comput. **25**(1), 159–171 (2021)
17. Ma, X., et al.: A multiobjective evolutionary algorithm based on decision variable analyses for multiobjective optimization problems with large-scale variables. IEEE Trans. Evol. Comput. **20**(2), 275–298 (2016)
18. Zhang, X., Tian, Y., Cheng, R., Jin, Y.: A decision variable clustering-based evolutionary algorithm for large-scale many-objective optimization. IEEE Trans. Evol. Comput. **22**(1), 97–112 (2016)
19. Ma, L., Huang, M., Yang, S., Wang, R., Wang, X.: An adaptive localized decision variable analysis approach to large-scale multiobjective and many-objective optimization. IEEE Trans. Cybern. (2021) https://doi.org/10.1109/TCYB.2020.3041212
20. Deb, K., Agrawal, R.: Simulated binary crossover for continuous search space. Complex Syst. **9**(4), 115–148 (1995)
21. Wang, Y., Liu, H., Long, H., Zhang, Z., Yang, S.: Differential evolution with a new encoding mechanism for optimizing wind farm layout. IEEE Trans. Industr. Inf. **14**(3), 1040–1054 (2018)
22. Cheng, R., Jin, Y., Olhofer, M., Sendhoff, B.: Test problems for large-scale multiobjective and many-objective optimization. IEEE Trans. Cybern. **47**(12), 4108–4121 (2017)
23. Nebro, A., Durillo, J., Garcia-Nieto, J., Carlos A.C.C., Luna, F., Alba, E.: SMPSO: a new PSO-based metaheuristic for multi-objective optimization. In: 2009 IEEE Symposium on Computational Intelligence in Multi-Criteria Decision-Making (MCDM), Nashville, USA, pp. 66–73 (2009)
24. Lin, Q., et al.: Particle swarm optimization with a balanceable fitness estimation for many-objective optimization problems. IEEE Trans. Evol. Comput. **22**(1), 32–46 (2018)
25. Lan, R., Zhu, Y., Lu, H., Liu, Z., Luo, X.: A two-phase learning-based swarm optimizer for large-scale optimization. IEEE Trans. Cybern. (2020) https://doi.org/10.1109/TCYB.2020.2968400
26. Yang, Q., Chen, W., Deng, J., Li, Y., Gu, T., Zhang, J.: A level-based learning swarm optimizer for large-scale optimization. IEEE Trans. Evol. Comput. **22**(4), 578–594 (2018)
27. Yang, Q., et al.: Segment-based predominant learning swarm optimizer for large-scale optimization. IEEE Trans. Cybern. **47**(9), 2896–2910 (2017)
28. Tian, Y., Zheng, X., Zhang, X., Jin, Y.: Efficient large-scale multiobjective optimization based on a competitive swarm optimizer. IEEE Trans. Cybern. **50**(8), 3696–3708 (2020)
29. Li, M., Yang, S., Liu, X.: Shift-based density estimation for Pareto based algorithms in many-objective optimization. IEEE Trans. Evol. Comput. **18**(3), 348–365 (2014)
30. Zille, H., Ishibuchi, H., Mostaghim, S., Nojima, Y.: A framework for large-scale multiobjective optimization based on problem transformation. IEEE Trans. Evol. Comput. **22**(2), 260–275 (2018)

31. Zhou, A., Jin, Y., Zhang, Q., Sendhoff, B., Tsang, E.: Combining model-based and genetics-based offspring generation for multi-objective optimization using a convergence criterion. In: Proceedings IEEE Transactions on Evolutionary Computation, Vancouver, BC, Canada, pp. 892–899 (2006)

32. Tian, Y., Xiang, X., Zhang, X., Cheng, R., Jin, Y.: Sampling reference points on the Pareto fronts of benchmark multi-objective optimization problems. IEEE Congr. Evol. Comput., 1–6 (2018)

33. Tian, Y., Cheng, R., Zhang, X., Jin, Y.: PlatEMO: a MATLAB platform for evolutionary multi-objective optimization [educational forum]. IEEE Comput. Intell. 12(4), 73–87 (2017)

KGT: An Application Mapping Algorithm Based on Kernighan–Lin Partition and Genetic Algorithm for WK-Recursive NoC Architecture

Hong Zhang[1]([✉])[iD] and Xiaojun Wang[2][iD]

[1] Henan University of Economics and Law, Zhengzhou 450046, Henan, China
[2] Beijing Institute of Technology, Beijing 100081, China

Abstract. Some previous researches have explored the application mappings for network-on-chip to reduce the power consumption and the network latency. However, some of these previous application mapping algorithms only find the local optimal result instead of a global best solution. To further save the power consumption and reduce the network latency, we propose a novel application mapping algorithm, called KGT mapping algorithm, for the triplet-based multi-core architecture (TriBA) topology which is WK-recursive based networks well conform to a modular design due to the properties of regularity and scalability. The KGT mapping algorithm exploits the advantage of both the Kernighan–Lin partitioning algorithm and genetic algorithm to reduce the overall power consumption and network latency. The KGT mapping algorithm generates a mapping solution by using KL partitioning algorithm. Next, to avoid the premature phenomena, we use a genetic algorithm to prevent the population trapped in the local optimal solution. Compared to the random mapping algorithm, the evaluation results demonstrate that the KGT mapping algorithm saves the power consumption by 22.75% and reduces the network latency by 17.5% on the average.

Keywords: WK-recursive network · Kernighan–Lin algorithm · Genetic algorithm · Application mapping · On-chip network

1 Introduction and Motivation

Some researches [1, 2] describe that on-chip multi-core processors such as the power consumption with respect to the whole on-chip power are 50%. In fact, there is several factors mainly affect the power consumption such as the network topology, the routing algorithm, application mappings. So a crucial challenge in NoC is how associate the IP cores implementing tasks of an application. This is application mapping which plays a key role to improve the performance of the overall on-chip multi-core architecture.

Several previous application mapping algorithms have been proposed [3–11]. A two steps Integer Linear Programming (ILP) mapping algorithm has been proposed by Ostler for process allocation and data mapping [3]. In [4], authors have proposed the Branch-and-Bound (BB) to topologically find the application mapping by searching the solution in tree branches and bounding unallowable solutions. Zhou et al. have

© Springer Nature Switzerland AG 2021
D.-S. Huang et al. (Eds.): ICIC 2021, LNCS 12836, pp. 86–101, 2021.
https://doi.org/10.1007/978-3-030-84522-3_7

proposed a genetic mapping algorithm which maps application onto NoC optimally with a minimum average delay [5]. Authors [8] have proposed SMAP algorithm which performs application mapping for 2DMesh-based NoC to minimize execution time. A Simulated Annealing (SA) mapping algorithm has been proposed for 2DMesh multi-core architecture topology which minimizes the on-chip area and the maximum network communication bandwidth [10].

In this paper, we propose a mapping heuristic algorithm (KGT mapping algorithm) that is based on Kernighan–Lin (KL) algorithm, genetic algorithm (GA) and the WK-recursive multicore architecture TriBA to reduce the overall network latency and power consumption. KL algorithm can reduce the fact network communication cost by placing frequently communicating cores closely. GA is a kind of mapping algorithm for exploring optimization and searching solutions. TriBA [12, 13] (Triplet-based architecture) is a novelty WK-recursive on-chip multicore architecture with the characteristic of scalability and locality.

2 Related Work

Several previous works have been proposed to use specially designed application mapping algorithms, for example Kernighan-Lin partitioning algorithm and genetic algorithm, to improve the different NoC architectures performance or reduce power consumption.

In [14, 15], Sahu et al. proposed three mapping algorithms LMAP, KLMAP and KMAP which apply for 2DMesh, Butterfly Fat Tree and Mesh of Tree structures respectively. The KL bi-partitioning mapping algorithms are designed to minimize the network latency and power consumption. Authors explored the opportunities in optimizing application mapping based on Kernighan-Lin algorithms for express channel-based on-chip network [16]. Manna et al. presented a KL bi-partitioning based approach to perform mapping the core graph of an application onto 2DMesh-based NoC architecture [17].

However, the KL mapping algorithm has its limitations and the resulted mappings generated by the KL algorithm may not be fully optimal. It differs from KL algorithm, the GA algorithm has been observed to perform better application mappings [18–21]. The key problem of the genetic algorithm is that the algorithm needs to avoid premature convergence to local optima.

A mapping algorithm based on genetic algorithm has been presented by Tosun to solve the energy and communication-aware mapping problems [18, 19]. In [20], authors proposed a novel logistic mapping strategy which employs an adaptive genetic algorithm for homogeneous 3D NoCs. In [21], Zang et al. designed an application mapping algorithm with genetic algorithm and searched for a low-power mapping solutions through adaptive crossover and mutation.

Meanwhile, the KL algorithm is bi-partitioning strategy which recursively partition the core graph to two sub-graphs. So the KL algorithm is only applied to like 2^n size Mesh or BFT architecture, not for new NoC architecture (For example TriBA is 3^n size NoC architecture).

Network-on-Chip (NoC) has been proposed to resolve the network latency and power consumption limitations of bus-based multi-core architecture [22, 23]. Generally, several factors affect the NoC performance and power consumption, such as the network topology, the routing algorithm, application mapping. So the design of network-on-chip (NoC) topology is an important field in the multi-core processor architecture. The triplet-based multi-core architecture (TriBA) is a kind of the multi-core WK-recursive [24, 25] network, which has some advantages such as scalability, regularity, locality and symmetry. The crucial design goal is to limit more the cores communication in local units, reducing the network latency and power consumption.

TriBA is a novel NoC multi-core architecture. Each core is connected with adjacent three neighbor cores. The triplet-based multi-core architecture explores more significantly locality than 2D mesh. Moreover, the remarkable characteristic of TriBA full interconnection flavor can reduce the network communication and network latency. The definition of TriBA topology is represented as follows.

Definition 1: Given a WK-recursive NoC topology with v_n ($n \geq 0$) cores (in here $v = 3$), the core's ID number is encoded in the sequence $a_{n-1}a_{n-2}...a_1a_0$, where $a_i\{1, 2,...,v\}(0 \leq i \leq n_-1)$ which contains the cluster number and the core number after partition at $level_i$ and the value of a_i means the position of the cluster number. The Fig. 1 shows TriBA topology network as n = 1, 2.

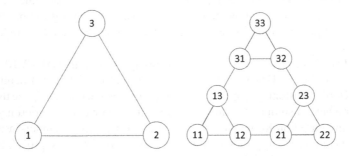

Fig. 1. TriBA multicore network topology with L = 1, 2

TriBA multi-core architecture has smaller degree, less number of total links, smaller network diameter and bisection width than other NoC topologies with the same number of cores [26]. It shows one of TriBA's cores can spend less time to send a message to other cores. Meanwhile, TriBA has a high bisection which means capacity of high network throughout. Table 1 shows the major criteria of TriBA and other NoC topologies.

Based on the mentioned reasons above, the novelty of our proposed KGT mapping algorithm employs the advantages of KL algorithm and GA algorithm for mapping application onto TriBA architecture. Firstly, we adopt tri-partitioning with a Kernighan-Lin algorithm which idea is come from the reference [27]. The modified Kernighan-Lin tri-partitioning algorithm which fits for the triplet-based characteristic of TriBA ensures the communication value among cores in the same partition is maximum value and the communication value among cores between partitions is minimum value. Secondly, we employ a GA algorithm to find the final global optimal mapping. To the best of our knowledge, the KGT mapping algorithm is the first work that employ the modified KL algorithm and GA algorithm onto TriBA, which satisfies the performance requirement of the application mapping and minimizes the average network latency and power consumption. Our experimental results show that, the KGT mapping algorithm reduces the network latency by an average of 17.5% and the power consumption gets decreased by an average of 22.75% respectively than the random mapping algorithm.

Table 1. Several factors of TriBA and general NoC topologies with n cores

Topology	Degree	Bisection width	Network diameter	Total links
TriBA	3	$\log_3 n$	$2^{\log_3 n} - 1$	$3(n-1)/2$
2DMesh	4	$\sqrt{n}$	$2(\sqrt{n}-1)$	$2(n - \sqrt{n})$
Torus	4	$2\sqrt{n}$	$\sqrt{n}$	$2n$
Hypercube	$\log_2 n$	$n/2$	$\log_2 n$	$n\log_2 n/2$

3 Problem Formulations

In this section we describe the power consumption model associated to the application mapping.

3.1 Energy Consumption Model

An energy consumption model has been proposed by Ye et al. for evaluating the power consumption of switch fabrics in network routers [28]. The equation of the bit energy (E_{bit}) is calculated as follows:

$$E_{bit} = E_{Sbit} + E_{Bbit} + E_{Wbit} \tag{1}$$

where E_{Sbit}, E_{Bbit} and E_{Wbit} represent the energy consumed by switch, buffering and interconnection wires, respectively.

However, for the on-chip multi-core architecture, the energy consumption should also include links between nodes. So a modified power consumption model has been proposed by Hu et al. for the on-chip multi-core architecture [4]. Thus, the energy consumption in sending one bit from a tile to its neighboring tile can be calculated as follows:

$$E_{bit} = E_{Sbit} + E_{Bbit} + E_{Wbit} + E_{Lbit} \tag{2}$$

where E_{Lbit} represent the energy consumed by link.

By evaluating the power consumption of all components on-chip multi-core architecture, Hu et al. found that the energy consumed by buffering and internal wires is negligible compared with switch and link. Thus, Eq. (2) can be reduced to:

$$E_{bit} = E_{Sbit} + E_{Lbit} \tag{3}$$

So, the energy consumption of sending one bit from node i to node j can be expressed as follows:

$$E_{bit}^{i,j} = n_{hops} \times E_{Sbit} + (n_{hops} - 1) \times E_{Lbit} \tag{4}$$

where n_{hops} is the number of routers the bit passes on its way along a path.

3.2 Definition of Application Mapping

In the application mapping for on-chip multi-core architecture, the communication bandwidth constraints between tasks are denoted with a core graph while the on-chip network topology is showed with a topology graph.

Definition 2: The task core graph is a directional graph, C(V, E). A vertex $v_i \in V$ indicates a task and the directional edge $e_{i,j} \in E$ represents the communication constraint between the cores v_i and v_j. $Comm_{i,j}$ denotes the weight of edge $e_{i,j}$, which indicates the bandwidth constraints of the communication from vertex v_i to vertex v_j.

Definition 3: The NoC topology graph is a multicore interconnects architecture graph T(U, F). A vertex $u_i \in U$ represents a node in multicore NoC topology and the directed edge $f_{i,j} \in F$ indicates a physical link for directed communicating between the vertices u_i and u_j. $Bw_{i,j}$ denotes the weighted value of the edge $f_{i,j}$, which shows the available communication bandwidth across the edge $f_{i,j}$.

The application mapping algorithm can be formulated by the mapping function:

Mapping algorithm: given a task core graph C(V, E) and the NoC topology graph T(U, F), find the function.

map: $V \rightarrow U$, such that, $map(v_i) = u_j, \forall v_i \in V, \exists u_j \in U, |V| \leq |U|$.

The communication between each pair of task cores is defined as a flow of single commodity [9], showed as $d^k, k = 1, 2, \ldots |E|$. The value of d^k indicates the communication bandwidth across a edge, which is denoted by $vl(d^k)$. The set of all commodities is represented by Dis and is defined as following:

$$Dis = \begin{cases} d^k : vl(d^k) = comm_{i,j}, \ k = 1, 2, \ldots |E|, \ \forall e_{i,j} \in E, \\ with \ source(d^k) = map(v_i), \ dest(d^k) = map(v_j) \end{cases}$$

The communication bandwidth constraints are indicated as following:

$$\forall link \ l_{n,m}, \sum_{k=1}^{|E|} vl(d^k) \times f(l_{n,m}, r_{source}(d^k), dest(d^k)) \leq BW(l_{n,m})$$

where $f(l_{n,m}, r_{i,j}) = \begin{cases} 1, \ if \ l_{n,m} \in L(r_{i,j}) \\ 0, \ otherwise \end{cases}$. If the communication bandwidth constraints are satisfied, the communication cost is given as following:

$$comm \cos t = \sum_{k=1}^{|E|} vl(d^k) \times hopcount(source(d^k), dest(d^k))$$

where $hopcount(a, b)$ is the minimum number of hops between nodes a and b.

4 The Proposed KGT Mapping Algorithm

We present the proposed heuristic-based KGT mapping algorithm in this section, which includes the Kernighan-Lin partitioning algorithm and genetic algorithm to minimize the overall communication cost among all of cores. The goal of KL algorithm is to partition a task graph into subsets recursively and obtain the minimum sum of the communication costs between the subsets. So we use the KL partition algorithm to obtain the first stage optimal mapping solution. At the second stage, we apply the genetic algorithm and obtain the optimal mapping solutions by inheritance, mutation, selection and crossover. When the offspring is not superior to the parent, the algorithm abandon the local best solution and regenerate another initial population from KL algorithm for searching the global optimal solution.

KL partitioning is a bi-partitioning strategy, however, Jesper Larsson [27] shows direct graph k-partitioning strategy. We can apply the tri-partitioning algorithm for dividing the number of cores to an exact power of 3 in core graph. Otherwise, we add some dummy nodes which are connected to all cores and between themselves. The communication costs of edges between cores to dummy node are the value 0 while the values of edges are infinity between dummy nodes.

Algorithm 1: KL_Tri-Partitioning alogrithm

KL_Tri-Partitioning(C)

Input: Core graph C=(V, E)

Output: Partition number of each core at each level of partitioning

begin

        If $|V| \leq 3$ then return

        best_tri-partiton = NIL

        best_cost = ∞

        For count = 1 to L

           tri-partition = KL_Tri(C)

           If ($gain_i$(tri-partition)<best_cost) then

              best_cost = $gain_i$ (tri-partiton)

              best_tri-partiton = tri-partition

           Generate graphs C1, C2 and G3 based on best_tri-partition

           KL_ Tri-Partitioning(C1)

           KL_ Tri-Partitioning(C2)

           KL_ Tri-Partitioning(C3)

end

_____
Algorithm 2: cost calculating algorithm
_____

KL_Tri(C)

    Generate random tri-partition of graph C

    Repeat

        Unlock all nodes

        For i = 1 to |V|/3

            Find three vertices whose exchange makes largest gain in cut cost

            Lock nodes with largest gain and store gain g_i

        Find such that $\sum_{i=1}^{|k|} g_i = Gain_k$ is maximized

        If $Gain_k > 0$ then

        Make first k tentative moves permanent

    Until $Gain_k \leq 0$
_____

The genetic algorithm is a heuristic search algorithm which is stem from the idea of nature evolution. As a result, all generated initial solutions are numbered as chromosomes and a group of chromosomes constitutes a population. The population uses three different genetic operators: (1) selection, (2) crossover and (3) mutation to evolve constantly and become to a better one. In every generation, the first pair of chromosomes is selected randomly from the current population based on the fitness function. The fitness value of every chromosome is calculated by the fitness function. Using the evaluation strategy, the chromosomes which generate better offspring will have higher probabilities to evaluation. Then these selected chromosomes in the first generation are considered as parents. The offspring resembles its parents and obtains the optimal genes from the parent generation. Moreover, to prevent the algorithm trapped in suboptimal solutions, the genetic algorithm uses the mutation operator after crossover operator. Every offspring has a chance to randomly swap several genes in the mutation stage. At last, the new generated population is used in the next iteration. So, a genetic algorithm constantly uses these steps until the termination criterion is satisfied or a maximum number of generations is reached.

The GA algorithm uses the following fitness function to evaluate the probability of an individual's survival in a chromosome. An individual's fitness value is larger which means its genes will more likely survive in the future generations. In KGT mapping algorithm, we adopt the reciprocal of energy consumption function as the fitness function (Table 2).

$$Fitness() = \frac{1}{Commcst} = \frac{1}{E_{total}} = \frac{1}{n_{hops} \times (E_{Sbit} + E_{Lbit})}$$

Algorithm 3: Gene Algotithm

GA function(KL_Tripartition(C))
Input: The new core graph C'= (V, E) by KL_Tripartitioning (C),
        NumIter, SizePop, ProCro, ProMut, TriBASize NoC, ProIni,
Output: The optimal chromosome (the best global solution)
begin
KL_Tripartitioning (C) and initialize the above parameters
for n=1to sizepop
        {find the optimal solution}
calculate the average fitness() of the optimal solution
for i=1 to maxgen
        {select operation of the optimal solution}
crossover operation of the optimal solution
reverse evolution operation of the optimal solution
for j=1 to sizepop
    {
        recalculate average fitness() of the optimal solution
        find the individual chromosome in the best fitness optimal solution
    }

Table 2. The definition of parameters

Parameter	Definition	Value
NumIter	number of iteration	50
SizePop	size of population	20
ProCro	probability of crossover	0.8
ProMut	probability of mutation	0.01
TriBASize	size of TriBA	27/81
ProIni	probability of initial solution	0.05

Algorithm 4: KGT Mapping Algorithm

Now the next task, each of these 3-core subsets is assigned to the appropriate basic unit of a 3^L TriBA architecture, L is the level of the on-chip multicore TriBA architecture. In here, the number of cores is 3^L. Although these 3-core subsets are attached to the nearby basic unit arbitrarily, it is still great opportunity to resolve an optimization solution by the proposed KGT mapping algorithm.

KGTMAP (G)

Input: Core Graph C'= (V, E), Topology Graph T= (U, F)

   Partition_ID (partition number) for each core at each level of KL_Tripartition

   algorithm and genetic algorithm

Output: Addressing number of each core in term of (level, C)

If |V|> |U|

   Mapping not possible

   Exit

For each core v in C

Level(c) ←0

   Position(c) ←0

L←$\lceil \log_3 |V| \rceil$

Mapping (L, C)

if L=1

   then c1=1 c2=2 c3=3

   return

   Mapping (L-1, C)

   {subset1=1 subset2=2 subset3=3

   L=L-1

}

Firstly, we produce a mapping by using KGT mapping algorithm. At each level of tri-partitioning, we assign a partition number 1, 2 and 3 to each subset by turn. These numbers have been utilized in the address assignment process in the KGT mapping algorithm. In the KGT mapping algorithm, these 3-core subsets are assigned according to the output results generated by KL tri-partitioning algorithm. After the mapping algorithm completed, each core has an assigned (level number, subset number) to identify its mapping position on the on-chip multi-core TriBA.

Secondly, the genetic algorithm encodes the population parameters into chromosomes, and uses the iterative way to cross and mutation to swap the chromosome coding numbers. The optimal solutions are close to the most dominant chromosomes step by step, and ultimately reach the optimization target (Figs. 2, 3, 4).

At last the KGT mapping completed, we obtain the best fitness solution. All of cores are mapped to the corresponding position of on-chip multi-core TriBA, meanwhile, those introduced dummy nodes will be removed included the related unnecessary routers and links.

5 Experimentation and Results

5.1 The Evaluation Model

In this paper, we evaluate our proposed KGT mapping algorithm in three aspects, the power consumption, network latency and instructions per cycle (IPC). In Sect. 3, we describe the energy model which is a key role to evaluate the communication cost

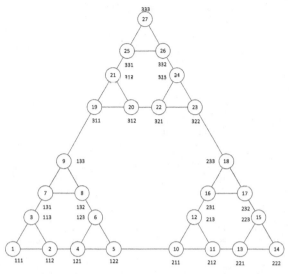

Fig. 2. The chromosome with 27 genes mapping onto TriBA NoC with 27 cores (the initial state)

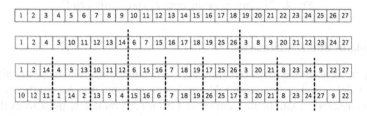

Fig. 3. The partition of 27 genes by KL algorithm

of KGT mapping algorithm. This mapping algorithm includes optimizing both power consumption and network latency. The formula of network latency is described as following:

$$LAT_{avg} = \frac{\sum_{i=1}^{N} LAT_i}{N}$$

where N represents the number of received message packets in certain cycles and LAT_i represents the network latency of the ith packet.

The average number of executed instructions for each certain clock cycle is represented IPC, which is used to evaluate the network throughput performance. The formula of IPC is described as following:

$$IPC = \sum_{i=0}^{n-1} instruction_i / cycle$$

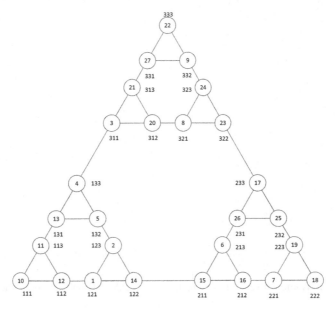

Fig. 4. 27 genes mapping onto 27 cores TriBA NoC

5.2 Simulator and Benchmarks

In this paper, we used Gem5 as our simulator to evaluate the KGT mapping algorithm, which is widely used as a configurable architecture simulator for multicore on-chip architecture-related research. In Gem5, the Orion [29] model is used to evaluate the power consumption of the various NoC topologies. Meanwhile, the benchmarks of PAR-SEC [30] are used in the following experiments. We use the WK-recursive NoC TriBA topology as the NoC topology, which is a regular topology with better NoC topology characteristics such as smaller network diameter, less total links and lower node degree than the 2DMesh topology. We compare the KGT mapping algorithm with several other algorithms on the TriBA NoC architectures: (1) BL_TriBA (the baseline): which maps the tasks onto the TriBA NoC topology randomly; (2) KL_TriBA: KL mapping algorithm on the TriBA structure; (3) GA_TriBA: which is the conventional genetic algorithm on TriBA NoC structure; (4) KGT: our proposed mapping algorithm on TriBA NoC structure.

5.3 Results and Analysis

The power consumption is normalized to the BL_TriBA random mapping algorithm. As shown in Fig. 5, the BL_TriBA random mapping algorithm consumes the highest power consumption while the KGT mapping algorithm has the least power, with an average of 5.21% than the random mapping. Figure 6 shows the experimental results of TriBA's power consumption with 81 cores. In this experimental result, the power savings of KGT mapping algorithm is more significant than in the 27core TriBA NoC architecture in Fig. 5. Overall, KGT mapping algorithm saves power consumption by an average of 22.75% compared to the baseline and achieves better performance compared to KL_TriBA and GA_TriBA.

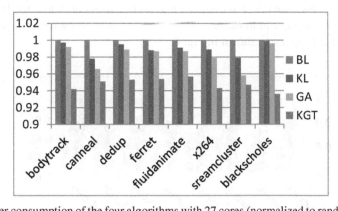

Fig. 5. Power consumption of the four algorithms with 27 cores (normalized to random mapping)

The reason is that KGT mapping algorithm has a smaller chance to get trapped in local optimum than conventional genetic algorithm because we add the jumping strategy in the KGT mapping algorithm. When the offspring are not as good as the parents, the initial population is abandoned to prevent the future iterative and evolutionary process which leads to worthless results. Moreover, KL_Tripartioning mapping algorithm combines the triplet-based characteristic of TriBA to make more communication transfer among three cores which have the characteristic of local full interconnect flavor. In consequence, the solution generated by the KGT mapping algorithm has less network communication cost and lower power consumption than the other mapping algorithm.

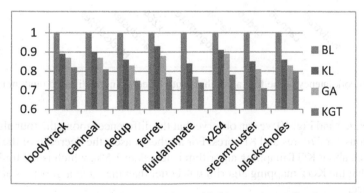

Fig. 6. Power consumption of the four algorithms with 81 cores (normalized to random mapping)

Figure 7 and Fig. 8 show the network latency of TriBA multi-core architecture normalized to the baseline case. The network latency of experimental results varies significantly, due to the various communications characteristics of these applications.

For 27 cores, KL_TriBA, GA_TriBA and KGT mapping algorithm decrease the network latency by the average 2.9, 4.6 and 10.2% respectively, compared to the baseline case. In the experimental result of 81 cores, the differences between four mapping algorithms are more significant because the communication loads between cores are greatly increased. The KGT mapping algorithm decreases the network latency by an average of 17.5% compared to the baseline as shown in Fig. 8.

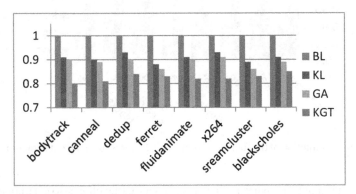

Fig. 7. Network latency of the four algorithms with 27 cores (normalized to random mapping)

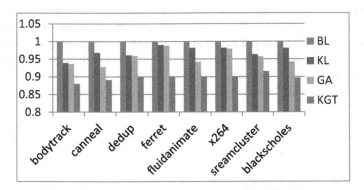

Fig. 8. Network latency of the four algorithms with 81 cores (normalized to random mapping)

From Fig. 9 and Fig. 10 we can observe that the difference among the four algorithms is not obvious for 27cores and 81 cores. For all these cases, the decrease of the network performance about KGT mapping algorithm is less than 1.5%, which is negligible. This indicates that the KGT mapping algorithm is better than the other algorithms at network performance.

Among the four mapping algorithms, KGT mapping algorithm achieves the best result in reducing the network communication cost and power consumption. When the size of the task graph is 27, it only takes several minutes to generate the best solutions, while it takes longer time to achieve the optimal solution with 81 nodes.

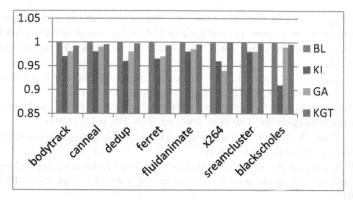

Fig. 9. IPC of the four algorithms with 27 cores (normalized to random mapping)

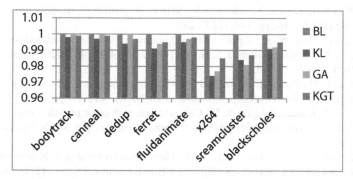

Fig. 10. IPC of the four algorithms with 81 cores (normalized to random mapping)

In summary, the KGT mapping algorithm remarkably decreases the power consumption and network latency by comparing with the random mapping algorithm. Comparing with the KL algorithm and the genetic algorithm, our KGT algorithm achieves better performance.

6 Conclusion

By combining the advantage of the KL algorithm and the genetic algorithm, we proposed a mapping algorithm on TriBA, called KGT mapping algorithm, to explore efficient mapping solutions. Both the power consumption and the network latency have a significant reduction in these experimental results. From the experimental results, we can see that the specific multi-core architecture plays its role to the greatest extent only when the mapping algorithm is designed in combination with the specific multi-core architecture.

References

1. Saeed, K., Raja, G.: Performance evaluation of modern broadwell architecture of a parallel & distributed microprocessor. In: ISWSN 2017, pp. 1–5. IEEE (2017)

2. He, F., Dong, X., Zou, N., Wu, W., Zhang, X.: Structured mesh-oriented framework design and optimization for a coarse-grained parallel CFD solver based on hybrid MPI/OpenMP programming. J. Supercomput. **76**(4), 2815–2841 (2019). https://doi.org/10.1007/s11227-019-03063-6
3. Ostler, C., Chatha, K.S.: An ILP formulation for system-level application mapping on network processor architectures. In: DATE 2007, pp. 99–104. IEEE/ACM (2007)
4. Hu, J., Marculescu, R.: Energy-aware mapping for tile-based NoC architectures under performance constraints. In: ASP-DAC 2003, pp. 233–239. IEEE (2003)
5. Zhou, W., Zhang, Y., Mao, Z.: An application specific NoC mapping for optimized delay. In: DTIS 2006, pp. 184–188. IEEE (2006)
6. Sarzamin, K., Sheraz, A.: An optimized hybrid algorithm in term of energy and performance for mapping real time workloads on 2d based on-chip networks. Appl. Intell. **48**(12), 4792–4804 (2018)
7. Sharma, P.K., Biswas, S., Mitra, P.: Energy efficient heuristic application mapping for 2-D mesh-based network-on-chip. Microprocess. Microsyst. **64**(1), 88–100 (2019)
8. Saeidi, S., Khademzadeh, A., Mehran, A.: SMAP: an intelligent mapping tool for network on chip. In: ISSCS 2007, pp. 1–4. IEEE (2007)
9. Cheng, C.-H., Chen, W.-M.: Application mapping onto mesh-based network-on-chip using constructive heuristic algorithms. J. Supercomput. **72**(11), 4365–4378 (2016). https://doi.org/10.1007/s11227-016-1746-3
10. Harmanani, H.M., Farah, R.: A method for efficient mapping and reliable routing for NoC architectures with minimum bandwidth and area. In: NEWCAS 2008, pp. 29–32. IEEE (2008)
11. Maljaars, J.M., Labeur, R.J., Trask, N.: Conservative, high-order particle-mesh scheme with applications to advection-dominated flows. Comput. Methods Appl. Mech. Eng. **348**(1), 443–465 (2019)
12. Chen, X., Shi, F., Yin, F., Wang, X.J.: A novel look-ahead routing algorithm based on graph theory for triplet-based network-on-chip router. IEICE Electron. Express **15**(4), 78–89 (2018)
13. Chen, X., Shi, F., Yin, F., Wei, Z.H., Wang, X.J.: An efficient distributed minimal routing algorithm for triplet-based WK-recursive network. In: HPCC 2018, pp. 547–554. IEEE (2018)
14. Sahu, P.K., Shah, N., Manna, K., Chattopadhyay, S.: A new application mapping algorithm for mesh based Network-on-Chip design. In: INDICON 2010, pp. 1–4. IEEE (2010)
15. Sahu, P.K., Shah, N., Manna, K., Chattopadhyay, S.: A new application mapping strategy for Mesh-of-Tree based Network-on-Chip. In: ICETECT 2011, pp. 518–523. IEEE (2011)
16. Zhu, D., Chen, L., Yue, S., Pedram, M.: Application mapping for express channel-based networks-on-chip. In: DATE 2014, pp. 1–6. IEEE/ACM (2014)
17. Manna, K., Choubey, V., Chattopadhyay, S., Sengupta, I.: Thermal variance-aware application mapping for mesh based network-on-chip design using Kernighan-Lin partitioning. In: PDGC 2014, pp. 274–279. IEEE (2015)
18. Tosun, S.: New heuristic algorithms for energy aware application mapping and routing on mesh-based NoCs. J. Syst. Archit. **57**(1), 69–78 (2011)
19. Elik, C., Bazlama, S., Neyt, F.: Energy and buffer aware application mapping for networks-on-chip with self-similar traffic. J. Syst. Archit. **59**(10-D), 1364–1374 (2013)
20. Wang, J., Li, L.I., Wang, Z., Zhang, R., Zhang, Y.: Energy-efficient mapping for 3D NoC using logistic function based adaptive genetic algorithms. Chin. J. Electron **23**(1), 254–262 (2014)
21. He, H., Fang, L.: Improved simulated annealing genetic algorithm based low power mapping for 3D NoC. In: EITCE 2018, pp. 402–408. IEEE (2018)
22. Mahendra, C., Gaikwad, A., Patrikar, R.M.: Review of XY routing algorithm for network-on-chip architecture. Int. J. Comput. Appl. **2**(1), 48–52 (2016)

23. Upadhyay, M., Shah, M., Bhanu, P.V.: Fault tolerant routing methodology for Mesh-of-Tree based Network-on-Chips using local reconfiguration. In: HPCS 2018, pp. 570–576. IEEE (2018)
24. Wang, Y.C., Juan, S.T.: Hamiltonicity of the basic WK-recursive pyramid with and without faulty nodes. Theoret. Comput. Sci. **562**(1), 542–556 (2015)
25. Lan, T., You, J.: Super spanning connectivity on WK-recursive networks. Theoret. Comput. Sci. **713**(2), 42–55 (2018)
26. Deng, N., Ji, W., Li, J., Zuo, Q.: A semi-automatic scratchpad memory management framework for CMP. In: Temam, O., Yew, P.-C., Zang, B. (eds.) APPT 2011. LNCS, vol. 6965, pp. 73–87. Springer, Heidelberg (2011). https://doi.org/10.1007/978-3-642-24151-2_6
27. Tr, L., Ff, J.: Direct graph k-partitioning with a Kernighan-Lin like heuristic. Oper. Res. Lett. **34**(6), 621–629 (2006)
28. Ye, T.T., Benini, L., Micheli, G.D.: Analysis of power consumption on switch fabrics in network routers. In: DAC 2002, pp. 524–529. IEEE (2002)
29. Xu, T.C., Leppänen, V.: Cache- and Communication-aware application mapping for shared-cache multicore processors. In: Pinho, L.M.P., Karl, W., Cohen, A., Brinkschulte, U. (eds.) ARCS 2015. LNCS, vol. 9017, pp. 55–67. Springer, Cham (2015). https://doi.org/10.1007/978-3-319-16086-3_5
30. Wang, Y., Zhang, D.: Data remapping for static NUCA in degradable chip multiprocessors. IEEE Trans. Very Large Scale Integr. Syst. **23**(5), 879–892 (2015)

Evolutionary Algorithms for Applications of Biological Networks: A Review

Gufeng Liu[1], Qunfeng Liu[2], Lijia Ma[1(✉)], and Zengyang Shao[1]

[1] College of Computer Science and Software Engineering, Shenzhen University, Shenzhen 518060, China
ljma1990@szu.edu.cn
[2] School of Computer Science and Technology, Dongguan University of Technology, Dongguan, China
liuqf@dgut.edu.cn

Abstract. With the rapid development of next-generation sequencing and high-throughput technologies, much biological data have been generated. The analysis of biological networks is becoming a hot topic in bioinformatics in recent years. However, many structure analyzing problems in biological networks are computationally hard, and most of heuristic algorithms cannot obtain good solutions. To solve this difficulty, many evolutionary algorithms have been proposed for analyzing the structures in biomedical fields. In this paper, we make a brief review of evolutionary algorithms for three common applications in biological networks such as protein complex detection, biological network alignment and gene regulatory network inference. Moreover, we give some discussions and conclusions of evolutionary algorithms for structure analyses in biological networks.

Keywords: Biological network · Evolutionary algorithm · Gene regulatory network · Protein-protein interaction

1 Introduction

In the past decades, with the continuous development of next-generation sequencing and high-throughput technologies, much biological data have been generated. With the completion of the sequencing of the human genome, the research of proteomic [1] has become an important area in bioinformatics. Proteins usually interact with each other to incorporate into a protein complex to accomplish biological functions. In recent years, biologists have found that cellular functions and biochemical events are coordinately carried out by groups of proteins interacting each other in functional modules. Detecting these functional modules or protein complexes in PPI networks is essential for understanding the structures and functions of fundamental cellular networks [2]. Moreover, understanding networks can help individuals understand the mechanism of cellular organization, process and functions [3]. Therefore, a comparative analysis of PPI networks across different species is necessary. Biological network alignment compares PPI networks over multiple species, aiming to find an optimal node mapping between PPI

© Springer Nature Switzerland AG 2021
D.-S. Huang et al. (Eds.): ICIC 2021, LNCS 12836, pp. 102–114, 2021.
https://doi.org/10.1007/978-3-030-84522-3_8

networks. The studies on biological network alignment be used to guide the knowledge transfer between conserved regions of molecular networks of different species [4]. On the other hand, with the development of DNA microarray technology, a large number of gene expression data have been generated [6]. Gene regulatory network inference plays an important role in biomedical research as the interactions among genes control most of the biological activities.

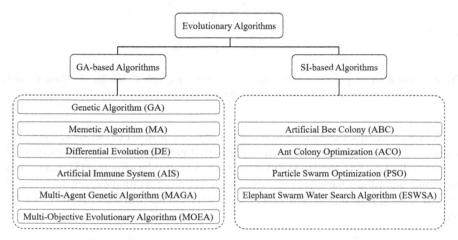

Fig. 1. Summary of the reviewed evolutionary algorithms

Although the above-mentioned biological tasks can be solve using high throughput experiments, these experiments are highly time-consuming. Evolutionary algorithms (EAs) have been proved to be efficient for many problems of engineering and computer science. EA is a subclass of evolutionary computation and belongs to the set of general stochastic search algorithms [6]. EAs can be classified from many perspectives, here we classify the EAs into two categories that are associated with Genetic Algorithm (GA-based) and Swarm Intelligence (SI-based) respectively. In this paper, we mainly review EAs for complexes detection, biological network alignment and gene regulatory network inference. A summary of related EAs is given in Fig. 1, while the related reviewed works are summarized as shown in Table 1.

The rest of this paper is organized as follows. Sections 2, 3 and 4 give a brief review of the applications of EAs in protein complex detection, biological network alignment and gene regulatory network inference, respectively. Sections 5 and 6 provide discussions and conclusions about the existing challenges and future directions of evolutionary algorithms for structure analyses in biological networks, respectively.

2   Applications in Protein Complex Detection

2.1   Definition of Protein Complex Detection

The detection of protein complexes in a PPI network is essentially a clustering problem, which illustrates in Fig. 2. A PPI network $\mathcal{G}$ is mathematically illustrated by an undirected graph $\mathcal{G} = (\mathcal{V}, \mathcal{E})$ with set of vertices $\mathcal{V}$ and edges $\mathcal{E} = (u, v) : u, v \in \mathcal{V}$. Each vertex is corresponded to a protein in PPI network while each edge connects two proteins that have interaction in biological networks. In some cases, the weights of edges describe the properties of communication among the PPI network.

Fig. 2. Illustration of protein complex detection. The subgraph $\mathcal{C}$ insides the dashed circle corresponds to a protein complex in an PPI network.

A protein complex corresponds to a subgraph $\mathcal{C} \in \mathcal{G}$ in a PPI network, formed by a dense cluster of proteins connected through multiple protein-protein interactions [7, 8]. Generally, protein complex in PPI networks is a molecular structure [8]. Many computational algorithms have been proposed to detect protein complexes, while network clustering is equivalent to the problem of subgraph isomorphism which is NP-Complete [9].

2.2 GA-Based Approaches

Pizzuti and Rombo [10] adopted Genetic Algorithm (GA) to discover clusters in PPI networks by different topological fitness functions, and experimental results demonstrate that GA is a competitive computational technique to deal with the problem of protein complex detection. Later they proposed a new clustering approach GA-RNSC [11] combining GA with the Restricted Neighborhood Search Clustering (RNSC) [12]. BiCAMWI [13] uses GA to extract the biclusters in the preprocessing phase and is able to detect protein complexes in dynamic PPI networks. GACluster [14] designs a new objective function to maximize intra-cluster cohesion and minimize inter-cluster coupling in a PPI network. It excludes the crossover operator to avoid disturbing the potentially good solutions. Besides, a modified version of the mutation operator from PROCOMOSS [15] has been adopted to reach a better solution with the obtained individuals. EGCPI [16] takes both topologies and protein attributes into consideration with an evolutionary clustering strategy. GA-PPI-Net [17] introduces a specific mutation operator to the GA for the detection of communities in the protein-protein or gene-gene interaction networks. Abduljabbar et al. [18] developed a heuristic operator based on Gene Ontology (GO) [19] for local optimization in order to explore the search space more effectively.

MGOC [20] is a bi-objective clustering method to detect protein complexes which adopts both topological properties of PPI networks and biological properties based on GO as objective functions. It adopts NSGA-II [21] as an underlying multi-objective framework to find the Pareto solutions. Similarly, bi-objective approaches [7, 15, 22] have been proposed for protein complex detection. Attea and Abdullah [23] as well as

Abdulateef et al. [24] modeled the complex detection problem as two optimization functions namely IntraComplex score and InterComplex score to measure the topological quality of complexes. On the other hand, MOBICLUST [25] presents three objective functions related to the bicluster properties to obtain dense biclusters. MOFPGA [26] is based on three network topological properties to optimize the density, size and characteristic path length of complexes. MODAPROC [27] consists of an objective function associated with disease to discover obscure disease specific gene and protein clusters.

In addition to MOEA approaches, MAE-FMD [2] is a global search algorithm based on a multi-agent evolutionary method. It first employs a group of agents as a population to carry out random walks from a start protein to other proteins in a PPI network and obtain their individual solution encodings. Then, it randomly places these agents into an evolutionary environment and performs three innovative agent-based operations to increase the energy levels of agents at each iteration. Fuzz-mod-DE [28] utilizes differential evolution (DE) and fuzzy membership induced ranking technique to predict protein complexes. The first phase mainly optimizes the individual objectives independently by DE while the second phase selects the unique solution to represent the protein complex in a PPI network.

2.3 SI-Based Approaches

IQ-Flow [29] is a functional-flow algorithm based on Quantum-behaved Particle Swarm Intelligence (PSO) to detect protein complexes. It is able to determine the optimum threshold when calculating the lowest similarity between modules. Sharafuddin et al. [30] developed a PSO based method which evaluates each subgraph with a cluster score as fitness function. Ji et al. proposed several Ant Colony Optimization (ACO) based algorithms for detecting functional modules in PPI networks. NACO-FMD [31] combines topological properties with functional information to conduct the ants searching effectively in the space of solutions. It is able to overcome noise to some extent by computing distances between each pair of proteins. ACO-MAE [32] is improved by combining ACO with a multi-agent concept which could avoid local optima to enhance search performance. HAM-FMD [33] is an improved version of ACO-MAE which adopts a different approach to produce clusters and applies a limited neighborhood into ACO to reduce the number of candidate nodes, which is more effective to detect the functional modules in PPI networks.

3 Applications in Biological Network Alignment

3.1 Definition of Biological Network Alignment

The goal of the Biological Network Alignment (BNA) problem is to align proteins across different biological networks based on their topological and biological similarity. We present the formal definition of the pairwise BNA in this section and it can be generalized to the multiple BNA problem. Given two undirected graph $\mathcal{G}_1 = (\mathcal{V}_1, \mathcal{E}_1)$ and $\mathcal{G}_2 = (\mathcal{V}_2, \mathcal{E}_2)$, where $\mathcal{G}_1$ and $\mathcal{G}_2$ consist of $m = |\mathcal{V}_1|$ and $n = |\mathcal{V}_2|$ vertices respectively. Each vertex $v \in \mathcal{V}_i$ is corresponded by a protein in PPI network and each

Table 1. Summary of the related evolutionary algorithms.

Protein Complex Detection				Genetic Regularity Network Inference			
Categories	Algorithms	Frameworks	Year	Categories	Algorithms	Frameworks	Year
GA-based	Abduljabbar et al. [18]	GA	2020	GA-based	Wang and Mohan [34]	GA	2019
	GA-PPI-Net [17]	GA	2019		GABNI [35]	GA	2018
	EGCPI [16]	GA	2016		HPGA [36]	GA	2018
	GACluster [14]	GA	2016		PCA-CMI [37]	GA	2017
	BiCAMWI [13]	GA	2016		RCGA$_D$ [38]	GA	2015
	GA-RNSC [11]	GA	2014		pithEBO-FCM [39]	MOEA	2020
	Pizzuti and Rombo [10]	GA	2012		MONET [40]	MOEA	2019
	Abdulateef et al. [24]	MOEA	2018		EMOEA$_{FCM}$-GRN [73]	MOEA	2019
	Attea and Abdullah [23]	MOEA	2018		Ren et al. [41]	MOEA	2016
	MODAPROC [27]	MOEA	2016		MAGARF$_{FCM}$-GRN [42]	MAGA	2018
	Bandyopadhyay et al. [7]	MOEA	2015		dMAGA-FCM$_D$ [43]	MAGA	2017
	MOEPGA [26]	MOEA	2015		dMAGA$_{FCM}$-GRN [44]	MAGA	2016
	Ray et al. [22]	MOEA	2013		MALASSO$_{RNN}$-GRN [45]	MA	2020
	MOBICLUST [25]	MOEA	2013		Fefelov et al. [46]	DE	2018
	PROCOMOSS [15]	MOEA	2012				
	MGOC [20]	MOEA	2012				
	Fuzz-mod-DE [28]	DE	2019				
	MAE-FMD [2]	MAGA	2014				
SI-based	Sharafuddin et al. [30]	PSO	2013				
	IQ-Flow [29]	PSO	2012				
	HAM-FMD [33]	ACO	2013				
	ACO-MAE [32]	ACO	2012				
	NACO-FMD [31]	ACO	2012				
Biological Network Alignment					Hurtado et al. [47]	PSO+MOEA	2020
GA-based	multiMAGNA++ [48]	GA	2018		RMPSO [49]	PSO	2019
	DynaMAGNA++ [50]	GA	2017		Sultana et al. [51]	PSO	2014
	MAGNA++ [52]	GA	2015		iGA-PSO [53]	PSO+GA	2014
	MAGNA [54]	GA	2014		DPSO [55]	PSO	2013
	GEDEVO-M [56]	GA	2014	SI-based	GA/PSO with DTW [57]	PSO+GA	2012
	GEDEVO [58]	GA	2013		hybrid GA-PSO [59]	PSO+GA	2012
	ImAlign [60]	AIS	2020		Mahfuz and Showkat [61]	ABC	2019
	Optnetalign [62]	MA+MAEA	2016		Forghany et al. [63]	ABC	2012
	MeAlign [64]	MA	2016		Kentzoglanakis and Poole [65]	ACO+PSO	2012
SI-based	PSONA [66]	PSO	2018		ACO$_{RD}$ [67]	ACO	2012
	NABEECO [68]	ABC	2013		ESWSA [69]	ESWSA	2017

pair of undirected edge (u, v): $u, v \in \mathcal{V}_i$ represents an interaction between two proteins u and v. We suppose that $m \leq n$ without loss of generality, therefore each node of the smaller network $\mathcal{G}_1$ can be aligned to a node of the larger network $\mathcal{G}_2$. An alignment of $\mathcal{G}_1$ to $\mathcal{G}_2$ is a total injective function $f : \mathcal{V}_1 \mapsto \mathcal{V}_2$, as shown in Fig. 3.

BNA problem aims at finding the optimal alignment between networks which can be measured through a scoring function S. Large number of evolutionary approaches have been proposed to solve this problem since its NP-Completeness.

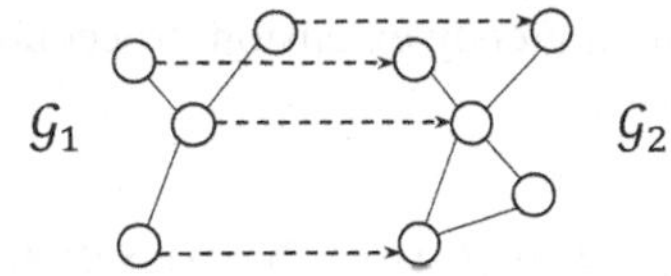

Fig. 3. Illustration of pairwise biological network alignment, dashed arrows represent the mapping from network $\mathcal{G}_1$ to network $\mathcal{G}_2$.

3.2 GA-Based Approaches

GEDEVO [58] is a Graph Edit Distance (GED) [70] based optimization model for BNA. It adopts GA to minimize the GED to obtain the optimal alignment of biological networks. GEDEVO-M [56] is an extension of GEDEVO which could align multiple PPI networks rather than two graphs. MAGNA [54] is a global network alignment approach based on GA, the presented crossover operator produces a child alignment that resembles both of its parents, while the selection and mutation operators are also utilized to meet the maximum number of iterations. Besides, it develops a new measure of alignment quality named S^3, by comparing with the drawbacks of the existing EC and ICS measures. Several improved versions of MAGNA have been proposed such as MAGNA++ [52], DynaMAGNA++ [50] and multiMAGNA++ [48]. MeAlign [64] is a Memetic Algorithm (MA) which combines GA with a local search refinement to solve the BNA problem. The objective function integrates both topological structure and sequence similarities of the biological network. The GA is used to search the regions of potential alignment solution, while the local search aims at finding the optimal solutions through a specific neighborhood heuristic strategy based on the regions. Optnetalign [62] is a multi-objective MA approach which balances the conflicting goals of topological and sequence similarity according to Pareto dominance and utilizes the crowded comparison operator first introduced in NSGA-II [21]. Several operators such as efficient swap-based local search, crossover and mutation have been adopted to create a population of alignments. Finally, it returns a set of solutions and each of which represents a possible alignment between biological networks. ImAlign [60] applies adaptive Artificial Immune System (AIS) for the BNA which combines the fitness and concentration of individuals by presenting an adaptive clonal section method. Extensive experiments demonstrate that it performs well in preserving the topological and biological properties of biological networks.

3.3 SI-Based Approaches

NABEECO [68] is the first Artificial Bee Colony (ABC) optimization approach for BNA problem. It focuses on the topology of biological networks and adopts GED [70] as optimization model to minimize. Non-topological nodes as pre-mappings can be used to accelerate convergence. PSONA [66] is a PSO based Network Aligner which redesigns the PSO in a discrete form to fit the BNA problem and combines the PSO with a swap-based local search to balance the exploration and exploitation processes of the algorithm. PSONA integrates both protein sequence similarity and interaction conservations to optimize a weighted GBA model. Furthermore, a seed-and-extend strategy is employed to initialize the alignments and a greedy extension procedure is designed to iteratively optimize the edge conservations.

4   Applications in Gene Regulatory Network Inference

4.1   Definition of Gene Regulatory Network Inference

The problem of Gene Regulatory Network (GRN) inference can be described and modeled efficiently by the Fuzzy Cognitive Maps (FCMs) [71] as illustrated in Fig. 4. Therefore, a GRN is represented as a signed directed graph that consists of N nodes. Each of the nodes corresponds to one gene. FCM can be denoted as an $N \times N$ weight matrix W, in which the weight $w_{ij} \in [-1, 1]$ represents the influence of node i to node j. GRN inference problem aims at inferring the complicated relationships among genes based on the gene expression data.

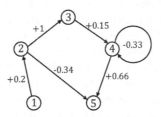

Fig. 4. Illustration of gene regulatory network inference modeled by the fuzzy cognitive maps, the weights represent the influence of different nodes.

However, as the number of identified genes increases, the GRN problem is becoming computational hardness. In this case, the EA based inference algorithms have been proposed to obtain good solutions efficiently.

4.2   GA-Based Approaches

RCGA$_D$ [38] first applies FCMs [71] to reverse engineer causal networks based on a decomposed framework and a Real-Coded Genetic Algorithm (RCGA). Three operators such as tournament selection, single point crossover and the non-uniform mutation are proposed to learn FCMs. PCA-CMI [37] combines GA and path consistency algorithms based on conditional mutual information to infer GRN. GA is performed to obtain the most optimal predictor set of each target gene. HPGA [36] consists of a parallel genetic algorithm part and a hill climbing part. A parallel genetic algorithm is used for global searches in the whole solution space, while the hill climbing is further utilized to refine the previous results. GABNI [35] is a method to infer Boolean network from time-series gene expression data. It applies a modified GA framework to achieve an efficient search, after the failure of the existing method to find the optimal solution. Wang and Mohan [34] developed a new variation of evolution strategy to permit both of the parents and offspring to survive into a new generation depending on the fitness score, which was utilized to learn sparse representations of GRNs.

Ren et al. [41] models a three-node enzyme network and optimizes two conflicting objectives simultaneously to find a robust adaptation in GRN. MONET [40] is a multi-objective algorithm to infer S-System [72] structures for GRNs by adopting the Pareto dominance concepts from NSGA-II [21]. Experiments indicate the competitive results when assessing robustness to noise. EMOEA$_{FCM}$-GRN [73] proposed by Liu et al. is a MOEA approach based on FCMs for GRN reconstruction. It first finds different optimal networks by applying MOEA and then combines the selected networks into the final solution through an efficient strategy. Later Shen et al. proposed another bi-objective approach pithEBO-FCM [39] based on the framework of MOEA/D [74] to reconstruct GRN. It can reduce the search space by decomposing the FCM learning task into inferring each local structure.

Several Multi-Agent Genetic Algorithm (MAGA) based approaches such as dMAGA$_{FCM}$-GRN [44], dMAGA-FCM$_D$ [43] and MAGARF$_{FCM}$-GRN [42] have been proposed successively to reconstruct GRN based on FCMs by Liu et al. in recent years. dMAGA$_{FCM}$-GRN is a dynamical MAGA in which agents are designed to represent the properties of GRN reconstruction and each generation of them is changed dynamically according to their energy in order to learn continuous states from data. Experimental results show that it is able to learn FCMs with 200 nodes. While dMAGA-FCM$_D$ can deal with a larger scale FCMs of 500 nodes, by integrating the decomposition-based approach with dynamical MAGA. MAGARF$_{FCM}$-GRN adopts random forests as the feature selection method to initialize the agents which accelerate the convergence speed of MAGA. Besides, it improves three genetic operators of MAGA and designs a new self-learning strategy for GRN inference. Apart from MAGA-based approaches, Fefelov et al. [46] proposed a hybrid algorithm combining clonal selection and trigonometric Differential Evolution (DE) for GRN reconstruction. Liu and liu developed a memetic algorithm MALASSO$_{RNN}$-GRN [45] based on the least absolute shrinkage and selection operator and recurrent neural network, which can obtain sparse solutions of GRN reconstruction.

4.3 SI-Based Approaches

Particle swarm optimization related approaches have been applied for GRN inference. DPSO [55] is an improved PSO method with a chaotic operator to model small networks based on the S-System, an L1 regularizer is adopted to avoid overshooting and increase the sparsity of the parameters in large networks. Sultana et al. [51] presented effective operators in PSO to infer the GRN based on the Linear Time Variant model. RMPSO [49] is also a modified version of PSO to infer GRN which can improve the exploration capability of particles in search space, as both repositories are maintained for storing personal and global best solutions of equal fitness, respectively. Lee and Hsiao proposed a hybrid GA-PSO [59] method to infer GRN, it first determines the network parameters through the search-based techniques and then reduces the task complexity by involving a network decomposition procedure. Later they developed a similar parallel evolutionary algorithm named iGA-PSO [53] which was executed in cloud computing environments to speed up the computation. GA/PSO with DTW [57] is another method that combines GA and PSO, the time delay relationship between two genes can be inferred by the dynamic time warping algorithm, then PSO is adopted to discrete the microarray dataset while

GA is used to generate a set of candidate GRNs. Hurtado et al. [47] recently proposed a multi-objective PSO approach to reconstruct GRN. It simultaneously optimizes the topology of a given network as well as tunes the kinetic order parameters in S-System and finally leads to the Pareto approximation front.

Forghany et al. [63] applied the Artificial Bee Colony (ABC) algorithm and its variant ABC* to capture the dynamics of GRN based on the S-System. The global searching capability of ABC* is enhanced when identifying model parameters. Mahfuz and Showkat [61] also proposed an ABC-based algorithm while they mainly focused on the accuracy of identified parameters of all 4 datasets. ACO_{RD} [67] is based on Ant Colony Optimization (ACO) and it is able to learn FCMs with more than 100 nodes with a decomposed approach which optimizes the whole weight matrix into small problems. Kentzoglanakis and Poole [65] proposed a PSO/ACO framework to learn parameters of each sub-structure of GRN. ESWSA [69] is another Swarm Intelligence approach to infer GRN based on the water searching strategy of elephants during drought, each of the regulatory parameters is calculated in separate search instances. ESWSA performs well with the presence of noise while it is not suitable for large-scale GRNs.

5   Discussions

EAs have increasingly gained attention in applications of biological networks due to their good properties in finding the global optimal solutions for large and complicated search spaces. However, there are several limitations for EAs to solve the above-mentioned biological analysis tasks. First, extensive experiments are needed to evaluate parameters of EAs to obtain good results. Moreover, it is difficult to determine the optimal trade-off between exploration and exploitation properties of EAs. In addition, the solutions are unstable due to the randomness of EAs and limited generations. Therefore, some prior knowledge of specific problems is needed to incorporate EAs to guide the search.

6   Conclusions

In this paper, we give a brief review of EAs for three common tasks such as protein complex detection, biological network alignment and gene regulatory network inference in biological networks. These tasks are computational difficulties due to the inherent nature of network structure analyses and the increasing complexity of biological data. Therefore, EAs are becoming a feasible technique to find good solutions for these tasks within a reasonable time.

In addition to the above three biological tasks, in recent years many other applications in biomedical and biochemistry have adopted EAs to provide reasonable solutions, which can be complementary to the previous classical approaches. The increasing number of successful applications of EAs make us believe that the demand for developing EAs will continue to grow in the near future.

References

1. Patterson, S.D., Aebersold, R.H.: Proteomics: the first decade and beyond. Nat. Genet. **33**(3), 311–323 (2003)

2. Ji, J., Jiao, L., Yang, C., Lv, J., Zhang, A.: MAE-FMD: multi-agent evolutionary method for functional module detection in protein-protein interaction networks. BMC Bioinformatics 15(1), 1–26 (2014)

3. Zhang, A.: Protein Interaction Networks: Computational Analysis. Cambridge University Press (2009)

4. Guzzi, P.H., Milenković, T.: Survey of local and global biological network alignment: the need to reconcile the two sides of the same coin. Briefings Bioinform. 19(3), 472–481 (2018)

5. Li, P.C.H., Sedighi, A., Wang, L. (eds.): Microarray technology. MMB, vol. 1368. Springer, New York (2016). https://doi.org/10.1007/978-1-4939-3136-1

6. Vikhar, P.A.: Evolutionary algorithms: a critical review and its future prospects. In: 2016 International Conference on Global Trends in Signal Processing, Information Computing and Communication (ICGTSPICC), pp. 261–265. IEEE (2016)

7. Bandyopadhyay, S., Ray, S., Mukhopadhyay, A., Maulik, U.: A multiobjective approach for identifying protein complexes and studying their association in multiple disorders. Algorithms Mol. Biol. 10(1), 1–15 (2015)

8. Hartwell, L.H., Hopfield, J.J., Leibler, S., Murray, A.W.: From molecular to modular cell biology. Nature 402(6761), C47–C52 (1999)

9. Cook, S.A.: The complexity of theorem-proving procedures. In: Proceedings of the Third Annual ACM Symposium on Theory of Computing, pp. 151–158 (1971)

10. Pizzuti, C., Rombo, S.: Experimental evaluation of topological-based fitness functions to detect complexes in PPI networks. In: Proceedings of the 14th Annual Conference on Genetic and Evolutionary Computation, pp. 193–200 (2012)

11. Pizzuti, C., Rombo, S.E.: An evolutionary restricted neighborhood search clustering approach for PPI networks. Neurocomputing 145, 53–61 (2014)

12. King, A.D., Pržulj, N., Jurisica, I.: Protein complex prediction via cost-based clustering. Bioinformatics 20(17), 3013–3020 (2004)

13. Lakizadeh, A., Jalili, S.: BiCAMWI: a genetic-based biclustering algorithm for detecting dynamic protein complexes. PLoS One 11(7), e0159923 (2016)

14. Ramadan, E., Naef, A., Ahmed, M.: Protein complexes predictions within protein interaction networks using genetic algorithms. BMC Bioinformatics 17(7), 481–489 (2016)

15. Mukhopadhyay, A., Ray, S., De, M.: Detecting protein complexes in a PPI network: a gene ontology based multi-objective evolutionary approach. Mol. BioSyst. 8(11), 3036–3048 (2012)

16. He, T., Chan, K.C.: Evolutionary graph clustering for protein complex identification. IEEE/ACM Trans. Comput. Biol. Bioinf. 15(3), 892–904 (2016)

17. Ben M'barek, M., Borgi, A., Ben Hmida, S., Rukoz, M.: GA-PPI-Net: A genetic algorithm for community detection in protein-protein interaction networks. In: van Sinderen, M., Maciaszek, L.A. (eds.) ICSOFT 2019. CCIS, vol. 1250, pp. 133–155. Springer, Cham (2020). https://doi.org/10.1007/978-3-030-52991-8_7

18. Abduljabbar, D.A., Hashim, S.Z.M., Sallehuddin, R.: An enhanced evolutionary algorithm for detecting complexes in protein interaction networks with heuristic biological operator. In: International Conference on Soft Computing and Data Mining, pp. 334–345. Springer (2020)

19. Ashburner, M., et al.: Gene ontology: tool for the unification of biology. Nat. Genet. 25(1), 25–29 (2000)

20. Ray, S., De, M., Mukhopadhyay, A.: A multiobjective go based approach to protein complex detection. Procedia Technol. 4, 555–560 (2012)

21. Deb, K., Pratap, A., Agarwal, S., Meyarivan, T.: A fast and elitist multiobjective genetic algorithm: NSGA-II. IEEE Trans. Evol. Comput. 6(2), 182–197 (2002)

22. Ray, S., Bandyopadhyay, S., Mukhopadhyay, A., Maulik, U.: Incorporating fuzzy semantic similarity measure in detecting human protein complexes in PPI network: a multiobjective approach. In: 2013 IEEE International Conference on Fuzzy Systems (FUZZ-IEEE), pp. 1–8. IEEE (2013)
23. Bara'a, A.A., Abdullah, Q.Z.: Improving the performance of evolutionary-based complex detection models in protein–protein interaction networks. Soft Comput. **22**(11), 3721–3744 (2018)
24. Abdulateef, A.H., Bara'a, A.A., Rashid, A.N., Al-Ani, M.: A new evolutionary algorithm with locally assisted heuristic for complex detection in protein interaction networks. Appl. Soft Comput. **73**, 1004–1025 (2018)
25. Maulik, U., et al.: Mining quasi-bicliques from hiv-1-human protein interaction network: a multiobjective biclustering approach. IEEE/ACM Trans. Comput. Biol. Bioinf. **10**(2), 423–435 (2012)
26. Cao, B., Luo, J., Liang, C., Wang, S., Song, D.: Moepga: a novel method to detect protein complexes in yeast protein–protein interaction networks based on multiobjective evolutionary programming genetic algorithm. Comput. Biol. Chem. **58**, 173–181 (2015)
27. Ray, S., Hossain, A., Maulik, U.: Disease associated protein complex detection: a multi-objective evolutionary approach. In: 2016 International Conference on Microelectronics, Computing and Communications (MicroCom), pp. 1–6. IEEE (2016)
28. Chowdhury, A., Rakshit, P., Konar, A., Atulya, K.N.: Prediction of protein complexes using an evolutionary approach. In: 2018 IEEE Symposium Series on Computational Intelligence (SSCI), pp. 1056–1063. IEEE (2018)
29. Lei, X., Huang, X., Shi, L., Zhang, A.: Clustering ppi data based on improved functional-flow model through quantum-behaved PSO. Int. J. Data Min. Bioinform. **6**(1), 42–60 (2012)
30. Sharafuddin, I., Mirzaei, M., Rahgozar, M., Masoudi-Nejad, A.: Protein-protein interaction network clustering using particle swarm optimization. In: IWBBIO, Citeseer, pp. 317–324 (2013)
31. Ji, J., Liu, Z., Zhang, A., Jiao, L., Liu, C.: Improved ant colony optimization for detecting functional modules in protein-protein interaction networks. In: Liu, C., Wang, L., Yang, A. (eds.) ICICA 2012. CCIS, vol. 308, pp. 404–413. Springer, Heidelberg (2012). https://doi.org/10.1007/978-3-642-34041-3_57
32. Ji, J., Liu, Z., Zhang, A., Jiao, L., Liu, C.: Ant colony optimization with multi-agent evolution for detecting functional modules in protein-protein interaction networks. In: Liu, B., Ma, M., Chang, J. (eds.) ICICA 2012. LNCS, vol. 7473, pp. 445–453. Springer, Heidelberg (2012). https://doi.org/10.1007/978-3-642-34062-8_58
33. Ji, J., Liu, Z., Zhang, A., Yang, C., Liu, C.: Ham-fmd: mining functional modules in protein–protein interaction networks using ant colony optimization and multiagent evolution. Neurocomputing **121**, 453–469 (2013)
34. Wang, Y., Mohan, C.K.: Gene regulatory network inference with evolution strategies and sparse matrix representation. In: 2019 IEEE International Conference on Bioinformatics and Biomedicine (BIBM), pp. 2105–2112. IEEE (2019)
35. Barman, S., Kwon, Y.K.: A Boolean network inference from time-series gene expression data using a genetic algorithm. Bioinformatics **34**(17), i927–i933 (2018)
36. Zheng, M., Zhang, S., Zhou, Y., Liu, G.: Inferring gene regulatory networks based on a hybrid parallel genetic algorithm and the threshold restriction method. Interdisc. Sci.: Comput. Life Sci. **10**(1), 221–232 (2018)
37. Iranmanesh, S., Sattari-Naeini, V., Ghavami, B.: Inferring gene regulatory network using path consistency algorithm based on conditional mutual information and genetic algorithm. In: 2017 7th International Conference on Computer and Knowledge Engineering (ICCKE), pp. 98–103. IEEE (2017)

38. Chen, Y., Mazlack, L.J., Minai, A.A., Lu, L.J.: Inferring causal networks using fuzzy cognitive maps and evolutionary algorithms with application to gene regulatory network reconstruction. Appl. Soft Comput. **37**, 667–679 (2015)
39. Shen, F., Liu, J., Wu, K.: A preference-based evolutionary biobjective approach for learning large-scale fuzzy cognitive maps: an application to gene regulatory network reconstruction. IEEE Trans. Fuzzy Syst. **28**(6), 1035–1049 (2020)
40. García-Nieto, J., Nebro, A.J., Aldana-Montes, J.F.: Inference of gene regulatory networks with multi-objective cellular genetic algorithm. Comput. Biol. Chem. **80**, 409–418 (2019)
41. Ren, H.P., Huang, X.N., Hao, J.X.: Finding robust adaptation gene regulatory networks using multi-objective genetic algorithm. IEEE/ACM Trans. Comput. Biol. Bioinf. **13**(3), 571–577 (2015)
42. Liu, L., Liu, J.: Inferring gene regulatory networks with hybrid of multi-agent genetic algorithm and random forests based on fuzzy cognitive maps. Appl. Soft Comput. **69**, 585–598 (2018)
43. Liu, J., Chi, Y., Zhu, C., Jin, Y.: A time series driven decomposed evolutionary optimization approach for reconstructing large-scale gene regulatory networks based on fuzzy cognitive maps. BMC Bioinformatics **18**(1), 1–14 (2017)
44. Liu, J., Chi, Y., Zhu, C.: A dynamic multiagent genetic algorithm for gene regulatory network reconstruction based on fuzzy cognitive maps. IEEE Trans. Fuzzy Syst. **24**(2), 419–431 (2015)
45. Liu, L., Liu, J.: Reconstructing gene regulatory networks via memetic algorithm and lasso based on recurrent neural networks. Soft. Comput. **24**(6), 4205–4221 (2020)
46. Fefelov, A., Lytvynenko, V., Voronenko, M., Babichev, S., Osypenko, V.: Reconstruction of the gene regulatory network by hybrid algorithm of clonal selection and trigonometric differential evolution. In: 2018 IEEE 38th International Conference on Electronics and Nanotechnology (ELNANO), pp. 305–309. IEEE (2018)
47. Hurtado, S., García-Nieto, J., Navas-Delgado, I., Nebro, A.J., Aldana-Montes, J.F.: Reconstruction of gene regulatory networks with multi-objective particle swarm optimisers. Appl. Intell. **51**, 1–20 (2020)
48. Vijayan, V., Milenković, T.: Multiple network alignment via multimagna++. IEEE/ACM Trans. Comput. Biol. Bioinform. **15**(5), 1669–1682 (2017)
49. Jana, B., Mitra, S., Acharyya, S.: Repository and mutation based particle swarm optimization (RMPSO): a new PSO variant applied to reconstruction of gene regulatory network. Appl. Soft Comput. **74**, 330–355 (2019)
50. Vijayan, V., Critchlow, D., Milenković, T.: Alignment of dynamic networks. Bioinformatics **33**(14), i180–i189 (2017)
51. Sultana, R., Showkat, D., Samiullah, M., Chowdhury, A.: Reconstructing gene regulatory network with enhanced particle swarm optimization. In: Loo, C.K., Yap, K.S., Wong, K.W., Teoh, A., Huang, K. (eds.) ICONIP 2014. LNCS, vol. 8835, pp. 229–236. Springer, Cham (2014). https://doi.org/10.1007/978-3-319-12640-1_28
52. Vijayan, V., Saraph, V., Milenković, T.: Magna++: mMaximizing accuracy in global network alignment via both node and edge conservation. Bioinformatics **31**(14), 2409–2411 (2015)
53. Lee, W., Hsiao, Y., Hwang, W.: Designing a parallel evolutionary algorithm for inferring gene networks on the cloud computing environment. BMC Syst. Biol. **8**(1), 1–19 (2014)
54. Saraph, V., Milenković, T.: Magna: maximizing accuracy in global network alignment. Bioinformatics **30**(20), 2931–2940 (2014)
55. Palafox, L., Noman, N., Iba, H.: Reverse engineering of gene regulatory networks using dissipative particle swarm optimization. IEEE Trans. Evol. Comput. **17**(4), 577–587 (2012)
56. Ibragimov, R., Malek, M., Baumbach, J., Guo, J.: Multiple graph edit distance: simultaneous topological alignment of multiple protein-protein interaction networks with an evolutionary algorithm. In: Proceedings of the 2014 Annual Conference on Genetic and Evolutionary Computation, pp. 277–284 (2014)

57. Lee, C., Leu, Y., Yang, W.: Constructing gene regulatory networks from microarray data using GA/PSO with DTW. Appl. Soft Comput. **12**(3), 1115–1124 (2012)
58. Ibragimov, R., Malek, M., Guo, J., Baumbach, J.: Gedevo: aAn evolutionary graph edit distance algorithm for biological network alignment. In: German Conference on Bioinformatics 2013. Schloss Dagstuhl-Leibniz-Zentrum fuer Informatik (2013)
59. Lee, W., Hsiao, Y.: Inferring gene regulatory networks using a hybrid ga–pso approach with numerical constraints and network decomposition. Inf. Sci. **188**, 80–99 (2012)
60. Wang, S., Ma, L., Zhang, X.: Adaptive artificial immune system for biological network alignment. In: Huang, D.-S., Jo, K.-H. (eds.) ICIC 2020. LNCS, vol. 12464, pp. 560–570. Springer, Cham (2020). https://doi.org/10.1007/978-3-030-60802-6_49
61. Mahfuz, O.B., Showkat, D.: Inference of gene regulatory network with s-system and artificial bee colony algorithm. In: 2018 Joint 7th International Conference on Informatics, Electronics & Vision (ICIEV) and 2018 2nd International Conference on Imaging, Vision & Pattern Recognition (icIVPR), pp. 117–122. IEEE (2018)
62. Clark, C., Kalita, J.: A multiobjective memetic algorithm for ppi network alignment. Bioinformatics **31**(12), 1988–1998 (2015)
63. Forghany, Z., Davarynejad, M., Snaar-Jagalska, B.E.: Gene regulatory network model identification using artificial bee colony and swarm intelligence. In: 2012 IEEE Congress on Evolutionary Computation, pp. 1–6. IEEE (2012)
64. Gong, M., Peng, Z., Ma, L., Huang, J.: Global biological network alignment by using efficient memetic algorithm. IEEE/ACM Trans. Comput. Biol. Bioinf. **13**(6), 1117–1129 (2015)
65. Kentzoglanakis, K., Poole, M.: A swarm intelligence framework for reconstructing gene networks: searching for biologically plausible architectures. IEEE/ACM Trans. Comput. Biol. Bioinf. **9**(2), 358–371 (2011)
66. Huang, J., Gong, M., Ma, L.: A global network alignment method using discrete particle swarm optimization. IEEE/ACM Trans. Comput. Biol. Bioinf. **15**(3), 705–718 (2016)
67. Chen, Y., Mazlack, L.J., Lu, L.J.: Inferring fuzzy cognitive map models for gene regulatory networks from gene expression data. In: 2012 IEEE International Conference on Bioinformatics and Biomedicine, pp. 1–4. IEEE (2012)
68. Ibragimov, R., Martens, J., Guo, J., Baumbach, J.: NABEECO: biological network alignment with bee colony optimization algorithm. In: Proceedings of the 15th annual conference companion on Genetic and evolutionary computation, pp. 43–44 (2013)
69. Mandal, S., Saha, G., Pal, R.K.: Recurrent neural network-based modeling of gene regulatory network using elephant swarm water search algorithm. J. Bioinform. Comput. Biol. **15**(04), 1750016 (2017)
70. Bunke, H., Riesen, K.: Graph edit distance–optimal and suboptimal algorithms with applications. Analysis of Complex Networks (2009)
71. Kosko, B.: Fuzzy cognitive maps. Int. J. Man Mach. Stud. **24**(1), 65–75 (1986)
72. Savageau, M.A.: Biochemical systems analysis. A Study of Function and Design in Molecular Biology, Addison Wesley Publ. (1976)
73. Liu, J., Chi, Y., Liu, Z., He, S.: Ensemble multi-objective evolutionary algorithm for gene regulatory network reconstruction based on fuzzy cognitive maps. CAAI Trans. Intell. Technol. **4**(1), 24–36 (2019)
74. Zhang, Q., Li, H.: MOEA/D: a multiobjective evolutionary algorithm based on decomposition. IEEE Trans. Evol. Comput. **11**(6), 712–731 (2007)

Computational Prediction of Protein-Protein Interactions in Plants Using Only Sequence Information

Jie Pan, Changqing Yu$^{(\boxtimes)}$, Liping Li, Zhuhong You, Zhonghao Ren, Yao Chen, and Yongjian Guan

School of Information Engineering, Xijing University, Xi'an 710123, China

Abstract. Protein-protein interactions (PPIs) in plants plays a significant role in plant biology and functional organization of cells. Although, a large amount of plant PPIs data have been generated by high-throughput techniques, but due to the complexity of plants cells, the PPIs pairs currently obtained by experimental methods cover only a small fraction of the complete plants PPIs network. In addition, the experimental approaches for identifying PPIs in plants are laborious, time-consuming, and costly. Hence, it is highly desirable to develop more efficient approaches to detect PPIs in plants. In this study, we present a novel computational method combining weighted sparse representation-based classifier (WSRC) with inverse fast Fourier transform (IFFT) representation scheme which was adopted in position specific scoring matrix (PSSM) to extract features from plant protein sequences. When performing the proposed method on the plant PPIs data set of *Maize*, we achieved excellent results with high accuracies of 89.12%. To further assess the prediction performance of the proposed approach, we compared it with the state-of-art support vector machine (SVM) classifier. Experimental results demonstrated that the proposed method has a great potential to become a powerful tool for exploring the plants cells function.

Keywords: Plant · Protein-protein interaction · Protein sequence · Inverse fast Fourier transform · Weighted sparse representation-based classifier

1 Introduction

In plants, the prediction of protein-protein interactions (PPIs) provides important information for understanding the molecular mechanisms underlying biological processes. Recently, a large number of high-throughput experimental approaches have been developed to identified PPIs, such as affinity-purification coupled to mass spectrometry (AP-MS) [1] and yeast two-hybrid (Y2H) [2] screens methods. Although we have accumulated a large amount of plants PPIs data [3, 4], these experimental approaches also have some inevitable drawbacks, which are not only costly, but also laborious and time-consuming. Moreover, these traditional biochemical experiments always suffer from high false positive rates and high false negative rates. And due to the complexity of plant growth and development systems, large-scale prediction experimental methods

© Springer Nature Switzerland AG 2021
D.-S. Huang et al. (Eds.): ICIC 2021, LNCS 12836, pp. 115–125, 2021.
https://doi.org/10.1007/978-3-030-84522-3_9

could not be adopted in plant domain, and now only a small fraction of the whole plants PPIs network can be detected. Therefore, it is very significance to develop the efficient computational approaches to identify PPIs in plants.

In recent years, much effort has been made to develop PPIs identification methods based on different data types, including literature mining knowledge [5], gene fusion [6], and protein structure information [7]. A large amount of PPIs dataset has been built, such as TAIR [8], PRIN [9], and MINT [10]. There are also some approaches that combine data and information from different sources to predict PPIs. However, without prior knowledge of corresponding proteins, these methods cannot be implemented.

Recently, the PPIs prediction methods, which extract information directly from amino acid sequences have received much attentions [11–18]. Many researchers have worked to provide sequences-based methods to detect novel PPIs, and experimental results indicated that PPIs in plants can be accurately identified using only sequence information [19–23]. For example, Sun et al. [24] presented a method that using a type of deep-learning algorithm called stacked autoencoder (SAE) to use sequence-based approaches for predicting PPIs. This model obtained the best results on 10-fold cross-validation which was based on protein sequence auto-covariance coding. You et al. [12] also developed a novel computational approach called PCA-EELM to predict PPIs. The main improvement of this study is that they adopted the PCA method to construct the most discriminative new feature set. In addition, many methods based on amino acid sequences have been developed in the literature [25–30].

In this study, we provided a novel computational method to detect the PPIs in plants from protein sequence information, which employing a novel position specific scoring matrix (PSSM) and combining the weighted sparse representation-based classifier (WSRC) with inverse fast Fourier transform (IFFT). This feature representation approach combined with the WSRC has remarkably performed in the prediction of the PPIs in plants. Furthermore, the main idea of our proposed model includes three steps. First, the plant protein sequence could be represented as a position specific scoring matrix so that we can obtain the biological evolutionary information between different types of amino acids. Second, utilizing the inverse Fast Fourier transform (IFFT) method to extracted a 400-dimensional vector from each plant proteins PSSM matrix. Thirdly, a powerful classifier, weighted sparse representation-based classifier, is employed to perform PPIs predictions on the *Maize* plant PPIs datasets. We also compared the proposed model with the state-of-the-art support vector machine (SVM) classifier to further evaluate the prediction performance. The experiments results demonstrated that our approach performs significantly well in predicting PPIs in plants.

2 Materials and Methodology

2.1 Data Collection and Data Set Construction

In this paper, we verify the proposed model on the *Maize* PPIs dataset. Maize is one of the most important food. We collected the *Maize* data set from *agriGO* [31] and *Protein-Protein Interaction Database for Maize* (*PPIM*) [32]. After the strict inclusion and exclusion screening, we select 14230 maize protein pairs as the positive dataset. For constructing the negative dataset, we selected 14230 additional maize protein pairs of different subcellular localizations. As a result, the whole *Maize* dataset is constructed by 28460 maize protein pairs.

2.2 Position-Specific Scoring Matrix (PSSM)

Through the Position-Specific Scoring Matrix (PSSM) which is reported by Gribskov [33] and it achieved great success in protein binding site prediction, protein secondary structure prediction and prediction of disordered regions [34–37]. The structure of PSSM can be represented as a matrix of N rows and 20 columns. Each protein sequence can be transformed as follows:

$$M = \{M_{\alpha,\beta}, \alpha = 1, \ldots, N, \ \beta = 1, \ldots, 20\} \tag{1}$$

where N denotes the length of a given plant protein sequence and column 20 represents the number of 20 amino acid. For each query sequence, the value $M_{\alpha,\beta}$, which could be described as β-th amino acid, will be set up by PSSM at the position of α. Thus, $M_{\alpha,\beta}$ can be calculated as:

$$M_{\alpha,\beta} = \sum_{k=1}^{20} p(\alpha, k) \times q(\beta, k) \tag{2}$$

Thus, the value of Dayhoff's mutation matrix between the β-th and k-th amino acids can be described as $q(\beta, k)$, and the occurrence frequency score of the k-th amino acid in the position of α with the probe can be represented by $p(\alpha, k)$. Hence, a high value means a strongly conservative position; otherwise, it will imply a weakly conservative position. In this study, we employed the *Position-Specific Iterated BLAST (PSI-BLAST)* tool [38] to generate the PSSM for each protein sequence. we assigned the e-value to 0.001 and selected 3 iterations in the process. In addition, all other parameters were set to default values to obtain highly and widely homologous sequences.

2.3 Inverse Fast Fourier Transform

In the fields of computational science and engineering, the Fast Fourier Transform (FFT) [39] is one of the most important algorithms. It is an indispensable algorithm in the field of Digital Signal Processing. However, FFT algorithm is not suitable in many practical applications when the data are not uniformly sampled. For this reason, we adopted the inverse fast Fourier Transform (IFFT) [40] method to obtain the transient response in time domain.

In the FFT, the irregularities of the twiddle factors can be solved by the Sine and Cosine transform of the signal. The Cosine and Sine transformations of the input signal are added together to obtain the FFT of the two-dimensional signal. As shown in Eq. (3) and Eq. (4), the required Sine matrix and Cosine matrix for FFT and IFFT can be defined as:

$$C(u + 1, x + 1) = \cos((pi/4)^*(u{*}x)) \tag{3}$$

$$S(u + 1, x + 1) = \sin((pi/4)^*(u{*}x)) \tag{4}$$

Hence, by the rules, the 2-D FFT can be yield by adding the Sine and Cosine transform as shown in Eq. (5):

$$F(u, v) = \sum_{u=0}^{M-1}\sum_{v=0}^{N-1}\left\{\begin{array}{l} F(u, v)(\cos(2\pi(ux + vy)/n) \\ +j\sin(2\pi(ux + vy)/N)) \end{array}\right\} \tag{5}$$

for $k = 0, 1, 2\ldots, N - 1$.

So, the IFFT for 2-D image can be described as follows:

$$f(x, y) = (1/MN)\sum_{u=0}^{M-1}\sum_{V=0}^{N-1}\left\{\begin{array}{l} F(u, v)(\cos(2\pi(ux + vy)/N) \\ -j\sin(2\pi(ux + vy)/N)) \end{array}\right\} \tag{6}$$

for $n = 0, 1, 2\ldots, N - 1$.

In our study, each protein sequence in the *Maize* dataset, will be converted into a 400-dimensional vector by means of an inverse fast Fourier transform.

2.4 Weighted Sparse Representation-Based Classification

Recently, with the improvement of linear representation methods (LRBM) and compressed sensing (CS) theory, sparse representation-based classification (SRC) [41, 42] algorithm has been proven to widely applied in signal processing, pattern recognition and computer vision. The SRC assumes that there is a training sample matrix $X \in R^{d \times n}$, which denotes n training set and d-dimensional feature vectors, and it also assumes that there are sufficient training samples belonging to the kth class and set up $X_k = [l_{k1}, \cdots l_{kn_k}]$, where n_k denotes the sample number of *kth* class and l_i represents the label of *ith* sample. Thus, the sample matrix X could be defined as $X = [X_1 \ldots X_K]$. The SRC algorithm can described the test sample $y \in R^d$ with the linear combination of kth-class training samples as:

$$y = \alpha_{k,1}l_{k,1} + \alpha_{k,2}l_{k,2} + \cdots + \alpha_{k,nk}l_{k,nk} \tag{7}$$

when the whole training set representation are taking into accounts, it can be further symbolized as follows:

$$y = X_{\alpha 0} \tag{8}$$

where $\alpha_0 = \left[0, \cdots, 0, \alpha_{k,1}, \alpha_{k,2} \cdots \alpha_{k,nk}, 0, \cdots, 0\right]^T$. It is well known that the nonzero entries in α_0 are only relevant to the *kth* class, so if the samples size is too large, α_0 will become sparse.

For SRC algorithm, the key of it is to search the α vector that formula (8) can satisfy and can minimize the l_0-norm of itself. It can be represented as:

$$\hat{\alpha_0} = \arg\min\|\alpha\|_0 \text{ subject to } y = X_\alpha \tag{9}$$

Since problem (9) is a NP-hard problem and it is difficult to be solved accurately. According to the CS theory, if the α is sparse enough, we can solve the related convex l_1-minimization problem instead of dealing with the solution of l_0-minimization problem directly:

$$\hat{\alpha_1} = \arg\min\|\alpha\|_1 \text{ subject to } y = X_\alpha \tag{10}$$

To deal with the occlusion, the Eq. (10) can be extended to the stable l_1-minimization problem:

$$\hat{\alpha_1} = \arg\min \|\alpha\|_1 \text{ subject to } \|y - X_\alpha\| \le \varepsilon \tag{11}$$

where $\varepsilon > 0$ represents the tolerance for reconstruction error. We can solve the Eq. (11) by using the standard linear programming methods.

After achieving the sparsest solution $\hat{\alpha_1}$, SRC can assign the test sample y to class k via the following rule:

$$\min_k r_k(y) = \left\| y - X \hat{\alpha}_1^k \right\|, \; k = 1 \ldots K \tag{12}$$

where $X \hat{\alpha}_1^k$ is the reconstruction which is built by training samples of class k and the class number of the whole samples can be defined as K. Then the SRC set a test sample as a sparse combination of training sample. Finally, we assigned it to the class which can minimizes the residual between itself and $X \hat{\alpha}_1^k$.

However, some studies [43–45] have reported that in some cases, locality structure of data is more important than sparsity. Moreover, the traditional SRC could not be guaranteed to be local. To solve this problem, Lu *et al.* [46] developed a novel variant of SRC called weighted sparse representation-based classifier (WSRC). The main improvement of this method is that it combines the locality structure of data with sparse representation. Through mapping the training data into a higher-dimensional kernel include feature space, it can yield a better performance of classification. Gaussian kernel-based distance was used in WSRC to calculate the weights:

$$d_G = (S_1, S_2) = e^{-\|s_1 - s_2\|^2 / 2\sigma^2} \tag{13}$$

where $s_1, s_2 \in R^d$ denotes two samples; σ is the Gaussian kernel width. In this way, WSRC can preserve the locality structure of data and it can address the following questions:

$$\hat{\alpha_1} = \arg\min \|W\alpha\|_1 \text{ subject to } y = X\alpha \tag{14}$$

and specifically,

$$diag(W) = \left[d_G(y, x_1^1), \ldots, d_G(y, x_{n_k}^k) \right]^T \tag{15}$$

where n_k is the sample number of training set in class k and W represents a block diagonal matrix about locality adaptor. Dealing with occlusion, we would finally solve the following stable l_1-minimization problem:

$$\hat{\alpha_1} = \arg\min \|W\alpha\|_1 \quad \text{subject to} \quad \|y - X\alpha\| \le \varepsilon \tag{16}$$

where $\varepsilon > 0$ denotes the tolerance value.

To summarize, the WSRC algorithm can be stated as follows:

Algorithm. Weighted Sparse Representation-based Classifier (WSRC)

1. Input: the matrix of training samples $X \in R^{d \times n}$ and a test sample $y \in R^d$.
2. Normalize the columns of X to have unit l_2 -norm.
3. Calculate the Gaussian distances between y and each sample in X and employ them to adjust the training samples matrix X to X'.
4. Solving the stable l_1 -minimization problem defined in Eq. (15).
5. Compute the residuals $r_k(y) = \left\| y - X \overset{\wedge k}{\alpha_1} \right\| (k = 1, 2, ..., K)$.
6. Output: the prediction label of y as $identify(y) = \arg\min(r_k(y))$.

---

3 Results and Discussion

3.1 Evaluation Criteria

To demonstrate the performance of the proposed approach, four evaluation criteria was used in this work, including accuracy (Acc.), sensitivity (Sen.), precision (Prec.), and Matthews correlation coefficient (MCC) [47–50]. Their corresponding calculating formulas are defined as follows:

$$Acc. = \frac{TN + TP}{TP + FP + FN + TN} \tag{17}$$

$$Sen. = \frac{TP}{TP + FN} \tag{18}$$

$$PR. = \frac{TP}{TP + FP} \tag{19}$$

$$MCC = \frac{TN \times TP - FN \times FP}{\sqrt{(TP + FP) \times (TN + FN) \times (TN + FP) \times (TP + FN)}} \tag{20}$$

where true positive (TP) denotes the number of plants protein-protein pairs classified as interacting correctly while true negative (TN) stands for the number of non-interacting PPIs pairs predicted correctly; false positive (FP) denotes the number of true non-interacting pairs predicted to be PPIs falsely, and false negative (FN) denotes the count of interacting plants PPIs pairs that predict to have no interaction. In addition, we also adopted the receiver operating characteristic (ROC) curves to assess the prediction performance of the proposed approach, and the area under the ROC curve (AUC) is calculated used for demonstrating the quality of prediction model.

3.2 Assessment of Prediction Ability

In this article, we used 5-fold cross-validation to evaluate the predictive ability of our model in *Maize* data sets. In this way, we can prevent overfitting and test the stability of the proposed method. More specifically, the whole data set is randomly divided into five parts, four of them were used to construct a training set and the rest one was adopted as a testing set. The cross validation has the advantages that it can minimize the impact of data dependency and improved the reliability of the results.

The five-fold cross validation results of the proposed approach on the *Maize* datasets are listed in Table 1. Form Table 1, we can observe that when applying the proposed method to the *Maize* dataset, we obtained best prediction results of average accuracy, precision, sensitivity, and MCC were 89.12%, 87.49%, 91.32%, and 80.59%, with corresponding standard deviations 0.59%, 1.38%, 0.64%, and 0.94%, respectively. Figure 1 shows the ROC curves for the proposed approach on *Maize* dataset. The significantly average AUC values suggesting that our method is fit well for our purposes to predict PPIs in plants from amino acid sequences.

Table 1. 5-fold cross-validation results yield on the *Maize* dataset using the proposed method.

Test set	Acc. (%)	PR. (%)	Sen. (%)	MCC (%)	AUC (%)
1	89.16	87.81	90.84	80.66	93.64
2	88.64	85.94	91.84	79.83	93.64
3	88.56	87.19	90.89	79.70	93.24
4	89.20	86.84	92.17	80.71	94.05
5	90.04	89.65	90.85	82.06	94.21
Average	**89.12 ± 0.59**	**87.49 ± 1.38**	**91.32 ± 0.64**	**80.59 ± 0.94**	**93.76 ± 0.38**

3.3 Comparison of the Proposed Model with Different Classifiert

Although the WSRC model obtained better performance in predicting PPIs of plants, we also need to further verify the prediction ability of the proposed method. We compared the prediction accuracy of the WSRC model with that of the state-of-art SVM model via the same feature extraction approach based on the Maize datasets. We applied the same feature extraction approach on the *Maize* datasets and compared the prediction accuracy of the WSRC model with the state-of-the-art SVM. We employed the LIBSVM tool to run this classification, and 5-fold cross-validation was also adopted in these experiments. In order to obtain better performance of SVM classifier, we should optimize several parameters of SVM classifier. In this study, the penalty parameter c and the kernel parameter g of SVM model was optimized by the gird search method. In the experiments of *Maize* dataset, we set $c = 5$, $g = 0.5$.

As shown in Table 2, it is clearly seen that when applied the SVM model to predict PPIs on *Maize* dataset, we yield good results with average accuracy, precision, sensitivity, MCC and AUC of 81.77%, 83.10%, 79.78%, 70.16% and 88.04%, respectively. For the

Maize datasets, the classification results yield by the SVM-based models are lower than those by the proposed approach. In summary, it is obvious that the overall prediction results of WSRC model is better than that of SVM-based approach. The ROC curves of SVM model are also shown in Fig. 1. All these experiments results indicated that the weighted sparse representation-based classifier is an effective and robust model for PPIs prediction in plants.

Table 2. 5-fold cross-validation results yield on the *Maize* dataset using the proposed method.

Test set	Acc. (%)	PR. (%)	Sen. (%)	MCC (%)	AUC (%)
IFFT + WSRC					
1	89.16	87.81	90.84	80.66	93.64
2	88.64	85.94	91.84	79.83	93.64
3	88.56	87.19	90.89	79.70	93.24
4	89.20	86.84	92.17	80.71	94.05
5	90.04	89.65	90.85	82.06	94.21
Average	**89.12 ± 0.59**	**87.49 ± 1.38**	**91.32 ± 0.64**	**80.59 ± 0.94**	**93.76 ± 0.38**
IFFT + SVM					
1	81.04	82.37	78.80	69.24	87.50
2	81.72	81.74	80.73	70.10	88.34
3	81.44	82.49	80.68	69.76	87.70
4	82.48	84.09	79.74	71.05	87.96
5	82.16	84.83	78.94	70.64	88.70
Average	**81.77 ± 0.57**	**83.10 ± 1.30**	**79.78 ± 0.92**	**70.16 ± 0.71**	**88.04 ± 0.48**

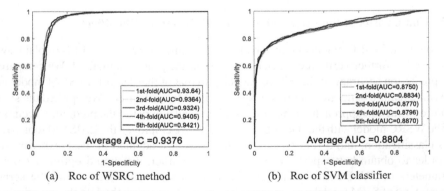

(a) Roc of WSRC method   (b) Roc of SVM classifier

Fig. 1. Comparison of the ROC curves obtained by WSRC and SVM-based method on *Maize* dataset (5-fold cross validation). (a) shows the ROC curves performed by WSRC model. (b) shows the ROC curves performed by SVM model.

4 Conclusion and Discussion

In this study, we present an effective and accurate computational method that utilize the information of amino acid sequences for predicting PPIs in plants. This method is based on a weighted sparse representation-based classifier combining with inverse fast Fourier transform and position-specific-scoring-matrix. The main point of this approach is to employ the unique of WSRC method including better generalization, simply and considering the sparsity and continuity of plants protein sequence data. The whole prediction model is composed of the following steps. Firstly, all the plant protein sequences were converted as the PSSM. Secondly, we employed the inverse fast Fourier transform to extract feature vectors from PSSM. Finally, weighted sparse representation-based classifier would be used as the machine learning classifier. The proposed approach performs significantly well on *Maize* PPIs datasets. In order to prove the efficient and reliability efficient of the proposed model, we also compare it prediction performance with the state-of-the-art SVM model. These experiments results indicates that our method can improve the accuracy of PPIs prediction in plants. In conclusion, the proposed method is a reliable and powerful prediction model for future proteomics research.

Acknowledgment. This work was supported by the National Natural Science Foundation of China under Grant NO. 61722212 and Grant NO. 62002297.

References

1. Chen, Y., Weckwerth, W.: Mass spectrometry untangles plant membrane protein signaling networks. Trends Plant Sci. **25**(9), 930–944 (2020)
2. Matiolli, C.C., Melotto, M.: A comprehensive Arabidopsis yeast two-hybrid library for protein-protein interaction studies: a resource to the plant research community. Mol. Plant-Microbe Interact. **31**, 899–902 (2018)
3. Di Silvestre, D., Bergamaschi, A., Bellini, E., Mauri, P.: Large scale proteomic data and network-based systems biology approaches to explore the plant world. Proteomes **6**, 27 (2018)
4. Waese, J., et al.: ePlant: visualizing and exploring multiple levels of data for hypothesis generation in plant biology. Plant Cell **29**, 1806–1821 (2017)
5. Hartmann, J., et al.: The effective design of sampling campaigns for emerging chemical and microbial contaminants in drinking water and its resources based on literature mining. Sci. Total Environ. **742**, 140546 (2020)
6. An, D., Cao, H.X., Li, C., Humbeck, K., Wang, W.: Isoform sequencing and state-of-art applications for unravelling complexity of plant transcriptomes. Genes **9**, 43 (2018)
7. Chou, K.-C., Shen, H.-B.: Plant-mPLoc: a top-down strategy to augment the power for predicting plant protein subcellular localization. PLoS One **5**, e11335 (2010)
8. Lamesch, P., et al.: The Arabidopsis Information Resource (TAIR): improved gene annotation and new tools. Nucleic Acids Res. **40**, D1202–D1210 (2012)
9. Gu, H., Zhu, P., Jiao, Y., Meng, Y., Chen, M.: PRIN: a predicted rice interactome network. BMC Bioinform. **12**, 1–13 (2011)
10. Licata, L., et al.: MINT, the molecular interaction database: 2012 update. Nucleic Acids Res. **40**, D857–D861 (2012)
11. Li, J.-Q., You, Z.-H., Li, X., Ming, Z., Chen, X.: PSPEL: in silico prediction of self-interacting proteins from amino acids sequences using ensemble learning. IEEE/ACM Trans. Comput. Biol. Bioinf. **14**, 1165–1172 (2017)

12. You, Z.-H., Lei, Y.-K., Zhu, L., Xia, J., Wang, B.: Prediction of protein-protein interactions from amino acid sequences with ensemble extreme learning machines and principal component analysis. In: BMC Bioinformatics, pp. 1–11. Springer (2013)

13. You, Z.-H., Lei, Y.-K., Gui, J., Huang, D.-S., Zhou, X.: Using manifold embedding for assessing and predicting protein interactions from high-throughput experimental data. Bioinformatics 26, 2744–2751 (2010)

14. Wang, Y.-B., et al.: Predicting protein–protein interactions from protein sequences by a stacked sparse autoencoder deep neural network. Mol. BioSyst. 13, 1336–1344 (2017)

15. You, Z.-H., Yu, J.-Z., Zhu, L., Li, S., Wen, Z.-K.: A MapReduce based parallel SVM for large-scale predicting protein–protein interactions. Neurocomputing 145, 37–43 (2014)

16. Hu, L., Wang, X., Huang, Y.-A., Hu, P., You, Z.-H.: A survey on computational models for predicting protein–protein interactions. Brief. Bioinform. (2021)

17. Lei, Y.-K., You, Z.-H., Dong, T., Jiang, Y.-X., Yang, J.-A.: Increasing reliability of protein interactome by fast manifold embedding. Pattern Recogn. Lett. 34, 372–379 (2013)

18. Li, Z.-W., You, Z.-H., Chen, X., Gui, J., Nie, R.: Highly accurate prediction of protein-protein interactions via incorporating evolutionary information and physicochemical characteristics. Int. J. Mol. Sci. 17, 1396 (2016)

19. Zhu, L., You, Z.-H., Huang, D.-S., Wang, B.: t-LSE: a novel robust geometric approach for modeling protein-protein interaction networks. PLoS One 8, e58368 (2013)

20. Wang, Y., You, Z.-H., Yang, S., Li, X., Jiang, T.-H., Zhou, X.: A high efficient biological language model for predicting protein–protein interactions. Cells 8, 122 (2019)

21. Huang, Y.-A., You, Z.-H., Chen, X., Chan, K., Luo, X.: Sequence-based prediction of protein-protein interactions using weighted sparse representation model combined with global encoding. BMC Bioinform. 17, 1–11 (2016)

22. Wang, L., et al.: An ensemble approach for large-scale identification of protein-protein interactions using the alignments of multiple sequences. Oncotarget 8, 5149 (2017)

23. Chen, Z.-H., You, Z.-H., Zhang, W.-B., Wang, Y.-B., Cheng, L., Alghazzawi, D.: Global vectors representation of protein sequences and its application for predicting self-interacting proteins with multi-grained cascade forest model. Genes 10, 924 (2019)

24. Sun, T., Zhou, B., Lai, L., Pei, J.: Sequence-based prediction of protein protein interaction using a deep-learning algorithm. BMC Bioinform. 18, 1–8 (2017)

25. Skoblov, M., et al.: Protein partners of KCTD proteins provide insights about their functional roles in cell differentiation and vertebrate development. BioEssays 35, 586–596 (2013)

26. Xia, J.-F., Zhao, X.-M., Huang, D.-S.: Predicting protein–protein interactions from protein sequences using meta predictor. Amino Acids 39, 1595–1599 (2010)

27. Song, X.-Y., Chen, Z.-H., Sun, X.-Y., You, Z.-H., Li, L.-P., Zhao, Y.: An ensemble classifier with random projection for predicting protein–protein interactions using sequence and evolutionary information. Appl. Sci. 8, 89 (2018)

28. Wang, Y.-B., You, Z.-H., Li, X., Jiang, T.-H., Cheng, L., Chen, Z.-H.: Prediction of protein self-interactions using stacked long short-term memory from protein sequences information. BMC Syst. Biol. 12, 107–115 (2018)

29. You, Z.-H., Li, S., Gao, X., Luo, X., Ji, Z.: Large-scale protein-protein interactions detection by integrating big biosensing data with computational model. BioMed. Res. Int. 2014 (2014)

30. Yi, H.-C., You, Z.-H., Guo, Z.-H., Huang, D.-S., Chan, K.C.: Learning representation of molecules in association network for predicting intermolecular associations. IEEE/ACM Trans. Comput. Biol. Bioinform. (2020)

31. Tian, T., et al.: AgriGO v2. 0: a GO analysis toolkit for the agricultural community, 2017 update. Nucleic Acids Res. 45, W122–W129 (2017)

32. Zhu, G., et al.: PPIM: a protein-protein interaction database for maize. Plant Physiol. 170, 618–626 (2016)

33. Gribskov, M., McLachlan, A.D., Eisenberg, D.: Profile analysis: detection of distantly related proteins. Proc. Natl. Acad. Sci. **84**, 4355–4358 (1987)
34. Li, Z.-W., et al.: Accurate prediction of protein-protein interactions by integrating potential evolutionary information embedded in PSSM profile and discriminative vector machine classifier. Oncotarget **8**, 23638 (2017)
35. Zhu, H.-J., You, Z.-H., Shi, W. L., Xu, S.-K., Jiang, T.-H., Zhuang, L.-H.: Improved prediction of protein-protein interactions using descriptors derived from PSSM via gray level co-occurrence matrix. IEEE Access **7**, 49456–49465 (2019)
36. Wang, L., et al.: Using two-dimensional principal component analysis and rotation forest for prediction of protein-protein interactions. Sci. Rep. **8**, 1–10 (2018)
37. Li, L.-P., Wang, Y.-B., You, Z.-H., Li, Y., An, J.-Y.: PCLPred: a bioinformatics method for predicting protein–protein interactions by combining relevance vector machine model with low-rank matrix approximation. Int. J. Mol. Sci. **19**, 1029 (2018)
38. Altschul, S.F., Koonin, E.V.: Iterated profile searches with PSI-BLAST—a tool for discovery in protein databases. Trends Biochem. Sci. **23**, 444–447 (1998)
39. Nussbaumer, H.J.: The fast Fourier transform. In: Fast Fourier Transform and Convolution Algorithms, pp. 80–111. Springer (1981)
40. Anitha, T., Ramachandran, S.: Novel algorithms for 2-D FFT and its inverse for image compression. In: 2013 International Conference on Signal Processing, Image Processing & Pattern Recognition, pp. 62–65. IEEE (2013)
41. Liao, B., Jiang, Y., Yuan, G., Zhu, W., Cai, L., Cao, Z.: Learning a weighted meta-sample based parameter free sparse representation classification for microarray data. PLoS One **9**, e104314 (2014)
42. Wright, J., Yang, A.Y., Ganesh, A., Sastry, S.S., Ma, Y.: Robust face recognition via sparse representation. IEEE Trans. Pattern Anal. Mach. Intell. **31**, 210–227 (2008)
43. Wang, J., Yang, J., Yu, K., Lv, F., Huang, T., Gong, Y.: Locality-constrained linear coding for image classification. In: 2010 IEEE computer society conference on computer vision and pattern recognition, pp. 3360–3367. IEEE (2010)
44. Sharma, A., Paliwal, K.K.: A deterministic approach to regularized linear discriminant analysis. Neurocomputing **151**, 207–214 (2015)
45. Roweis, S.T., Saul, L.K.: Nonlinear dimensionality reduction by locally linear embedding. Science **290**, 2323–2326 (2000)
46. Lu, C.-Y., Min, H., Gui, J., Zhu, L., Lei, Y.-K.: Face recognition via weighted sparse representation. J. Vis. Commun. Image Represent. **24**, 111–116 (2013)
47. Wong, L., You, Z.-H., Li, S., Huang, Y.-A., Liu, G.: Detection of protein-protein interactions from amino acid sequences using a rotation forest model with a novel PR-LPQ descriptor. In: International Conference on Intelligent Computing, pp. 713–720. Springer (2015)
48. Lei, Y.-K., You, Z.-H., Ji, Z., Zhu, L., Huang, D.-S.: Assessing and predicting protein interactions by combining manifold embedding with multiple information integration. In: BMC Bioinformatics, pp. 1–18. Springer (2012)
49. Zhu, L., You, Z.-H., Huang, D.-S.: Increasing the reliability of protein–protein interaction networks via non-convex semantic embedding. Neurocomputing **121**, 99–107 (2013)
50. An, J.-Y., et al.: Identification of self-interacting proteins by exploring evolutionary information embedded in PSI-BLAST-constructed position specific scoring matrix. Oncotarget **7**, 82440 (2016)

Image and Signal Processing

A Diabetic Retinopathy Classification Method Based on Novel Attention Mechanism

Jinfan Zou[1,2,3](✉), Xiaolong Zhang[1,2,3], and Xiaoli Lin[1,2,3]

[1] School of Computer Science and Technology, Wuhan University of Science and Technology, Wuhan, China
{xiaolong.zhang,linxiaoli}@wust.edu.cn
[2] Institute of Big Data Science and Engineering, Wuhan University of Science and Technology, Wuhan, China
[3] Hubei Key Laboratory of Intelligent Information Processing and Real-Time Industrial System, Wuhan University of Science and Technology, Wuhan, China

Abstract. The convolutional neural network (CNN) has been effectively used to do feature extraction in medical image analysis including fundus images for diabetic retinopathy. However, some important detail features are easily ignored, which causes low prediction accuracy. To solve this problem, a CNN feature extraction method based on a novel pixel-level attention mechanism is proposed to achieve an efficient and accurate identification of diabetic retinopathy. Firstly, data normalization and data augmentation are used to improve the quality of datasets. Then, according to the structure characteristics of the two backbone networks Inception- ResnetV2 and EfficientNet-B5, the corresponding attention modules are introduced respectively. Finally, attention networks are trained as feature extractors of fundus images, and complementary deep attention feature descriptors are formed through feature fusion, which can effectively improve the accuracy of the classification model. The experimental results on EyePACS, Messidor and OIA datasets show that the proposed method outperforms the previous ones.

Keywords: Diabetic retinopathy · Attention mechanism · Convolutional neural network · Feature fusion · Image classification

1 Introduction

Diabetic retinopathy (DR) is a retinal disease caused by hyperglycemia in diabetics with symptoms such as hemorrhages, microaneurysms, and exudates. DR is one of the important causes of blindness in adults worldwide [1]. DR is one of the most common fundus complications, which seriously affects vision and even causes blindness [2]. Early standardized treatment can effectively delay the progression of DR and prevent the onset of blindness [3]. Professional ophthalmologists complete the diagnosis of DR by analyzing fundus images, which can be time-consuming and labor-intensive. Therefore, it is necessary to implement automatic DR detection based on computer-aided diagnosis.

© Springer Nature Switzerland AG 2021
D.-S. Huang et al. (Eds.): ICIC 2021, LNCS 12836, pp. 129–142, 2021.
https://doi.org/10.1007/978-3-030-84522-3_10

In previous studies, image processing techniques were mainly used to extract local lesion features (e.g., hemorrhages, microaneurysms, and hard exudates, etc.) from fundus images, and then binary classifiers were used to predict DR grading [4, 5]. However, these methods are based on small datasets, which are prone to some problems such as model overfitting and difficulty in being widely used.

With the development of deep learning, many researchers have applied it to DR detection tasks with fruitful results. Xie et al. used data labeled with the lesioned regions to train CNN [6], which requires a large amount of manually labeled sample information, so the practical application is limited. Zhou et al. used generative adversarial networks (GAN) to synthesize high-resolution fundus images to increase the diversity of data [7], but the network structure was not improved. Zeng et al. used binocular fundus images as input to train the InceptionV3 model with the Siamese-like structure based on transfer learning [8], but the method lacks a focus on fine-grained features. Fundus images are typical fine-grained images. Therefore, more attention should be paid to the important details. Wang et al. proposed a zoom-in network based on attention mechanism, in which the suspicious regions generated by attention maps are zoomed in for details to classify DR images [9], but the method lacks channel attention. Bajwa et al. integrated multiple CNN models and combining fine- and coarse-grained classifiers to evaluate the performance of the models [10], but the method sacrifices a significant amount of training time to slightly improve the classification accuracy. Although these methods perform well, the lack of more attention to detailed features makes them still inadequate in extracting discriminative features.

By introducing attention mechanism into CNN, the network can pay more attention to the key information in the image [11]. Hu et al. proposed Squeeze-and-Excitation Nework (SENet) to model the channel relationship [12]. Mei et al. Proposed a CNN model based on the attention mechanism for focusing on the correlation between neighboring pixels in the spatial dimension [13].

Inspired by the above work, we propose a novel pixel-level attention mechanism. Through spatial attention and channel attention, key features are strengthened and secondary features are weakened to strengthen the discrimination of features. Our main work is as follows: 1) Solve the problems of data noise and unbalanced category distribution through data normalization and data augmentation. 2) Select Inception-ResNetV2 and EfficientNet-B5 as the backbone networks based on transfer learning to extract deep features. 3) The corresponding pixel-level attention module is introduced into the backbone network to extract effective deep attention features by processing each pixel point on the image strongly or weakly. 4) Fuse features of the attention networks to form complementary deep attention feature descriptors to improve the information expression ability of images.

2 Methodology

We propose a DR classification model based on a pixel-level attention mechanism (PATT). The overall framework is shown in Fig. 1, where PATT consists of a spatial attention module and a channel attention module. ESA denotes the spatial attention module corresponding to EfficientNet-B5, ISA denotes the spatial attention module corresponding to Inception-ResNetV2, and CA denotes the channel attention module corresponding to the two backbone networks. Firstly, as the backbone networks based on transfer learning [14], Inception-ResNetV2 and EfficientNet-B5 are trained using preprocessed images. Then, the corresponding pixel-level attention modules including spatial attention and channel attention are introduced into the backbone networks. Finally, the fused deep attention features are input to the LightGBM classifier for training after global average pooling (GAP) and dropout layers (Drop), and the predicted results are output.

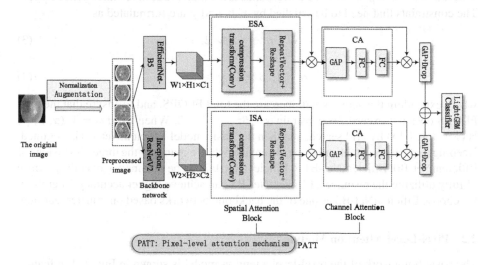

Fig. 1. The overall framework of the DR classification model

2.1 Backbone Networks Based on Transfer Learning

Inception-ResNetV2 was proposed based on the InceptionV3 and ResNet models [15]. We use the inception structure to expand the local receptive field to obtain more spatial information, and use the residual structure in ResNet to train a deeper network to obtain deep features.

EfficientNet is an adaptive optimization network that balances accuracy and speed [16]. Traditional CNN models improve performance by simply enlarging the network width, network depth, or image resolution. While when the model is large, blindly deepening and widening the network and increasing the image resolution will instead degrade its performance. EfficientNet achieves a good scaling balance of the three elements.

EfficientNet is called N and is defined as

$$N = \bigotimes_{i=1,2,\ldots,s} F_i^{L_i}\left(X_{<H_i, W_i, C_i>}\right). \tag{1}$$

Where $\otimes$ denotes the convolution operation; i denotes the number of convolutional layers; s denotes the number of convolutional layers with different structures; F_i denotes the convolutional layer; L_i denotes the network depth; and $X_{<H_i, W_i, C_i>}$ denotes the input tensor. Under the condition of resource maximization, the optimal combination of network depth L_i, network width C_i and input image resolution size (H_i, W_i) needs to be found to obtain the highest model accuracy, and the optimal corresponding scaling of these three dimensions Parameter (d, w, r) is formulated as

$$N(d, w, r) = \underset{i=1,2,...,s}{\otimes} \hat{F}^{d \cdot \hat{L}_i}\left(X_{<r \cdot \hat{H}_i, r \cdot \hat{W}_i, w \cdot \hat{C}_i>}\right). \tag{2}$$

Where $d = \alpha^\varphi, w = \beta^\varphi, r = \gamma^\varphi, \varphi$ corresponds to the magnitude of resource consumption. The constraints that need to be satisfied by α, β, and γ are formulated as

$$\alpha \cdot \beta^2 \cdot \gamma^2 \approx 2, \tag{3}$$

$$\alpha \geq 1, \beta \geq 1, \gamma \geq 1. \tag{4}$$

where expanding the network will increase the total FLOPS, and to ensure that the total FLOPS does not exceed 2^φ, constrain $\alpha \cdot \beta^2 \cdot \gamma^2 \approx 2$. When fixed $\varphi = 1$, (α, β, γ) is (1.2, 1.1, 1.15) by grid search, and then the basic model EfficientNet-B0 is obtained; Keeping (α, β, γ) constant and using different φ while expanding the three dimensions of EfficientNet-B0, EfficientNet-B1 to EfficientNet-B7 can be obtained. Under the premise of fully utilizing the resources, EfficientNet-B5 can achieve higher accuracy. Therefore, we choose EfficientNet-B5 as one of the backbone networks based on transfer learning.

2.2 Pixel-Level Attention Module

The basic framework of the pixel-level attention model is shown in Fig. 2. The feature maps obtained through the backbone network are used as the input of the spatial attention module, which can be adaptively and selectively focused on the regions of interest through the attention of a series of convolutional layers. The focused feature map is multiplied point by point with the original feature map to obtain a feature map with key pixels strengthened for better learning of key information. After the spatial attention module, the channel attention module is introduced to perform weighted aggregation through feature recalibration. The larger the weight, the more useful information is contained in the channel, so that the key points can be located more accurately.

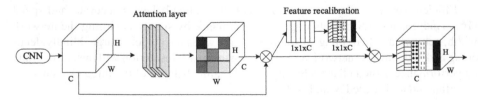

Fig. 2. The basic framework of the pixel-level attention model

Considering that the size of the feature map output via EfficientNet-B5 is (7, 7, 2048) and the size of the feature map output via Inception-ResNetV2 is (5, 5, 1536), we design the corresponding spatial attention modules separately. The basic structure of the spatial attention module for EfficientNet-B5 is shown in Fig. 3. The feature map $X \in R^{H \times W \times C}$ output from the last convolutional layer of EfficientNet-B5 is used as the input of the attention module. Then select some 1×1 convolutional layers as the mapping function for training, and add the corresponding activation function ReLu or Sigmoid after each convolutional layer, and finally output the feature map $F(X) \in R^{H \times W \times 1}$. $F(X)$ corresponds to the weight distribution on a single spatial plane, where the larger the weight, the more key information is contained in the spatial region. To be consistent with the X size, we transform $F(X)$ into $F'(X) \in R^{H \times W \times C}$ by repeating the vector and reshaping. Finally, the multiplication is used to process the strength of each pixel of the feature map X. The output Y can be expressed as

$$Y = X \times F'(X). \tag{5}$$

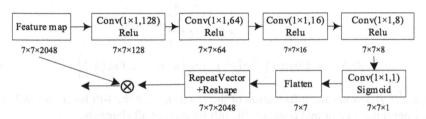

Fig. 3. The spatial attention module for EfficientNet-B5 (ESA module)

The basic structure of the spatial attention module for Inception-ResNetV2 is shown in Fig. 4. The input feature map $X \in R^{H \times W \times C}$ is convolved to obtain a non-linear expression through the ReLu activation function, and then Sigmoid activation function is to obtain the weight distribution $G(X) \in R^{H \times W \times 1}$ on a single spatial plane. Transform the size of $G(X)$ to obtain the feature map $G'(X) \in R^{H \times W \times C}$. Finally, multiply X by $G'(X)$ to obtain the feature map of the pixels with strong or weak processing for the purpose of spatial attention. The output Z can be expressed as

$$Z = X \times G'(X). \tag{6}$$

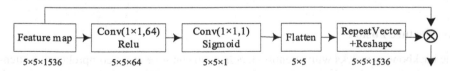

Fig. 4. The spatial attention module for Inception-ResNetV2 (ISA module)

After spatial attention mask filtering of the original feature maps, the channel attention module is added. The principle of the channel attention module used by the two backbone networks is the same. They perform squeeze and excitation training on the

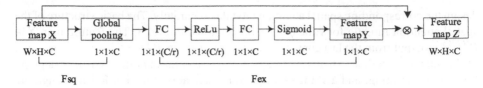

Fig. 5. The channel attention module (CA module)

feature map to obtain the weights of different channels for channel feature selection. The basic structure of the channel attention module is shown in Fig. 5.

The feature mapping processes in Fig. 5 are set up as following. *Fsq* is a feature mapping process where the feature map X is squeezed into a feature map with a size of *1 × 1 × C* through the global pooling, and the global feature of the channel is obtained corresponding to a scalar. The *Fsq* operation can be expressed as

$$Fsq(u^l) = \frac{1}{H \times W} \sum_{i=1}^{H} \sum_{j=1}^{W} u^l(i,j), \qquad (7)$$

$$D(X) = Concat\left[Fsq(u^1), Fsq(u^2), \ldots, Fsq(u^c)\right]. \qquad (8)$$

Where $u^l(i, j)$ denotes the information of the point (i, j) on the l-th layer and $D(X)$ is a one-dimensional vector that connects the information of all channels.

Another feature mapping process is *Fex*, whose input is *D(X)* in formula (8) that is passed through two fully connected layers and ReLu and Sigmoid, which reduce the number of network parameters and obtain more complex Non-linear expression. Then obtain a one-dimensional feature vector *Y* containing the channel weight distribution. After experimental debugging, setting the channel scaling parameter r to 16 performs better. The *Fex* operation can be expressed as

$$Fex(z, W) = \sigma(W_2\delta(W_1z)). \qquad (9)$$

Where z denotes the feature map obtained after the *Fsq* operation; δ denotes the ReLu activation function; σ denotes the Sigmoid activation function; W_1, W_2 denote fully connected layer.

Finally, the attention feature map Z is obtained by multiplying X and Y, where the key features on the channel are captured and useless channel features are filtered out.

2.3 Feature Fusion

The backbone networks with attention mechanism can give more comprehensive attention to the fine-grained lesion features from the space and channel. To further make full use of the global information of fundus images, the features extracted by the attention networks are filtered though the global average pooling layer and the dropout layer, and then fused to achieve feature complementarity and expand feature representations. The feature vectors are concated to achieve feature fusion. Input the fused features into the LightGBM classifier for training, and then output predicted result. The traditional gradient boosting decision tree (GBDT) needs to scan all features, so its computational complexity is relatively large and the efficiency is slow when processing big data. To solve the problem, LightGBM was proposed, where Gradient-based one-side sampling (GOSS) and exclusive feature bundling (EFB) are used to improve training efficiency while ensuring accuracy [17].

2.4   Training Algorithm

Figure 6 shows the classification model training algorithm. The training set is input into the backbone networks with the attention module, where ESA and ISA correspond to the spatial attention modules of EfficientNet-B5 and InceptionResNetV2, respectively. CA is the channel attention module adopted by the backbone networks. The feature maps obtained by the attention module are sequentially input to the global average pooling

---

Algorithm 1: The proposed model training

D: Preprocessed Dataset
L: Dataset Label
ESA: Spatial Attention Block for EfficientNet-B5
ISA: Spatial Attention Block for Inception-ResNetV2
CA:Channel Attention Block with Squeeze and Excitation
Model one:

$A_1 \leftarrow EfficientNet(input\_shape)(D)$

$A_2 \leftarrow ESA(A_1)$

$A_3 \leftarrow CA(A_2)$

$A_4 \leftarrow GlobalAveragePooling2D()(A_3)$

$A_5 \leftarrow Dropout(0.5)(A_4)$

Return A_5

Model two:

$B_1 \leftarrow Inception\,\mathrm{Re}\,sNetV2(input\_shape)(D)$

$B_2 \leftarrow ISA(B_1)$

$B_3 \leftarrow CA(B_2)$

$B_4 \leftarrow GlobalAveragePooling2D()(B_3)$

$B_5 \leftarrow Dropout(0.5)(B_4)$

Return B_5
1: Start
2: $F \leftarrow Concat(A_5, B_5)$
3:Traing LightGBM using D, L
4. End

Fig. 6. The proposed model training algorithm description

layer and dropout layer, which can prevent networks overfitting while reducing the amount of calculations. Among them, the dropout value is 0.5, and the one-dimensional feature vector is output. The fused features are input into the LightGBM classifier for training to extract deep attention features with better characterization capability. In order to save training time, the early stopping method is used. In each iteration of the network training process, cross entropy [18] is used as the loss function of DR grading, and the network parameters are continuously updated according to the loss value of each iteration. The Adam [19] optimizer is used to avoid the local optimization and speed up the training speed during the training process.

3   Datasets

There are three public datasets including EyePAC, Messidor, and OIA datasets in the work. There are 35126 fundus images marked with DR grades in the EyePACS dataset, which are categorized into 0 to 4 grades according to the severity of DR. There are 1200 fundus images in TIF format in the Messidor dataset, which are categorized into 0 to 3 grades. There are 13673 fundus images for DR grading in the OIA dataset, which are categorized into 0 to 5 grades.

Considering grade 0 (normal) and grade 1 (mild non-proliferative DR) have little effect on health, but grade 2 (moderate non-proliferative DR) and above must be treated with corresponding measures, Fundus images are categorized into two grades in the work. Among them, grade 0 and grade 1 are categorized and re-marked as 0, grade 2 and above are categorized and re-marked as 1.

3.1   Data Normalization

There are some problems in the datasets, such as color interference and category imbalance, so we use data normalization and data augmentation to preprocess the images.

The specific steps of image normalization are as follows:

1) Eliminate useless images (overexposure, underexposure or shooting failure, etc.).
2) Crop the black borders around the image.
3) The method of equalizing brightness and contrast is used to normalize the image. This process is described as

$$M_{out} = M \times \alpha + \beta \times G(\rho) * M + \gamma. \tag{10}$$

Where $G(\rho)$ denotes the Gaussian smoothing function with standard deviation ρ; * denotes the convolution operation; M and M_{out} are input and output images, respectively. According to experimental debugging, $(\alpha, \beta, \rho, \gamma)$ is set to $(4, -4, 10, 128)$.

4) Resize the image to 224 * 224 pixels.

Some images are randomly selected from the EyePACS dataset. The comparison before and after preprocessing are shown in Fig. 7.

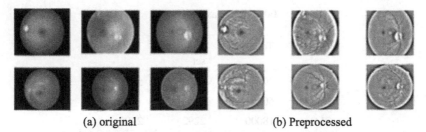

(a) original   (b) Preprocessed

Fig. 7. Comparison before and after preprocessing

3.2 Data Augmentation

Since the proportion of grade 1 images in the datasets is small, in order to avoid overfitting the classification model, data augmentation is used. Augment grade 1 images through operations such as rotation, mirroring, and flipping to achieve category balance.

In addition, the Keras Image Data Generator is used for the training set to augment samples and improve the generalization ability of the model. The related augmentation methods and corresponding parameters are shown in Table 1.

Table 1. Data augmentation parameters

Method	Parameter
Rotation	10
Width_shift	0.2
Height_shift	0.2
Shear	0.2
Zoom	0.1
Horizontal_flip	FALSE
Brightness	[0.1, 2]
Fill_mode	nearest

4 Experimental and Results

4.1 Experimental Setup

We uses EyePACS, Messidor and OIA datasets to verify the performance of the classification model. The preprocessed datasets are randomly divided into training set, validation set and test set, where the number distribution is shown in Table 2.

Table 2. Dataset division

Dataset	Training set	Validation set	Test set
EyePACS	30000	5000	5000
Messidor	6000	1000	1000
OIA	18000	2292	2292

4.2 Evaluation Metrics

To measure the model performance, we use accuracy (Acc), sensitivity (Sen), specificity (Spe), and AUC value, which are formulated as

$$\text{Acc} = \frac{TP + TN}{TP + FP + TN + FN}, \tag{11}$$

$$\text{Sen} = \frac{TP}{TP + FN}, \tag{12}$$

$$\text{Spe} = \frac{TN}{FP + TN}. \tag{13}$$

Where TP, FP, TN, and FN denote true positive, false positive, true negative, and false negative, respectively.

4.3 Model Performance Analysis

To select better CNNs and a classification model, some CNNs are tested, including InceptionV3, ResNet-50, DenseNet, Inception-ResNetV2 and EfficientNet-B5. Some classification models are used in CNNs, including SVM, RandomForest, XGBoost, and LightGBM. The experiments show that the performance of LightGBM is slightly improved by 0.02% compared to the others. Table 3 shows the test results of CNNs with LightGBM on the EyePACS dataset. It can be seen that the Inception-ResNetV2 and EfficientNet-B5 have the most significant performance.

Table 3. Experimental results of different CNNs model on the EyePACS dataset

Experiment	Models	Acc/%
1	Inception-V3	84.45
2	ResNet-50	85.3
3	DenseNet	87.23
4	Inception-ResNetV2	89.84
5	EfficientNet-B5	91.04

Table 4 shows the test results of five methods designed in the work on the EyePACS dataset. Experiments 1 to 5 denote Inception-ResNetV2, EfficientNet-B5, Inception-ResNetV2 with the PATT module, EfficientNet-B5 with the PATT module, and attention network fusion (Ensembles), respectively. It can be seen from the Table 4 that the performance of the backbone network with the pixel-level attention module is better than that of the original backbone network, and the performance of feature fusion is better than that of a single network. These show that the pixel-level attention module and feature fusion can effectively extract key features, which is helpful to improve the performance of the model.

Table 4. Experimental results of different model on the EyePACS dataset

Experiment	Methods	Acc/%	Sen/%	Spe/%
1	Inception-ResNetV2	89.84	84.87	93.85
2	EfficientNet-B5	91.04	85.64	95.53
3	Inception-ResNetV2 + PATT	94.1	90.23	97.14
4	EfficientNet-B5 + PATT	96.02	92.92	98.41
5	Ensembles	**96.36**	**93.45**	**98.59**

Table 5 shows the collected results on the EyePACS and the Messidor datasets for the DR classification task. On the EyePACS dataset, Bajwa et al. [10] achieved a good result by integrating ResNet and DenseNet networks and combining coarse and fine-grained classifiers, which is slightly higher in sensitivity and AUC value. However, our accuracy is 1.05% higher, the specificity is much better, and our method is superior on the Messidor dataset. On the Messidor dataset, Wang et al. [9] proposed zoom-in Net to zoom in for important details, which is extremely poor in specificity. On the OIA dataset, our method achieves an accuracy of 94.28%, a sensitivity of 90.86%, a specificity of 97.93%, and a AUC values of 0.94. Through a comprehensive analysis of the performance of these methods, it can be concluded that our method has good effectiveness and generalizability.

Table 5. Comparison of experimental results of different methods

Method	Acc/%		Sen/%		Spe/%		AUC/%	
	EyePACS	Messidor	EyePACS	Messidor	EyePACS	Messidor	EyePACS	Messidor
Seound et al. [20]	N/A	74.5	N/A	N/A	N/A	N/A	N/A	91.6
Vo et al. [21]	N/A	89.7	N/A	89.3	N/A	90	N/A	89.1
Wang et al. [9]	N/A	91.1	N/A	97.8	N/A	50	N/A	95.7

(*continued*)

Table 5. (*continued*)

Method	Acc/%		Sen/%		Spe/%		AUC/%	
	EyePACS	Messidor	EyePACS	Messidor	EyePACS	Messidor	EyePACS	Messidor
Zeng et al. [8]	N/A	N/A	80.7	N/A	95	N/A	94.9	N/A
Nguyen et al. [22]	82	N/A	80	N/A	82	N/A	90.4	N/A
Bajwa et al. [10]	95.34	89.25	95	89	81	91	99.23	96.45
Ensembles	96.46	93.7	93.87	91.97	98.42	95.6	97.3	96.8

In order to illustrate the reliability of the model more intuitively, we use the Grad-CAM algorithm to generate a heat map [23]. In Fig. 8, we can clearly see that there are obvious bleeding points in the red frame of image (a). The heat map generated by the Grad-CAM algorithm is fused with the original image (a) to obtain the image (b). It can be observed that our method has learned the lesion information very well, and improved the capability of the network to extract effective features.

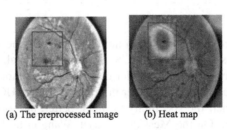

(a) The preprocessed image   (b) Heat map

Fig. 8. Heat map visualization

5   Conclusion

In this work, we propose a novel pixel-level attention mechanism and feature fusion method for DR grading. Considering the importance of fine-grained features for DR grading, we design corresponding pixel-level attention modules for different network structures to maximize the advantages of individual networks, and improve the expression of key features and weaken the interference of irrelevant information by processing the strength and weakness of each pixel point on the image with the pixel-level attention mechanism. Meanwhile, the effective feature information is expanded by fusing the features of the attention network. Finally, the LightGBM classifier is used to train the fused deep attention features and extract more discriminative feature representations to complete the DR classification task. Finally, the experiments on DR datasets verify the effectiveness and generalization of the classification model. In the future work, we will improve the classification model so that it can accurately extract more discriminative features to expand the gap between classes and improve the accuracy of DR grading.

Acknowledgments. The authors thank the members of Machine Learning and Artificial Intelligence Laboratory, School of Computer Science and Technology, Wuhan University of Science and Technology, for their helpful discussion within seminars. This work was supported by, National Natural Science Foundation of China (61972299, U1803262, 61702381), Hubei Province Natural Science Foundation of China (No. 2018CFB526).

References

1. Cho, N.H., Shaw, J.E., Karuranga, S., et al.: IDF diabetes atlas: global estimates of diabetes prevalence for 2017 and projections for 2045. Diabetes Res. Clin. Pract. **138**, 271–281 (2018)
2. Zhang, G., Chen, H., Chen, W., et al.: Prevalence and risk factors for diabetic retinopathy in China: a multi-hospital-based cross-sectional study. Br. J. Ophthalmol. **101**(12), 1591–1595 (2017)
3. Kourgialis, N.: Diabetic retinopathy-silently blinding millions of people world-wide. Invest Ophtalmol. Vis. Sci. **57**, 6669–6682 (2017)
4. Deka, A., Sarma, K.K.: SVD and PCA features for ANN based detection of diabetes using retinopathy. In: Proceedings of the CUBE International Information Technology Conference, pp. 38–41 (2012)
5. Kumar, M., Nath, M.K.: Detection of microaneurysms and exudates from color fundus images by using SBGFRLS algorithm. In: Proceedings of the International Conference on Informatics and Analytics, pp. 1–6 (2016)
6. Xie, Y., Huang, H., Hu, J.: Staging and lesion detection of diabetic retinopathy based on deep convolution neural network. J. Comput. Appl. **40**(8), 2460–2464 (2020)
7. Zhou, Y., Wang, B., He, X., et al.: DR-GAN: conditional generative adversarial network for fine-grained lesion synthesis on diabetic retinopathy images. arXiv preprint arXiv:1912.04670 (2019)
8. Zeng, X.L., Chen, H.Q., Luo, Y., et al.: Automated detection of diabetic retinopathy using a binocular siamese-like convolutional network. In: 2019 IEEE International Symposium on Circuits and Systems (ISCAS), pp. 1–5. IEEE (2019)
9. Wang, Z., Yin, Y., Shi, J., et al.: Zoom-in-net: deep mining lesions for diabetic retinopathy detection. In: International Conference on Medical Image Computing and Computer-Assisted Intervention, pp. 267–275 (2017)
10. Bajwa, M.N., Taniguchi, Y., Malik, M.I., et al.: Combining fine-and coarse-grained classifiers for diabetic retinopathy detection. In: Annual Conference on Medical Image Understanding and Analysis, pp. 242–253 (2019)
11. Fukui, H., Hirakawa, T., Yamashita, T., et al.: Attention branch network: learning of attention mechanism for visual explanation. In: Proceedings of the IEEE/CVF Conference on Computer Vision and Pattern Recognition, pp. 10705–10714 (2019)
12. Hu, J., Shen, L., Sun, G.: Squeeze-and-excitation networks. In: Proceedings of the IEEE conference on computer vision and pattern recognition, pp. 7132–7141 (2018)
13. Mei, X., Pan, E., Ma, Y., et al.: Spectral-spatial attention networks for hyperspectral image classification. Remote Sens. **11**(8), 963 (2019)
14. Russakovsky, O., Deng, J., Su, H., et al.: Imagenet large scale visual recognition challenge. Int. J. Comput. Vision **115**(3), 211–252 (2015)
15. Szegedy, C., Ioffe, S., Vanhoucke, V., et al.: Inception-v4, inception-resnet and the impact of residual connections on learning. arXiv preprint arXiv:1602.07261 (2016)
16. Tan, M., Le, Q.V.: Efficientnet: rethinking model scaling for convolutional neural networks. arXiv preprint arXiv:1905.11946 (2019)

17. Ke, G., Meng, Q., Finley, T., et al.: Lightgbm: a highly efficient gradient boosting decision tree. In: Advances in Neural Information Processing Systems, pp. 3146–3154 (2017)
18. Hussien, E., Elsayed, Z.: Improving error back propagation algorithm by using cross entropy error function and adaptive learning rate. Int. J. Comput. Appl. **161**(8), 5–9 (2017)
19. Kingma, D.P., Ba, J.: Adam: a method for stochastic optimization. arXiv preprint arXiv:1412. 6980 (2014)
20. Seoud, L., Hurtut, T., Chelbi, J., et al.: Red lesion detection using dynamic shape features for diabetic retinopathy screening. IEEE Trans. Med. Imag. **35**(4), 1116–1126 (2016)
21. Vo, H.H., Verma, A.: New deep neural nets for fine-grained diabetic retinopathy recognition on hybrid color space. In: 2016 IEEE International Symposium on Multimedia (ISM), pp. 209–215. IEEE (2016)
22. Nguyen, Q.H., Muthuraman, R., Singh, L., et al.: Diabetic retinopathy detection using deep learning. In: Proceedings of the 4th International Conference on Machine Learning and Soft Computing, pp. 103–107 (2020)
23. Selvaraju, R.R., Cogswelli, M., Das, A., et al.: Grad-CAM: visual explanations from deep networks via gradient-based localization. Int. J. Comput. Vision **128**(2), 336–359 (2020)

A Comparable Study on Dimensionality Reduction Methods for Endmember Extraction

Guangyi Chen[1(✉)] and Wenfang Xie[2]

[1] Department of Computer Science and Software Engineering, Concordia University, Montreal, QC H3G 1M8, Canada
[2] Department of Mechanical, Industrial and Aerospace Engineering, Concordia University, Montreal, QC H3G 1M8, Canada
wfxie@encs.concordia.ca

Abstract. Endmember extraction is widely used to detect spectrally unique signatures of pure ground materials in hyperspectral imagery (HSI). In this paper, we study which method is good for reducing the dimensionality of the input data cubes for endmember extraction. In existing works, either principal component analysis (PCA) or minimum noise fraction (MNF) is used to reduce the dimensionality of the input data cubes. For the first time in literature, we combine locally linear embedding (LLE) with pixel purity index (PPI) and N-FINDR for endmember extraction. In addition, we compare LLE with PCA and MNF for endmember extraction. We find that MNF is better than PCA and LLE, and PCA is comparable to LLE when combined with PPI and N-FINDR for endmember extraction.

Keywords: Hyperspectral imagery (HSI) · Dimensionality reduction · Locally linear embedding (LLE) · Principal component analysis (PCA) · Minimum noise fraction (MNF) · Endmember extraction

1 Introduction

Linear spectral unmixing is a widely adopted method to classify mixed pixels in hyperspectral imagery (HSI). It requires to find spectrally unique signatures of pure ground components, usually known as endmembers, and needs to express mixed pixels as linear combinations of endmembers. We briefly review several existing spectral unmixing methods published in the literature here. Keshava and Mustard [1] studied spectral unmixing methods in a systematically way. Chang and Du [2] estimated the number of spectrally distinct signal sources in HSI data cubes. They developed three Neyman–Pearson detection theory-based eigen thresholding techniques to solve such problems that model the virtual dimensionality (VD) estimation as a binary composite hypothesis testing problem and the VD estimation error can be determined by receiver operating characteristic (ROC) analysis. Boardman et al. [3] studied mapping target signatures via partial unmixing of AVIRIS data, which is called pixel purity index (PPI). The PPI is widely utilized in hyperspectral image analysis for endmember extraction due to its publicity and availability in the Environment for Visualizing Images (ENVI) software.

© Springer Nature Switzerland AG 2021
D.-S. Huang et al. (Eds.): ICIC 2021, LNCS 12836, pp. 143–150, 2021.
https://doi.org/10.1007/978-3-030-84522-3_11

Chang and Plaza [4] proposed a fast-iterative algorithm for implementation of pixel purity index (FIPPI). Instead of randomly generating vectors as initial endmembers, the FIPPI determines an appropriate initial set of endmembers to speed up its process. Furthermore, it calculates the number of endmembers by VD. Winter [5] invented the N-FINDR: an algorithm for fast autonomous spectral endmember determination in hyperspectral data. This technique is based on the geometry of convex sets to find a unique set of purest pixels in a data cube. The N-volume contained by a simplex from the purest pixels is larger than any other volume from any other combination of pixels. The method inflates a simplex inside the data, starting from a random set of pixels. For each pixel and each endmember, the endmember is replaced with the spectrum of the pixel and the volume is recalculated. If it increases, the spectrum of the new pixel replaces that endmember. This process is iterated until no more replacements are done. Plaza et al. [6] proposed a novel method for unsupervised pixel purity determination and endmember extraction from multidimensional data cubes. It used both spatial and spectral information in a combined manner. It was based on mathematical morphology, a classic image processing method that can be applied to the spectral domain while keeping its spatial characteristics. The method was evaluated by a specifically designed framework that utilized both simulated and real hyperspectral data cubes. Ozkan et al. [7] developed a new endmember extraction and hyperspectral unmixing method, which is a two-step autoencoder network. It was entirely enhanced and restructured by introducing additional layers and a projection metric. They introduced a new loss function that is a Kullback-Leibler divergence term with spectral angle distance (SAD) similarity and additional penalty terms to improve the sparsity of the estimates. These modifications set the common properties of endmembers, such as nonlinearity and sparsity for autoencoder networks. Their method is suitable for large-scale data and it can speed up on graphical processing units (GPU). Nascimento and Dias [8] proposed a novel method for unsupervised endmember extraction from hyperspectral data, which is called vertex component analysis (VCA). The method utilizes two points: (a) the endmembers are the vertices of a simplex and (b) the affine transformation of a simplex is also a simplex. Wu et al. [9] studied simplex growing algorithm (SGA) and extended it to a real-time processing algorithm that can solve for four problems in N-FINDR, (a) use of random initial endmembers that introduces inconsistent results, (b) high computational complexity that is from an exhaustive search for finding all endmembers simultaneously, (c) dimensionality reduction due to big data volumes, and (d) lack of real-time capability.

In this paper, we investigate different dimensionality reduction methods and combine them with existing endmember extraction methods such as PPI and N-FINDR. We compare locally linear embedding (LLE) with principal component analysis (PCA) and minimum noise fraction (MNF), and we find that MNF performs the best for endmember extraction. Furthermore, PCA is comparable to LLE for endmember extraction as demonstrated in our experiments. Even though LLE is a nonlinear dimensionality reduction method, it does not outperform linear dimensionality reduction method (MNF) for endmember extraction, which is undesirable.

The organization of this paper is given as follows. Section 2 studies LLE, PCA and MNF as dimensionality reduction techniques for endmember extraction. Section 3 performs experiments to show the effectiveness of different dimensionality reduction methods when combined with PPI and N-FINDR. Finally, Sect. 4 concludes the paper and proposes future research directions.

2   The Proposed Method

Hyperspectral data analysis is an important job in identification, detection, estimation, and discrimination of earth surface materials. It is closely related to dimensionality reduction, endmember extraction, atmospheric correction, spectral unmixing and classification phases. One important aim in hyperspectral data processing is to yield high classification rate, which relies on the extracted endmembers. An endmember is a spectrally unique, idealized, and pure signature of a surface material. Endmember extraction is an important goal to obtain the high accuracy for hyperspectral data classification and spectral unmixing.

Hyperspectral data analysis needs to determine certain basis spectra called endmembers. After these spectra are determined, the HSI data cube can be unmixed into the fractional abundance of each material in each pixel. The requirement to study a huge amount of multivariate data introduces the major problem of dimensionality reduction. The LLE [10] calculates low-dimensional, neighborhood-preserving embeddings of high-dimensional data. The LLE maps its high dimensional inputs into a single global coordinate system in a much lower dimension, and its optimizations only involve global minima. By studying the local structure of linear reconstructions, the LLE finds the global structure of nonlinear manifolds.

In this paper, we study PCA, MNF and LLE to reduce the dimensionality of the input HSI data cube and combine them with existing endmember extraction methods such as PPI and N-FINDR. Experiments demonstrate that the dimensionality reduction using MNF is the best for endmember extraction for all testing cases. Also, PCA and LLE are comparable for endmember extraction. In our previous work [11], we have already proposed an improved LLE for hyperspectral imagery analysis including endmember extraction.

Our proposed study for endmember extraction can be described as follows:

Input: A - An HSI data cube with spatial size of $M \times N$ pixels and P spectral bands. K – The number of neighborhoods used in LLE (50).
Output: The endmember spectra extracted from the HSI data cube A.

1. Find the number of distinct endmembers [2] in data cube A, denoted as Num.
2. Set $d = 2 \times$ Num for LLE + PPI and $d =$ Num for all other methods.
3. Perform PCA, MNF and LLE to the given HSI data cube A with K neighbours and retain d output images, denoted the output as B $(M \times N \times d)$.
4. Extract endmember spectra of dimension $d \times 1$ from B by means of PPI.
5. Extract endmember spectra of dimension $d \times 1$ from B by means of N-FINDR.
6. Find the spatial locations of those endmembers for both PPI and N-FINDR and output the endmember spectra of dimension $P \times 1$ from the original data cube A.
7. Compute the spectral angle mapper (SAM) between ground truth endmember spectra and extracted endmember spectra for both PPI and N-FINDR.

The major contribution of this paper can be given as follows. We investigate different dimensionality reduction methods such as LLE, PCA and MNF for endmember extraction. For the first time in history, we combine LLE with PPI and N-FINDR for endmember extraction. Our experiments demonstrate that MNF is better than PCA and LLE, and PCA is comparable to LLE for endmember extraction. Although LLE is a nonlinear dimensionality reduction method, it cannot beat linear dimensionality method MNF for endmember extraction, which is undesirable. Another limitation of LLE is that it requires a big amount of memory to run and it is slower than PCA and MNF. Therefore, LLE can only be applied to small and intermediate sized HSI data cubes.

3 Experimental Results

In our experiments, we test the performance of different dimensionality reduction methods for endmember extraction when combined with PPI and N-FINDR. We test two widely used data cubes in our experiments. The first data cube is the Indian pines data cube with spatial size of 145×145 pixels and 220 spectral bands. The second data cube is the jasperRidge data cube with spatial size of 100×100 pixels and 198 spectral bands. Both data cubes are taken from Matlab without much detail about them. We use the following commands in Matlab to load them:

$$A = \text{hypercube}('\text{indian\_pines.dat}');$$

$$A = \text{hypercube}('\text{jasperRidge2\_R198.hdr}');$$

In this paper, we choose MATLAB R2020b, which contains hyperspectral imaging Add-Ons. Figure 1 show the #50 spectral band for the Indian pines data cube and the jasperRidge data cube, respectively. We choose $K = 50$ as the number of neighbours in the LLE algorithm. We use the following Matlab command to find the number of endmembers in the data cube A:

$$d = \text{countEndmembersHFC}(A);$$

Figures 2 and 3 show the endmember spectra for the Indian pines and jasperRidge data cubes, respectively. For the Indian pines data cube, MNF + PPI and MNF + N-FINDR performs the best. For the jasperRidge data cube, MNF + PPI and MNF + N-FINDR also obtain the best results for endmember extraction. Nevertheless, LLE is comparable to PCA and it needs more memory than PCA or MNF to run on a computer.

Tables 1 and 2 show the SAM between the ground truth endmembers and the extracted endmembers by using LLE, PCA and MNF for the Indian pines data cube and the jasperRidge data cube, respectively. The numbers of distinct endmembers for both data cubes are determined by [2]. We use FIPPI to approximately estimate the ground truth endmembers because we do not know the ground truth endmembers for these two data cubes. From both tables, we can see that MNF performs better than PCA and LLE, and PCA is comparable to LLE for endmember extraction.

(a)   **(b)**

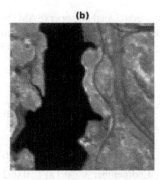

Fig. 1. The #50 spectral bands for (a) the Indian pines data cube; (b) the jasperRidge data cube.

Table 1. The spectral angle mapper (SAM) between the ground truth endmember spectra and the extracted endmember spectra for the Indian pines data cube for different dimensionality reduction methods. The number of distinct endmembers is determined by [2]. The best method is highlighted in bond font.

	PPI			N-FINDR		
	PCA	MNF	LLE	PCA	MNF	LLE
Endmember 1	**0**	**0**	0.1385	0.0405	**0**	0.0331
Endmember 2	0.0405	**0**	**0**	0.1182	**0.0613**	0.1385
Endmember 3	0.1385	**0.0405**	0.1288	0.0655	**0.0256**	0.0766
Endmember 4	**0**	0.0163	**0**	0.0617	0.0454	**0**
Endmember 5	0.1224	**0.0788**	0.1093	0.1077	**0**	0.1053
Endmember 6	**0.0703**	0.0740	0.1224	**0.0408**	0.0761	0.1288
Endmember 7	0.0647	**0**	0.1233	**0**	0.0725	**0**
Endmember 8	0.1233	0.1288	**0.0766**	0.0851	**0.0786**	0.1175
Endmember 9	0.0676	**0**	0.1112	**0.0543**	0.0997	0.0880
Endmember 10	0.0632	**0.0256**	0.1175	**0.0656**	0.0846	0.0703
Endmember 11	**0.0301**	0.0470	0.1318	0.1004	0.1224	**0.0653**
Endmember 12	0.0987	**0.0441**	0.0961	0.1247	0.0740	**0.0591**
Endmember 13	0.1195	**0**	0.0653	0.0642	**0.0390**	0.0812
Endmember 14	0.0676	**0**	0.1422	0.0870	**0.0808**	0.1311
Endmember 15	0.0650	**0**	0.0890	0.1320	0.1122	**0.0975**
Endmember 16	0.1084	**0**	0.0859	**0.0384**	0.0766	0.1316
Endmember 17	0.0535	**0**	**0**	0.0744	**0.0163**	0.0583
Endmember 18	0.0383	**0.0307**	0.0405	0.0997	**0.0676**	0.0697

Table 2. The spectral angle mapper (SAM) between the ground truth endmember spectra and the extracted endmember spectra for the jasperRidge data cube for different dimensionality reduction methods. The number of distinct endmembers is determined by [2]. The best method is highlighted in bond font.

	PPI			N-FINDR		
	PCA	MNF	LLE	PCA	MNF	LLE
Endmember 1	**0**	**0**	**0**	0.0392	**0**	0.3534
Endmember 2	0.0294	0.0363	**0**	**0.0600**	0.0923	0.1245
Endmember 3	0.1814	0.0923	**0.0297**	0.3691	0.1231	**0**
Endmember 4	0.0392	**0**	0.1053	**0**	**0**	**0**
Endmember 5	0.0233	**0**	**0**	**0.0190**	0.0195	0.0531
Endmember 6	0.0998	**0.0249**	0.3278	0.0998	**0.0784**	0.1199
Endmember 7	0.0156	**0**	0.1542	0.0794	**0**	0.0769
Endmember 8	**0**	0.1529	**0**	0.0645	**0**	0.1540
Endmember 9	0.1582	0.1542	**0.0156**	0.0745	**0**	0.0450
Endmember 10	0.0303	**0.0195**	0.1245	**0.0350**	0.0392	0.0785

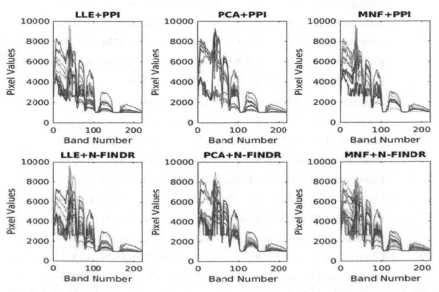

Fig. 2. Endmember spectra for PPI and N-FINDR when combined with LLE, PCA and MNF for the Indian pines data cube.

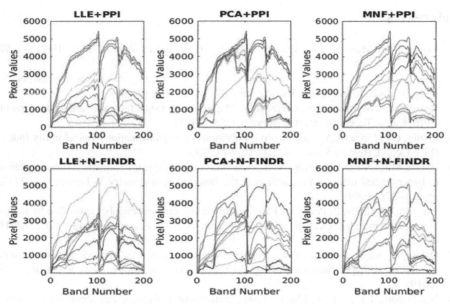

Fig. 3. Endmember spectra for PPI and N-FINDR when combined with LLE, PCA and MNF for the jasperRidge data cube.

4 Conclusions

An endmember in mineralogy is a mineral which is at the extreme end of a mineral series in purity. Endmember extraction is a main problem in unmixing mineral analysis and accurate unmixing relies on precise endmember determination. It is not easy to find completely pure pixels that contain only one endmember spectrum in the HSI data cube. Since we are considering subpixel content, it is difficult to know whether a pixel is pure or not. Even if there exist such pixels, finding them in high-dimensional HSI data cube is exceedingly difficult.

In this paper, we have studied different dimensionality reduction methods before an endmember extraction method is applied. In existing works, only PCA or MNF is combined with endmember extraction methods such as PPI and N-FINDR. In this paper, we study LLE as a dimensionality reduction technique for endmember extraction as well. We find that MNF is better than PCA and LLE, and PCA is comparable to LLE for endmember extraction for two testing data cubes. Nevertheless, the limitation of LLE is that it needs a lot of memory to run and it is slower than PCA and MNF. Consequently, LLE can only be applied to small and intermediate sized HSI data cubes.

Future research will be done in the following ways. We may apply LLE to hyperspectral imagery denoising as an extension to our previous work [12] (PCA + Wavelet Shrinkage). We may also study deep convolutional networks (CNN) for endmember extraction and hyperspectral imagery denoising as our next research projects.

References

1. Keshava, N., Mustard, J.F.: Spectral unmixing. IEEE Signal Process. Mag. **19**(1), 44–57 (2002)
2. Chang, C.I., Du, Q.: Estimation of number of spectrally distinct signal sources in hyperspectral imagery. IEEE Trans. Geosci. Remote Sens. **42**(3), 608–619 (2004)
3. Boardman, J.W., Kruse, F.A., Green, R.O.: Mapping target signatures via partial unmixing of AVIRIS data, Technical report, California, USA (1995)
4. Chang, C.I., Plaza, A.A.: Fast iterative algorithm for implementation of pixel purity index. IEEE Geosci. Remote Sens. Lett. **3**(1), 63–67 (2006)
5. Winter, M.E.: N-FINDR: an algorithm for fast autonomous spectral endmember determination in hyperspectral data. In: Proceedings at the Imaging Spectrometry V, Denver, vol. 3753, pp. 266–275 (1999)
6. Plaza, A., Martinez, P., Perez, R., Plaza, J.: Spatial/spectral endmember extraction by multi-dimensional morphological operations. IEEE Trans. Geosci. Remote Sens. **40**(9), 2025–2041 (2002)
7. Ozkan, S., Kaya, B., Akar, G.B.: EndNet: Sparse autoEncoder network for endmember extraction and hyperspectral unmixing. IEEE Trans. Geosci. Remote Sens. **57**(1), 482–496 (2019)
8. Nascimento, J.M.P., Dias, J.M.B.: Vertex component analysis: a fast algorithm to unmix hyperspectral data. IEEE Trans. Geosci. Remote Sens. **43**(4), 898–910 (2005)
9. Wu, C., Chang, C., Ren, H., Chang, Y.L.: Real-time processing of simplex growing algorithm. In: IEEE International Geoscience and Remote Sensing Symposium, Cape Town, South Africa (2009)
10. Roweis, S., Saul, L.: Nonlinear dimensionality reduction by locally linear embedding. Science **290**(5500), 2323–2326 (2000)
11. Qian, S.E., Chen, G.Y.: A new nonlinear dimensionality reduction method with application to hyperspectral image analysis. In: Proceedings of IEEE International Geoscience and Remote Sensing Symposium (IGARSS), Barcelona, Spain, 23–27 July 2007
12. Chen, G.Y., Qian, S.E.: Denoising of hyperspectral imagery using principal component analysis and wavelet shrinkage. IEEE Trans. Geosci. Remote Sens. **49**(3), 973–980 (2011)

Hyperspectral Image Classification with Locally Linear Embedding, 2D Spatial Filtering, and SVM

Guang Yi Chen[1(✉)], Wen Fang Xie[2], and Shen-En Qian[3]

[1] Department of Computer Science and Software Engineering, Concordia University, Montreal, QC H3G 1M8, Canada
[2] Department of Mechanical, Industrial and Aerospace Engineering, Concordia University, Montreal, QC H3G 1M8, Canada
wfxie@encs.concordia.ca
[3] Space Science and Technology, Canadian Space Agency, St-Hubert, QC J3Y 8Y9, Canada
shen-en.qian@canada.ca

Abstract. In this paper, we propose a novel algorithm for hyperspectral image (HSI) classification. We use locally linear embedding (LLE) to reduce the dimensionality of the data cube, 2D spatial filtering on the LLE output band images, and support vector machine (SVM) to classify HSI pixels. In this way, both spatial information and spectral information are taken into consideration for HSI classification. Experiments demonstrate that this new algorithm is extremely competitive when compared to several existing methods.

Keywords: Hyperspectral image classification · Locally linear embedding (LLE) · 2D spatial filtering · Support vector machine (SVM)

1 Introduction

Hyperspectral image (HSI) classification is especially important in analyzing land cover for remotely sensed hyperspectral images, and it has been investigated widely in geoscience, environmental science, and computer vision. If we have knowledge on true classes, the main goal in HSI classification is to assign a class label to each pixel of the observed hyperspectral image. According to machine learning and feature extraction theories, many HSI classification methods have been developed to improve the classification accuracy.

We briefly review several existing methods here. Liu et al. [1] proposed a new method for dimensionality reduction of hyperspectral images based on improved spatial–spectral weight manifold embedding. F. Zhou et al. [2] studied hyperspectral image classification via spectral-spatial LSTMs. Chen et al. [3] investigated hyperspectral image classification by kernel sparse representation. Melgani and Bruzzone [4] worked on the classification of hyperspectral remote sensing images with support vector machines. Fauvel et al. [5] investigated a spatial–spectral kernel-based approach for the classification of remote-sensing images. Camps-Valls and Bruzzone [6] studied kernel-based methods

© Springer Nature Switzerland AG 2021
D.-S. Huang et al. (Eds.): ICIC 2021, LNCS 12836, pp. 151–159, 2021.
https://doi.org/10.1007/978-3-030-84522-3_12

for hyperspectral image classification. Li et al. [7] proposed a generalized composite kernel framework for hyperspectral image classification. Chen et al. [8] studied hyperspectral image classification using dictionary-based sparse representation. Li et al. [9] worked on spectral-spatial classification of hyperspectral data using loopy belief propagation and active learning. Kang et al. [10] studied spectral-spatial hyperspectral image classification with edge-preserving filtering. Cheng et al. [11] investigated semisupervised hyperspectral image classification via discriminant analysis and robust regression. Chen et al. [12] worked on deep learning-based classification of hyperspectral data with extremely good results.

In this paper, we develop a new algorithm for HSI classification. We choose locally linear embedding (LLE) [13] to reduce the dimensionality of the input data cube, 2D spatial filtering [14] to the LLE output bands, and support vector machine (SVM) [15] to classify HSI data cube. As a result, both spatial information and spectral information are taken into consideration for HSI classification. Experiments show that this new algorithm is extremely competitive when compared to several existing methods.

The organization of this paper is given as follows. Section 2 proposes a new method for HSI classification with LLE and SVM. Section 3 performs experiments to show the effectiveness of our proposed method. Finally, Sect. 4 concludes the paper.

2 The Proposed Method

The requirement to study a huge amount of multivariate data brings in the basic problem of dimensionality reduction (DR): how to discover compact representations of high-dimensional data? The LLE [13] is an unsupervised learning method, which can calculate low-dimensional, neighborhood-preserving embeddings of high-dimensional data. Instead of clustering methods for local DR, the LLE maps its high dimensional inputs into a single global coordinate system in a much lower dimension, and its optimizations only solve global minima. By investigating the local symmetries of linear reconstructions, the LLE learn the global structure of nonlinear manifolds.

The huge need for two-dimensional (2D) digital filters [14] can be derived from different reasons: high efficiency for image processing and analysis, big application flexibility, good expansion of standard computers and minicomputers or special type processors at decreasing cost. Efficient 2D digital filters of finite impulse response (FIR) type via suitable window functions is also available in the literature.

The purpose of an SVM algorithm [15] is to find a hyperplane in a higher dimensional space that correctly classifies the data points. To separate the two classes of data samples, there exist many possible hyperplanes that can be chosen. The aim is to find a plane that has the maximum margin, i.e., the maximum distance between data points of both classes. By maximizing the margin, we can introduce some reinforcement so that future data points can be classified with more confidence.

We propose a new method for HSI classification in this section. We choose LLE to reduce the dimensionality of the input HSI data cube, 2D spatial filtering to convolve the LLE output band images, and SVM to classify the HSI data cube. Our new method considers both spatial information and spectral information at the same time in the process of classification. Experiments demonstrate that our method is extremely competitive when compare to existing methods.

Our proposed method for HSI classification can be described as follows:

Input: A - An HSI data cube with spatial size of M×N pixels and P spectral bands.

K – The number of neighborhoods used in LLE (60).

d - The number of spectral band images to be retained after dimensionality reduction using LLE (60).

Output: The class of each spatial pixel in the HSI data cube A.

1. Perform LLE to the given HSI data cube A and retain d output images, denoted as B (M×N×d).
2. Set 2D filer $f = 1/49 \times ones(7,7)$, where $ones(7,7)$ is a matrix of size 7×7 with all values 1.
3. Spatially convolve B with filter f, denoted as C (M×N×d).
4. Classify each hyperspectral pixel vector of size d×1 in C as one of the known classes by SVM classifier. Let the output be *Map*.

The main contributions of this paper are as follows. We select to use nonlinear dimensionality reduction method LLE instead of principal component analysis (PCA), which is a linear method. Our new method takes into consideration of both spatial information and spectral information at the same time in the process of classification. As a result, better classification results can be obtained. Our new method is simple, fast, and easy to implement with particularly good classification accuracy. The motivation of this study is to find a better way to classify HSI images with better classification results. Our experiments demonstrate that this new method compares favorably to several existing methods for HSI classification.

3 Experimental Results

In our experiments, we use LIBSVM [16] as a classifier to classify every spatial pixel in HSI data cube. We select radial basis function as the kernel for SVMs. We randomly pick 5% of HSI pixels as validation data and the rest of pixels as testing data. We find the best parameter C in SVM and we use this parameter C to classify pixels in the whole data cube. We choose two widely used data cubes in our experiments, which can be described as follows.

(1)**Indian Pines.** This scene was acquired by the Airborne Visible/Infrared Imaging Spectrometer (AVIRIS) over the Indian Pine test site in northwestern Indiana, USA, on June 12, 1992. This data cube has 145×145 pixels with 200 spectral bands. Figure 1 depicts the false-colour composite image and the ground-truth map. Table 1 tabulates ground truth classes and the pixel number for every class in this data cube.

(2) **Salinas.** This scene was also acquired by the AVIRIS sensor over Salinas Valley, California. The spatial size of the HSI is 512×217 pixels. AVIRIS data cubes have 224 spectral bands. We discard the 20 water absorption bands: [108–112], [154–167], and 224. Salinas ground truth contains 16 classes. Figure 2 demonstrates a false-colour composite image and the ground truth map. Ground truth classes and the total pixel numbers for all classes are shown in Table 2. Since LLE needs a lot of memory to run, we down sample this scene by a factor of 2 in the spatial domain.

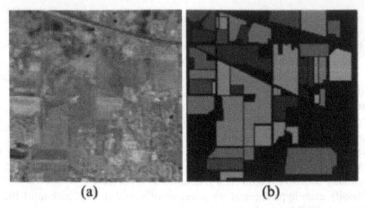

(a)   (b)

Fig. 1. Indian Pines data cube. (a) False-color composite image (b) Ground-truth map with 16 classes.

Table 1. Ground truth classes and the total pixel number for each class in Indian Pines data cube.

Class names	Total samples
Alfalfa	46
Corn notill	1428
Corn mintill	830
Corn	237
Grass pasture	483
Grass trees	730
Grass pasture mowed	28
Hay windrowed	478
Oats	20
Soybean notill	972
Soybean mintill	2455
Soybean clean	593
Wheat	205
Woods	1265
Buildings Grass Trees Drives	386
Stone Steel Towers	93

We use the following commands in Matlab to train and test SVMs:

$$model = svmtrain(Y\_Train\_Val, X\_Train\_Val, ['-s\,0\,-t\,2\,-c\,C']);$$

$$[label, acc, dec] = svmpredict(Y\_Train\_Val, X\_Train\_Val, model);$$

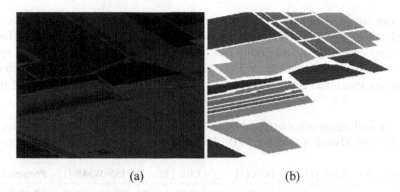

(a)   (b)

Fig. 2. Salinas data cube. (a) False-color composite image (b) Ground-truth map with 16 classes.

Table 2. Ground truth classes and the total pixel number for each class in Salinas data cube.

Class names	Total samples
Broccoli green weeds 1	2009
Broccoli green weeds 2	3726
Fallow	1976
Fallow rough plow	1394
Fallow smooth	2678
Stubble	3959
Celery	3579
Grapes untrained	11271
Soil vinyard develop	6203
Corn senesced green weeds	3278
Lettuce romaine 4wk	1068
Lettuce romaine 5wk	1927
Lettuce romaine 6wk	916
Lettuce romaine 7wk	1070
Vinyard untrained	7268
Vinyard vertical trellis	1807

We randomly pick k percent of the pixels from each class for training and the rest of the pixels for testing. We run our program for ten times and obtain the mean overall accuracy and standard deviation (STD) from it. We normalize each pixel's spectral vector Vec as follows:

$$outVec = (Vec - mean(Vec))/STD(Vec).$$

We run our Matlab code for ten times and obtain the mean overall accuracy and standard deviation (STD). The overall accuracy and STD are shown in Table 3 for the Indian Pines data cube. We randomly pick 10%, 20%, 30%, 40%, and 50% of spatial pixels as training data set and the rest of spatial pixels as testing data set for each class in the Indian Pines data cube. We randomly pick 2%, 4%, 6%, 8%, and 10% of pixels

Table 3. Overall classification results of different methods for the Indian Pines data set with K = 60 and d = 60 (Overall Accuracy ± STD). The best results are highlighted in bold font.

Percentage (%)	RAW [1]	PCA [1]	LLE [1]	ISS-WME [1]	Proposed
10	49.82 ± 1.94	68.40 ± 1.14	65.93 ± 1.71	75.38 ± 1.47	**97.45 ± 0.44**
20	51.86 ± 1.59	72.34 ± 1.47	71.42 ± 1.27	81.25 ± 1.51	**99.04 ± 0.22**
30	53.11 ± 1.35	74.82 ± 1.69	74.37 ± 1.48	83.83 ± 1.73	**99.39 ± 0.11**
40	54.28 ± 1.81	76.07 ± 1.44	75.98 ± 1.21	84.80 ± 1.80	**99.63 ± 0.10**
50	54.77 ± 1.39	76.65 ± 1.76	76.67 ± 1.27	84.71 ± 0.93	**99.77 ± 0.06**

Table 4. Classification results of different methods for each class for the Indian Pines data set (Overall Accuracy) with K = 60, d = 60 and 50% training pixels. The best results are highlighted in bold font.

Class names	RAW [1]	PCA [1]	LLE [1]	ISS-WME [1]	Proposed
Alfalfa	13.04	52.17	30.77	86.96	**99.57**
Corn notill	38.42	69.37	40.06	72.17	**99.48**
Corn mintill	25.06	48.76	44.34	56.47	**99.86**
Corn	14.41	77.11	26.27	53.95	**99.92**
Grass pasture	59.06	90.87	62.38	94.65	**99.96**
Grass trees	86.48	97.81	97.90	98.86	**99.89**
Grass pasture mowed	35.71	76.19	50.28	83.81	**100**
Hay windrowed	88.70	99.86	97.13	99.68	**100**
Oats	11,24	43.33	30.00	86.67	**100**
Soybean notill	25.17	63.51	94.24	75.17	**99.90**
Soybean mintill	71.15	83.32	73.62	79.52	**99.85**
Soybean clean	56.41	64.75	52.70	74.07	**99.36**
Wheat	74.51	94.12	75.21	**99.87**	99.80
Woods	94.57	97.68	86.71	97.66	**99.98**
Buildings Grass Trees Drives	29.02	45.77	34.20	52.85	**100**
Stone Steel Towers	91.30	90.58	88.70	**98.41**	95.22
Overall Accuracy	54.77	76.65	76.67	84.71	**99.77**

as training data set and the rest of pixels as testing data set for each class in the Salinas data cube. For the Indian Pines data cube, the overall classification accuracy and STD are shown in Table 3; the individual class accuracies with 50% training pixels and the rest as testing pixels are shown in Table 4; the individual class accuracies with 10% training pixels and the rest as testing pixels are shown in Table 5. For the Salinas data cube, the overall classification accuracy and STD are shown in Table 6. We choose K = 60 and d = 60 in all our experiments for our LLE transform. The results of all compared methods (RAW, PCA, LLE, and ISS-WME) in Tables 3, 4 and 6 are copied from [1], and the results of all compared methods (RAW, PCA, LDA, and SSLSTMs) in Table 5 are taken from [2]. The best results are highlighted in bold font. Our proposed method in this paper improves upon existing methods significantly in classification rates for all cases in our experiments.

Table 5. Classification results of different methods for each class for the Indian Pines data set (Overall Accuracy) with K = 60, d = 60 and 10% training pixels. The best results are highlighted in bold font.

Class names	RAW [2]	PCA [2]	LDA [2]	SSLSTMs [2]	Proposed
Alfalfa	56.96	59.57	63.04	88.78	**90.73**
Corn notill	79.75	68.75	72.04	93.76	**95.48**
Corn mintill	66.60	53.95	57.54	92.42	**97.20**
Corn	59.24	55.19	46.58	86.38	**96.29**
Grass pasture	90.31	83.85	91.76	89.79	**97.01**
Grass trees	95.78	91.23	94.41	97.41	**99.32**
Grass pasture mowed	80.00	82.86	72.14	84.80	**100**
Hay windrowed	97.412	93.97	98.74	99.91	**100**
Oats	35.00	34.00	26.00	74.44	**96.67**
Soybean notill	66.32	64.18	60.91	95.95	**95.82**
Soybean mintill	70.77	74.96	76.45	96.93	**97.75**
Soybean clean	64.42	41.72	67.45	89.18	**96.74**
Wheat	95.41	93.46	96.00	98.48	**99.18**
Woods	92.66	89.45	93.79	98.08	**99.50**
Buildings Grass Trees Drives	60.88	47.77	65.54	92.85	**97.18**
Stone Steel Towers	87.53	88.17	83.66	87.86	**92.86**
Overall Accuracy	77.44	72.58	76.67	95.00	**97.45**

Table 6. Overall classification results of the different methods for the Salinas data set with K = 60 and d = 60 (Overall Accuracy ± STD). The best results are highlighted in bold font.

Percentage (%)	RAW [1]	PCA [1]	LLE [1]	ISS-WME [1]	Proposed
2	63.34 ± 1.13	75.12 ± 2.18	89.23 ± 2.42	88.53 ± 1.23	**96.16 ± 1.04**
4	66.47 ± 1.51	77.25 ± 1.34	89.38 ± 2.94	90.22 ± 0.99	**98.06 ± 0.36**
6	68.01 ± 1.27	77.56 ± 1.84	91.61 ± 2.32	91.90 ± 1.29	**98.73 ± 0.34**
8	68.96 ± 2.22	77.96 ± 1.87	91.92 ± 3.21	91.16 ± 1.04	**99.08 ± 0.14**
10	69.13 ± 1.21	79.02 ± 2.29	92.16 ± 2.73	92.19 ± 1.02	**99.36 ± 0.13**

4 Conclusions

In this paper, we have proposed a novel method for HSI classification by means of LLE and SVM. We choose nonlinear dimensionality reduction method LLE instead of PCA, which is a linear method. Our novel method takes into consideration of both spatial information and spectral information at the same time in the process of classification, so better classification results can be achieved. Experimental results show that this novel method compares favorably to several existing methods for HSI classification.

References

1. Liu, H., Xia, K., Li, T., Ma, J., Owoola, E.: Dimensionality reduction of hyperspectral images based on improved spatial–spectral weight manifold embedding. Sensors **20**(16), 4413 (2020)
2. Zhou, F., Hang, R., Liu, Q., Yuan, X.: Hyperspectral image classification using spectral-spatial LSTMs. Neurocomputing **328**, 39–47 (2019)
3. Chen, Y., Nasrabadi, N.M., Tran, T.D.: Hyperspectral image classification via kernel sparse representation. IEEE Trans. Geosci. Remote Sens. **51**(1), 217–231 (2013)
4. Melgani, F., Bruzzone, L.: Classification of hyperspectral remote sensing images with support vector machines. IEEE Trans. Geosci. Remote Sens. **42**(8), 1778–1790 (2004)
5. Fauvel, M., Chanussot, J., Benediktsson, J.A.: A spatial–spectral kernel-based approach for the classification of remote-sensing images. Pattern Recogn. **45**(1), 381–392 (2012)
6. Camps-Valls, G., Bruzzone, L.: Kernel-based methods for hyperspectral image classification. IEEE Trans. Geosci. Remote Sens. **43**(6), 1351–1362 (2005)
7. Li, J., Marpu, P.R., Plaza, A., Bioucas-Dias, J.M., Benediktsson, J.A.: Generalized composite kernel framework for hyperspectral image classification. IEEE Trans. Geosci. Remote Sens. **51**(9), 4816–4829 (2013)
8. Chen, Y., Nasrabadi, N.M., Tran, T.D.: Hyperspectral image classification using dictionary-based sparse representation. IEEE Trans. Geosci. Remote Sens. **49**(10), 3973–3985 (2011)
9. Li, J., Bioucas-Dias, J.M., Plaza, A.: Spectral-spatial classification of hyperspectral data using loopy belief propagation and active learning. IEEE Trans. Geosci. Remote Sens. **51**(2), 844–856 (2013)
10. Kang, X.D., Li, S., Benediktsson, J.A.: Spectral-spatial hyperspectral image classification with edge-preserving filtering. IEEE Trans. Geosci. Remote Sens. **52**(5), 2666–2677 (2014)
11. Cheng, G., Zhu, F., Xiang, S., Wang, Y., Pan, X.: Semisupervised hyperspectral image classification via discriminant analysis and robust regression. IEEE J. Sel. Topics Appl. Earth Observ. Remote Sens. **9**(2), 595–608 (2016)

12. Chen, Y., Lin, Z., Zhao, X., Wang, G., Gu, Y.: Deep learning-based classification of hyperspectral data. IEEE J. Sel. Topics Appl. Earth Observ. Remote Sens. **7**(6), 2094–2107 (2014)
13. Roweis, S., Saul, L.: Nonlinear dimensionality reduction by locally linear embedding. Science **290**(5500), 2323–2326 (2000)
14. Damelin, S., Miller, W.: The Mathematics of Signal Processing. Cambridge University Press, Cambridge (2011). ISBN 978-1107601048
15. Cortes, C., Vapnik, V.N.: Support-vector networks. Mach. Learn. **20**(3), 273–297 (1995)
16. Chang, C.C., Lin. C.J.: LIBSVM: a library for support vector machines. ACM Trans. Intell. Syst. Technol. **2**(27), 1–27 (2011)

A Hierarchical Retrieval Method Based on Hash Table for Audio Fingerprinting

Tianhao Li, Maoshen Jia$^{(\boxtimes)}$, and Xuan Cao

Beijing Key Laboratory of Computational Intelligence and Intelligent System, Faculty of
Information Technology, Beijing University of Technology, Beijing 100124, China
jiamaoshen@bjut.edu.cn

Abstract. The focus of audio retrieval research is to find the target audio faster
and more accurately in the audio database according to a query audio. In this
paper, a low-dimensional audio fingerprint extraction method based on local linear
embedding (LLE) and an efficient hierarchical retrieval method are proposed. In
the fingerprint extraction part, the audio fingerprint is computed by the energy
comparison. The proposed method reduces the dimensionality of the energy vector
and the number of energy comparisons by introducing the LLE algorithm, which
results in a low-dimensional audio fingerprint. The retrieval part is divided into
two stages, which are hash value retrieval for single-frame audio fingerprints and
fingerprint block retrieval for consecutive multi-frame audio fingerprints. In the
first stage, the reference audios with the same hash value as the query audio are
filtered out as candidates. In the second stage, the exact retrieval result is found by
calculating the similarity between the query fingerprint block and the reference
fingerprint block. The proposed method reduces the computational complexity
of fingerprint matching by narrowing the scope of retrieval, thus improving the
retrieval speed. In the experimental part, the effectiveness of the proposed method
is evaluated and compared with some state-of-the-art methods. The experiments
prove that the retrieval accuracy and computation speed can reach a high level
after using the proposed method.

Keywords: Audio retrieval · Audio fingerprint · Hash table

1 Introduction

With the development of information technology, the information production speed
has also increased rapidly in recent years [1]. This has resulted in the accumulation
of large amounts of information data. Especially, along with the advancement of the
internet industry, it has made people exposed to huge amount of multimedia data every
day. Therefore, higher technical demands are required on information compression,
recognition and retrieval [2]. In the era, audio retrieval, as an important component of
multimedia retrieval, has become an emerging research area.

The accuracy and speed for audio retrieval have been the key points considered in
research. An audio retrieval system can be divided into two parts: feature extraction and
feature retrieval.

© Springer Nature Switzerland AG 2021
D.-S. Huang et al. (Eds.): ICIC 2021, LNCS 12836, pp. 160–174, 2021.
https://doi.org/10.1007/978-3-030-84522-3_13

In terms of feature extraction, audio fingerprint is favored because of its simple computation, small data size and less susceptibility to signal distortion [3]. Audio fingerprint can represent an original audio signal consisting of a large amount of data with less data size. It is derived from the most important part of audio data perceived by the human ear, which has perceptual similarity [4]. And it also outperforms traditional audio features such as linear prediction coefficients (LPC) [5], short time zero-crossing rate, and Mel frequency cepstral coefficient (MFCC) [6] in audio retrieval works. Audio fingerprinting is used in a wide range of applications such as music recognition and copyright protection [7].

There are various algorithms for audio fingerprinting. The most classic one is Philips audio fingerprinting based on the energy difference of adjacent frequency bands proposed by Haitsma et al. [8]. The algorithm has a high robustness and can maintain good retrieval performance when the signal is disturbed. It requires only 3 s of query audio length as its retrieval granularity (retrieval of the shortest audio segment length required). However, its resistance to linear speed changes of the audio signal is low and more errors appear in the retrieval results in this case [9, 10]. Another representative algorithm is called Shazam audio fingerprinting proposed by Wang which uses spectral peak pairs for matching [11]. This method first extracts a series of peaks on the time-frequency domain of the audio signal. Then it constructs an audio fingerprint by using the time and frequency information of every two peaks. The method performs very well in dealing with the effects of noise, filtering and compression. However, the number of the fingerprints generated by the algorithm and the overall data size is large. Its retrieval performance degrades significantly as the database size increases.

In recent years, Anguera combined the two methods mentioned above and proposed an audio fingerprint called Masked Audio Spectral Keypoints (MASK) [12]. In this method, it uses the spectral peaks as the center to construct the fingerprint computation region and computes the audio fingerprint by the energy difference. The method has higher robustness and is applicable to the case of multiple audio mixtures, such as music and speech. Although the performance of this fingerprint is superior, there is still room to reduce the data size of the fingerprint. In the completed work, we introduce local linear embedding algorithm (LLE) [13] to reduce the dimension of the band energy, so as to extract an audio fingerprint with lower dimension and reduce the overall data size [14].

In feature retrieval, the most common method is to use the similarity of features as a metric to discriminate the matching results. In order to improve the efficiency of retrieval, Haitsma builds a look up table (LUT) [15]. It can quickly find the location in database where the fingerprint is same as the query. Then the location is used as a starting point for audio segment matching. However, due to the low feasibility and practicality of LUT, a hash table is used as an alternative. Although this method has a significant improvement on the retrieval speed, it has the problems of high memory consumption and low space utilization because of the uneven data distribution in hash table [16, 17].

In addition, Gupta proposes a method for counting the total number of fingerprint matches [18, 19]. It counts the maximum number of fingerprint matches for each fingerprint as the starting point for the reference audio in the database. The fingerprint with the highest count is selected as the final starting point. The purpose of the counting is to reduce unnecessary matches by selecting the fingerprint segments that are most similar to the query for further computation. This method requires a large amount of computation and is more complicated when applied to large-sized database.

In this paper, a hierarchical retrieval model based on hash table is proposed to address the problems of high computational complexity and high memory consumption in current audio retrieval systems. The method creates hash tables for each reference audio in the database, which reduces the amount of data loaded into memory and consumes less memory space. In first stage of retrieval, reference audios are filtered with the hash value of a single query fingerprint. Then the whole audio segment is matched in second stage. The retrieval process reduces the number of matches and the amount of computation. The proposed method saves valuable memory resources and effectively improves the speed of retrieval.

The remainder of paper is organized as follows: Sect. 2 introduces the audio fingerprint extraction method used in this paper. Section 3 presents the proposed retrieval strategy and fingerprint matching method. Section 4 shows the experimental results of the retrieval system in terms of accuracy and speed. Finally, Sect. 5 summarizes the proposed method and its performance.

2 Audio Fingerprint Extraction Based on LLE

The amount of original audio data is large, so the audio fingerprint features need to be extracted for the purpose of reducing the computation. In this paper, a low-dimensional audio fingerprint extraction method based on LLE is proposed. The process of the proposed method is as follows.

2.1 Pre-processing

In the framing step, the frame length is set to 100 ms overlapped with an interval of 10 ms, considering that the extraction of audio fingerprints requires high frequency resolution and wide coverage of time. The framing strategy is mainly to reduce the error caused by the inconsistency between the starting points of the query and reference audio. With this strategy, even in the worst case (5 ms difference between the starting points), the retrieval performance is still good.

2.2 Maximum Spectral Points Extraction

The pre-processed audio data is converted from time domain to frequency domain by discrete Fourier transform (DFT), and the spectrum is divided into 18 bands. The peak which has the maximum amplitude of each frame in the frequency domain is selected, and its frame number and band number are recorded as index information. To ensure the selection has good coverage over the whole audio segment, the number of the peaks should be between 70 and 100 per second.

2.3 Sub-region Division

After the peak selection step, the MASK region is constructed with the peak of each frame as the center. The region consists of 5 bands and 19 frames symmetrically distributed around the central peak. In the MASK region construction part, overlap is allowed between the regions belonging to different peaks. In order to prevent the peaks from appearing at the boundaries of the band, the first and last two bands are reserved in the peak selection process. In this way, no boundary crossing occurs when creating MASK regions.

Each MASK region is again divided into four groups of sub-regions: horizontal, vertical, central and marginal groups. Each sub-region is calculated to obtain an energy value, and the energy values in the same group form an energy vector. The energy vectors are used in the subsequent computation of fingerprint generation.

2.4 Audio Fingerprint Generation

The LLE-based audio fingerprint extraction is applied in this paper to downscale the energy vector of each frame in the horizontal and central group. The audio fingerprint is then obtained by comparing the differences in energies between sub-regions. The flow of the method is shown in the following figure.

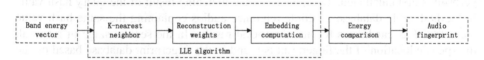

Fig. 1. Block diagram of audio fingerprint extraction based on LLE.

As shown in Fig. 1, the energy vector of each frame is downscaled by using LLE to obtain a low-dimensional vector. Next, the novel energy vector is used to obtain a low-dimensional audio fingerprint by comparing the energy of adjacent sub-regions to achieve the purpose of fingerprint dimensionality reduction. If the sign of the energy difference is positive then it is recorded as 1, and the opposite is 0. The final length of the audio fingerprint is 16 bits. The downscaled audio fingerprints have smaller data size, occupy less storage space, and are retrieved more efficiently. Since the amount of information contained in a single fingerprint is small, it is not sufficient to match the whole audio segment. Therefore, the method uses an audio fingerprint block composed of multiple consecutive frames of audio fingerprints for feature comparison to complete the audio retrieval work.

3   Hierarchical Retrieval Based on Hash Table

In order to reduce the computational complexity of retrieval and improve the retrieval speed, this paper proposes a hierarchical retrieval method based on hash table. The retrieval process in this paper is divided into two stages: the first stage is the hash retrieval of single-frame audio fingerprints; the second stage is the fingerprint block retrieval of consecutive multi-frame audio fingerprints. The implementation process of the proposed method is as follows.

3.1 Retrieval Model Overview

The general hierarchical retrieval method is to categorize the stored data during the database establishment process [20]. Then, in the retrieval process, the category of the query audio is judged first to initially narrow the scope of the search, and then the similarity between features is calculated to find the exact result. This method is a retrieval from the perspective of stored data content, which requires a certain degree of differentiation between data. It cannot cope with the situation of storing the same kind of data and data types that cannot be specifically distinguished. At the same time, in the retrieval process, all the required data and data indexes have to be loaded into the memory, which takes up a large amount of memory space. To solve the above problems, this paper proposes a hierarchical retrieval method based on hash table from the perspective of retrieval structure. Unlike other methods that build a hash table uniformly for all reference fingerprints, the proposed method builds separate hash tables for fingerprints of each reference audio. A single hash table stores less data, making it take up less memory when it is called in a retrieval.

In the first stage of retrieval, the hash value corresponding to the query audio fingerprint is first calculated. Then, based on the matching result of the query hash value in the hash table database, multiple reference audios containing the same hash value are selected as candidates for audio retrieval results. In the second stage of retrieval, the specific location of the fingerprint is found in the fingerprint database based on the candidate hash table. Finally, the location is used as the starting point to compute the similarity between the query and the reference audio fingerprint block. If the computation result is lower than the similarity threshold, the reference audio is output as a retrieval result. The proposed method first selects multiple candidates from the reference audio by matching the hash value of a single fingerprint, narrowing the scope of retrieval. Then the accurate retrieval results are obtained by similarity computation of audio fingerprint blocks. The retrieval process reduces the computational effort of fingerprint matching and improves the retrieval performance of the system. The proposed method includes two parts, database construction and retrieval of the query audio. More details are presented in the following content.

3.2 Database Construction

The database used in this paper is as follows:

Audio database. It stores the information of the original reference audio.

Audio fingerprint database. The audio fingerprint data of each reference audio is stored in this database.

Hash table database of audio fingerprints. It stores the hash table data of each reference audio, including the hash value and frame index.

The latter two databases are constructed on the basis of the audio database after the computation.

Creation of Hash Table. In this paper, separate hash tables are created for the finger-print data of each reference audio. The hash table is to map the original audio fingerprint into a definite integer, namely the hash value, by means of a hash function. With the hash value as the table header, the frame index (the label of the frame where it is located) of the fingerprint corresponding to the hash value is stored in the table. The search first calculates the hash value of the query audio fingerprint, and finds the frame index of the fingerprint stored in the hash table by this value. Then find the location of the fingerprint which is same as the query in the fingerprint database. It realizes the fast location of the query fingerprint in the fingerprint database.

The most important thing in the design of the hash function is to ensure that there is a unique correspondence between the audio fingerprint and the hash value. Therefore, this paper adopts binary-to-decimal conversion for the hash calculation of audio fingerprints, as shown in Fig. 2.

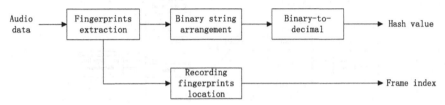

Fig. 2. Block diagram of the calculation process of hash value.

The audio fingerprint used in this paper consists of 16 binary bits, which can be equated to a binary sequence of 16 bits in length. In Fig. 2, the method converts the binary sequence into its corresponding decimal number which is used as a hash value. The frame indexes of the audio fingerprints corresponding to the hash values are stored in tables, and these frame indexes are used to quickly locate the starting point of a fingerprint block match during retrieval. For a reference audio with M frames of audio fingerprint, the fingerprint of its m-th frame is denoted as f_m. f_m is a 16-dimensional audio fingerprint vector, which is represented as follows:

$$f_m = [f(m, 1), f(m, 2), \ldots \ldots, f(m, 16)] \tag{1}$$

The hash value of f_m is calculated by the following formula:

$$H_m = \sum_{n=1}^{16} 2^{n-1} f(m, n) \tag{2}$$

where $f(m, n)$ is the reference audio fingerprint of n-th bit in m-th frame, $n = 1, 2 \ldots 16$, corresponding to the 16-bit audio fingerprint. H_m is the hash value calculated from the m-th frame of the reference fingerprint.

During the conversion, if one bit of the binary sequence changes, its corresponding decimal number will be different. The hash function satisfies the requirement that different fingerprints correspond to different hash values, and also avoids hash conflicts to a certain extent. Once the hash values and corresponding frame indexes are obtained, the entire hash table can be created for the reference audio. The schematic diagram of the hash table structure is shown in Fig. 3.

In Fig. 3, the left side shows the hash table of the reference audio, including the table header consisting of hash values and the stored frame index. On the right side is the fingerprint block of the reference audio, which contains the audio fingerprint data of each frame in the reference audio. It can be seen that there may be more than one frame index corresponding to the hash value, which indicates that the fingerprint is repeated in the fingerprint block. The multiple frame indexes corresponding to the duplicate fingerprints are all stored in the hash table to prevent any missing during fingerprint retrieval. The structure of the above hash table is used to quickly find the location of the audio fingerprint in the fingerprint database, which improves the retrieval speed.

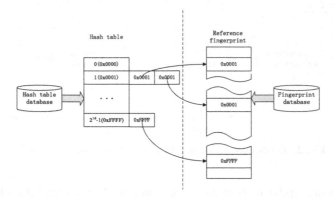

Fig. 3. Structure of hash table.

Processing of Reference Audio. The Fingerprint Extraction and Hash Table Creation of the Reference Audio is Shown in Fig. 4.

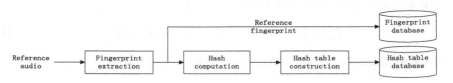

Fig. 4. Block diagram of the reference audio processing.

According to the process shown in Fig. 4, given an input audio signal S, the fingerprint features of S are first extracted using the audio fingerprint extraction method in Sect. 2. Then, the M-frame reference audio fingerprint f_m is obtained and stored in the audio fingerprint database. The hash value H_m of each frame of reference fingerprint is calculated according to Eq. (2). Finally, the hash table of the reference audio is created and stored in the hash table database.

3.3 Retrieval of Query Audio

Hierarchical Retrieval. In this paper, a two-stage retrieval model is established: the first stage is hash matching of single fingerprint, and the second stage is matching of fingerprint blocks.

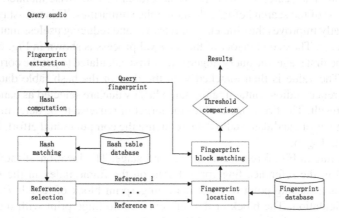

Fig. 5. Block diagram of hierarchical search. (Color figure online)

Given a piece of audio to be examined, its fingerprint is first extracted. A frame of the query fingerprint is selected and its hash value is calculated according to Eq. (2). Through the first stage of hash matching, multiple reference audios with the same hash value are found in the hash table database as candidates for the retrieval result. Then, through the second stage of fingerprint block matching, the location of the fingerprint is found in the reference audio fingerprint block using the frame index in the hash table of the reference audio. The location is used as a starting point to calculate the similarity between fingerprint blocks. Finally, the computation result is compared with the similarity threshold to determine whether the candidate audio is the retrieval result. The flow of the retrieval part is shown in Fig. 5.

According to Fig. 5, the retrieval of the query audio can be divided into the following steps.

Step 1: Given a query audio S', the fingerprint features are computed by using the audio fingerprint extraction method in Sect. 2. A total of R frames of audio fingerprint F'_r is obtained, where the n-th bit of the r-th frame of the query audio fingerprint is denoted as $F'(r, n)$.

Step 2: From query fingerprint block, a frame of audio fingerprint is selected to calculate its hash value according to Eq. (2), and the hash value H'_r is obtained.

Step 3: The first stage of retrieval (red box part in Fig. 5). Using H'_r. to match in the hash table database to find multiple hash tables of reference audio with the same hash value. The corresponding reference audio (Reference 1…Reference n) will be selected as candidates for the retrieval result.

Step 4: The second stage of retrieval (blue box part in Fig. 5). Based on the frame index in the hash table, the location of the fingerprint is found in the fingerprint database.

The fingerprint blocks with frame length T are selected sequentially from the location, and the similarity between the fingerprint blocks of the query audio and the reference audio is computed. The similarity is compared with the threshold, and if the similarity is less than the threshold, the reference audiis output as a retrieval result. On the contrary, it means there is a large difference between them.

The hierarchical search in this paper is structured in a stepwise manner, narrowing down the scope of the search before calculating the similarities. This two-stage retrieval structure greatly improves the efficiency of retrieval and reducing useless matches in the retrieval process. The actual layout of the retrieval process is shown in Fig. 6.

During the first stage, the query fingerprint is first calculated to get the corresponding hash value. The value is then matched with the data in the hash table database, and multiple reference audios containing the same hash value are selected as candidates for the retrieval result. The first stage enables the target of retrieval to be determined within a limited number of candidate audio data, reducing the computational effort. It is shown as a red line in Fig. 6.

The blue line in Fig. 6 represents the second stage. The location of the fingerprint is first found in the reference fingerprint based on the frame index in the hash table. Then use it as a starting point to construct a fingerprint block of length T frames for similarity calculation. The block of the reference audio fingerprint with the length of T frames from the m-th frame is called $\mathbf{F_m}$. $\mathbf{F_m}$ contains the 16-dimensional reference audio fingerprints of T frames, whose matrix representation is shown in Eq. (3).

$$\mathbf{F_m} = \begin{bmatrix} f_m \\ \vdots \\ f_{m+T} \end{bmatrix} = \begin{bmatrix} f(m,1) & \cdots & f(m,16) \\ \vdots & \ddots & \vdots \\ f(m+T,1) & \cdots & f(m+T,16) \end{bmatrix} \tag{3}$$

where $f_m \cdots f_{m+T}$ are the T frames of audio fingerprints in the fingerprint block, and each frame of audio fingerprints can be written in the form of a 16-dimensional vector according to Eq. (1). The reference audio fingerprint block $\mathbf{F_m}$ in matrix form is constructed with fingerprint dimension and fingerpri frame number as rows and columns. In the same way, the query audio fingerprint block $\mathbf{F'_r}$ of length T frames from r-th frame can be constructed.

Bit Error Rate (BER) is used in this paper to determine the similarity between the query audio and the reference audio. The BER calculates the ratio of the number of fingerprint bits in query fingerprint block and reference fingerprint block that differ in the same position to the total number of fingerprint bits in the block. The number of bits that differ between fingerprints is realized by calculating the Hamming distance between them. The formula for calculating a single fingerprint is as follows:

$$d(m,r) = \sum_{n=1}^{16} f(m,n) \oplus f'(r,n) \tag{4}$$

where $d(m,r)$ denotes the Hamming distance between the m-th frame of the reference fingerprint and the r-th frame of the query fingerprint, $\oplus$ is XOR operation, and the parameter n is the number of bits of the fingerprint. Therefore the matrix form of Eq. (4) can be represented as follows:

$$\mathbf{D} = \mathbf{F_m} \oplus \mathbf{F'_r}$$

$$= \begin{bmatrix} F(m,1) & \cdots & F(m,16) \\ \vdots & \ddots & \vdots \\ F(m+T,1) & \cdots & F(m+T,16) \end{bmatrix} \oplus \begin{bmatrix} F'(r,1) & \cdots & F'(r,16) \\ \vdots & \ddots & \vdots \\ F'(r+T,1) & \cdots & F'(r+T,16) \end{bmatrix} \quad (5)$$

where $\mathbf{D}$ is the matrix obtained after the XOR operation of fingerprint blocks $\mathbf{F_m}$ and $\mathbf{F'_r}$, whose size is the same as the fingerprint block matrix ($T \times 16$). The elements stored in the matrix $\mathbf{D}$ are the Hamming distances of the fingerprints at the same positions in the fingerprint blocks $\mathbf{F_m}$ and $\mathbf{F'_r}$.

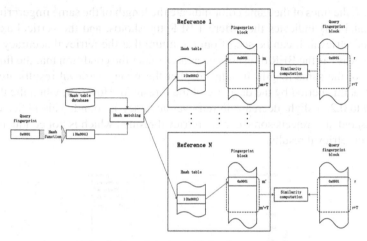

Fig. 6. Search layout. (Color figure online)

The total Hamming distance between the reference and the query fingerprint block is obtained by summing all the elements in $\mathbf{D}$.

$$d_{ham} = \sum_{i=1}^{T} \sum_{j=1}^{16} D_{ij} \quad (6)$$

where D_{ij} is the element of the i-th row and j-th column of the matrix $\mathbf{D}$, d_{ham} is the Hamming distance between fingerprint blocks $\mathbf{F_m}$ and $\mathbf{F'_r}$. Combining the fingerprint block length T and the number of bits of the fingerprint (16 bits), the BER between fingerprint blocks can be calculated as follows:

$$E_{BER} = \frac{d_{ham}}{T \times 16} \quad (7)$$

By comparing E_{BER} with BER threshold, if it is lower than the threshold, it means that the query audio is more similar to the reference audio, and vice versa, it means that the two are more different. Finally, the retrieval results with BER below the threshold among the candidate reference audio are output which is sorted in ascending order according to E_{BER}. When the selected frame of the query fingerprint has no matching result in the audio database, the next frame of the query is to be selected in turn to continue the hierarchical retrieval. The retrieval process continues until the length of the remaining query fingerprint block is less than T frames, which means that the length for fingerprint block matching is not enough.

Setting of BER Threshold and T. The parameters used for fingerprint block matching are determined by experiments. Four different fingerprint block lengths and four different BER thresholds are selected for the experiments. By comparing the retrieval accuracy under each parameter condition and considering the complexity of retrieval, the best set of parameters is selected to be applied to the final retrieval system. The retrieval accuracy with each combination of parameters is shown in the figure below.

In Fig. 7, the lines of the same color indicate the length of the same fingerprint block, the horizontal axis indicates the different BER thresholds, and the vertical axis is the accuracy of retrieval. It can be seen from the figure that the retrieval accuracy initially tends to increase as the BER threshold decreases under the condition that the fingerprint blocks are of the same length. It indicates that the wrong retrieval results are mainly eliminated at this period by the decrease of the threshold. However, when the threshold is reduced to 0.2, a slight decrease in retrieval accuracy is found. This indicates that a small threshold also rejects some correct retrieval results, which is not conducive to the screening of retrieval results.

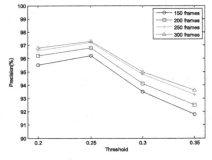

Fig. 7. Retrieval accuracy under different BER thresholds and fingerprint block lengths. (Color figure online)

Under the condition of the same threshold, the retrieval accuracy tends to increase with the increase of fingerprint block length. And after the length reaches 250 frames, the rise of retrieval accuracy tends to level off. Although the longer the length of fingerprint blocks, the more accurate the calculation of similarity between fingerprint blocks, the computational complexity of fingerprint block matching also increases. Therefore, the length of fingerprint blocks cannot be increased endlessly.

After considering the effects of both parameters on the retrieval results, a BER threshold of 0.25 and a fingerprint block length of 250 frames are experimentally selected to be applied in the final retrieval system. Under this parameter condition, the similarity calculation results are more accurate and the retrieval results can be effectively filtered. At the same time, the complexity of the calculation is moderate and the retrieval efficiency of the proposed method is relatively high.

4 Experimental Results and Analysis

4.1 Experimental Data

The experiments in this paper are performed on data sets with different signal-to-noise ratios to verify the retrieval performance of the proposed method. The data sampling rate is 8 kHz. At this sampling rate, the audio data contains enough information and covers the more sensitive frequency bands of the human ear to meet the requirements of the audio retrieval task.

The experimental data is introduced as follows:

Database 1 consists of 5000 audio files, including the data collected in the laboratory and the data collected from the internet. The length of each audio file is between 3 s and 5 min. The storage size of the data is 12.3 GB and the total length is 230 h.

Database 2 is the dataset generated by adding 40 dB white noise to database 1.

Database 3 is the dataset generated by adding 30 dB white noise to database 1.

Database 4 is the dataset generated by adding 20 dB white noise to database 1.

Database 5 is the dataset generated by adding 10 dB white noise to database 1.

Database 6 is a random interception of 1000 audio segments with the length of 3 s from database 1 and it is used as the query audio set.

4.2 Performance Evaluation

In this paper, recall and precision rate which commonly used in the field of information retrieval, are selected as the metrics to evaluate the performance of the algorithm. They are defined as follows:

Recall rate = the number of correct targets detected from the retrieval source/the number of targets that should have been retrieved × 100%

Precision rate = the number of correct targets detected from the retrieval source/the number of targets actually retrieved × 100%

4.3 Results and Analysis

The experiments include both accuracy and speed of retrieval, and are analyzed separately. In this paper, audio retrieval experiments are conducted in a pure audio database and several database with different SNR, respectively. The aim is to verify the robustness and the overall stability of the method in this paper.

To evaluate the performance of the proposed method, the experiments compare the retrieval results with two classical methods, Philips and Shazam. They are widely used in the field of audio retrieval and have good retrieval performance. Philips fingerprinting is done by dividing the spectrum into 33 bands and calculating the 32-dimensional audio fingerprint by comparing the differences in adjacent bands. Shazam fingerprinting uses the peak information of the audio in the frequency domain to create a hash with a storage space of 32 bits as the audio fingerprint. The MASK fingerprint is also used as a reference method for performance comparison in the experiments. The MASK fingerprint is the original fingerprint before dimensionality reduction using the method in this paper, and its size is 22 dimensions. The retrieval performance of each method with different SNR is shown in Fig. 8.

Figure 8(a) and (b) show the results of retrieval recall and precision, respectively. It can be seen that the proposed method is basically consistent with the reference method in terms of retrieval performance. As the SNR decreases, the retrieval performance of the proposed method also decreases slowly and smoothly, which indicates that it is more resistant to noise.

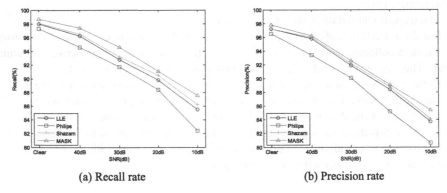

(a) Recall rate   (b) Precision rate

Fig. 8. Comparison of the retrieval performance of each method.

To show more clearly the size of the audio fingerprints extracted by each method, they are summarized in Table 1 in this paper.

Table 1. Audio fingerprint size of different methods.

Methods	Audio fingerprint size
Philips	32 bits
Shazam	32 bits
MASK	22 bits
LLE	16 bits

The fingerprint extraction methods in Table 1 include the method proposed in this paper and the three comparison methods. Among them, the LLE audio fingerprint has smaller size, occupies less storage space, and has higher retrieval efficiency. In summary, this paper achieves the purpose of reducing the amount of fingerprint data and saving storage space by reducing the dimensions of audio fingerprints. And the method performs well in overall retrieval performance and stability.

To verify the improvement of the retrieval efficiency of the methods in this paper, the average time consumed by each method in the retrieval process is experimentally counted. The results are shown in Fig. 9.

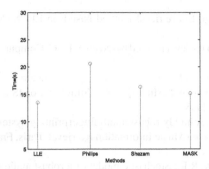

Fig. 9. Comparison of retrieval time by method.

As can be seen in Fig. 9, the proposed method (LLE) has the shortest retrieval time. It shows that the proposed method effectively improves the retrieval speed and has a low retrieval complexity. In conclusion, the proposed method achieves fast retrieval speed and good robustness while ensuring high retrieval accuracy.

5 Conclusions

In this paper, an LLE-based audio fingerprint extraction method is applied and a hierarchical retrieval structure based on hash table is proposed. The proposed method achieves good performance in both retrieval accuracy and speed by reducing the data size of audio fingerprints and the computation of the retrieval process. While ensuring that the retrieval work has high retrieval accuracy, a fast retrieval speed is obtained. Experiments prove that the method is robust to changes in the query audio and is not inferior to the state-of-the-art methods in terms of retrieval performance. To sum up, the method in this paper has less fingerprint data amount and lower retrieval complexity, and achieves higher retrieval performance.

Acknowledgments. This work has been supported by the National Natural Science Foundation of China (No. 61971015) and the Cooperative Research Project of BJUT-NTUT (No. NTUT-BJUT-110–05).

References

1. Zhang, X., Zou, X., Hu, Q., Zhang, P.: Audio retrieval method based on weighted DNA. China Sciencepaper **13**(20), 2295–2300 (2018)
2. Li, G., Wu, D., Zhang, J.: Concept framework for audio information retrieval: ARF. J. Comput. Sci. Technol. **18**(5), 667–673 (2003)
3. Li, W., Li, X., Chen, F., Wang, S.: Review of digital audio fingerprinting. J. Chin. Comput. Syst. **29**(11), 2124–2130 (2008)
4. Cano, P., Batle, E., Kalker, T., Haitsma, J.: A review of algorithms for audio fingerprinting. In: Proceedings of the Multimedia Signal Processing, St. Thomas, VI, USA, pp. 169–173, 9–11 December 2002

5. Jiang, X., Li, Y.: Audio data retrieval method based on LPCMCC. Comput. Eng. **35**(11), 246–247 (2009)
6. Jiang, X.: An audio data retrieval method based on MFCC. Comput. Digital Eng. **36**(9), 24–26 (2008)
7. Doets, P.J.O., Lagendijk, R.L.: Extracting quality parameters for compressed audio from fingerprints. In: Proceedings of the International Conference on ISMIR, London, UK, pp. 498–503, 11–15 September 2005
8. Haitsma, J., Kalker, T.: A highly robust audio fingerprinting system. In: Proceedings of the International Conference on Music Information Retrieval, Paris, France, pp. 107–115, 13–17 October 2002
9. Doets, P.J.O., Lagendijk, R.L.: Stochastic model of a robust audio fingerprinting system. In: Proceedings of the 5th International Conference on Music Information Retrieval, Barcelona, Spain, pp. 2–5, 10–14 October 2004
10. Park, M., Kim, H., Yang, S.: Frequency-temporal filtering for a robust audio fingerprinting scheme in real-noise environments. ETRI J. **28**(4), 509–512 (2006)
11. Wang, A., Li, C.: An industrial strength audio search algorithm. In: Proceedings of the International Conference on Music Information Retrieval, Baltimore, MD, USA, pp. 7–13, 27–30 October 2003
12. Anguera, X., Garzon, A., Adamek, T.: MASK: robust local features for audio fingerprinting. In: Proceedings of the IEEE International Conference on Multimedia and Expo, Melbourne, Australia, pp. 455–460, 9–13 July 2012
13. Chen, J., Zheng, M.: Locally linear embedding: a review. Int. J. Pattern Recogn. Artif. Intell. **25**(7), 985–1008 (2011)
14. Jia, M., Li, T., Wang, J.: Audio fingerprint extraction based on locally linear embedding for audio retrieval system. Electronics **9**(9), 1483 (2020)
15. Haitsma, J., Kalker, T.: A highly robust audio fingerprinting system with an efficient search strategy. J. New Music Res. **32**(2), 211–221 (2003)
16. Chen, M., Xiao, Q., Matsumuto, K., Yoshida, M., Kita, K.: A fast retrieval algorithm based on fibonacci hashing for audio fingerprinting systems. In: Proceedings of the 2013 International Conference on Advanced Information Engineering and Education Science, Beijing, China, pp. 219–222, 19–20 December 2013
17. Yao, S., Wang, Y., Niu, B.: An efficient cascaded filtering retrieval method for big audio data. IEEE Trans. Multimedia **9**(17), 1450–1459 (2015)
18. Gupta, V., Boulianne, G., Cardinal, P.: CRIM's content-based audio copy detection system for TRECVID 2009. Multimedia Tools Appl. **66**(2), 371–387 (2012)
19. Ouali, C., Dumouchel, P., Gupta, V.: A robust audio fingerprinting method for content-based copy detection. In: 2014 12th International Workshop on Content-Based Multimedia Indexing (CBMI), Klagenfurt, Austria, pp. 1–6 (2014)
20. Borjian, N., Kabir, E., Seyedin, S., Masehian, E.: A query-by-example music retrieval system using feature and decision fusion. Multimedia Tools Appl. **77**(5), 6165–6189 (2017). https://doi.org/10.1007/s11042-017-4524-1

Automatic Extraction of Document Information Based on OCR and Image Registration Technology

Shen Ran[✉], Hu Ruoyun, Ding Qi, and Jin Liangfeng

State Grid Zhejiang Marketing Service Center, 138 Yunlian Road, Cangqian Street, Yuhang District, Hangzhou 311100, China

Abstract. Recent have witnessed significant advances in scene text detection and recognition. In this paper, we propose a novel pipeline for real-life State Grid image OCR, named Rectification two-branch OCR, that guarantees the robustness and performance of the real scene recognition. Specially, we first consider the influence of image quality and adopt specific preprocessing techniques for different kinds of data, containing both traditional image processing methods and deep learning-based methods. Then, we propose a simple but effective end-to-end neural network based trainable model to further locate and recognize scene text. At last, we combine the image rectification module and OCR module as a new pipeline for the State Grid image OCR. Extensive experiments on several real scene datasets demonstrate the effectiveness of our proposed methods and also prove it can handle more irregular input image such as rotate and crop.

Keywords: Image rectification · OCR · Neural network

1 Introduction

With the development of informatization and digitization, paperless office has become the new formal of society. An increasing number of documents and forms are scanned and processed in the form of electronic images, which not only saves the cost, improve the efficiency, but also promotes the development of environmental protection. With the combination of digital and tradition office mode, there is an urgent need for a batch of optical character recognition (OCR) tools that can transform digital images and formatted documents to automatically identify and enter the paper documents.

The output of OCR systems is heavily influenced by the quality of the input image. Here image preprocessing comes into play to improve the quality of input image so that the OCR engine can produce more accurate results. It is a data preparation step for contrast enhancement, noise reduction of filtering. Segmentation steps always follow, with the goal to isolate the regions of interest (ROIs) from the background. As for the business image data, ROIs are different for different types of data. For example, when processing ID card images, the name, date of birth and address on the screen should be segmented first and then the segmented characters can be recognized by the OCR

© Springer Nature Switzerland AG 2021
D.-S. Huang et al. (Eds.): ICIC 2021, LNCS 12836, pp. 175–185, 2021.
https://doi.org/10.1007/978-3-030-84522-3_14

system. In this work, we adopt specific preprocessing techniques for different kinds of data, containing both traditional image processing methods and deep learning-based methods, and aim at segmenting original images into several ROIs to prepare data for the OCR modules.

2 Related Work

2.1 Image Rectification and Registration

Image rectification is mainly divided into two categories: geometric rectification and gray level rectification. For the traditional paper geometric rectification technique, we first get the gradient information of the image by Canny [1] and other filter operators based on the second-order information of the image, and then extract the boundary information after removing the noise by Gaussian filter. By changing the size of the Gaussian filter kernel, we can get the edge information of different scales. Then we get the more robust edge information by dilation and erosion algorithms, and further remove the isolated noise independent of the main part of the image. Finally, the image distortion is determined by finding the edge position of the outermost contour and the standard position difference, and the original image is obtained by using the back projection transformation.

At present, the mainstream image matching schemes are based on feature points. Among the, non-deep learning features such as SIFT, SURF and ORB features are widely used in the industry [2]. In recent years, as the deep learning develops rapidly, a number of feature point detection and descriptors based on deep learning are gradually mature, which surpass all non-deep learning feature schemes in performance. However, these methods mainly focus on the matching relationship at the patch level, and do not use the image-level prior to eliminate the mismatching. In 2015, Zagoruyko and Komodakis proposed DeepCompare [3], which takes image pairs as the input and learns the similarity of them directly through the Siamese network. Han et al. proposed MatchNet [4], which can learn descriptors and measurement methods at the same time. Simo-Serra et al. proposed DeepDesc [5] that measures the distance between features extracted by image patches learned by hinge loss, and achieved better performance by balancing positive and negative examples through hard example mining. In 2016, Kurmar et al. [6] proposed to use global loss to expand the distance between positive and negative examples. Balntas et al. proposed TFeat network [7] by using shallow convolutional network and faster hard example mining strategy. Balntas et al. made supervision generalize to positive and negative examples and proposed a new supervision method. In 2017, Tian et al. proposed L2Net [8] through an improved unbalanced sampling strategy of positive and negative examples and a new loss function based on the relative distance in Euclidean space, which achieved the best performance beyond the predecessors under deep supervision. Mishchuk et al. [9] broke the best performance of descriptor learning again through a hard example mining strategy and a simple triplet loss function. The network not only achieved better results than all traditional methods in the experimental dataset, but also showed its powerful generalization in the following datasets. Zhang et al. [10] added the global orthogonal regularization term to the original triplet loss, hoping that the descriptors could use as wide a parameter space as possible. In 2018, Tian et al. [11] referred to the graph matching problem based on the first-order similarity of HardNet and added the second-order similarity regularization, which achieved better results on the HPatches dataset. In 2019, Luo et al. [12] introduced the spatial constraints between key points into the network to further improve the performance. In addition, after Yi et al. first proposed the end-to-end detection and description scheme in 2016, DeTone, Ono, Shen and Dusmanu et al. [13] proposed the end-to-end feature point detection and description framework based on CNN, which further improved the effect and generalization of image matching. Sarlin et al. [14] modeled matching problem by using Transform's encoder and optimal transmission problem, to obtain the best results at present.

2.2 OCR

In recent years, the detection and recognition of scene text (optical character recognition) has aroused more and more research interest in the computer vision community, especially after the revival of neural networks and the growth of image data sets. Scene text detection and recognition provides an automatic and fast way to access text information contained in natural scenes, thereby benefiting various practical applications, such as contract detection, instant translation, and assistance for the blind.

At present, the research on optical character recognition is relatively mature. We can divide the general optical character recognition methods into two main categories, namely, staged detection and end-to-end detection.

Staged Detection. Phased detection is to divide the entire process into two phases: text detection and character detection. The text area of the input image is first detected, and then the text area is sent to the detector to detect specific text, and finally the output result is obtained. For the text detection stage, many methods have been proposed to detect scene text. Jaderberg et al. Use edge boxes [15] to generate candidate regions and refine candidate boxes through regression. Some methods use the symmetry of the text to detect scene text. Adapted from Faster R-CNN and SSD, with carefully designed modifications. There are also some segmentation methods. These methods first detect text segments, and then link them to text instances through spatial relationships or link prediction. In addition, there is a direct return to the text box from the dense segmentation map. Lyu et al. [16] proposes to detect and group corners of text to generate text boxes. Liao et al. [17] proposed a rotation-sensitive regression for orientation-oriented scene text detection. Compared with the popularity of horizontal or multi-directional scene text detection, almost no work has focused on text instances of arbitrary shapes. Recently, due to application requirements in actual scenes, the detection of text with arbitrary shapes has gradually attracted the attention of researchers.

End-to-End Detection. The text detection problem and recognition are completely separated in two stages, but in fact, these two tasks are actually complementary and mutually reinforcing. If it is alone, it may cause sub-optimization problems. So some end-to-end detection methods have been proposed, the most famous of which is the Mask TextSpotter series. Among them, Mask TextSpotter V1 [18] is based on mask RCNN and performs end-to-end text recognition based on segmentation. The mask branch can not only predict the segmentation map to segment the text area, but also predict the character probability map. Mask TextSpotter V2 adds spatial attention to the recognition part on the basis of v1 to improve the text recognition ability of the frame.

Mask TextSpotter V3 adds a segmentation network on the basis of v2 to replace RPN to generate candidate frames, which are more accurate. However, end-to-end detection often has the disadvantages of low accuracy and complex network structure.

3   Text Detection and Recognition Based on Two-Branch Mask OCR Model

3.1   Framework

In view of the reliability and flexibility of the end-to-end method, based on actual application scenarios, our method chooses the end-to-end method to minimize network parameters and increase network flexibility while ensuring accuracy.

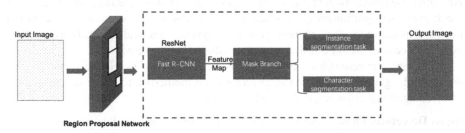

Fig. 1. The overall framework of the method in this section.

The overall framework of our article is shown in Fig. 1. Our method consists of four parts: first we use the feature pyramid network as the skeleton, the regional candidate box network generates text candidate boxes, Fast R-CNN is used as the box regression method, and finally a mask branch is used for text scene detection and character recognition. In the training process, first we use the feature pyramid network to generate text candidate boxes, and then we input these candidate box features into Fast R-CNN to generate more accurate text candidate boxes, and finally our mask branch outputs text scene graphs and Font segmentation diagram. Let's introduce these parts separately.

Network Architecture. Due to the different sizes of text in natural pictures, in order to build a feature map that can integrate high-order semantic information at various scales, we use a feature pyramid structure based on a residual neural network with a depth of 50. The Feature Gold Pagoda uses a bottom-up architecture to fuse feature maps of different resolutions to achieve better results.

Regional Candidate Box Network. We use the regional candidate box network to generate text candidate boxes for the subsequent Fast R-CNN and mask branch. We allocate anchor points at different stages according to the size of the anchor points. Specifically, the area of the anchor is set to {322, 642, 1282, 2562, 5122} pixels in five stages {P2, P3, P4, P5, P6} respectively. Each stage also uses a different aspect ratio {0.5, 1, 2}. In this way, the regional candidate box network can handle texts of various sizes and aspect ratios.

Fast R-CNN. Then we will use Fast R-CNN to do the classification task regression task. The role of this stage is to provide more and more accurate candidate frames for the next stage. The input of this stage is a 7 * 7 feature map generated by the regional candidate network.

Mask Branch. There are two tasks in the mask branch, one is the global text instance segmentation task and the other is the character segmentation task. This branch passes through four convolutional layers and one deconvolutional layer. The size of a given input is fixed to 16 * 64. The mask branch can predict 38 mappings (size 32 * 128), including global text instance maps, 36 Character images and character background images. Regardless of the shape of the text instance, the global text instance graph can give accurate positioning of the text area. The character mapping table is a 36-character mapping table, including 26 letters and 10 Arabic numerals. The character background image of the character area also needs to be removed in the post-processing stage.

3.2 Label Generation

This section introduces how to generate the label form required by the above steps given a picture I and its corresponding label. We first convert the polygon into a horizontal rectangle that covers the smallest area of the rectangle. Then we generate the target of the region candidate box network and Fast R-CNN. For the mask branch of a given label P, C (which may not exist), two types of target images will be generated, as well as the candidate boxes output by the regional candidate box network: the global image for text instance segmentation and the semantic segmentation for characters Character illustration. Given a label r with text, we first use the matching mechanism to obtain the best matching horizontal rectangle. The corresponding polygons and characters can be further obtained. Next, move and adjust the size of the matching polygon and character frame, and finally get the final semantic segmentation character map.

3.3 Objective Function

We define our objective function as:

$$L = L_{rpn} + \partial L_{RCNN} + \partial_2 L_{mask} \tag{1}$$

The first two are defined as the loss function of the candidate box network and the loss function of Fast R-CNN, which are consistent with the original paper. L_{mask} is divided into two parts: the global loss function and the character loss function. The global loss function is defined as the average. The cross entropy loss function of, and the character loss function is a weighted spatial soft-max loss function. In our article, ∂ and ∂_2 are set to 1.

3.4 Forward Propagation

In the training process, the input candidate frame of the mask branch is different from the training process of the regional candidate network. In the inference phase, we use the output of Fast R-CNN as the candidate frame for generating the predicted global mapping and character mapping, because Fast R-CNN The output is more accurate. The specific reasoning process is as follows: First, enter the test image, we obtain the output of Fast R-CNN, and filter out redundant candidate frames through non-maximum suppression. Then, input the reserved candidate boxes into the mask branch to generate a global map and a character map. Finally, the contour of the text area is calculated on the global map and the predicted polygon is directly obtained. In addition, the character sequence can be generated through the pixel voting algorithm on the character map we proposed.

Pixel Voting Algorithm. We decode the predicted character map into a character sequence through the proposed pixel voting algorithm. We first binarize the background image, the value ranges from 0 to 255, and the threshold is 192. Then obtain all character areas according to the connected areas in the binarization graph. We calculate the average value of each region for all character images, and these values can be regarded as the character category probability of the region. The character class with the largest average value will be assigned to this area. After that, we group all characters from left to right according to the writing habits of Chinese characters.

4 Experiments

4.1 Data of Image Rectification

The test data we used are the photos taken from realistic business scenarios provided by the State Grid, including images of the property title certificate, both sides of ID card, contracts and electricity meters. Figure 2 illustrates some raw data to process. The exact size of the testing set is shown in Table 1.

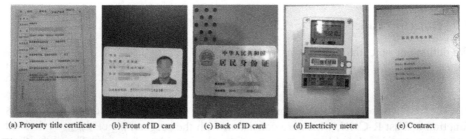

(a) Property title certificate   (b) Front of ID card   (c) Back of ID card   (d) Electricity meter   (e) Contract

Fig. 2. Examples of some raw data to process. Each column from (a) to (e) represents one example of the corresponding type of data. Some sensitive privacy information is blocked by mosaic.

Due to the shooting angle and varying illumination, most data cannot be presented with high quality as shown in Fig. 2, making it difficult to recognize the key information. Therefore, the first step is to rectify the raw data and segment the ROIs to prepare clean data for OCR.

Table 1. The size of testing set.

Data	Size
Property title certificate	230
Front of ID card	688
Back of ID card	746
Contract	518
Electricity meter	895

4.2 Image Rectification Evaluation

To evaluate the effectiveness of the proposed method, we conduct extensive experiments on the above dataset and compare the following methods: (a) traditional rectification method (denoted by T), (b) the combination of T and SIFT-based matching method, (c) the combination of T with deep learning-based matching method, (d) the fine-tuned deep learning model with extra data, and (e) our proposed hierarchical multi-scale feature aggregation image matching algorithm.

The results are shown in Table 2. The accuracy means the proportion of rectified images that contains the key information to recognize.

Table 2. The rectification accuracy on different testing data. These five columns, from the 2nd one to the 6th one, report the accuracy on the property title certificate, front of ID card, back of ID card, contract, and electricity.

Method	Accuracy				
	Property	Front	Back	Contract	Meter
(a) Traditional	73.32%	80.75%	79.84%	75.38%	79.32%
(b) T + SIFT	78.50%	84.04%	84.82%	81.02%	83.40%
(c) T + DL	84.38%	87.20%	86.90%	85.63%	85.60%
(d) Fine-tuning	92.24%	95.45%	94.30%	91.02%	92.53%
(e) Ours	**93.45%**	**97.88%**	**97.02%**	**93.41%**	**94.27%**

Our proposed method achieves excellent performance on various dataset of different tasks. Among these methods, the rectification method with image matching can improve the success rate of traditional method. The deep learning-based matching method achieves the significant improvement by replacing the traditional matching technique. Besides, the self-supervised method based on data augmentation improves the performance of the pre-trained model, and the success rate is more than 95% on the specified dataset. Finally, with the improved deep learning-based image matching algorithm, the success rate on the specified dataset is more than 97%.

Some rectification results are shown in Fig. 3. The proposed method works well with various conditions, including normal shooting, shadow, severe screen glare, bad illumination and rotation.

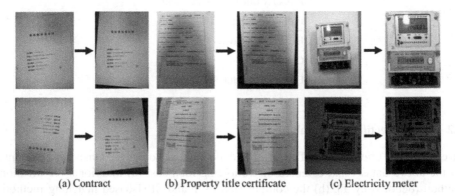

(a) Contract   (b) Property title certificate   (c) Electricity meter

Fig. 3. Examples of the rectification results. Each part, from left to right, shows the original image and the rectified key regions.

4.3 Data of Text Detection and Recognition

The test data we used are the photo of the contract, the front photos of the ID card, and the back photos of the ID card provided by the power grid. Figure 4 shows an example of the photos of the contract. The specific quantities of the three types of test data are shown in Table 3. After these images are corrected, they generally have a relatively straight angle before being imported into the OCR model. Only a few images will become very distorted due to the bad filming angle, which will result in failure to recognize successfully.

4.4 Information Extraction and Denoising

Input the picture to be recognized into the OCR model to get the text recognition result in the picture. The OCR model will give the recognized text, the confidence of the recognized text result and the pixel coordinates of the text box in the original image, as shown in Fig. 4. Take the contract as an example, as shown in Fig. 5.

In order to extract only the required information, for example, only the name and ID number on the ID card are needed to identify the ID card, we chose to use the approximate coordinates of the name and ID number obtained after correcting the picture and the pixel coordinates of the text box in the original image. For comparison, if the coordinates of the text box are within the range of the corrected coordinates, leave this message. In addition, we have set specific denoising sentences for many situations. For example, when identifying the number of the contract, the information of the province and city that is in the same position as the number is often extracted together, and we search in the extracted results to find the characters "province", "city", and "district". Then leave only the results after these characters, so that you can accurately filter out the information you want.

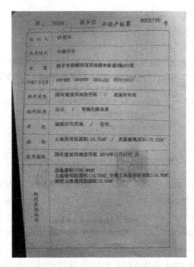

Fig. 4. An example of the contract photos.

Table 3. Test dataset.

Dataset	Quantity
Contract	230
IDcard_front	688
IDcard_back	746

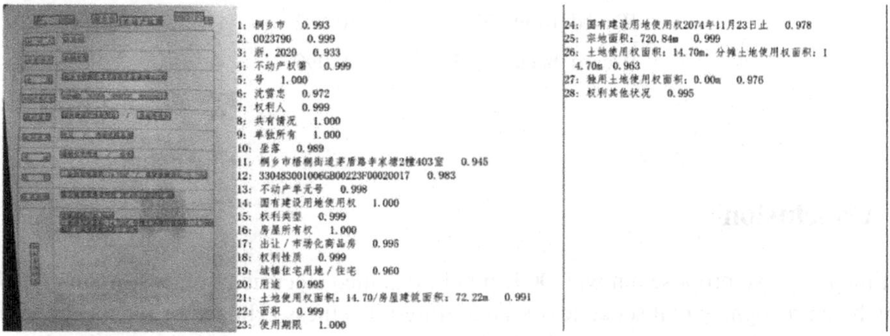

1: 桐乡市 0.993	24: 国有建设用地使用权2074年11月23日止 0.978
2: 0023790 0.999	25: 宗地面积: 720.84m 0.999
3: 新，2020 0.933	26: 土地使用权面积: 14.70m，分摊土地使用权面积: 1
4: 不动产权第 0.999	4.70m 0.983
5: 号 1.000	27: 独用土地使用权面积: 0.00m 0.976
6: 沈雪忠 0.972	28: 权利其他状况 0.995
7: 权利人 0.999	
8: 共有情况 1.000	
9: 单独所有 1.000	
10: 坐落 0.989	
11: 桐乡市梧桐街道茅盾路2幢403室 0.945	
12: 330483001006G800223F00020017 0.983	
13: 不动产单元号 0.998	
14: 国有建设用地使用权 1.000	
15: 权利类型 0.999	
16: 房屋所有权 1.000	
17: 出让 / 市场化商品房 0.996	
18: 权利性质 0.999	
19: 城镇住宅用地 / 住宅 0.960	
20: 用途 0.995	
21: 土地使用权面积: 14.70/房屋建筑面积: 72.22m 0.991	
22: 面积 0.999	
23: 使用期限 1.000	

Fig. 5. The recognition result of contract.

After information extraction and denoising, we output the results to a TXT file to save. Take the contract shown in Fig. 4 as an example, and the output effect is shown in Fig. 6.

不动产权号:[['0023790']]
地址:['桐乡市']
权利人:['沈雪忠']
坐落:桐乡市梧桐街道茅盾路李家塘2幢403室
使用期限:['国有建设用地使用权2074年11月23日止']

Fig. 6. An example of the results obtained by extracting information and denoising after the recognition of the contract.

4.5 Results

Because there are a few test images of the contract that are very slanted after correction and cannot be recognized by human eyes, the recognition accuracy of our model on the test data of the contract is only about 85%. In contrast, because ID card photos are basically normal images with no major abnormalities, the recognition accuracy on ID card test images can reach about 95%. From this point of view, our model has excellent performance and can get very good recognition results as long as it is in general conditions.

Since our model can be used on the GPU, the recognition speed can be greatly improved. The time to identify a contract is about 1 s, the time to identify the front of the ID card is about 0.08 s, and the time to identify the back of the ID card is about 0.09 s. The specific experimental results are shown in Table 4.

Table 4. Recognition accuracy and recognition time of three kinds of test data.

Dataset	Accuracy(%)	Time(s)
Contract	85.6%	1.112
IDcard_front	97.2%	0.082
IDcard_back	95.5%	0.099

5 Conclusion

In this paper, we propose a novel OCR pipeline named Rectification two-branch OCR, which can recognize real scene text image more robustness. The experiments show the effectiveness of our proposed methods. In future work, we consider integrating image correction and recognition into a network to make the entire module more flexible and easy to use.

Acknowledgement. This paper is supported by Science and Technology Project of State Grid Corporation of China (No. 5211DS190032).

References

1. Canny, J.: A computational approach to edge detection. IEEE Trans. Pattern Anal. Mach. Intell. **6**, 679–698 (1986)
2. Castleman, K.R.: Digital Image Processing. Pearson Education, Prentice Hall (2007)
3. Zagoruyko, S., Komodakis, N.: Learning to compare image patches via convolutional neural networks. In: Proceedings of the IEEE Conference on Computer Vision and Pattern Recognition, pp. 4353–4361 (2015)
4. Han, X., Leung, T., Jia, Y., et al.: MatchNet: unifying feature and metric learning for patch-based matching. In: Proceedings of the IEEE Conference on Computer Vision and Pattern Recognition, pp. 3279–3286 (2015)
5. Simo-Serra, E., Trulls, E., Ferraz, L., Kokkinos, I., Fua, P., Moreno-Noguer, F.: Discriminative learning of deep convolutional feature point descriptors. In: Proceedings of the IEEE International Conference on Computer Vision, pp. 118–126 (2015)
6. Kumar, B.G.V., Carneiro, G., Reid, I.: Learning local image descriptors with deep Siamese and triplet convolutional networks by minimising global loss functions. In: Proceedings of the IEEE Conference on Computer Vision and Pattern Recognition, pp. 5385–5394 (2016)
7. Balntas, V., Riba, E., Ponsa, D., et al.: Learning local feature descriptors with triplets and shallow convolutional neural networks. BMVC **1**(2), 3 (2016)
8. Tian, Y., Fan, B., Wu, F.: L2-Net: deep learning of discriminative patch descriptor in Euclidean space. IN: Proceedings of the IEEE Conference on Computer Vision and Pattern Recognition, pp. 661–669 (2017)
9. Mishchuk, A., Mishkin, D., Radenovic, F., et al.: Working hard to know your neighbor's margins: local descriptor learning loss. arXiv preprint arXiv:1705.10872 (2017)
10. Zhang, X., Yu, F.X., Kumar, S., et al.: Learning spread-out local feature descriptors. In: Proceedings of the IEEE International Conference on Computer Vision, pp. 4595–4603 (2017)
11. Tian, Y., Yu, X., Wu, B., Fan, B., Heijnen, H., Balntas, V.: SOSNet: second order similarity regularization for local descriptor learning. In: Proceedings of the IEEE/CVF Conference on Computer Vision and Pattern Recognition, pp. 11016–11025 (2019)
12. Luo, Z., Zhou, L., Bai, X., et al.: ASLFeat: learning local features of accurate shape and localization. In: Proceedings of the IEEE/CVF Conference on Computer Vision and Pattern Recognition, pp. 6589–6598 (2020)
13. DeTone, D., Malisiewicz, T., Rabinovich, A.: Superpoint: self-supervised interest point detection and description. In: Proceedings of the IEEE Conference on Computer Vision and Pattern Recognition Workshops, pp. 224–236 (2018)
14. Sarlin, P.E., DeTone, D., Malisiewicz, T., et al.: Superglue: learning feature matching with graph neural networks. In: Proceedings of the IEEE/CVF Conference on Computer Vision and Pattern Recognition, pp. 4938–4947 (2020)
15. Zitnick, C.L., Dollár, P.: Edge boxes: locating object proposals from edges. In: Fleet, D., Pajdla, T., Schiele, B., Tuytelaars, T. (eds.) ECCV 2014. LNCS, vol. 8693, pp. 391–405. Springer, Cham (2014). https://doi.org/10.1007/978-3-319-10602-1_26
16. Lyu, P., Yao, C., Wu, W., Yan, S., Bai, X.: Multi-oriented scene text detection via corner localization and region segmentation. In: Proceedings of CVPR, pp. 7553–7563 (2018)
17. Liao, M., Zhu, Z., Shi, B., Xia, G.S., Bai, X.: Rotation-sensitive regression for oriented scene text detection. In: Proceedings of CVPR, pp. 5909–5918 (2018)
18. Lyu, P., Liao, M., Yao, C., Wu, W., Bai, X.: Mask textspotter: an end-to-end trainable neural network for spotting text with arbitrary shapes. In: Ferrari, V., Hebert, M., Sminchisescu, C., Weiss, Y. (eds.) Computer Vision – ECCV 2018. LNCS, vol. 11218, pp. 71–88. Springer, Cham (2018). https://doi.org/10.1007/978-3-030-01264-9_5

Using Simplified Slime Mould Algorithm for Wireless Sensor Network Coverage Problem

Yuanye Wei[1], Yongquan Zhou[1,2(✉)], Qifang Luo[1,2], and Jian Bi[1]

[1] College of Artificial Intelligence, Guangxi University for Nationalities,
Nanning 530006, China
zhouyongquan@gxun.edu.cn
[2] Guangxi Key Laboratories of Hybrid Computation and IC Design Analysis,
Nanning 530006, China

Abstract. Wireless sensor network (WSN) coverage control is a study of how to maximize the network coverage to provide reliable monitoring and tracking services with guaranteed quality of service. The application of optimization algorithm is helpful to effectively control the network nodes energy, improve the perceived quality of services, and extend the network survival time. This paper presents a simplified slime mould algorithm (SSMA) for optimization WSN coverage problem. We mainly conducted thirteen groups of WSNs coverage optimization experiments, and compared with several well-known metaheuristic algorithms. From the experimental results and Wilcoxon rank sum test results demonstrate that SSMA is generally competitive, outstanding performance and effectiveness.

Keywords: WSN · Slime mould algorithm (SMA) · Simplified slime mould algorithm (SSMA) · Effective coverage rate · Metaheuristic

1 Introduction

Wireless sensor network (WSNs) is composed of a large number of densely distributed sensor network node, each node has limited computing, storage and wireless communication capabilities, and can sense the surrounding environment at close range. It has the characteristics of small size, low cost, and low power consumption. It can assist in sensing, collecting and processing the information of the monitored object in real time, so with the rapid development of science and technology in various fields, WSNs has been widely used in military affairs, environmental monitoring, medical and health care, safety monitoring and other fields [1].

In various applications of wireless sensor networks, coverage has always been a crucial issue, which determines the monitoring ability of the target area. An appropriate node deployment strategy can not only improve the service quality of WSNs, but also effectively promote its energy utilization. However, searching for the optimal node deployment scheme is a difficult task, especially for large-scale sensor networks [2]. As this context, scholars choose to focus on meta-heuristic algorithm, which has made a good contribution to the coverage optimization of WSNs. Ref. [3] used PSO Optimized

© Springer Nature Switzerland AG 2021
D.-S. Huang et al. (Eds.): ICIC 2021, LNCS 12836, pp. 186–200, 2021.
https://doi.org/10.1007/978-3-030-84522-3_15

sink node path location in WSNs; Ref. [4] a novel FFO was introduced to the coverage optimization. Ref. [5] used PSO and Voronoi diagram to optimize sensor coverage in WSNs. Ref. [6] used a novel DE applied to optimize the coverage problem of WSNs. Ref. [7] used ACO based sensor deployment protocol for wireless sensor networks. Ref. [8] proposed a Hybrid Red Deer SSA for Optimal Routing Protocol in Wireless Multimedia Sensor Networks. Ref. [9] proposes FPA to solve MEB problem in WSNs. Ref. [10] used FA for power management in (WSNs) et al.

The Slime Mould Algorithm [11] is a novel metaheuristic algorithm, which was proposed by Seyedali M. et al. in 2020. It to simulate the positive and negative feedback generated by the propagation wave of biological oscillator during the foraging process of Slime mould in nature, and to guide their behavior and morphological changes. Because of its simple concept and strong searching ability, it has been paid attention to and studied by many scholars since it was put forward. Ref. [12] used HSMA_WOA for COVID-19 chest X-ray images; Ref. [13] dynamic structural health monitoring based on SMA; Ref. [14] proposed hybrid ANN model with SMA for Prediction of Urban Stochastic Water Demand; Ref. [15] based on SMA to extract the optimal model parameters of solar PV panel; Ref. [16] SMA for feature selection; Ref. [17] SMA-AGDE for solving various optimization problems; Ref. [18] using SMA to Mitigating the effects of magnetic coupling et al.

In this paper, in order to improve the search accuracy and population diversity of the con SMA in solving the coverage optimization problem of wireless sensor networks, we propose a simplified slime mould algorithm (SSMA). The SSMA has three highlights:

(1) A simplified slime mould algorithm is proposed to solving WSNs coverage optimization.
(2) Simplify the position update formula and a new adaptive oscillation factor, for balance the development and exploration capacity, and improve the coverage capacity of WSNs.
(3) SSMA will be experimentally experimented in a small-scale and large-scale WSNS coverage problems, and comparative analysis of 6 state-of-the-art algorithms.

The rest of this paper is structured as follows: Sect. 2 briefly reviews the original SMA algorithm. Section 3 details the simplified SMA (SSMA). Section 4 gives the practical application of SSMA in solving WSNs coverage optimization problem. Section 5 carries out experimental analysis and discussion. Section 6 provides the conclusions and future directions of this work.

2   Wireless Sensor Network Coverage Optimization

Assumed that the monitoring area of WSNs is a two-dimensional plane, which is digitized into $L \times M$ grids, and the size of each grid is set to unit 1. N homogeneous sensors are deployed in this area, and the node set can be expressed as $Z = \{Z_1, Z_2, Z_3, ... Z_N\}$ having the same sensing radius R. In this paper, Boolean model is used as the node sensing model, so long as the target is within the node sensing range, it can be successfully sensed. Assuming that the coordinates of a node z_i in the detected area are (x_i, y_i) and the position coordinates D_j of the target point are (x_j, y_j), the distance between the node and the target point is:

$$d(z_i, D_j) = \sqrt{(x_i, y_i)^2 - (x_j, y_j)^2}$$

(2.1)

where $p(z_i, D_j)$ represents the perceptual quality of node z_i to D_j. When the position D_j of node z_i is within the circle of perceived range, the perceived quality is 1, otherwise it is 0, and the mathematical expression is:

$$p(Z_i, D_j) = \begin{cases} 1, & \text{if } d(z_i, D_j) \leq R \\ 0, & \text{otherwise} \end{cases}$$

(2.2)

Generally, the sensor's perception probability of the target is less than 1. In order to improve the perception probability of the target, multiple sensors need to cooperate to detect, so the sensor's perception probability of a certain target is:

$$p(Z, D_j) = 1 - \prod_{i=1}^{N} \left[1 - p(z_i, D_j)\right]$$

(2.3)

The coverage rate of the node set Z over the entire monitoring area is:

$$f = \frac{\sum\limits_{j=1}^{L \times M} p(Z, H_j)}{L \times M}$$

(2.4)

Equation (2.4) is the objective function of WSN coverage optimization problem.

3   Slime Mould Algorithm

Slime mould algorithm [11] mainly simulates the behavior and morphological changes of slime mould in the foraging process in nature. SMA can effectively simulates the information transmission mechanism of feeding back the food concentration in the search space through the propagation wave generated by the oscillator, and simulates the threat to the existence in the foraging process with the weight factor, which improves the effectiveness of the algorithm.

Slime mould can judge the position according to the concentration of surrounding food during foraging, and adjust their foraging mode according to the influence of environment.

3.1 Mathematical Model

When slime mould foraging, they can sense the concentration and position of food through odour in the air and then approach it constantly. This process can be expressed by the following equation:

$$\vec{S}(t+1) = \begin{cases} rand * (ub - lb) + lb, \ rand < pc \\ \vec{S^*}(t) + \vec{kb} * (\vec{W} * \vec{S}_{R1}(t) - \vec{S}_{R2}(t)), \ r < q \\ \vec{kc} * \vec{S}(t), \ r \geq q \end{cases} \tag{3.1}$$

where ub and lb are represents the upper and lower boundaries of the search space, $rand$ and r are random values between $[0,1]$, and pc is a constant, which is taken same as stander SMA. Parameters $\vec{kb}$ and $\vec{kc}$ are oscillation factor, t represents the current iteration times, $\vec{S^*}$ represent the current optimal individual, $\vec{S}$ represent the current individual, $\vec{S}_{R1}$ and $\vec{S}_{R2}$ represents two individuals randomly selected in the slime mould population, $\vec{W}$ represent weight.

The q mathematical formula of is as follows:

$$q = \tanh|f(i) - DF| \tag{3.2}$$

where $i \in 1, 2, ...n$, $f(i)$ indicated fitness of individual $\vec{S}$, DF represents the best fitness value obtained in all iterations. $\vec{kb}$ and $\vec{kc}$ mathematical formula as follows:

$$\vec{kb} = [-a, a] \tag{3.3}$$

$$a = \arctan h\left(-\frac{it}{T_{max}} + 1\right) \tag{3.4}$$

$$\vec{kc} = \left(1 - \frac{it}{T_{max}}\right) \tag{3.5}$$

$\vec{W}$ mathematical formula as follows:

$$W(smell(i)) = \begin{cases} 1 + r \cdot \log(\dfrac{BF - f(i)}{BF - WF + \varepsilon} + 1) \\ 1 - r \cdot \log(\dfrac{BF - f(i)}{BF - WF + \varepsilon} + 1), \ other \end{cases} , i \leq \dfrac{2}{N} \tag{3.6}$$

$$smell = sort(f) \tag{3.7}$$

where r is a randomly values interval of $[0,1]$, T_{max} indicates the maximum number of iterations, BF indicates the best fitness in the current iteration, WF indicates the worst fitness in the current iteration, $smell$ represents the corresponding values sorted by fitness values. is the fitness of the whole population.

4 Simplified Slime Mould Algorithm

Since the SMA [11] was put forward, it has been paid attention to and studied by many scholars because of its simple concept and good effect in solving various practical problems. However, like many meta-heuristic algorithms, it also has the disadvantages of premature convergence and easy to fall into local optimum. When solving the coverage optimization problem of wireless sensor networks, the standard shape memory alloy has obvious defects such as local optimum location, low coverage rate and insufficient convergence speed. Therefore, we propose a simplified slime mold optimization algorithm (SSMA) to solve the optimization problem of wireless sensor networks. The position update in Eq. (3.1(3)) of the SMA is that individuals search around themselves, which leads to poor exploration ability. Individuals cannot find the global optimal solution quickly, and easily fall into local optimal solution. The position update in Eq. (3.1(2)), individuals update their position according to the global optimal solution and two random solutions is random choose in the population. This mechanism can balance the exploration and development ability of the algorithm well. In addition, the oscillation factor has a vital impact on the optimization performance of SMA. When the oscillation factor is chosen as trigonometric function, it not only improves the convergence speed of optimization, but also accords with the actual theory of propagation wave. Therefore, we propose a simplified slime mould algorithm (SSMA) to solve the coverage optimization problem of WSNs.

The update formula is as follows:

$$\vec{S}(t+1) = \begin{cases} rand * (ub - lb) + lb, \, rand < pc \\ \vec{S}^*(t) + \vec{kb} * (\vec{W} * \vec{S_{R1}}(t) - \vec{S_{R2}}), \, otherwise \end{cases} \tag{4.1}$$

The oscillation factor formula is as follows:

$$\vec{kb} = [-a, a] \tag{4.2}$$

$$a = \gamma(\cos(\pi \cdot \frac{it}{T_{max}}) + \lambda) \tag{4.3}$$

where a is the value of the change range of γ, $\gamma = 0.5$; λ is the value step of the slime mould, $\lambda = 1$.

The SSMA mainly depends on the optimal solution and two random solution in the population. Equation (4.1) is the core of the SSMA. When z is less than 0.03, slime mould searches randomly in the whole search space, and Eq. (4.2) is executed; otherwise, Eq. (4.3) is executed. The best individual can lead the development stage of slime mould, while two random individuals in the population mainly dominate the exploration stage of the SSMA. This formula can balance the exploration and development of the algorithm. The pseudo-code of the SMA is shown in Algorithm 1.

Algorithm 1. SSMA pseudo-code
1.   Initialize the parameter, population, ub, lb, Max_iter;
2.   Initialize the position of slime mould $S_i(1,2,...,n)$;
3.   **While** ($t < Max\_ter$)
4.   Calculate the fitness of all slime mould;
5.   Update $DF, BF, WF, \vec{S}$
6.   Calculate the W by Eq. (3.6);
7.   **for** each search portion
8.   Update q, kb, kc;
9.   Update position
10.   **If** $rand < pc$
11.   Update position by Eq. (4.1)
12.   **Else**
13.   Update position by Eq. (4. 2)
14.   **End if**
15.   **End for**
16. $t = t+1$;
17.   **End while**
18.   **Return** $DF, \vec{S}$

5 Simulation Result

5.1 Problem Definition

As mentioned in the second section, we will define it as follows: the coverage of sensor nodes is a circle with a fixed radius. Whether a certain position is in the coverage area of the sensor node can be determined by calculating the distance between the target point and the node. In order to evaluate the coverage of WSNs in two-dimensional area, the whole monitoring area is divided into $L \times M$ grids. If WSNs can cover points, then the coverage rate is 1. In addition, we assume that all sensors are the same monitoring area, and ignore the boundary effect on the sensor network.

During initialization, all sensor nodes are randomly scattered in a given monitoring area, and the initial coordinates of these sensor nodes are the initial input values of the algorithm. Each search agent in the algorithm represents a placement scheme of sensor nodes. In the two-dimensional monitoring area, the dimension of search agent is twice the number of sensor nodes; In other words, the $2x - 1$ dimensional data represents the abscissa of the x-th sensor node, while the $2x$ dimensional data represents its ordinate. The algorithm takes WSNs coverage as the fitness function that is Eq. (2.4), as mentioned in the second section, and maximizes the fitness function as the optimization goal. Finally, the optimized coverage and the positions of all sensor nodes are output.

We compare the experimental results of SSMA with SMA [11], PSO [19], GWO [20], WOA [21], MPA [22] and FPA [23]. Table 1 summarizes the experimental parameter settings for comparison algorithms. In order to eliminate the experimental error caused by chance, the average value of 20 independent runs is used as the comparison result. We are mainly divided into two kinds of experiments: low dimension and large-scale dimension. Other parameter settings: population size N, maximum iterations T; The monitoring radius R, the monitoring area $L \times M$, the sensor of node: Node, and the dimension Dim.

Table 1. Parameter settings for each algorithm

Algorithm	Year	Parameters and values
PSO [19]	1995	$C_2 = 1.5, C_2 = 2, W_{max} = 0.9, W_{min} = 0.4$
FPA [23]	2012	$p = 0.8$
GWO [20]	2014	$a = [2, 0]$
WOA [21]	2016	$a = [2, 0], b = 1, A = [2, 0]$
MPA [22]	2020	$Fads = 0.2, P = 0.5$
SMA [11]	2020	$z = 0.03$
SSMA	~	$z = 0.03, \lambda = 1, \gamma = 0.5$

5.2 Comparative Analysis of Experiments

In this sector, we compared the SSMA with some competitive MAs on WSNs coverage optimization. Furthermore, all of the experimental series were performed on MATLAB 2017b and were run on a CPU Core i3–6100 v4 (3.70GHz) with 8 GB RAM in this paper.

In the experimental part, we will set up 3 type groups of coverage optimization experiments, among which there are two groups of low-dimensional ones, which are divided into Case1-Case3 and C1-C4, the number of nodes is unequal between 5 to 80 nodes, moreover, there are third groups of large-dimensional experiments, which name C5 to C10, mainly from 100 to 1200 nodes. The parameter settings of three groups of experiments are shown in Table 2. The experiments mainly compared the experimental results of SSMA with six state-of-the-art algorithms to analyze the robustness and convergence speed.

5.2.1 Low-Dimension Experimental Analysis

In this section, we mainly test and analyze the number of nodes in low dimension. There are two types of cases in low dimension: Case1-Cas3; C1-C4. Examples Case1-Case3 come from literature [24], C1-C4 are self-defined cases in this paper, third groups for large dimension. We used the above two types with different radio for experimental comparison. Case 1-Case 3 and C1-C4 compares these two groups of seven cases through Tables 3 and Figs.1 to Fig. 3. Moreover, Tables 3 list results for independent runs for 20 times, where "Ave" and "Std" represent the average coverage and standard deviation,

Table 2. Parameter settings

Case	R(m)	Dim	Node	L*M(m)	Maximum iteration	Population size
Case 1	1	20	10	5*5	250	30
Case 2		40	40	10*10		
Case 3		160	80	15*15		
C 1	10	10	5	50*50	100	30
C 2		40	20	80*80		
C 3		120	60	100*100		
C 4		160	80	150*150		
C5	45	200	100	800*800		
C6	30	600	300			
C7	25	1000	500			
C8	20	1400	700			
C9	18	2000	1000			
C10	15	2400	1200			

respectively. Black bold numbers represent the optimal solution, blue bold represents the secondly best value, and green bold represents the third best.

As Case1 to Case3. It can be clearly seen from Table 3 about the average coverage rate (mean fitness) of SSMA is significantly higher than that of the other six methods. C1 is reached 63.05%, ranking first, which was 1.61%, 4.54%, 0.56%, 3.07%, 1.66% and 10.36% higher than PSO, FPA, GWO, WOA, MPA and SMA, respectively. As C2 is reached 86.43%, which was 13.97%, 15.07%, 2.55%, 11.44%, 7.9% and 19.35% higher than PSO, FPA, GWO, WOA, MPA and SMA respectively. The average coverage rate of SSMA in C3 reaches 99.47%, which is 13.46%, 8.76%, 6.89%, 5.79%, 3.2% and 12.41% higher than PSO, FPA, GWO, WOA, MPA and SMA respectively. The average coverage rate of SSMA in C4 is 85.13%, which is 21.58%, 13%, 10.64%, 9.64%, 4.14% and 15.15% higher than PSO, FPA, GWO, WOA, MPA and SMA respectively. The average coverage rate of C1-C4, SSMA in C1 reached 63.05%, ranking first, which was 1.61%, 4.54%, 0.56%, 3.07%, 1.66% and 10.36% higher than PSO, FPA, GWO, WOA, MPA and SMA respectively. The C2 is reached 86.43%, which was 13.97%, 15.07%, 2.55%, 11.44%, 7.9% and 19.35% higher than PSO, FPA, GWO, WOA, MPA and SMA respectively. Case for C3 reaches 99.47%, which is 13.46%, 8.76%, 6.89%, 5.79%, 3.2% and 12.41% higher than PSO, FPA, GWO, WOA, MPA and SMA respectively. While mean coverage C4 is 85.13%, which is 21.58%, 13%, 10.64%, 9.64%, 4.14% and 15.15% higher than PSO, FPA, GWO, WOA, MPA and SMA respectively. From the above data display and analysis, it can be seen that SSMA has obvious advantages in solving the coverage optimization problem of wireless sensor networks, and has relatively strong robustness and fast convergence speed.

Table 3. Results for low-dimension experimental

case	Algorithm	PSO	FPA	GWO	WOA	MPA	SMA	SSMA
Case1	Ave	0.7960	**0.97**	**0.962**	0.922	**0.962**	0.84	**0.984**
	Std	0.0733	**0.0178**	**0.0275**	0.0378	0.0330	0.0486	**0.0239**
Case2	Ave	0.6815	0.8515	**0.8925**	0.8405	**0.922**	0.7685	**0.9715**
	Std	0.0613	**0.0081**	0.0481	0.0228	0.0337	**0.0157**	**0.0139**
Case3	Ave	0.5842	0.7633	**0.7909**	0.7507	**0.8589**	0.7067	**0.9044**
	Std	0.0514	**0.0076**	0.0748	0.0254	0.0270	**0.0107**	**0.0193**
C1	Ave	**0.6144**	0.5851	**0.6249**	0.5998	0.6139	0.5269	**0.6305**
	Std	0.0343	**0.0106**	0.0112	0.0135	**0.0085**	0.0176	**0.0039**
C2	Ave	0.7246	0.7136	**0.8388**	0.7499	**0.7853**	0.6708	**0.8643**
	Std	0.0383	**0.0136**	**0.0165**	0.0230	0.0199	0.0169	**0.0124**
C3	Ave	0.8601	0.9071	0.9258	**0.9368**	0.9627	0.8706	**0.9947**
	Std	0.0427	**0.0056**	0.0493	0.0115	0.0126	**0.0083**	**0.0025**
C4	Ave	0.6355	0.7213	0.7449	**0.7549**	0.8099	0.6963	**0.8513**
	Std	0.0196	**0.0054**	0.0614	0.0124	0.0130	**0.0115**	**0.0068**

The convergence curve of Case1 and C2 show in Fig. 1 that the convergence curve of SSMA is much faster than the other six algorithms, and other algorithms are easy to fall into the local optimum, and loss the optimization ability. In Fig. 2, The standard deviation of SSMA is more stable and higher than other algorithms. As can be seen from the coverage in Fig. 3, the coverage of SSMA is a superior than other algorithms. As the dimensionality increases, the coverage performance is stronger.

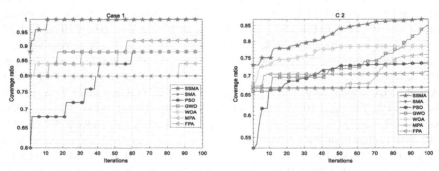

Fig. 1. Coverage curve on case1 and C2

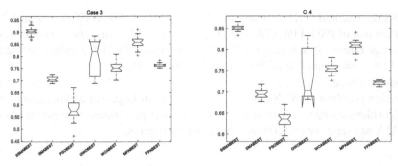

Fig. 2. Box plot on Case3 and C4

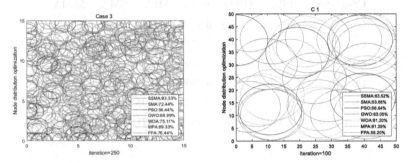

Fig. 3. Nodes distribution optimization on case 3 and C1

5.2.2 Large-Scale Experimental Analysis

In this section, we applied the proposed SSMA on a large-scale wireless sensor node, and compare the results with some methods. The experimental parameters are list in Table 2, which are same as previous mention.

Other algorithms may adapt in low scale dimension problem, but they may lose performance in large-scale dimension. However, the SSMA is still applicable in the large-scale nodes. Results of six large-scale cases show in Table 4. It is obviously that the coverage SSMA of C5, C6, C7 and C9 are ranked first among the six large-scale cases of C5-C10, and the SSMA of C8 and C10 are ranked second. For C8, the coverage rate of SSMA was 0.18% less than that of MPA ranked first, and C10, the coverage rate of SSMA was 0.05% less than that of MPA ranked first. It can be seen that the SSMA ranked second in C8 and C10 is not much different from the MPA ranked first. C5(200 dimensions), the coverage rate of SSMA reached 78.35%, which was 25.44%, 10.89%, 12.87%, 8.39%, 2.82% and 13.41% higher than that of PSO, FPA, GWO, WOA, MPA and SMA.C6(600 dimensions), the coverage rate of SSMA reached 83.32%, which was 29.67%, 6.94%, 8.54%, 5.41%, 0.49% and 8.68% higher than that of PSO, FPA, GWO, WOA, MPA and SMA. In C7(1000 dimensions), the coverage rate of SSMA reached 86.29%, which was 30.44%, 5.46%, 6.95%, 4.32%, 0.21% and 6.92% higher than that of PSO, FPA, GWO, WOA, MPA and SMA. In experiment C8(1400 dimensions), the coverage rate of SSMA reached 81.02%, which was 29.58%, 4.45%, 5.47%, 3.43% and 5.54% higher than that of PSO, FPA, GWO, WOA and SMA. In C9(2000 dimension), the coverage rate of SSMA reached 85.15%, which was 31.64%, 3.94%, 5%, 0.1%, 3.22% and 4.83% higher than that of PSO, FPA, GWO, WOA, MPA and SMA. In C10(2400 dimensions), the coverage rate of SSMA reached 78.16%, which was 28.21%,3.19%,4.05%,2.82%, and 4.612% higher than that of PSO,FPA,GWO,WOA,SMA.

From the above results and discussions, it shows that SSMA can solve the coverage optimization problem of WSNs. SSMA do still used in large-scale experiments, while other algorithms compared with SSMA have lost their performance. Experiments show that SSMA has superior performance and competitiveness.

Table 4. Average values results on large-scale experimental

Case	PSO	FPA	GWO	WOA	MPA	SMA	SSMA
C5	0.5291	0.6746	0.6548	**0.6997**	**0.7553**	0.6494	**0.7835**
C6	0.5365	0.7638	0.7478	**0.7791**	**0.8283**	0.7464	**0.8332**
C7	0.5585	0.8083	0.7934	**0.8197**	**0.8608**	0.7937	**0.8629**
C8	0.5144	0.7657	0.7555	**0.7759**	**0.8120**	0.7548	**0.8102**
C9	0.5351	0.8123	0.8015	0.8193	**0.8505**	0.8032	**0.8515**
C10	0.4995	**0.7497**	0.7411	0.7534	**0.7821**	0.7404	**0.7816**

Figure 4 shows the coverage convergence curve from C7 and C8. We seen that the convergence curve of SSMA is faster than other algorithms, and the optimization speed is stronger. With the increase of dimensions, the ability of optimization is less obvious and the difficulty of optimization increases. Figure 5 shows the graph of standard deviation, from which it can be seen that SSMA is the best, and Fig. 6 shows the optimization of node distribution of randomly selected samples in each case. It can be observed that with the increase of iteration times, the network coverage gradually rises to the optimal value. However, the increase of the dimension of optimization problem brings great challenges to the algorithm. It can be seen that with the increase of dimensions, the performance of optimization becomes a challenge, and the performance of optimization gradually weakens on high-dimensional tasks.

5.3 Analysis of Statistical Significance

Wilcoxon Sum-Rank Test (WSRT), as a non-parameter test, can effectively evaluate the statistically significant difference between the two optimization algorithms. Table 5 shows the P-values of Wilcoxon test [25] obtained by different SSMA for 13 cases, which are statistically significant when the significance level is 0.05. When we calculate the P-value higher than 0.05, it means that there is no significant difference between the two methods. On the contrary, when the P-value is less than 0.05, we can see that there are great differences between them. We use Wilcocxon test to compare the performance difference between the two methods. ' +' indicates that SSMA is superior to comparison methods, '-' indicates that SSMA is second only to competitors, and ' =' indicates that there is no difference in performance between SSMA and comparison methods for each

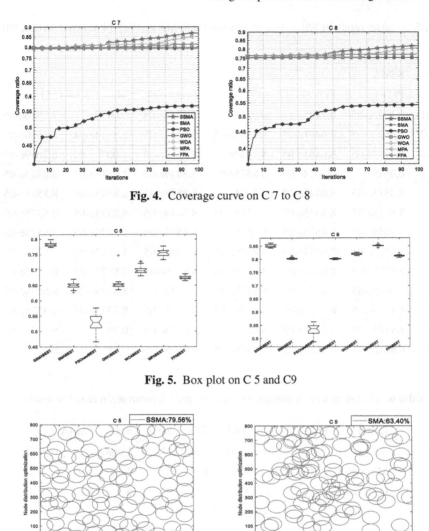

Fig. 4. Coverage curve on C 7 to C 8

Fig. 5. Box plot on C 5 and C9

Fig. 6. Node distribution optimization of SSMA and SMA on C 5

example in 20 runs independently. The results highlight the obvious advantages of SSMA over all other competitors in two different dimensional examples.

It can be seen from Table 5 that the performance of our proposed SSMA is better than that of other algorithms. In addition, in Table 6, we use Friedman test [26] to rank several algorithms. According to the Mean rank obtained by Friedman test, the maximum mean rank of SSMA variables is 6.95, which indicates that the overall performance of the proposed SSMA in wireless sensor network coverage optimization is the best. The rank of other test algorithms is MPA > GWO > FPA > WOA > SMA > PSO.

Table 5. p-values results of Wilcoxon rank sum test for SSMA vs. PSO, FPA, GWO, WOA, MPA, SMA

Case	SSMA					
	PSO	FPA	GWO	WOA	MPA	SMA
Case1	8.1993e-05	0.1185	0.0239	3.4338e-04	0.0437	7.8386e-05
Case2	8.4627e-05	7.6461e-05	8.6955e-05	8.6584e-05	2.8314e-04	8.4506e-05
Case3	8.7949e-05	8.6461e-05	8.7949e-05	8.7699e-05	1.0078e-04	8.7699e-05
C 1	0.0645	8.8074e-05	8.8575e-05	8.8449e-05	1.1112e-04	8.8324e-05
C 2	8.8575e-05	8.8449e-05	4.4934e-04	8.8575e-05	8.8575e-05	8.8575e-05
C 3	8.8575e-05	8.8575e-05	8.8575e-05	8.8449e-05	8.8575e-05	8.8575e-05
C 4	8.8449e-05	8.8575e-05	8.8575e-05	8.8575e-05	8.8575e-05	8.8575e-05
C 5	8.8575e-05	8.8074e-05	8.8575e-05	8.8449e-05	1.1112e-04	8.8324e-05
C 6	8.8575e-05	8.8449e-05	4.4934e-04	8.8575e-05	8.8575e-05	8.8575e-05
C 7	8.8575e-05	8.8575e-05	8.8575e-05	8.8449e-05	8.8575e-05	8.8575e-05
C 8	8.8575e-05	8.8575e-05	8.8575e-05	8.8575e-05	8.8575e-05	8.8575e-05
C 9	8.8575e-05	8.8575e-05	8.8575e-05	8.8575e-05	0.390	8.8575e-05
C 10	8.8575e-05	8.8575e-05	8.8575e-05	8.8575e-05	0.6813	8.8575e-05

Table 6. Overall wilcoxon sum test rank results and friedman mean rank test results

Result	PSO	FPA	GWO	WOA	MPA	SMA	SSMA
+ / = /-	12/1/0	13/0/0	12/1/0	13/0/0	11/2/0	13/0/0	~
Mean rank	1.20	4.08	4.25	3.65	5.85	2.23	6.95
Over rank	7	4	3	5	2	6	1

6   Conclusions

In this paper, a SSMA is proposed and applied to solve the coverage optimization problem of WSNs. For SSMA, we have mainly made two improvements. To improved update position mathematical formula, mainly to improve the optimization performance of the SSMA hence improve the coverage rate; the second is the improvement of oscillation factor. The second part of the improvement was mainly improved the optimization speed in the early stage of search and the convergence speed. For experimental part, we mainly conducted 13 cases of experiments: low-dimensional experiments and high-dimensional experiments, and compared with six state-of-the-art algorithms. In addition, Wilcoxon rank sum test and Friedman text are used to determine the significant difference between the results of SSMA and other competitors. Experimental results show that these improvements to SSMA can improve search efficiency and speed up convergence. The proposed SSMA was superior to most comparisons methods. For the future work, first, the proposed SSMA can be applied to other real-world problems, such as route path planning, transportation safety management, job shop scheduling, and graph coloring problem. Second, other versions of SSMA can be extended, such as multi-objective version, complex version, binary version, quantum coding version, etc. Finally, combining SSMA with other algorithms may be a promising aspect. Fourth, the updating formula and parameters of SSMA can be improved, and the correctness can be verified by experiments on benchmark problem.

Acknowledgment. This work is supported by National Science Foundation of China under Grant 62066005, and by the Project of Guangxi Natural Science Foundation under Grants No. 2018GXNSFAA138146.

References

1. Singh, A., Sharma, S., Singh, J.: Nature-inspired algorithms for wireless sensor networks: a comprehensive survey. Comput. Sci. Rev. **39**, 100342 (2021)
2. Wang, S., Yang, X., Wang, X., Qian, Z.: A virtual force algorithm-lévy-embedded grey wolf optimization algorithm for wireless sensor network coverage optimization. Sensors **19**(12), 2735 (2019)
3. Mendis, C., Guru, S.M., Halgamuge, S., Fernando, S.: Optimized sink node path using particle swarm optimization. In: 20th International Conference on Advanced Information Networking and Applications, 2006, AINA 2006,. IEEE Computer Society (2006)
4. Song, R., Xu, Z., Liu, Y.: Wireless sensor network coverage optimization based on fruit fly algorithm. Int. J. Online Eng. (Ijoe) **14**(6), 58–70 (2018)
5. Aziz, N.A., Alias, M.Y., Mohemmed, A.W.A.: wireless sensor network coverage optimization algorithm based on particle swarm optimization and Voronoi diagram. In: International Conference on Networking. IEEE (2009)
6. Kuila, P., Jana, P.K.: A novel differential evolution based clustering algorithm for wireless sensor networks. Appl. Soft Comput. J. **25**, 414–425 (2014)
7. Liao, W.H., Kao, Y., Wu, R.T.: Ant colony optimization based sensor deployment protocol for wireless sensor networks. Expert Syst. Appl. **38**(6), 6599–6605 (2011)
8. Ambareesh, S., Madheswari, A.N.: HRDSS-WMSN: a multi-objective function for optimal routing protocol in wireless multimedia sensor networks using hybrid red deer salp swarm algorithm. Wireless Pers. Commun. **119**(1), 117–146 (2021). https://doi.org/10.1007/s11277-021-08201-z
9. Rajeswari, M., Thirugnanasambandam, K., Raghav, R.S., Prabu, U., Saravanan, D., Anguraj, D.K.: Flower pollination algorithm with powell's method for the minimum energy broadcast problem in wireless sensor network. Wireless Pers. Commun. **119**, 1111–1135 (2021)
10. Pakdel, H., Fotohi, R.: A firefly algorithm for power management in wireless sensor networks (WSNs). J. Supercomputing 1–22 (2021). https://doi.org/10.1007/s11227-021-03639-1
11. Li, S., Chen, H., Wang, M., Heidari, A.A., Mirjalili, S.: Slime mould algorithm: a new method for stochastic optimization. Future Gener. Comput. Syst. **111**, 300–323 (2020) aliasgharheidari.com
12. Abdel-Basset, M., Chang, V., Mohamed, R.: Hsma_woa: a hybrid novel slime mould algorithm with whale optimization algorithm for tackling the image segmentation problem of chest x-ray images. Appl. Soft Comput. **95**, 106642 (2020)

13. Tiachacht, S., Khatir, S., Thanh, C.L., Rao, R.V., Mirjalili, S., Wahab, M.A.: Inverse problem for dynamic structural health monitoring based on slime mould algorithm. Eng. Comput. 1–24. (2021) https://doi.org/10.1007/s00366-021-01378-8
14. Zubaidi, S. L., et al.: Hybridised artificial neural network model with slime mould algorithm: a novel methodology for prediction of urban stochastic water demand. Water 12(10), 2692 (2020)
15. Mostafa, M., Rezk, H., Aly, M., Ahmed, E.M.: A new strategy based on slime mould algorithm to extract the optimal model parameters of solar PV panel. Sustain. Energ. Technol. Assess. 42, 100849 (2020)
16. Abdel-Basset, M., Mohamed, R., Chakrabortty, R.K., Ryan, M.J., Mirjalili, S.: An efficient binary slime mould algorithm integrated with a novel attacking-feeding strategy for feature selection. Comput. Indus. Eng. 153, 107078 (2021)
17. Houssein, E.H., Mahdy, M.A., Blondin, M.J., Shebl, D., Mohamed, W.M.: Hybrid slime mould algorithm with adaptive guided differential evolution algorithm for combinatorial and global optimization problems. Expert Syst. Appl. 174, 114689 (2021)
18. Djekidel, R., et al.: Mitigating the effects of magnetic coupling between HV transmission line and metallic pipeline using slime mould algorithm. J. Magn. Magn. Mater. 529, 167865 (2021)
19. Kennedy, J., Eberhart, R.: Particle swarm optimization. In: Proceedings of IEEE 1995 International Conference on Neural Networks, vol. 4, pp. 1942–1948 (2002)
20. Mirjalili, S., Mirjalili, S.M., Lewis, A.: Grey wolf optimizer. Adv. Eng. Softw. 69(3), 46–61 (2014)
21. Mirjalili, S., Lewis, A.: The whale optimization algorithm. Adv. Eng. Softw. 95(95), 51–67 (2016)
22. Faramarzi, A., Heidarinejad, M., Mirjalili, S., Gandomi, A.H.: Marine Predators Algorithm: A nature-inspired metaheuristic. Expert Syst. Appl. 152, 113377 (2020)
23. Yang, X.-S.: Flower pollination algorithm for global optimization. In: Durand-Lose, J., Jonoska, N. (eds.) UCNC 2012. LNCS, vol. 7445, pp. 240–249. Springer, Heidelberg (2012). https://doi.org/10.1007/978-3-642-32894-7_27
24. Miao, Z., Yuan, X., Zhou, F., Qiu, X., Song, Y., Chen, K.: Grey wolf optimizer with an enhanced hierarchy and its application to the wireless sensor network coverage optimization problem. Appl. Soft Comput. 96, 106602 (2020)
25. Herrmann, D.: Wahrscheinlichkeitsrechnung und Statistik — 30 BASIC-Programme. Vieweg+Teubner Verlag, Berlin (1984) https://doi.org/10.1007/978-3-322-96320-8_25
26. Ashcroft, S., Pereira, C.: The friedman test: comparing several matched samples using a non-parametric method. In: Ashcroft, S., Pereira, C. (eds.) Practical Statistics for the Biological Sciences: Simple Pathways to Statistical Analyses, pp. 105–108. Macmillan Education, London (2003). https://doi.org/10.1007/978-1-137-04085-5_12

Super-Large Medical Image Storage and Display Technology Based on Concentrated Points of Interest

Jun Yan, Yuli Wang, Haiou Li, Weizhong Lu, and Hongjie Wu[✉]

School of Electronic and Information Engineering, Suzhou University of Science and Technology, Suzhou 215009, China

Abstract. In the field of medical image processing today, there are more and more medical image categories, such as cell images, tissue images, etc. The wide variety of images is of great help in medical diagnosis, not only for visual observation but also for precise analysis of various causes of disease. Due to the development of medicine, the requirements for images are also higher and the amount of data is becoming larger, and the images have reached tens of thousands of pixels, for the current computer, the current environment can no longer meet the needs of image loading display. In response to the above problems, this paper proposes a method for storing and displaying oversized medical images based on centralized points of interest, which achieves fast loading and displaying of oversized cell images, and has been practically applied in relevant medical institutions, achieving certain results in compressed storage and real-time display of cell images, showing the effectiveness and advancement of the method.

Keywords: Medical image · Cell image · Image pyramid · Loading and display

1 Introduction

With the continuous development of modern image technology, image processing is widely used in various fields of people's life, promoting the progress of today's society. In this paper, a technique is proposed to be applied to the fast loading and display of medical images in the medical field. With the development of information technology and medical level, medical images are more demanding, using traditional processing methods, the development of computer hardware can no longer meet the need for medical information processing [1–3]. It is also a big challenge for memory [4]. The loading display of high-resolution images has become an important issue that needs to be addressed in the medical industry [5, 6]. In response to the above problems, this paper proposes a fast storage and display technology for super-large medical images. The implemented technologies mainly include super-large medical image block layered image pyramid technology and fast image loading and display technology.

D.-S. Huang et al. (Eds.): ICIC 2021, LNCS 12836, pp. 201–208, 2021.
https://doi.org/10.1007/978-3-030-84522-3_16

2 Principles and Methods

2.1 Problem Description

For very large image data processing, the first problem is how to compress and store very large images, and the second problem is the multi-resolution display of images. In view of these two main problems, we need to consider that the image data will always be larger than the computer's memory, which means that it is impossible for the image data to be stored in the memory for processing. In this case, you need to use compression storage technology [7, 8]. How to organize the data in the hard disk with the saved data file has become a very critical issue. For multi-resolution fast display images, we use a technique similar to image pyramid [9, 10]. The image pyramid finally needs to record the image information of each layer and each block.

2.2 Image Preprocessing

This paper uses a total of more than 20 sets of image slice data sets. Each set of image sets can be merged into a complete image. Each set of data sets has at least 300 slice images, and can contain up to 800 slice images. Each slice image the resolution is 1920 × 1080 or 2448 × 2048, so the original image is currently 37562 × 37000, 39878 × 38955 and 39221 × 38740, and higher resolutions may appear later. These image data sets are directly provided by relevant medical institutions.

As far as the first problem is concerned, the image compression storage problem, associated with the image scanning method, directly scanned images, multi-resolution can not meet the demand. Therefore, we are constantly exploring and find a way to digitally slice. There are hundreds, thousands or even tens of thousands of microscopic images of digital slice images [11, 12]. Each sliced image has a high resolution, and after multiple magnification, it can fully meet our expectations. Direct splicing way as Fig. 1a, the effect is not very good, splicing results as Fig. 1b.

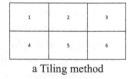

a Tiling method

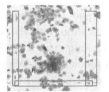

b Practical results

Fig. 1. Splicing method and actual results

The microscope scans eventually formed in two ways, as shown in Fig. 2, and the edges of the slices were often offset to varying degrees due to the influence of tilt angle and whether the scan was bidirectional, while the overlap relationship among them was mainly influenced by the bidirectional scan [13].

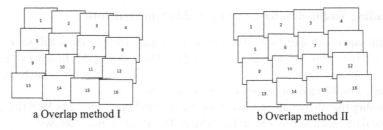

a Overlap method I   b Overlap method II

Fig. 2. Overlap method

2.3 Image Stitching and Cutting Based on Spatial Coordinates

After our research, in the field of image stitching, predecessors have done a lot of research, and some have proposed a feature-based method [14]. This method extracts feature points (edge points, corner points, inflection points, etc.), estimates transformation parameters for feature points, and finally stitches them, or extracts feature curves and surfaces, or based on pixels and voxels., The method that all pixels share a feature subset. In this method, due to translation transformation and rotation transformation, as well as the unevenness of the platform and the error caused by the focus adjustment, the distance between the objective lens and the scanning plane will change, thereby causing the image Small changes in size [15]. There are proposed area-based approaches [16], Simply put, it is to compare the template of a certain area of image 1 with the search area of comparison image 2 to obtain the similarity and perform stitching. This method will cause the search space to become larger due to the uncertainty of the initial position, thereby reducing the speed [17]. A method based on block matching, also known as template matching, is proposed. The main operation is to use the template to translate on the graph and calculate the correlation value. The largest correlation value is the best match [18]. Based on this, a spatial coordinate based approach is proposed for seamless stitching with the upper left corner of the slice image as the starting coordinate. For the edge jaggedness generated after stitching, the inner joining rectangle method was used.

Take the maximum value of the first column, take the maximum value of the first row, take the minimum value of the last column + slice width, take the minimum value of the last row + height, the maximum and minimum values refer to the horizontal and vertical of each slice Coordinates, the specific formula can be:

$$
\begin{cases}
\max(x_{(j-1)}) \times ColumnCount + 1 & (j = 1, 2, \cdots, RowCount) \\
\max(y_i) & (i = 1, 2, \cdots, ColumnCount) \\
\min(x_{j \times ColumnCount}) + ImageWidth & (j = 1, 2, \cdots, RowCount) \\
\min(y_i) + ImageHeight & (i = totalCount - ColumnCount, \cdots, totalCount)
\end{cases}
$$
$$(1)$$

Among them: ColumnCount is the number of slices in each row, RowCount is the number of slices in each column, and totalCount is the total number of slices.

2.4 Loading Mechanism Based on Pyramid Data Structure

The minimum unit of operating system memory management is 4K or 8K, so the size of the slice is $256 \times 256, 512 \times 512, 1024 \times 1024, 2048 \times 2048$ [19]. In order to make the enlarged image clearer, the system takes 256×256. To implement hierarchical access techniques, a multi-resolution pyramid model is used [20].

Assuming that the width and height of the original layer image is W0*H0, and the size of the data block is BlockSize*BlockSize, The formula is as follows:

$$Layers = log_2^{max(W_0,H_0)/BlockSize} + 1 \tag{2}$$

max(W0, H0) means to take the maximum value of the parameters W0, H0.

Assuming that the resolution of the image of the i-th layer is Wi*Hi, the calculation rule of the image resolution of the i + 1-th layer Wi + 1*Hi + 1 is:

$$W_{i+1} = floor(W_i/2) \tag{3}$$

$$H_{i+1} = floor(H_i/2) \tag{4}$$

In the formula, floor() means to round down the result in parentheses.

In each image layer, for a specified pixel (x, y), it is assumed that the pixel coordinates start from 0, and the data blocks in the horizontal and vertical directions also start from 0. Then the following formula can be derived:

Horizontal and block number:

$$T_x = floor\left(x/BlockSize\right) \tag{5}$$

$$T_y = floor\left(y/BlockSize\right) \tag{6}$$

Number of horizontal and vertical chunks:

$$N_x = ceil\left(W_i/BlockSize\right) \tag{7}$$

$$N_y = ceil\left(H_i/BlockSize\right) \tag{8}$$

Total number of image blocks per layer:

$$N = N_x * N_y \tag{9}$$

ceil() means to round up the result in parentheses.

2.5 Quick Display Mechanism Based on Concentrated Points of Interest

Since the number of sliced images is increasing and the final generated image pyramid structure file is getting larger, an approach based on concentrated points of interest is proposed below. Load display scheduling algorithm based on concentrated points of interest: Open the image data file, read the index items and the memory block index table, and get the number of pyramid levels of the image. Calculate the layer of the displayed picture according to the current display resolution. According to the relative position of the picture and the display area, calculate the index of the taken picture in the layer. Calculate the position of the image block in the file according to the layer number, horizontal and vertical index of the taken picture. Read the corresponding image data block and load it into the screen display area. According to the underlying image block, the edges and surrounding images are calculated in real time, and the overall image is complemented. The formula is as follows (Fig. 3):

$$W^* = W_0 \Big/ 2^i (i = 0, \cdots , layer) \tag{10}$$

$$H^* = H_0 \Big/ 2^i (i = 0, \cdots , layer) \tag{11}$$

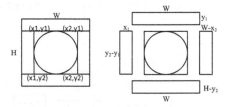

Fig. 3. Edge image decomposition

After the above optimization, the size of the pyramid structure file is greatly reduced, the display is basically the same as the previous speed, and the time consumption during compression and loading is also greatly reduced.

The following Table 1 is a time-space comparison between this method and the normal method. It is mainly used for experimental evaluation. This method uses a method based on interest points, but the general method does not use this method. In Table 1, three sets of image data are used Set, compare the two methods from different slice numbers and resolution of slice maps. It can be intuitively found from Table 1 that the optimized method has a certain improvement in time and space compared with the general method, which confirms the superiority of this method.

3 Experimental Realization

The main process of the system is to process the medical digital slice map acquired by automatic microscope scanning, cut the original map, create index, store the file and compress the resolution (Fig. 4).

According to the above flowchart and Table 1, the processing process not only reduces the time, but also reduces the space size of the generated files. Figure 5 is the final complete image taken from the image set of 3–45% of the cells.

The system consists of two main modules, compression and browsing. The compression module is designed to compress the slices acquired by automatic microscope

Table 1. Time and space comparison between this method and normal method

Image set	Number of slices	Resolution	Method	Generation time	Generated file size
0923–6336-1	340	2448 × 2048	normal method	1206.96 s	274 MB
0923–6336-1	340	2448 × 2048	This method	1003.83 s	212 MB
0926–6336	340	2448 × 2048	normal method	1250.6 s	281 MB
0926–6336	340	2448 × 2048	This method	998.05 s	225 MB
cell3–45%	792	1920 × 1080	normal method	1925.12 s	486 MB
Cell3–45%	792	1920 × 1080	This method	1523.07 s	406 NB

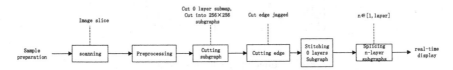

Fig. 4. Image stitching flow chart

Fig. 5. Final image of 3–45% of cells

scanning. Image browsing module is the main module. The main function of the browsing module is to load the content that needs to be displayed in real time according to the user's operation. Among them, the user's basic operations on the medical image mainly include translation, enlargement, reduction, and rotation.

4 Conclusion

This system greatly improves the display speed of super-large medical images by using the above image data processing methods, so that the amount of image data that can be processed by the computer system is not limited by the computer memory capacity, and the processing and display speed is also not affected by the amount of image data. The influence of, can realize the real-time display of super-large image data. This system has been applied in individual medical institutions in Suzhou, providing help for rapid diagnosis of medical institutions.

Acknowledgement. This paper is supported by the National Natural Science Foundation of China (62073231, 61772357, 61902272, 61876217, 61902271), National Research Project (2020YFC2006602) and Anhui Province Key Laboratory Research Project (IBBE2018KX09).

References

1. Wang, Y., Chuan-Fu, L.I.: The new research progress of artificial intelligent methods in medical image processing. Chin. J. Med. Phys. (2013)
2. Ghule, A.G., Deshmukh, P.R.: Image segmentation available techniques, open issues and region growing algorithm. J. Signal Image Process. (2012)
3. Amintoosi, M., Fathy, M., Mozayani, N.: A fast image registration approach based on SIFT key-points applied to super-resolution. J. Photogr. Sci. **60**, 185–201 (2012)
4. Liu, P.: Research on super-large image segmentation technology and its application in medical images. Dalian University of Technology (2010)
5. Zhou, G., Li, Y., Meng, Q.: Application of fuzzy cluster analysis in medical image processing. Chin. J. Health Inf. Manag. **08**, 69–73 (2011)
6. Schadt, E., Linderman, M., Sorenson, J., et al.: Computational solutions to large-scale data management and analysis. Nat. Rev. Genet. **11**, 647–657 (2010)
7. Ping, Z., Jian, Z., Wu, S., et al.: Image processing system and method for in vitro culture and expansion of umbilical cord mesenchymal stem cells (2019)
8. Bansal, N.: Image compression using hybrid transform technique. J. Glob. Res. Comput. Sci. **4** (2013)
9. Jiang, D.: GIS map dynamic roaming algorithm based on image pyramid. Comput. Eng. Des., 199–203 (2013)
10. Ravisankar, P., Sree Sharmila, T., Rajendran, V.: Acoustic image enhancement using Gaussian and laplacian pyramid – a multiresolution based technique. Multimedia Tools Appl. **77**(5), 5547–5561 (2017). https://doi.org/10.1007/s11042-017-4466-7
11. Xu, R.: Application of Distributed Parallel Computing in Accelerated Processing of Digital Human Image Resampling. Third Military Medical University (2007)
12. Mahalanobis, A., Kubala, K.S., Ashok, A., et al.: The algorithm stitching for medical imaging. Comput. Imag., 98700M (2016)
13. Jiang, M., Tao, Z., Jiang, T., et al.: Development of laser confocal scanning microscope based on continuous scanning mode. Opt. Instru., 60–65 (2013)
14. Cheng, Q.: Research and implementation of feature-based image mosaic fusion technology. Dig. Technol. Appl., 116 (2016)
15. Yue, Y., Miao, L., Peng, S.: Automatic stitching method of multi-layer microscopic images. J. Comput. Aided Des. Graph., 000959–000964 (2006)

16. Li, Y., Li, G., Gu, S., Long, K.: Image stitching algorithm based on re-gion segmentation and scale-invariant feature transformation. Opt. Precis. Eng. **24**, 1197–1205 (2016)
17. Liu, L.: Endoscope image stitching. Xidian University (2012)
18. Li, X., Lu, S.: Research on panoramic image stitching algorithm based on block matching. Technol. Econ. Mark., 27–28 (2019)
19. Li, L.: Research on Multi-scale Expression and Fast Display Technology of Embedded GIS Spatial Data. Southeast University (2018)
20. Wang, J., Wang, B., Yang, G., et al.: Medical image fusion technology based on morphological pyramid. Ord. Ind. Autom., 82–84 (2014)

Person Re-identification Based on Hash

Bo Song$^{(\boxtimes)}$, Xinfeng Zhang, Tianyu Zhu, Bowen Ren, and Maoshen Jia

Beijing University of Technology, Beijing, China
{zxf,jiamaoshen}@bjut.edu.cn

Abstract. Person re-identificaton aims to retrieve interested person objects from the person image database in cross camera scene, which has a wide range of application value in the field of video surveillance and security. With the generation of massive monitoring data, the retrieval speed of person re-identificaton is required to be higher. Facing the problem of the slow retrieval speed of person re-identification in large-scale monitoring data, we propose a person re-identification method based on hash. When training hash mapping function, we innovatively adds a batch hash code learning (BHL) module in the network to generate hash code as supervision information, which contributes greatly in retaining similarity information between person image pairs. Although the retrieval speed gets improved due to the concise binary hash features from hash mapping function above, we still need to retain the high accuracy at the same time and a coarse-to-fine (CF) retrieval strategy is proposed. Experiments on two public person re-identification datasets, Market-1501 and DukeMTMC-ReID, show the effectiveness of the proposed method. Compared with the benchmark model, the performance of our method is only 0.5% lower in mAP, but higher by 0.3% in Rank-1 on Market-1501 dataset. And on DukeMTMC-ReID dataset, Rank-1 and mAP are only 0.3% and 0.4% lower. However, the retrieval speed of the two datasets is increased by 12 times and 16 times respectively.

Keywords: Person re-identificaton · Hash · Deep feature · Retrieval speed

1 Introduction

Person re-identification (Re-ID) aims to retrieve interested person from the image database collected by different cameras, which has a wide range of application prospects and values in the field of video surveillance and security. Most of current person re-identification methods only focus on how to improve the retrieval accuracy of models. Both the traditional methods based on manual features [1–3] and the deep learning-based methods [4–7] expect to extract a discriminative feature of person image. These features are continuous real values. In the retrieval stage, we calculate the Euclidean distance or cosine similarity between the image to be queried and all the image features in the image database, and then compare them to get the returned results. Because calculating Euclidean distance or cosine similarity requires multiple multiplication operations, it takes a long time. With the increasing amount of data in the image database, each retrieval will take longer time, which is obviously not conducive to the better implement

© Springer Nature Switzerland AG 2021
D.-S. Huang et al. (Eds.): ICIC 2021, LNCS 12836, pp. 209–222, 2021.
https://doi.org/10.1007/978-3-030-84522-3_17

of person re-recognition technology. Therefore, it is necessary to study the fast retrieval of person re-identification.

There are two ways to realize the fast retrieval of person re-identification. One is to reduce the dimension of person feature representation. That is, the original 512-dimensional features used to represent person are now 256-dimensional, 128-dimensional, 64-dimensional, etc. However, when features are not redundant, reducing their dimensions will greatly reduce the accuracy of retrieval. In addition, multiple multiplication operations in Euclidean distance or cosine similarity cannot be avoided. Therefore, this method is not feasible. The other method is to map the person features into the form of binary hash features and calculate the Hamming distance when comparing the person hash features. Compared with the traditional cosine similarity calculation method, the Hamming distance has lower time complexity because it does not need multiple multiplication operations. Table 1 shows a comparison of the time taken to calculate the cosine similarity and the Hamming distance with different feature length on the Intel Xeon E5–2620 CPU. It can be seen that the Hamming distance calculation is one order of magnitude faster than the cosine similarity calculation. Therefore, this paper studies how to map the deep feature into a binary hash feature without reducing the much accuracy so as to achieve the purpose of improving the retrieval speed.

Table 1. Time consuming comparison of cosine similarity and Hamming distance under different feature lengths

Feature length	Computation time (s)	
	Cosine similarity	Hamming distance
128	2.0×10^{-5}	1.6×10^{-6}
256	4.3×10^{-5}	2.7×10^{-6}
512	8.9×10^{-5}	4.8×10^{-6}
1024	1.6×10^{-4}	1.0×10^{-5}
2048	3.5×10^{-4}	2.2×10^{-5}

This paper mainly studies fast retrieval of person re-identification. In general, the main contributions of this paper are as follows: 1) We propose a method to map person deep features into binary hash features, which consists of two stages. In the first stage, ResNet50 [8] is used to train a model that is of extracting person deep features. The second stage is to add two modules, which are batch hash code learning (BHL) and hash map learning (HML), on the basis of the model trained in the first stage. The main function of BHL module is to generate the hash code that retains the pair similarity of a batch image deep feature, and then use the hash code as the supervision information to better help the learning of the hash mapping function. The function of HML is to map the deep feature of the image into a continuous hash code, and then map it into a discrete form by a binary function. Using hash feature as person feature can effectively improve the retrieval speed. 2) We propose a coarse-to-fine (CF) retrieval strategy, which further improves the accuracy of retrieval. 3) We verify the effectiveness of the method proposed in this paper through experiments on two classic datasets.

2 Related Work

Hashing Algorithm. At present, the research on hash algorithm is mainly in the field of large-scale image retrieval, which can be divided into data-independent method (LSH [9]) and data-dependent method (SH [10], ITQ [11]) according to whether it is dependent on data. For the data-independent hash method, the hash function is often generated randomly and independent of any data, and more hash bits are often needed to obtain better results. The data-dependent method is to overcome the limitation of data-independent algorithm, hoping to use data information to generate a relatively short encoding length and better retrieval accuracy of hash code. The data-dependent algorithms can be divided into supervised hashing (KSH [12], SDH [13], CNNH [14], NINH [15], DPSH [16]) and unsupervised hashing (SH [10], ITQ [11], SADH [17]) according to whether the label information of data is used. Supervised hashing mainly extracts the similarity relationship between original data and label information into the obtained hash code, while unsupervised hashing does not use the label information of dataset, but only learns the hash function from image data.

Person Re-identification. Person re-identification aims to solve the problem of person retrieval in cross-camera scenes. The main problem it faces is the small intra-class distance and large inter-class distance caused by the large changes in illumination, occlusion, posture and perspective. The existing person re-identification methods mainly study how to extract a robust and discriminative feature. According to the different feature extraction methods, the current person re-identification methods can be divided into the traditional manual features-based approach [1–3] and the deep learning-based approach [4–7]. However, they are all to extract continuous real value features of person, and the use of these features requires the calculation of cosine similarity or Euclidean distance in the retrieval stage, which will lead to slow retrieval when the image database becomes large. Under the same feature length, the Hamming distance calculation based on hash feature is much faster than the cosine similarity and Euclidean distance calculation. Therefore, some hashing methods (CSBT [18], PDH [19], SADH [17], ABC [20]) have been proposed. The main idea of these methods is to extract the binary hash feature of the image to replace the continuous real value feature. However, these methods generally require longer hash code length to make the retrieval results have better accuracy, and the increase of hash code length will increase the retrieval time, which deviates from the original intention of using hash method. This paper tries to solve these problems to a certain extent.

3 Method

In order to improve the person re-identification retrieval speed, this paper proposes a method which can map the deep feature to hash code. Firstly, a hash mapping function is learned to map the deep feature into a continuous hash code while maintaining semantic similarity and distance similarity. Then, a binary function is used to discretize the hash code. Finally, a concise binary hash code is obtained. In addition, in order to improve the retrieval speed and give consideration to the accuracy at the same time, this paper proposes a coarse-to-fine retrieval strategy.

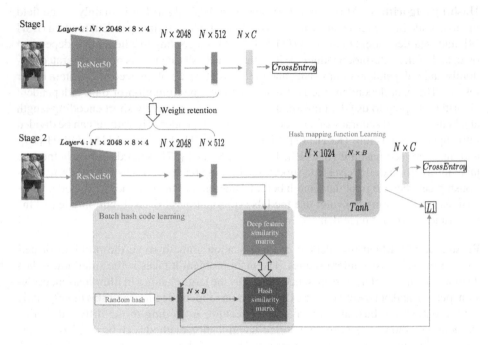

Fig. 1. Overall network structure

3.1   Overall network structure

Figure 1 shows the overall network structure of the method proposed in this paper, which includes two stages. Stage 1 is the deep feature learning stage. The main purpose of which is to train a convolutional neural network model to extract deep features from person images. Stage 2 is hash mapping function learning stage. The main goal of this stage is to train a function that can map deep features into continuous hash codes.

3.2   Deep Feature Learning

In the stage of deep feature learning, we directly used ResNet-50 [8] as the basic network. The network structure and network parameters were consistent with those before the last global average pooling layer (GAP) of ResNet-50. At the same time, two layers of full connection layer were added after the last global average pooling layer (GAP), which are FC1 layer and FC2 layer shown in Table 2. FC1 layer mainly plays the role of feature dimension reduction, that is, the 2048-dimension feature is reduced to 512-dimension after global average pooling. This layer can reduce the influence brought by feature redundancy to a certain extent. FC2 layer is the classification layer. The number of neuron nodes c in this layer is consistent with the number of person identities in the dataset. The detailed network structure and parameters of Stage 1 are shown in Table 2. In this part, the loss function is cross entropy loss based on identity.

Table 2. Stage 1 network structure and configuration

Layer Name	Feature Map Size	Configuration
Conv1	128×64	filter 7×7, 64, stride 2
		max pooling 3×3, stride 2
Conv2_x	64×32	$\begin{bmatrix} 1 \times 1, 64 \\ 3 \times 3, 64 \\ 1 \times 1, 256 \end{bmatrix} \times 3$
Conv3_x	32×16	$\begin{bmatrix} 1 \times 1, 128 \\ 3 \times 3, 128 \\ 1 \times 1, 512 \end{bmatrix} \times 4$
Conv4_x	16×8	$\begin{bmatrix} 1 \times 1, 256 \\ 3 \times 3, 256 \\ 1 \times 1, 1024 \end{bmatrix} \times 6$
Conv5_x	8×4	$\begin{bmatrix} 1 \times 1, 512 \\ 3 \times 3, 512 \\ 1 \times 1, 2048 \end{bmatrix} \times 3$
GAP	1×1	global average pooling 8x4
FC1	-	512, dropout, tanh
FC2	-	c-dim, softmax

3.3 Hash Mapping Function Learning

After the training in Stage 1, we can get a model that can extract the deep features of person. In Stage 2, we discard the classification layer FC2 and take the 512-dimension output of FC1 layer as the extracted deep features of person. We use a two-layer full connection layer to learn the hash mapping function in order to increase the non-linear expression ability of the hash mapping function. Its parameter configuration is shown in Table 3.

ReLU is used in the activation function of Hash_FC1 layer to alleviate the gradient disappearance to a certain extent. Since we ultimately hope to obtain a hash code in the form of $\{-1, 1\}$ binarization, Tanh is used in the activation function configuration of Hash_FC2 to facilitate binarization. The final output of the hash_FC2 layer can be expressed in the form of Formula (1), where $v = [v_1, v_2, \cdots, v_{512}]$ represents the deep features extracted from a person image. W_1 and W_2 represent the full connection weights of the Hash_FC1 and Hash_FC2 layer. b_1 and b_2 represent the bias parameters of the two layers respectively. The deep feature of an image can be mapped into $[-1, 1]$ through formula (1). Since it is a continuous real value, we use the binary function formula (2) to obtain a binary hash code. Because the binary function is not derivative, we do not binarize during training.

$$h(v) = tanh\left(W_2^T(W_2^T v + b_1) + b_2\right) \tag{1}$$

$$b_k(v) = sgn(h_k(v)), \quad k = 1, 2, \cdots, q \tag{2}$$

In order to make the trained hash mapping function retain part of the semantic information, we add a classification layer after the output of Hash_FC2.The number of neurons in the classification layer was the same as the number of person categories in the dataset. The cross entropy loss based on identity was still used for the classification layer.

In order to make the trained hash mapping function retain the similarity information between the deep feature pairs of person images, we present a batch hash code learning process. The main goal of this process is to generate a corresponding binary hash code for each deep feature in a batch, so that this hash code can retain similarity information between the deep feature pairs. That is if the deep feature pairs are similar in Euclidean space, their corresponding hash codes should also be similar in Hamming space. The learning process of this batch hash code is as follows. 1) Calculate the cosine similarity matrix S between the normalized n person deep features $\{v_1, v_2, \cdots, v_n\}$ in a batch firstly. 2) Randomly generate n binary hash codes of length B, with the value of -1 or 1, denoted as H. And calculate the Hamming similarity matrix between them. 3) Draw on the experience of reference [14], H is optimized by Formula (3) to minimize its value through the method of coordinate descent.

$$\min_{H} ||S - \frac{1}{q}HH^T||_F^2 \tag{3}$$

The binary hash code obtained at the end of batch hash code learning can retain the similarity information between sample pairs. Using this hash code as the supervision information to conduct a L1 Loss with the output of Hash_FC2, the hash mapping function can retain part of the similarity information when mapping the deep features.

Table 3. Hash mapping function configuration

Layer name	Configuration
Hash_FC1	1024, ReLU
Hash_FC2	q-dim, Tanh

3.4 Search Strategies from Coarse to Fine

The deep features of person images contain more high-level semantic information, and it is more accurate for person retrieval than binary hash feature. In order to further improve the retrieval accuracy, a coarse-to-fine retrieval strategy is proposed in this paper. As shown in Fig. 2, the returned person image with green border in the figure are the correctly retrieved image, while the ones with red border are the incorrectly retrieved image.

Coarse Retrieval. Given a person image I_q to be retrieved and an image library $G = \{g_1, g_2, \cdots, g_m\}$ containing m person images, their deep features are extracted firstly and denoted as V_q and $V = \{v_1, v_2, \cdots, v_m\}$ respectively. Then, these deep features are mapped into continuous hash features by using the hash mapping function learned in stage 2 and binarized by Eq. (2), denoted as H_q and $H = \{h_1, h_2, \cdots, h_m\}$. $G_h = \{g_1^h, g_2^h, \cdots, g_m^h\}$ can be obtained by calculating the Hamming distance between H_q and H and sorting from small to large. Under coarse retrieval, G_h is the result returned by the system.

Fine Retrieval. In the experiment, it is found that under coarse retrieval, the images of person different from the one to be retrieved will sometimes come to the front in the returned results. That is, the Hamming distance between those images and the person image to be retrieved is also relatively small. In order to improve the retrieval accuracy, we propose the fine retrieval on the basis of the coarse retrieval to optimize the results of the coarse retrieval. Specifically, a threshold H_t is set at first. In the result of coarse search G_h, there are l retrieval results $G_{ht} = \{g_1^h, g_2^h, \cdots, g_l^h\}$ which are smaller than the threshold H_t. We recalculate the cosine distance by using the deep feature and sorted from largest to smallest to get the result $G_{ct} = \{g_1^c, g_2^c, \cdots, g_l^c\}$. The final result returned by the system is $G_c = \{g_1^c, g_2^c, \cdots, g_l^c, g_{l+1}^h, \cdots, g_m^h\}$.

Fig. 2. Retrieval strategy from coarse to fine

4 Experiment

4.1 Datasets and Evaluation Indicators

This paper evaluates the proposed method on two common public datasets, Market-1501 and DukeMTMC-ReID. These two datasets and the evaluation criteria used in this paper will be introduced in detail.

Market-1501 [21] dataset was collected in front of a supermarket in Tsinghua University, using six cameras, including five high-resolution cameras and one low-resolution camera. There are 1501 person identities in this dataset, among which 12936 images in the training set contain 751 person identities, 19732 images in the test part contain 750 person identities and some interference items, and there are 3368 person images to be retrieved.

DukeMTMC-ReID [22] was collected from Duke University. It is a subset of DukeMTMC, a person tracking dataset, and uses eight outdoor cameras. The dataset contains 1812 person identities of which 1402 appear in more than two cameras. Among 1402 person identities, there are 702 person identities and 16522 images in total for training. Another 702 person identities and 408 interfering person images are selected as candidate sets and 2228 person images are to be retrieved.

In this paper, we use cumulative match characteristic (CMC) curve and mean average precision (mAP) to evaluate the accuracy of our method. CMC mainly calculates Rank-1.

4.2 Experimental Environment and Parameter Setting

The proposed method is implemented in Pytorch deep learning framework and is implemented on Ubuntu16.04 operating system with Intel Xeon E5–2620 having RAM of 16 GB memory and GTX TitanX GPU with 12 GB of GPU memory.

According to reference [5], the size of each input image is adjusted to 256×128 in the two stages of training and augmented by horizontal flip and random erasing [23]. The batch size is set as 32, the learning rate of FC1 and FC2 in the full connection layer is set as 0.05 in the stage 1 training. The learning rate of Conv1 ~ Conv5_x is set as 0.005. In stage 2, the deep feature extraction part uses the result of stage 1 as the pre-training weight. The learning rate of hash map function learning network is set to 0.05, and we use stochastic gradient descent SGD algorithm to optimize the network weight parameters. We train a total of 60 iterations on the dataset, and the learning rate is reduced 10 times after 30 iterations. In the retrieval strategy from coarse to fine, when the feature length mentioned in this paper is 512 and the threshold is set to 200.

4.3 Experimental Results and Analysis

4.3.1 Experimental Comparison with Related Methods

Tables 4 and 5 show the comparison of this paper with relevant methods on Market-1501 and DukeMTMC-reID datasets. R in the table represents the deep feature of real values, and B represents the hash feature of binary values. The non-hashing correlation methods proposed in this paper in Tables 4 and 5 can be divided into two groups. One is the non-hashing method for extracting person deep features, and the other is the hashing correlation method. From Tables 4 and 5 we can analyze the following three points: 1) Although this method does not work hard on the person deep feature extraction network, a simple ResNet-50, to extract deep features and learn a hash mapping function for mapping the extracted deep features into hash codes, it has still greatly surpassed the existing hash methods in accuracy, which even can be comparable with the methods based on deep features, such as IDE [4] and GLAD [5]. 2) Compared with the existing non-hashing method using deep feature directly, the method in this paper has lower retrieval time. 3) The proposed coarse-to-fine strategy, which first uses hash feature to sort the person images whose Hamming distance is less than the threshold of 200, and then uses the deep feature to calculate cosine similarity to reorder the sorted results. The Rank-1 and mAP of the results on the Market-1501 dataset are improved by 3.0% and 5.1% respectively. At the same time, the retrieval time is about 0.03 s longer than that of only calculating Hamming distance, which fully reflects the good performance of this retrieval strategy.

Table 4. Comparison with related methods on market-1501 dataset (%, s)

Method	Feature type and length	Market-1501		
		Rank-1	mAP	Retrieval time consuming (s)
IDE [4]	R 2048	88.1	72.8	6.4×10^0
GLAD [5]	R 4096	89.9	73.9	1.3×10^1
PCB [6]	R 12288	93.8	81.6	3.2×10^1
CSBT [18]	B 512	42.9	20.3	1.1×10^{-1}
PDH [19]	B 512	44.6	24.3	1.1×10^{-1}
SADH [17]	B 512	42.3	22.8	1.1×10^{-1}
ABC [20]	B 512	69.4	48.5	1.1×10^{-1}
Ours	B 512	84.8	65.6	1.1×10^{-1}
Ours w/CF	B + R 512	87.8	70.7	1.4×10^{-1}

Table 5. Comparison with related methods on DukeMTMC-ReID dataset (%, s)

Method	Feature type and length	DukeMTMC-ReID		
		Rank-1	mAP	Retrieval time consuming (s)
IDE [4]	R 2048	69.4	55.4	6.0×10^0
GLAD [5]	R 4096	–	–	–
PCB [6]	R 12288	83.3	69.2	2.9×10^1
CSBT [18]	B 512	47.2	33.1	8.0×10^{-2}
PDH [19]	B 512	–	–	–
SADH [17]	B 512	–	–	–
ABC [20]	B 512	69.9	52.6	8.0×10^{-2}
Ous	B 512	75.9	57.4	8.0×10^{-2}
Ous w/CF	B + R 512	78.9	60.7	1.0×10^{-1}

4.3.2 Hyperparameters Setting

1) *Influence of different learning rates and batch sizes on model performance.*

As a hyperparameter, learning rate will generally bring different model results when set differently in training. Because too large or too small learning rate setting may lead to the network falling into local optimization instead of global optimization. Figure 3 shows the performance results of the model with different learning rates when the length of the hash code is fixed at 512 and the batch size is 32 on the Market-1501 dataset. It can be seen that when the learning rate is set to 0.05, the performance indexes of the model achieve relatively good results.

Batch size is also a hyperparameter which has a great impact on the performance results of the model. In the learning module batch hash code in stage 2 of the method proposed in this paper, the deep feature of a batch is used to calculate its cosine similarity matrix. This matrix plays a guiding role in the optimization of generating random hash codes. The optimized hash code serves as the supervision information to guide the learning of the hash mapping function. So the batch size has an important impact on the performance of the model. Figure 4 shows the results of the model under batch size of 8, 16, 32 and 64 respectively when the length of the hash code is fixed at 512 and the learning rate is 0.05 on Market-1501 dataset. It can be seen that the model achieves better results when the batch size is set at 32.

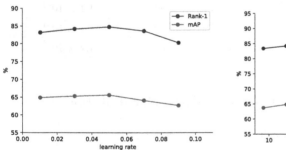
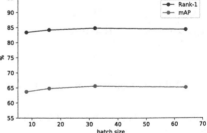

Fig. 3. Different learning rates   **Fig. 4.** Different batch sizes

2) *Influence of hash code length on model performance.*

The length of the hash code is an important factor affecting the performance indexes of the model. Because the short hash code may lose features, while the long hash code will make features redundant, which will affect the accuracy of model retrieval. In addition, the longer hash code will bring the problem of longer retrieval time. Table 6 shows the model under different length of hash code performance indexes. We can see from Table 6: 1) when the set length of hash code from 128 to 256, the Rank-1 of the model and mAP increased by 5.3% and 6.2% respectively. When the set length of hash code from 256 to 512, the model of the Rank-1 and mAP increased by 2.4% and 4.4% respectively. Indicators improved a lot. However, when the hash length is set from 512 to 1024, the Rank-1 and mAP of the model are only increased by 0.3% and 0.8% respectively. When the hash length is set from 1024 to 2048 and from

2048 to 4096, the Rank-1 and mAP of the model are improved little. This indicates that the feature is lost when the dimension is lower than 512, while the feature is redundant when the dimension is higher than 512. 2) With the increase of the hash code setting, the retrieval time is also increasing. Considering the accuracy of the retrieval and the retrieval time, it is more appropriate to set the hash code length to 512 dimensions.

Table 6. Performance comparison of different length hash codes (%, s)

Hash code length	Rank-1	mAP	Retrieval time consuming
128	77.1	55.0	4.7×10^{-2}
256	82.4	61.2	8.6×10^{-2}
512	84.8	65.6	1.1×10^{-1}
1024	85.1	66.4	2.0×10^{-1}
2048	85.7	67.3	4.7×10^{-1}
4096	86.2	67.5	1.0×10^{0}

4.3.3 Network Ablation Experiment

In order to demonstrate the effectiveness of each module of the method proposed in this paper, a network ablation experiment is carried out in this paper. The experimental results are shown in Tables 7 and 8, where R represents the real value deep feature and B represents the binary value hash feature. The baseline model in the table is the model trained in Stage 1 of Fig. 1, on which 512-dimension deep features are directly extracted for retrieval test. The direct hashing(DH) means that the deep features are directly discretized into hash features using Hamming distance for retrieval. HMF is hash map learning module, BHL is batch hash code learning module and CF stands for coarse-to-fine (CF) retrieval strategy. It can be seen from Tables 7 and 8: 1) the use of deep feature has a higher retrieval accuracy than the use of hash feature. 2) The direct binary discretization of deep features has poor performance in terms of accuracy. 3) Using the deep features to relearn a hash mapping function (excluding hash code learning module) improves more than deep features binary discretization directly on Rank-1 and mAP for 5.8% and 6.3% respectively on the Market-1501 dataset. Rank-1 and mAP increased by 4.4% and 2.4% on DukeMTMC-ReID dataset. It shows that the output of the hash mapping function can retain the semantic information in the deep feature well. 4) Batch hash code learning module is designed to retain the similarity information between feature pairs. It can be seen from Tables 7 and 8 that the network with this module is 4.7% and 4.9% higher than the network without this module in Rank-1 and mAP on Market-1501 dataset. It is 4.7% and 3.7% higher in Rank-1 and mAP on the DukeMTMC-ReID dataset. It shows that the output of the hash mapping function can retain the similarity information between deep features well. 5) The proposed coarse-to-fine retrieval strategy takes a little retrieval time, but can further greatly improve the performance index of the model. Compared with the benchmark model based on deep feature to calculate cosine similarity, the mAP on Market-1501 dataset has only lower by 0.5%, and exceeds the Rank-1. On the DukeMTMC-ReID dataset, the difference between Rank-1 and mAP was only lower by 0.3% and 0.4%. However, the retrieval speed is increased by 12 times and 16 times respectively. So the method proposed in this paper has more advantages from the comprehensive consideration of accuracy and retrieval time.

Table 7. Retrieval performance under different network structures on Market-1501 dataset

Baseline	DH	HMF	BHL	CF	Feature Type and Length	Market-1501		
						Rank-1	mAP	Retrieval Time Consuming
√	×	×	×	×	R 512	87.5	71.2	1.7×10^{0}
√	√	×	×	×	B 512	74.3	54.4	1.1×10^{-1}
√	×	√	×	×	B 512	80.1	60.7	1.1×10^{-1}
√	×	√	√	×	B 512	84.8	65.6	1.1×10^{-1}
√	×	√	√	√	B+R 512	87.8	70.7	1.4×10^{-1}

Table 8. Retrieval performance under different network structures on DukeMTMC-ReID dataset

Baseline	DH	HMF	BHL	CF	Feature Type and Length	DukeMTMC-ReID		
						Rank-1	mAP	Retrieval Time Consuming
√	×	×	×	×	R 512	79.2	61.1	1.6×10^{0}
√	√	×	×	×	B 512	66.8	51.3	8.0×10^{-2}
√	×	√	×	×	B 512	71.2	53.7	8.0×10^{-2}
√	×	√	√	×	B 512	75.9	57.4	8.0×10^{-2}
√	×	√	√	√	B+R 512	78.9	60.7	1.0×10^{-1}

5 Conclusion

In order to solve the problem of slow retrieval speed in large-scale image data of person re-identification, we propose a method that can map deep features into hash codes. By learning a hash mapping function, this method can map a deep feature to a continuous hash code under the condition of keeping semantic similarity and distance similarity unchanged, and then discretize it through a binary function to get a concise binary hash code. In addition, in order to improve the retrieval speed and give consideration to the accuracy, we propose a retrieval strategy from coarse to fine. We have carried out experiments in two common data sets, Market-1501 and DukeMTMC-ReID, and analyzed the experimental results, and demonstrated the effectiveness of this method.

Acknowledgements. This work is supported by the project (NTUT-BJUT-110–05) of China.

References

1. Gray, D., Tao, H.: Viewpoint invariant pedestrian recognition with an ensemble of localized features. In: European Conference on Computer Vision. Springer, pp. 262–275 (2008). https://doi.org/10.1007/978-3-540-88682-2_21
2. Pedagadi, S., Orwell, J., Velastin, S., Boghossian, B.: Local fisher discriminant analysis for pedestrian re-identification. In: Proceedings of the IEEE Conference on Computer Vision and Pattern Recognition, pp. 3318–3325 (2013)
3. Liao, S., Hu, Y., Zhu, X., Li, S.Z.: Person re-identification by local maximal occurrence representation and metric learning. In: Proceedings of the IEEE Conference on Computer Vision and Pattern Recognition, pp. 2197–2206 (2015)
4. Zheng, L., Yang, Y., Hauptmann, A.G.: Person re-identification: past, present and future. arXiv preprint arXiv:1610.02984 (2016)
5. Wei, L., et al.: GLAD: global-local-alignment descriptor for pedestrian retrieval. In: Proceedings of the 25th ACM International Conference on Multimedia (2017)
6. Sun, Y., Zheng, L., Yang, Y., Tian, Q., Wang, S.: Beyond part models: person retrieval with refined part pooling (and a strong convolutional baseline). In: Ferrari, V., Hebert, M., Sminchisescu, C., Weiss, Y. (eds.) ECCV 2018. LNCS, vol. 11208, pp. 501–518. Springer, Cham (2018). https://doi.org/10.1007/978-3-030-01225-0_30
7. Wang, G., et al.: Learning discriminative features with multiple granularities for person re-identification. In: Proceedings of the 26th ACM International Conference on Multimedia (2018)
8. He, K., et al.: Deep residual learning for image recognition. In: Proceedings of the IEEE Conference on Computer Vision and Pattern Recognition (2016)
9. Datar, M., et al.: Locality-sensitive hashing scheme based on p-stable distributions. In: Proceedings of the Twentieth Annual Symposium on Computational Geometry (2004)
10. Weiss, Y., Torralba, A., Fergus, R.: Spectral hashing. In: NIPS, vol. 1(2) (2008)
11. Gong, Y., et al.: Iterative quantization: a procrustean approach to learning binary codes for large-scale image retrieval. IEEE Trans. Patt. Anal. Mach. Intell. **35**(12), 2916–2929 (2012)
12. Liu, W., et al.: Supervised hashing with Kernels. In: 2012 IEEE Conference on Computer Vision and Pattern Recognition. IEEE (2012)
13. Shen, F., et al.: Supervised discrete hashing. In: Proceedings of the IEEE Conference on Computer Vision and Pattern Recognition (2015)
14. Xia, R., et al.: Supervised hashing for image retrieval via image representation learning. In: Proceedings of the AAAI Conference on Artificial Intelligence, vol. 28(1) (2014)
15. Lai, H., et al.: Simultaneous feature learning and hash coding with deep neural networks. In: Proceedings of the IEEE Conference on Computer Vision and Pattern Recognition (2015)
16. Li, W.-J, Wang, S., Kang,, W.-C.: Feature learning based deep supervised hashing with pairwise labels. arXiv preprint arXiv:1511.03855 (2015)
17. Shen, F., et al.: Unsupervised deep hashing with similarity-adaptive and discrete optimization. IEEE Trans. Patt. Anal. Mach. Intell. **40**(12), 3034–3044 (2018)
18. Chen, J., et al.: Fast person re-identification via cross-camera semantic binary transformation. In: Proceedings of the IEEE Conference on Computer Vision and Pattern Recognition (2017)
19. Zhu, F., et al.: Part-based deep hashing for large-scale person re-identification. IEEE Trans. Image Process. **26**(10), 4806–4817 (2017)
20. Liu, Z., et al.: Adversarial binary coding for efficient person re-identification. In: 2019 IEEE International Conference on Multimedia and Expo (ICME). IEEE (2019)
21. Zheng, L., et al.: Scalable person re-identification: a benchmark. In: Proceedings of the IEEE International Conference on Computer Vision (2015)

22. Ristani, E., et al.: Performance measures and a data set for multi-target, multi-camera tracking. In: European Conference on Computer Vision. Springer, Cham (2016). https://doi.org/10.1007/978-3-319-48881-3_2
23. Zhong, Z., et al.: Random erasing data augmentation. In: Proceedings of the AAAI Conference on Artificial Intelligence, vol. 34(07) (2020)

A Robust and Automatic Recognition Method of Pointer Instruments in Power System

Jian-Xun Mi[1,2]($\boxtimes$), Xu-Dong Wang[1,2], Qing-Yun Yang[1,2], and Xin Deng[2]

[1] Chongqing Key Laboratory of Image Cognition, Chongqing University of Posts and Telecommunications, Chongqing 400065, China
[2] College of Computer Science and Technology, Chongqing University of Posts and Telecommunications, Chongqing 400065, China

Abstract. Currently, the research on substation automation inspection relies on inspection robots and computer vision technologies. However, due to limitations of existing methods, some existing methods cannot meet the requirements of identifying instruments of wide varieties. Additionally, the applications of these methods are hindered by interferences, such as illumination and shadow. This paper presents an automatic reading method for pointer instruments. First, it proposes an instrument location method based on correlation filtering, which exhibits a good tolerance of severe lighting changes. Second, to identify the pointer angle, which is the major task in inspection, a human vision-inspired method is proposed, which directly reads the orientation of the pointer by Gabor-based filtering. This one-stage pointer angle identification algorithm is superior compared with a popular method based on Hough-transform in terms of robustness and accuracy. Because the latter includes several steps which lead to be fragile in practical applications. The experimental results show that the proposed method exhibits strong robustness against illumination variations and shadows, which is verified in different kinds of instruments.

Keywords: Machine vision · Automatic recognition method · Correlation filtering · Gabor filter · Pointer instrument

1 Introduction

The pointer-type instruments have several advantages: simple structure, strong anti-electromagnetic interference capability, and low price, so it has wide applications in the power system. At present, the substation mainly relies on human resources to collect the readings of the instruments. However, the inspection task has a high labor intensity and security risks, and its reliability is greatly affected by human factors. Therefore, it is necessary to develop a system that automatically reads the instrument indication. Many researchers use robotics and computer vision technology to achieve automatic power inspection.

At present, the substation automation inspection program collects real-time data of instrument images by intelligent inspection robots [1]. The automatic recognition

© Springer Nature Switzerland AG 2021
D.-S. Huang et al. (Eds.): ICIC 2021, LNCS 12836, pp. 223–237, 2021.
https://doi.org/10.1007/978-3-030-84522-3_18

method of the instrument includes two main steps: (1) positioning an instrument; (2) recognizing the instrument reading.

The positioning step detects an instrument's coordinates in the image collected with a robot and then extracts the instrument's dial area. The current mainstream methods include Hough transform circle detection, template matching, and feature point matching. Lin Zhang et al. [4] proposed a method based on visual saliency, which uses the local pixel inhomogeneity factor model to improve the instrument area's visual saliency and locate the instrument by Hough transform circle detection.

The recognition step identifies the reading of the pointer in the dial area. This step can be implemented by two methods: distance-based one and angle-based one. The distance-based process uses [1] the distance relationship between the pointer and the scale to calculate the instrument's reading. Jianna Chi et al. [3] proposed a method that uses the central projection method to determine the position of the scale. The angle-based method recognizes the pointer angle to calculate the reading of the instrument. The core technology for the angle-based method involves the Hough transform line detection [5]. The key factor which affects the performance of the method is the extraction of lines corresponding to the pointer (skeleton or edge). But for instruments with complex surfaces, the thinning algorithm could produce unexpected straight lines, which makes it hard to distinguish the true pointer skeleton.

Before extracting the pointer skeleton, the pointer was extracted by the silhouette method (gray subtraction) to remove objects other than the pointer [5]. These methods can achieve good results for images acquired in the laboratory under the ideal condition. Since image noise affects the extraction of the pointer skeleton, a preprocessing step, such as median filtering [6], is required. There are also some problems in extracting the pointer skeleton, such as the skeleton is discontinuous. Han Jiale [7] use corrosion expansion to solve it after extracted the pointer skeleton. Because the substation has a complex spatial structure, the camera could not face the instrument surface vertically when captures images, which may cause errors in instrument readings. To solve this problem, some scholars introduced binocular vision to reduce the interference of the shooting angle on the recognition of the reading [8]. Besides, the environmental factors of the substation have a great influence on the recognition of the instrument reading, especially the illumination changes. Chao Zheng et al. [9] believed that the existing methods are most sensitive to light, so the retinal cortical theory is used to reduce the influence of illumination changes, and the perspective transformation is used to correct the camera angle. Because the Hough transform cannot cope with various complicated situations, these researches are all about how to accurately and robustly extract pointers and enable them to be correctly detected by Hough transforms.

Our contribution to this paper is that we proposed a new combination of the instrument positioning method and the instrument reading recognition method. It shows robust performance, especially when the instrument is in the outdoor environment. Due to the work situation, our instruments are posed outside and need a robot to take a picture for reading the instrument. To optimize our method under these interferences under the outdoor environment, a novel idea is proposed to recognize instrument reading, which is inspired by human vision. Without modeling or identifying the pointer in a dial instrument, we directly acquire the pointer's orientation by introducing direction-sensitive filtering (DSF). To manipulate the features of a straight pointer, i.e., length and width, DSF with a designed Gabor function shows good performance.

2 Related Works

According to our investigations, the current research still has some limitations for the instrument positioning in the outside environment. Due to the diversity of instrument types, the instrument positioning methods based on the Hough transform circle detection are only applicable in circular instruments. [13] Moreover, the Hough transform circle detection may easily detect multiple circles of similar size, making it impossible to determine the instrument position accurately. [2] Also, the feature-point-based methods, such as ORB, SIFT, and SURF, are with insufficient feature points and feature point matching errors due to the instability of the detail information in the dial.

For the recognition step, the Hough transform line detection utilizes only the pointer's linear characteristics regardless of the pointer width information for the recognition of instrument reading. Therefore, the result of Hough transform line detection depends on the production of extracting the pointer line. However, those silhouette-based methods detecting the pointer's outline are of poor performance in real environment applications. Several interferences on the surface of instruments cause more noisy residuals, which also leads to a more significant impact on the Hough transform's accuracy. Besides, the pointer shape also influences results by the Hough transform-based pointer detection. Moreover, the projection-based methods [12] also face the same challenges. [3] We propose a new automatic instrument reading method in view of the existing problems, which include the instrument's positioning and the pointer angle recognition.

In addition to the Hough transform-based method, other efforts are made to read the pointer instrument. Guan Yu Dong et al. [10] suggested fitting a curve using the correspondence between the pointers of different angles and the intersection of the same line and the display. P.A. Belan et al. [11] used the radial projection method and the Bresenham algorithm to determine the pointer position. Zhi-Juan Yang et al. [12] proposed a system, which used the circle-based Regional cumulative Histogram method to calculate the direction of the pointer.

3 Automatic Instrument Reading Method

Automatic instrument reading methods mainly involve four steps: instrument [1] positioning, image preprocessing, pointer angle recognition, and instrument reading calculation. The major efforts are put into the object instrument positioning and the recognition of the angle of a pointer.

For this positioning problem, one needs to find where the target is. Since positioning is the first step in the recognition, its accuracy significantly influences the method selection of succeeding steps and finally affects recognition of the pointer angle. For the Pointer angle recognition problem, the current methods always focus on lines which is not stable for our situation of the outdoor environment, and there are a lot of stains or some inferences to obstruct our reading process. So, we proposed a new way to focus on changing the pointer's shape and direction. It will reduce the risk of being affected by obstacles and perform more robust.

3.1 Instrument Positioning

For the automatic instrument reading method, the first step is to locate the instrument's position in the image captured by the robot. At present, the substation inspection robot collects instrument image data by tracking the pre-planned inspection path. A rehearsal inspection of all the target instruments is conducted. At each inspection spot, the camera parameter settings are recorded for acquiring the best image quality of target instruments and other information according to the requirements of the following steps. The images acquired in this rehearsal procedure are called preset images. Ideally, the position of an instrument in each captured image is fixed. However, the systematic errors of the robot, such as positioning error, the mechanical error of the gimbal, could cause deviation between the acquired image and the preset image. This deviation contains observable translational transformations, negligible scaling transformations, and negligible rotation transformations. To eliminate such transformations by image processing technique usually is not a tough task. However, substations are generally in an outdoor environment and have a complex spatial structure, so locating a target instrument can seriously suffer with lighting factors, rain, and shadow factors.

In our method, positioning an instrument is solved by positioning the area of an image containing this instrument. Since the spatial relationship among the objects in the image is fixed, the target object's position can be obtained by locating the position of the partial region. Our idea is quite different from most existing methods that use only the instrument's information to handle the positioning problem [2–4]. Intuitively, in the instrument positioning problem, the image information around the instrument can be used to obtain more auxiliary features because there is no strict distinction between foreground and background in this problem. Therefore, this paper proposes an instrument positioning method based on correlation filtering, which can effectively improve the accuracy and robustness of instrument positioning. This method utilizes template features, such as HOG features, to reduce the effects of illumination changing and image detail instability. At the same time, the correlation filtering further reduces the influence of local distortion of the image data matrix on the positioning process. The specific framework is shown in Fig. 1.

Feature Extraction. In the substation environment, the influence of illumination factors is more significant, mainly due to the change of illumination intensity, uneven illumination on the instruments, and local reflection. According to our investigation, we find that the form feature and spatial relationship of the image are stable in this problem, so the features belonging to the template class, such as HOG (Histogram of Oriented Gradient), are suitable to describe the image. We pick HOG as the feature descriptor, which is believed to be robust to illumination variations. Besides, we do not need to consider rotation transformation and scale transformation, so the computational cost of extracting the HOG feature is much smaller than the previously proposed method [14].

Correlation Filter Model and Training. In the field [2] of signal processing, correlation is used to describe the correlation of two signals. Later, it was introduced into the field of image processing. The more similar the two images are, the larger the correlation is. The correlation calculation on an image can be generally represented by $f = x \otimes h$,

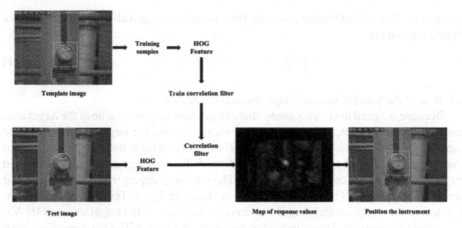

Fig. 1. Framework of proposed algorithm.

where x is the input image, h is the filtering template, and f is the filtering response result, and $\otimes$ is the convolution operation [15].

For the instrument positioning problem, since our target is rigid, the target has no distinct shape and scale change, which means the template feature is relatively stable. The correlation filtering using the template features is suitable for solving instrument positioning problems. To use the correlation filter to locate the target, we need to find a filter template that can describe the appearance of the target well, which can get the maximum correlation in the target position when the input image is filtered by the sliding window method, as shown in Fig. 2.

Fig. 2. Denotation of correlation filtering.

In some early studies, a simple artificial template image is severed as a correlation filter, but the robustness of positioning was poor. Actually, the shapes of the instruments in the substation are various, so if we only design simple filter templates, the filters cannot adapt to all instruments. Therefore, it is necessary to obtain a filter template for each instrument adaptively. Since the correlation filter's computational form is equivalent to the linear regression, the correlation filtering problem can be transformed into the linear regression problem [16]. The linear regression is used to generate the correlation filter templates for each instrument adaptively. The form of the correlation filter can be expressed as $f = w^T x$, where w is the correlation filter template and is the input image. To increase the robustness of the model, this paper uses the L1 norm for regularization constraints. The task of finding a suitable filter template is equivalent to making the loss function minimal:

$$\min_{w} \sum_{i} (f(x_i) - y_i)^2 + \lambda \|w\|_1^2 \qquad (1)$$

where x_i is the training sample image and λ is the regular term.

Because we need to obtain a model that can get max response value in the target area and a much smaller response value in other areas, we take the target area as a positive sample and other areas as negative samples. The target area (blue box) and the position of the instrument (black box) are manually calibrated. The training sample set is generated by the template image collected in advance. The training samples (red box) are collected around the target area's position (blue box), as shown in Fig. 3. The closer to the target, the denser the sample collection. If the center coordinate of the target area is (M, N); the sample is represented as x_i, its center coordinate is (m_i, n_i). The corresponding label value is:

$$y_i = \exp\left(-\frac{(m_i-M)^2-(n_i-N)^2}{2\lambda_{label}^2}\right) \qquad (2)$$

where λ_{label} is the standard deviation. As the sample position close to the target area, the label value will be close to 1.

Fig. 3. Collect training sample set by Template image. The red boxes are the training samples. The blue box is target area. The black box is the instrument. (Color figure online)

After the training sample set is obtained, the HOG feature of the training samples is extracted to train the model. Since Eq. 1 has no closed-form solution, we use the coordinate descent method [17] to solve it.

Finally, after obtaining the correlation filter template, the correlation of each area of the test image is calculated by the sliding window method, as shown in Fig. 2. The point with the highest correlation value is the center point of the target area. Because the spatial relationship between the instrument image and the target area belongs to the prior knowledge, the coordinate of the instrument image is obtained.

3.2 Image Preprocessing

Since the actual environment of the substation belongs to the wild environment, there are often disturbances such as stains and shadows on the dial. Moreover, there are many kinds of instruments. For example, some dials have more letters or other interference, as shown in Fig. 4(a). In this paper, the morphological closing operation is used to process the image of the dial area to reduce the interference of strip noise. The morphological closing operation essentially performs an expansion operation on the image before performing an etching operation. It can remove some small hollow areas, as shown in Fig. 4(b). However, when the thickness of the pointer is similar to the scale, it is risky to perform the closing operation. It should be noted that the closing operation is not necessary, but it does improve the accuracy of pointer angle recognition in some instances.

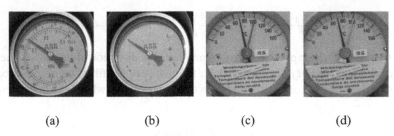

(a)   (b)   (c)   (d)

Fig. 4. The example of closed operation. (a) the original image. (b) the image after closed operation (c) The original image. (d) R channel image (Color figure online)

For some unique instruments, some additional preprocessing is required, as shown in Fig. 4(c) and (d). For such instruments, the red pointer is the upper limit of the alert value, and the black pointer indicates the current display of the instrument. In this paper, the image of the R channel is used to perform the subsequent steps by separating the RGB channels of the image.

3.3 Pointer Reading Recognition Based on DSF

There are two types of methods for instrument reading recognition: angle method and distance method. At present, most of the plans for reading the indication of instruments are the angle methods. In the existing research, the pointer reading recognition method obtains the instrument's reading by fitting the line corresponding to the pointer and calculating the angle of the line. For example, the Hough transform and the various projection methods judge whether there is a straight line in the image by counting the number of pixels on a certain line, but it isn't easy to judge the length of the line. These methods do not fully utilize the shape characteristics of the pointer. When there are many pixels with discontinuous stains in a straight line, such as shadow, it may be judged that such a line exists in the image.

Due to the problems [1] of the current methods, we consider from the perspective of human vision. When identifying a pointer, humans distinguish between pointers and other objects by shape and direction rather than just the pointer's line. Since few objects on the instrument's surface are similar in shape to the pointer, humans do not misidentify instruments. So, we need a method that is sensitive to direction and shape. Inspired by human vision, we propose a one-stage pointer angle identification method, which introduces direction-sensitive filtering (DSF) with a designed Gabor function. It can competitively select shapes to calculate the pointer's angle and avoid other interference such as shadows and scales.

The Gabor transform is a windowed short-time Fourier transform whose window function is a Gaussian function. The two-dimensional Gabor function is the only function that can reach the lower bound of the frequency domain and the spatial uncertainty principle [18]. The two-dimensional Gabor filter is similar to the receptive field of mammalian visual cortex simple cells. It has excellent spatial locality and direction selectivity and can enhance image features such as edge, peak, valley, and ridge contour [19, 20]. The form of the two-dimensional Gabor filter is:

$$\phi_{u,v}(z) = \frac{\|k_{u,v}\|^2}{\sigma^2} \exp\left(-\frac{\|k_{u,v}\|^2 \|z\|^2}{2\sigma^2}\right) \times \left[\exp(ik_{u,v}z) - \exp\left(-\frac{\sigma^2}{2}\right)\right] \qquad (3)$$

where $z(x, y)$ is the image pixel point coordinate; u is the direction of the Gabor filter; v is the scale of the Gabor filter; $k_{u,v}$ is the center frequency of the filter, which describes the response of the Gabor filter in different directions and scales, which can be represented for:

$$k_{u,v} = \begin{pmatrix} k_v cos\varphi_u \\ k_v sin\varphi_u \end{pmatrix} \qquad (4)$$

where $k_v = \frac{K_{max}}{f^v}$, k_v is the sampling scale; K_{max} is the maximum frequency; φ_u is the filter direction selectivity; $\frac{\|k_{u,v}\|^2}{\sigma^2}$ is used to compensate for the attenuation of the energy spectrum determined by the frequency; $\exp\left(-\frac{\|k_{u,v}\|^2 \|z\|^2}{2\sigma^2}\right)$ is the Gaussian function; $\exp(k_{u,v}z)$ is the oscillation function, the real part for the cosine function, the imaginary part is a sinusoidal function; $\exp\left(-\frac{\sigma^2}{2}\right)$ is a DC component, which is used to reduce the interference of the absolute value of the image grayscale on the filter, so that the sensitivity of the filter to the illumination change of the image is reduced.

Before introducing the specific method, we first verify the feasibility of using the Gabor filter for pointer recognition. The image is processed by Gabor filtering in different directions, as shown in Fig. 5. It can be seen that an observable response occurs when the scale and direction of the object in the image simultaneously conform to the parameters of the Gabor filter. This selectivity reduces the need for preprocessing and makes the algorithm robust to complex dials. However, there is still some noise in the response image, as shown in Fig. 5(b), (c). These noises are primarily responses to edges and scales that are similar to a pointer. To reduce such interference, we can narrow down the image range processed by the Gabor filter to the dial or constrain the statistical region range when counting each response image's response value. The algorithm proposed in this paper adds the mask to constrain the statistical area in the step of statistical response value, as shown in Fig. 5(d). It should be noted that there is not only one mask here. Each angle of the Gabor filter has a mask that is the same as its angle. For each angle's mask, we set a correction area for the central part of the plate, which means we want to avoid reading the lines around the edge to reduce the kind of interference for reading the pointer. After obtaining the pointer's angle, the pointer reading is calculated by the angular relationship between the pointer and the scale. The pointer reading R calculation can be obtained by:

$$R = R_{min} + \frac{\theta_{ac}}{\theta_{ab}} \times Rang \qquad (5)$$

where R_{min} is the value of the starting scale, θ_{ac} is the angle between the starting scale and the pointer, and θ_{ab} is the angle between the starting and ending scales, and Rang is the indication range of the scale.

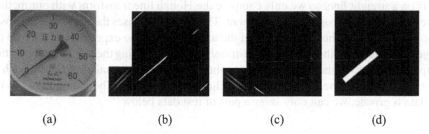

(a)   (b)   (c)   (d)

Fig. 5. A test result about Gabor filter. (a) The grayscale of an instrument. (b) The kernel of Gabor filter and the result of image processed by it. (c) The kernel of Gabor filter and the result of image processed by it. (d) The mask.

Therefore, this paper proposes a pointer angle recognition method based on Gabor filtering, which uses a multi-angle Gabor filter to process the instrument image and then counts the response image of each angle. The angle corresponding to the image with the largest cumulative response value is the angle of the pointer.

Finally, the pointer reading is calculated by Eq. 5. The method framework is shown in Fig. 6.

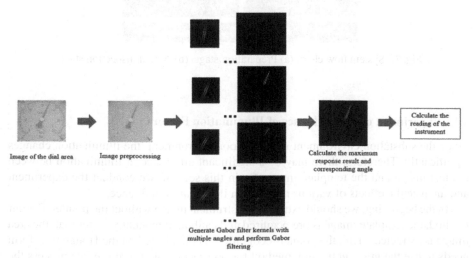

Fig. 6. Framework of proposed algorithm. The range of angles of the Gabor filter covers the range of the pointer.

4 Experiment

In this section, we conduct the experiment to test our proposed method. Following the layout of the real substation, a standard test site is established. The instrumentation automatic reading system is divided into two stages: 1. Preparation stage. 2. Actual inspection stage. The specific process is shown in Fig. 7. It should be mentioned that the projection method and the Hough transform both use the cumulative number of the pixels on a straight line, so we only compare the Hough line transform with our method in the pointer angle recognition experiment. The experiment uses the algorithm proposed in Sect. 3.1 to locate the instrument and the test images of the experimental part are the images extracted after the instrument positioning. After finishing the test, our cooperation company embeds the method to its robots and makes them for real application, such as in the substations at the Zigong City of Sichuan Province and Chongqing City. As the test data is private, we can only show a part of test data below.

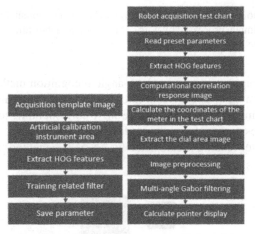

Fig. 7. System flow chart. (a) Preparation stage. (b) Actual inspection stage.

4.1 Experiment on the Influence of Illumination Factors

Since the substation environment is an outdoor environment, the illumination changes significantly. Therefore, there may be a significant difference in illumination between the test image and the template image. So in this section, we conduct the experiment and analyze the effects of various methods in illumination influence.

In the beginning, we should extract the instrument image without the pointer. For our method, the template image is pre-acquired and calibrated manually. After that, the area image is extracted. Here, the pointer recognition method based on the Hough transform needs to use the gray subtraction method for preprocessing. Therefore, we process the area image through the software Photoshop to get the dial image without a pointer.

The test images collected by the inspection robot are shown in Fig. 8(a). These images are processed to indicate the pointer angle by the Hough line transform method.

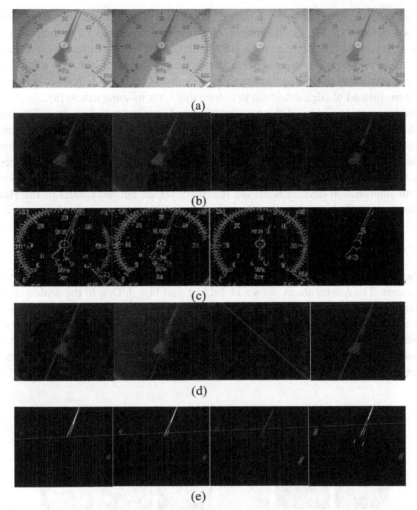

Fig. 8. Test of different lighting conditions. (a) Images under different lighting conditions. (b) a result image of the gray subtraction method between (a) and the target area image extracted from the pre-acquired template image. (c) The result image of edge extraction of (b) using the canny operator. (d) The result image of using Hough transform line detection to process (c). (e) The maximum response value images in the result of processing (a) by the algorithm proposed in this paper

From Fig. 8(b), it can be found that the effect of the gray subtraction method is not observable when there is a significant difference between the illumination environment of the test image and the illumination environment of the template image. When Fig. 8(b) are edge-extracted by the Canny edge detector, the interference such as the scale cannot be stably removed. The pointer image can be better extracted only when the test image and the template image are in the same lighting environment. Additionally, a situation where the light is too dark can also cause the grayscale subtraction method to remove the pointer image incorrectly. So the edge detection is used to extract pointer features, as shown in Fig. 8(c). We use the Hough transform to detect the edge of the pointer but not the center axis of the pointer, as shown in Fig. 8(d). When the difference in the angle between the two edges of the pointer becomes larger, the error between the detected angle and the actual angle also becomes larger. Some researchers have used refinement algorithms instead of edge detection to reduce this error to some extent [9].

The recognition results of the proposed method are shown in Fig. 8(e). The image shown here is the resulting image with the largest cumulative response value obtained by the proposed algorithm. It can be seen that the maximum response result image obtained is the filtered result image when the Gabor filter kernel is in the same direction as the central axis of the pointer. The recognition results of the proposed method in different illumination environments are relatively stable, but the results of the Hough line transform will change between the two edges of the pointer.

4.2 Error Discussion

We choose some samples at random to test. The sample-set includes various types of instruments. The results are as shown in Table 1 and Fig. 9. Due to the scale accuracy limitation, there are errors in personal readings. Except for the human error, the character of the algorithm and system error can also lead to the relative error. Because the error is lower than the minimum calibration precision, the relative error is acceptable. The proposed method in this paper can recognize the reading of the instrument accurately, in a result, it can effectively improve the maintenance efficiency of substations.

Fig. 9. The results of the sample set by this paper proposed method. The sequence of images is consistent with the sequence in Table 1, from left to right, from top to bottom, and the serial Numbers correspond to 1–20 in turn.

Table 1. The results of the reading of the instruments by the proposed method.

Number of instruments	Reading by human	Reading by Hough transform based methods	Reading by proposed methods	Relative error
1	1	1.32	1.08	0.0833
2	56	83.82	55.84	0.0028
3	90	94.25	89.62	0.0041
4	0.5	0.53	0.495	0.01
5	67	62.32	65.5	0.0223
6	20	16.45	20.25	0.0125
7	25	25.74	25.31	0.0124
8	32	31.82	32.12	0.0038
9	0.179	0.19	0.17	0.0503
10	0.1	0.06	0.096	0.04
11	6.9	6.18	6.84	0.0087
12	0.24	0.21	0.244	0.0167
13	0.14	0.08	0.137	0.0214
14	58	11.32	57.38	0.0106
15	30	28.43	30.61	0.0203
16	24.2	33.67	24.19	0.0004
17	2.5	0.23	2.6	0.04
18	26	28.25	25.86	0.0053
19	32	41.21	33.07	0.0334
20	0.26	0.17	0.266	0.023

5 Conclusion

This paper presents a computer vision-based automatic reading method for the substation instruments. Firstly, the instrument localization algorithm accurately locates the instrument by a method based on correlation filtering. Secondly, the image preprocessing section utilizes a closing operation to reduce the effect of the scale on the recognition pointer angle. Then, the pointer recognition algorithm uses DSF to recognize the pointer angle and overcomes the influence of scale, shadow, and other interference on the result. Finally, the instrument reading is calculated by the correspondence between the angle and the reading. We collected test data sets by establishing test sites that met the applicable standards. According to the experimental results, the instrument automatic reading method proposed in this paper has good robustness and accuracy, and it can meet the monitoring needs of the actual substation. Furthermore, the pointer angle recognition module in our method can also be replaced with a processing module for other types of instruments, such as an oil level gauge, etc. In the future, we will study multi-scale information and spatial transformation to improve the accuracy and reduce the influence of early human participation.

References

1. Shengfang, L., Xingzhe, H.: Research on the AGV based robot system used in substation inspection. In: International Conference on Power System Technology, pp. 1–4 (2006). https://doi.org/10.1109/ICPST.2006.321495
2. Jian, Y., Xin, W., Xue, Z., Zhenyou, D.: Cloud computing and visual attention-based object detection for power substation surveillance robots. In: Canadian Conference On Electrical and Computer Engineering, pp. 337–342 (2015). https://doi.org/10.1109/CCECE.2015.7129299
3. Chi, J., Liu, L., Liu, J., Jiang, Z., Zhang, G.: Machine vision based automatic detection method of indicating values of a pointer gauge. Math. Probl. Eng. **2015**, 1–19 (2015)
4. Zhang, L., Fang, B., Zhao, X., Zhang, H.: Pointer-type meter automatic reading from complex environment based on visual saliency. In: International Conference on Wavelet Analysis and Pattern Recognition, pp. 264–269 (2016). https://doi.org/10.1109/ICWAPR.2016.7731651
5. Alegria, F.C., Serra, A.C.: Automatic calibration of analog and digital measuring instruments using computer vision. IEEE Trans. Instrum. Measur. **49**(1), 94–99 (2000)
6. Yue, X., Min, Z., Zhou, X., Wang, P.: The research on auto-recognition method for analogy measuring instruments. In: International Conference on Computer, Mechatronics, Control and Electronic Engineering, pp. 207–210 (2010). https://doi.org/10.1109/CMCE.2010.5609721
7. Jiale, H., En, L., Bingjie, T., Ming, L.: Reading recognition method of analog measuring instruments based on improved Hough transform. In: IEEE 2011 10th International Conference on Electronic Measurement and Instruments, Chengdu, China, pp. 337–340 (2011). https://doi.org/10.1109/ICEMI.2011.6037919
8. Yang, B., Lin, G., Zhang, W.: Auto-recognition method for pointer-type meter based on binocular vision. J. Comput. **9**(4), 787–793 (2014)
9. Zheng, C., Wang, S., Zhang, Y., Zhang, P., Zhao, Y.: A robust and automatic recognition system of analog instruments in power system by using computer vision. Measurement **9**(2), 413–420 (2016)
10. Yudong, G., Yang, Z., Bowen, H., Hong, Z., Dayu, S.: Pointer-type meter reading method research based on image processing technology. In: Second International Conference on Networks Security, Wireless Communications and Trusted Computing, Wuhan, China, pp. 107–110 (2010). https://doi.org/10.1109/NSWCTC.2010.33
11. Belan, P.A., Araujo, S.A., Librantz, A.F.H.: Segmentation-free approaches of computer vision for automatic calibration of digital and analog instruments. Measurement **46**(1), 177–184 (2013)
12. Yang, Z., Niu, W., Peng, X., Gao, Y., Qiao, Y., Dai, Y.: An image-based intelligent system for pointer instrument reading. In: 4th IEEE International Conference on Information Science and Technology, Shenzhen, China, pp. 780–783 (2014). https://doi.org/10.1109/ICIST.2014.6920593
13. Mukhopadhyay, P., Chaudhuri, B.B.: A survey of Hough transform. Pattern Recogn. **48**(3), 993–1010 (2015)
14. Felzenszwalb, P.F., Girshick, R.B., Mcallester, D.A., Ramanan, D.: Object detection with discriminatively trained part-based models. IEEE Trans. Pattern Anal. Mach. Intell. **32**(9), 1627–1645 (2010)

15. Bolme, D.S., Beveridge, J.R., Draper, B.A., Lui, Y.M.: Visual object tracking using adaptive correlation filters. In: 2010 IEEE Computer Society Conference on Computer Vision and Pattern Recognition, San Francisco, CA, USA, 2010, pp. 2544–2550 (2010). https://doi.org/10.1109/CVPR.2010.5539960.
16. Henriques, J.F., Caseiro, R., Martins, P., Batista, J.: High-speed tracking with kernelized correlation filters. IEEE Trans. Pattern Anal. Mach. Intell. **37**(3), 583–596 (2015)
17. Friedman, J.H., Hastie, T., Hofling, H., Tibshirani, R.: Pathwise coordinate optimization. Ann. Appl. Stat. **1**(2), 302–332 (2007)
18. Daugman, J.: Complete discrete 2-D Gabor transforms by neural networks for image analysis and compression. IEEE Trans. Acoust. Speech Signal Process. **36**(7), 1169–1179 (1988)
19. Lyons, M., Akamatsu, S., Kamachi, M., Gyoba, J.: Coding facial expressions with Gabor wavelets. In: Proceedings Third IEEE International Conference on Automatic Face and Gesture Recognition, Nara, Japan, pp. 200–205 (1998). https://doi.org/10.1109/AFGR.1998.670949
20. Vukadinovic, D., Pantic, M.: Fully automatic facial feature point detection using Gabor feature based boosted classifiers. In: IEEE International Conference on Systems, Man and Cybernetics, Waikoloa, HI, USA, 2005, vol. 2, pp. 1692–1698 (2005). https://doi.org/10.1109/ICSMC.2005.1571392.

Partial Distillation of Deep Feature for Unsupervised Image Anomaly Detection and Segmentation

Qian Wan[1], Liang Gao[1], Lijian Wang[2], and Xinyu Li[1(✉)]

[1] School of Mechanical Science and Engineering, Huazhong University of Science and Technology, Wuhan, China
wanqian19@hust.edu.cn, {gaoliang,lixinyu}@mail.hust.edu.cn
[2] AVIC China Aero-Polytechnology Establishment, Beijing, China

Abstract. Unsupervised image anomaly detection and segmentation is challenging but important in many fields, such as the defect of product inspection in intelligent manufacturing. The challenge is that, the labeled anomalous data is few and only normal data is available, causing the distribution of the anomaly unknowable. Unsupervised methods based on image-reconstruction and feature-embedding have been recently studied for anomaly detection and segmentation, while there still exists the problems in term of the reconstruction quality and testing performance. Inspired by the application of pre-trained feature-embedding in anomaly detection, this paper proposes a novel framework called Partial Distillation of Deep Feature (PDDF) for unsupervised image anomaly detection and segmentation. Only normal images are used to partially distill the knowledge of the deep features extracted by the pre-trained network at training. The difference (dissimilarity) between the distilled feature with the pre-trained one is applied to score the uncertainty of anomaly at testing. The PDDF achieves the comparing results compared with the state-of-the-art methods on MVTec AD dataset, with the area under the receiver operating characteristic curve (ROCAUC) of 96.7% for unsupervised anomaly detection, 96.9% and the area under the per-region-overlap curve (PROAUC) of 91.8% for unsupervised anomaly segmentation.

Keywords: Anomaly detection · Anomaly segmentation · Knowledge distillation

1 Introduction

Anomaly Detection and Segmentation is an important and long-standing problem faced by the intelligent manufacturing system, in which such as the defect inspection [1], fault diagnose [2]. The well-known challenge is that the labeled anomalous samples are few and only normal samples are available at training, causing the distribution of the anomaly unknowable [3]. In this paper, unsupervised image anomaly detection and segmentation have been studied.

© Springer Nature Switzerland AG 2021
D.-S. Huang et al. (Eds.): ICIC 2021, LNCS 12836, pp. 238–250, 2021.
https://doi.org/10.1007/978-3-030-84522-3_19

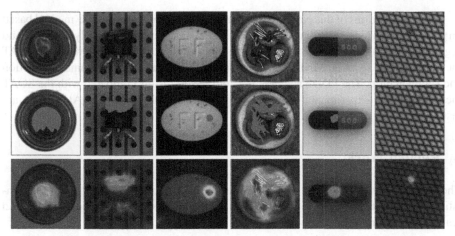

Fig. 1. Qualitative results of the proposed PDDF on the MVTec AD dataset [4]. **Top Row:** The testing image with anomalies (defects). **Mid Row:** The ground-truth of anomalies in red color. **Bottom Row:** The anomaly scoring maps detected in pixel-wise by the proposed PDDF.

Recently, some researches based on image-reconstruction [4–8] and feature-embedding [9–14] methods have been studied for unsupervised image anomaly detection and segmentation due to the few labeled anomalous data. Try to prevent the anomaly being reconstructed and then measure the difference between the input and the reconstruction among image-reconstruction based methods. The quality of the reconstruction decides the performance of this type method, but the anomalous regions always lead to a worse reconstruction [8]. The feature-embedding based methods utilize the high discriminative feature in embedding space to distinguish the normal features from the abnormal ones. The embedded space can be constructed by the pre-trained neural networks [9] or self-supervised training [12]. Due to the large amount of retrievals of the similar features for testing, there are computationally expensive [9, 12]. Also in another type of feature-embedding, distilling the knowledge from the teacher network to the student network has been studied for anomaly scoring [13]. This largely reduces the expensive computation while the accuracy has some drop compared with [9, 12].

To solve the above issue, this paper proposed the Partial Distillation of Deep Feature (PDDF) for unsupervised image anomaly detection and segmentation inspired by [13]. The PDDF directly utilize the difference between the distilled feature with the pre-trained features to detect the anomaly at testing stage. To preserve the well quality of the extraction of the low-level features, the head of pre-trained network does not participate in the knowledge distillation and is shared to the student network. Only the deep layers of the pre-trained teacher network are required to distill the knowledge to the student network during training in the PDDF, thus called the partial knowledge distillation. The PDDF achieves a comparing results compared with the sate-of-the-art (SOTA) methods on MVTec AD dataset [4]. The Fig. 1 shows some qualitative results of the proposed PDDF on the MVTec AD dataset, and the anomalies contained in testing images are scored with higher values.

The contributions of this paper are summarized as: 1) The novel PDDF method based on partially knowledge distilling is proposed for the unsupervised anomaly detection and segmentation. 2) The extensive experiments are conducted compared with the SOAT methods on MVTec AD, MNIST and Fashion-MINST datasets for anomaly detection and segmentation. The PDDF achieves the comparing results compared with the SOAT methods. 3) The experimental ablation study of the PDDF on MVTec AD dataset are discussed in this paper.

2 Related Work

The previous methods studied for unsupervised image anomaly detection and segmentation can be categorized into two types as image-reconstruction based methods and feature-embedding based methods.

2.1 Image-Reconstruction Based Methods

These type methods have been mostly studied with auto-encoder network. The main idea behinds them is that the distribution of normal samples can be learned by the reconstruction of an encoder-decoder manner. The anomaly could not be reconstructed similar as the input, thus the difference between the reconstruction with the input can be viewed as the score for anomaly detection.

Auto-encoder trained with the structural similarity loss rather than L2 loss, making a better performance on anomaly segmentation [5]. AnoGAN [6] is proposed to learn a manifold of normal latent with a deep convolutional generative adversarial network. The convolutional adversarial variational auto-encoder incorporates the Grad-CAM [17] to locate the anomalies from the learned latent space [7]. Iterative energy-based projection with gradient decent is proposed to enhance the reconstruction quality of the variational auto-encoder [8]. The issue that the anomalous regions deteriorate the reconstruction quality can be reduced in some extent and improve the performance on anomaly segmentation. Despite these type methods are very interpretable and understandable, the reconstruction error may lead to a bad performance [13].

2.2 Feature-Embedding Based Methods

The feature-embedding based methods map the image into the discriminative embedding space to distinguish the anomaly from the normal embedding.

The pre-trained ResNet18 [15] is used to embed the image to the feature space, and cluster the features with k-means [10]. The distance between the feature with the clustering centers is scored as the uncertainty of anomaly. Student-Teacher framework is proposed to locate the anomaly by that the student network makes worse regression for anomalous region when trained under the supervision of the teacher network [11]. Patch SVDD [12] learns the embedded space in patch-wise by the self-supervised learning manner and retrieves the similar embedding features from normal images for the ones of testing images for anomaly detection and segmentation. The sub-image anomaly detection with deep pyramid correspondences (SPADE) is proposed to retrieves the similar feature in the embedded space constructed by the pre-trained network. Multiresolution knowledge distilling (MKD) method uses the pre-trained VGG network [16] to train a random initialized network by knowledge distilling [13], and locates the anomaly by the testing loss backpropagation. The student-teacher feature pyramid matching (STFPM) [14] is proposed to distill the whole student network and measure the discrepancy between the teacher network and student network to detect the anomaly.

Unsupervised methods based on above two types have been mainly studied for anomaly detection and segmentation, but there still exists the problems in term of the reconstruction quality and testing performance. The proposed PDDF achieves the comparing performance on unsupervised image anomaly detection and segmentation compared with the SOAT methods.

3 Method

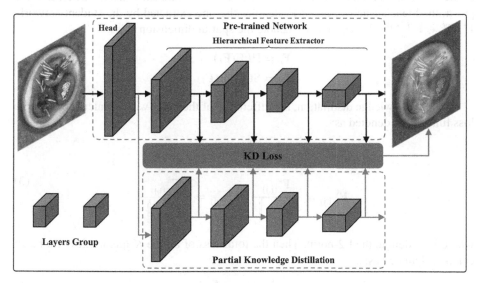

Fig. 2. The framework of the proposed PDDF.

The Fig. 2 shows the framework of the proposed PDDF. The pre-trained network is divided into two parts, the head and the hierarchical feature extractor. During the training, only the normal images are used to distill the knowledge from the hierarchical feature extractor (HFE) of the pre-trained teacher network to train the random initialized student network. The head of the pre-trained teacher network is shared with the student network. For the testing, the discrepancy between the HFE of pre-trained teacher network with the well-distilled student network is scored for the uncertainty of the anomaly. The head is the group of the first few layers of the network. The hierarchical feature extractor is consisted with the groups of the deep layers of the pre-trained network. Such as the pre-trained ResNet [15], the group of first four layers is set as the head and the subsequent three groups of residual layers is set as the HFE. The pre-trained network has been trained by supervised learning on ImageNet dataset [18].

3.1 Partial Knowledge Distillation

Trying to preserve the well quality of the low-level feature extracted from the network, the head of the pre-trained teacher does not participate the distillation procedure. And the head is shared to the distilling-required student network both during the training and testing. This ensures the same rich low-level features extracted by both networks. Given the input image $\mathbf{X}$ with size $H \times W \times C$, the features $\mathbf{F}_l$ extracted by the head of the network is denoted as following, $\mathbf{F}_l \in \mathbb{R}^{H_l \times W_l \times C_l}$, H_l, W_l, C_l denote the spatial dimension sizes:

$$\mathbf{F}_l = \text{Head}(\mathbf{X}) \tag{1}$$

After feature extraction at the head of the pre-trained teacher, the deep hierarchical features are extracted by the HFE. The HFE consists of the several groups of deep layers of the network. In each hierarchy, the $\mathbf{F}_t$ denotes the feature extracted by the HFE of the pre-trained teacher network, and $\mathbf{F}_s$ denotes the one extracted by the student network. $\mathbf{F}_t^k$, $\mathbf{F}_s^k \in \mathbb{R}^{H_k \times W_k \times C_k}$, H_k, W_k, C_k denote the spatial dimension sizes.

$$\mathbf{F}_t = \text{HFE}(\mathbf{F}_l)$$
$$\mathbf{F}_s = \text{Student}(\mathbf{F}_l) \tag{2}$$

For the knowledge distillation, the difference of the pixel-wise is calculated and the loss function is denoted as:

$$loss^k = \frac{1}{H_k \times W_k} \left\| \hat{\mathbf{F}}_t^k - \hat{\mathbf{F}}_s^k \right\|_2^2$$
$$\hat{\mathbf{F}}_{t\,(i,j)}^k = \frac{\mathbf{F}_{t\,(i,j)}^k}{\left\| \mathbf{F}_{t\,(i,j)}^k \right\|_2}, \quad \hat{\mathbf{F}}_{s\,(i,j)}^k = \frac{\mathbf{F}_{s\,(i,j)}^k}{\left\| \mathbf{F}_{s\,(i,j)}^k \right\|_2} \tag{3}$$

where $\|\cdot\|_2$ denote the L2-norm. Then the total loss of all the K hierarchies of a batch can be calculated as:

$$loss = \frac{1}{K} \sum_{batch} \sum_{k=1}^{K} loss^k \tag{4}$$

During the training, given a batch of normal images from the training set, the parameters of the student network are once updated by the *loss* back-propagation in every epoch. After training of all the setting epochs, the well-distilled student network is obtained and can be used for testing.

3.2 Anomaly Scoring

Anomaly Segmentation for Testing Image. Through the partial knowledge distillation, the student network can be obtained by distilling the knowledge of the deep features from the pre-trained teacher. For the pixel-wise anomaly segmentation of a testing image, in each hierarchy, the discrepancy between the HFE of the pre-trained network with the student network is computed in position-wise as:

$$\mathbf{A}_{(i,j)}^{k} = \left\| \hat{\mathbf{F}}_{t(i,i)}^{k} - \hat{\mathbf{F}}_{s(ij)}^{k} \right\|_{2}^{2} \tag{5}$$

By bilinear interpolation, the anomaly scoring map $\mathbf{A}^k$ is resized to the same spatial size of the testing image. And the final anomaly scoring map $\hat{\mathbf{A}}$ can be obtained by over all the hierarchies as:

$$\hat{\mathbf{A}}^{k} = \text{resize}(\mathbf{A}^{k})$$
$$\hat{\mathbf{A}} = \frac{1}{K} \sum\nolimits_{k=1}^{K} \hat{\mathbf{A}}^{k} \tag{6}$$

Anomaly Detection for Testing Image. The anomalous region obtains the much higher discrepancy value than the normal region in the anomaly scoring map. Then a maximum value of each anomaly scoring map can be viewed as the score of uncertainty of anomaly as:

$$S = \max(\hat{\mathbf{A}}) \tag{7}$$

4 Experiments

4.1 Experimental Setup

MVTec AD Dataset. The MVTec AD dataset [4] is published by MVTec Software GmbH. The dataset has been widely studied for the unsupervised image anomaly detection and segmentation. It consists of 15 different categories collected from the real-world scenarios, totally containing 3629 normal images for training and 467 normal images and 1258 abnormal images for testing. The abnormal images contain different scale anomalies (defect) compared with the normal image. The anomalies are fine-grained and this further increases the difficulty for unsupervised anomaly detection and segmentation. The PDDF has been compared with the SOAT methods on the dataset.

MNIST and Fashion-MNIST Datasets. MNIST dataset [19] consists of 10 categories, 60k training images and 10k testing 28×28 gray-scale handwritten digit images. Fashion-MNIST dataset [20] also consists of 10 categories, 70k training images and 10k testing 28×28 gray-scale fashion product images. The anomaly detection task is also conducted on both two datasets in this paper. The image from the other categories are viewed as the anomalous images and the images of the same categories are viewed as the normal ones.

Implement Details. The images of MVTec AD dataset are resized to 256×256 both for training and testing, normalized by the mean and standard deviation value of ImageNet dataset. The images of MNIST and Fashion-MNIST datasets are resized to 32×32 both for training and testing while not normalized. The student network is trained by the stochastic gradient descent (SGD) with the learning rate of 0.5, the momentum of

0.9, the weight decay of 0.0001. The batch size is set as 32 and totally trained with 200 epochs. The structure of student network is set as the same as the hierarchical feature extractor of the pre-trained teacher, while is initialized randomly. The default setting of the pre-trained teacher network is ResNet18 and the corresponding student is the randomly initialized structure same as the HFE of the teacher in this paper.

Evaluation Metrics. For anomaly detection, the area under the receiver operating characteristic curve (ROCAUC) is commonly adopted to evaluate the performance of a model. For anomaly segmentation, except the ROCAUC metrics, the area under the per-region-overlap curve (PROAUC) is proposed to evaluate the model fairly for different scale anomalies [11].

4.2 Anomaly Detection

For the anomaly detection task which is inferring whether a testing is abnormal, the PDDF is compared with the SOAT methods. The compared methods are AE$_{SSIM}$ [5], CAVGA-Du [7], AnoGAN [6], VAE-grad [8], SPADE [9], Patch SVDD [12], MKD [13], and STFPM [14].

As reported in Table 1, the PDDF achieves the comparing results over 15 categories on MVTec AD dataset and gets a higher average ROCAUC result compared with the SOAT methods. This demonstrates that the PDDF is effective for the fine-grained anomaly detection from the same category. The capability of the PDDF on the anomaly detection is obtained by measuring the maximum discrepancy between the student network and the HFE, since that the student network is well distilled from the HFE of the pre-trained network only on normal images in the same categories and it can regress much more discrepancy on the anomalous regions which unseen during distilling stage.

The experiments on the class-level anomaly detection of the PDDF method are conducted on the MNIST [19] and Fashion-MNIST [20] datasets. On these two datasets, the compared methods are OCSVM [21], DSVDD [22], AnoGAN [6], DSEBM [23], DAGMM [24], ARAE [25], LSA [26]. The results of corresponding methods are average on 10 categories for the both two datasets. As showed in Table 2, the PDDF also achieves comparing results on the class-level anomaly detection on the MNIST and Fashion-MNIST datasets compared with the SOAT methods. The class-level anomaly detection shows that the student network generalizes bad on these categories which unseen during the distilling stage, so the maximum discrepancy for anomaly detection can be measured for these categories.

4.3 Anomaly Segmentation

For the anomaly segmentation task, the PDDF is compared with the SOAT methods on MVTec AD dataset. The compared methods are AE$_{SSIM}$ [5], AnoGAN [6], VAE-grad [8], SPADE [9], ST [11], Patch SVDD [12], MKD [13], and STFPM [14].

As showed in Table 3, the PDDF also achieves the comparing results compared with the SOAT methods over 15 categories under the ROCAUC metrics, despite the total average ROCAUC is slightly lower than the best STFPM method [14].

Table 1. The anomaly detection results comparing with the SOAT methods on MVTec AD dataset. (ROCAUC%). Best in bold and Second in underline bold.

Category	AE$_{SSIM}$	CAVGA-D$_u$	AnoGAN	VAE-grad	Patch SVDD	STFPM	MKD	SPADE	Proposed PDDF
Carpet	67.0	73.0	49.0	67.0	**92.9**	-	79.3	92.8	**98.7**
Grid	69.0	75.0	51.0	83.0	**94.6**	-	78.1	47.3	**99.6**
Leather	46.0	71.0	52.0	71.0	90.9	-	95.1	**95.4**	**100.0**
Tile	52.0	70.0	51.0	81.0	**97.8**	-	91.6	96.5	**99.0**
Wood	83.0	85.0	68.0	89.0	**96.5**	-	94.3	95.8	**99.5**
Bottle	88.0	89.0	69.0	86.0	98.6	-	**99.4**	97.2	**100.0**
Cable	61.0	63.0	53.0	56.0	**90.3**	-	89.2	84.8	**97.2**
Capsule	61.0	83.0	58.0	86.0	76.7	-	80.5	**89.7**	**95.7**
Hazelnut	54.0	84.0	50.0	74.0	92.0	-	**98.4**	88.1	**100.0**
Metal Nut	54.0	67.0	50.0	78.0	**94.0**	-	73.6	71.0	**99.3**
Pill	60.0	**88.0**	68.0	80.0	86.1	-	82.7	80.1	**96.8**
Screw	51.0	77.0	35.0	71.0	81.3	-	83.3	66.7	**88.8**
Toothbrush	74.0	91.0	57.0	89.0	**100.0**	-	**92.2**	88.9	86.4
Transistor	52.0	73.0	67.0	70.0	**91.5**	-	85.6	90.3	**92.6**
Zipper	80.0	87.0	59.0	67.0	97.9	-	93.2	96.6	**97.8**
Average	63.0	78.0	55.0	77.0	92.1	**95.5**	87.7	85.5	**96.7**

Table 2. The anomaly detection results comparing with the SOAT methods on MNIST and fashion-MNIST datasets. (ROCAUC%). Best in bold and Second in underline bold.

	OCSVM	DSVDD	AnoGAN	DSEBM	DAGMM	ARAE	LSA	Proposed PDDF
MNIST	96.0	94.8	91.3	-	-	**97.5**	**97.5**	97.9
Fashion-MNIST	92.8	92.8	-	85.5	41.7	**93.6**	92.2	**93.0**

Table 3. The anomaly segmentation results comparing with the SOAT methods on MVTec AD dataset. (ROCAUC%). Best in bold and Second in underline bold.

Category	AE_{SSIM}	AnoGAN	VAE-grad	Patch SVDD	STFPM	MKD	SPADE	Proposed PDDF
Carpet	87.0	54.0	73.5	92.6	**98.8**	95.6	**97.5**	**98.8**
Grid	94.0	58.0	96.1	**96.2**	**99.0**	91.8	93.7	**99.0**
Leather	78.0	64.0	92.5	97.4	**99.3**	98.1	97.6	**99.2**
Tile	59.0	50.0	65.4	91.4	**97.4**	82.8	87.4	**95.3**
Wood	73.0	62.0	83.8	90.8	**97.2**	84.8	88.5	**94.4**
Bottle	93.0	86.0	92.2	98.1	**98.8**	96.3	98.4	**98.5**
Cable	82.0	78.0	91.0	96.8	95.5	82.4	**97.2**	**96.3**
Capsule	94.0	84.0	91.7	95.8	98.3	95.9	**99.0**	**98.6**
Hazelnut	97.0	87.0	97.6	97.5	98.5	94.6	**99.1**	**98.6**
Metal Nut	89.0	76.0	90.7	**98.0**	97.6	86.4	**98.1**	96.5
Pill	91.0	87.0	93.0	95.1	**97.8**	89.6	96.5	**98.0**
Screw	96.0	80.0	94.5	95.7	98.3	96.0	**98.9**	**98.7**
Toothbrush	92.0	90.0	98.5	98.1	**98.9**	96.1	97.9	**98.7**
Transistor	90.0	80.0	91.9	**97.0**	82.5	76.5	**94.1**	84.5
Zipper	88.0	78.0	86.9	95.1	**98.5**	93.9	**96.5**	**98.5**
Average	87.0	74.3	89.3	95.7	**97.0**	90.7	96.5	**96.9**

Due to that the ROCAUC metrics is in favor of large anomalous region, the PROAUC metrics is introduced to fairly treat the anomalous regions with different scale [11]. The PROAUC is calculated under the 30% false-positive rate [11]. This paper further evaluates the PDDF under the PROAUC metrics. As reported in Table 4, the PDDF also achieves the comparing results compared with the SOAT methods over 15 categories under the PROAUC metrics. The PDDF can well segment the anomalies with different scales compared with the SOAT methods. The PDDF equips with the well capability for anomaly segmentation because the position-wise feature of student network is far away from the feature of the HFE of the pre-trained network when evaluated on the anomalous regions. The reason of it is that the student network only trained with the features of

Table 4. The anomaly segmentation results comparing with the SOAT methods on MVTec AD dataset. (PROAUC%). Best in bold and Second in underline bold.

Category	AE$_{SSIM}$	CNN_Dict	AnoGAN	VAE	ST	STFPM	SPADE	Proposed PDDF
Carpet	64.7	46.9	20.4	50.1	69.5	**95.8**	94.7	**94.8**
Grid	84.9	18.3	22.6	22.4	81.9	**96.6**	86.7	**96.7**
Leather	56.1	64.1	37.8	63.5	81.9	**98.0**	97.2	**97.9**
Tile	17.5	79.7	17.7	87.0	**91.2**	92.1	75.9	83.8
Wood	60.5	62.1	38.6	62.8	72.5	**93.6**	87.4	**88.7**
Bottle	83.4	74.2	62.0	89.7	91.8	**95.1**	95.5	94.9
Cable	47.8	55.8	38.3	65.4	86.5	87.7	**90.9**	92.4
Capsule	86.0	30.6	30.6	52.6	91.6	92.2	**93.7**	92.5
Hazelnut	91.6	84.4	69.8	87.8	93.7	94.3	**95.4**	**95.2**
Metal Nut	60.3	35.8	32.0	57.6	89.5	**94.5**	**94.4**	91.7
Pill	83.0	46.0	77.6	76.9	93.5	**96.5**	94.6	**95.7**
Screw	88.7	27.7	46.6	55.9	92.8	93.0	**96.0**	**93.7**
Toothbrush	78.4	15.1	74.9	69.3	86.3	**92.2**	93.5	91.4
Transistor	72.5	62.8	54.9	62.6	70.1	69.5	**87.4**	**72.3**
Zipper	66.5	70.3	46.7	54.9	93.3	**95.2**	92.6	**95.0**
Average	69.4	51.5	44.3	63.9	85.7	**92.1**	91.7	**91.8**

the normal images which extracted by the HFE of the pre-trained network during the distilling stage. Then the student network outputs the bad features compared with the one of the HFE when tested on the anomalous regions. So the distance of them can be measured in position-wise manner and the anomalies which measured with much large distances can be segmented out.

Table 5. Anomaly detection and segmentation results of different pre-trained network in the PDDF on MVTec AD dataset. (AUCROC%)

	VGG16	VGG19	ResNet18	ResNet34
Detection	93.7	93.6	**96.7**	96.6
Segmentation	95.0	95.2	96.9	**97.0**

4.4 Ablation Study

Influence of Pre-trained Network. The partial knowledge distillation of different pre-trained networks has been studied in this section. The pre-trained VGG16, VGG19 [16],

Table 6. Anomaly detection and segmentation results of different hierarchical feature in the PDDF on MVTec AD dataset. (AUCROC%)

	1	2	3	1&2	1&3	2&3	1&2&3
Detection	83.1	92.5	96.5	93.1	96.2	**96.8**	96.7
Segmentation	92.7	95.0	96.2	95.7	96.7	96.7	**96.9**

ResNet18, ResNet34 [15] have been studied as the pre-trained teacher network. In VGG network, the group of first down-sample layers is set as the head and the subsequent three groups of down-sample layers is set as the hierarchical feature extractor. In ResNet network, the group of first four layers is set as the head and the subsequent three groups of residual layers is set as the hierarchical feature extractor.

Table 5 shows the results of four different pre-trained networks for anomaly detection and segmentation. The VGG network shows the slightly lower performance on the two tasks compared with ResNet and the ResNet18 performs best. The much deeper ResNet34 does not shows the better performance than the shallower ResNet18. The deeper network requires large amount normal images to make the knowledge distilling from the HFE of the pre-trained network and may not generalize well on category which only has the several hundred images.

Influence of Hierarchical Feature. The different hierarchical discrepancy has been studied to show the hierarchical influence in partial knowledge distillation for anomaly detection and segmentation. The pre-trained teacher network is set as the ResNet18 and the first three groups of residual layers are set as the different hierarchies.

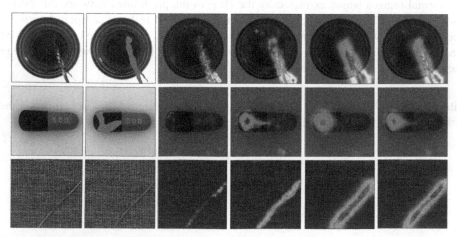

Fig. 3. The anomaly scoring maps of different hierarchies. First columns are the input images with anomalies. Second columns are the ground-truth in red color. The last four columns are respectively the anomaly scoring maps of hierarchy 1, 2, 3 and the combination of 1&2&3.

As showed in Table 6, the average results among 15 categories on MVTec AD dataset for anomaly detection and segmentation are reported. The combination of the hierarchies performs well on the two tasks. And the deeper hierarchy has the more discriminability on anomaly detection, but may damage the performance for anomaly segmentation due the reduce of the resolution of the feature map. The combination of the hierarchies 2 and 3 shows the best performance on the anomaly detection on MVTec AD dataset. The combination of deeper hierarchies can be well applied to the anomaly detection. The shallow hierarchies can be applied to make up the issue of the reduce of resolution of the deeper hierarchies when segmentation required. The Fig. 3 shows some qualitative results of different hierarchical scoring maps for testing images.

5 Conclusion

This paper proposes the PDDF method for unsupervised image anomaly detection and segmentation. Only the normal images are used to partially distill the knowledge of the features extracted by the pre-trained network at training. The head of the pre-trained teacher does not participate in the knowledge distillation and is shared to the student network to extract the low-level features. The discrepancy between the distilled feature with the pre-trained one is applied to detect anomaly at testing. With the experiments on MVTec AD, MNIST and Fashion-MNIST datasets, the PDDF achieves the comparing results compared with the state-of-the-art methods.

Acknowledgement. This work was supported by the National Key R&D Program of China under Grant No. 2018AAA0101704, and the Program for HUST Academic Frontier Youth Team 2017QYTD04.

References

1. Gao, Y., Gao, L., Li, X.: A generative adversarial network based deep learning method for low-quality defect image reconstruction and recognition. IEEE Trans. Industr. Inf. **17**, 3231–3240 (2021)
2. Wen, L., Li, X., Gao, L., Zhang, Y.: A new convolutional neural network-based data-driven fault diagnosis method. IEEE Trans. Industr. Electron. **65**, 5990–5998 (2018)
3. Ruff, L., et al.: A unifying review of deep and shallow anomaly detection. In: Proceedings of the IEEE, pp. 1–40 (2021)
4. Bergmann, P., Fauser, M., Sattlegger, D., Steger, C.: MVTec AD — a comprehensive real-world dataset for unsupervised anomaly detection. In: IEEE Conference on Computer Vision and Pattern Recognition (CVPR), pp. 9592–9600 (2019)
5. Bergmann, P., Lowe, S., Fauser, M., Sattlegger, D., Steger, C.: Improving unsupervised defect segmentation by applying structural similarity to autoencoders. In: 14th International Joint Conference on Computer Vision, Imaging and Computer Graphics Theory and Applications (VISAPP), pp. 372–380 (2019)
6. Schlegl, T., Seeböck, P., Waldstein, S.M., Schmidt-Erfurth, U., Langs, G.: Unsupervised anomaly detection with generative adversarial networks to guide marker discovery. In: Niethammer, M., et al. (eds.) IPMI 2017. LNCS, vol. 10265, pp. 146–157. Springer, Cham (2017). https://doi.org/10.1007/978-3-319-59050-9_12

7. Venkataramanan, S., Peng, K.-C., Singh, R.V., Mahalanobis, A.: Attention guided anomaly localization in images. In: Vedaldi, A., Bischof, H., Brox, T., Frahm, J.-M. (eds.) ECCV 2020. LNCS, vol. 12362, pp. 485–503. Springer, Cham (2020). https://doi.org/10.1007/978-3-030-58520-4_29
8. Dehaene, D., Frigo, O., Combrexelle, S., Eline, P.: Iterative energy-based projection on a normal data manifold for anomaly localization. In: International Conference on Learning Representations (ICLR) (2020)
9. Cohen, N., Hoshen, Y.: Sub-image anomaly detection with deep pyramid correspondences. arXiv preprint arXiv: 2005.02357 (2020)
10. Napoletano, P., Piccoli, F., Schettini, R.: Anomaly detection in nanofibrous materials by CNN-based self-similarity. Sensors **18**, 209 (2018)
11. Bergmann, P., Fauser, M., Sattlegger, D., Steger, C.: Uninformed students: student-teacher anomaly detection with discriminative latent embeddings. In: IEEE Conference on Computer Vision and Pattern Recognition (CVPR) (2020)
12. Yi, J., Yoon, S.: Patch SVDD: patch-level SVDD for anomaly detection and segmentation. In: Ishikawa, H., Liu, C.-L., Pajdla, T., Shi, J. (eds.) ACCV 2020. LNCS, vol. 12627, pp. 375–390. Springer, Cham (2021). https://doi.org/10.1007/978-3-030-69544-6_23
13. Salehi, M., Sadjadi, N., Baselizadeh, S., Rohban, M., Rabiee, H.: Multiresolution knowledge distillation for anomaly detection. arXiv preprint arXiv: 2011.11108 (2020)
14. Wang, G., Han, S., Ding, E., Huang, D.: Student-teacher feature pyramid matching for unsupervised anomaly detection. arXiv preprint arXiv: 2103.04257 (2021)
15. He, K., Zhang, X., Ren, S., Sun, J.: Deep residual learning for image recognition. In: IEEE Conference on Computer Vision and Pattern Recognition (CVPR), pp. 770–778 (2016)
16. Simonyan, K., Zisserman, A.: Very deep convolutional networks for large-scale image recognition. In: International Conference on Learning Representations (ICLR) (2015)
17. Selvaraju, R.R., Cogswell, M., Das, A., Vedantam, R., Parikh, D., Batra, D.: Grad-CAM: visual explanations from deep networks via gradient-based localization. In: IEEE International Conference on Computer Vision (ICCV), pp. 618–626 (2017)
18. Russakovsky, O., et al.: ImageNet large scale visual recognition challenge. Int. J. Comput. Vision **115**(3), 211–252 (2015). https://doi.org/10.1007/s11263-015-0816-y
19. LeCun, Y., Corinna Cortes.: The mnist database of handwritten digits. (2005).
20. Xiao, H., Rasul, K., Vollgraf, R.: Fashion-MNIST: a novel image dataset for benchmarking machine learning algorithms. arXiv preprint arXiv: 1708.07747 (2017)
21. Chen, Y., Zhou, X., Huang, T.: One-class SVM for learning in image retrieval. In: Proceedings 2001 International Conference on Image Processing, vol. 1, pp. 34–37 (2001)
22. Ruff, L., et al.: Deep one-class classification. In: ICML (2018)
23. Zhai, S., Cheng, Y., Lu, W., Zhang, Z.: Deep structured energy based models for anomaly detection. In: ICML (2016)
24. Zong, B., et al.: Deep autoencoding gaussian mixture model for unsupervised anomaly detection. In: ICLR (2018)
25. Salehi, M., et al.: ARAE: adversarially robust training of autoencoders improves novelty detection. arXiv preprint arXiv: 2003.05669 (2020)
26. Abati, D., Porrello, A., Calderara, S., Cucchiara, R.: Latent space autoregression for novelty detection. In: IEEE/CVF Conference on Computer Vision and Pattern Recognition (CVPR), pp. 481–490 (2019)

Speech Recognition Method for Home Service Robots Based on CLSTM-HMM Hybrid Acoustic Model

Chenxin Zhao, Xiaohua Wang$^{(\boxtimes)}$, and Lei Zhang

School of Electronic Information, Xi'an Polytechnic University, Xi'an 710048, China
wangxiaohua@xpu.edu.cn

Abstract. In this paper, we propose a speech recognition method based on CLSTM-HMM hybrid acoustic model for home service robots that cannot recognize speech commands effectively in domestic noise. Hybrid speech features characterize dynamic and static features, which can characterize speech features well, CNN has advantages in selecting good features, LSTM and HMM are usually used for speech recognition, which both methods have demonstrated strong capabilities. The experimental results prove that the recognition rate of home service robots based on CLSTM-HMM speech recognition method can reach 91.44%, 91.36% and 90.92%, respectively. Over all, the home service robots based on CLSTM-HMM speech recognition method have ideal recognition performance in domestic noise.

Keywords: Speech recognition · Acoustic model · Home service robot · Domestic noise

1 Introduction

Nowadays robots have penetrated into our daily life [1]. Affected by the diversity of domestic noise, the home service robots cannot effectively recognize voice commands. Therefore, the speech recognition performance of home service robots needs to be effectively improved [2]. Effective speech features and acoustic models are the key problems in speech recognition, which have important effects on the speech recognition performance of home service robots in the domestic noise.

Meir frequency coefficient (MFCC) [3] and Gammatone frequency cepstral coefficient (GFCC) [4] are two commonly used methods for speech feature extraction. MFCC cannot effectively extract speech features in noise environments, and MFCC features can only characterize the static properties of speech signals. GFCC cannot obtain comprehensive speech information. Although the hybrid features of MFCC and GFCC proposed improve the speech recognition performance to certain extent [5], the high computation complexity of the system is caused by the multiple feature dimensions. The method based on MFCC and Teager energy operator (TEO) [6] proposed in the literature extracts a mixture of static and dynamic features.

© Springer Nature Switzerland AG 2021
D.-S. Huang et al. (Eds.): ICIC 2021, LNCS 12836, pp. 251–263, 2021.
https://doi.org/10.1007/978-3-030-84522-3_20

Acoustic modeling is an important part of speech recognition systems. The traditional mainstream approach is represented by GMM-HMM [7], but with the diversification of sound environments, the traditional acoustic models can no longer effectively recognize speech signals in complex environments. Novoa proposed a DNN-HMM acoustic model to avoid the problem of unrecognized speech signals during actual human-computer interaction [8]. Liu proposed a long short-term memory recurrent neural network with HMM combined with a deep bidirectional model, and this system achieved better results than DNN-HMM in speech recognition [9]. Tian et al. used a hybrid acoustic model based on CNN, RNN and HMM [10], and reduced the recognition error rate of the speech system by using RNN to process the contextual information between adjacent speech frames. In order to im-prove the performance of speech recognition systems, researchers have explored various aspects such as acoustic model training scale and feature dimensionality. Bukhari et al. proposed a hybrid acoustic model combining CNN, RNN and HMM to process the state related to upper and lower speech frames using RNN and achieve good recognition results [11]. Graves et al. proposed an acoustic model combining CTC and LSTM to achieve end-to-end speech recognition [12]. However, LSTM is subject to overfitting during training, and Billa et al. solved the overfitting problem during training well by adding Dropout regularization [13]. Zhang proposed the SCBAMM model, which solves the problem of gradient explosion and gradient disappearance to some extent by using the spatial information of the features [14]. Toktam proposed an adaptive windowed multi-depth residual network (AMRes), where the residual network (ResNet) is combined with an acoustic model to effectively reduce the speech recognition error rate by exploiting the depth information of the speech signal [15].

In this paper, we adopt the method of Yao [16] to extract hybrid speech features according to the characteristics of domestic noise. To better overcome the diversity of speech signals, we use the CLSTM-HMM hybrid acoustic model for recognition, while adding Dropout to reduce the overfitting problem. We will verify whether the speech recognition method based on CLSTM-HMM hybrid acoustic model has good recognition effect in real home life through noise experiments in domestic environment. The final experiment results indicated that the recognition rate of the method reached 91.44%, 91.36% and 90.92 under different domestic noise environments, respectively. Over all, the recognition effect of home service robots based on the speech recognition method in this paper in domestic noise is the focus of this study.

The remainder of this paper is organized as follows. Hybrid speech features TEOGFCC + ΔTEOGFCC extraction method is described in Sect. 2. CLSTM-HMM hybrid acoustic model is proposed in Sect. 2. The experiment results of home service robots based on this paper method in domestic noise are shown in Sect. 3. Conclusion and future works are provided in Sect. 4.

2 Hybrid Speech Features TEOGFCC + ΔTEOGFCC Extraction Method in Domestic Environment

2.1 Related Basic Theories

Gammatone Filter

The gammatone filter has some noise immunity [17] and is able to analyze speech commands efficiently. In a home noisy environment, the speech commands are passed through a 64-channel gammatone filter bank and the short-time energy is calculated for each frame. Its step response in the time domain can be expressed as follows.

$$g_i(t) = \alpha t^{n-1} exp(-2\pi b_i t) cos(2\pi f_i + \phi_t) U(t), t \geq 0, 1 \leq i \leq N \tag{1}$$

where α denotes the amplitude of the Gammatone filter, n denotes the order of the filter, N denotes the number of filters, f_i denotes the centre frequency of the filter, ϕ_i denotes the initial phase of the filter, usually $\phi_i = 0$, $U(t)$ denotes the unit step function, b_i denotes the attenuation factor of the Gammatone filter.

$$E_{ERB}(f_i) = 24.7 \times \left(4.37 \times \frac{f_i}{1000} + 1\right) \tag{2}$$

Applying the Laplace transform to Eq. (4).

$$G_i(s) = \frac{A}{2}\left[\frac{(n-1)!}{(s+b-jw)^n} + \frac{(n-1)!}{(s+b+jw)^n}\right] \tag{3}$$

According to the experience of previous experiments, the Gammatone filter order is $n = 4$, so 64 Gammatone filters are taken and superimposed into a Gammatone filter bank, the impact response function of the Gammatone filter bank as follow.

$$g_i(n) = \frac{1}{2\pi j}\int_{-\infty}^{+\infty} G_i(z)z^{n-1}dz \tag{4}$$

Teager Energy Operator (TEO)

TEO is a nonlinear difference operator proposed by Teager et al. [18], which serves to characterize the energy transform of the signal and the instantaneous energy value. In a domestic noisy environment, TEO provides a good estimate of the "real" source of energy and guarantees that the system will be more robust in domestic noise environment.

For a discrete speech signal, Teager defined it as.

$$T[x(n)] = x^2(n) - x(n+1)x(n-1) \tag{5}$$

Where $T[x(n)]$ is the output of the Teager and $x(n)$ is the sampled value of the speech signal at the point. In a noisy environment, the voice command signal $x(n)$ is the sum of the pure voice signal $s(n)$ and the noisy voice signal $w(n)$, i.e.

$$x(n) = s(n) + w(n) \tag{6}$$

TEO of a voice signal can be expressed as:

$$T[x(n)] = T[s(n)] + T[w(n)] + 2T[s(n), w(n)] \qquad (7)$$

Where $T[s(n), w(n)] = s(n)w(n) - 0.5s(n-1)w(n+1) - 0.5s(n+1)w(n-1)$ is defined as the mutual Teager energy of $s(n)$ and $x(n)$, since $s(n)$ and $w(n)$ are independent of each other and are zero-mean, which yields.

$$T[x(n)] = T[s(n)] + T[w(n)] \qquad (8)$$

We add the TEO to the speech features extraction method, which can not only reflect the energy change of the speech signal, but also achieve the effect of speech enhancement.

2.2 Hybrid Speech Features TEOGFCC + ΔTEOGFCC Extraction Based on GFCC and TEO

We adopt GFCC and TEO hybrid speech features extraction method to extract hybrid speech features TEOGFCC + ΔTEOGFCC. The steps are as follows.

The steps for extracting hybrid speech features TEOGFCC + ΔTEOGFCC are as follows.

Step 1. Pre-emphasis, framing and windowing are performed on the captured speech signal $x(n)$.

$$y(n) = x(n) - \alpha(n-1) \qquad (9)$$

Where $x(n)$ represents the sampled value of the speech signal at n moments of the speech signal and the pre-emphasis factor $\alpha = 0.98$.

The speech signal is smooth in a short time, so we are required to frame it, and also to reduce the edge effect of the speech frame, we add the Hamming window to the framed speech signal.

Step 2. The Teager energy value is obtained for each frame of the pre-processed speech signal.

Step 3. The output of step 2 is FFT [19] transformed to convert the time domain signal to frequency domain signal to obtain the energy spectrum value $Y(t, i)$ for each frame of the speech signal.

Step 4. The $Y(t, i)$ is filtered through a Gammatone filter bank, and the output of each filter bank is logarithmically compressed to obtain a set of logarithmic energy spectra, further simulating the non-linear nature of human perception of speech.

Step 5. The above logarithmic results are subjected to the DCT [20] to reduce the correlation between the dimensional feature parameters due to filter overlap and obtain the speech features TEOGFCC, which are calculated as follows.

$$C_{TEOGFCC}(i) = \sum_{j=1}^{N} m_j cos\left[\frac{\pi i}{N}(j - 0.5)\right], i = 1, 2, L, \ldots, M \qquad (10)$$

Where M denotes the feature dimension and N is the number of Gammatone filters to obtain the feature distribution of each speech signal frame.

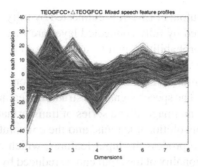

Fig. 1. Speech signal 'Backward' in Mandarin Hybrid speech features TEOGFCC + △TEOGFCC.

Step 6. Speech features TEOGFCC for first-order difference to get speech features △TEOGFCC, and mix TEOGFCC and △TEOGFCC by PCA algorithm [21] to get hybrid speech features TEOGFCC + △TEOGFCC.

In the process of extracting hybrid speech features, the speech feature TEOGFCC can only reflect static features, therefore, we added the feasible region parameter △TEOGFCC characterizing dynamic features. we used PCA algorithm to perform principal component analysis on the high-dimensional original matrix for the purpose of dimensionality reduction [22], and according to the obtained new TEOGFCC and △TEOGFCC the contribution rate of each inverse spectral coefficient, and the speech features with higher contribution rate are formed into the hybrid speech features TEOGFCC + △TEOGFCC.

3 Speech Recognition Based on CLSTM-HMM

3.1 Network Structure of the CLSTM-HMM Hybrid Speech Acoustic Model

We establish the following hybrid acoustic model and the network structure shown in Fig. 2.

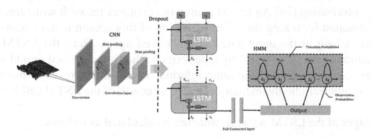

Fig. 2. Network structure of the CLSTM-HMM hybrid acoustic model.

The input of acoustic model is 24-dimensional hybrid speech features TEOGFCC + △TEOGFCC. CNN extracts spatial information through weight sharing in convolutional layer and max-pooling in pooling layer for the purpose of reducing the feature parameters in CNN. The output of LSTM is connected by fully connected layer, then the feature parameters extracted by fully connected layer are fused and normalized to output various classification probabilities, finally softmax outputs the classification results. hmm performs forced alignment on the output results.

CNN can be divided into three parts such as convolutional layer, pooling layer and fully connected layer [23]. The speech signal has structural features in the time-frequency spectrum and is viewed as an image after a series of transformations. The convolutional layer contains multiple convolutional kernels, and the convolutional layer applies a set of convolutional kernels to process the input features, which can share weights in the network [24]. The dimensionality of the input can be reduced by maximum pooling [25]. The maximum pooling takes the maximum value of the pixels in the area covered by the pooling window to obtain the pixel values of the output feature maps. The structure of the CNN is shown in Fig. 3.

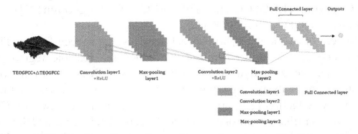

Fig. 3. Convolutional neural network structure diagram.

The formula is as follows.

$$B = f(A * \mu + b) \tag{11}$$

Where B represents the voice feature map after convolution, f is the activation function, A is the original voice feature matrix, μ is the convolution kernel, $*$ represents the convolution operation, and b is the bias amount.

LSTM is a special kind of recurrent neural network that learns long-term deprogramming information [26] An LSTM neuron is a complex network structural unit that stores information for a long time. The activation of this neuron is not overwritten by new inputs as long as the input gate remains closed again. Thus, the LSTM "gates" the information of the "unit state" by time series so that it is removed or added. These three gates selectively control the input information and the stored information from the previous moment into the memory cell. The structure of the LSTM cell is shown in Fig. 4.

The output of the LSTM network structure is calculated as follows.

$$i_t = \sigma(W_{xi}x_t + W_{hi}h_{t-1} + W_{ci}c_{t-1} + b_i) \tag{12}$$

$$f_t = \sigma(W_{xf}x_t + W_{hf}h_{t-1} + W_{cf}c_{t-1} + b_f) \tag{13}$$

$$c_t = f_t c_{t-1} + i_t tanh(W_{xc}x_t + W_{hc}c_{t-1} + b_c) \tag{14}$$

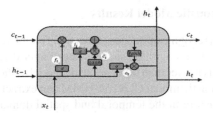

Fig. 4. Structure of LSTM units.

$$o_t = \sigma(W_{xo}x_t + W_{ho}h_{t-1} + W_{co}c_{t-1} + b_o) \tag{15}$$

$$h_t = o_t \tanh(c_t) \tag{16}$$

Where $x = \{x_1, x_2, \cdots, x_i, \cdots, x_N\}$, N is the number of samples of the input speech signal, σ is the sigmoid activation function, b_i, b_f, b_o and b_c, are the biases of the corresponding gates and W is the matrix of weights between the gates.

3.2 Training of the CLSTM-HMM Hybrid Acoustic Model

In this paper, we build a hybrid acoustic model CLSTM-HMM on Windows system using TensorFlow framework and Python. The hybrid acoustic model is experimentally validated on the THCHS-30 Chinese dataset. The THCHS-30 Chinese dataset is 30 h of speech data from 40 people in a quiet office environment [27], including 30 people's speech sample training set and 10 persons' speech sample test set.

We train the proposed CLSTM-HMM hybrid acoustic model to obtain the corresponding acoustic model. The training steps of CLSTM-HMM hybrid acoustic model are as follows.

Step 1: Extract the hybrid speech features TEOGFCC + ΔTEOGFCC. Initialize the model parameters. Set the training number epoch to 50 times, and set the learning rate to 0.001.

Step 2: CNN consists of 2 convolutional layers, 2 pooling layers and 2 fully connected layers. Set the number of convolutional kernels of CNN to 64, the size and number of convolutional kernels to 256, the step size to 1, and choose Relu as the activation function. Set the pooling layer to max-pooling.

Step 3: The pooled feature parameters are input to the LSTM through the fully connected layer to extract the temporal features, the number of nodes of the LSTM is set to 256, and the weight of Dropout is set to 0. 5.

Step 4: Dropout is added between the nodes of the LSTM to prevent overfitting. Update the weights and biases in the model, while adding 1 to the number of iterations.

Step 5: Judge whether the error value reaches the set threshold or whether the number of iterations for acoustic model optimization reaches the preset value.

Step 6: HMM models the timing of the speech signal and forces the alignment of the output values.

3.3   Comparison of Acoustic Model Results

Table 1 shows the speech recognition success rates of different acoustic models. According to the data in Table 1, the recognition success rate of hybrid acoustic model CLSTM-HMM is 92.16%, which is higher than other acoustic models. The following conclusions are drawn: the hybrid acoustic model CLSTM-HMM can extract deeper speech features by virtue of its deep structure in the temporal and spatial domains, while the LSTM is used to process the contextual information between adjacent speech frames to effectively solve the problems of gradient disappearance and gradient explosion, and the Dropout is added during the training process to reduce overfitting, which makes the hybrid acoustic model has better recognition effect than other acoustic models.

Table 1.   Comparison of speech recognition rates for different acoustic models in the THCHS-30 dataset.

Acoustic models	Identification rate/%
LSTM-HMM	89.73%
GMM-HMM	86.38%
CNN-HMM	88.47%
RNN-HMM	89.29%
DNN-HMM	87.26%
CLSTM-HMM	92.16%

The following are the results of the effect of training time on different acoustic models.

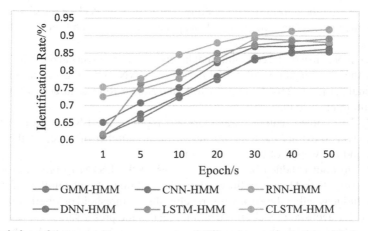

Fig. 5.   Variation of the recognition success rate of different acoustic models with the number of iterations (Epoch).

The changes of recognition rates after several iterations of different acoustic models are shown in Fig. 5. The results show that the recognition success rate of CLSTM-HMM is 6.73% higher than that of GMM-HMM, which indicates that CLSTM-HMM has better recognition performance. The addition of Dropout in CLSTM-HMM reduces the overfitting during training, ensures the rapid convergence of the model after several iterations, and greatly improves the recognition performance of the model.

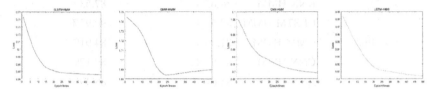

Fig. 6. Acoustic model loss curves on the THCHS-30 dataset.

For evaluating the computation performance of the method, we compare the loss function curves of different acoustic models during the training process. As shown in Fig. 6, it demonstrates the training time of 50 epochs for four typical acoustic models. CLSTM-HMM converges the fastest, converges the best, and has the highest recognition accuracy for speech commands.

4   Experiment of Home Service Robots

4.1   Experimental Results and Analysis

In the noise experiment, the speech recognition effect of home service robots based on hybrid acoustic model CLSTM-HMM speech recognition method is experimented by comparing the recognition rates of different acoustic models.

In modern homes, according to the investigation of indoor noise sources, indoor noise mainly includes traffic noise, household appliance noise and industrial machinery noise. In order to simulate the actual noise environment, white noise, pink noise and volvo noise from the Noisex-92 standard library [28] are used to simulate the noise in the home environment, and the speech messages with signal-to-noise ratios of 10 dB and 20 dB are set for all speech commands.

White noise is the noise with equal noise power spectral density contained in each equal-bandwidth frequency band over a wide range of frequencies, so we simulate domestic appliances noise in a domestic environment with white noise. Pink noise has equal intensity at each octave, i.e., it has the same or similar energy in a certain range (octave), so it is used to simulate industrial machinery noise. Volvo noise is directly defined as car noise, and it is used to simulate traffic noise.

The experiment results are shown in Table 2. The recognition effect of the home service robot based on the speech recognition method of CLSTM-HMM model is better than that of the home service robot based on other methods under different signal-to-noise ratios. The recognition rates reach 90.92% under white noise, 91.44% under pink noise, and 91.36% under Volvo noise respectively.

Table 2. Recognition rate of different acoustic models in home noise environment

Noise type	Acoustic model	Signal-to-noise ratio/dB	
		10 dB	20 dB
White	GMM-HMM	81.34%	84.75%
	CNN-HMM	85.75%	88.16%
	RNN-HMM	84.83%	87.53%
	CLSTM-HMM	88.51%	90.92%
Pink	GMM-HMM	81.86%	84.91%
	CNN-HMM	86.54%	89.17%
	RNN-HMM	85.64%	88.49%
	CLSTM-HMM	88.73%	91.44%
Volvo	GMM-HMM	75.91%	83.47%
	CNN-HMM	85.62%	88.06%
	RNN-HMM	84.35%	87.86%
	CLSTM-HMM	88.97%	91.36%

At the meantime, a comparison with previous studies indicates that our method still has good results. The recognition rate of HMM/ANN [29] is 79.2% at 10 db and 89.1% at 20 db. Compared with its method, CLSTM-HMM hybrid acoustic model-based speech recognition method we studied is more noise resistant and accurate in the domestic environment.

The reason is that the hybrid speech feature TEOGFCC + ΔTEOGFCC extraction method can extract the hybrid speech features that can better represent the speech characteristics under certain noise environment. the ability of CNN to extract deep features and the LSTM speech network structure not only reduce the training time of acoustic models, but also overcome the diversity of speech signals well.

4.2 Simulation Results

In this chapter, the home service robots based on CLSTM-HMM hybrid acoustic model speech recognition method is simulated in domestic noise. We set up a 14.04 LTS Ubuntu system and an Indigo ROS system for this simulation.

The communication mechanism of ROS is used to connect the individual modules together to complete the voice commands control system. The communication relationship between the nodes is shown in the Fig. 7.

According to Fig. 7, the box indicates the message, the ellipse indicates the node, and the rectangle indicates the topic. /audio_capture to get the voice commands control signal captured through microphone and publish the message to /microphone topic; /wakeup_node subscribes to //microphone topic, detecting whether the input voice commands control signal is a valid voice commands to start the speech recognition system, and then, sends the valid voice commands control signal as a message to the /asr_topic

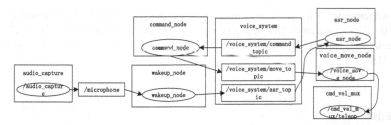

Fig. 7. The node relationship diagram of speech control system in ROS system.

topic; /asr_node subscribes to the /asr_topic topic, sends the detected valid voice commands control signal to the language library for matching, and publishes the message to the /command_topic topic; node /command_node subscribes to the /command_topic topic, and after The node /voice_node subscribes to the /command_topic topic, receives voice commands and transmits them to the robot to control the movement of the robot.

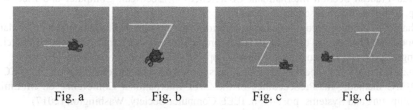

| Fig. a | Fig. b | Fig. c | Fig. d |

Fig. 8. Simulation results of the speech-controlled robot.

The robot recognizes the voice commands and completes the corresponding actions by the speech recognition method in this paper. The results are shown in Fig. 8. Fig. a, b, c, and d show the results of the robot recognizing voice commands "Forward", "Turn right", "Turn left", and "Backward". And complete the corresponding actions. According to Fig. 8, it can be shown for different voice commands, the robot can complete the corresponding actions.

5 Conclusion

We propose a speech recognition method based on a hybrid CLSTM-HMM acoustic model to address the inability of home service robots to recognize speech commands from domestic noise. Hybrid speech features TEOGFCC + △TEOGFCC well represent the speech characteristics, CNN has excellent feature extraction ability, and LSTM and HMM show powerful ability in speech recognition. The experiment in domestic noise shows that the speech recognition method of home service robot based on CLSTM-HMM hybrid acoustic model has satisfying recognition effect in domestic noise.

In this paper, the response time is also one of the factors that affect the effectiveness of speech control for home service robots, so this study will be the focus of our future work.

Acknowledgements. This work was supported by Natural Science Foundation of China [51905405]; Key research & Development plan of Shaanxi province China [2019ZDLGY01-08]; Ministr of Education Engineering Science and Technology Talent Training Research Project of China [18JDGC029].

References

1. Chivarov, N., Chikurtev, D., Pleva, M., Ondas, S.: Exploring human-robot interfaces for service mobile robots. In: 2018 World Symposium on Digital Intelligence for Systems and Machines (DISA), Kosice, pp. 337–342 (2018)
2. Liu, C.-Y., Hung, T.-H., Cheng, K.-C., Li, T.-H.S.: HMM and BPNN based speech recognition system for home service robot. In: 2013 International Conference on Advanced Robotics and Intelligent Systems (2013)
3. Kumar, A., Rout, S.S., Goel, V.: Speech mel frequency cepstral coefficient feature classification using multi level support vector machine. In: Proceedings of the 2017 4th IEEE Uttar Pradesh Section International Conference on Electrical, Computer and Electronics, pp. 134–138. IEEE, Piscataway (2017)
4. Shahin, I.: Emotion recognition based on third-order circular suprasegmental hidden Markov model. In: Proceedings of the 2019 IEEE Jordan International Joint Conference on Electrical Engineering and Information Technology, pp. 800–805. IEEE, Poscataway (2019)
5. Tazi, E.B.: A robust speaker identification system based on the combination of GFCC and MFCC methods. In: Prcoceedings of the 2017 International Conference on Multimedia Computing and Systems, pp. 54–58. IEEE Computer Society, Washington (2017)
6. Can, G., Akabas Cetin, A.E.: Recognition of acoustic signatures using non-linear teager energy based features. In: Proceedings of the 2016 International Workshop on Computational Intelligence for Multimedia Understanding, p. 78011900. IEEE, Piscataway (2016)
7. Van Hai, D., Chen, N.F., Lim, B.P.: Multi-task learning for phone recognition of under-resourced languages using mismatched transcription. IEEE/ACM Trans. Audio Speech Lang. Process. **PP**(99), 1 (2017)
8. Novoa, J., Wuth, J., Escudero, J.P., Fredes, J., Mahu, R., Yoma, N.B.: DNN-HMM Based Automatic Speech Recognition for HRI Scenarios (2018). 9781450349536
9. Liu, Y.: Research on continuous speech recognition with large vocabulary based on deep learning. Chongqing University of Posts and Telecommunications (2018)
10. Ying, T.: Research on acoustic event detection method based on CRNN-HMM. Shenyang University of Technology (2019)
11. Bukhari, D., Wang, Y., Wang, H.: Multilingual convolutional, long short-term memory, deep neural networks for low resource speech recognition. Procedia Comput. Sci. **107**, 842–847 (2017)
12. Graves, A., Jaitly, N.: Towards end-to-end speech recognition with recurrent neural networks. In: International Conference on Machine Learning, pp. 1764–1772 (2014)
13. Billa, J.: Dropout approaches for LSTM based speech recognition systems. In: 2018 IEEE International Conference on Acoustics, Speech and Signal Processing (ICASSP), Calgary, AB, pp. 5879–5883 (2018)
14. Zhang, H., Huang, H., Han, H.: Attention-based convolution skip bidirectional long short-term memory network for speech emotion recognition. IEEE Access **9**, 5332–5342 (2021). https://doi.org/10.1109/ACCESS.2020.3047395
15. Zoughi, T., Homayounpour, M.M., Deypir, M.: Adaptive windows multiple deep residual networks for speech recognition. Expert Syst. Appl. **139**, 112840 (2020), ISSN 0957-4174

16. Xiaohua, W., Pengchao, Y., Liping, M., Wenjie, W., Lei, Z.: Feature extraction algorithm for robot speech control in workshop environment. J. Xi'an Univ. Electron. Sci. Technol. **47**(02), 16–22 (2020)
17. Lian, H.L., Zhou, J., Hu, Y.T., Zheng, W.M.: Conversion of whispered speech to normal speech using deep convolutional neural networks. J. Acoustics **45**(01), 137–144 (2020)
18. Zhang, Q.Q., Liu, Y., Pan, J.L., et al.: Continuous speech recognition based on convolutional neural networks. J. Eng. Sci. **37**(09), 1212–1217 (2015)
19. Peng, Y.L., Li, R., Ma, X.H., Li, Y.B.: Harmonic detection algorithm by Hanning dual-window full-phase FFT tri-spectral line interpolation. Power Sci. Eng. **37**(04), 25–29 (2021)
20. Li, X.: Research on speech feature parameter extraction method. Xi'an University of Electronic Science and Technology (2006)
21. Bai, J., Shi, Y.Y., Xue, P.Y., et al.: Fusion of nonlinear power function and spectral subtraction method for CFCC feature extraction. J. Xi'an Univ. Electron. Sci. Technol. **46**(1), 86–92 (2019)
22. Bandela, S.R., Kumar, T.K.: Stressed speech emotion recognition using feature fusion of teager energy operator and MFCC. In: Proceedings of the 2017 8th International Conference on Computing, Communications and Networking Technologies, Piscataway, p. 8204149. IEEE (2017)
23. Monteiro, R.L.D.C., Pererra, V., Costa, H.G.: Analusis of the better life index trough a cluster algorithm. Soc. Indicators Res. **142**(2), 477–506 (2019)
24. Liu, J.H.: Chinese Speech Recognition based on Deep Convolutional Neural Network. Taiyuan University of Technology, Taiyuan (2019)
25. Fan, Y.Y.: A Deep Learning Approach for Speaker Recognition . Nanchang University of Aviation, Nanchang (2019)
26. Huang, R.: Experimental design of Chinese speech recognition based on Bi-RNN. Modern Comput. (Professional Edition) **10**, 92–95 (2019)
27. Sun, L., Du, J., Dai, L.R., et al.: Multiple-target deep learning for LSTM-RNN based speech enhancement. In: 2017 Hands-free Speech Communications and Microphone Arrays (HSCMA). IEEE (2017)
28. Zhang, Z.T.: Research on speech recognition technology based on wavelet and PNCC feature parameters. Chongqing University (2018)
29. Feng, Y.: Speech recognition technology based on the study of HMM and DNN hybrid model. Hebei University of Science and Technology (2020)

Serialized Local Feature Representation Learning for Infrared-Visible Person Re-identification

Sizhe Wan[1,5]([⊠]), Changan Yuan[2,5], Xiao Qin[3,5], and Hongjie Wu[4,5]

[1] Institute of Machine Learning and Systems Biology, School of Electronics and Information Engineering, Tongji University, Shanghai 201804, China
[2] Guangxi Academy of Science, Nanning 530025, China
[3] School of Computer and Information Engineering, Nanning Normal University, Nanning 530299, China
[4] School of Electronic and Information Engineering, Suzhou University of Science and Technology, Suzhou 215009, China
[5] Institute of Science and Technology for Brain Inspired Intelligence (ISTBI), Fudan University, Shanghai 200433, China

Abstract. Infrared-Visible person re-identification is a kind of cross-modality person re-identification. The purpose of the task is that given a person image we need to find another image on the same person from gallery. The query images and gallery images are not only in RGB modality but in infrared modality as well. The cross-modality person ReID task can deal with the limitation of single modality because we usually can get images in more than one modality. In our work, we take advantage of both global feature and local feature. We use a dual-path structure to extract features from RGB images and infrared images respectively. Besides, we add the LSTM structure in each path to learn the serialized local features. The loss function consists of cross-entropy loss and hetero-center loss so that the model can bridge the cross-modality and intra-modality gaps to capture the modality-shared features and improve the cross-modality similarity. Finally, we do experiments on two datasets including SYSU-MM01 and RegDB, then compare with other methods in recent studies.

Keywords: Infrared-Visible person re-identification · Cross-modality · Dual-path structure · Serialized local feature learning

1 Introduction

Person re-identification (PReID) is a technology that uses computer vision in machine learning to determine whether the person images captured by different cameras are same. This technology is widely used in public security and intelligent video surveillance [1, 2]. In recent years, PReID has aroused great interest in the computer vision and it has made great progress. A person re-identification task can be divided into two parts, firstly, person tracking and image capture, and secondly person identification and retrieval.

© Springer Nature Switzerland AG 2021
D.-S. Huang et al. (Eds.): ICIC 2021, LNCS 12836, pp. 264–273, 2021.
https://doi.org/10.1007/978-3-030-84522-3_21

With the rapid development of deep learning and neural network technology, especially computer vision technology [3–5, 9, 22, 23, 25, 26, 31–33, 36], more and more works focus on identify the specific pedestrian on visible camera module [6–8]. However, with the deepening of the research, the researchers found that the image quality in real life is not really good. In addition, in real life, not all images taken by visible light cameras are all, and there are many other types of data describing pedestrians, especially infrared images taken by infrared cameras. When the recognition effect of visible light images was not satisfactory, the researchers thought of a method of combining infrared image. The infrared images are single-channel modality which have no color features. Because of the difference between RGB images and infrared images, the traditional models that trained on RGB images can not address the problem on cross-modality images, which is a challenging task.

Some methods have been proposed for the cross-modality PReID. Wu et al. contributed a dataset named SYSU-MM01 [10] and proposed a deep zero-padding model to address the cross-modality issue. Dai et al. proposed cmGAN [11] jointed triplet loss and cross-entropy loss. Ye et al. proposed TONE which use two-stream CNN architecture and hierarchical cross-modality metric learning (HCML) [12]. Ye et al. proposed a two-stream deep model for feature extracting with a bi-directional dual constrained top-ranking (DCTR) loss [13].

Obviously, we can see that most of methods ignore the discriminative local feature representations. Due to the differences among camera equipment, the appearance, which is a global feature, is easily affected by wearing, scale, occlusion, posture and viewing angle, etc. But the local characteristics of the same pedestrian are similar, which contains many visible details such as the type of clothes or the style of the shoes. Therefore, the local feature representations have significant discriminability and are not affected by the cross-modality modes [14].

In our work, we use a dual-path model which contains two individual branches to extract each modality features respectively, one is visible stream and the other is infrared stream. We use the ResNet50 [15] model as the backbone of the model. We use the attention mechanism to make the model extract local feature representations instead of global information from cross-modal images. We split the feature maps learned by the two branches into several stripes and use LSTM or Transformer to learn the local information. Finally, we do experiments on two public datasets SYSU-MM01 and RegDB and achieve great performance.

2 Related Works

Many approaches that deal with the cross-modality retrieval problem were proposed under such condition that there is a gap between RGB domains and infrared domains. Most of current studies present a cross-modality matching deep model in RGB domains.

Kai et al. proposed a pedestrian identification method in infrared images that only relied on local image features, combining re-identification with pedestrian detection and tracking methods [18]. Mφgelmose et al. proposed a pedestrian re-identification method that combines visible light image, depth information and thermal infrared image data [19]. Wu et al. found that both the dual-stream structure and the fully-connected structure can be represented by a single-stream network structure under special circumstances, and proposed a single-stream network using a deep zero-filling method [20]. In addition, Wu also created a benchmark dataset SYSU-MM01 [10] for pedestrian re-recognition in infrared images. Ye et al. proposed a dual-stream network learning identifiable feature representation with high-order loss with bidirectional constraints [12], taking into account the feature changes between and within modalities, and then combining feature loss and contrast loss to improve the dual-stream network, and proposed a hierarchical cross-modal learning method called HCML(Hierarchical cross-modality metric learning) [12]. Aiming at the problem of insufficient identification information for the same pedestrian between visible light and infrared images, Dai et al. designed a cross-modal generation confrontation network [21] and the CMC and mAP on the SYSU-MM01 dataset are 12.17% and 11.85% higher than the work of Wu et al., respectively. Wang et al. proposed a dual-level discrepancy reduction learning scheme, D2RL (Dual-level Discrepancy Reduction Learning scheme) [16]. They were the first to use the GAN network to synthesize a multi-spectral image of two types of images of visible light and infrared images for person re-identification under infrared conditions. Zhu et al. design a novel loss function (named HC Loss) [17] to constrain the distance between two centers of heterogenous modality. Based on that, Wu et al. propose a novel attention mechanism named Position Attention-guided Learning Module (PALM) to capture long-range dependencies and enhance the discriminative local feature representations of heterogenous modality.

3 Proposed Method

We will introduce the model of our proposed framework and methods, as shown in Fig. 1. The proposed model mainly composed of three components: (1) Dual-path structure, (2) Serialized local feature learning, (3) Cross-entropy loss and hetero-center loss.

3.1 Dual-Path Structure

To extract the person features in the cross-modality problem, we use a dual-path local information structure as the basic architecture. As its name, there are two paths in this structure. One is a visible path for the RGB images and the other is an infrared path for the infrared images. And we use two ResNet-50 networks as the backbone for both paths respectively to extract person features in each modality. The ResNet-50 is pre-trained model, which has four res-convolution modules, we can also call them four stages. These modules are independent the two paths so that they can extract some specific person features in different modality independently.

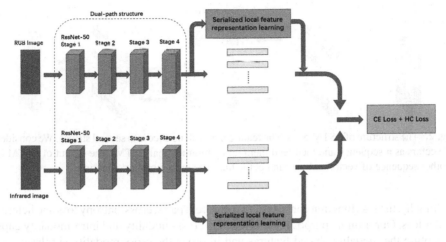

Fig. 1. The overall structure of the model. We use the dual-path structure as the basic architecture. We use pre-trained ResNet-50 networks as the backbone to extract modality feature. The bidirectional LSTM is used in the side brunch on each path to learn the local feature representation by part-based method. The loss function is composed of cross-entropy loss and hetero-center loss.

We get the global feature from the backbone but it is not enough. The feature map, which is the output from backbone, will be divided into several stripes horizontally and use average pooling and fully connected layer to get feature vectors. Then the vectors are merged with the features outputted from the side brunch, which is describe in next section. Finally, the merged feature vectors are feed into two weight-shared FC layers. At this step, the feature vectors from two different modalities are cross connected, benefits the following operation that bridge the gap between two modalities, which is reflected in the loss function.

3.2 Serialized Local Feature Representation Learning

In order to learn the serialized local feature representation, we use the part-base approach. An image of a person can be divided into several horizontal stripes. So that the person can be divided into stripes according to his or her body parts. The horizontal stripes which contain parts of the person body as well as local feature can be considered as a sequence. To learn the sequential feature representation, we finally decide to choose RNN which base on LSTM units. By using RNN model, the feature represent can be learned with spatial contexts. In the LSTM cell, there are three different gates and they can forget the information in contexts which is not important and remain the information in important contexts. The structure of LSTM cell shows in the Fig. 2.

Specifically, we use two LSTM models that do not share their parameters on the low-level feature in each path respectively. For example, the feature extracted from the RGB path will be split into sequence and feed into a side brunch, which is built as bidirectional LSTM. The result feature vector will be merged into the feature vector that result from original path, and then calculate the loss. The infrared path has the same structure.

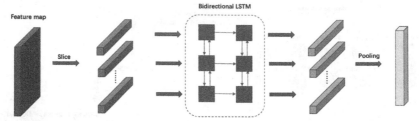

Fig. 2. The structure of LSTM parts. The feature map will be slice into several stripes. We consider all vectors as a sequence and then feed them into a bidirectional LSTM. The output of LSTM is another sequence of vector and we can get the final vector by pooling operation.

Finally, the loss function consists of two different parts, cross-entropy loss and hetero-center loss. Our training purpose is to bridge the cross-modality and intra-modality gaps to capture the modality-shared features and improve the cross-modality similarity. So that we use the cross-entropy loss (CE loss) for each path to learn the modality-specific features for classification. The CE loss function can be formulated as:

$$L_{CE} = -\sum\nolimits_{n=1}^{N} \log \frac{e^{W_{y_n}^T x_n + b_{y_n}}}{\sum_{i=1}^{I} e^{W_i^T x_n + b_i}} \qquad (1)$$

where N denotes the batch size, and W_i denotes the ith column of the weights, and b denotes the bias, and parameter I denotes the number of identities, and x_n denotes the nth visible and infrared features in the yth class.

Besides, we use the Hetero-Center (HC) loss to improve the cross-modality similarity. Hetero-Center (HC) loss depends on the center distance and the difference between two modality feature distributions. Hetero-Center (HC) loss is formulated as:

$$L_{HC} = \sum\nolimits_{m=1}^{M} \left[\left\| C_{m,1} - C_{m,2} \right\|_2^2 \right] \qquad (2)$$

where $C_{m,1} = \frac{1}{V} \sum_{n=1}^{V} X_{m,1,n}$ and $C_{m,2} = \frac{1}{I} \sum_{n=1}^{I} X_{m,2,n}$ is the centers of feature representations of visible images and infrared images in the ith class. V and I denote the numbers of visible images and infrared images in the ith class respectively, and M is the number of classes.

4   Experiments

4.1   Dataset Description

SYSU-MM01[10] is the largest dataset for cross-modality person Re-ID. There are 287,628 RGB images captured by four visible cameras and 15,729 infrared images captured by two infrared cameras. There 491 identities in the dataset and each of them is captured by at least one visible camera and one infrared camera. The whole dataset can be divided into two parts: training set and test set. In the training set, there are 22,258 RGB images and 11,909 infrared images, which contains 395 different identities. As for test set, there are 301 RGB images and 3,803 infrared images, which contains 96 different identities.

RegDB is captured by dual-camera systems [24]. There are 8,240 images in the dataset, which contains 412 identities. There are 206 identities in the training and 206 identities in the test set. There are 10 different RGB images and 10 different infrared images for each identity respectively.

The evaluation metrics which including Cumulative Matching Characteristics (CMC) and mean Average Precisions (mAP) are applied in all experiments.

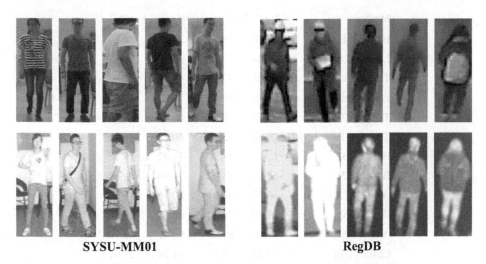

SYSU-MM01   **RegDB**

Fig. 3. Examples from the SYSU-MM01 and RegDB benchmarks.

4.2 Implementation Details

The experiments are deployed on a TITAN XP GPU. The deep learning framework is PyTorch. The backbone network is the pre-trained ResNet-50. The size of input person images are 288×144. We use data augmentation mechanism such as random erasing and horizontal random flipping methods. Learning rate is initially set to 0.01 and the momentum is set to 0.9. The batch size is set to 64. The number of person identities is set to 4 in a batch and there are 8 RGB images and 8 infrared images for each identity respectively.

4.3 Comparison with Other Methods

We do experiments on our method and compare the performance with other methods, including HOG [25], LOMO [26], Two-stream [10], Zero-Padding [10], TONE [12], TONE + HCML [12], BCTR [13], BDTR [13], cmGAN [11], D^2RL [16], AlignGAN [40], CMGN [43], HC Loss [34], JSIA-ReID [44], XIV [45].

Table 1. Comparisons on SYSU-MM01

Methods	Rank-1	Rank-10	Rank-20	mAP
HOG [25]	2.76	18.25	31.91	4.24
LOMO [26]	3.64	23.18	37.28	4.53
Two-stream [10]	11.65	47.99	65.50	12.85
Zero-Padding [10]	14.80	54.12	71.33	15.95
TONE [12]	12.52	50.72	68.60	14.42
TONE + HCML [12]	14.32	53.16	69.17	16.16
BCTR [13]	16.12	54.90	71.47	19.15
BDTR [13]	17.01	55.43	71.96	19.66
cmGAN [11]	26.97	67.51	80.56	27.80
D^2RL [16]	28.90	70.60	82.40	29.20
AlignGAN [29]	42.40	85.00	93.70	40.70
CMGN [30]	27.21	68.19	81.76	27.91
HC Loss [17]	56.96	91.50	96.82	54.95
JSIA-ReID [44]	38.10	80.70	89.90	36.90
XIV [45]	49.92	89.79	95.96	50.73
Ours	**54.88**	**90.27**	**95.87**	**54.10**

We evaluate our model on SYSU-MM01 dataset. Table 1 shows the comparative results on Rank-1, Rank-10, Rank-20 accuracy and mAP value of our model and other methods. In our model, Rank-1 accuracy is 54.88%, and Rank-10 accuracy is 90.27%, and Rank-20 accuracy is 95.87%, and mAP is 57.21%. Most of the results are higher than other methods that we list in the table.

We also evaluate our model on RegDB dataset. The results are showed in Table 2. In our model, Rank-1 accuracy is 85.05%, and Rank-10 accuracy is 96.94%, and mAP 72.87% performance. It can be seen that all the results is higher than other methods which we list in the table.

Table 2. State-of-the-art comparisons on RegDB

Methods	Rank-1	Rank-10	mAP
HOG [25]	13.49	33.22	10.31
LOMO [26]	0.85	2.47	2.28
Two-stream [10]	12.43	30.36	13.42
Zero-Padding [10]	17.75	34.21	18.90
TONE [12]	16.87	34.03	14.92
TONE + HCML [12]	24.44	47.53	20.80
BCTR [13]	32.67	57.64	30.99
BDTR [13]	33.47	58.42	31.83
D^2RL [16]	43.40	66.10	44.10
AlignGAN [29]	57.90	–	53.60
CMGN [30]	35.13	61.07	32.14
HC Loss [17]	83.00	–	72.00
JSIA-ReID [34]	48.50	–	49.30
XIV [35]	62.21	83.13	60.18
Ours	**85.05**	**96.94**	**72.87**

5   Conclusions

In our work, we adopt a dual-path structure as the basic structure, which consists of two independent brunches for each modality respectively. The backbone of each brunch is a pre-trained ResNet-50 network, which can extract features. The outputs of ResNet-50 are divided into several stripes to learn local feature. What's more, we add a side brunch for each path respectively to learn serialized local feature representation, which contains a bidirectional LSTM. We consider the split local features as a sequence so that the connection among the local features can be extracted. Finally, we conduct the joint supervision of cross-entropy loss and hetero-center loss. The proposed methods get nice performance on the two benchmark datasets including SYSU-MM01 and RegDB.

Acknowledgements. This work was supported by the grant of National Key R&D Program of China, No. 2018AAA0100100; in part by supported by National Natural Science Foundation of China, Nos. 61861146002, 61772370, 61732012, 61932008, 61772357, 62073231, and 62002266; in part by the Scientific & Technological Base and Talent Special Program of the Guangxi Zhuang Autonomous Region, GuiKe AD18126015; and in part by "BAGUI Scholar" Program of Guangxi Province of China.

References

1. Ye, M., et al.: Deep learning for person re-identification: a survey and outlook (2020)

2. Wu, D., et al.: Deep learning-based methods for person re-identification: a comprehensive review. Neurocomputing **337**, 354–371 (2019)
3. Wang, X.F., Huang, D.S., Xu, H.: An efficient local Chan-Vese model for image segmentation. Pattern Recogn. **43**(3), 603–618 (2010)
4. Huang, D.S., Du, J.-X.: A constructive hybrid structure optimization methodology for radial basis probabilistic neural networks. IEEE Trans. Neural Netw. **19**(12), 2099–2115 (2008)
5. Li, B., Huang, D.S.: Locally linear discriminant embedding: an efficient method for face recognition. Pattern Recogn. **41**(12), 3813–3821 (2008)
6. Qian, X., Fu, Y., Xiang, T., Jiang, Y.-G., Xue, X.: Leader-based multi-scale attention deep architecture for person re-identification. IEEE Trans. Pattern Anal. Mach. Intell. **42**, 371–385 (2019)
7. Wu, Y., Zhang, K., Wu, D.: Person re-identification by multi-scale feature representation learning with random batch feature mask. IEEE Trans. Cogn. Dev. Syst. (2020). https://doi.org/10.1109/TCDS.2020.3003674
8. Wu, D.: Omnidirectional feature learning for person re-identification. IEEE Access **7**, 28402–28411 (2019)
9. Wang, X.F., Huang, D.S.: A novel density-based clustering framework by using level set method. IEEE Trans. Knowl. Data Eng. **21**(11), 1515–1531 (2009)
10. Wu, A.: RGB infrared cross modality person re identification. In: Proceedings of the IEEE International Conference on Computer Vision, pp. 5380–5389 (2017)
11. Dai, P., Ji, R.: Cross modality person re identification with generative adversarial training. In: Proceedings of International Joint Conference on Artificial Intelligence, pp. 677–683 (2018)
12. Ye, M., Lan, X., Li, J.: Hierarchical discriminative learning for visible thermal person re identification. In: Thirty Second AAAI Conference on Artificial Intelligence (2018)
13. Ye, M., Wang, Z., Lan, X., Yuen, P.C.: Visible thermal person re-identification via dual-constrained top-ranking. In: Proceedings of International Joint Conference on Artificial Intelligence, pp. 1092–1099 (2018)
14. Bai, X., Yang, M., Huang, T., et al.: Deep-person: learning discriminative deep features for person re-identification. Pattern Recogn. (2017)
15. He, K., Zhang, X.: Deep residual learning for image recognition. In: Proceedings of the IEEE Conference on Computer Vision and Pattern Recognition, pp. 770–778 (2016)
16. Wang, Z., Wang, Z., Zheng, Y.: Learning to reduce dual level discrepancy for infrared visible person re identification. In: Proceedings of the IEEE Conference on Computer Vision and Pattern Recognition, pp. 618–626 (2019)
17. Zhu, Y., Yang, Z., Wang, L.: Hetero-center loss for cross-modality person re-identification. Neurocomputing **386**, 97–109 (2019)
18. Kai, J.L., Arens, M.: Local feature based person reidentification in infrared image sequences. In: Proceedings of the 7th IEEE International Conference on Advanced Video and Signal Based Surveillance, Boston, USA, pp. 448–455. IEEE (2010)
19. Møgelmose, A., Bahnsen, C., Moeslund, T.B., Clapes, A., Escalera, S.: Tri-modal person re-identification with RGB, depth and thermal features. In: Proceedings of the 26th IEEE Conference on Computer Vision and Pattern Recognition Workshops, Portland, USA, pp. 301–307. IEEE (2013)
20. Wu, A.C., Zheng, W.S., Yu, H.X., Gong, S.G., Lai, J.H.: RGB-infrared cross-modality person re-identification. In: Proceedings of the 2017 IEEE International Conference on Computer Vision, Venice, Italy, pp. 5390–5399. IEEE (2017)
21. Dai, P.Y., Ji, R.R., Wang, H.B., Wu, Q., Huang, Y.Y.: Cross-modality person re-identification with generative adversarial training. In: Proceedings of the 2018 International Joint Conference on Artificial Intelligence, Stockholm, Sweden, pp. 677–683 (2018)
22. Huang, D.S., Ip, H.H.S., Chi, Z.-R.: A neural root finder of polynomials based on root moments. Neural Comput. **16**(8), 1721–1762 (2004)

23. Huang, D.S.: A constructive approach for finding arbitrary roots of polynomials by neural networks. IEEE Trans. Neural Netw. **15**(2), 477–491 (2004)
24. Nguyen, D.T., Hong, H.G., Kim, K.W.: Person recognition system based on a combination of body images from visible light and thermal cameras. Sensors **17**(3), 605 (2017)
25. Dalal, N., Triggs, B.: Histograms of oriented gradients for human detection. In: Proceedings of the IEEE Conference on Computer Vision and Pattern Recognition, pp. 886–893 (2005)
26. Liao, S., Hu, Y., Zhu, X.: Person re-identification by local maximal occurrence representation and metric learning. In: Proceedings of the IEEE Conference on Computer Vision and Pattern Recognition, pp. 2197–2206 (2015)
27. Huang, D.S., Chi, Z., Siu, W.-C.: A case study for constrained learning neural root finders. Appl. Math. Comput. **165**(3), 699–718 (2005)
28. Huang, D.S., Ip, H.H.S., Chi, Z., Wong, H.S.: Dilation method for finding close roots of polynomials based on constrained learning neural networks. Phys. Lett. A **309**(5–6), 443–451 (2003)
29. Wang, G., Zhang, T., Cheng, J.: RGB-infrared cross-modality person re-identification via joint pixel and feature alignment. In: Proceedings of the IEEE International Conference on Computer Vision, pp. 3623–3632 (2019)
30. Jiang, J., Jin, K., Qi, M., Wang, Q., Wu, J., Chen, C.: A Cross-modal multi-granularity attention network for RGB-IR person re-identification. Neurocomputing **406**, 59–67 (2020)
31. Huang, D.S., Ip, H.H.S., Law Ken, C.K., Chi, Z.: Zeroing polynomials using modified constrained neural network approach. IEEE Trans. Neural Netw. **16**(3), 721–732 (2005)
32. Huang, D.S., Ip, H.H.S., Law Ken, C.K., Chi, Z., Wong, H.S.: A new partitioning neural network model for recursively finding arbitrary roots of higher order arbitrary polynomials. Appl. Math. Comput. **162**(3), 1183–1200 (2005)
33. Huang, D.S., Zhao, W.-B.: Determining the centers of radial basis probabilistic neural networks by recursive orthogonal least square algorithms. Appl. Math. Comput. **162**(1), 461–473 (2005)
34. Wang, G.-A., Zhang, T., Yang, Y.: Cross-modality paired images generation for RGB–infrared person re–identification. In: Thirty-Fourth AAAI Conference on Artificial Intelligence (2020)
35. Li, D., Wei, X., Hong, X., Gong, Y.: Infrared-visible cross-modal person re-identification with an X modality. In: Thirty-Fourth AAAI Conference on Artificial Intelligence (2020)
36. Zhao, Z.Q., Huang, D.S., Sun, B.-Y.: Human face recognition based on multiple features using neural networks committee. Pattern Recogn. Lett. **25**(12), 1351–1358 (2004)

A Novel Decision Mechanism for Image Edge Detection

Junfeng Jing[1] 📷, Shenjuan Liu[1] 📷, Chao Liu[1] 📷, Tian Gao[1] 📷,
Weichuan Zhang[1,2(✉)] 📷, and Changming Sun[2] 📷

[1] School of Electrical and Information, Xi'an Polytechnic University, Xi'an 710048, China
[2] CSIRO Data61, PO Box 76, Epping, NSW 1710, Australia

Abstract. Edge detection plays an important role in image processing and computer vision tasks such as image matching and image segmentation. In this paper, we argue that edges in image should always be detected even under different image rotation transformations or noise conditions. Then a novel edge detection method is presented for improving the robustness of image edge detection. Firstly, we perform the following operations on the input image to obtain a set of images: random destruction operation, random rotation operation, and randomly addition of Gaussian noise with zero mean and random variance. Secondly, multi-directional Gabor filters with multiple scales are used as a tool to smooth the set of images and obtain candidate edges from the set of images. Thirdly, a novel edge decision mechanism is designed with a 0.95 confidence interval for selecting true edges from the candidate edges. Finally, two edge detection evaluation criteria (i.e., the aggregate test receiver-operating-characteristic and the Pratt's figure of merit) are utilized to evaluate the proposed edge detection method against five state-of-the-art methods. The experimental results show that our proposed method outperforms all the other tested methods.

Keywords: Multi-directional Gabor filters with multiple scales · Edge detection · Edge decision mechanism

1 Introduction

Edge detection is one of the most fundamental operations in the field of image analysis and image processing [13]. The extracted edge contours from input images are widely used as critical cues for various image understanding tasks such as image segmentation [23], object detection [15], and corner detection [28, 30]. The existing edge detection methods can be roughly divided into three categories [9]: hand-crafted based methods [9, 16, 19, 31], classical learning based methods [6, 14, 25], and deep learning based methods [4, 10, 11]. Hand-crafted based methods mainly utilize gradient information of images for extracting edges. In [16], gravity field intensity operator is used to replace the image intensity operator in the Canny method for improving the accuracy performance of edge detection. In [20], Hough transformation is utilized to replace the operation of hysteresis threshold setting in the Canny method for reducing the impact of threshold on

© Springer Nature Switzerland AG 2021
D.-S. Huang et al. (Eds.): ICIC 2021, LNCS 12836, pp. 274–287, 2021.
https://doi.org/10.1007/978-3-030-84522-3_22

edge detection. In [31], anisotropic Gaussian filters with automatic scales are designed for detecting edges from input images and a novel edge connection strategy is proposed for extracting cross edges from images. Classical learning-based methods mainly utilize object level supervision models [22] or hand-crafted feature information [1] to learn edge features for extracting edges from images. In [6], a structured random forest [8] is trained for extracting edges from images. Meanwhile, principal component analysis technique [26] is employed for improving the speed of the training process. In [14], edge detection is mapped as a curve searching task in a 2D discrete space. The hierarchically construct difference filters are applied to complete this process of search. Deep learning based methods mainly employ multi-level hierarchical features in convolutional neural network models for extracting edges from images. In [11], the feature information extracted by each convolutional layer of the VGG network is merged to locate image edges. In [10], the ground truth of image edges and the edges detected by the Canny detector with different scale parameters are utilized to train each convolutional layer of VGG-16 for improving robustness of image edge detection. Deng et al. [4] designed an end-to-end network architecture to effectively utilize hierarchical feature information and output precise boundary masks for detecting edges.

In [3], Canny proposed three criteria for image edge detection: signal-to-noise ratio, localization accuracy, and single edge accuracy. Meanwhile, Canny presented a famous Canny edge detection method. Following the three edge detection criteria, many edge detection methods [7, 9, 14, 27, 31] were proposed for detecting edges from images. There are two main problems for the existing edge detection method: (1) they are sensitive for edge detection with image rotation; (2) the results of edge detection greatly depend on the choice of a threshold. In this paper, we argue that edges in images should always be detected even under different image rotation transformations or noise conditions.

In this work, we present a new edge detection method which has the ability to robustly extract edge contours from input images with different image rotation transformations or different thresholds with a given confidence intervals (e.g., 0.95). Firstly, we perform serval operations on the input image to obtain a set of images. Secondly, multi-directional Gabor filters [12, 17] with multiple scales are used as a tool to smooth the set of images and obtain candidate edges from the set of images. Thirdly, a novel edge decision mechanism is designed with a 0.95 confidence interval for selecting true edges from the candidate edges. Finally, two edge detection evaluation criteria (i.e., the aggregate test ROC [2] and the FOM [24]) are utilized to evaluate the proposed edge detection method against five state-of-the-art methods. The experimental results show that our proposed method outperforms all the other tested methods.

2   Proposed Method

In this section, the imaginary part of Gabor filters (IPGFs) is briefly introduced, and then a novel edge detection method with a novel edge decision mechanism is proposed.

2.1   The Imaginary Part of Multi-directional Gabor Filters

It was proved that the IPGFs have high response value to linear parts of an image and are robust to different image scales and rotation transformations [12, 17]. In this paper, the multi-directional IPGFs are applied to extract the local intensity variation information of an input image I(x, y) for detecting edges. A two-dimensional discrete IPGF with a rotation angle $\theta k = \frac{\pi k}{K}$, k = 0, 1,..., K-1, can be expressed as

$$\varphi(m, n; \epsilon, \alpha, \beta, k) = \frac{\epsilon^2}{\pi \alpha \beta} \exp\left(-\left(\left(\frac{\epsilon^2}{\alpha^2}\right)p^2 + \left(\frac{\epsilon^2}{\beta^2}\right)q^2\right)\right) \sin(2\pi \epsilon p),$$

$$p = m \cos\theta_k + n \sin\theta_k,$$

$$q = -m \sin\theta_k + n \cos\theta_k.$$

(1)

where α and β are the sharpness along the major and minor axis respectively which control the shape of the IPGF, ϵ is the central frequency of the IPGF, and m, n represents the pixel coordinate in the integer lattice Z^2. An IPGF with $\theta_k = 0$ is shown in Fig. 1.

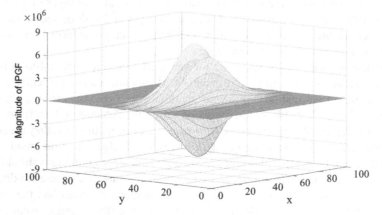

Fig. 1. An IPGF with $\theta_k = 0$, $\epsilon = 0.25$, $\alpha = 1$ and $\beta = 1.2$.

For an input image $I(m,n)$, the magnitude response of the discrete IPGFs along orientation θ_k is calculated as

$$\xi\left(m_x, n_y; \epsilon, \alpha, \beta, k\right) = I\left(m_x, n_y\right) \otimes \varphi\left(m_x, n_y; \epsilon, \alpha, \beta, k\right),$$

(2)

where $\otimes$ denotes a convolution operation.

It is proved in [29] that multi-directional anisotropic filters have the ability to accurately depict the difference between a step edge and other types of edges. Taking a step edge pixel and an L-type edge as an example, the magnitude response of anisotropic IPGFs (AIPGFs) and isotropic IPGFs (IIPGFs) on these two type edges are shown in Fig. 2. It can be seen from Fig. 2 that the magnitude response of the AIPGFs with multi-directional filtering can depict the difference between step edge and an L-type edge while the magnitude response of the IIPGFs cannot. Therefore, the anisotropic multi-directional IPGFs are employed in this paper for smoothing the input image and detecting edges.

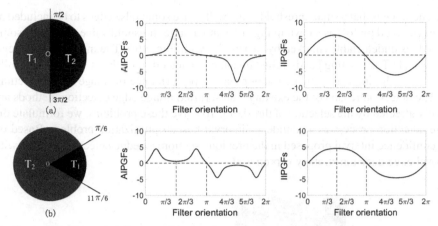

Fig. 2. A step edge and an L-type edge are shown in (a) and (b) in the first column, the corresponding magnitude responses of AIPGFs and IIPGFs are shown in the second column and the third column respectively.

2.2 Confidence Interval

Conceptually, a confidence interval [5] is a range that indicates the degree which the true value is likely to fall around the measured value. For a given set of sample data, we defined Ω as the observed object, W as all possible observations, and X as the actual observations. X is actually a random variable defined on Ω with a range of values on W. In this case, for some observed value $X = t$, the confidence interval is $(\mu(t), \nu(t))$. In fact, if the true value is ω, then the confidence level is the probability ς

$$c = Pr(u(X) < w < v(X)), \tag{3}$$

The calculation of the confidence interval depends on the statistics used. The confidence interval is calculated at a predetermined significance level δ, which is set to 0.05 in most cases. It means that the confidence is 0.95. The confidence interval is calculated by

$$c = Pr(c_1 \leq \mu \leq c_2) = 1 - \delta, \tag{4}$$

where P_γ is the probability, δ is the significance level, and μ is the mean value of sample data.

2.3 New Edge Detection Measure

Our investigation shows that the edge detected by the existing hand-crafted edge detection algorithm in the image under different rotation transformation and threshold are both greatly different from that detected in the original image. Take Fig. 3 and Fig. 4 as examples, the edge detection results from the Canny method [3] for the original image and the image rotated are shown in Fig. 3. In this test, the parameter settings for the Canny detector remain the same. It can be seen that some obvious edge contours in the window area (marked by 'o') cannot be detected in the rotated image as shown in Fig. 3(b) and (d). The same is that a small threshold value will cause some false edges to be included as edges (marked by 'o') as shown in Fig. 4(b) and a large threshold value will make some real edges undetectable (marked by 'o') as shown in Fig. 4(c). The main reasons are as follows: (1) The existing hand-crafted image edge detection methods have not carefully considered the impact of image rotation on the performance of image edge detection. (2) The detection results of the existing hand-crafted image edge detection methods are largely affected by the selection of threshold. To solve these problems, we formulate the edge detection process as a confidence interval based optimization problem. Based on the confidence interval proposed in the previous section, an edge detection method with a novel edge decision strategy is proposed.

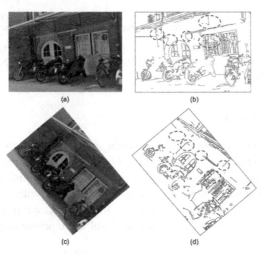

Fig. 3. Examples of changes of edge detection results for the original image and the rotated image. (a) An original image, (b) Edge detection result for the original image, (c) The original image is rotated by $\pi/4$ clockwise, (d) Edge detection result after the original image is rotated by $\pi/4$ clockwise.

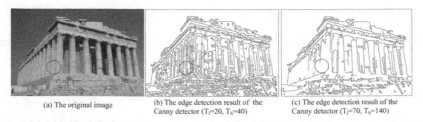

(a) The original image   (b) The edge detection result of the Canny detector (T_l=20, T_h=40)   (c) The edge detection result of the Canny detector (T_l=70, T_h=140)

Fig. 4. The influence of threshold on image edge detection.

Three Operations on an Input Image. In this paper, we argue that edges in an image should always be detected even under different image transformations or noise conditions. Follow this concept, a series of operations are performed on the input image to obtain a robust edge detector: *Random destruction*: Given an input image I, we uniformly partition the image into $R * S$ sub-regions as shown in Fig. 5(a). *Random rotation*: Given an input image I, it is randomly rotated by θ_k clockwise as shown in Fig. 5(b). *Random Gaussian noise*: Zero-mean white Gaussian noise with random variance is added to the input image I to generate noisy images with different variances as shown in Fig. 5(c).

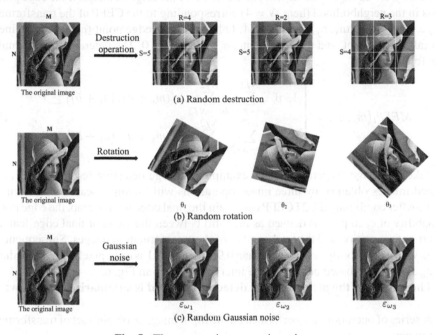

Fig. 5. Three operations on an input image.

Candidate edges extracted by the AIPGFs with multiple scales. For an input image $I(m, n)$, a series of images $\Lambda(m, n)$ are obtained based on the image operations we designed. Subsequently, the multi-directional AIPGFs with multiple scales are utilized to smooth the images $\Lambda(m, n)$ and $I(m, n)$ respectively as follows

$$\xi_{\sigma,k}\left(m_x, n_y\right) = I\left(m_x, n_y\right) \otimes \varphi\left(m_x, n_y; \in, \alpha, \beta.k\right),$$
$$\in = 0.25 \times \left(\sqrt{2}\right)^{\sigma-1}. \tag{5}$$

where $\xi_{\sigma,k}(m_x, n_y)$ represents the magnitude response of AIPMGFs with scale σ along direction k.

The discrete edge strength map (ESM) at each scale is constructed by the maximum magnitude response of AIPMGFs as follows

$$\mathrm{ESM}\left(I_\sigma\left(m_x, n_y\right)\right) = \max_{k=0,1,\ldots,K-1}\left\{\left|\xi_{\sigma,k}\left(m_x, n_y\right)\right|\right\}, \tag{6}$$

Edge Extraction Strategy Based on the Confidence Interval. After obtaining the multi-scale ESMs of each image of image sets $\Lambda(m, n)$, the non-maximum suppression and double threshold technique proposed in the Canny algorithm [3] are applied to obtain the multi-scale candidate edge feature map (CEFP) sets from each image of image sets $\Lambda(m, n)$.

Finally, for an edge point (m_x, n_y) in the CEFP of the original image, if the edge point exists in the neighborhood (here, $N = 4$) corresponding to the CEFP of the transformed image, the $MEFP_i(m_x, n_y)$ is equal to 1. Otherwise, the edge point (m_x, n_y) is defined as a false edge point, and the $MEFP_i(m_x, n_y)$ is equal to 0. The determination formula is as follows

$$
MEFP_i\left(m_x, n_y\right) = \begin{cases} 1, & \text{if } \sum_{a=-N/2}^{N/2} \sum_{b=-N/2}^{N/2} \Lambda\left((m_x + a), (n_y + b)\right) \geq 1 \\ 0, & \text{if } \sum_{a=-N/2}^{N/2} \sum_{b=-N/2}^{N/2} \Lambda\left((m_x + a), (n_y + b)\right) = 0 \end{cases} \tag{7}
$$

Take the image 'Vegetables' as an example, the edge detection results of 27 transformed images obtained by three image operations with multiple scales are shown in Fig. 6. After combining the 27 CEFPs to obtain the final edge feature map, the edge pixel probability of each pixel is defined as the ratio between the value of final edge feature map in same position and the total number of the transformation images. Subsequently, pixels with a probability value higher than 0.95 are defined as edge pixels. The procedure of edge extraction based on confidence interval is shown in Fig. 6.

The outline of the proposed edge detection method is summarized as follow:

1. A series of operations are performed on the input image to obtain a set of transformed images.
2. Multi-directional IPGFs with multiple scales are used to smooth the series of images and obtain multi-scale ESMs of each transformed image and original image.
3. The non-maximum suppression and double threshold technique proposed in the Canny algorithm [3] are applied to obtain the multi-scale CEFP sets for each image.
4. Combine 27 CEFPs to obtain the final edge feature map and calculate the confidence probability of candidate edge points.
5. Obtain the final edge pixels by the confidence threshold.

3 Experiment Results and Performance Evaluation

In this section, the aggregate test ROC [2] and the FOM [24] are applied to evaluated the edge detection performance for six state-of-the-art detectors (the ANDD [31], the AMDD [18], the IAGKs [19], the Canny [3], the edge drawing (ED) [21] and the proposed method).

The codes for these methods are from authors. The parameter settings for the proposed detector are: $K = 8$, $\epsilon = 0.25$, $\alpha = 1$, $\beta = 1.2$, $\sigma1 = 1$, $\sigma2 = 2$ and $\sigma3 = 3$. Three angles are randomly selected from $[0, 2\pi]$ to rotate the image clockwise, and

(a) The original image

(b) The CEPF from the original image

(c) The result of edge detection with σ_1

(d) The result of edge detection with σ_2

(e) The result of edge detection with σ_3

(f) The final edge detection result from the original image

Fig. 6. Edge extraction schematic drawing based on confidence interval. (a) The original image, (b) The CEPF from the original image, (c) The result of edge detection with σ_1, (d) The result of edge detection with σ_2, (e) The result of edge detection with σ_3, (f) The final edge detection result from the original.

the random variance of Gaussian noise is random selected from [1, 15], R and S are randomly selected from [1, 8]. The percentage of not-edge pixels ζ is 0.8, and the factor of threshold ratio τ is 0.15.

3.1 The Evaluation of ROC

For the detected edge pixel sets, if there exists an edge pixel that belongs to the detected edge set which is not far away from the ground truth, a true positive (TP) is determined. Conversely, when an edge point is near the non-edge region of the ground truth, it is recorded as a false positive (FP). Let nTP and nFP be the number of pixels of true positive and false positive respectively. Let ne and nne be the number of edge pixels and non-edge pixels in the ground truth. Therefore, the detection result on ROC curve can be expressed by the unmatched metric and the FP metric:

$$\gamma_{Unmatch} = 1 - \frac{n_{TP}}{n_e},$$

$$\gamma_{FP} = \frac{n_{FP}}{n_{ne}}.$$

(8)

In order to avoid the influence of parameter selection on the edge detection performance of a detector, the ROC curve on the optimal parameters is trained on a single image at first. And then the test is carried out on the image set. The average value of all the ROC curves of the images is used to obtain the overall test ROC curve. The image database contains a set of fifty object images and another set of ten aerial images with the specified ground truths.

The Canny detector [3] has the scale factor σ, the percent of not-edge pixels ζ, and the factor of threshold ratio τ to adjust. The parameter settings are taken in the set σ = 1, 1.2,..., 5, ζ = 0.6, 0.61,..., 0.95, and τ = 0.2, 0.21,...,0.5. The ANDD detector [31] and the AMDD detector [18] both have five adjustable parameters: the number of orientations, the anisotropic factor $\rho = 1/\sqrt{(\frac{\pi}{2K})}$, scale factor σ, the percent of not-edge pixels ζ, and the factor of threshold ratio τ. The parameter settings are K = 4, 6,..., 12, $\sigma2 = \rho2, \rho2 + 1,..., \rho2 + 10$, ζ = 0.6, 0.61,..., 0.95 and τ = 0.1, 0.15,..., 0.6. The adjustable parameters of the ED [21] detector include the scale factor σ, the gradient threshold Gth, the percentage of not-edge pixels ζ, and the factor of threshold ratio τ. The allowable setting range of these parameters are σ = 1, 1.2..., 0.6, Gth = 2, 4,...,48, ζ = 0.6, 0.61,..., 0.95 and τ = 0.1. The IAGKs [19] needs to set four parameters. the anisotropic factor $\rho2$ = 2, 3,..., 12, the scale factor $\sigma2$ = 2, 3,..., 12, the percent of not-edge pixels ζ = 0.6, 0.61,..., 0.95 and the factor of threshold ratio τ = 0.1, 0.15,..., 0.6.

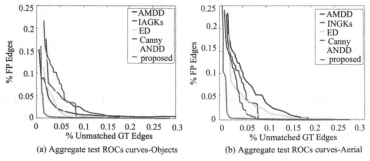

(a) Aggregate test ROCs curves-Objects   (b) Aggregate test ROCs curves-Aerial

Fig. 7. Comparison of the aggregate test ROC of the six detectors for the set of object images and aerial images. (a) Aggregate test ROC curves-Objects, (b) Aggregate test ROC curves-Aerials.

The results shown in Fig. 7 are the ROCs curve for the noise-free case. Compared with other methods, the proposed detector achieves the best performance on edge detection. The ANDD [31] and the AMDD [18] detectors achieve the sub-optimal performance. The performance of the IAGKs [19], the Canny [3], and the ED [21] detectors are relatively weak. The main reason is that these methods apply the isotropic filter to obtain the edge response, which cannot accurately exactly the intensity variation information on edge.

3.2 FOM Evaluation

Let n_e be the number of edge pixels in the ideal edge and n_d be the number of edge pixels in the detected edge. The FOM metric can be defined as:

$$FOM = \frac{1}{\max(n_e, n_d)} \sum_{k=1}^{n_d} \frac{1}{1 + \omega d(k)^2}, \tag{9}$$

where d(k) is the distance from the kth detected edge pixel to the ideal edge pixel, and ω is set to 1/4. The value range of FOM is [0.1]. An excellent FOM detection result should be 1, which means that all edge pixels are detected in the ideal edge and no false edge pixels appear. The closer FOM is to 1, the better edge detection performance is.

Fig. 8. Five test images for the FOM evaluation.

The images with five different scenes used for FOM evaluation are shown in Fig. 8. Firstly, by adjusting the high-low threshold ratio to 0.5, three edge maps with almost the same number of edge pixels were obtained from three detectors (the Canny detector, the ANDD method, and the proposed detector). Then the ideal edge is formed by two of the three edge maps whose edge pixels exist at least. Secondly, based on the ideal edge map, the noise robustness of the three detectors is evaluated by the change of their FOM with the noise level. Edge maps by the three detectors for noise-free 'block' image is shown in Fig. 9 and the edge map by the three detectors at $\varepsilon\omega = 15$ is shown in Fig. 10. Obviously, the proposed detector is more robust to noise than the other detectors. That is, as the noise level increases, its FOM decreases more slowly than the other two detectors. The reason is that the method in this paper selects the confidence set up with many noise levels inside the space of intervention, which help produce fewer false edges, which is consistent with our previous ROC assessment. At noise levels $\varepsilon\omega = 5$, 10 and 15, the

FOM of the three detectors are summarized in Table 1 and the algorithm complexity is obtained by calculating the average time of the proposed algorithm being executed 100 times is shown in Table 2.

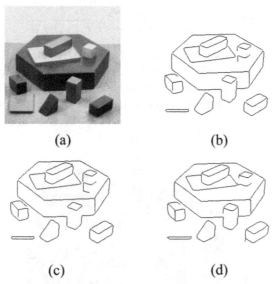

(a)   (b)

(c)   (d)

Fig. 9. Edge maps by the three detectors for the noise free 'Blocks' image. (a) Noise free 'Blocks' image, (b) Edge map by the Canny detector with the percentage of non-edge pixels 0.6, (c) Edge map by the ANDD detector with the percentage of non-edge pixels 0.64, (d) Edge map by the proposed detector with the percentage of non-edge pixels 0.71.

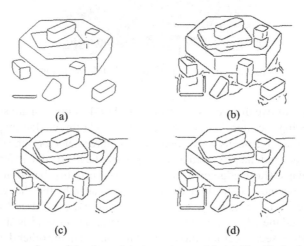

(a)   (b)

(c)   (d)

Fig. 10. Visual comparison of the three detectors at $\varepsilon_\omega = 10$. (a) The ideal edge map, (b) Edge map by the Canny detector, (c) Edge map by the ANDD detector, (d) Edge map by the proposed detector.

Table 1. FOM comparison of the three detectors at three noise levels, where each cell lists the FOM and the number of detected edge pixels (NDEPs)

Image/ ideal edge pixels		Noise level								
		Canny			ANDD			Proposed		
		$\varepsilon_\omega = 5$	$\varepsilon_\omega = 10$	$\varepsilon_\omega = 15$	$\varepsilon_\omega = 5$	$\varepsilon_\omega = 10$	$\varepsilon_\omega = 15$	$\varepsilon_\omega = 5$	$\varepsilon_\omega = 10$	$\varepsilon_\omega = 15$
Sofa	FOM	0.8601	0.8436	0.8174	0.8905	0.8633	0.8158	**0.9088**	**0.9723**	**0.8207**
/15,139	NDEPs	15181	15154	15157	15144	15133	15110	15189	15201	15144
Lena	FOM	0.8792	0.8480	0.7992	0.8854	0.8461	0.8001	**0.9217**	**0.8902**	**0.8211**
/17,758	NDEPs	18020	17921	17978	18101	17680	17830	18031	18013	17893
Block	FOM	0.9480	0.9341	0.9357	0.9601	0.9578	0.9398	**0.9758**	**0.9586**	**0.9573**
/1973	NDEPs	1981	1963	1980	1989	1970	1960	1976	1972	1972
Shave	FOM	0.8337	0.7710	0.7394	0.8494	0.8123	0.7613	**0.8732**	**0.8152**	**0.7915**
/6277	NDEPs	6234	6197	6221	6208	6215	6278	6293	6208	6270
House	FOM	0.9394	0.9287	0.9172	0.9401	0.9350	0.9189	**0.9672**	**0.9520**	**0.9386**
/2996	NDEPs	3005	2982	2986	2980	2983	30024	2989	2987	3013

Table 2. Mean run-time of the proposed edge detector.

Task	Times(s)				
	Sofa	Lena	Block	Shave	House
Three operations	0.23	0.28	0.16	0.36	0.22
Candidate edges extracted	4.64	5.69	3.26	6.52	3.87
Edge extraction on the confidence interval	0.36	0.65	0.29	0.74	0.28

4 Conclusions

In this paper, we argue that edges in an image should be always detected even under different image rotation transformations or noise conditions. Then, we present a new edge detection method which has the ability to robustly extract edge contours from input images in different image rotation transformations or different thresholds with a given confidence intervals (e.g.,0.95). Firstly, we perform serval operations on the input image to obtain a set of images. Secondly, multi-directional Gabor filters with multiple scales are used as a tool to smooth the set of images and obtain candidate edges from the set of images. Thirdly, a novel edge decision mechanism is designed with a 0.95 confidence interval for selecting true edges from the candidate edges. Finally, two edge detection evaluation criteria (i.e., the aggregate test receiver-operating-characteristic (ROC) and the Pratt's figure of merit (FOM)) are utilized to evaluate the proposed edge detection method against five state-of-the-art methods. The experimental results show that our proposed method outperforms all the other tested methods. Furthermore, the proposed edge detection method has a great potential to be applied in image segmentation and many other image processing and computer vision tasks such as object detection and image matching.

Acknowledgments. This work was supported by the Shaanxi Innovation Ability Support Program (2021TD-29)- Textile Intelligent Equipment Information and Control Innovation Team and Shaanxi innovation team of universities –Textile Intelligent Equipment Information and Control Innovation Team.

References

1. Arbelaez, P., Maire, M., Fowlkes, C., Malik, J.: Contour detection and hierarchical image segmentation. IEEE Trans. Pattern Anal. Mach. Intell. **33**(5), 898–916 (2010)
2. Bowyer, K., Kranenburg, C., Dougherty, S.: Edge detector evaluation using empirical ROC curves. Comput. Vis. Image Underst. **84**(1), 77–103 (2001)
3. Canny, J.: A computational approach to edge detection. IEEE Trans. Pattern Anal. Mach. Intell. **8**(6), 679–698 (1986)
4. Deng, R., Shen, C., Liu, S., Wang, H., Liu, X.: Learning to predict crisp boundaries. In: Proceedings of the European Conference on Computer Vision, pp. 562–578 (2018)
5. DiCiccio, T.J., Efron, B., et al.: Bootstrap confidence intervals. Stat. Sci. **11**(3), 189–228 (1996)
6. Dollár, P., Zitnick, C.L.: Structured forests for fast edge detection. In: Proceedings of the IEEE International Conference on Computer Vision. pp. 1841–1848 (2013)
7. Kimia, B.B., Li, X., Guo, Y., Tamrakar, A.: Differential geometry in edge detection: accurate estimation of position, orientation and curvature. IEEE Trans. Pattern Anal. Mach. Intell. **41**(7), 1573–1586 (2018)
8. Kontschieder, P., Bulo, S.R., Bischof, H., Pelillo, M.: Structured class-labels in random forests for semantic image labelling. In: 2011 International Conference on Computer Vision, pp. 2190–2197 (2011)
9. Liu, Y., Xie, Z., Liu, H.: An adaptive and robust edge detection method based on edge proportion statistics. IEEE Trans. Image Process. **29**, 5206–5215 (2020)
10. Liu, Y., Lew, M.S.: Learning relaxed deep supervision for better edge detection. In: Proceedings of the IEEE Conference on Computer Vision and Pattern Recognition, pp. 231–240 (2016)
11. Liu, Y., Cheng, M.M., Hu, X., Wang, K., Bai, X.: Richer convolutional features for edge detection. In: Proceedings of the IEEE Conference on Computer Vision and Pattern Recognition, pp. 3000–3009 (2017)
12. Luan, S., Chen, C., Zhang, B., Han, J., Liu, J.: Gabor convolutional networks. IEEE Trans. Image Process. **27**(9), 4357–4366 (2018)
13. Marr, D., Hildreth, E.: Theory of edge detection. Proc. R. Soc. London Ser. B Biol. Sci. **207**(1167), 187–217 (1980)
14. Ofir, N., Galun, M., Alpert, S., Brandt, A., Nadler, B., Basri, R.: On detection of faint edges in noisy images. IEEE Trans. Pattern Anal. Mach. Intell. **42**(4), 894–908 (2019)
15. Qin, X., Zhang, Z., Huang, C., Gao, C., Dehghan, M., Jagersand, M.: BasNet: boundary-aware salient object detection. In: Proceedings of the IEEE Conference on Computer Vision and Pattern Recognition, pp. 7479–7489 (2019)

16. Rong, W., Li, Z., Zhang, W., Sun, L.: An improved Canny edge detection algorithm. In: Proceedings of the IEEE International Conference on Mechatronics and Automation, pp. 577–582 (2014)
17. Sharifi, O., Mokhtarzade, M., Beirami, B.A.: A deep convolutional neural network based on local binary patterns of gabor features for classification of hyperspectral images. In: Proceedings of the International Conference on Machine Vision and Image Processing, pp. 1–5 (2020)
18. Shui, P.L., Wang, F.P.: Anti-impulse-noise edge detection via anisotropic morphological directional derivatives. IEEE Trans. Image Process. **26**(10), 4962–4977 (2017)
19. Shui, P., Zhang, W.: Noise-robust edge detector combining isotropic and anisotropic Gaussian kernels. Pattern Recogn. **45**, 806–820 (2012)
20. Song, R., Zhang, Z., Liu, H.: Edge connection based Canny edge detection algorithm. Pattern Recognit Image Anal. **27**(4), 740–747 (2017). https://doi.org/10.1134/S1054661817040162
21. Topal, C., Akinlar, C.: Edge drawing: a combined real-time edge and segment detector. J. Vis. Commun. Image Represent. **23**(6), 862–872 (2012)
22. Wang, L., et al.: Learning to detect salient objects with image-level supervision. In: Proceedings of the IEEE Conference on Computer Vision and Pattern Recognition, pp. 136–145 (2017)
23. Wei, Y., et al.: STC: A simple to complex framework for weakly-supervised semantic segmentation. IEEE Trans. Pattern Anal. Mach. Intell. **39**(11), 2314–2320 (2016)
24. Pratt, W.K.: Digital Image Processing (1978)
25. Xiaofeng, R., Bo, L.: Discriminatively trained sparse code gradients for contour detection. In: Proceedings of the Advances in Neural Information Processing Systems, pp. 584–592 (2012)
26. Yang, J., Zhang, D., Frangi, A.F., Yang, J.y.: Two-dimensional PCA: a new approach to appearance-based face representation and recognition. IEEE Trans. Pattern Anal. Mach. Intell. **26**(1), 131–137 (2004)
27. Yu, Z., Feng, C., Liu, M.Y., Ramalingam, S.: CaseNet: deep category-aware semantic edge detection. In: Proceedings of the IEEE Conference on Computer Vision and Pattern Recognition, pp. 5964–5973 (2017)
28. Zhang, W., Shui, P.: Contour-based corner detection via angle difference of principal directions of anisotropic Gaussian directional derivatives. Pattern Recogn. **48**(9), 2785–2797 (2015)
29. Zhang, W., Sun, C.: Corner detection using multi-directional structure tensor with multiple scales. Int. J. Comput. Vision **128**(2), 438–459 (2020)
30. Zhang, W., Sun, C., Breckon, T., Alshammari, N.: Discrete curvature representations for noise robust image corner detection. IEEE Trans. Image Process. **28**(9), 4444–4459 (2019)
31. Zhang, W., Zhao, Y., Breckon, T.P., Chen, L.: Noise robust image edge detection based upon the automatic anisotropic Gaussian kernels. Pattern Recogn. **63**, 193–205 (2017)

Rapid Earthquake Assessment from Satellite Imagery Using RPN and Yolo v3

Sanjeeb Prasad Panday[1](✉), Saurav Lal Karn[1], Basanta Joshi[1](✉), Aman Shakya[1](✉), and Rom Kant Pandey[2](✉)

[1] Department of Electronics and Computer Engineering, Pulchowk Campus, Institute of Engineering, Tribhuvan University, Kathmandu, Nepal
{sanjeeb,basanta,aman.shakya}@ioe.edu.np
[2] Sanothimi Campus, Tribhuvan University, Bhaktapur, Nepal
Romkant.Pandey@sac.tu.edu.np

Abstract. Nepal suffers from earthquakes frequently as it lies in a highly earthquake prone region. The relief that is to be sent to the earthquake affected area requires rapid earliest assessment of the impact in the area. The number of damaged buildings provides us with the necessary information and can be used to assess the impact. Disaster damage assessment is one of the most important parts in providing information about the impact to the affected areas after the disaster. Rapid earthquake damage assessment can be done via the satellite imagery of the affected areas. This research work implements the Region Proposal Network (RPN) and You only look once (Yolo) v3 for generating region proposals and detection. Sliding window approach has been implemented for the method to work on large satellite imagery. The obtained detection has been com-pared with the ground truth. The proposed method achieved the overall F1 score of 0.89 as well as Precision of 0.94 and Recall of 0.86.

Keywords: Disaster · Earthquake · RPN · Yolo v3 · Satellite imagery · Remote sensing

1 Introduction

Earthquakes are more frequent in Nepal due to its fragile geography and location between the Eurasian and Indian tectonic plates [1]. In such conditions, planning the relief works is a huge task for the government. Planning relief works require information about the damage caused by disaster. Multiple factors are responsible for determining the damage caused by the disaster. Some of these factors are damage to structures, land mass changes, number of deaths and impact on people's life. Over the past few decades, researchers have tried multiple ways to assess the damage from the disaster.

One of the early research papers published in IEEE was building detection from high-resolution satellite image by Wei Liu and V. Prinet [2] which deals with the building detection using probability model. These authors worked on probabilistic model to extract buildings images from satellite imagery. However, this work cannot be generalized i.e. model created for one scene cannot be used for another scene. Another

D.-S. Huang et al. (Eds.): ICIC 2021, LNCS 12836, pp. 288–301, 2021.
https://doi.org/10.1007/978-3-030-84523-3_23

interesting approach was implemented by Amy Zhang, Xianming Liu, Andreas Gros and Tobias Tiecke was building detection from satellite images on a Global Scale [3]. Their approach used the image segmentation and classification model. They employed the use of pixel wise segmentation to detect buildings. In the same manner, another work on damage assessment was done by L. Chiroiu and G. Andre [4]. These authors employed the use of high-resolution satellite imagery for the assessment of 2001, Bhuj, India earth-quake by using the optical and radar satellite imagery of the affected area as basis for assessment. Similarly, M.R. Archana et al. [5] gave an interesting approach for earthquake assessment using pre-event and post-event imagery by focusing on similarity between both images by assuming that the buildings must have rectangular footprint and isolated.

Another interesting work is done by Chandan Dinesh Parape and Masayuki Tamura of Kyoto University Japan [6] by employing the usage of the morphological operators for segmenting images while ISODATA for the feature extraction and classification. This method used the pre-disaster and post-disaster images for processing and comparing with the ground truth. One of the major flaws of this work is that this method cannot be generalized to segment images as the reflectance changes from image to image. In the same field, Facebook AI Research group [7] has been doing some advanced research in recent times. They proposed a new approach to segment the buildings by utilizing the two-stage procedure with Region Proposal Network in the first stage and predicting the class and box offset. This method works by extracting proposed multiple regions of interest and classification is done in a second step. This method has considerable accuracy on smaller images but fails to predict considerably in larger images. Faster R-CNN based method has been applied in the multiple types of damage assessment [8] but further research is necessary to apply it for damage assessment from satellite images. YOLO (You only look once) [9], a one stage detector algorithm has been applied for detection of col-lapsed buildings in post-earthquake remote sensing images [10]. This method detects well in images with a smaller number of buildings but further evaluation in larger images with many buildings is necessary.

In this regard, this research work incorporates the backbone of two stage detectors (Faster R-CNN) i.e. Region Proposal Network along with the (One stage detectors) Yolo v3 classifier due to its great classification accuracy. In addition, the sliding window [11] approach to process on large satellite imagery is also proposed in this research so that processing can be done even on a small computing machine. The segmentation approach implemented in this research work is highly generalized and can be used to detect objects other than **"building"** by training with a dataset having required labels.

2 Proposed Methodology

The current research work implements the usage of Region Proposal Network (RPN) and Yolo v3 to detect the buildings from pre-disaster and post-disaster satellite imagery. The proposed method combines both methods for the optimal detection of buildings. The block diagram for the proposed method is shown in Fig. 1.

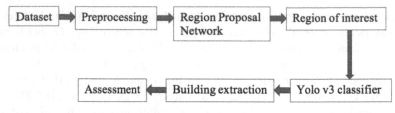

Fig. 1. Block diagram

2.1 Dataset

For the initial implementation, two datasets were used namely Crowd AI Map-ping Challenge dataset [12] and AWS space net challenge dataset [13]. Both of these datasets are available online and can be downloaded from their official websites. The proposed model was trained using these online available dataset. For the case of Nepal, the model trained on the above-mentioned dataset fails due to the different orientation, shapes and colors of the buildings. To solve this issue, satellite imagery of rural areas of Nepal was acquired and annotated for the ground truth.

Satellite imagery has been acquired from Bing satellite images [14] (due to best possible image present) via the SAS Nightly Software. Satellite imagery has been acquired for the areas in the Sindhupalchok district, Nepal. The acquired imagery was of the zoom level of 20. The higher zoom only increased the size of image but was not clear. The zoom level 20 provided the most affordable resolution. These acquired images are quite large in size. For the ease of annotation, the images are cropped in the size of 512 * 512. These slices are smaller in size and can be used for the training. The generated slices of original images are annotated via the VGG annotator tool [15]. This tool provides the annotations in the MS COCO format.

2.2 Preprocessing

The acquired dataset is preprocessed before proceeding with the training process. For Crowd AI Mapping Challenge Dataset, no preprocessing was required as the dataset was well balanced and the annotations were in MS COCO format which is suitable for the training. For the AWS spacenet dataset, the images were over 0.5 MB and the building footprints were in the form of latitude and longitude. The footprints were converted to bounding boxes using the utilities provided by the spacenet [13]. To handle large images, masks were generated using the obtained bounding box and training was done only using these masks. This method although decreased the time for training but due to loss of features showed very poor detection. To increase the accuracy, the large images were resized to 515 * 512. This reduced the size of the image without significant loss of the features. After resizing, the annotation was converted to the Pascal VOC format which is suitable for the Yolo v3 classifier training.

Custom dataset has been used to train both RPN and YOLO v3. For RPN, custom dataset annotations were already in MS COCO format thus no preprocessing was required. For the YOLO v3, the annotations were converted in the Yolo v3 annotation format.

2.3 Region Proposal Network

The Region Proposal Network (RPN) has proven to be very efficient to detect objects in image and is considered as the backbone of Faster R-CNN [16, 17]. To detect regions of images where the object lies, a small network is slide over a convolutional feature map that is the output by the last convolution layer. In default configuration for the RPN, number of anchors per pixel is taken as 9 i.e., 9 proposals are generated for every pixel in the image. However, in this work, 21 anchors i.e., 21 proposals are generated for every pixel in the image have been taken due to various sizes of buildings in satellite imagery. The total number of proposals for an image is given by Eq. 1.

$$p = w * h * k \tag{1}$$

Where p, w, h, k are the number of proposals, width, height and proposals for the pixel per image respectively. The proposals are assigned labels based on the intersection over union. The anchors with IOU score over 0.7 are assigned to be ones containing buildings. In this work, with image size of 512 * 512 and anchor size of 21, the number of proposals can be calculated using Eq. 1 as 5,505,024 per image. Anchor placement has been shown for a single pixel in Fig. 2.

As it can be seen, the number of proposals is very large. To reduce the number of proposals, RPN has been used. RPN can be built over the Resnet, VGG, AlexNet and DeepNet. In this work, Resnet 101 [18] has been used as the convolutional network to generate feature maps from the image. Furthermore, in this work, Mask R-CNN [7] which is an extension of Faster R-CNN by adding a branch for predicting an object mask (Region of Interest) in parallel with the existing branch for bounding box recognition will be used.

2.4 Region of Interest

The region proposal network provides us with the so called **"proposals"**. These proposal regions are extracted from the original image using Python pillow library for the extraction of region of interests. The obtained proposals are used as regions of interests and are sent as input to the Yolo v3 classifier.

2.5 Yolo V3

Yolo (You only look once) v3 [19] is a state-of-the-art object detection algorithm which is able to detect objects in real time and can be used to detect objects in video. This implementation of Yolo v3 uses Dark net architecture. For the activation function, Yolo v3 uses the softmax activation function. Softmax function converts numbers into probabilities that sums up to 1. It is generally used to map the non-normalized output of a network to a probability distribution over predicted output classes.

In this work, the feed ROIs are passed to the Yolo v3 classifier. Yolo v3 classifier divides the images into regions and predicts bounding boxes and prob-abilities for each region. A threshold of 0.8 is set for building detection. Regions with probabilities over the threshold are treated as buildings.

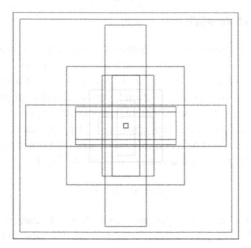

Fig. 2. Anchor placement for a single pixel.

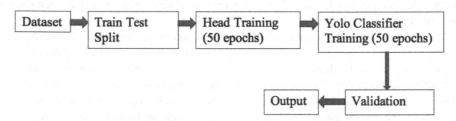

Fig. 3. Anchor placement for a single pixel.

Yolo v3 detects a large number of objects in images but doesn't work in the case of satellite imagery. The default Yolo v3 classes don't include building let alone in satellite imagery. So, in order to accomplish our task of making a Yolo v3 classifier to work on our method, a custom class of building is added to train the model using Spacenet dataset and annotations converted to Pascal VOC format. The dataset consisted of 6000 images. The dataset was split in 60% training, 20% validation and 20% test data. The training is concluded in two stages. On the first stage only the heads of the network are trained. This is done to decrease the validation loss in a short span of time. The heads are trained for 50 epochs. Once the heads are trained, weights are saved. In the second stage, the overall network is trained for the next 50 epochs. The block diagram of training Yolo v3 shown in Fig. 3.

The training was performed on Inter(R) Core(TM) i5-7200 CPU @2.50 GHz. The training was completed in 8 days. The training resulted in validation loss of 11.5%. The model is coarse due to being not trained on large dataset. Dataset has been acquired for multiple cities from the official site of AWS spacenet.

The same training method has been applied to the Custom dataset training with Yolo v3. The heads are trained for 50 epochs to decrease the validation loss in a short span of time. Once the heads are trained, weights are saved. In the second stage, the overall network is trained for the next 50 epochs. The training was done on the Google Colab Notebook with GPU environment and 12 GB RAM. The training resulted in validation loss of 13.46%.

2.6 Building Extraction

Building extraction is one of the important steps of this work. The ROIs which are said to be buildings are counted as buildings. The combination of these methods works quite well to detect buildings in small imagery but fails to detect buildings in large satellite imagery. In normal case, a single satellite imagery of an area is over 4000 * 4000 pixels. During experimentation, it was found out that the integrated method performs poorly in large imagery. In order to solve this issue, the sliding window approach was integrated. This approach passes only a section of image through the process and accumulates the results. As the complete image has been passed through the process, the accumulated results are compiled to give the predictions. The sliding window method greatly improved the accuracy of our method to detect buildings in large imagery. The algorithm of the pipeline of the modified methodology with sliding window approach reduces the memory consumption for processing larger images but also increases the time required to perform detection in large satellite imagery.

2.7 Assessment

Once the predictions are accumulated, the number of buildings detected in pre-disaster and post-disaster images are compared. Based on this comparison, rapid damage assessment has been done. This method provides initial assessment of damage in the area. Assessment for Indonesian earthquake was done. The im-ages for this assessment were obtained from Digital Globe website. A sample assessment has been done for the Pipal data, Sindhupalchok, Nepal using im-ages obtained from the Google Earth Pro.

3 Results and Analysis

3.1 Morphological Operators

At first method, the building extraction was done using morphological operators over the image and segmentation using the surface reflectance. This method incorporated multiple use of the opening and closing operators to reduce the noise. Surface reflectance was incorporated to segment the buildings from the other objects. This method relied heavily on the surface reflectance. As the surface reflectance of the building differs from scene to scene as well as building to building, one value of surface reflectance used to segment in one image may not be best suited for another image. Due to this problem, this method was rejected.

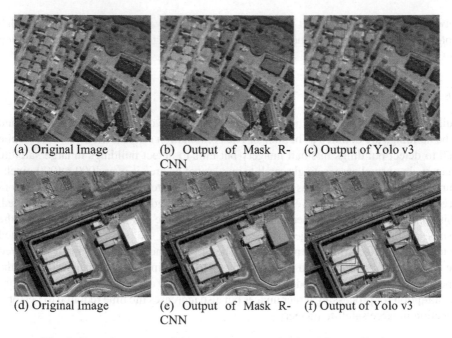

(a) Original Image   (b) Output of Mask R-CNN   (c) Output of Yolo v3

(d) Original Image   (e) Output of Mask R-CNN   (f) Output of Yolo v3

Fig. 4. Detection output of the Mask R-CNN and Yolo v3 for smaller images

3.2 Mask R-CNN

The next method used was the R-CNN algorithm i.e. the two stage procedure with Region Proposal Network in the first stage and predicting the class and box o set in the next stage. The output of Mask R-CNN for the satellite imagery with size of 438 * 406 pixels along with the original image is shown in Fig. 4.

In order to train the RPN, crowdAI mapping challenge dataset was used. The dataset was split in 60% training data, 20% validation data and 20% test data. The training was done on Inter(R) Core(TM) i5-7200 CPU @2.50 GHz. 30 epochs were run for training. The training was completed in 14 days. The training obtained the validation loss of 5%. For the training of RPN with custom dataset of Nepal, the dataset was split in 60% training data, 20% validation data and 20% test data. The training was done on the Google Colab with GPU environment and 12 GB ram. Overall 160 epochs were run for training. The training obtained the validation loss of 18%.

As we can see in Fig. 4, Mask R-CNN gives the classification with significant accuracy in smaller images. As seen in Fig. 4a and Fig. 4b, masks are predicted over 25 out of 35 buildings present. While in Fig. 4d and Fig. 4e out of 6 buildings, masks are predicted over 5 buildings. However, for the larger satellite images, this method fails to correctly predict as this algorithm needs to resize the image for prediction. The output for large satellite imagery (990 * 660) is shown in Fig. 5. As we can see in Fig. 5b, the masks only overlap over the 6 buildings over a large area. This method is slower to train and predict. The accuracy of this method is great in smaller images but fails to predict on larger images. To increase the accuracy, we shifted towards much more sophisticated algorithm.

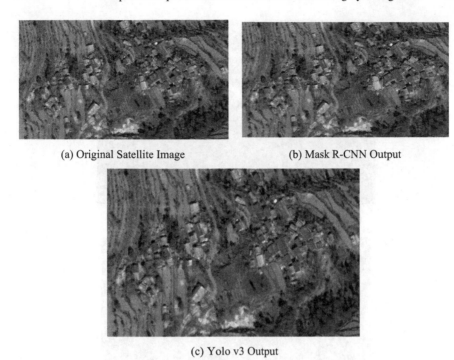

(a) Original Satellite Image   (b) Mask R-CNN Output

(c) Yolo v3 Output

Fig. 5. Detection results of Mask R-CNN and Yolo v3 on larger image

3.3 Yolo V3

The next method used is Yolo v3 algorithm. The algorithm is applied over the whole image. The network divides images into regions and predicts bounding boxes and probabilities for each region. The output of this algorithm for smaller images is compared with RPN and the result is shown in Fig. 4. This method classifies much better in satellite imagery over the Mask R-CNN as shown in Fig. 5c. However, it can be seen that, number of wrong areas have also been classified as buildings. This method has great accuracy for the smaller images but fails on larger images.

3.4 Combined Approach

Due to shortcomings of the above approach, a new integrated approach is pro-posed to improve the process for the building extraction. This approach uses a combination of Region Proposal Network to detect regions containing specified objects in our case 'building' and Yolo v3 classifier to classify if the proposed regions are buildings or not. The combined approach, with model trained in international standard dataset, is applied to the satellite image of Kunchok, Sindhupalchok. The result was not so promising as the number of buildings detected was very bad. To solve this issue, sliding window approach is integrated in the pipeline. Under this method, a section of image is passed via the process at once and the output is accumulated. Once all the sections of images are processed, accumulated output is mapped to the original image and the output of

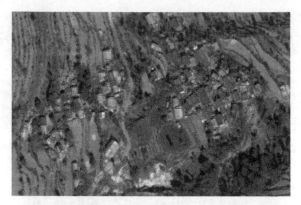

(a) Result of Combined approach with sliding window

b) Result of Custom Dataset trained model

Fig. 6. Detection results for Kunchok, Sindhupalchok

the combined approach with sliding window method is shown in Fig. 6a. The combined approach along with sliding window approach has produced better results than other methods implemented above.

Further, both RPN and Yolo v3 were trained on manually annotated images of the Sindhupalchok district. The detection result for the Kunchok, Sindhupalchok by the network trained on custom dataset is shown in Fig. 6b. By comparing Figs. 6a and 6b, it can be clearly seen that the model trained on manually an-notated images predicts even those buildings located at the right and left corner of the image which were not predicted by the model trained on an international standard dataset. One other prediction was done to test how the results vary for the large satellite imagery. The implemented approach was tested on Pipal Danda image acquired via the Google Earth pro and it's output is shown in Fig. 7. We can see that the bounding boxes are detected over a large number of buildings. The output of the combined approach trained on a custom dataset which showed better results for multiple areas in Nepal are shown in Fig. 8. It can be seen that separate buildings in the busy areas are also predicted.

Fig. 7. Combined approach detection result on larger images

3.5   Custom Dataset Assessment

The satellite imagery for Pipal Danda, Nepal of multiple timelines was obtained from Google earth pro to highlight the possibility of rapid disaster assessment. The images of the same location captured in 2018, 2016 and 2015 A.D. after the earthquake respectively are obtained. The assessment of building detection was done on those images to show the changes the specified area has gone over the interval of 3 years. An attempt was made to capture the image just before the 2015 A.D. earthquake but was not possible due to unavailability of day time images.

The buildings detected for the Pipal Danda, Nepal are shown in Fig. 9, Fig. 10 and Fig. 11. At first assessment was done on the imagery of 2018 A.D. to show the present buildings in the area. Then 2016 A.D. images were analyzed and the building count is compared to calculate the percentage change. The combined approach detected 145 buildings for the 2018 A.D. image. The combined approach detected almost 115 buildings for the 2016 A.D. image. As we can see, the number of buildings has changed by 26.08% in 2 years. The combined approach detected almost 55 buildings for the image captured in 2015 A.D. just after the earthquake. This shows that the number of buildings has changed by 62.06%. This approach can be used to assess the damage area by analyzing the images captured before and after the earthquake.

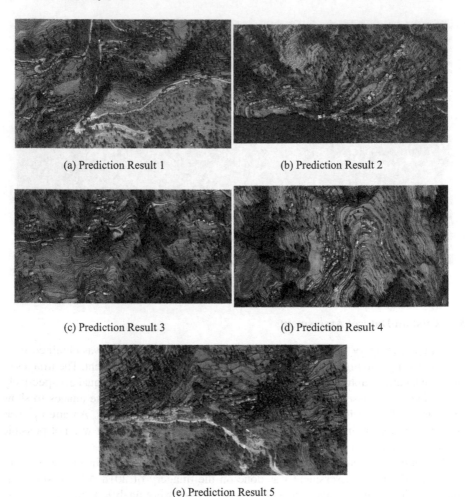

(a) Prediction Result 1

(b) Prediction Result 2

(c) Prediction Result 3

(d) Prediction Result 4

(e) Prediction Result 5

Fig. 8. Various detection results on custom dataset trained model

3.6 Evaluations

After the detection of the buildings with the combined approach, the detected buildings are compared with the ground truth images i.e. manually annotated satellite images to identify number of the absent buildings. On the basis of absent buildings, the assessment was made and confusion matrix was obtained as shown in Table 1. The method achieved the overall F1 score of 0.89 as well as Precision of 0.94 and Recall of 0.86.

Fig. 9. 2018 A.D. image of Pipal Danda, Nepal

Fig. 10. 2016 A.D. image of Pipal Danda, Nepal

Fig. 11. 2015 A.D. image of Pipal Danda, Nepal

Table 1. Building assessment evaluation

Actual/Predicted	Yes	No	Total
No	6	5	11
Yes	6	75	81
Total	**12**	**80**	**92**

4 Conclusion

In this research, multiple methods like Morphological Operators, Mask R CNN and Yolo v3 were applied for identifying a suitable method for building detection. It was observed that hybrid approach with Mask R-CNN along with Yolo v3 performed better. The model was trained with an international dataset and pretty good results were obtained with standard images. However, the detection for multiple satellites imagery of Nepal acquired from Google Earth Pro was only improved after training with manually annotated custom dataset for Nepal. For the building damage/change assessment, the images of the same area acquired over multiple time ranges were acquired. The detection of buildings was done on the acquired image of different timelines. The number of absent buildings were identified and considered as a basis for the damage/change assessment. The pro-posed method achieved the overall F1 score of 0.89 as well as Precision of 0.94 and Recall of 0.86. At present, the custom dataset is small so the predictions are still missing. Increasing the data set will help to solve this problem. Currently, assessment is done based on the number of missing buildings which is not a very good factor for assessment. Using the difference in the pixel values of building over the pre-disaster and post-disaster imagery will make a good assessment. The method doesn't predict well on the dark images thus giving the wrong predictions. Using gray scale images for the training could result in better predictions in dark illuminated images.

References

1. Dey, S.: A devastating disaster: a case study of nepal earthquake and its impact on human beings. IOSR J. Hum. Soc. Sci. **20**, 28–34 (2015)
2. Liu, W., Prinet, V.: Building detection from high-resolution satellite image using probability model. In: Proceedings, 2005 IEEE International Geoscience and Remote Sensing Symposium 2005. IGARSS 2005, vol. 6, pp. 3888–3891. Citeseer (2005)
3. Fries, B., et al.: Measuring the accuracy of gridded human population density surfaces: a case study in bioko island, equatorial guinea. BioRxiv (2020)
4. Chiroiu, L., Andre, G.: Damage assessment using high resolution satellite imagery: application to 2001 Bhuj, India, earthquake. In: Proceedings of the 7th National Conference on Earthquake Engineering (2001)
5. Jenis, A.: Earthquake damage assessment of buildings using pre-event and post-event imagery, vol. 1, March 2012
6. Parapayalage, C.D.K.: Building extraction in hazardous areas using extended morphological operators with high resolution optical imagery (2014)
7. He, K., Gkioxari, G., Dollár, P., Girshick, R.: Mask R-CNN. In: Proceedings of the IEEE International Conference on Computer Vision, pp. 2961–2969 (2017)
8. LuqmanAli, W.K., Chaiyasarn, K.: Damage detection and localization in masonry structure using faster region convolutional networks. Int. J. **17**(59), 98–105 (2019)
9. Redmon, J., Farhadi, A.: Yolov3: an incremental improvement. arXiv preprint arXiv:1804.02767 (2018)
10. Ma, H., Liu, Y., Ren, Y., Yu, J.: Detection of collapsed buildings in post-earthquake remote sensing images based on the improved yolov3. Remote Sens. **12**(1), 44 (2020)
11. Glumov, N., Kolomiyetz, E., Sergeyev, V.: Detection of objects on the image using a sliding window mode. Opt. Laser Technol. **27**(4), 241–249 (1995)
12. Crowd AI Mapping Challenge Dataset. https://www.crowdai.org/challenges/mapping-challenge. Accessed 19 Oct 2020
13. AWS Spacenet challenge dataset. https://spacenetchallenge.github.io/datasets/datasetHomePage.html. Accessed 19 Oct 2020
14. Bing Satellite Images. https://www.bing.com/maps/aerial. Accessed 19 Oct 2020
15. VGG Image Annotator, Visual Geometry Group - University of Oxford. https://www.robots.ox.ac.uk/~vgg/software/via/. Accessed 19 Oct 2020
16. Ren, S., He, K., Girshick, R., Sun, J.: Faster R-CNN: towards real-time object detection with region proposal networks. arXiv preprint arXiv:1506.01497 (2015)
17. Karmarkar, T.: Regional proposal network (RPN)—backbone of faster R-CNN, 18 August (2018). 6
18. He, K., Zhang, X., Ren, S., Sun, J.: Deep residual learning for image recognition. In: Proceedings of the IEEE Conference on Computer Vision and Pattern Recognition, pp. 770–778 (2016)
19. Rastiveis, H., Samadzadegan, F., Reinartz, P.: A fuzzy decision making system for building damage map creation using high resolution satellite imagery. Nat. Hazards Earth Syst. Sci. **13**(2), 455–472 (2013)

Attention-Based Deep Multi-scale Network for Plant Leaf Recognition

Xiao Qin[1], Yu Shi[1], Xiao Huang[1], Huiting Li[1], Jiangtao Huang[1(✉)],
Changan Yuan[2(✉)], and Chunxia Liu[3]

[1] Nanning Normal University, Nanning 530299, China
Jiangtao@nnnu.edu.cn
[2] Guangxi Academy of Science, Nanning 530025, China
[3] Guangxi Technological College of Machinery and Electricity, Nanning 530007, China

Abstract. Plant leaf recognition is a computer vision task used to identify plant species. To address the problem that current plant leaf recognition algorithms have difficulty in recognizing fine-grained leaf classification between classes, this paper proposes a DMSNet (Deep Multi-Scale Network) model, a plant leaf classification algorithm based on multi-scale feature extraction. In order to improve the extraction ability of different fine-grained features of the model, the model is improved on the basis of Multi-scale Backbone Architecture model. In order to achieve better plant leaf classification, a visual attention mechanism module to DMSNet is added and ADMSNet (Attention-based Deep Multi-Scale Network), which makes the model focus more on the plant leaf itself, is proposed, essential features are enhanced, and useless features are suppressed. Experiments on real datasets show that the classification accuracy of the DMSNet model reaches 96.43%. In comparison, the accuracy of ADMSNet with the addition of the attention module reaches 97.39%, and the comparison experiments with ResNet-50, ResNext, Res2Net-50 and Res2Net-101 models on the same dataset show that DMSNet improved the accuracy by 4.6%, 18.57%, 3.72% and 3.84%, respectively. The experimental results confirm that the DMSNet and ADMSNet plant leaf recognition models constructed in this paper can accurately recognize plant leaves and have better performance than the traditional models.

Keywords: Plant leaf classification · Multi-scale backbone · ECANet

1 Introduction

The classification of plant images is an important research interest in plant taxonomy. The traditional features extracted from plant leaves are mainly color, shape, size, texture, etc. However, plant leaves have complex variation, and there will be differences in size and texture for the same category of leaves, and different categories of leaves may have extremely similar shapes or textures. Therefore, plant leaf classification is a great challenge. Convolutional neural networks can learn the intrinsic pattern of sample data and obtain the hidden feature information in the image. Therefore, convolutional neural networks are widely used in various image recognition tasks in the field of computer

© Springer Nature Switzerland AG 2021
D.-S. Huang et al. (Eds.): ICIC 2021, LNCS 12836, pp. 302–313, 2021.
https://doi.org/10.1007/978-3-030-84522-3_24

vision [1–7]. Furthermore, convolutional neural networks have shown good classification results on the task of plant leaf image classification. Uzal et al. [8] used convolutional neural networks for extracting features and subsequently used support vector machines to classify soybean leaves. Sun et al. [9] designed a 26-layer deep learning model consisting of eight residual modules to classify plant images on the campus of Beijing Forestry University and achieved outstanding results. Hu et al. [10] proposed an MSF-CNN model with input images downsampled as multiple low-resolution images. Then, these input images with different scales are progressively fed into the MSF-CNN architecture, and this model has good classification results on the MalayaKew Leaf dataset and LeafSnap Plant Leaf dataset. Bodhwani et al. [11] made it more applicable to ResNet-50 [12] by improving it for plant leaf classification and achieved a classification accuracy of 93% on the 185-class plant leaf dataset.

The important features of plant leaf images are generally fine-grained, and the current convolutional neural networks mostly applied in plant leaf classification pay less attention to fine-grained features. In order to make the convolutional neural network model pay more attention to the fine-grained features of plant leaves, this paper tries to take the Multi-scale Backbone Architecture model Res2Net [13] as the base architecture, adding multi-scale feature extraction, so that the model can capture more fine-grained features, and experiments prove the effectiveness of the model.

The main work of this paper is as follows:

(1) DMSNet, an algorithm applicable to plant leaf classification, is proposed, and multi-scale feature extraction is added to the Multi-scale Backbone Architecture model to increase the diversity of features and capture fine-grained features from different perspectives, making the model more applicable to plant leaf classification.
(2) In response to the fact that the DMSNet model does not pay enough attention to the fine-grained features of plant leaves, a channel attention module is added on the basis of DMSNet. The ADMSNet model is constructed, and the attention module allows the model to extract the features of plant leaves better and have better classification effects.
(3) Experiments using DMSNet and ADMSNet models on real plant leaf dataset, the accuracy reached 96.43% and 97.39%, respectively.
(4) The DMSNet and ADMSNet models were compared experimentally with the convolutional neural network model Res2Net. There is a significant improvement in the accuracy rate. It proves the effectiveness of the model proposed in this paper and outperforms the traditional model.

2 Related Work

2.1 Res2Net

Res2Net [13] was proposed in 2019 by Gao et al. They constructed a hierarchical residual class link Basic2Block module on a single residual block. In the Basic2Block module of Res2Net, a hierarchical residual linkage in a single residual block enables changes in the sensory field at a finer granularity level to capture details and global properties.

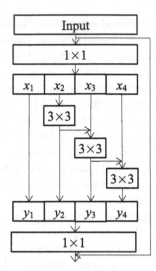

Fig. 1. Basic2Block module of Res2Net ($s = 4$)

Figure 1 shows the Basic2Block module in Res2Net, which replaces the n-channel convolution kernels with s-w-channel convolution kernels, where $n = s*w$. It can have stronger multi-scale feature extraction capability with similar computational effort. After $1*1$ convolution, the input is sliced into s feature subsets, defined as x_i, $i \in \{1,2,...,s\}$, each of which has the same scale size, except for x_1, the other feature subsets which are convolved with $3*3$ convolution kernels, where these convolutions are defined as $F_i(x_i)$, and the output is defined as y_i. Thus y_i can be written as Eq:

$$y_i = \begin{cases} x_i, i = 1; \\ F_i(x_i), i = 2; \\ F_i(x_i + y_{i-1}), 2 < i \le s. \end{cases} \tag{1}$$

Each $3*3$ convolution operation can potentially accept all the feature information on its left side, and each output can increase the perceptual field, so each Basic2Block can acquire a different number of feature combinations with different perceptual field sizes.

2.2 ECANet

Wang et al. [14] analyzed the channel attention module of SENet [15] and its variants (SE-Var1, SE-Var2 and SE-Var3). They demonstrated that avoiding dimensionality reduction is important for weight learning of channel attention and that proper cross-channel interaction can maintain performance while significantly reducing model complexity. Therefore, the authors propose an efficient channel attention ECA module for convolutional neural network models. This module can efficiently implement local cross-channel feature information interaction without dimensionality reduction by one-dimensional convolution, and the adaptively sized one-dimensional convolution kernel determines the coverage of local cross-channel feature information interaction.

The channel attention ECA module is shown in Fig. 2. First, the module first pools the input features globally averaged to obtain a feature information y of $1*1*C$, C being the number of channels of the input features. For each channel y_i, the ECA module only considers y_i to interact with the k adjacent channels, and to further improve the performance of the module, the ECA will let all channels share the weight information, thus the weight ω_i of each y_i is calculated as:

$$\omega_i = \sigma(\sum_{j=1}^{k} \alpha^j y_i^j), y_i^j \in \Omega_i^k \qquad (2)$$

Where: $\Omega^k{}_i$ is the set of k neighboring channels of y_i and σ is the activation function. To be able to achieve the above interaction between channels, the module uses a convolution kernel of size k to achieve the following equation:

$$\omega = \sigma(C1D_k(y)) \qquad (3)$$

$C1D_k$ is a fast one-dimensional convolution of kernel size, while for the kernel size k represents the coverage of the local cross-channel interaction, i.e., channel y_i will have k neighboring channels involved in its attention prediction. And this idea of obtaining cross-channel information interactions uses only k parameters three main steps of the ECA module: (1) global average pooling of the input features to obtain $1*1*C$ features. (2) Obtain a one-dimensional convolution kernel of adaptive size k by computation. (3) One-dimensional convolution of the $1*1*C$ features to obtain the weights of each channel. It can be seen that the channel attention module ECA can be used on the convolutional neural network model with just a small increase in parameters to give a significant improvement in the model's effectiveness.

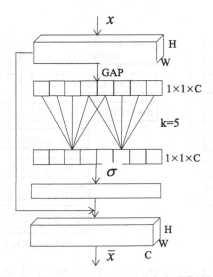

Fig. 2. Channel attention module in ECANet

3 DMSNet (Deep Multi-scale Network)

As mentioned above, the Res2Net module has good performance in object recognition and segmentation type tasks, thus this paper proposes a plant leaves classification module DMSNet (Deep Multi-Scale Network) based on the Res2Net module as the basic architecture. The features of plant leaves are complex which means plant leaves of different categories may be extremely similar, while the same category of plant leaves may also have large differences, which requires the module to have good access to the fine-grained features of plant leaves. The fine-grained features of the image can be acquired through the Multi-scale Backbone Architecture of Res2Net very well. This paper maintains that there will be different fine-grained features in the images of different sizes obtained after downsampling the same image. In order to get these different fine-grained features, this paper adds a new process to the Res2Net network based on the multi-scale grain size feature extraction process of the input image after downsampling. The specific procedures are (1) using the original size image to obtain the features x_2, (2) downsampling the original image to obtain the features of different grain size x_2', (3) conducting channel splicing of x_2 and x_2'. The main architecture of Res2Net-50 and DMSNet are shown in Fig. 3, respectively. Inside the red grid lines in Fig. 3 are the improvement of DMSNet to Res2Net-50 by replacing the original four Basic2Block module stacked layers with a downsampled convolutional module. The features acquired by this module are fused with the features outputted by the stacked layers after three Basic2Block modules. This part of the module is called Multi-Scale Feature Extraction (MSFE) in this paper. The second half is consistent with the architecture of Res2Net-50, where DMSNet inputs the different angular features obtained by MSEF to the stacked layer of 6 Basic2Blocks and the stacked layer of 3 Basic2Blocks. Finally, pooling, full connectivity, and classification are performed.

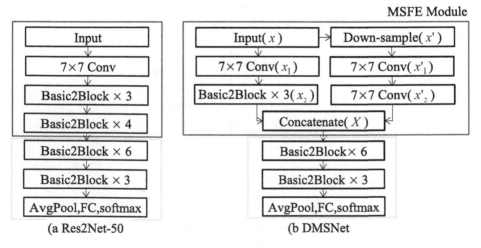

Fig. 3. Architecture of Res2Net-50 and DMSNet

The main algorithm description of the MSFE module is shown in Table 1: (1) The image data x first undergoes the convolution of 7*7 for initial feature extraction to obtain the feature x_1. (2) Further feature extraction is then performed using three stacked layers of the Basic2Block module (noted in the algorithm description as $H()$) to obtain a combination of features with different receptive field sizes. (3) The original image data x is downsampled to obtain smaller image data x'. (4) Similarly, the image data x' undergoes the convolution of 7*7 for initial feature extraction to obtain the feature x_1'. (5) The feature x_1' are then convolved by a second 7*7 convolution to obtain the feature x_2'. The purpose of this convolution is to extract the features that are different from those in step (2) while making the channel size consistent with them to facilitate the subsequent channel splicing. (6) The final stage is to splice the acquired feature x_2 with the feature x_2' to obtain the final output feature X.

Table 1. Algorithm procedure of the MSFE module.

Multi-scale feature extraction procedure under DMSNet
Input: images x
Output: feature map X
1 : $x_1 = Conv\_7(x)$ 2 : $x_2 = H(x_1)$ 3 : $x' = \text{downsample}(x)$ 4 : $x_1' = Conv\_7(x')$ 5 : $x_2' = Conv\_7(x_1')$ 6 : $X = cat(x_2, x_2')$
Return X

4   ADMSNet (Attention-Based Deep Multi-scale Network)

There are features that are not useful for model classification which increasing the diversity of features and affecting the classification effectiveness of the module among the features extracted by the DMSNet module. And the attention mechanism based on computer vision proposed in recent years can make the module pay more attention to the beneficial features. In this paper, the ECA module (inside the yellow grid line) is added to the Basic2Block module of Res2Net to form the ECA-Basic2Block (ECA-B) module. The purpose is to conduct a better model algorithm to extract useful fine-grained features from the complex textures of plant leaves, allowing the module to have better classification performance. And the ADMSNet plant leaves classification model is constructed based on the (ECA-B) module.

Figure 4 shows the overall architecture of ADMSNet. The overall algorithm procedure is basically the same as that of DMSNet, and the main improvement is that the

ECA-B module is used to replace all the Basic2Block modules. Compared with the DMSNet model, the feature extraction capability of ADMSNet is greatly improved due to the increase of a tiny amount of parameters.

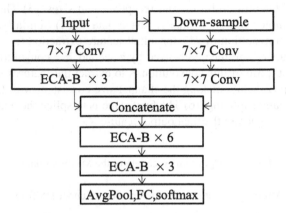

Fig. 4. Architecture of ADMSNet module

Figure 5 shows the framework of ECA-B module.

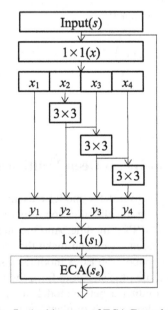

Fig. 5. Architecture of ECA-B module

Table 2 shows the algorithm description of ECA-B: (1) ECA-B takes the feature s obtained in the previous layer as the input to the current module, and subsequently obtains the feature x after the first 1*1 convolution. (2) The feature x are segmented averagely in the channel into feature $x_1, x_2,..., x_i$, where $i = 4$. The segmentation process of the features is noted in the algorithm description as $G(x)$. (3) Feature $y_1, y_2,..., y_i$ can be obtained according to Eq. 1 mentioned above. (4) Subsequently, feature y can be obtained through channel splicing of the above features. (5) The feature s_1 is then obtained by a second 1*1 convolution. (6) Then, input the feature s_1 to the ECA module and conduct weight calculation of each channel to obtain the feature s_e with attention weights. (7) Finally, the acquired feature s_e are added with the initial input feature s to obtain the feature S as the input of the next module.

Table 2. Procedure of ECA-B module

Procedure of ECA-B nodule
Input: feature map s
Output: feature map S
$1 : x = Conv\_1(s)$
$2 : x_1, x_2, x_3, x_4 = G(x)$
$3 : y_1 = x_1; y_2 = F_2(x_2); y_3 = F_3(x_3 + y_2); y_4 = F_4(x_4 + y_3)$
$4 : y = cat(y_1, y_2, y_3, y_4)$
$5 : s_1 = Conv\_1(y)$
$6 : s_e = eca(s_1)$
$7 : S = s_e + s$
Return S

5 Experiment

5.1 Experimental Environment and Parameter Setting

All the experiments in this paper were done on a plant leaves dataset created by the author. The hardware includes an Intel i7–7700 processor, NVIDIA GTX1080Ti GPU are used, with learning-rate set to 0.1, momentum set to 0.9, weight-decay set to 1e-4, and iteration times set to 300. The image size input initially is 128*128.

5.2 Experimental Dataset

The plant leaf dataset used in this paper contains a total of 15207 pieces belong to 201 categories of plant leaves, and each category has at least 60 images. Also, 70% of the images are used as the training set and 30% as the validation set. Figure 6 shows some images of the plant leaves dataset to belong to 8 categories.

Fig. 6. Part of images from the plant leaves dataset

5.3 Classification Experiments with Real Data

To solve the classification disturbs such as noise and similarity in plant leaf image data, initialization modules of ResNet-50 [12], ResNext-50 [16], Res2Net-50 [13], and Res2Net-101 [13] are used to perform predictive classification on the small sample plant leaves dataset we produced. Table 3 shows the experimental results, and their correct rates are 77.86%, 91.83%, 92.71%, and 92.95%, respectively. Among which ResNext-50 is proposed by borrowing the idea of ResNet-50, but it is not very effective in plant leaf classification. The reason is that its improved residual module does not focus on the fine-grained features of plant leaves images, which are very important for plant leaf classification. ResNet-50 and Res2Net-50 are better at obtaining the fine-grained features of plant leaves, and thus show better accuracy on our small sample plant leaf dataset. The reason why Res2Net-50 is precise than ResNet-50 is its improvement of the residual module of ResNet-50, which allows the network to acquire a combination of features with different receptive field sizes. The proposed module DMSNet improves the classification accuracy by 3.72% to 96.43% compared to Res2Net-50 after adding multi-scale feature extraction. This proves that the method proposed in this paper of multi-scale feature extraction to diversify feature information is effective. The ADMSNet module obtained by adding the ECA module based on DMSNet can focus more on the plant leaves themselves, enhancing beneficial features and suppressing useless features according to the importance of their features. With the addition of the ECA module, the accuracy of the ADMSNet module is improved by 0.96% compared with DMSNet, which proves that the addition of the attention module is effective. And ADMSNet has the best classification effect in the current experiments.

5.4 Ablation Experiments of ADMSNet

To further verify that both the multi-scale feature extraction and the addition of channel attention in the model proposed in this paper can improve the model's classification accuracy for small data plant leaves, the ablation experiments of ADMSNet are compared. Table 4 shows the results of the ablation experiments of ADMSNet, where the Res2Net-50-ECA model refers to replacing all the Basic2Block of Res2Net-50 with ECA-B. As shown in Table 4, the two models ADMSNet and Res2Net-50-ECA, which

Table 3. Accuracy comparison among DMSNet, ADMSNet and other module

Method	Test set accuracy
ResNext [16]	77.86%
ResNet-50 [12]	91.83%
Res2Net-50 [13]	92.71%
Res2Net-101 [13]	92.95%
DMSNet	**96.43%**
ADMSNet	**97.39%**

used the ECA-B module, showed varying degrees of improvement in classification accuracy on the small plant leaf dataset, with ADMSNet being the most pronounced. This demonstrates that the channel attention mechanism can indeed improve the classification accuracy of the models.

At the same time, it can be found that the accuracy of DMSNet is almost the same as that of Res2Net-50-ECA. On the one hand, it shows that the addition of the channel attention module to the Res2Net-50 model enhances the attention to the fine-grained features of the plant leaf images, and the accuracy improvement is very obvious. On the other hand, the DMSNet proposed in this paper can also make the model pay attention to more fine-grained features by increasing the diversity of features through multi-scale feature extraction, and the effect is obvious. ADMSNet combines the above two methods and thus works best. It is also further demonstrated that there are different fine-grained features on different image data sizes, and a small number of these different features are helpful for model classification. All the experimental results fully validate that the proposed method in this paper has good classification performance on real datasets.

Table 4. Results of the ablation experiments of ADMSNet

Method	Test set accuracy
Res2Net-50 [13]	92.71%
DMSNet	**96.43%**
Res2Net-50-ECA	96.05%
ADMSNet	**97.39%**

6 Conclusions

In this paper, based on the characteristics of the plant leaf image dataset taken in the natural environment, which has a large amount of noise and leads to difficult extraction of plant image features and difficult classification, we propose the DMSNet algorithm and ADMSNet algorithm based on Res2Net. In the ADMSNet algorithm, by adding multi-scale input and ECA module and increasing the diversity of image features and the feature extraction ability of the network, the image classification with more noise of plant leaves is effectively solved.

In future work, this paper will increase the sample size of the real sampled dataset and apply the method proposed in this paper to the classification task of a more complex real plant leaf image dataset to improve further the ability of the model to extract plant leaf features and improve the accuracy of plant leaf classification in a natural environment.

Acknowledgements. This paper was supported by the National Natural Science Foundation of China (Grant No. 61962006, 61802035, 61772091); the Project of Science Research and Technology Development in Guangxi (Grant No. AA18118047, AD18126015, AB16380272); the BAGUI Scholar Program of Guangxi Zhuang Autonomous Region of China (Grant No. 2016 [21], 2019 [79]); the Natural Science Foundation of Guangxi (Grant No. 2018GXNSFAA138005) and the Sichuan Science and Technology Program (Grant No. 2018JY0448, 2019YFG0106, 2020YJ0481, 2019YFS0067, 2020YFS0466, 2020JDR0164, 2020YJ0430).

References

1. Zoran, D., Chrzanowski, M., Huang, P., et al.: Towards robust image classification using sequential attention models. arXiv: Computer Vision and Pattern Recognition (2019)
2. Xie, Q., Hovy, E., Luong, M., et al.: Self-training with noisy student improves ImageNet classification. arXiv: Learning (2019)
3. Bertinetto, L., Muller, R., Tertikas, K., et al.: Making better mistakes: leveraging class hierarchies with deep networks. arXiv: Computer Vision and Pattern Recognition (2019)
4. Wu, Y., Zhang, K., Wu, D., et al.: Person re-identification by multi-scale feature representation learning with random batch feature mask. IEEE Tran. Cogn. Dev. Syst. (2020)
5. Chen, Z., Wei, X., Wang, P., et al.: Multi-label image recognition with graph convolutional networks. In: Computer Vision and Pattern Recognition, pp. 5177–5186 (2019)
6. Wang, G., Wang, K., Lin, L., et al.: Adaptively connected neural networks. In: Computer Vision and Pattern Recognition, pp. 1781–1790 (2019)
7. Yuan, C., Wu, Y., Qin, X., et al.: An effective image classification method for shallow densely connected convolution networks through squeezing and splitting techniques. Appl. Intell. **49**(10), 3570–3586 (2019)
8. Uzal, L.C., Larese, M.G., et al.: Deep learning for plant identification using vein morphological patterns. Comput. Electron. Agric. **127**, 418–424 (2016)
9. Sun, Y., Liu, Y., et al.: Deep learning for plant identification in natural environment. Comput. Intell. Neurosci. **2017**(4), 7361042–7361042 (2017)
10. Hu, J., Chen, Z., Yang, M., et al.: A multi-scale fusion convolutional neural network for plant leaf recognition. IEEE Sig. Process. Lett. **25**, 853–857 (2018)
11. Bodhwani, V., Acharjya, D.P., Bodhwani, U.: Deep residual networks for plant identification. Procedia Comput. Sci. **152**, 186–194 (2019)
12. He, K., Zhang, X., Ren, S., et al.: Deep residual learning for image recognition. In: Proceedings of the IEEE Conference on Computer Vision and Pattern Recognition, pp. 770–778(2016)
13. Gao, S., Cheng, M.M., Zhao, K., et al.: Res2Net: a new multi-scale backbone architecture. IEEE Trans. Pattern Anal. Mach. Intell. **43**, 652–662 (2019)

14. Wang, Q., Wu, B., Zhu, P., et al.: ECA-Net: efficient channel attention for deep convolutional neural networks. In: 2020 IEEE/CVF Conference on Computer Vision and Pattern Recognition (CVPR). IEEE (2020)
15. Jie, H., Li, S., Gang, S.: Squeeze-and-excitation networks. In: 2018 IEEE/CVF Conference on Computer Vision and Pattern Recognition (CVPR). IEEE (2018(
16. Xie, S., Girshick, R., Dollár, P., et al.: Aggregated residual transformations for deep neural networks. In: Proceedings of the IEEE Conference on Computer Vision and Pattern Recognition (2016)

Information Security

Short Video Users' Personal Privacy Leakage and Protection Measures

Haiyu Wang[1,2]([✉])

[1] School of Economics and Management, Zhengzhou Normal University, Zhengzhou, China
[2] School of Journalism and Communication, Zhengzhou University, Zhengzhou, China

Abstract. Information security has long been a concern of the people. The widespread use of short videos has raised the academia and industry's attention to the leakage of users' personal privacy and protection measures. In short videos, there are three main situations of unconscious active disclosure of users' personal privacy, conscious partial protection, and consented information collection. When users disclose personal information, there are optimistic biases and the need to build group identity. The privacy agreement of the short video APP obscures the user's wishes. Based on this, users need to increase their awareness of personal privacy protection, and platform companies need to strengthen the protection of users' personal information in terms of short video download permissions, protocol security, and industry self-discipline.

Keywords: Short videos · Privacy breach · Privacy protection

1 Introduction

Information security has long been a concern for people. As early as 1973, some Western researchers proposed that computers and the Internet will become an important threat to personal information privacy in the future [1]. The concept of privacy has been widely present in the American legal academic literature since the 19th century. The "information privacy security" we are talking about now refers more to a series of issues such as data protection that have emerged in the society since the Internet era. The development of the Internet makes people realize that network data may be monitored by others. Privacy not only refers to avoiding others' prying eyes in the legal sense, but also includes network data privacy. With the widespread use of media platforms, this contradiction has been further magnified in Douyin short videos.

As a new social method, mobile short video has been popular and sought after by netizens for adapting to the needs of expression and viewing in fragmented time. Driven by technology and commercial propaganda, mobile short video has become a tool for people's daily life. But it is also the unique dissemination advantage of mobile short video that makes it easy to contribute to the spread of online false information, online violence, and online pornography. Under the glamorous appearance of the short video industry, the hidden factors that threaten the healthy and stable development of society cannot be ignored [2].

© Springer Nature Switzerland AG 2021
D.-S. Huang et al. (Eds.): ICIC 2021, LNCS 12836, pp. 317–326, 2021.
https://doi.org/10.1007/978-3-030-84522-3_25

Most of the short videos produced content for users. The "protagonist" of the video is usually oneself. Based on the psychology of "recording life", the content shot is mostly daily life. Personal living environment, income, emotional status, etc. is spied on. Short videos are spread based on weak relationships [3]. Generally, all users can watch and download the content that is shot locally. Even if the posted content is deleted, there is no guarantee that the content will not be recorded and saved. This also means that their login information, password forms, browsing behavior, identity information and other personal privacy can be written, stored, read and used in the server in the form of data, and the information security risks faced by users have suddenly increased. Although security vendors and government organizations have made unremitting efforts at different levels of protection technology, strategies, and laws and policies, from the current point of view, the adverse effects of information privacy and data leakage are increasing.

According to the 47th "Statistical Report on China's Internet Development Status" issued by the China Internet Network Information Center (CNNIC), as of December 2020, the number of Internet users in my country has reached 989 million, an increase of 85.4 million from March 2020, and the Internet penetration rate has reached 70.4%. Among them, the number of short video users is 873 million, accounting for 88.3% of the total netizens. A total of 980 questionnaires and 840 valid questionnaires were distributed in this survey.

Table 1. Gender Proportion of short video users

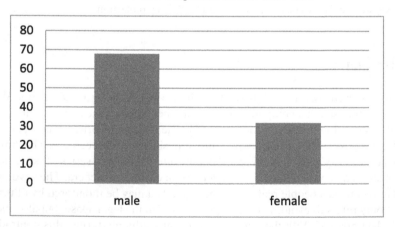

Table 1 shows the proportion of user gender attributes, short video users accounted for 68% of male users, and the ratio of men to women has changed from a relatively balanced ratio to nearly 7:3.

Table 2 shows that users aged 0–17 accounted for 14.77%, users aged 18–24 accounted for 35.57%, users aged 25–34 accounted for 29.75%, and users over 35 accounted for 17.9%.

Table 2. Age ratio of short video users

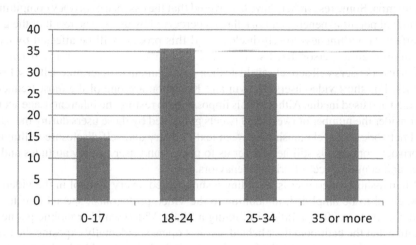

The rise in the proportion of young users has increased the risk of privacy leakage to a certain extent. Because young people's perception of privacy risks is relatively weak, they cannot effectively avoid risks. This article takes the current short video apps with a large number of users, high frequency of use, and a wide range of influences as examples, summarizes the types of users' personal privacy leaks in the use of short videos, and proposes to enhance the awareness of users' personal information protection, strengthen the platform's self-discipline and industry standards The perspective of user information protection, enrich the theoretical results of information security, and provide suggestions for the sound development of the short video industry [4].

In the following second part, three types of privacy leakage are summarized: unconscious information leakage, conscious partial protection and passive information collection. The third part points out the reasons for information leakage: personal optimism bias, group identity needs, and privacy agreements to obscure users' wishes. The fourth part proposes information protection measures.

2   Types of Privacy Leaks

2.1   Unconscious Disclosure

Self-disclosure willingness, as an important aspect of measuring the disclosure of users' private information, often appears in empirical research on information privacy. Although researchers usually regard it as the personal information sharing behavior of social users, only a few studies have obtained user push data. In most cases, self-disclosure is still obtained through attitude scales. The lower the willingness to self-disclose, it is greater the emphasis on protecting public and private privacy. In addition to the strong positive correlation between self-disclosure willingness and final information sharing behavior, researchers also found that privacy concerns negatively affect self-disclosure. However, follow-up studies have also found that the relationship between the two is not always negatively correlated, which is the so-called "privacy paradox" phenomenon. Some researchers have discovered that the essence of privacy computing is the game of potential benefits and risk threats perceived by users. The result of the game will affect the willingness to self-disclose, and this process will be affected by external stimuli such as personalized services and economic compensation [5]. Therefore, social network users will actively evaluate their environment to determine self-disclosure behavior. For short video users, Douyin and Kuaishou are one of the most frequently contacted and used media. Although it is impossible to restore the inherent usage scenarios, consider the number of tweets and tweets generated by these users during the use of social networks. The relationship between text delivery and self-disclosure willingness and privacy concerns is still helpful for us to understand user privacy attitudes and the causes and consequences of privacy behaviors.

When a short video user is recording a short video, every symbol in the video has the potential to be amplified, in addition to the video protagonist, special effects, and music. On 19 June 2018, a little girl filming a Shake Video with her mobile phone did not notice that the bathroom door behind her was open, accidentally exposing the image of her mother taking a shower. Within a short period of time the video's play count rose sharply, after which Jitterbug took down the video and responded that it would further improve the platform's audit mechanism [6].

The author learned in the process of research that some short video users inadvertently revealed their city and unit information in addition to showing the amount of salary in the salary slip during the live broadcast, which caused a lot of trouble for themselves. Another user told the researcher that when she was chatting with her grandmother, her grandmother accidentally picked up her clothes, which was considered by the network supervision to be too revealing, and the user was banned as a result.

Table 3 shows that 63.52% of personal information is often unconsciously leaked, indicating that more than half of users have a weak awareness of personal privacy protection.

Table 3. Proportion of unconscious disclosure

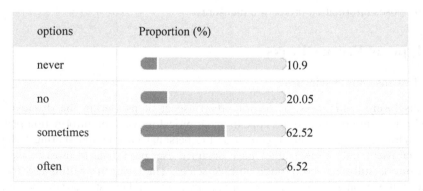

options	Proportion (%)
never	10.9
no	20.05
sometimes	62.52
often	6.52

2.2 Conscious Local Protection

According to Goffman's "self-presentation" theory, society is a big stage, and everyone is performing themselves. In interpersonal communication, people conduct marketing management on themselves in different ways, hoping to get more people's attention and praise. In the intelligent age, users have more and more opportunities to communicate with strangers and show themselves through short video platforms. In order to gain traffic and attention, and even become "net celebrities," they often consciously protect their privacy locally. Some short video users express themselves to others, share details of daily life, and show their cars and houses, but they will mosaic the car logos and license plates [7]. These contents often imply personal interests, family members and economic conditions, etc., and a lot of user information can be obtained intuitively. The conscious protection of key information shows that netizens' awareness of self-protection and privacy protection is increasing.

When uploading personal information, users obscure their names, units and ID numbers; salary amounts, geographic regions and occupations are often not obscured. The trending hash tag #WageYeah has 8,894 videos with a total of 7,758,000 views. The user named "Buddha Girl" uploaded a text message of her salary arrival with the "Salary ah" effect, which received 16,000 likes and 9,623 comments, with the amount of salary and account balance visible.

Table 4 shows that only about 27% of respondents are aware of the need to proactively protect personal information.

Table 4. Proportion of personal information protection

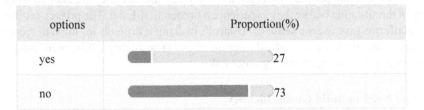

options	Proportion(%)
yes	27
no	73

2.3 Consented Information Collection

The current Internet era is the era of data. While data brings us convenience, problems such as the collection and leakage of Internet user information have become increasingly prominent. Short video users register for a Douyin account. There are five options for associated mobile phone number, headline number, Weibo number, QQ number, and WeChat account. No matter which method you choose to register, Douyin will unknowingly collect user address book information, But the next step is to pass mobile phone verification, set a password, and then real-name verification, that is, need to bind ID card, email, Weibo account, etc. If you want to share the video on other platforms, you can share it with one click. Then it will be pushed to users with the sign of "may know people", and at the same time, I will be recommended to others. After searching for related keywords several times to find videos, video recommendations from previous related searches often appear on the homepage of video sites, such as searching for game commentary, and the next two or three days of push and recommendation are all related game commentary videos. It can be seen that browsing on video websites will leave traces, while Internet operators use the information left by users for analysis and then make recommendations to users. After turning off the "Allow recommend me to friends" function in the privacy settings section, other users cannot find you through their address book, but you can still find people you may know through friend recommendations. In addition, Douyin users can choose to fill in labels such as gender, age, school, and region. If they are not filled in completely, the system will prompt, "You haven't filled in your profile yet, click to add…" No matter whether the user produces on the Douyin platform or not Or watch the video, and regardless of the frequency of the user's production or viewing of the video, a user portrait based on individual attribute information will be generated [8].

3 Analysis of the Causes of Privacy Leakage in Short Videos

3.1 Optimistic Bias of Information Subjects

An empirical study by the University of California on the privacy risks of Face book showed that people generally have an optimism bias about privacy risks, i.e. they believe they are at lower risk than the average person. Optimism bias is also known as the 'third person effect', where people tend to overestimate the impact on others and underestimate the impact on themselves when faced with the negative effects of mass media. Based on optimism bias, many people's perceptions of privacy and security are skewed from the reality of the situation before they experience a breach first-hand. The privacy settings on some platforms give users a sense of 'control', making it difficult for them to recognize that their control over personal information is a 'false reality', thus increasing their confidence and exposing them to greater risk.

3.2 The Need to Build Group Identity

In 1998, British sociologists Abercrombie and Lyndhurst introduced the concept of 'dispersed audiences' in their monograph Audience: A Sociological Theory of Performance and Imagination, in which people are performers while they are spectators. In addition, they propose the spectator/performer paradigm, which suggests that people engage in imaginative performances in the media and see themselves as performers in the spotlight, and that when identity construction is complete, this narcissistic state of mind triggers their next media consumption, in which the audience's identity is constantly reconstructed. While open, Jitter bit uses big data to divide its users into invisible "circles". Each user is interested in a different kind of content, but those who are interested in a certain kind of content and get pushed have some common characteristics. Big data technology crawls through keywords, categorizes the content and pushes it to users who are likely to be interested. If the content posted is not recognized, it cannot be interacted with, it cannot be 'liked' or 'commented on', and the producers of information cannot build group identity.

3.3 Privacy Agreements Obscures Users' Wishes

According to the "2019–2020 Public Opinion Monitoring Report on the Illegal Collection of Personal Information by Apps" published by Ai Media Consulting, 63.3% of respondents did not read the privacy terms carefully when using apps, and 24.3% of respondents never read the privacy terms. Jitterbug's privacy agreement is over 10,000 words long, with a lot of jargon, and many people click to agree to it in a hurry or without ever reading it. By agreeing to Jitter bit's privacy agreement, users are tacitly acknowledging that the Jitter bit platform can share user information with software service providers, smart device providers and advertising service providers. In addition, there are three situations that are not covered by the Privacy Agreement: information collected by third-party services accessed through Jitterbug; information collected by other companies or organizations that serve ads in Jitterbug; and information collected by third-party service providers.

4 Responses to Privacy Protection in Short Videos

4.1 Raising Awareness of Information Security Among Short Video Users

The short video user group is huge, the quality is also uneven, and the understanding of privacy and privacy protection is not in place, so there are many short video works that intentionally or unintentionally leak privacy. Some short video users lack the awareness of respecting the privacy of others, the method of shooting is improper, and the videos are uploaded without permission after tracking and secret shooting, even for legitimate purposes, the method is not appropriate. Due to their inadequate understanding of privacy, some users cannot distinguish whether short video content infringes other people's privacy or leaks personal privacy. They lack the necessary checks before information is released, resulting in inadvertent infringement of others or privacy leakage. In addition, there are users who use the short video platform as their social space, record and share their personal lives through the platform, and put information such as home address, work unit, income, and family members in the video, thereby inadvertently leaking personal privacy. Due to the low threshold of the short video platform, the platform has many young users. The young people lack the ability to discern privacy issues and have no concept of privacy protection. The short videos produced and disseminated often lead to privacy leaks.

In the era of intelligent communication, there are many users of online short videos. Due to the development of technology, users are no longer the one-dimensional audiences in the era of mass communication, but become consumers, producers and disseminators of short videos. Based on the changes in the role of short video users, users should assume more social responsibilities in the management of short video privacy leakage. For huge short video users, there is a phenomenon that some short video users have low media literacy. In order to gain eyeballs and attention, they deliberately spread vulgar, violent, and private content that involves the privacy of others, so as to satisfy the viewer's desire for private voyeurism and curiosity. For this behavior of some short video users, it is necessary to cultivate their sense of social responsibility and improve their awareness of information ethics and legal responsibilities. Users should be the supervisors of the mobile short video industry, do not disseminate short videos with privacy leaks, and actively participate in supervision and reporting to purify the short video dissemination space. Short video users need to stay awake while enjoying the convenience brought by short videos, protect their privacy, and also not be a producer or disseminator of the privacy of others. Users are the key link in short video privacy protection. Whether ordinary users have privacy protection awareness is related to the content quality of short videos and the healthy development of the short video industry.

4.2   Short Video Platform Governance

The governance of short video privacy leakage needs to rely on multiple subjects, multiple methods, and collaborative governance to achieve results. The short video platform must give full play to the role of the main body, and legal norms, ethical restrictions, and technical means must be multi-pronged.

The short video platform has financial and technological advantages, and is a powerful governance entity. It should assume greater social responsibility in the governance of privacy leaks, rather than just focusing on economic benefits. Many short video platforms are business-oriented. When criticized by public opinion, they often attribute the problem to non-human factors such as algorithm neutrality and intelligent recommendation. In fact, "this kind of misunderstanding and choice of information judgment is derived from its business The fundamental shortcoming of standardized operation." The short video platform should strengthen supervision and manual review, improve the level of technical governance, and strengthen the synergy between manual review and technical monitoring. The huge number of short video users, explosive growth in the number of video works, complex content, insufficient content review and information control are important reasons for personal privacy leakage. The platform itself must strengthen supervision and manual review of content, and improve content review standards. At present, most short video applications use algorithms for machine review of keywords or key frames. Such review is not comprehensive and there will be omissions. On the one hand, the platform must continue to innovate technology and maximize the advantages of technology; on the other hand, the platform should continue to strengthen manual operations, strengthen human investment, expand the size of the review team, and strictly control the release of online short videos. In addition to its own efforts, the short video platform can also use the power of users to improve the user complaint mechanism. The discovery of short video privacy leakage, users can play an important role in it. As a platform party, we must make full use of these forces to establish and improve the complaint department, respond in a timely manner, delete the illegal content of the complaint, and avoid the negative social impact caused by the proliferation of short videos.

By default, short videos on the Jitterbug platform can be shared and downloaded by all viewers. If a link is simply shared to another social media platform and the content uploader removes the posting, then the link is also invalid. However, viewers can download videos locally to their mobile devices, which can enhance the risk of privacy breaches if saved by someone with ulterior motives. In this regard, a link could be added when users upload content, allowing them to choose whether they want others to download their content locally. This does not avoid the risk of privacy breaches, but it respects the user's privacy and puts the decision in the user's hands.

4.3 Short Video Industry Self-discipline

Industry self-discipline refers to that information managers take the initiative to protect information. All commercial websites such as Sina, 360, Google and Baidu have a "Privacy Protection" policy on their homepages. The representative of the industry's self-discipline model is the United States. In the United States, there are mainly two self-regulatory models used to protect personal information. Consult industry guidelines and online privacy certification programs. For example, the Online Privacy Alliance of America (OPA) announced in 1998 that online privacy guidelines belong to the recommended industry guide model, while the United States Truste certification belongs to the online privacy certification program model. However, industry groups have no right to punish. Companies that violate the regulations.

Through combing the network industry specifications, it is found that the number of industry specifications formulated by the Internet industry and the short video industry is very limited, and the specifications formulated are also relatively vague, unable to adapt to the actual complex conditions. For a long time, the Internet industry has not been supervised, and under the new situation, it is impossible to effectively protect the privacy rights in mobile short video applications. Therefore, the existing Internet industry associations and short video industry associations should have more functions to strengthen the guiding role of the industry associations. Government departments should urge short video related industry associations to formulate corresponding industry regulations based on the prominent privacy information protection issues in the short video development process to strengthen the guidance of the short video industry. In order to ensure the effective implementation of the regulations, government departments can transfer some of their powers, give short video related industry associations the right to penalize, and take measures such as warnings and fines for platforms that violate industry regulations. For serious violations of laws and regulations, they can be prohibited from entering the short video industry.

5   Conclusion

With the new ecology of the Internet, the use of short videos has penetrated deeply into ordinary individuals and all walks of life. While enjoying the convenience brought by short videos, we also deeply feel that under the open Internet environment, the work related to user information privacy protection has a long way to go. The research on this issue from the theoretical perspective of privacy protection in this paper will provide ideas for platform information protection schemes, and provide strategic guidance for short video users to improve their privacy protection, and provide decision-making reference for the government's privacy protection legislation in the big data environment At the same time, it also provides research support for related industries to carry out industry self-discipline or privacy protection certification, and strives to promote the construction of a strong guarantee system in related fields, and make useful attempts and explorations.

References

1. Khan, M.L.: Social media engagement: what motivates user participation and consumption on YouTube? Comput. Hum. Behav. **66**(1), 236–247 (2017)
2. Ferris, A.L., Hollenbaugh, E.E.: A uses and gratifications approach to exploring antecedents to Facebook dependency. J. Broadcast. Electron. Media **62**(1), 51–70 (2018)
3. Rauniar, R., Rawski, G., Yang, J., Johnson, B.: Technology acceptance model (TAM) and social media usage: an empirical study on Facebook. J. Enterp. Inf. Manag. **27**(1), 3–6 (2014)
4. Metzger, M.J., Suh, J.J.: Comparative optimism about privacy risks on Facebook. J. Commun. **67**(2), 203–232 (2017)
5. Hsu, C.L., Lin, J.C.C.: Effect of perceived value and social influences on mobile app stickiness and in-app purchase intention. Technol. Forecast. Soc. Chang. **108**, 42–53 (2016)
6. Haili, P., Minxue, H.: The cultivation and the difference effect of consumer emotional relationships: an interactive compensatory effect of satisfaction, attachment, and identification the customer's behaviors. Nankai Bus. Rev. **20**(4), 16–26 (2017)
7. Hutson, E., Kelly, S., Mitello, L.K.: Systematic review of cyber bullying interventions for youth and parents with implications for evidence- based practice. World Evid. Based Nurs. **15**(1), 72–79 (2018)
8. Peled, Y.: Cyber bullying and its influence on academic, social, and emotional development of undergraduate students. Heliyon, **5**(3), e01393 (2019)

An Efficient Video Steganography Method Based on HEVC

Si Liu, Yunxia Liu$^{(\boxtimes)}$, Cong Feng, and Hongguo Zhao

College of Information Science and Technology,
Zhengzhou Normal University, Zhengzhou, China
liuyunxia0110@hust.edu.cn

Abstract. As one of the most popular communication media today, digital video is generally regarded as an ideal covert communication carrier. Therefore, video steganography technology has attracted the attention of researchers in the field of data hiding and has become one of the research hotspots in this field. This paper presents a HEVC video steganography method based on QDST coefficient. The secret message is embedded into the multi-coefficients of the selected 4 × 4 luminance QDST blocks to avert the intra-frame distortion drift. And the matrix encoding technology is used to reduce the modification of these embedded blocks. The experimental results show that the proposed algorithm can effectively increase the visual quality and get good embedding capacity.

Keywords: HEVC · Video steganography · Matrix encoding · Multi-coefficients · Intra-frame distortion drift

1 Introduction

Steganography mainly studies how to hide secret messages in digital media (such as images, audio, video, etc.) for covert communication [1]. With the popularization of video in the Internet and the gradual promotion of 5G mobile Internet, using video as a carrier of steganography is conducive to the realization of covert communication. Compared with images and audio, video contains more information and more complex coding modes, so video is an ideal carrier for steganography [2]. Many modules in the video compression coding process can be used as the carrier of video steganography, such as intra prediction, motion estimation, DCT/DST transform, etc. [3, 4].

With the rapid development of the digital video industry, high-definition video is becoming more and more popular in daily applications, and ultra-high-resolution formats with higher resolution are gradually being produced on the market. At the same time, due to the development of services such as video dialogue and video-on-demand, the demand for efficient video transmission is increasing. These new demands have put forward higher performance standards for the definition, frame rate and compression rate of video coding. HEVC is the abbreviation of High Efficiency Video Coding [5]. It is a new video compression standard used to replace the H.264/AVC coding standard. On January 26, 2013, HEVC officially became an international standard. HEVC has a good parallel

© Springer Nature Switzerland AG 2021
D.-S. Huang et al. (Eds.): ICIC 2021, LNCS 12836, pp. 327–336, 2021.
https://doi.org/10.1007/978-3-030-84522-3_26

processing architecture, which greatly improves the ability to process high-resolution video. In addition, HEVC has a higher compression rate and supports data loss recovery, reducing the difficulty of transmission. And Compared with H.264/AVC, HEVC is more adaptable to the increasingly diverse high-definition network video services. It can be expected to replace HAVC in the future and become the new mainstream video compression coding standard and be widely used. Therefore, research on steganography methods based on this standard has very high application prospects.

For the intra-frame QDST (Quantized DST) coefficients of the HEVC standard, some literatures have proposed corresponding video steganography algorithms [6–12]. Chang et al. [6] employed a three-coefficients to solve the QDST coefficient distortion drift problem for 4 × 4 luminance blocks. Liu et al. [9] proposed a robust and improved visual quality steganography method for HEVC in 4 × 4 luminance QDST blocks. To improve the robustness of data hiding, the embedded data are first encoded into the encoded data by using the BCH syndrome code (BCH code) technique. To improve the visual quality of data hiding, three groups of the prediction directions are provided to limit the intra-frame distortion drift. Liu et al. [10] employed a multi-coefficients to realize reversible steganography in 4 × 4 luminance QDST blocks, but each 4 × 4 luminance block can only be embedded in 1 bits of information, so the embedding capacity is limited. Liu et al. [12] proposed a data hiding method for H.265/HEVC video streams without intra-frame distortion drift, the message is embedded into the multi-coefficients of the 4 × 4 luminance DST blocks of the selected frames which meet the specific conditions.

In this paper, two sets of multi-coefficients are used to avert the intra-frame distortion drift in 4 × 4 luminance QDST blocks. In order to further improve the visual quality of the proposed algorithm, the matrix encoding technology is used to reduce the modification of these embedded blocks. Experimental results show that the proposed algorithm has both good visual quality and high embedding capacity.

The rest of the paper is organized as follows. Section 2 describes the theoretical framework of the proposed algorithm. Section 3 describes the proposed algorithm. Experimental results are presented in Sect. 4 and conclusions are in Sect. 5.

2 Theoretical Framework

2.1 Intra-frame Prediction

HEVC uses transform coding of the prediction error residual in a similar manner as in H.264. The residual block is partitioned into multiple square transform blocks. The supported transform block sizes are 4 × 4, 8 × 8, 16 × 16, and 32 × 32. For the transform block size of 4 × 4, an integer transform derived from a DST (Discrete Sine Transform) is applied to the luma residual blocks for intra prediction modes. In terms of complexity, the 4 × 4 DST-style transform is not much more computationally demanding than the 4 × 4 DCT-style transform, and it provides approximately 1% bit-rate reduction in intra-frame coding.

A prediction block of HEVC intra prediction method is formed based on previously encoded adjacent blocks. The sixteen pixels in the 4 × 4 QDST block are predicted by using the boundary pixels of the adjacent blocks which are previously obtained, which use a prediction formula corresponding to the selected optimal prediction mode, as shown in Fig. 1. Each 4×4 block has 33 angular prediction modes (mode 2–34).

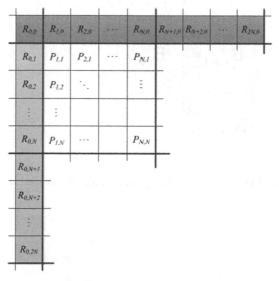

Fig. 1. Labeling of prediction samples

2.2 Intra-frame Distortion Drift

Distortion drift refers to that embedding the current block not only causes the distortion of the current block, but also causes the distortion of its adjacent blocks. As illustrated in Fig. 2, we assume that current prediction block is $B_{i,j}$, then each sample of $B_{i,j}$ is the sum of the predicted value and the residual value. Since the predicted value is calculated by using the samples which are gray in Fig. 2. The embedding induced errors in blocks $B_{i-1,j-1}$, $B_{i,j-1}$, $B_{i-1,j}$, and $B_{i-1,j+1}$ would propagate to $B_{i,j}$ because of using intra-frame prediction.

For convenience, we give several definitions, the 4×4 block on the right of the current block is defined as right-block; the 4×4 block under the current block is defined as under-block; the 4×4 block on the left of the under-block is defined as under-left-block; the 4×4 block on the right of the under-block is defined as under-right-block; the 4×4 block on the top of the right-block is defined as top-right-block, as shown in Fig. 3. The 4×4 block embedding induced errors transfer through the boundary pixels to these five adjacent blocks.

2.3 Matrix Encoding

Matrix encoding is a method of selecting a specific carrier with a specific function and embedding more information while modifying fewer carriers. Matrix encoding was first

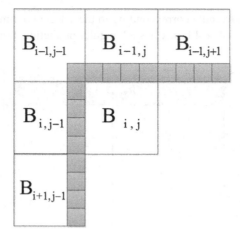

Fig. 2. The prediction block $B_{i,j}$ and the adjacent encoded blocks

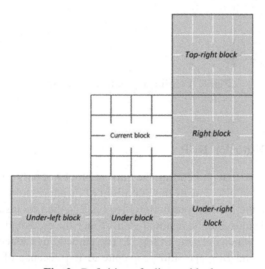

Fig. 3. Definition of adjacent blocks

proposed by Crandall [13]. The embedding efficiency is an indicator of the data hiding algorithm, which represents the ratio of the embedded bit to the modified bit. Matrix encoding can reduce the modification of the carrier when embedding the same number of bits, which can improve the embedding efficiency.

This paper uses the (1, 2, 3) matrix encoding mode. Matrix encoding of (1, 2, 3) mode requires three positions to be embedded with secret information, and two bits can be embedded with one bit modified.

For example, there are three positions a_1, a_2, and a_3 where secret information is to be embedded, and $a_1 \oplus a_3$ ($\oplus$ stands for XOR operation) represents the secret information bit w_1, and $a_2 \oplus a_3$ represents the secret information bit w_2.

If $a_1 \oplus a_3$ and w_1 match, and $a_2 \oplus a_3$ and w_2 match, then no modification will be made. At this point, two bits of secret information have been embedded.

If $a_1 \oplus a_3$ and w_1 match, and $a_2 \oplus a_3$ and w_2 do not match, then modify a_2 to make $a_2 \oplus a_3$ and w_2 match, so that two bits of information are embedded.

If $a_1 \oplus a_3$ and w_1 do not match, and $a_2 \oplus a_3$ and w_2 match, then modify a_1 to make $a_1 \oplus a_3$ and w_1 match, so that two bits of information are embedded.

If $a_1 \oplus a_3$ and w_1 do not match, and $a_2 \oplus a_3$ and w_2 do not match, then modify a_3 to make $a_1 \oplus a_3$ and w_1 match, and $a_2 \oplus a_3$ and w_2 match, so that two bits of information are embedded.

In this paper, we consider even numbers as 0 and odd numbers as 1, so in the actual embedding operation, even number XOR even number is equal to 1, even number XOR even number is equal to 0, and odd number XOR odd number is equal to 0.

3 Description of Algorithm Process

3.1 Embedding

Because the prediction block uses the boundary pixels of its adjacent blocks, if the embedding error just changed the other pixels of the current block instead of the boundary pixels used for intra-frame angular prediction, then the distortion drift can be avoided. Based on this idea, two conditions are proposed to prevent the distortion drift.

Condition 1: Right-mode $\in \{2$–$25\}$, Under-right-mode $\in \{11$–$25\}$, Top-right-mode $\in \{2$–$9\}$.

Condition 2: Under-left-mode $\in \{27$–$34\}$, Under-mode $\in \{11$–$34\}$.

If the current block meets Condition 1, the pixel values of the last column should not be changed in the following intra-frame prediction. If the current block meets Condition 2, the pixel values of the last row should not be changed in the following intra-frame prediction. If the current block meets the Condition 1 and 2 at the same time, the current block should not be embedded. If both the Condition 1 and 2 cannot be satisfied, the current block can be arbitrarily embedded where the induced errors won't transfer through the boundary pixels to the five adjacent blocks, that means the distortion drift won't happen, but in this paper we don't discuss this situation, the current block should also not be embedded.

Two sets of multi-coefficients are proposed to meet the above conditions when embedded in. The multi-coefficients can be defined as a three-coefficient combination (C_1, C_2, C_3), C_1 is used for bit embedding, and C_2, C_3 are used for distortion compensation. The specific definitions are as follows, multi-coefficients of VS applies to condition 1, and multi-coefficients of HS applies to condition 2:

VS (Vertical Set) $= (C_{i0} = 1, C_{i2} = -1, C_{i3} = 1), (C_{i0} = -1, C_{i2} = 1, C_{i3} = -1)$ $(i = 0,1,2,3)$.

HS (Horizontal Set) $= (C_{0j} = 1, C_{2j} = -1, C_{3j} = 1), (C_{0j} = -1, C_{2j} = 1, C_{3j} = -1)$ $(j = 0,1,2,3)$.

For example, add 1 to a_{00} in a 4×4 QDST block, then subtract 1 from a_{02}, and add 1 to a_{03}. This modification will not change the pixel value in the last column of this block, as shown in Fig. 4(a).

Similarly, subtract 1 from a_{00} in a 4×4 QDST block, then add 1 to a_{20}, and subtract 1 from a_{30}. This modification will not change the pixel value of the last row of the block, as shown in Fig. 4(b).

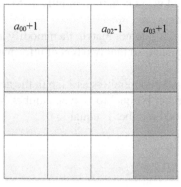

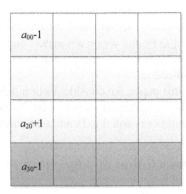

(a) Examples of VS   (b) Examples of HS

Fig. 4. (a) Examples of VS (b) Examples of HS

Assume

$$\begin{pmatrix} a0 & a1 & a2 & a3 \\ a4 & a5 & a6 & a7 \\ a8 & a9 & a10 & a11 \\ a12 & a13 & a14 & a15 \end{pmatrix}$$ is the 4×4 QDST coefficients matrix used to be embedded.

If the QDST block meets condition 1, then the coefficients a_0, a_4, a_8 and a_{12} are used to embed secret information. If the QDST block meets condition 2, then the coefficients a_0, a_1, a_2 and a_3 are used to embed secret information.

Take the coefficients a_0, a_1, a_2 applicable to condition 2 as an example. Suppose the 2-bit secret information to be embedded are w_1 and w_2, respectively.

If $a_0 \oplus a_2 = w_1$, and $a_1 \oplus a_2 = w_2$, then no modification is required, and the 2-bit data has been embedded.

If $a_0 \oplus a_2 \neq w_1$, and $a_1 \oplus a_2 = w_2$, then modify a_0 to make $a_0 \oplus a_2 = w_1$. Specifically, if $a_0 \geq 0$, then $a_0 = a_0 + 1$, $a_8 = a_8 - 1$, $a_{12} = a_{12} + 1$; if $a_0 < 0$, then $a_0 = a_0 - 1$, $a_8 = a_8 + 1$, $a_{12} = a_{12} - 1$. As mentioned above, a_8 and a_{12} are used to prevent distortion drift.

If $a_0 \oplus a_2 = w_1$, and $a_1 \oplus a_2 \neq w_2$, then modify a_1 to make $a_1 \oplus a_2 = w_2$. Specifically, if $a_1 \geq 0$, then $a_1 = a_1 + 1$, $a_9 = a_9 - 1$, $a_{13} = a_{13} + 1$; if $a_1 < 0$, then $a_1 = a_1 - 1$, $a_9 = a_9 + 1$, $a_{13} = a_{13} - 1$.

If $a_0 \oplus a_2 \neq w_1$, and $a_1 \oplus a_2 \neq w_2$, then modify a_2 to make $a_0 \oplus a_2 = w_1$ and $a_1 \oplus a_2 = w_2$. Specifically, if $a_2 \geq 0$, then $a_2 = a_2 + 1$, $a_{10} = a_{10} - 1$, $a_{14} = a_{14} + 1$; if $a_2 < 0$, then $a_2 = a_2 - 1$, $a_{10} = a_{10} + 1$, $a_{14} = a_{14} - 1$.

By using the aforementioned matrix encoding technique, embedding 2-bit secret information only needs to modify at most 3 coefficients. If the conventional line-by-line embedding method is used, at most 6 coefficients need to be modified to embed the 2-bit secret information.

The coefficient a_3 is not used in this embedding process, it will be combined with the first two embedding coefficients of the next embeddable block to embed 2-bit secret information.

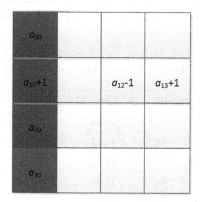

Fig. 5. Examples of embedding

As shown in Fig. 5, the coefficients a_{00}, a_{10}, and a_{20} of the first QDST block form a matrix encoding unit, where a_{10} is used to embed secret information; the coefficient a_{30} and the coefficients b_{00}, b_{01} of the second QDST block form a matrix encoding unit, where b_{01} is used to embed secret information.

After the original video is entropy decoded, we get the intra-frame prediction modes and QDST block coefficients. By using the matrix encoding technique, we embed the secret information by the multi-coefficients into the selected 4×4 luminance QDST blocks which meet the conditions. Finally, all the QDST block coefficients are entropy encoded to get the carrier video.

3.2 Data Extraction

After entropy decoding of the HEVC carrier video, two exclusive-OR operations are performed on the coefficients of the first row or first column in the selected 4×4 luminance QDST blocks which meet the condition 1 and condition 2. The exclusive-OR operation is repeated until the number of extraction bits is equal to the secret information bits notified in advance, thereby extracting the embedded secret information.

Take the previously mentioned coefficients a_0, a_1, a_2 applicable to condition 2 as an example, calculate $a_0 \oplus a_2$ to extract the first bit of this matrix encoding unit, and calculate $a_1 \oplus a_2$ to extract the second bit of this matrix encoding unit. This operation is repeated until the number of extraction bits is equal to the number of secret information bits.

4 Case Study

The proposed method has been implemented in the HEVC reference software version HM16.0. In this paper we take "Basketball" (416 * 240), "Keiba" (416 * 240), "Race-Horses" (832 * 480) and "SlideShow" (1280 * 720) as test video. The GOP size is set to

1 and the values of QP (Quantization Parameter) are set to be 16, 24 and 32. The method in [12] is used for performance comparisons.

As shown in Table 1, the PSNR (Peak Signal to Noise Ratio) of our method is better than the method proposed in [12] in each video sequences. Because compared to the method proposed in [12], our method uses matrix encoding technology to further reduce the modification of the QDST embedding block.

Table 1. PSNR (dB) of embedded frame in each video sequences

Sequences	Method	QP = 16	QP = 24	QP = 32
Basketball	In this paper	47.63	42.54	36.88
	In [12]	46.25	40.59	35.12
Keiba	In this paper	48.02	42.36	37.01
	In [12]	46.53	40.78	35.58
RaceHorses	In this paper	47.64	42.52	36.43
	In [12]	46.10	41.34	35.10
SlideShow	In this paper	47.77	41.88	37.51
	In [12]	46.28	40.28	36.09

In terms of embedding capacity, as shown in Table 2, the embedding capacity of our method is lower than the method in [12] of average per frame. Because the method proposed in [12] can embed 1-bit secret information in each row or column, our method is to embed 0.66 bit secret information in each row or column. Although the embedding capacity of our method is slightly lower than the other one, it is still acceptable.

Table 2. Embedding capacity (bit) of embedded frame in each video sequences

Sequences	Method	QP = 16	QP = 24	QP = 32
Basketball	In this paper	3501	3978	3582
	In [12]	5304	6027	5248
Keiba	In this paper	3651	3770	3580
	In [12]	5532	5713	5576
RaceHorses	In this paper	9624	11578	9657
	In [12]	14582	17543	14632
SlideShow	In this paper	13321	15462	13962
	In [12]	20184	23428	21156

Figure 6 to Fig. 7 are the video frame screenshots of two steganography methods. Since our method reduces the modification to the QDST blocks, it has a higher visual quality.

(a) Method in this paper   (b) Method in [12]

Fig. 6. (a) Method in this paper (b) Method in [12].

(a) Method in this paper   (b) Method in [12]

Fig. 7. (a) Method in this paper (b) Method in [12].

5   Conclusion

This paper proposed a novel HEVC video steganography method based on QDST coefficient. Two sets of multi-coefficients are used to avert the intra-frame distortion drift in 4×4 luminance QDST blocks. In order to further improve the visual quality of the proposed algorithm, the matrix encoding technology is used to reduce the modification of these embedded blocks. Experimental results demonstrate the feasibility and superiority of the proposed method.

Acknowledgment. This paper is sponsored by the National Natural Science Foundation of China (NSFC, Grant 61572447).

References

1. Liu, Y., Liu, S., Wang, Y., et al.: Video steganography: a review. Neurocomputing **335**(28), 238–250 (2019)
2. Liu, Y., Liu, S., Wang, Y., et al.: Video coding and processing: a survey. Neurocomputing **408**, 331–344 (2020)
3. Liu, Y., Li, Z., Ma, X., et al.: A robust without intra-frame distortion drift data hiding algorithm based on H. 264/AVC[J]. Multimedia Tools Appl. **72**(1), 613–636 (2014)
4. Liu, Y., Hu, M., Ma, X., et al.: A new robust data hiding method for H. 264/AVC without intra-frame distortion drift. Neurocomputing **151**, 1076–1085 (2015)

5. Sze, V., Budagavi, M., Sullivan, G.J. (eds.): High efficiency video coding (HEVC). ICS, Springer, Cham (2014). https://doi.org/10.1007/978-3-319-06895-4
6. Chang, P.-C., Chung, K.-L., Chen, J.-J., Lin, C.-H.: An error propagation free data hiding algorithm in hevc intra-coded frames. In: Signal & Information Processing Association Summit & Conference, pp. 1–9 (2013)
7. Liu, S., Liu, Y., Lv, G., Feng, C., Zhao, H.: Hiding bitcoin transaction information based on HEVC. In: Qiu, M. (ed.) SmartBlock 2018. LNCS, vol. 11373, pp. 1–11. Springer, Cham (2018). https://doi.org/10.1007/978-3-030-05764-0_1
8. Gaj, S., Kanetkar, A., Sur, A., Bora, P.K.: Drift-compensated robust watermarking algorithm for H. 265/HEVC video stream. ACM Trans. Multimedia Comput. Commun. Appl. (TOMM). **13**(1), 11 (2017)
9. Liu, Y., et al.: A robust and improved visual quality data hiding method for HEVC. IEEE Access **6**(2018), 53984–53997 2018
10. Liu, S., Liu, Y., Feng, C., Zhao, H.: A reversible data hiding method based on HEVC without distortion drift. In: Huang, D.-S., Hussain, A., Han, K., Gromiha, M.M. (eds.) ICIC 2017. LNCS (LNAI), vol. 10363, pp. 613–624. Springer, Cham (2017). https://doi.org/10.1007/978-3-319-63315-2_53
11. Zhao, H., Liu, Y., Wang, Y., Wang, X., Li, J.: A blockchain-based data hiding method for data protection in digital video. In: Qiu, M. (ed.) SmartBlock 2018. LNCS, vol. 11373, pp. 99–110. Springer, Cham (2018). https://doi.org/10.1007/978-3-030-05764-0_11
12. Liu, Y., Liu, S., Zhao, H., Liu, S.: A new data hiding method for H. 265/HEVC video streams without intra-frame distortion drift. Multimedia Tools Appl. **78**(6), 6459–6486 (2019)
13. Crandall, R.: Some notes on steganography. http://os.inf.tu-dresden.de/westfeld/crandall.pdf. (1998)

Analysis on the Application of Blockchain Technology in Ideological and Political Education in Universities

Shanshan Gu[1,2(✉)]

[1] School of Marxism, Wuhan University of Technology, Wuhan, China
[2] Zhengzhou Normal University, Zhengzhou, China

Abstract. As a prominent representative of the innovation and development of the information technology, block chain technology has been gradually recognized and applied to various fields. However, the application of blockchain technology is still in the exploration and research stage. Further researches and consideration about the blockchain technology applications can promote social changes, especially in the university ideological and political education. It can provide one new solution to the problems of ideological and political education activities. First, we elaborate the characteristics and social significance about educational application with adopting blockchain technology. Then, by analyzing the characteristics of block chain technology, we discuss the feasibility of its application in ideological and political education in universities, and put forward the construction idea and scheme of block chain based ideological and political education in universities. Finally, we analyze the existing challenges and solutions to the implementation of the proposed scheme.

Keywords: Blockchain · Universities · Ideological and political education

1 Introduction

In the information society, the new media and technologies should be applied to make work more effectively, and the traditional advantages of ideological and political work should be highly integrated with information technology to enhance the appealing and anachronistic [1]. The application of new technology is the innovation requirement of ideological and political work in colleges and universities, as well as the inevitable choice of exploring new ways and methods and improving the effectiveness. We believe that block chain technology plays an important role in promoting the innovation of ideological and political education in universities and we should make use of its principle and characteristic effectively to enrich the ways of ideological and political education in universities.

Blockchain is a kind of distributed data recording technology which records all transactions or behaviors of electronic record [2]. Block chain is composed of "block" and "chain". Each block contains all the transaction data that occurred when they are created and manipulated in the entire blockchain network.

© Springer Nature Switzerland AG 2021
D.-S. Huang et al. (Eds.): ICIC 2021, LNCS 12836, pp. 337–345, 2021.
https://doi.org/10.1007/978-3-030-84522-3_27

All blocks are linked in chronological order and with hash values to form a block chain. Each block in the block chain can share all transactions data of the whole system within a specific time, and the sharing of content makes each block equal. Since each block shares all information of the system, the information can be cross-verified to ensure the authenticity of the information. Meanwhile, the asymmetric encryption principle of the block chain enables each node to reach a consensus with other nodes without mutual trust. The exploration on the application of block chain technology has been started, and applications based blockchain is gradually expanding to finance, justice, medical care, logistics and other fields [3]. In the financial field, global banking giants including HSBC, bank of America and more than 40 international financial institutions have formed an R3 consortium to jointly develop block chain technology. In the medical field, Gem (a block chain enterprise) provided a global blockchain based health-care scheme which aimed to provide more personal and low-cost service for people. In the food sector, Walmart is also trying to use block chain technology to record food sources, in an attempt to make consumers have more food information, so as to improve the current mess in the food industry. Blockchain technology is a new development direction of information technology, and there is huge development space for the development of education [4].

The remainder of this paper is organized as follows: Sect. 2 reviews the background of blockchain based ideological and political education in universities, especially provides a detailed consideration about feasibility when utilizing blockchain technology. Section 3 proposes the blockchain based scheme of ideological and political education in universities and the challenges and solutions to the implementation of this scheme is presented in Sect. 4. Finally, the conclusion is shown in Sect. 5.

2 Background of Blockchain Based Ideological and Political Education

Ideological and political education in colleges and universities must adhere to the correct political direction in a new era. As the main chain, the ideology of socialism forms a stable data structure with the joint efforts and consensus of all nodes, which can be taken advantage of such erroneous thoughts as democratic socialism, historical nihilism, western constitutional democracy. The principle of accounting with distributed technology ensures its stability and security, promotes the flow of socialist core values based on the principle of trust mechanism, promotes the positive influence on mainstream ideology, and consolidates the guiding position of Marxism in the field of ideology.

As a part of ideological work, ideological and political education in colleges and universities should pay more attention to the innovation of ideological and political education theory, teaching and management means, and education methods. Block chain technology principle of application for ideological and political education in colleges and universities present dynamic development [5], the participation of all the nodes embodies the characteristics of overall education, block chain structure to make the way of all-round education and whole-process education more systematic, "cultivating what kind of person" "how to cultivate people" and "cultivate people" for who the consensus of the mechanism for the working group of ideological and political education goal is more clear, "three full" synergy mechanism make the ideological and political education more effective [6].

At present, the ideological and political education in colleges and universities generally remains single mode, one is the teaching as the main channel of ideological and political theory education mode, another is based on the daily education such as counselor of education mode, which in the implementation of the two single intersection and fusion point is less, mostly play a role in their respective fields, and in a narrow range, as for the effect of ideological and political education of the overall effectiveness is not strong. And the block chain technology brings new ideas. In block chain technology, every node shares all data information and stores complete account books. Data information is backed up among different nodes, and data maintenance and update need to be completed by multiple nodes of the system. According to the characteristics of block chain technology, it can be envisaged to construct the chain resource network of ideological and political education in colleges and universities: through the recording block of online learning resources uploaded by ideological and political educators, the blocks are sequentially linked by time axis, and all the data on the chain are entered into the database server of ideological and political education in colleges and universities [7]. Students can access the learning resources in the server and will be granted the right to record blocks on the learning chain after completing the learning tasks. The academic affairs office can give students corresponding credits according to the credit system according to their learning resource blocks, record the blocks and link them. Each node shares all the information in the whole chaining resource network, and once the information of a block is changed, all users can grasp it in real time.

3 Proposed Scheme of Blockchain Based Ideological and Political Education

Education is a social activity with developmental characteristics, a long-term process of continuous improvement and a structural feature of "chain". Ideological and political education in colleges and universities is a sub-category of education that the block chain technology can be used to study and solve the problem is feasible. The block chain technology is used in ideological and political education in colleges and universities. The educational elements or links in the ideological and political education of college students, which are linked, interdependent and restricted, are connected into a chain relationship based on specific connections or rules, and are further integrated into a holistic education network system.

3.1 Elements

Node

In block chain technology, a node can simply be understood as a computer participating in the connection, in other words, a user. Then in the construction of ideological and political education chain in colleges and universities, the nodes can be positioned as the participants of ideological and political education in colleges and universities, that is, ideological and political educators, college students and relevant administrators.

Block

In block chain technology, a block generally refers to a block that stores transaction data for a specific period of time. Therefore, in the application of ideological and political education chain in colleges and universities, each block is a module used to record information when educators, learners and managers complete tasks in accordance with relevant rules in a specific period.

Chain

In the block chain technology, the links between the blocks are generally in the form of times tamps, that is, according to the time sequence as the connection condition, for example, a node (miner) first obtains the calculation results and passes the verification to create a new block, and so on. In the application of ideological and political education chain in colleges and universities, time can also be used as the connection condition to link the blocks.

3.2 Chain Resource Network Assumption

In general, the assumption of resource network structure of ideological and political education chain in colleges and universities (Fig. 1) is based on block chain technology, forming a new learning system in which ideological and political educators update Shared resources, students' learning and feedback on relevant resources. To be clear, this chain of network structure should exist in one university, not all universities. Each university can set up its own chain resource network.

First, the first block chain is a resource chain mainly composed of ideological and political educators. Ideological and political education workers with the achievements of teaching or scientific research during the period are recorded as a block. This is similar to the block chain technology in which miners calculate the results by calculating the formula, and the first miner who gets the results gets the right to record the block at that time. Ideological and political educators will constantly produce results in their own learning and research process. On this block chain, the teaching and scientific research achievements of ideological and political educators are recorded in chronological order in the form of time axis. For example, suppose in January of 2018, a Mooc course "Introduction to Mao Zedong thought and the theoretical system with Chinese characteristics" shot by an ideological and political education teacher in a university was put online for students to choose, and the teaching results of the Mooc course could be marked by the online time. Then in February of 2018, a campus cultural activity organized by an ideological and political education instructor of the university won the award at the university level or above, and the planning and implementation of this campus cultural activity was marked by the time of winning the award. Then, in March of 2018, a team of teachers of a professional course of the university completed the provincial curriculum ideological and political project, and the time of completion was taken as the marking block for all relevant contents in the project. All the block contents on this block chain will be linked together in the order of timeline and uploaded to the ideological and political education database server of the university, which will be automatically stored according to the three categories of ideological and political courses, ideological and political courses and daily ideological and political education for learners to learn by political courses and daily ideological and political education for learners to learn by classification.

Second, the second block chain is a learning chain with college students as the main body. College students can access the ideological and political education database server of the university to obtain learning resources. When students complete a certain resource at a certain time, they can obtain the right to record the block. What needs to be made clear here is that each student's learning trajectory is different, and the learning content is also of individual selectivity. Therefore, in this block chain, each block recorded by students is a separate chain. Encryption technology can be used to protect the privacy of learners and the trust problem can be eliminated by default consensus mechanism. Thus, a new personal online learning environment is created to meet students' personalized learning needs. For example, suppose that in September 2018, student A completes the Mooc course "Introduction to Mao Zedong thought and the theoretical system of socialism with Chinese Characteristics", and the learning progress, homework completion, and online examination results are marked by the completion time. In November of 2018, student A watched and studied the contents related to the campus cultural activities awarded by the ideological and political education instructor, and the learning record was marked by time, so that the learning trajectory of student A was linked by the relationship of time axis. For another example, if student B completes the two learning contents of student A in October and December of 2018, the learning record of student B will be marked with its own learning time and linked. In addition, in this block chain, students can encrypt the data related to their privacy and personal information, and student information is only visible to students and academic administrators, so as to guarantee students' personal rights and interests. Each student can only see their own learning block content and learning trajectory, and cannot consult others' data. However, the learning block contents and learning records of all students in this block chain will also be uploaded to the ideological and political education database server of universities. All data in the database server are open and transparent, and students have no right to access other students' learning records.

Finally, the third block chain is a management chain with educational administration department as the main body. According to the consensus mechanism of block chain technology, universities can set up the credit system of network resources. The department in charge of academic affairs (Office of Academic affairs) can check the ideological and political education database server of the university in the whole process, and obtain the right to record students' credit blocks through students' learning records, which are also marked by time. For example, in the database server of ideological and political education, student A completed the "Introduction to Mao Zedong thought and the theoretical system with Chinese Characteristics". In terms of course learning, the academic affairs office can record the credit blocks for student A according to the study completion of student A and the credit system of learning from network resources, marked by time, and finally still based on the timeline Links the credit blocks together to count toward student A's total credit. Similarly, student B forms its own chain of credits according to the chronological order.

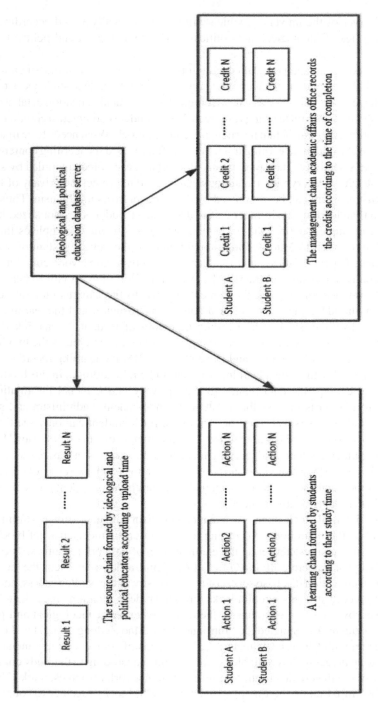

Fig. 1. Network structure of Blockchain based ideological and political education universities.

According to the principle of sharing content in block chain technology, data on all chains can also be shared in the internal chain network of universities. In the resources chain of ideological and political education, the ideological and political education workers can upload their own data to the ideological and political education of the database server, and at the same time can also share student' learning resources on record and the records given by the educational administration department. According to student's learning, ideological and political education workers can adjust timely to improve education effect. In the student learning chain, students can share ideological and political ideas, The achievement resources of the educators, as well as know their own learning achievements and access to the recognition information of the administrative department of education, which have played a certain auxiliary effect on students' own learning arrangement and planning. In the management chain of the educational administration department, the academic affairs office can grasp the real-time and specific situation of the ideological and political teaching, scientific research and learning of all teachers and students, which is of great significance to improve the overall quality of ideological and political teaching.

4   Challenges and Solutions to the Implementation of the Proposed Scheme in Ideological and Political Education

In recent years, colleges and universities have made a lot of efforts in industry-university-research cooperation, but the research on the application of blockchain technology is still in its infancy, and the real application needs more time, especially in Ideological and political education in colleges and universities. The idea of building a chain resource network by using block chain technology in ideological and political education in colleges and universities is actually the idea of combining block chain technology with innovative learning methods in the current information age. Learning in the information age must be based on abundant resources, and network learning resources play an important role in learning in the information age. Network learning resources and network learning are rapidly changing the content and way of learning. At present, the approaches of ideological and political education in colleges and universities still focus on classroom teaching and daily ideological and political education. The use of block chain technology to establish a chain resource network in colleges and universities, for students, to provide students with rich and high-quality learning resources of the university, and through the study of the content of resources to obtain the corresponding credit, innovation of learning forms, improve the positive learning. For ideological and political educators, there is a more convenient way to record their own research results, a broader platform for learning interaction with students, and a more direct channel to understand students' cognitive status to timely adjust the teaching content and methods. However, there are still some obstacles that need to be solved to construct the chain resource network by block chain technology in the current ideological and political education in colleges and universities.

4.1 Construction of Ideological and Political Education Database Server in Colleges and Universities

In the imagined chain resource network, the database server of ideological and political education in colleges and universities is a very important link. The construction assumption of the database server of ideological and political education in colleges and universities is based on the server concept in the block chain technology. Any server in the block chain shares all the data in the whole block chain. In the chain resource network of ideological and political education in colleges and universities, the database server of ideological and political education in colleges and universities is also a system that stores all the resource data. The resource chain, learning chain and management chain all share the data of this database server. As far as the specific situation of each university is concerned, there are basically no similar facilities and platforms, and no corresponding functional departments to manage and maintain them. Block chain technology requires higher computing capacity of servers, which requires universities to increase investment and strengthen hardware construction to meet the needs of ideological and political education. At the same time, colleges and universities should strengthen the talent training of block chain technology application and set up special departments to manage and maintain the database server.

4.2 Enthusiasm of Ideological and Political Educators in Universities

Basic current college ideological and political education is based on classroom teaching and daily ideological and political education as the main way. The teachers, counselors and other ideological and political education workers have to undertake heavy teaching and administrative tasks, so they may not be willing to take the time to make the network learning resources, also may not be willing to share their research results. A small number of teachers are accustomed to the traditional mode of education and find it difficult to accept the challenges brought by the new technology in terms of concepts and abilities. On one hand, colleges and universities should take the production and uploading of online learning resources as the basic work of ideological and political educators, stipulate the workload of each ideological and political educator, and take this as an indicator to be included in the assessment of professional title evaluation or job promotion. On the other hand, colleges and universities can regularly evaluate the online learning resources that ideological and political educators are responsible for and award the winners.

4.3 College Students' Initiative and Consciousness in Learning Resources

Under the traditional education model, college students lack initiative and consciousness in learning. Students form a strong dependence on the traditional method of "teachers teach, students learn", and the concept of active learning is relatively weak. In the assumption of ideological and political education chain resource network in colleges, students need to actively study the learning resources in the database server of ideological and political education in colleges and universities to obtain corresponding credits, but students lack initiative of learning resources content. They may take the form of hanging up the phone or substitute learning to complete the learning task, so that students cannot achieve the effect of improving the learning effect. Therefore, colleges and universities should optimize the conditions of network learning, increase investment in hardware, try to meet the needs of students' independent learning, and provide a good hardware environment for students' independent learning. At the same time, in the process of learning resources for students, universities can set up a stage assessment link and may adopt the face recognition quasi-examination access system. In addition, the learning resources can be added in each segment of the question and answer session to improve the initiative of students to learn resources, and also make students understand their own knowledge or skills through the response to questions and stage assessment.

5 Conclusion

In this paper, the characteristics and social significance about educational application based on blockchain are taken into consideration firstly. Then, with the analysis of characteristics between blockchain and ideological and political education filed, the scheme of block chain based ideological and political education in universities is proposed. Finally, the existing challenges and solutions to the implementation of the proposed scheme are also analyzed and discussed.

References

1. Zhang, S.: Xi Jinping stressed that ideological and political work should be integrated into the whole process of education and teaching to create a new situation in the development of China's higher education. People's Daily 9 Dec 2016 (2016)
2. Li, D., Wei, J.W.: Block chain technology principle, application field and challenges. Telecommun. Sci. **12**, 20–25 (2016)
3. Yuan, Y., Wang, F.Y.: Development status and prospect of block chain technology. J. Autom. **4**, 481–494 (2016)
4. Xu, T.: Application and challenges of block chain technology in education and teaching. Mod. Educ. Technol. **1**, 108–114 (2017)
5. Yang, X.M., Li, X., Wu, H.Q.: Application model and realistic challenge of block chain technology in the field of education. Res. Mod. Distance Educ. **2**, 34–45 (2017)
6. Xu, R.H.: A Study on the Education Chain of the Theoretical System of Socialism with Chinese Characteristics for College Students. School Marxism University of Electronic Science and Technology of China, Chengdu (2017)
7. Tang, Y.M., Xie, Z.J.: A brief discussion on network learning resources and their management. Mod. Educ. Technol. **4**, 12–14 (2007)

Parallel Security Video Streaming in Cloud Server Environment

Mi-Young Kang$^{(\boxtimes)}$

Honam University, 120, Honamdae-gil, Gwangsan-gu, Gwangju, Republic of Korea
kmy2021@honam.ac.kr

Abstract. With the rapid development of various multimedia, an increasing volume of multimedia data are being generated. Such data may include sensitive medical, commercial, and military imaging information. As the technology used to provide image information in a cloud based web system is being developed, important images that require protection are provided through internal intra networks without using the cloud server; the demand for the service is increasing. In terms of video security, several encryption algorithms have been applied to secure transmissions in recent years. A large number of multimedia encryption schemes are also used in real products. However, there is a problem where a third party can acquire an image by exploiting a security vulnerability in a cloud-based image transmission. We have focused our research on a cloud server environment that provides real-time service to transmit secure video with desirable encryption speed. However, security levels and video streaming quality should also be considered. Real-time streaming technology that provides secure real-time service for encrypted and unencrypted video without image degradation has not yet been reported. We obtained satisfactory results in the parallel security streaming development environment. The results of this study are evaluated and analyzed in the manuscript.

Keywords: Cloud server · Parallel security streaming · Real time service · Video content

1 Introduction

The multimedia communication by wired/wireless integration has become possible. In addition, because modern Internet speeds can provide ubiquitous service, various developments are being made, which makes users want to use multimedia services more frequently. As a result, companies that provide services have started strengthening their supply with the demand. However, if security is neglected, sales will not increase based on demand alone. Therefore, companies need strengthened security to protect their interests and content. Encryption for data containing text, images, audio, and video is performed to make the data unreadable, and thus, more secure. The primary objective is to encrypt or transmit data invisibly. Data encryption consists mostly of content scrambling.

In this study, we analyze existing data encryption processing technology. In addition, we propose an optimal method for encrypting an image in real-time when applying intranet.

© Springer Nature Switzerland AG 2021
D.-S. Huang et al. (Eds.): ICIC 2021, LNCS 12836, pp. 346–357, 2021.
https://doi.org/10.1007/978-3-030-84522-3_28

2   Related Works

Generally, the enterprise first tries to block image access after confirming the IP of the external access client for secure video processing.

Recently, the application range and frequency of video technology have become more active in many areas. As a result, the need for information security is also increasing.

Various types of image encryption technology have been developed in the past, but the performance required for image encryption and decryption is still insufficient. In particular, it is difficult to provide performance suitable for a real-time internet service to a large number of users.

To date, many articles have been published, and these articles can be divided into three research types. The following describes a method of encrypting image pixels:

1) Before video coding techniques [1–6]: Pixel-based encryption
2) Within video coding: Transform-based or a combined approach involving coding and encryption
3) After video coding: Bit-level encryption (Table 1).

Table 1. Video image pixel encryption method.

	Characteristic	Disadvantages
Before video encryption technique is applied	Provides high security without compromising compression efficiency and image quality	Encrypts at the bit level within pixels prior to image compression. The amount of data required for encryption is very less
Within video encryption technique	Has the advantage of encrypting while performing compression algorithms, such as MPEG; this can speed up encryption	Perceptual encryption method: Has the disadvantage of lost compression efficiency or image quality Selective encryption method: Weak security strength
After video encryption technique is applied	How to encrypt this file with traditional encryption algorithms, such as DES, RSA, IDEA, and AES, after the entire video has been compressed	It is advantageous in that security stability is high and compression efficiency and image quality are not lost However, owing to the heavy computation speed, there is a problem with real-time performance

Various methods, such as pure permutation [7], Zig-zag permutation [8], Huffman codeword permutation [9], and compression logic-based random permutation [10], have been developed. The advantage of most of these approaches is that image compression and encryption can be handled together to reduce processing time. The disadvantage is that there is some visual perturbation in the MPEG stream as it performs encryption before or during compression.

We aim to provide a real-time service for secure video along with desirable encryption speed in a cloud server system. We are aiming to achieve security stability within post-processing video encryption technology, along with desirable real-time performance with an internal video encryption method based on a compromise-type permutation (selective encryption) scheme (Table 2).

Table 2. Video image pixel encryption method.

Image encryption method	Characteristic
Pure permutation	• Video files are mostly composed of numerical data • Data modulation: Secures data by mixing large amounts of it in various ways • Security is achieved by means of scrambling through the exchange of bytes within the frame of the MPEG stream • When encrypting a generic document, it is vulnerable to known-plaintext attacks
Zig-zag permutation	• Zig-zag permutation is combinatorial mathematics that sequentially sets set $\{1,2,3,\ldots,n\}$ to be larger or smaller than the previous entry • Develops a variety of ways to create exchange lists where all possible exchanges have a uniform distribution
Huffman codeword permutation	• Light MPEG video encryption method that performs encryption and MPEG compression simultaneously • Combines encryption and MPEG compression to save computation time • The use of arbitrary Huffman codewords may reduce compression efficiency • Using codewords of the same size as the standard Huffman codeword to minimize the impact on compression efficiency
Compression logic-based random permutation	• Applies random exchange to multiple exchange groups instead of converting 8 * 8 coefficients of one DCT I*64 vector data using random exchange list • This method is safe from vulnerability to DCT compression efficiency

A wide variety of selective encryption techniques have been developed but the techniques developed by Mayer and Gagegast [11] are representative technologies.

A number of variants of selective encryption techniques similar to this technology have been developed and are in use. However, during the investigation, it was discovered that the target of our study was not met.

First, according to the security level, an important part of the MPEG video stream is encrypted with the conventional encryption algorithms, DES and RSA.

The security level increases when the amount of data of the important part of the MPEG video stream to be encrypted increases, while the processing speed decreases inversely.

Second, the encryption ratio varies depending on which parameters are encrypted.

Encrypting only the stream header allows for a real-time service with a very low encryption ratio. However, there is a concern that it can be deciphered easily.

Third, encrypting all bit streams provides an encryption ratio of 100% and allows for very high security. However, because the processing time is long, real-time service is difficult to achieve.

The selective encryption method does not change image quality to the level of encrypting the header. Various methods of encrypting the frame result in image quality variations.

In this paper, we aim to provide secure and encrypted video services to users in real time by applying techniques to parallel streaming servers that can minimize service latency without deformation of image quality.

3 Real-Time Characteristics of Video Security Services

In the video security service of the proposed parallel security streaming (PSS) system, the real-time stream image must be high quality and be able to provide a stable stream based on the Internet connection speed.

In this service environment, we have designed the basis of the transmission/reception structure to guarantee the real-time performance of the video security service based on the efficiency of through parallel processing.

Table 3 shows the resolution, frame rate, or bit rate of relatively high-quality moving images for the future.

The video transmission rate range is presumed to be compressed by the H.264 method [12] and requires a very high transmission rate.

If we make it possible to service videos of the quality shown in the table, it can be confirmed that the lower quality service that is currently being provided is not a problem for the real-time service at all.

The process by which a user uses the video security service of a PSS system.

1. The administrator distributes the video data to the security streaming server (SSS).
2. The user accesses the main server and requests the service for the desired video through the media information created for the connection.
3. When the user requests the video service, the SSSs where the actual data is stored from the main server, the size of the storage block data, the piece size of the block

Table 3. Video frame, resolution, and transfer rate relationship in H.264

Pixels and frame	Resolution	Video bitrate range
1440p @60fps	2560 * 1440	9,000–18,000 kbps
1440p @30fps	2560 * 1440	6,000–13,000 kbps
1080p @60fps	1920 * 1080	4,500–9,000 kbps
1080p @30fps	1920 * 1080	3,000–6,000 kbps
720p @60fps	1280 * 720	2,250–6,000 kbps
720p @30fps	1280 * 720	1,500–4,000 kbps

for encryption, the size of the data, the encryption method, and receives the meta key information necessary to do so.

4. Through the transmitted meta key information, the user client establishes a connection with each SSSs. Next, the transmission data block is transmitted according to the SSS and block information of the meta key. The video streaming data are restored before the encryption through decryption and reassembling processes according to the encryption information of the meta key.

5. When the receiving module of the user client transfers the restored video streaming data to the source filter of the existing media player, it automatically detects all the standard format data supported by the media player and plays the video.

Owing to this structure, the transmission load between a single server and user is efficiently distributed to the transmission load between a plurality of SSSs and users. Real-time service is provided using the time obtained from the distribution for decryption. At this time, the security of the meta key information itself is very important.

Therefore, meta key information should be protected by the highest level of traditional encryption algorithms, such as DES, RSA, IDEA, and AES [13].

However, the size of the meta key information is not large and it is necessary to decode the video before the video playback starts. Therefore, we confirmed that it does not affect the real-time video service.

The detailed flow of the communication protocol designed in this study is shown in Figure. Figure 1 shows the process of importing encrypted video data from SSSs, starting from requesting a desired video after the user logs in to the main server at the client in a packet flow.

1. If the user wishes to request a desired video from the main server:
   The client process sends a PSSvideoReq message to the main server. When the main server receives this message, it sends a PSSvideoResp message to the client.
2. The PSSvideoResp message includes the SSSs storing of encrypted data of a desired moving picture, a storage block data size, a fragment size of a block for encryption, an encryption method, and meta key information.
3. If the client process receives the meta key information from the main server through the PSSvideoResp message on the main server:
   TCP connection with SSSs holding encrypted video files.
4. The meta key information is received encrypted. Therefore, it is necessary to first decode the meta key information and then extract the addresses of the SSSs from it.
5. In the case of Fig. 1, the IP address of three servers (SSS0, SSS1, SSS2) can be extracted from the meta key information. Therefore, a TCP connection is established between the connect function of the client and accept function of the SSS through a three-way handshake.
6. Then, the user client decrypts the encrypted video blocks through three connection-oriented channels established in three servers (SS0, SS1, and SS2). Then, the decoded video data is transferred to the source filter of the media filter.
7. The SSSopenFile message transmitted from the client includes the filename in which encrypted video blocks stored in the respective servers (SSS0, SSS1, SSS2) are stored.
8. Each server opens this file and sends it to the client in a piece unit packet of the block.

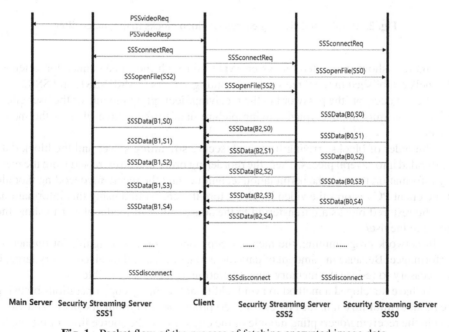

Fig. 1. Packet flow of the process of fetching encrypted image data

The three servers (SSS0, SSS1, and SSS2) perform packet transmission in parallel and the time taken to transmit the packet in the Internet situation is relatively longer than the time it takes for the client to receive the packet.

Therefore, if the number of servers is increased, the effect of parallel processing can be expected to improve.

For example, at University, the transmission rate of FTP and video streaming was measured from the outside on weekdays, and a transmission rate between 23 and 74 Mbps could be measured.

In addition, if the network card can handle 1 Gbps, increasing the number of servers to 12 can improve performance, even when considering the overhead.

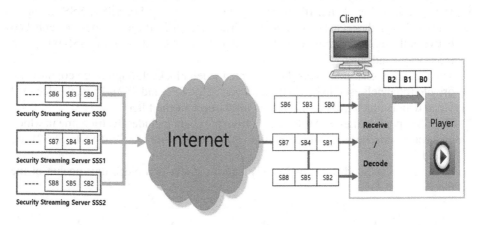

Fig. 2. Process of delivering encrypted video blocks to the source filter

Figure 2 shows the video blocks encrypted through the three connection-oriented channels established in the three secure streaming servers (SSS0, SSS1, and SSS2).

Then, it decrypts the password in the receiving/decrypting module of the user client PC and transmits the decrypted moving picture data to the source filter of the media player.

The order of blocks arriving from each secure streaming server and the block data are divided into several pieces. Then, the reordering order can be extracted from the meta key information through the permutation technique, and the receiving/decoding module of the client PC restores the video blocks before the encryption using this information.

The restored blocks are transferred to the source filter of media player to show the movie to the user.

In network programming, the memory copy operation has a significant impact on performance. Because the amount of data to be transmitted and received is very large, it is necessary to reduce the memory copy operation as much as possible.

We have developed a method to avoid additional memory copy operations by using the point to minimize it in the process of decrypting and decrypting the decrypted video data in the receiving/decrypting module of the client PC to the source filter of the media player.

Rearranging the pieces in the received block is accomplished by rearranging the pointers. Then, because the video data is transferred to the source filter of the media player along with the reordered pointer, only an overhead for changing the pointer value occurs. This method plays a role in providing a real-time streaming service, which was the goal of this study.

4 Performance Evaluation of PSS Service

To use a typical server as a video security service solution, hardware and media transmission and encryption software components are required. To provide real-time secure video service, hardware, such as network transmission/reception devices for real-time video data transmission, is a key element for storage devices with a large capacity and high transmission speed. In particular, the efficient arrangement of video data and user management are essential to support a large number of users.

This is due to the characteristics of the Internet service, such as the real-time service requirement of a very large number of users at the same time, the inconsistency in the processing speed between transmission and reception systems, the concentration of partial data flow in the Internet network, variable exists.

In existing video security processing technology, it is not appropriate to use encryption technology that causes lack of real-time property or weak image quality in the Internet service, considering the importance of the image in the smart sensor network system. In the current Internet environment, high processing speed, large storage capacity for image data storage, scalability for additional storage space required for additional users, high-speed network resources, and the distribution and utilization of resources are required to provide the necessary services.

Moreover, after a small amount of buffering time, the user can provide uninterrupted video. Even though the importance of security in smart network systems is taken into account, the real-time service of moving images should be maintained.

It is also essential to maintain the quality of the original image.

The security video service in this study is intended to develop real-time Internet service technology without any change in original image quality. The service is called PSS. The performance evaluation of the video security service collected the basic data in the communication network environment of University and used it as basic data for performance evaluation. After confirming the IP address of the external access client of the sensor network management system from the DB in the internal network image data processing support algorithm (for secure image processing), the program serving the client whose IP address was accessible is tested 100 times. It was confirmed that the service provided to the client stored the IP address in the DB.

To evaluate the real-time performance of a secure video service, we focused on the fact that slowness of bandwidth provided to leased line of server connected to Internet of server and sensor disk access bandwidth of server respectively.

When a small number of users use the service, there is no problem providing real-time service because the bandwidth requirement is small.

However, when a large number of users request a service, the bandwidth increases proportionately, and if the bandwidth is not sufficient, a bottleneck occurs where small bandwidth is provided among the two bandwidths.

P_{disk} is the bandwidth of the server's disk access; P_{out}, the bandwidth of the transmission line connecting the Internet;

P_{user}, the bandwidth demanded by each user for real-time video service on the Internet; and $P_{user} \geq P_{user} \times N_{user}$ if the bandwidth is N_{user}.

N_{user}, it is possible to provide a real-time service to the user.

Increasing the bandwidth of leased lines more than the disk access bandwidth is technically problematic and costly.

Therefore, this performance evaluation is performed considering $P_{disk} \geq P_{user} \times N_{user}$ and assuming that $P_{disk} \leq P_{out}$, $P_{disk} = 1$ Gbps, $P_{disk} = 3$ Gbps, and $P_{disk} = 5$ Gbps.

In system, the bandwidth required by the user for the real-time secure video service depends on the quality of the video provided by the server ($P_{user} =$ bandwidth). According to a report provided by YouTube, the video transmission rate range when compressed using the H.264 method.

The reason why the video transmission rate required for the specific quality video is different when the video is compressed by the H.264 method is that the compression rate changes depending on the characteristics of the video. The maximum value of each range is applied for the performance evaluation of the maximum number of users that can provide real-time service.

Then, the bandwidth P_{user} required for real-time service is as follows (Table 4).

Table 4. Real-time service demand bandwidth-based image quality (source: youtube)

Resolution	Number of frames	Bit rate range
2560 * 1440	@60fps	18 Mbps
2560 * 1440	@30fps	13 Mbps
1920 * 1080	@60fps	9 Mbps
1920 * 1080	@30fps	6 Mbps
1280 * 720	@60fps	6 Mbps
1280 * 720	@30fps	4 Mbps

The because $P_{disk} \geq P_{user} \times N_{user}$, the maximum number of users that can perform real-time service on one server is $N_{max,user} = P_{disk}/P_{user}$.

If the encrypted video is distributed to N_{sss} SSSs, the bandwidth $P_{sssuser}$ required for real time on each SSS is $P_{sssuser} = P_{user}/N_{sss}$.

Therefore, the maximum number of users that can perform real-time service in each SSS is as follows.

$$N_{max,user} = P_{disk}/P_{sssuser} = (N_{sss}P_{disk})/P_{user} \tag{1}$$

If the disk access bandwidth of the server is $P_{disk} = 1$ Gbps, when the bandwidth required for the real-time service of the moving picture is $P_{user} = 13$ Mbps, $P_{user} = 9$ Mbps, $P_{user} = 6$ Mbps, and $P_{user} = 4$ Mbps,

The performance evaluation results for the maximum number of users capable of real-time service are as follows (Fig. 3 and 4).

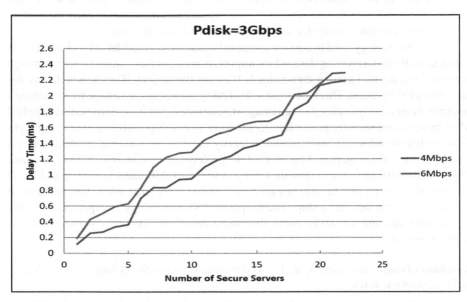

Fig. 3. Maximum number of users available for real-time service (P_{disk} = 3 Gbps).

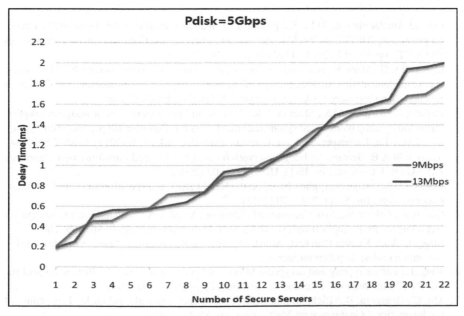

Fig. 4. Maximum number of users available for real-time service (P_{disk} = 5 Gbps).

According to these results, five SSSs can provide real-time video service with P_{user} = 4 Mbps to 1250 simultaneous users.

5   Conclusions

In this paper, we have started to investigate whether it is possible to provide a real-time secure video service with desirable encryption speed in a smart sensor network system. The methods currently in service and being researched are analyzed.

As the technology used to service image information in cloud-based web systems is advancing, there is a growing demand for important services that require security through internal intra-networks rather than a cloud. This was the subject of study, which was distressed by enlargement. This is because a third party can acquire an image by exploiting a security vulnerability in cloud-based image transmission. We have designed and applied PSS-based video security processing technology to develop real-time streaming technology that provides safe and real-time security services for video encryption without compromising image quality. As a result of the performance evaluation, it was confirmed that five security servers can provide 4 Mbps video to 1,250 users in real time. In case of multiple servers, two security servers can provide more than 190% performance. It proved that it can become a ship. The proposed PSS technology is expected to secure important video and service by controlling the external and internal connections in the intranet network.

Disclosure Policy.   The author(s) declare(s) that there is no conflict of interests regarding the publication of this article.

References

1. George Amalarethinam, D.I.., Sai geetha, J.: Enhancing security level for public key cryptosystem using MRGA. In: World Congress on Computing and Communication Technologies (WCCCT), pp. 98–102 (2014). ISBN: 978-1-4799-2876-7
2. Ganapathy, G., Mani, K.: Add-on security model for public key cryptosystem based on magic square implementation. In: Proceedings of the world congress on Engineering and Computer Science , San Fransisco, USA, vol. 1 (2009). ISBN 978-988-7012-6-8
3. Rathod, H., Sisodia, M.S., Sharma, S.K.: Design and implementation of image encryption algorithm by using block based symmetric transformation algorithm (hyper image encryption algorithm). Int. J. Comput. Technol. Electron. Eng. (IJCTEE) 1(3) (2011). ISSN 2249–6343
4. Younes, M.A.B., Jantan, A.: Image encryption using block-based transformation algorithm. IAENG Int. J. Comput. Sci. **35**(1), IJCS_35_1_03 (2008)
5. Kester, Q.-A.: Image encryption based on the RGB PIXEL transposition and shuffling. Int. J. Comput. Netw. Inf. Secur. **7**, 43–50 (2013)
6. Gutub, A., Ankeer, M., Abu-Ghalioun, M., Shaheen, A., Alvi, A.: Pixel indicator high capacity technique for RGB Applications(WoSPA 2008). University of Sharjah, Sharjah (2008)
7. Slaggel, A.J.: Known-Plaintext Attack Against a Permutation Based VideoEncryption Algorithm (2004). http://eprint.iacr.org.
8. Tang, L.: For encrypting and derypting MPEG video data efficiently. In: Proceedings of the Forth ACM International Multimedia Conference, pp. 219–230 (1996)
9. Shi, C., Bhargava, B..: Light weight MEPG video encryption algorithm. In: Proceeding of the International Conference on Multimedia, pp. 55–61 (1998)
10. Wang, H., Xu, C.: A new lightweight and scalable encryption algorithm for streaming video over wireless networks. In: International Conference on Wireless Network, pp. 180–185 (2007)

11. Meyer, J., Gadegast, F.: Security Mechanisms for Multimedia Data with the Example MPEG-1 video. Technical University of Berlin, Project Description of SECMPEG (1995)
12. "Draft ITU-T Recommendation and Final Draft International Standard of Joint Video Specification (ITU-T Rec. H.264/ISO/IEC 14496–10 AVC)," Joint Video Team(JVT), Doc. JVT-G050, Technical Report (2003)
13. Daemen, J., Rijmen, V.: AES proposal: the rijndael block cipher. Protom World Int.l, Katholieke Universiteit Leuven, ESAT-COSIC, Belgium, Technical Report (2002)
14. Abomhara, M., Zakaria, O., Khalifa, O.O.: An overview of video encryption techniques. Int. J. Comput. Theory Eng. 2(1), 1793–8201 (2010)
15. Shah, J., Saxena, V.: Video encryption: a survey. IJCSI Int. J. Comput. Sci. Issues 8(2) (2011). ISSN(Online): 1694–0814
16. Dongare, A.S., Alvi, A.S., Tarbani, N.M.: An efficient technique for image encryption and decryption for secured multimedia application. Int. Res. J. Eng. Technol. (IRJET) 04 (2017)
17. Merlyne Sandra Christina, C., Karthika, M., Vasanthi, M., Vinotha, B.: Video encryption and decryption using RSA algorithm. IJETT 33(7), 328–332 (2016)

An Efficient Video Steganography Scheme for Data Protection in H.265/HEVC

Hongguo Zhao[1], Menghua Pang[2], and Yunxia Liu[1(✉)]

[1] College of Information Science and Technology,
Zhengzhou Normal University, Zhengzhou, China
liuyunxia0110@hust.edu.cn
[2] College of Mathematics and Statistics, Zhoukou Normal University, Zhoukou, China

Abstract. An efficient and novelty video steganography scheme with adopting averting intra prediction distortion drift technique is proposed in this paper for the enhanced protection of crucial data related to H.265/HEVC digital video. With adoption of intra prediction modes selection and DCT/DST coefficients correlation, the intra prediction distortion drift can be totally prevented in intra frame (I frame). With adoption of selection of prediction patterns of multiple adjacent prediction blocks, the intra prediction distortion drift in inter frames (B frame or P frame) can be totally prevented. Compared to the previous related works, the proposed scheme applied the averting intra prediction distortion drift technique not only intra but also inter frames, and further improve the visual quality (imperceptibility) of carrier video about H.265/HEVC. Moreover, larger embedding capacity can also be achieved rather than embedding only manipulated on I frames. The experimental results have been proven the superiority of efficiency and performance about the proposed scheme.

Keywords: Video steganography · Intra distortion drift · Prediction modes · Intra/inter prediction

1 Introduction

Recently video steganography technology is becoming a powerful tool for the protection of prominent (or sensitive) data related to digital videos. Especially for the illegal distribution and propagation of movie products, video steganography can achieve a strong protection level by embedding watermark into movie products, and trace back the distribution trajectories through the network broadcasting [1]. Another supporting foundation is the continuous development of video coding technology. Since the primary video coding standards have been developed by the well-known ITU-T and ISO/IEC organizations (where H.261, H.263 is released by ITU-T, MPEG-1, MPEG-4 visual by ISO-IEC, and H.262/MPEG-2, H.264/MPEG-4 by the combination), H.265/HEVC was finalized in 2013, and now has been a key technology standard for multi-application scenarios, including scalable video coding, 3-D/stereo/multi-view video coding and the most important, the high compression efficiency for captured movies, especially for HD

© Springer Nature Switzerland AG 2021
D.-S. Huang et al. (Eds.): ICIC 2021, LNCS 12836, pp. 358–368, 2021.
https://doi.org/10.1007/978-3-030-84522-3_29

format products [2]. As claimed in [2, 3], for the identical perceptual video, H.265/HEVC can achieve approximately 50% bit-rate reduction compared to the proceeding standard H.264/AVC. While considering the strong needs of video products against the illegal distribution, protection of video contents [4], and the compression capability, compu tational resources for practical use, video steganography technology can be a desirable tool to tackle these protection problems of privacy data referred to video content or application [5].

Video steganography technology provides a potential path to protect video products from malicious spreading and legitimate rights tracing through network. Video steganog raphy researches always searches the most optimized video signal redundancy to embed secret data into video contents (carrier videos), the changes to the carrier video intro duced by embedding secret data are imperceptible for video observers except for data extractors. The existing video steganography methods can be classified into spatial and transform domain based researches according to the embedding domains. Also, based on video specific features, the video steganography researches can be classified to pre diction modes [6], motion vector [7] (MV), and other specific features [8] (e.g., variable length code-VLC).

While considering the scenario of transferring video through network, transform domain based video steganography methods provides more practical application mean ings. The main reason is that embedding data into spatial domains (always LSB substi tution on pixels) can be easily lost after loss compression (e.g., H.264/AVC encoder). However, in the transform domain based researches, discrete cosine transform (DCT) coefficient is a basic and renowned carrier due to its majority occupying the bitstream and reversibility [4, 9–11]. Moreover, based on video specific features, intrapicture pre diction modes are also a hot research field for embedding due to its fundamental role for I frame prediction and reconstruction (reference samples or decoding) process. More over, different from MV based methods and VLC based methods which will introduce large visual distortion for carrier videos, embedding based on prediction modes limit the distortion drift and guarantee a comparative large embedding capacity due to more textural features in I frame [7]. Based on above consideration, the combination between DCT coefficients and intrapicture prediction modes would be a potential tool to design an efficient video steganography scheme for digital video protection, especially for HD or beyond HD format videos compressed by H.265/HEVC.

In this paper, we focus on the issues of video security protection, by designing an efficient and video steganography scheme to promote the security level of the digital video transmitted on network. Based on the relevant researches of DCT/DST (discrete sine transform) domain and intrapicture prediction modes, we propose to combine the DCT/DST coefficients and intrapicture prediction modes in 4 × 4 DCT/DST block to embed secret data. In order to minimize the visual distortion, we design a module to totally avert intra distortion drift (prediction modes and DCT/DST coefficient in I frame, prediction modes groups in B and P frames). Experimental evaluation has been proven that the proposed method can achieve high visual quality and security performance for carrier videos and sufficient embedding capacity for embedding.

The remainder of this paper is organized as follows: Sects. 2 reviews the related technical backgrounds of DCT/DST transformation process, prediction modes and intrapicture distortion drift. Section 3 proposes the scheme of our video steganography research based on the combination of DCT coefficients and intrapicure prediction modes in I frame and prediction modes groups in B and P frames. In Sect. 4, the experimental results are presented and evaluated our scheme. Finally, the conclusion is shown in Sect. 5.

2   Related Technical Backgrounds

The intrapicture prediction is used in I frames and B or P frames to reduce the signal spatial redundancy in H.265/HEVC. In the prediction process of I frames, the intrapicture prediction is the only prediction pattern and in B or P frames, there's probably intrapicture prediction in prediction units as the extension and supplementary for inter motion estimation. As shown in Fig. 1(a), when predicting the samples of current PU, the reference samples marked as gray will be used as reference to generate the predicted samples for current PU. Moreover, the adjacent blocks containing the reference sample is indexed in this paper as Top-Left, Top, Top-Right, Left, and Down-left blocks with their corresponding locations on current PU. The prediction angles are depicted in Fig. 2(b), where there are 33 angular prediction modes for current PU. The best prediction mode can be selected from these 33 angular prediction directions, or planar and DC modes, which is determined by the comparison of calculation of distortion (measured by SAD) and encoded bits number (entropy with CABAC). After prediction modes have been confirmed, the residual samples can be acquired by the subtraction between the original samples and the predicted samples in current PU.

The H.265/HEVC prediction modes provides higher precision than its proceeding standard H.264/AVC due to its more angular prediction directions in small prediction blocks, e.g., 4 × 4 prediction unit.

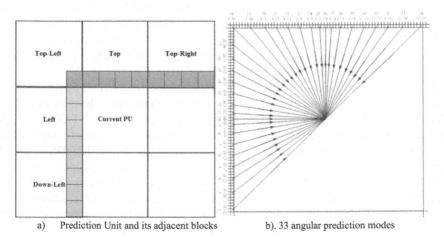

a)   Prediction Unit and its adjacent blocks   b). 33 angular prediction modes

Fig. 1. Intrapicture prediction modes and process

2.1 Intrapicture Distortion Drift in H.265/HEVC

The intrapicture distortion drift is always introduced by embedding secret data into carrier videos. Here the generation process is depicted for embedding data into DCT/DST coefficients. As shown in Fig. 1(a), if we embed secret data into the current PU's adjacent block, such as Top-Left, Top, Top-Right, Left or Down-Left. Then the reference samples, which are marked as gray region might be changed due to the embedding manipulation. However, these errors would be accumulated, and propagated to the current PU prediction process. As a result, the predicted samples would be not identical to the original predicted samples for the same current PU. The embedding errors accumulation process is defined as intrapicture distortion drift in video steganography and it would bring a considerable distortion for the carrier video.

2.2 Transformation and Inverse Transformation in H.265/HEVC

Due to the actual embedding is manipulated on DCT/DST coefficients, the main transformation and inverse transformation in H.265/HEVC is elaborated in this section. The transformation process can translate the residual samples from spatial domain to transform domain for higher precision and lower dynamic range. In H.265/HEVC, 4×4 transform block use DST transformation matrix, and other dimension blocks use DCT transformation matrix. The one-dimensional DST transformation process of 4×4 transform block can be formulated as following:

$$Y = AX \qquad (1)$$

Where Y presents the transformed coefficients, X presents the residual samples after prediction process, and A indicates the transformation matrix. If the dimension of current transform block is 4×4, the transformation matrix A can be depicted as following:

$$A = \frac{2}{3} \begin{bmatrix} \sin\frac{\pi}{9} & \sin\frac{2\pi}{9} & \sin\frac{3\pi}{9} & \sin\frac{4\pi}{9} \\ \sin\frac{3\pi}{9} & \sin\frac{3\pi}{9} & 0 & -\sin\frac{3\pi}{9} \\ \sin\frac{4\pi}{9} & -\sin\frac{\pi}{9} & -\sin\frac{3\pi}{9} & \sin\frac{2\pi}{9} \\ \sin\frac{2\pi}{9} & -\sin\frac{4\pi}{9} & \sin\frac{3\pi}{9} & -\sin\frac{\pi}{9} \end{bmatrix} \qquad (2)$$

Rounding and scaling above transformation matrix A, we can acquire the integer transformation matrix H and overwrite the two-dimensional DST transformation as follows:

$$H = \begin{bmatrix} 29 & 55 & 74 & 84 \\ 74 & 74 & 0 & -74 \\ 84 & -29 & -74 & 55 \\ 55 & -84 & 74 & -29 \end{bmatrix} \qquad (3)$$

$$Y = HXH^T \qquad (4)$$

Above Eq. (4) depicts the actual DST transformation process in H.265/HEVC. After transformation, the coefficients Y will go through post-scaling and quantization process as follows:

$$\widetilde{Y} = (Y. \times MF)/2^{(qbits+T\_Shift)} \qquad (5)$$

Where $qbits = 14+floor(QP/6)$ $MF = 2^{qbits}/Q_{step}$, and QP is the quantization parameter and Q_{step} presents the quantization step, which is determined by coding configuration and rate-distortion optimization (RDO) process for bit-rate restriction scenario.

The inverse transformation always occurs in decoding or reconstruction process. The re-scaling and inverse quantization process (Eq. 6), inverse transformation process (Eq. 7) are depicted as following:

$$Y' = \tilde{Y} . \times Qstep. \times 2^{6-shift} \tag{6}$$

$$X' = H^T Y' H \tag{7}$$

Where $shift = 6 + floor(QP/6) - IT\_Shift$, Y' depicts the transformation coefficients after re-scaling and inverse quantization, and X' presents the acquired residual samples after inverse DST transformation process and would be used for future reconstruction samples with predicted samples.

3 Proposed Prediction Modes and DCT/DST Coefficients Based Video Steganography Scheme

The proposed video steganography scheme based on intra prediction modes and DCT/DST coefficients is illustrated in Fig. 2. The scheme can be divided into two components, including embedding and extraction process. In embedding section, appropriate 4 × 4 embedded blocks can be selected based on the intra prediction modes of the adjacent blocks corresponding to 4 × 4 current block in I frame. Meanwhile, in B and P frames, the embedded block is selected based on its adjacent blocks' prediction patterns (intra or motion estimation). Then the specific coefficients are selected for embedding according to embedding mapping rules, where different frame types meet different embedding rules. After entropy encode (CABAC or CALVC), the carrier video will be encoded to bitstream and transmitted through external network. The extraction section is an inverse loop of embedding. To guarantee the security of secret data, essential encryption and decryption are also manipulated before and after video steganography process.

3.1 Prediction Modes Selection for Embedding Blocks

The intrapicture prediction modes selection process of 4 × 4 prediction unit can be divided into two sections according different frame types (I frame, P frame and B frame). The main goal to do prediction modes selection is to totally avert intra distortion drift introduced by embedding secret data into DCT/DST coefficients. In I frames, we define three categories about adjacent blocks prediction modes, where some components need to jointly combine the specific DCT/DST coefficients to avert intra distortion drift. In B and P frames, we define one specific prediction unit where its adjacent prediction block patterns are all motion estimation. If we embedding secret data into this specific prediction unit, the distortion introduced by embedding will not accumulate to its adjacent block prediction process, which will totally avert intra distortion drift in interpicture frames.

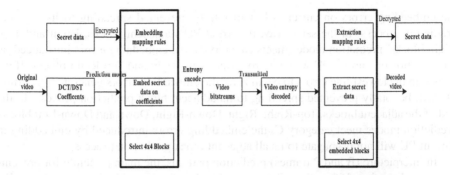

Fig. 2. Proposed video Steganography Scheme based on intra prediction modes and DCT/DST coefficients

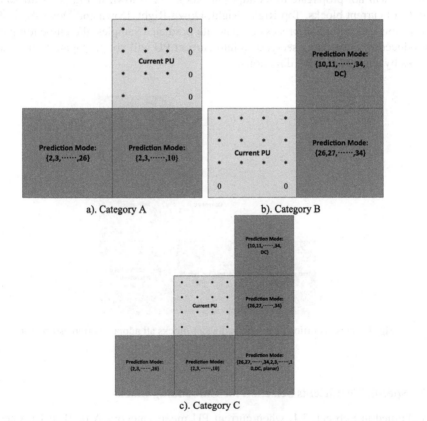

a). Category A   b). Category B

c). Category C

Fig. 3. Selection of prediction modes for embedding block in intrapicture frame (I frame).

As shown in Fig. 3, prediction modes can be classified into three categories based on adjacent block prediction modes in I frame. When the current PU meets category A, the embedding errors introduced by modification on current PU region will not propagated to its Down and Down-Left blocks. If we constrain the right column errors to be zero, the embedding errors on current PU will totally prevented spreading to its adjacent prediction process. In the same way, if current PU's adjacent blocks, Right and Top-Right blocks' prediction modes meets category B, the embedding errors introduced by modification on current PU will not propagate to its Right and Top-Right blocks. If we constrain the lowest line errors to be all zero, the embedding errors generated in current PU will be totally prevented spreading to its adjacent block prediction process. In the end, if the adjacent blocks, Top-Right, Right, Down-Right, Down and Down-Left blocks prediction modes meet category C, the embedding errors introduced by embedding in current PU will not propagate to its all adjacent block prediction process.

In interpicture (B and P frames) prediction patterns, the main principle for preventing intra distortion drift is that all adjacent blocks of current block all adopts motion estimation process to generate predicted samples, then embedding secret data into current PU will not propagate to its adjacent blocks. As shown in Fig. 4, if all adjacent blocks of current blocks, Top-Right, Right, Down-Right, Down and Down-Left blocks utilize motion estimation process to generate predicted samples, the embedding errors introduced by embedding secret data into current PU will not propagate to its adjacent blocks by angle prediction directions.

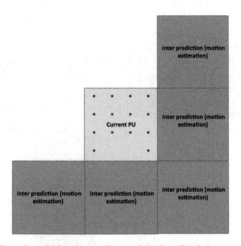

Fig. 4. Intra distortion drift when adjacent blocks all adopts motion estimation.

3.2 Specific Coefficients Selection for Embedding Blocks

As claimed in Subsect. 3.1, when current PU meets category A or B in I frames, the constrains of embedding errors to be zero should be imposed to avert intra distortion drift, which can be achieved by the selection operation on specific coefficients on current embedding block. The operation of embedding data is done on quantized DCT/DST coefficients, so the embedding error E can be presented according to formulas (6) and (7) as follows:

$$E = H^T (\Delta. \times Qstep. \times 2^{6-shift})H \tag{8}$$

Where Δ presents the actual modification on quantized DCT/DST coefficients. When current PU meet category A, the rightest column intensity values of embedding error E should be enforced to all zero, which can be represented as

$$E = \begin{bmatrix} e_{00} & e_{01} & e_{02} & 0 \\ e_{10} & e_{11} & e_{12} & 0 \\ e_{20} & e_{21} & e_{22} & 0 \\ e_{30} & e_{31} & e_{32} & 0 \end{bmatrix} \quad (9)$$

Combining Eq. (8) and (9), we can find the modification on DCT/DST coefficients should meet the following principle:

$$\Delta = \begin{bmatrix} \delta_{00} & 0 & -\delta_{00} & \delta_{00} \\ \delta_{10} & 0 & -\delta_{10} & \delta_{10} \\ \delta_{20} & 0 & -\delta_{20} & \delta_{20} \\ \delta_{30} & 0 & -\delta_{30} & \delta_{30} \end{bmatrix}$$

In which, if we embed data '1' into δ_{00}, we should correspondingly embed data '-1' and '1' into coefficients δ_{02} and δ_{03}.

Similarly, When current PU meet category B, the lowest row intensity values of embedding error E should be enforced to all zero, which can be represented as

$$E = \begin{bmatrix} e_{00} & e_{01} & e_{02} & e_{03} \\ e_{10} & e_{11} & e_{12} & e_{13} \\ e_{20} & e_{21} & e_{22} & e_{23} \\ 0 & 0 & 0 & 0 \end{bmatrix} \quad (10)$$

Combining Eq. (8) and (10), we can find the modification on DCT/DST coefficients should meet the following principle:

$$\Delta = \begin{bmatrix} \delta_{00} & \delta_{01} & \delta_{02} & \delta_{03} \\ 0 & 0 & 0 & 0 \\ -\delta_{00} & -\delta_{01} & -\delta_{02} & -\delta_{03} \\ \delta_{00} & \delta_{01} & \delta_{02} & \delta_{03} \end{bmatrix}$$

In which, if we embed data '1' into δ_{00}, we should correspondingly embed data '-1' and '1' into coefficients δ_{20} and δ_{30}.

Extraction module is an inverse process compared to embedding module. During the decoding process, in I frame, the current PU's quantized DCT/DST coefficients and its adjacent blocks prediction modes should be obtained from the selection category types with the same constraints as embedding. However, in interpicture frames (B and P frames), we only need to know the current PU and its adjacent blocks prediction patterns (intra prediction modes or motion estimation). Based on above pre-works, we can exactly extract the embedded secret data from corresponding DCT/DST coefficients with the same operation as embedding module.

4 Experimental Evaluation

The proposed video steganography scheme is manipulated and evaluated with the H.265/HEVC reference software HM16.0, including several public test video samples from resolution in the range of 416×240 to 1920×1080. The coding parameters are set as follows: frame-rate is set to be 30 frames/s, quantization parameter is set to be 32, and the test video sequence is set to be all intra frames, B and P frames with the interval 4. The main evaluation includes PSNR, embedding capacity and bit-rate increase, and the actual experiment results of this work is represented as following:

The subjective visual quality about our proposed scheme is depicted in Fig. 5, where the first column are the original video samples resoluations in the range of 416×240 to 1280×720 (BasketballPass: 416×240, RaceHorses: 832×480 and KristenAndSara: 1280×720), the second column are the compressed video samples without video steganography and the third column are the proposed video steganography scheme. It can be seen that our proposed scheme has achieved good visual quality on carrier videos compared to the reference samples without embedding. Figure 6 provides comparisons about PSNR and bitrate when we embed various volumes of secret data in test video sample BasketballPass. The experimental results have proved a good embedding performance of our proposed scheme on PSNR and bit-rate increase.

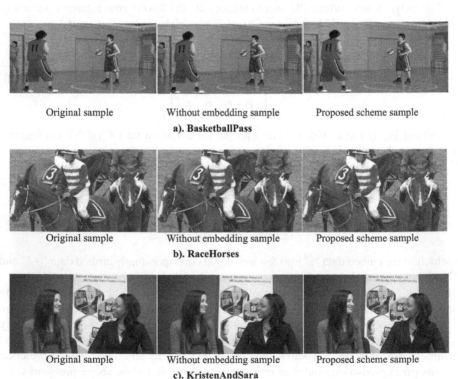

Original sample   Without embedding sample   Proposed scheme sample
a). BasketballPass

Original sample   Without embedding sample   Proposed scheme sample
b). RaceHorses

Original sample   Without embedding sample   Proposed scheme sample
c). KristenAndSara

Fig. 5. Subjective visual quality of the proposed video steganography scheme

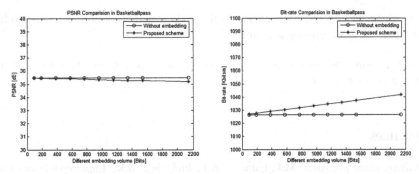

Fig. 6. Comparisons of PSNR and bit-rate with embedding different volume secret data.

Table 1. Performance of the proposed method

Video sequence	PSNR	Proposed method		
		PSNR	Capacity (bits)	Bit-rate increase
BasketballPass	35.48	35.36	2144	0.63%
RaceHorses	34.7	34.4	10092	1.20%
KristenAndSara	46.38	45.11	10028	1.41%

Table 1 provides the embedding performance of our proposed video steganography method, where 20 frames are used to test the PSNR, embedding capacity and bit-rate increase. For the visual quality evaluation, PSNR is the average value of all tested frames, and bit-rate increase is the comparison between the compressed video bitsteams with and without embedding secret data. It can be seen that for visual quality, the PSNR of our proposed scheme is 35.36 dB, 34.4 dB, and 45.11 dB for tested videos BasketballPass, RaceHorses and KristenAndSara. However, the original PSNR values are 35.48 dB, 34.7 dB and 46.38 dB, respectively. For embedding capacity, our proposed scheme has achieved 2144 bits, 10092 bits, and 10028 bits, respectively. For bit-rate increase, our proposed scheme is 0.63%, 1.2% and 1.41%, respectively. It can be seen from Table 1 that our proposed scheme can achieve a good performance on visual quality, embedding capacity and bit-rate increase.

5 Conclusion

In this paper, an effective video steganography scheme based on selection of intrapicture prediction modes is proposed for video secret data protection. The proposed scheme mainly utilize the three categories of prediction modes in I frames, prediction patterns in B and P frames, and corresponding DCT/DST coefficients to embed secret data. The experimental results show that our proposed scheme can achieve a good embedding performance on visual quality, embedding capacity and bit-rate increase. The proposed video steganography can provide a strong support for the protection of copyright of digital videos, an effective trace back tool for malicious distribution of video products based on H.25/HEVC coding standard.

Acknowledgment. This paper is sponsored by the National Natural Science Foundation of China (NSFC, Grant No. 61572447).

References

1. Asikuzzaman, M., Alam, M.J., Lambert, A.J., Pickering, M.R.: Imperceptible and robust blind video watermarking using chrominance embedding: a set of approaches in the DT CWT domain. IEEE Trans. Inf. Forensics Secur. **9**(9), 1502–1517 (2014)
2. Sullivan, G.J., Ohm, J.R., Han, W.J., Wiegand, T.: Overview of the high efficiency video coding (HEVC) standard. IEEE Trans. Circuits Syst. Video Technol. **22**(12), 1649–1668 (2012)
3. Ohm, J.R., Sullivan, G.J., Schwarz, H., Tan, T.K., et al.: Comparison of the coding efficiency of video coding standards-Including high efficiency video coding (HEVC). IEEE Trans. Circuits Syst. Video Technol. **22**(12), 1669–1684 (2012)
4. Liu, Y.X., Zhao, H.G., Liu, S.Y., et al.: A robust and improved visual quality data hiding method for HEVC. IEEE Access. **6**, 53984–53987 (2018)
5. Zhang, X.P.: Reversible data hiding with optimal value transfer. IEEE Trans. Multimedia. **15**(2), 316–325 (2012)
6. Liu, Y., Jia, S., Hu, M., et al.: A reversible data hiding method for H.264 with Shamir's (t, n)-threshold secret sharing. Neurocomputing **188**, 63–70 (2016)
7. Yang, J., Li, S.: An efficient information hiding method based on motion vector space encoding for HEVC. Multimedia Tools Appl. **77**(10), 11979–12001 (2017). https://doi.org/10.1007/s11 042-017-4844-1
8. Zhao, H.G., Liu, Y.X., Wang, Y.H., et al.: A video steganography method based on transform block decision for H.265/HEVC. IEEE Access. **9**, 55506–55521 (2021). https://doi.org/10. 1109/ACCESS.2021.3059654
9. Yao, Y.Z., Zhang, W.M., Yu, N.H.: Inter-frame distortion drift analysis for reversible data hiding in encrypted H.264/AVC video bitstreams. Signal Process. **128**, 531–545 (2016)
10. Zhao, H., Pang, M., Liu, Y.: Intra-frame adaptive transform size for video steganography in H.265/HEVC bitstreams. In: Huang, D.-S., Premaratne, P. (eds.) Intelligent Computing Methodologies: 16th International Conference, ICIC 2020, Bari, Italy, 2–5 October 2020, Proceedings, Part III, pp. 601–610. Springer International Publishing, Cham (2020). https://doi.org/10.1007/978-3-030-60796-8_52
11. Zhao, H.G., Liu, Y.X., Wang, Y.H., Wang, X.M., Li, J.X.: A blockchain-based data hiding method for data protection in digital video. In: International Conference on Smart Blockchain: SmartBlock 2018, Tokyo, Japan, pp. 99–110 (2018)

A Robust Lossless Steganography Method Based on H.264/AVC

Shuyang Liu$^{(\boxtimes)}$

School of Statistics, East China Normal University, Shanghai, China
51204404004@stu.ecnu.edu.cn

Abstract. This paper presents a robust lossless steganography H.264/AVC method. First, the embedded message is distributed by a polynomial and obtains a series of sub- messages and BCH is used to encode each sub- message. Then, the message is embedded into the frames of the video which meet the specific prediction modes. Last, the hidden message will be recovered by the t sub- messages of the video frames. This method can recover the original video completely while extracting the embedded message, and have a good robustness in frame and bit error. The experiment results show that the method can reach a good effect and have a good visual quality.

Keywords: Steganography · H.264/AVC · Robust · Lossless · BCH · Secret sharing

1 Introduction

Steganography, a technology which can embed message into cover media contents, has received extensive concern with the rapid development of internet and digital media, and has also been widely noted for it can be used in many fields such as copy right protection, user identification, access control, etc. The technique has reached a good result for some previous video compression standards, such as MPEG1–4 [1, 2, 16–18], and the existing H.265 video data hiding schemes are studied by few scholars [3, 22]. H.264/AVC (advanced video coding) is the most widely application video coding standard with high compression efficiency published by ITU-T VCEG and ISO/IEC MPEG [7] and is suit for network transmission. To embed message, extant steganography methods need to choose a video coding structure, such as DCT/DST (discrete cosine/sine transform) coefficient [4, 21], motion vector [5, 20], intra prediction [6, 19], histogram column diagram [10] to embed message.

The intra-frame distortion drift is a problem for steganography methods with H.264, since embedding message may change the frame coefficients, and then the reconstructed pixels of related frames may be influenced [7].There have been some H.264 steganography method without distortion drift by embedding message into 4×4 DCT blocks, which have reached a good effect [8, 9]. Though coefficients-change in large-scale has been avoided, still, some coefficients in [8, 9] have been changed permanently so there may be some loss in video quality. That is, the methods in [8, 9] are loss, and cannot be

© Springer Nature Switzerland AG 2021
D.-S. Huang et al. (Eds.): ICIC 2021, LNCS 12836, pp. 369–378, 2021.
https://doi.org/10.1007/978-3-030-84522-3_30

qualified. In some fields such as medical and legislation which have a high standard in the robustness of original video, the loss in the carrier cannot be permitted. Consequently, further study and investigation are required.

Beside the man-made change of coefficients in embedding, the data are always face with attack, such as the bad weather, hack attack and even equipment aging during the transaction, intentionally or unintentionally. That is, the video may sustain some random damage in changed-coefficients, which may further cause the loss of embedded message. In some fields with a high standard in the accuracy of embedded message, algorithms can fix the embedded message are needed. But many traditional researches didn't focus on the lossless or only can achieve semi-lossless of the original video carrier while recovering the embedded message [11, 12]. Therefore, research on H.264 (H.264/AVC) lossless steganography methods which can fix the embedded message is very valuable.

Some researchers have reached lossless steganography method, but they cannot fix frame loss or fix bit loss [13, 14]. Some researchers focus on the robust of steganography since the embedded message sometimes cannot survive from packet loss, video-processing operations, and so on. In [13], the BCH is used to recover error bit. In [14], secret sharing is used to recover the lost or error of package or frame. In [11], both two methods are used, and the algorithm has a good robustness in bit error, bit loss, frame error and frame loss. However, [11] isn't a lossless method.

In this paper, we provide a new robust lossless steganography method based on H.264. This method can recover the original video completely while extracting the embedded message, and have a good robustness in frame and bit error. The experiment results show that the method can reach a good effect and have a good visual quality.

2   Theoretical Frame-Work

2.1   Prevention of Intra-frame Distortion Drift

The intra-frame distortion drift happens when the bits are embedded into I frames. The prediction block's calculation is based on the current block and coded block. Define upper-left block $B_{i-1, j-1}$, under-left block $B_{i, j-1}$, upper block $B_{i-1, j}$ and upper-right block $B_{i-1, j+1}$, then prediction of $B_{i, j}$ is calculated by the gray part of $B_{i-1, j-1}$, $B_{i, j-1}$, $B_{i-1, j}$ and $B_{i-1, j+1}$ (Fig. 1) with the intra-frame prediction mode of $B_{i, j}$, the error in the gray part will be calculated and transmitted into current block $B_{i, j}$, then the error becomes larger and larger with the procession of calculation and causes a loss in video visual quality, i.e., the intra-frame distortion drift. If we don't embed bits into the edge (the gray part) of the block, the intra-frame distortion drift can be avoided.

There are nine 4×4 block prediction modes (nominated by 0-8) in H.264 (Fig. 2). Then we define:

$$\text{Condition1} : \text{upper} - \text{right block} \in \{0, 3, 7\}$$
$$\text{Condition2} : \text{under} - \text{left block} \in \{0, 1, 2, 4, 5, 6, 8\}$$
$$\text{and under} - \text{block} \in \{0, 8\}$$
$$\text{Condition3} : \text{under} - \text{right block} \in \{0, 1, 2, 3, 7, 8\}$$

When current block $B_{i, j}$ satisfies condition1, the embedded error in $B_{i, j}$ will not be transmitted into the right block from $B_{i, j}$'s right edge.

When current block Bi, j satisfies condition2, the embedded error in Bi, j will not be transmitted into the under and left-under block from Bi, j's under edge.

When current block Bi, j satisfies condition3, the embedded error in Bi, j will not be transmitted into the under-right block from Bi, j's lower-right corner.

Then the intra-frame distortion drift can be controlled by the 3 conditions.

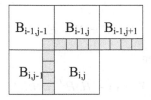

Fig. 1. The prediction block $B_{i,j}$ and the adjacent blocks

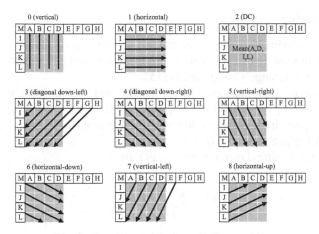

Fig. 2. 4×4 luma block prediction modes.

2.2 BCH Code

Define n the codeword length, k the code dimension, BCH (n, k, t) can correct t frame errors at most. Define BCH (n, k, t)'s generalized parity-check matrix H:

$$H = \begin{bmatrix} 1 & \Delta^2 & \ldots & \Delta^2 \\ 1 & \Delta^2 & \ldots & \Delta^2 \\ 1 & \Delta^2 & \ldots & \Delta^2 \\ 1 & \Delta^2 & \ldots & \Delta^2 \end{bmatrix} \tag{1}$$

Define Δ as $GF(2^m)$'s primitive element. The current data stream is $M = \{m_0, m_1, ..., m_{n-1}\}$, embedded data is $N = \{n_0, n_1, ..., n_{n-1}\}$.Then

$$N = M H^{\mathrm{T}} \tag{2}$$

If the received data stream is $S = \{s_0, s_1, \ldots, s_{n-1}\}$, then

$$N(x) = S_0 + S_1 x + S_2 x^2 + S_3 x^3 + \ldots + S_{n-1} x^{n-1}$$

$$S(x) = s_0 + s_1 x + s_2 x^2 + s_3 x^3 + L\, s_{n-1} x^{n-1},$$

Define E, E satisfies:

$$S = N + E \tag{3}$$

Using Eqs. (2) and (3), then

$$Y = (S - H)H^T = EH^T \tag{4}$$

Then embedded message can be calculated by (3) and can be extracted by (2).

2.3   Secret Sharing

Secret sharing is a technology which can sent sub-secrets to participants and the recovery of secret needs at least several participants. Shamir proposed a threshold secret sharing method based on Lagrange's interpolation [15].

Assume the divided message is $a_0 = k$, and the polynomial $p_{t-1}(x) = a_0 + a_1 x + a_2 x^2 + \cdots a_{t-1} x^{t-1} \bmod p$. (the prime p > 0, $a_i \in Z_p (i = 1, \ldots, n)$).

Assume the participant number is $n(n > p)$, y_i is calculated at $x_i (i = 1, 2 \ldots, i\text{-}1)$, then send it to n participants as (x_i, y_i) $(i = 1, \ldots, n)$.

If $p_{i-1}(x_i)$ has t values (x_i, y_i) $(i = 1, \ldots, t)$, then

$$p_{t-1}(x) = \sum_{i=1}^{t} y_i \prod_{j=1, j \neq i}^{t} \frac{x - x_j}{x_i - x_j} \tag{5}$$

and $k = p_{t-1}(0) = \sum\limits_{i=1}^{t} y_i \prod\limits_{j=1, j \neq i}^{t} \frac{-x_j}{x_i - x_j} = \sum\limits_{i=1}^{t} b_i y_i.$

Where $b_i = \prod\limits_{j=1, j \neq i}^{t} \frac{-x_j}{x_i - x_j}$ is the embedded message.

3   Proposed Algorithm

3.1   Embedding

The 4×4 DCT blocks are chosen to embed message.

As shown in Fig. 3. The positive integer N and coefficients $\tilde{Y}_{ij}(i, j = 0, 1, 2, 3)$ are chosen as an example to describe the embedding of algorithm:

(1) The embedded message are distributed by a polynomial and obtain a series of sub-messages.
(2) The sub- messages are encoded with BCH encoding method.

(3) The blocks are chosen with the DC coefficient larger than user-defined parameter threshold to insert message;
(4) The blocks satisfied (3) and met condition1, 2, 3 are chosen to insert the sub-message modulated by BCH technique.
(5) The DCT coefficients are encoded and the embedded H.264 video is got.

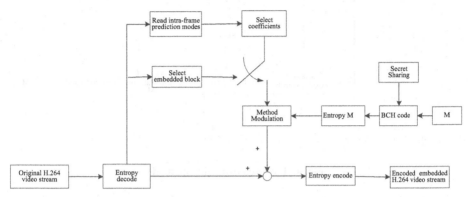

Fig. 3. Proposed embedding method

Modulation Method:
If $\left|\tilde{Y}_{ij}\right| = N + 1$ or $\left|\tilde{Y}_{ij}\right| \neq N$, the DCT coefficients $\tilde{Y}_{ij}$ is modified with (6).

If the embedded bit is 1 and $\left|\tilde{Y}_{ij}\right| = N$, the DCT coefficients $\tilde{Y}_{ij}$ is modified with (7).

If the embedded bit is 0 and $\left|\tilde{Y}_{ij}\right| = N$, $\tilde{Y}_{ij} = \tilde{Y}_{ij}$.

$$
\tilde{Y}_{i,j} = \begin{cases} \tilde{Y}_{i,j} - 1 & if \quad \tilde{Y}_{i,j} \geq 0 \, and \; \left|\tilde{Y}\right|_{i,j} > N \\[2mm] \tilde{Y}_{i,j} + 1 & if \quad \tilde{Y}_{i,j} < 0 \, and \; \left|\tilde{Y}\right|_{i,j} > N \\[2mm] \tilde{Y}_{i,j} & if \qquad \left|\tilde{Y}\right|_{i,j} < N \end{cases} \tag{6}
$$

$$
\tilde{Y}_{i,j} = \begin{cases} \tilde{Y}_{i,j} + 1 & if \quad \tilde{Y}_{i,j} \geq 0 \, and \; \left|\tilde{Y}\right|_{i,j} = N \\[2mm] \tilde{Y}_{i,j} - 1 & if \quad \tilde{Y}_{i,j} < 0 \, and \; \left|\tilde{Y}\right|_{i,j} = N \end{cases} \tag{7}
$$

3.2 Extracting

The extracting method is shown in Fig. 4. The received H.264 video is decoded, and the decoded DCT coefficients and 4 × 4 intra-frame prediction modes are got. Then the 4 × 4 block is choose with absolute value of DC coefficients, self-defined threshold and adjacent blocks whose prediction mode satisfy condition1, 2, 3. And the decoded message M is extracted from the chosen blocks with extract method.

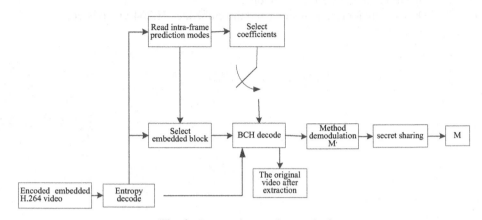

Fig. 4. Proposed extraction method

Extracting Method

If $\left|\tilde{Y}_{ij}\right| = N + 2$ or $\left|\tilde{Y}_{ij}\right| \neq N + 2$ or $\left|\tilde{Y}_{ij}\right| \neq N + 1, \tilde{Y}_{ij}$ modify the coefficients as Eq. (8).

If $\left|\tilde{Y}_{ij}\right| = N + 1$, extract bit 1 and $\tilde{Y}_{ij}$ modify the coefficients as Eq. (9).

If $\left|\tilde{Y}_{ij}\right| = N$, extract bit 0 and keep $\tilde{Y}_{ij}$.

$$\tilde{Y}_{i,j} = \begin{cases} \tilde{Y}_{i,j} - 1 & if \quad \tilde{Y}_{i,j} \geq 0 \, and \, \left|\tilde{Y}\right|_{i,j} > N + 1 \\ \tilde{Y}_{i,j} + 1 & if \quad \tilde{Y}_{i,j} < 0 \, and \, \left|\tilde{Y}\right|_{i,j} > N + 1 \\ \tilde{Y}_{i,j} & if \quad \left|\tilde{Y}\right|_{i,j} < N \end{cases} \tag{8}$$

$$\tilde{Y}_{i,j} = \begin{cases} \tilde{Y}_{i,j} - 1 & if \quad \tilde{Y}_{i,j} \geq 0 \, and \, \left|\tilde{Y}\right|_{i,j} = N + 1 \\ \tilde{Y}_{i,j} + 1 & if \quad \tilde{Y}_{i,j} < 0 \, and \, \left|\tilde{Y}\right|_{i,j} = N + 1 \end{cases} \tag{9}$$

The M′ is dealt with BCH decode and secret sharing after the extraction, and then the message M is got.

4   Experiment Results

The proposed algorithm has been implemented based on JM16.0 the standard H.264 software. Each experimental video encodes 300 frames at 30 frames/s, with quantize parameter 28 and an intra-period of 15 and QCIF of 176 × 144. The series of experiment videos are Akiyo, News, Container, Basketball, Coastguard, etc. In this paper, the PSNR (peak signal to noise ratio) is calculated by original YUV videos and embedded videos(include I, B, P frames) and take the average of all the frames (including I, B, and P frames) of a video sequence. Define frame-drop rate to describe the percentage of the lost frames in the simulate attack, and the survival rate is defined to describe the proportion of bits that this algorithm has correctly extracted.

The experiment result of the algorithm in this paper with BCH (63,7,15) and Secret Sharing (3,8) is listed in Table 1, and the experiment result with BCH (7,4,1) and Secret Sharing (3,8) is listed in Table 2. From the table we know the embedding capacity is 5040 bits and 4704 bits respectively and the PSNR is 39.89 in average and 39.92 in average respectively. When the frame loss rate is 5% or 10%, the survival rate can reach 100% with BCH (63,7,15). It can be concluded from the results the method can get a good embedding capacity and visual quality and has a good robustness. We can use the huge number of video frames to get a larger embedding capacity, and can influence embedding capacity and survival rate by the adjustment in (t,n). And the algorithm is very save for message transformation, since the common loss rate in internet transformation is about 10%.

In Table 3, we compare the experiment result of this algorithm with BCH (63,7,15) and (3,8) and the result of algorithm in [14] with (3,8).From the table, we can see our algorithm is obviously better than that in [14] in survival rate. That's because our algorithm can correct the bit error with BCH while the algorithm in [14] cannot.

The Figs. 5, 6, 7 and 8 show the original frame of video Akiyo and Container and extracted frame of Akiyo and Container. We can see our algorithm has a good visual quality, that's because our method can recover the original video completely.

Table 1. Embedding performance with BCH (63,7,15) and Secret Sharing (3,8)

Video sequence	Loss rate (%)	PSNR (dB)	Capacity (bits)	Survival rate (%)
Akiyo	5	40.54	5040	100
	10	40.54	5040	100
	20	40.54	5040	95.83
News	5	39.51	5040	100
	10	39.51	5040	100
	20	39.51	5040	91.26
Container	5	39.64	5040	100
	10	39.64	5040	100
	20	39.64	5040	92.54

Table 2. Embedding performance with BCH (7,4,1) and Secret Sharing (3,8)

Video sequence	Loss rate (%)	PSNR (dB)	Capacity (bits)	Survival rate (%)
Akiyo	5	40.53	4704	100
	10	40.53	4704	98.16
	15	40.53	4704	95.36
News	5	39.55	4704	97.55
	10	39.55	4704	93.72
	15	39.55	4704	90.96
Container	5	39.68	4704	93.04
	10	39.68	4704	95.54
	15	39.68	4704	91.51

Table 3. Embedding performance comparison between our method and [14]

Video sequence	Loss rate (%)	Survival rate (%) algorithm in this paper	Survival rate (%) algorithm in [14]
Akiyo	5	100	100
	10	100	99.79
	15	95.84	96.35
	20	95.33	79.84
News	5	100	100
	10	100	87.72
	15	94.58	88.64
	20	89.76	78.78
Container	5	100	100
	10	100	86.66
	15	98.05	89.17
	20	92.54	90

Fig. 5. The original frame of video Akiyo

Fig. 6. The extracted frame of Akiyo

Fig. 7. The original frame of video Container

Fig. 8. The extracted frame of Container

5 Conclusion

In this paper, we provide a lossless robust lossless steganography method based on H.264/AVC. During the video transformation, there are some attacks which may cause the loss of embedding message, and we use BCH and Secret Sharing to correct the mistake. The experiment result shows that our algorithm has a good robustness and reaches a good visual quality. Besides, our method can recover the original video completely, which means that the method can be used in some fields which is sensitive to video's accuracy and visual quality. In the future, we will look for some other methods to get a better result.

References

1. Shanableh, T.: Data hiding in MPEG video files using multivariate regression and flexible macroblock ordering. IEEE Trans. Inf. Forensics Secur. **7**(2), 455–464 (2012)
2. Sarkar, A.A., Madhow, U., Chandrasekaran, S., et al.: Adaptive MPEG-2 video data hiding scheme. In: Electronic Imaging. International Society for Optics and Photonics (2007)
3. Nightingale, J., Wang, Q., Grecos, C., et al.: The impact of network impairment on quality of experience (QoE) in H265/HEVC video streaming. IEEE Trans. Consum. Electron. **60**(2), 242–250 (2014)

4. Chang, P.C., Chung, K.L., Chen, J.J., et al.: A DCT/DST-based error propagation-free data hiding algorithm for HEVC intra-coded frames. J. Vis. Commun. Image Represent. **25**(2), 239–253 (2014)
5. Yao, G., Feng, P.: Information Hiding for H.264 in Video Stream Switching Application (2010). IEEE
6. Lee, N.S.: Method and apparatus for intra prediction of video data, US (2006)
7. Sullivan, G.J., Ohm, J.-R., Han, W.-J., Wiegand, T.: Overview of the high efficiency video coding (HEVC) standard. IEEE Trans. Circuits Syst. Video **22**, 1649–1668 (2012)
8. Ma, X.: A data hiding algorithm for H.264/AVC video streams without intra-frame distortion drift. IEEE Trans. Circ. Syst. Video Technol. **20**, 1320–1330 (2010)
9. Ma, X., Li, Z., Lv, J., et al.: Data hiding in H.264/AVC streams with limited intra-frame distortion drift. In: International Symposium on Computer Network & Multimedia Technology. IEEE (2010)
10. Zhao, J., Li, Z.-T., Feng, B.: A novel two-dimensional histogram modification for reversible data embedding into stereo H.264 video. Multimedia Tools Appl. **75**(10), 5959–5980 (2016)
11. Shuyang, L., Liang, C.: A new reversible robust steganography method based on H.264/AVC. Appl. Res. Comput. **36**(04), 1144–1147 (2019)
12. Liu, Y., Leiming, J., Mingsheng, H., Zhao, H., Jia, S., Jia, Z.: A new data hiding method for h.264 based on secret sharing. Neurocomputing **188**, 113–119 (2016)
13. Liu, A.Y., et al.: A robust reversible data hiding scheme for H.264 without distortion drift. Neurocomputing 151(Pt.3), 1053–1062 (2015)
14. Liu, Y., Chen, L., Hu, M., et al.: A reversible data hiding method for H.264 with Shamir's (t, n)-threshold secret sharing. Neurocomputing **188**(may5), 63–70 (2016)
15. Shamir, A.: How to share a secret. Commun. ACM **22**(11), 612–613 (1979)
16. Idbeaa, T., Jumari, K., Samad, S.A.: Data hiding based on quantized AC-coefficients of the I-Frame for MPEG-2 compressed domain. Int. Rev. Comput. Softw. **7**(4), 1458–1462 (2012)
17. Mobasseri, B.G., Marcinak, M.P.: Data hiding in MPEG-2 bitstream by creating exceptions in code space. Signal Image Video Process. (2014)
18. Liu, H., Huang, J., Shi, Y.Q.: DWT-based video data hiding robust to mpeg compression and frame loss. Int. J. Image Graph. **5**(01), 111–133 (2011)
19. Bouchama, S., Hamami, L., Aliane, H.: H.264/AVC data hiding based on intra prediction modes for real-time applications. In: Lecture Notes in Engineering & Computer Science, vol. 2200, no. 1 (2012)
20. Ke, N., Yang, X., Zhang, Y.: A novel video reversible data hiding algorithm using motion vector for H.264/AVC. Tsinghua Sci. Technol. **22**(5), 489–498 (2017)
21. Lin, T.J., Chung, K.L., Chang, P.C., et al.: An improved DCT-based perturbation scheme for high-capacity data hiding in H.264/AVC intra frames. J. Syst. Softw. **86**(3), 604–614 (2013)
22. Zhao, H., Liu, Y., Wang, Y., Liu, S., Feng, C.: A video steganography method based on transform block decision for H.265/HEVC. IEEE Access **9**, 55506–55521 (2021). https://doi.org/10.1109/ACCESS.2021.3059654

Research on Application of Blockchain Technology in Higher Education in China

Cong Feng[1,2](✉) and Si Liu[1]

[1] College of Information Science and Technology,
Zhengzhou Normal University, Zhengzhou, China
[2] Faculty of Social Sciences and Liberal Arts, UCSI University,
Kuala Lumpur, Malaysia

Abstract. The rise of new technology usually has a great impact on education. Blockchain, as an emerging technology, has developed rapidly in recent years and has become a hot topic of education researchers. This technology is initially applied in the field of higher education in China now. In view of this, this paper first introduces the concept of blockchain technology, and expounds the characteristics and types of its application in the field of education in detail. Then combined with the existing typical application cases of blockchain technology in higher education field, the impact and value of blockchain technology on higher education are discussed, and possible challenges are analyzed. It is expected that this paper can provide reference for the innovation of comprehensive and deep integrated education concept and education mode based on blockchain in China.

Keywords: Blockchain · Higher education · Application · Challenge

1 Introduction

Technological developments will unavoidably lead to changes in the field of education. Blockchain technology is an emerging technology which has developed rapidly in recent years and has been included in national strategies by some developed countries such as Britain and the United States. Besides, the Chinese government also attaches great importance to the development of this technology. In the "13th Five-Year" National Information Planning in 2016, the importance of "strengthening the advanced layout of strategic frontier technology was emphasized", and the key frontier technology status of blockchain has been determined form then on [1].

In 2018, the Ministry of Education issued "the education informatization 2.0 action plan", which points out that based on new technologies such as blockchain and large data, intelligent learning records, transfer, exchange, certification and other effective methods should be actively explored, so as to ubiquitous and intelligent learning system, and promote deep integration of the information technology and intelligent technology during the whole teaching process [2].

In 2019, on the 18th of the political bureau of the central collective learning, Chairman Xi Jinping proposed to accelerate the blockchain technology innovation and development, regarding blockchain as an important breakthrough of the core technology

© Springer Nature Switzerland AG 2021
D.-S. Huang et al. (Eds.): ICIC 2021, LNCS 12836, pp. 379–389, 2021.
https://doi.org/10.1007/978-3-030-84522-3_31

independent innovation, and put forward to actively promote the application of this technology in areas such as education, employment so as to provide more intelligent, convenient and high quality public services for the people [3].

In the "Key Points of Educational Informatization and Network Security in 2020" issued by the Ministry of Education in 2020, it is mentioned that the application of blockchain technology in students' online learning and teachers' online teaching behavior recording and identification should be explored, and a new model of online teaching evaluation that is both extensible and credible should be established [4].

From the release of a series of national documents, it can be seen that the education model based on blockchain has become a new development trend of future education. There is no doubt that blockchain technology has great application potential in the field of education, but this technology is still in its preliminary exploration stage, and has only been tried in the field of higher education in China at present. Based on this, this paper will introduce the concept and core technology of blockchain technology, discuss the ideas and value of educational application of this technology in combination with existing typical educational cases, and analyze the possible challenges it faces, in order to promote the application process of it in the higher education filed in China.

2   Blockchain Technology

2.1   Concept of Blockchain

At present, blockchain has not formed a unified definition in the academic world, but it is widely believed that blockchain is a kind of accounting technology that is jointly maintained by multiple parties, uses cryptography to ensure the security of transmission and access, and can realize the consistent storage of data, which is difficult to tamper, and prevents denial, which also known as Distributed Ledger Technology [5]. As shown in Fig. 1, the block generates and records the transaction information to be processed in chronological order. It is composed of a block header and a block body. The block header is responsible for connecting to the next block through the main chain, and the block body is responsible for storing data information. When blocks and chains are formed, the system automatically generates a timestamp and time labels the data information.

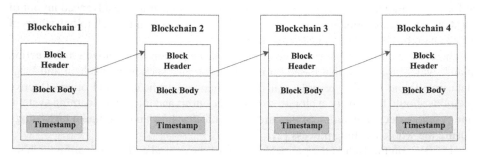

Fig. 1. General structure of blockchain

2.2 Main Features of Blockchain

According to the data structure and operation principle of blockchain, combined with its application in the field of education, blockchain technology has the following characteristics [6–8].

Decentralization: It is the most essential feature of blockchain technology. Blockchain technology is based on the Peer-to-Peer protocol in the network protocol, and according to its decentralized characteristics, to achieve the high reliability of distributed nodes and computer servers, which can well solve the Byzantine problem and has a strong fault tolerance. Since each node has a copy of the data, the distributed database is not affected by a problem on a single node.

Openness: This feature is based on the consensus mechanism of blockchain technology. To achieve trust, no third party is needed, only all participants need to agree with the consensus mechanism. Once a consensus is reached, it is recorded on the blockchain. Therefore, no one can modify and delete.

Immutable: Once transactions have been entered in the database and the account has been updated, the records cannot be changed because they are linked to every transaction record that preceded them. Various algorithms are used to ensure that the records in the database are permanent, chronologically sorted, and accessible to all other nodes on the network.

Security: As long as 51% of all data nodes cannot be controlled, network data cannot be manipulated and modified willfully, which makes the blockchain itself relatively safe and avoids subjective and artificial data changes [9].

Traceability: Each transaction on the ledger is recorded with a timestamp and validated through a verified hash reference. Therefore, you can trace the data information of any block by the correlation between arbitrary blocks.

2.3 Types of Blockchain

With the continuous development of technology and application, Blockchain has derived from the original narrow sense of "decentralized distributed verification network" into three types with different characteristics. According to the different implementation methods, it can be divided into Public Blockchain, Consortium Blockchain and Private Blockchain [10].

Public Blockchain: A blockchain in which everyone can participate equally, which is close to the original design sample of the blockchain. Everyone on the chain can freely access, send, receive, and authenticate transactions. It is a "decentralized" blockchain. The bookkeeper of the Public Blockchain is all the participants.

Consortium Blockchain: A "partially decentralized" blockchain composed of a limited number of organizations that can be accessed within the consortium. Each consortium member still adopts a centralized form, and the consortium members realize data co-verification in the form of a blockchain, so it is a "partially decentralized" blockchain. The bookkeepers of the consortium blockchain are determined by the members of the consortium through consultation and are usually representatives of the institutions.

Private Blockchain: A fully centralized blockchain, in which all members of the chain need to submit data to a central organization or a central server for processing. They only have the right to initiate transactions without the right to verify them. The bookkeeper is the only one, that is, the owner of the chain.

3   Multi-scenario Application of Blockchain in Higher Education

Nowadays, many types of blockchain applications in higher education have been investigated, as shown in Fig. 2. In this paper, some typical application scenarios of blockchain in higher education are discussed.

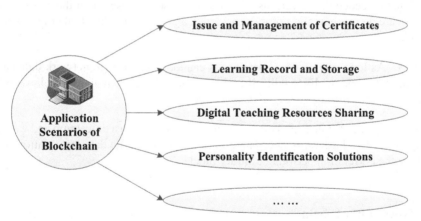

Fig. 2. Blockchain-based application scenarios in higher education

3.1   Issue and Management of Certificates

During their university career, students may be issued many certificates, such as diploma certificates, degree certificates, transcripts, award certificates, English level certificates, professional qualification certificates, etc. It has been widely proved that the security, transparency, and immutability of blockchain could bring benefits to the issuance and management of those certificates [8]. And this is also the field in which the blockchain technology is most widely applied in higher education. Through blockchain technology, students can directly access all kinds of certificate information they have obtained any-time. What's more, it enables education institutes to manage students' academic certificates securely and efficiently. And the certificate information recorded on the blockchain will also be permanently stored in the cloud server to effectively solve the authentication problem and enhance the credibility.

During the past years, an increasing number of applications of blockchain were proposed to address this problem. In 2014, when the University of Nicosia, Cyprus (UNIC) commenced to apply this technology on an official basis to store and confirm the diplomas; it was also the first university that began to accept fees for studies as bitcoins [11]. In 2016, the Sony Global Education supported by IBM developed a blockchain platform to store, protect, and exchange information related to students' performance and progress [12]. In 2017, the Malta College of Arts, Science & Technology began to issue Blockcerts-based digital certificates. The same year, Chinese researchers developed the ECBC blockchain to control certificates of education. Its performance rate is higher than the previous standard [13].

3.2 Learning Record and Storage

One of the core technologies of blockchain is to provide immutable distributed data records. When applied in the teaching field, blockchain technology can be used to create a learning information recording and query platform, which can record the learning situation and exam results of learners in the software system, as well as provide services such as query results. On the one hand, these records are helpful for teachers and researchers to analyze students' learning needs and situations and design corresponding teaching models, so as to realize personalized teaching. On the other hand, it is also convenient for students to self-monitor their learning situation and adjust their status to enter the next stage of independent learning. This platform, which records student learning information data, can provide convenience for learners, educational institutions and recruiters to inquire about the required information.

With the recorded learning information stored, the blockchain can be expanded to evaluate the students' professional ability, which provides recruitments and job selections with great convenience. Zhao proposed a design scheme of student ability evaluation system in which K-means clustering algorithm was applied to analyze students' academic performance and achievements that were stored on the blockchain [14]. Objective evaluation of students' professional ability is beneficial to students' employment. Wu studied a block chain-based skill evaluation model and applied it to measure students' professional knowledge and expertise. The evaluation process was simplified significantly, and high efficiency was achieved [15]. Thus it can be seen that it is feasible to use blockchain to record and query learning information, and there is still a great space for development in the future.

3.3 Digital Teaching Resources Sharing

At present, there are some problems in teaching resources in higher education, such as lack of high-quality teaching resources, low utilization rate of teaching resources, unbalanced allocation of teaching resources, high cost of platform operation and management, and difficult guarantee of resource safety [16]. Using the characteristics of blockchain such as decentralization, de-trust, collective maintenance, security, openness, anonymity and autonomy to build digital resource sharing can well solve these problems. The sharing of teaching resources need the coordination and support of all participants involved, including teachers and students, departments within universities and university alliances. With different participants and the equal status of the same types of participants in the sharing of teaching resources in consideration, the Public blockchain is constructed based on the teachers and students of the university, the Private blockchain based on the internal departments of the university and the Consortium blockchain based on among universities [17].

The Public blockchain takes teachers and students of all universities as participants. Due to the completely decentralized characteristics of the public chain, all university teachers and students can acquire or create resources according to their own preferences, so it can greatly meet the diversity, complexity and personalized needs of users. At the same time, based on the financial and monetary functions of the blockchain, a virtual currency mechanism used in the resource co-construction and sharing system is developed, and relevant incentive mechanisms that meet the needs of teachers and students in universities are developed. For example, teachers and students need to consume virtual currency when using relevant resources, and can earn corresponding virtual currency when sharing and creating resources. By introducing market mechanism, the virtual currency pegged to the relevant rights and interests of the college teachers and students. With more virtual currency earned, more resources are acquired and the users enjoy more rights and interests, so as to arouse the enthusiasm of teachers and students to use more resources, to encourage more college teachers and students to participate into the construction of teaching resources, and to promote the benign development of the teaching resource management. In addition, blockchain can guarantee the integrity and immutability of data, which provides the most favorable technical support for protecting the relevant private data of teachers and students as well as the originality and copyright of teaching resources.

The private chain is based on the internal departments of the university. The Private blockchain has the characteristics of complete centralization, which is more suitable for the internal data management of a specific organization. The central organization controls the write permissions, and selectively opens the read permissions to the outside. In view of the characteristics of scattered storage and diversified platforms of teaching resources in various departments in colleges and universities, the writing rights of teaching resources in various departments are controlled within colleges and universities. The use of blockchain technology can achieve cross-platform access to resources without intermediary platform. At the same time, through the link consensus conditions are achieved. With the use of smart contract technology, various departments in universities can not only manage their own teaching resources in the previous way, but also intelligently identify and overall view the resource data of other departments, so as to avoid repeated construction of resources. Similarly, the smart contract of the blockchain can be used to formulate corresponding reward rules to encourage participants. Using private chain to construct teaching resources in universities can not only conducts information docking at the fastest transaction speed, but also intelligently achieves cross-platform resource management and docking among various departments to ensure that digital teaching resources are not damaged by the outside world.

The Consortium blockchain has the characteristics of partial decentralization. Who can be the participants of the consortium blockchain are decided by organizations or individuals constructing the blockchain. Only after confirmation, can new nodes join the consortium blockchain. In the process of co-constructing and sharing teaching resources among universities based on the consortium blockchain, the participants of the university consortium are decided by the organizers. In view of the characteristics of the blockchain, the platform is intelligently implemented in accordance with the consensus mechanism. In addition, the reward rules are applied in the same way. After realizing the organization and management mechanism, performance evaluation mechanism and rights distribution mechanism of the university consortium through the smart contract of the consortium blockchain, the coordination and management cost of the university consortium will be greatly saved, and the co-construction and sharing of teaching resources can be guaranteed.

3.4 Personality Identification Solutions

The identity authentication method based on blockchain has the characteristics of modularization of identity verification, multi-service linkage authentication model, authenticity of data, user privacy security and so on. Different from the traditional identity authentication methods, blockchain identity authentication can transmit data to the identity authentication party for verification through the secure distribution channel [18]. The identity authentication party does not need to directly connect with the system of identity dependent party, but only needs to verify certificates for the identity authentication applicant. Identity Dependency can realize multi-service and multi-scene linkage of user identity information access and authenticity verification with no need to connect with each authenticator. The campus network can use blockchain technology to carry out reliable, credible and efficient privacy protection, data protection and access control for sensitive information generated in the authentication process of teachers, students and other identity subjects, so as to reduce the risk of information leakage. Various application platforms of universities are composed of application consortium blockchain, and each platform jointly maintains the consortium blockchain through online collaboration. When each platform member's own data changes, it automatically notifies and synchronizes other application platform members through the distributed structure of blockchain. When students and teachers log in successfully on one platform, other platforms synchronize the credit authorization data. There is no need for duplicate authentication when a user accesses other platforms within the consortium blockchain.

4 Issues and Challenges of Blockchain in Education

Blockchain technology now is still in its infancy stage of development, and most of the research is conducted in the financial field. Compared with the financial field, problems in the education field have stronger uniqueness and complexity. So far, the application of blockchain technology in the education field faces many challenges, such as limited data storage space, difficult promotion and operation, disputes over the property rights of educational data, and privacy protection risks of teachers and students caused by the hidden security of blockchain technology itself [19, 20].

4.1 The Data Storage Space Factor

Blockchain technology is a technology innovation of Internet finance, but many technical characteristics, especially network capacity, are still in the initial stage of development. The blockchain database records all the data information of each transaction from the beginning to the present. Any user who wants to store data needs to download and store the CreationBlock that carries all resource information. With the application of big data technology in the education field, the amount of data generated by teachers, students, and education management departments will show a blowout growth, result in more and more data information carried by the blocks in the blockchain, which puts forward higher requirements on the storage space of blockchain database. With the increasing amount of various types of data, on the one hand, the data storage space will be limited, which will affect the uploading and updating of data information, and on the other hand, the efficiency of data transmission will be reduced significantly, so as to affect the demand for real-time acquisition of data.

4.2 The Safety Factor

Security is another important aim that blockchain education has to obtain. Even though the blockchain is inherently capable of providing a secure solution, threats from within and without still exist [21]. In theory, only when more than 51% of nodes are attacked and controlled by hackers at the same time can data information be leaked or tampered with. However, with the development of mathematics, cryptography and computing technology, it is difficult to guarantee that the algorithm will not be cracked in the future, causing the leakage of teachers' and students' information [22]. Moreover, all transactions are open and transparent, and any information can be traced and inquired. Then some conclusions can be inferred, and the behaviors of teachers and students may be predicted, which is not conducive to the protection of teachers' and students' privacy.

4.3 The Policy and Empirical Factors

At present, there is no universal standard in the application field of blockchain technology, so its promotion and operation in the field of education face the dual challenges of lack of policy protection and practical experience. First of all, due to the lack of policy protection and guidance, the decentralization will have a strong impact on traditional education institutions, causing the traditional education platform in the distribution of interests suffer heavy losses. Take management department for instance, decentralization reduces expenses for the university administrative staff, so university administrations may resist implementation of this technology. It is not conducive to large-scale popularization and application of blockchain technology. Secondly, due to the few application cases of blockchain technology in the field of education, experts and scholars take a wait-and-see attitude towards its promotion and application in education field, and lack of motivation to promote its further application.

4.4 The Property Rights Disputes Factor

Due to the decentralization, the data on the blockchain is distributed recorded and stored on the blockchain, which makes the property rights of students' data blurred. Currently, the school affairs office is generally responsible for data management. With the application of blockchain, all data were stored on the blockchain, making the responsibilities of the entity management department downplayed. Then new problems appear. Who owns the right of attribution and use of these virtual data? Who owns the results based on data analysis? The series of the issues caused by the data property rights have to be studied and addressed to in the further application of blockchain in education field.

To sum up, in the next few years, it will take unremunerated investment from enterprises, gradual intervention from governments, continuous demonstration from experts and selfless cooperation from educational institutions to truly apply blockchain technology to education.

5 Conclusion

This paper focuses on some typical application scenarios of blockchain in higher education, including issue and management of certificates, learning record and storage, digital teaching resources sharing and personality Identification solutions. However, there are still many problems and challenges, such as limited data storage space, difficult promotion and operation, disputes over the property rights of educational data, privacy protection risks of teachers and students caused by the hidden security of blockchain technology itself, etc. Therefore, future research will focus on solving these problems. It is expected that educational institutions, teachers, educational researchers and learners will hold an open and positive attitude to the impact and changes and made corresponding preparations for the extensive and in-depth application of blockchain technology. In the next few years, it will take unremunerated investment from enterprises, gradual intervention from governments, continuous demonstration from experts and selfless cooperation from educational institutions to truly apply blockchain technology to education.

Acknowledgment. This paper is sponsored by the National Natural Science Foundation of China (NSFC, Grant 61572447).

References

1. No. 73 of China's Development and Reform Commission. Notice of the State Council on the issuance of the "13th Five-Year" National Informatization Plan (2016)
2. The Ministry of Education of the People's Republic of China. Notice of the Ministry of Education on the issuance of the "Action Plan for Educational Informatization 2.0". http://www.moe.gov.cn/srcsite/A16/s3342/201804/t20180425_334188.html

3. Jinping, X.: Take blockchain as an important breakthrough for independent innovation of core technology and accelerate the innovation and development of blockchain technology and industry. http://paper.people.com.cn/rmrb/html/2019-10/26/nw.D110000renmrb_2 0191026_2-01.htm

4. The General Office of the Ministry of Education of China: Key Points of Educational Informatization and Network Security in 2020. Education Science and Technology Department (2020). No. 1 Document. http://www.ict.edu.cn/news/jrgz/xxhdt/

5. China Academy of Information and Communications Technology Trusted Blockchain Promotion Plan: Blockchain White Paper, October 2019. http://www.ceweekly.cn/2019/1112/274880.shtml

6. Yumna, H., Khan, M.M., Ikram, M., Ilyas, S.: Use of blockchain in education: a systematic literature review. In: Nguyen, N.T., Gaol, F.L., Hong, T.-P., Trawiński, B. (eds.) ACIIDS 2019. LNCS (LNAI), vol. 11432, pp. 191–202. Springer, Cham (2019). https://doi.org/10.1007/978-3-030-14802-7_17

7. Kamišalić, A., Turkanović, M., Mrdović, S., Heričko, M.: A preliminary review of blockchain-based solutions in higher education. In: Uden, L., Liberona, D., Sanchez, G., Rodríguez-González, S. (eds.) Learning Technology for Education Challenges: 8th International Workshop, LTEC 2019, Zamora, Spain, July 15–18, 2019, Proceedings, pp. 114–124. Springer, Cham (2019). https://doi.org/10.1007/978-3-030-20798-4_11

8. Alammary, A., Alhazmi, S., Almasri, M., Gillani, S.: Blockchain-based applications in education: A systematic review. Appl. Sci. **9**(12), 2400 (2019)

9. Yao, Z.J., Ge, J.G.: A review of the principle and application of blockchain. Inf. Technol. Appl. Sci. Res. **8**(2), 3–17 (2017)

10. Tan, X.J.: Blockchain technology development review, status quo and prospects. China Social Science Network, 9 December 2019

11. UNIC: Publication of Academic Certificates (2014). https://www.unic.ac.cy/iff/blockchain-certificates/

12. Sony Global Education: Sony Global Education Develops Technology Using Blockchain for Open Sharing of Academic Proficiency and Progress Records (2016). https://www.sony.net/SonyInfo/News/Press/201602/16-0222E/

13. Fedorova, E.P., Skobleva, E.I.: Application of blockchain technology in higher education. Eur. J. Contemp. Educ. **9**(3), 552–571 (2020)

14. Zhao, W., Liu, K., Ma, K.: Design of student capability evaluation system merging blockchain technology. J. Phys. Conf. Ser. **1168**, 032123 (2019)

15. Wu, B., Li, Y.: Design of evaluation system for digital education operational skill competition based on blockchain. In: IEEE 15th International Conference on e-Business Engineering (ICEBE), Xi'an, China, 12–14 October 2018, pp. 102–109 (2018)

16. Tian, Z.G.: Exploration on co-construction and sharing of digital teaching resources in colleges and universities in the new era. High. Educ. BBS **11**, 74–77 (2017)

17. Luo, M.R., Yuan, X.Y., Cui, Y.: The co-construction and sharing of university digital teaching resources based on "blockchain." Univ. Libr. Work **2**, 34–38 (2020)

18. Kangkang, S., Shuhong, Y., Feng, X., Hua, Z.: Research on application of blockchain in informationization construction of colleges and universities. China Education Informationization (2021)

19. Ma, Y., Fang, Y.: Current status, issues, and challenges of blockchain applications in education. Int. J. Emer. Technol. Learn. (iJET) **15**(12), 20 (2020)

20. Yang, X.M., Li, X., Wu, H.Q., Zhao, K.Y.: The Application model and challenges of blockchain technology in education. Mod. Distance Educ. Res. **2**, 34–35 (2017)
21. Top five blockchain security issues in 2019. https://ledgerops.com/blog/2019/03/28/top-five-blockchain-security-issues-in-2019
22. Lin, I.C., Liao, T.C.: A survey of blockchain security issues and challenges. Int. J. Netw. Secur. **19**(5), 653–659 (2017)

Neural Networks

Multi-class Text Classification Model Based on Weighted Word Vector and BiLSTM-Attention Optimization

Hao Wu[1,2], Zhuangzhuang He[1,2], Weitao Zhang[1,2], Yunsheng Hu[1,2], Yunzhi Wu[1,2(✉)], and Yi Yue[1,2]

[1] Anhui Provincial Engineering Laboratory of Beidou Precision Agriculture, Anhui Agricultural University, Hefei, China
WUyzh@ahau.edu.cn

[2] School of Information and Computer, Anhui Agricultural University, Hefei 230036, Anhui, China

Abstract. To address the problems that traditional multi-category text classification algorithms generally have high dimensionality of text vectorization representation, do not consider the importance of words to the overall text, and weak semantic feature information extraction. A multi-category text classification model based on Weighted Word2vec, BiLSTM and Attention mechanism (Weight-Text-Classification-Model, WTCM) is proposed. First, the text is vectorized by the Word2vec model; then the weight value of each word is calculated by the TF-IDF algorithm and multiplied with the word vector to construct a weighted text vector representation; then the semantic feature information is extracted by the context-dependent capability of BiLSTM; the Attention mechanism layer is incorporated after the BiLSTM layer to assign weights to the output of each moment After the BiLSTM layer, an Attention mechanism layer is incorporated to assign weights to the sequence information output at each moment; finally, it is input to the softmax classifier for multi-category text classification. The experimental results show that the classification accuracy, recall and F-value of the WTCM model are as high as 91.26%, 90.98% and 91.12%, which can effectively solve the multi-category text classification problem.

Keywords: Word2vec · TF-IDF algorithm · BiLSTM unit · Attention mechanism · Multi category text classification

1 Introduction

Automated text classification [1] has long been a hot problem in the field of Natural Language Processing (NLP) research. Quality text classification algorithms can effectively reduce information redundancy, improve the efficiency of information retrieval, and facilitate users to quickly access effective information. It has been widely used in question and answer systems [2], sentiment analysis [3], spam filtering [4] and other fields.

© Springer Nature Switzerland AG 2021
D.-S. Huang et al. (Eds.): ICIC 2021, LNCS 12836, pp. 393–400, 2021.
https://doi.org/10.1007/978-3-030-84522-3_32

The current mainstream text classification algorithms are divided into shallow machine learning algorithms and deep learning algorithms based on neural networks. Shallow machine learning algorithms include statistical classification methods such as Naïve Bayes (NB) [5], Support Vector Machines (SVM) [6], and K-Nearest Neighbor (KNN) [7]. Min Zhang et al. [8] fused sentiment dictionaries based on the plain Bayesian algorithm to classify text for sentiment, but the rule-based fused dictionaries have difficulties in feature extraction. Zheng Fei et al. [9] combined deep learning with LDA model to complete text classification, but LDA still has the defect of semantic neglect in feature representation. Jun Deng et al. [10] used Word2vec to solve the semantic divide problem and combined with SVM to complete microblog comment sentiment binary classification. And deep learning algorithms mainly include Convolutional Neural Networks (CNN) and its variants and Recurrent Neural Network (RNN) and its variants. With the development of neural network technology, researchers use RNN algorithms to capture contextual semantic information and complete text classification, but RNN algorithms have problems such as gradient explosion and disappearance. In order to solve the defects of the original RNN algorithm, Hochreiter [11] et al. modified the RNN structure by adding three gate structures and proposed the Long Short Term Memory (LSTM) neural network. Kim et al. [12] proposed the convolutional neural network with the help of its ability to better capture local relevance, and proposed the Kim et al. [12] proposed a TextCNN model to apply it to a text classification task by taking advantage of the ability of convolutional neural networks to better capture local relevance. Wang et al. [13] combined CNN and LSTM for text classification and improved the accuracy of text classification. With Bahdanau et al. [14] proposed the attention mechanism. Yang et al. [15] et al. in 2016 used the attention mechanism for the sentiment classification task of review text. Researchers have also tried to combine neural networks with attention mechanisms. Wenfei Lan [16] et al. incorporated the Attention mechanism behind the LSTM layer to complete the classification of short Chinese news texts and achieved good results. Yunshan Zhao [17] et al. used a convolutional neural network text classification model based on the Attention mechanism, which has high accuracy and strong generality in the text classification task.

2   Proposed Model

In order to overcome the shortcomings of existing comment sentiment analysis methods, a sentiment analysis method based on Weighted Word2Vec-Bilstm-Attention is proposed. Sentiment analysis task is essentially a text classification task, and most of the current sentiment analysis studies use distributed word representation. However, in the two sentences "I bought an apple phone" and "I ate an apple", the meanings and semantic information of "apple" are different, and an improved word representation method is proposed to integrate the generated word vectors into the traditional TF-IDF algorithm to generate weighted word vectors. It is then fed into a bidirectional long and short term memory neural network (BiLSTM) for contextual information feature extraction. Then the Attention mechanism is fused to adjust the weights of the sequence information output at each moment to better represent the comment vector. Finally, the sentiment tendency of the comments is obtained by a feedforward neural network classifier. Figure 1 shows the overview of our network.

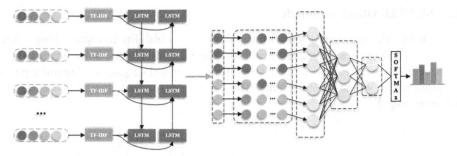

Fig. 1. WTCM model

2.1 The Construction of the Weighted Word Vector

The CBOW model in Word2vec is used to be able to map text into low-dimensional vectors and to overcome problems such as the semantic divide, which is based on contextual distribution to predict target words. For the word w_k, the context is expressed as follows:

$$context(w_k) = \{w_{k-t}, w_{k-(k-1)}, \ldots, w_{k+t}\}$$

However, Word2vec's word vector is not able to determine the importance of words to the whole text, so the TF-IDF algorithm is used to calculate the weight value of words and portray the importance of words to the text. TF-IDF is a common weighting technique used in Information Retrieval and Text Mining. In this paper, TF-IDF is used to calculate the weight value of each word. The formulas of TF-IDF are as follow:

$$tf_{i,j} = \frac{n_{i,j}}{\sum_k n_{k,j}}$$

where $n_{i,j}$ is the number of occurrences of the word in the text, and $\sum_k n_{k,j}$ is the sum of the occurrences of all words in the text. IDF is the inverse document frequency, which is used to indicate the importance of words. The calculation formula is as follows.

$$idf_i = \log \frac{|D|}{|\{j : t_i \in d_j\}|}$$

Where $|D|$ is the total number of documents in the corpus, $|\{j : t_i \in d_j\}|$ means the number of documents containing the word t_i. Where the TF-IDF value of a word is the product of $tf_{i,j}$ and idf_i, calculated as

$$w_{tf-idf} = tf_{i,j} * idf_i$$

The weights of the weighted word vectors $\vec{e}$ are calculated as follows:

$$\vec{e} = w_{tf-idf} * \vec{w}$$

$\vec{w}$ is defined as the distributed word vector trained by word2vec.

2.2 BiLSTM-Attention Module

To overcome the traditional recurrent neural network's inability to capture long-range semantic information and gradient explosion, the long-short term memory neural network overcomes the related problems by introducing update gates i, forgetting gates f, output gatesc o, and memory units. The LSTM network structure is shown in the following Fig. 2.

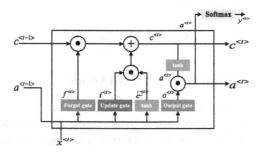

Fig. 2. The diagram of LSTM network structure.

The formula for each gate in the LSTM model is shown below.

$$\Gamma_f = \sigma(W_f[a^{<t-1>}, x^{<t>}] + b_f); \Gamma_u = \sigma(W_u[a^{<t-1>}, x^{<t>}] + b_u)$$
$$\tilde{c}^{<t>} = \tanh(W_c[a^{<t-1>}, x^{<t>}] + b_c); \Gamma_o = \sigma(W_o[a^{<t-1>}, x^{<t>}] + b_o)$$
$$c^{<t>} = \Gamma_u * \tilde{c}^{<t>} + \Gamma_f * c^{<t-1>}; a^{<t>} = \Gamma_o * \tanh c^{<t-1>}$$

where the BiLSTM is a combination of two LSTMs forward and backward, calculated as

$$\overrightarrow{h_t} = LSTM(x_t, \overrightarrow{h_{t-1}}); = LSTM(x_t, \overleftarrow{h_{t-1}}); h_t = w_t\overrightarrow{h_t} + v_t\overleftarrow{h_t} + b_t$$

The attention mechanism was first applied in the field of image and first applied in the field of natural language processing in the field of machine translation. Similar to the attention allocation mechanism of human brain, it improves the quality of extracting semantic feature information by calculating the probability weight value of output sequence information at each moment. The calculation formula is as follows.

$$\tilde{h_t} = \tanh(w_v h_t + b_v); \alpha_t = soft\max(\tilde{h_t}); \hat{h_t} = \sum_i \alpha_t h_t$$

2.3 Feedforward Neural Network Classifier

Finally, a normalization operation is performed by the softmax layer to output the predicted values, which are calculated as follows.

$$Y = soft\max(\hat{h_t})$$

3 Experiments

3.1 Experiments Environment and Dataset

The experimental environment is Python 3.7.0, Inter Core i5-8250U 1.80 GHz, and 8 GB of memory, and the related libraries and their versions are shown in Table 1 below.

Table 1. Library and its version

Third party library	Versions
Jieba	0.42.1
Keras	2.3.1
Numpy	1.16.4
Pandas	0.23.4
Scikit-learn	0.19.2
Tensorflow	1.14.0
Gensim	3.8.1

The experimental dataset is Today's Headline News dataset, which contains news texts from April to May 2018, with a total of 382,688 news texts distributed among 15 categories, namely, people's livelihood, culture, entertainment, sports, finance and economics, real estate, automobile, education, science and technology, military, tourism, international, securities, agriculture, and e-sports. In order to make the test results more convincing, all 10,000 texts were selected and divided according to the training set, test set and validation set 8:1:1.

3.2 Comparison of the Performance of Weighted Word Vectors

In order to verify the effectiveness of the weighted word vector algorithm, the original word vector and the weighted word vector are compared separately for text classification accuracy experiments. Different Window_size and Vector Dimension are set. The window size is 0, 1, 2…10; the word vector dimension is 0, 50, 100…300. The experimental results are shown in Fig. 3 and 4.

Through experiments, it is found that after introducing the TF-IDF algorithm to weight the Word2vec word vectors, the algorithm performance is improved. Word2vec can effectively overcome the problems such as the existence of high-dimensional sparsity in one-hot coding, and the weight values introduced by the TF-IDF algorithm are the feature enhancement of the word vectors.

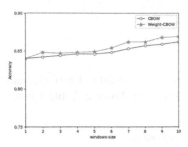

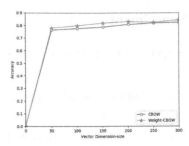

Fig. 3. Window_ Size curve   **Fig. 4.** Vector dimension curve

3.3   Network Comparison

In this subsection, four classical models, namely, support vector machine (SVM), plain Bayesian (NB), convolutional neural network (TextCNN), and long and short term memory neural network (LSTM), are selected for comparison experiments using the control variables method. The superparameters of the proposed model in this paper are determined, as shown in the Table 2.

Table 2. Parameter setting of classification model

Parameters	Values
max_len	30
hidden_size	128
att_size	64
dropout	0.2
epochs	20
batch_size	128
loss function	categorical_crossentropy
optimizer	Adam

In this paper, precision, recall, and f-value are used to evaluate our model. The results are shown in the table, where the accuracy and loss value iteration curves of the model proposed in this paper are shown in Fig. 5 and Fig. 6, respectively (Table 3).

We can see that when the WTCM model is trained using the gradient descent method, the loss value of the function is gradually decreasing and eventually converges to a state of convergence with a value of 0.0257. The value of the accuracy rate is finally as high as 91.26% as the model is continuously optimized.

Table 3. Comparison of experimental results

Method	Precision	Recall	F-score
Naïve Bayesian	84.57%	85.59%	85.08%
SVM	83.39%	84.43%	83.91%
TextCNN	85.74%	86.15%	84.94%
LSTM	89.57%	88.73%	89.15%
Proposed model	**91.26%**	**90.98%**	**91.12%**

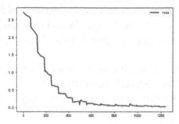

Fig. 5. Loss iterative curve of classification model

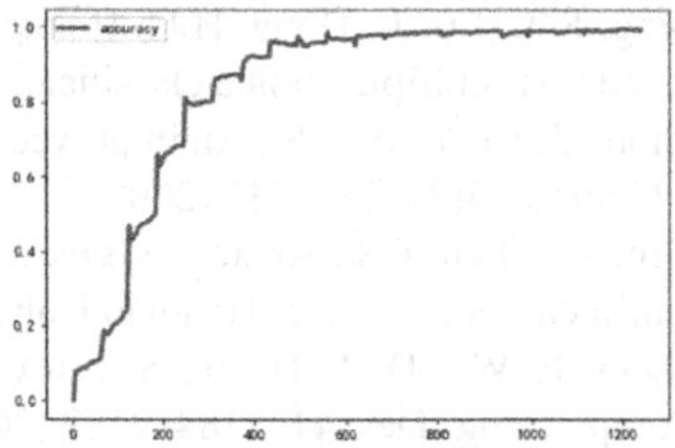

Fig. 6. Accuracy iterative curve of classification model

4 Conclusion

In this paper, we propose the WTCM model for multi-category text classification to address the problems that traditional multi-category text classification algorithms generally have high dimensionality of text vectorization representation, do not consider the importance of words to the overall text, and have weak semantic feature information extraction. Word2vec is used for text vectorization to generate a low latitude dense word vector with semantic information. The TF-IDF algorithm is introduced for word weight values, and the word vector is used to construct a weighted word vector representation with the word vector, which reflects the importance of words to the text. Using BiLSTM can effectively solve the gradient disappearance and explosion problem and can well capture the contextual semantic information, deep learning of text and complete deep semantic feature extraction. Finally, the Attention mechanism layer is fused to assign probability weight values to the high-level feature vectors output by BiLSTM to highlight key information.

Acknowledgement. This work was supported in part by Major Research and Development Projects of Anhui Province Key R&D Program Project (201904e01020015); Qinghai Province Key R&D and Transformation Program (2020-QY-213); Qinghai Province Basic Research Program Project (2020-ZJ-913); National Key R&D Program Subproject (2017YFD0301303).

References

1. Thompson, P.: Looking back: on relevance, probabilistic indexing and information retrieval. Inf. Process. Manage. **44**, 963–970 (2008)
2. Tang, Y., Lin, L., Luo, Y.M., Pan, Y.: Advances in information retrieval technology based on automatic question and answer systems. Comput. Appl. **28**, 2745–2748 (2008)
3. Hecai, L., Yonghong, L.: A framework design for sentiment disposition analysis service in natural language processing. Fujian Comput. **35**, 76–77 (2019)
4. Lu, W., Li, Z., Zhu, C.: Research on spam filtering based on plain Bayesian algorithm. Sens. Microsyst. **39**, 46–48,52 (2020)
5. Sharma, N., Singh, M.: Modifying Naive Bayes classifier for multinomial text classification. In: International Conference on Recent Advances & Innovations in Engineering (2017)
6. Ding, S.T., Lu, J., Hong, H.F., Huang, A., Guo, Z.Y.: Design and implementation of SVM-based text multiple choice classification system. Comput. Digit. Eng. **48**, 147–152 (2020)
7. Zhou, Y., Li, Y., Xia, S.: An improved KNN text classification algorithm based on clustering. J. Comput. **4**(3), 230–237 (2009)
8. Min, Z., Chen, F.S.: An active Bayesian text sentiment classification method combined with sentiment dictionary. J. Huaqiao Univ. (Nat. Sci. Edn.) **39**, 623–626 (2018)
9. Zheng, F., Wei, D.-T., Huang, S.: A text classification method based on LDA and deep learning. Comput. Eng. Des. **41**, 2184–2189 (2020)
10. Deng, J., Sun, S., Wang, R., Song, X., Li, H.: Sentiment evolution analysis of microblog public opinion based on Word2Vec and SVM. Intell. Theor. Pract. **43**, 112–119 (2020)
11. Hochreiter, S., Schmidhuber, J.: Long short-term memory. Neural Comput. **9**, 1735–1780 (1997)
12. Kim, Y.: Convolutional Neural Networks for Sentence Classification. Eprint Arxiv (2014)
13. Wang, H., Song, W., Wang, F.: A text classification method based on a hybrid model of LSTM and CNN. Small Microcomput. Syst. **41**, 1163–1168 (2020)
14. Bahdanau, D., Cho, K., Bengio, Y.: Neural machine translation by joint learning to align and translate. arXiv:1409.0473 (2014)
15. Yang, Z., Yang, D., Dyer, C., He, X., Hovy, E.: Hierarchical attention networks for document classification. In: Conference of the North American Chapter of the Association for Computational Linguistics: Human Language Technologies (2017)
16. Lan, W.F., Xu, W.W., Wang, D.Z., Pan, P.C.: LSTM-attention based Chinese news text classification. J. Zhongnan Univ. Nationalities (Nat. Sci. Edn.) **37**, 129–133 (2018)
17. Zhao, Y., Duan, Y.: A text classification model based on attention mechanism for convolutional neural networks. J. Appl. Sci. **37**, 541–550 (2019)

Fault Diagnosis Based on Unsupervised Neural Network in Tennessee Eastman Process

Wei Mu, Aihua Zhang$^{(\boxtimes)}$, Zinan Su, and Xing Huo

School of Control Science and Engineering, Bohai University, Jinzhou 121013, China

Abstract. In the industrial process, in order to solve the problem that the supervised neural network used for fault diagnosis always needs to compare with the corresponding output value constantly and takes a long time. In this paper, the competitive neural network and the improved self-organizing feature mapping neural network in unsupervised neural network is proposed for fault diagnosis. In the learning process, the active neighborhood between neurons can be gradually reduced without obtaining output values, so as to enhance the activation degree of central neurons, and then the weights and thresholds can be automatically adjusted. In this way, maintenance personnel can get more time for timely maintenance and reduce losses to a great extent. Finally, the feasibility of the proposed method is verified by the simulation of Tennessee Eastman process.

Keywords: Fault diagnosis · Competitive neural network · Self-organizing feature mapping neural network · Tennessee Eastman process

1 Introduction

Fault diagnosis has always been a hot topic. With the increasing reliability and security of the underlying system, it is very important to find the fault situation in the system as soon as possible. Due to the increasingly mature fault diagnosis technology of models, many industrial application methods have been rapidly developed in the past few years [1–3]. Tennessee Eastman Process (TEP) is a simulation platform developed by Eastman Company to simulate real chemical processes [4, 5]. Towns first introduced TEP as a benchmark for academic research in 1993 [6]. Since then, TEP has been widely used in the simulation and verification of various control and process monitoring methods.

In recent years, fault diagnosis methods based on data-driven technology have become a subject of extensive research [7–10]. Due to the low accuracy of traditional fault diagnosis methods, more and more scholars adopt neural networks for fault diagnosis. Zhou et al. [11] studied the fault diagnosis technology based on BP network information fusion, and applied the information fusion technology to the fault diagnosis of the MFCS comparative amplifier board. Zheng et al. [12] established the bearing fault diagnosis model of coal mine hoist based on wavelet neural network to diagnose the bearing fault of coal mine hoist. Tang et al. [13] used an adaptive neural fuzzy network (ANFIS) to model the fault diagnosis system of lithium iron phosphate battery pack, and then completed the training and construction of the parameters of the diagnosis system.

© Springer Nature Switzerland AG 2021
D.-S. Huang et al. (Eds.): ICIC 2021, LNCS 12836, pp. 401–414, 2021.
https://doi.org/10.1007/978-3-030-84522-3_33

Based on Labview and BP neural network, Yu et al. [14] performed fault diagnosis for rotating machinery. Ouhibi R et al. [15] proposed a fault diagnosis method for induction motors based on a probabilistic neural network (PNN). Aiming at the frequent occurrence of elevator faults, Zhang et al. [16] proposed an elevator fault diagnosis method that combined fault tree analysis, improved particle swarm optimization (PSO), and probabilistic neural network (PNN). Yang et al. [17] proposed a transformer fault diagnosis method based on dissolved gas analysis (DGA), BAT algorithm (BA), and optimized probabilistic neural network (PNN). Through the data of aircraft accidents, Khan et al. [18] showed that when the flight crew failed to correctly monitor the flight path of the aircraft, the unsupervised neural network method was used to better understand why the pilots failed to correctly monitor, and analyzed the accident and event data. Khan et al. [19] developed an Spoken term detection (STD) method based on acoustic signal processing that combines several techniques of noise removal, dynamic noise filtering, and evidence combination to improve existing speaker-dependent STD methods and, in particular, to improve the reliability of query discourse recognition.

All the above scholars will have supervised neural networks widely used in fault diagnosis. The supervised neural network is a radial basis function (RBF) feedforward neural network based on Bayesian decision theory, although it has strong fault tolerance and significant advantages in pattern classification. However, there is still a challenge: for example, the performance of probabilistic neural networks in supervised neural networks is greatly affected by the element smoothing factor of its hidden layer, which affects the classification performance; and you need a set of input and output values before you can do any further modeling.

Therefore, considering the above reasons, this paper uses two typical neural networks in unsupervised learning neural network, namely competitive neural networks and self-organizing feature mapping (SOM) neural networks, to classify and identify step faults, random faults, and valve faults in TEP. The accuracy is also verified by simulation. Finally, it makes up for the deficiency of supervised neural networks under the condition of ensuring accuracy. He et al. [20] proposed a modeling method for motor bearing fault identification using wavelet singularity entropy (WSE) and self-organizing feature mapping (SOM) neural network. The results show that the model can effectively identify the end position of the motor fault bearing and its internal pitting failure position. Compared with the traditional support vector machine (SVM) and BP neural network (BP) recognition models, the proposed model has higher accuracy and better stability in fault identification, and is more suitable for multi-classification problems such as fault identification.

The rest of this article is organized as follows. The structure and algorithm of competitive neural networks in unsupervised neural networks are described in Sect. 2. In Sect. 3, the structure and algorithm of self-organizing feature mapping neural networks in an unsupervised neural network are introduced in detail. Section 4 illustrates the effectiveness of competitive neural networks algorithm and SOM algorithm through the simulation experiment of TEP. Finally, we summarize the conclusions in Sect. 5.

2 Competitive Neural Network

2.1 The Structure of Competitive Neural Networks

The Competitive neural network is a typical and widely used unsupervised learning neural network, its structure is shown in Fig. 1. The Competitive neural network is generally composed of input layer and competition layer. Similar to RBF and other neural networks, the input layer only realize the transmission of the input mode and does not participate in the actual operation. All neurons in the competitive layer compete with each other to win the response to the input mode. Finally, only one neuron wins and makes the connection weights and thresholds related to the winning neuron develop in a direction more conducive to its competition, while the corresponding weights and thresholds of other neurons remain unchanged.

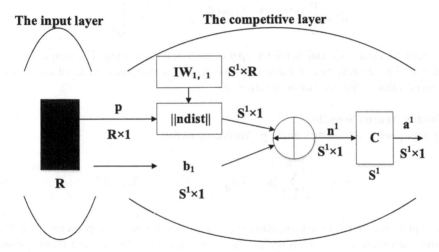

Fig. 1. Neural network structure

2.2 Algorithm of Competitive Neural Network

Initialization Process

The competitive neural network is composed of R neurons, and the competitive layer is composed of S1 neurons. Let the input matrix of the training sample be

$$P = \begin{pmatrix} p_{11} & p_{12} & \cdots & p_{1Q} \\ p_{21} & p_{22} & \cdots & p_{2Q} \\ \vdots & \vdots & & \vdots \\ p_{R1} & p_{R2} & \cdots & p_{RQ} \end{pmatrix}_{R \times Q} \tag{1}$$

Where, Q is the number of training samples; P_{ij} represents the i input variable of the j training sample, and $p_i = [p_{i1}, p_{i2}, \ldots, p_{iQ}], i = 1, 2, \ldots, R$. The initial connection weight of the network is

$$IW^{1,1} = [\omega 1, \omega 2, \ldots \omega_R]_{S^1 \times R} \tag{2}$$

Where,

$$\omega_i = \left[\frac{min(p_i) + max(p_i)}{2} \frac{min(p_i) + max(p_i)}{2} \cdots \frac{min(p_i) + max(p_i)}{2}\right]'_{S^1 \times 1},$$

i = 1,2,...,R. The initial threshold of the network is

$$b^1 = \left[e^{1 - \log(\frac{1}{S^1})}, e^{1 - \log(\frac{1}{S^1})} \ldots, e^{1 - \log(\frac{1}{S^1})}\right]'_{S^1 \times 1} \tag{3}$$

The network needs to initialize relevant parameters before learning. The weight learning rate is α, The learning rate of the threshold is β, The maximum number of iterations is T, initial value of the number of iterations N = 1.

Winning Neuron Selection
Randomly select a training sample p, According to the

$$n_i^1 = -\sqrt{\sum_{j=1}^{R} (p^j - IW_{ij}^{1,1})} + b_i^1, i = 1, 2, \ldots, S^1 \tag{4}$$

It computes the input of the competitive layer neurons. Where, n_i^1 represents the output of the i neuron in the competition layer; p^j represents the value of the I input variable of sample P; $IW_{ij}^{1,1}$ represents the connection weight of the i neuron in the competition layer and the j neuron in the input layer; b_i^1 represents the threshold value of the i neuron in the competition layer. If the k neuron in the competitive layer is the winning neuron,

$$n_k^1 = max(n_i^1), i = 1, 2, \ldots, S^1, k \in [1, S^1] \tag{5}$$

Weight and Threshold Updates
The weights and thresholds corresponding to the winning neuron k are modified as follows:

$$IW_k^{1,1} = IW_k^{1,1} + \alpha(p - IW_k^{1,1}) \tag{6}$$

$$b^1 = e^{1 - \log[(1-\beta)e^{1 - \log(b^1)} + \beta \times \alpha^1]} \tag{7}$$

The weights and thresholds of the remaining neurons remain unchanged. Among them, $IW_K^{1,1}$ is the k row of $IW^{1,1}$. It is the weight corresponding to the winning neuron k; a^1 is the output of neurons in the competitive layer,

$$a^1 = \left[a_1^1, a_2^1, \ldots, a_{S^1}^1\right], a_i^1 = \begin{cases} 1, i = k \\ 0, i \neq k \end{cases}, i = 1, 2, \ldots, S^1 \tag{8}$$

End of Iteration Judgment

If the sample is not finished, another sample is selected at random and return to Step 2). If N < T, set N = N + 1, return to step 2); Otherwise, the iteration ends. The algorithm flow chart is shown in Fig. 2.

3 Self-organizing Feature Mapping Neural Network

3.1 The Structure of SOM Neural Network

Self-organizing feature mapping (SOM) neural network is improved on the basis of competitive neural network, aiming at the situation that only one neuron wins each time, that is, only one neuron's weight and threshold value is worthy of modification. SOM neural network, not only the weight and threshold corresponding to the winning neurons are worth adjusting, but also other neurons within the adjacent range have the opportunity to adjust their weight and threshold, which greatly improves the learning ability and generalization ability of the network. As shown in Fig. 3, SOM neural network is similar in structure to competitive neural network. It is also a two-tier network composed of an input layer and a self-organizing feature mapping layer (competition layer).The input layer is used to receive information from the outside world, and the output layer is used for simulation, comparison, and response. One of the most important features of the SOM network architecture is the ability to extract the pattern characteristics of the input signals.

3.2 Algorithm of SOM Neural Network

The learning algorithm of SOM neural network is similar to the algorithm of competitive neural network, with great difference only in the weight adjustment part. The specific algorithm process is divided into the following parts:

Initialization Process

In SOM neural network, the input layer is composed of R neurons and the competition layer is composed of S^1 neurons. A small random number is assigned to each neuron in the competition layer as the initial value. $IW_{ij}^{1,1}(i = 1, 2, ..., S^1; j = 1, 2, ...R)$ $(IW_{ij}^{1,1}$ represents the connection weight between the i neuron in the competition layer and the j neuron in the input layer). And the initial neighborhood is $N_c(t)$ the initial threshold learning rate is set as the η. The maximum number of iterations is T, and the initial value of the number of iterations is N = 1.

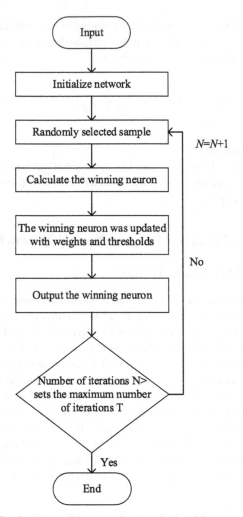

Fig. 2. Competitive neural network algorithm process

Winning Neuron Selection

Similar to the competitive neural network, a sample is randomly selected and the Formula (4) and (5) is used to calculate the winning neuron k.

Weight Update

$$\begin{cases} IW_j^{1,1} = IW_j^{1,1} + \eta(t)(p - IW_j^{1,1}), \ j \in N_c(t) \\ IW_j^{1,1} = IW_j^{1,1}, \qquad\qquad\qquad j \notin N_c(t) \end{cases} \tag{9}$$

According to Eq. (9), the weight of the winning neuron k and all neurons in its neighborhood are updated $N_c(t)$.

Update the Learning Rate and Neighborhood
Before the next iteration, the weight values of the neurons in the winning neuron and the neighborhood should be updated first, and the learning rate and neighborhood should be updated by using Eqs. (10) and (11).

$$\eta = \eta(1 - \frac{N}{T}) \tag{10}$$

$$N_c = \left\lceil N_c(1 - \frac{N}{T}) \right\rceil \tag{11}$$

Where, the symbol $\lceil \rceil$ denotes round up.

End of Iteration Judgment
If the learning sample is not finished, select another sample at random and return to Step 2). If N < T, let N = N + 1, return to step 2); Otherwise, the iteration ends. The algorithm flow chart is shown in Fig. 4.

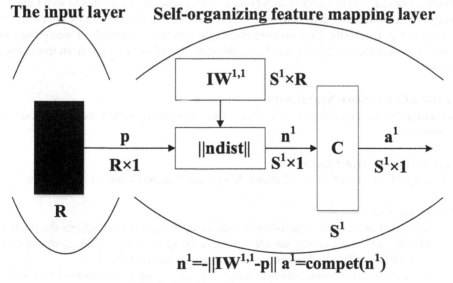

Fig. 3. SOM neural network structure

4 Unsupervised Learning Neural Network Simulation of TEP

The unsupervised neural network is an algorithm of an artificial intelligence network, whose purpose is to classify the original sample to understand the internal structure of the sample. Unlike supervised learning networks, unsupervised learning networks do not know whether their classification results are correct or not, that is, they are not supervised reinforcement (telling them which learning is correct). But its advantage is that it can provide input paradigms for this kind of network, and it will automatically pull out the underlying class rules from those paradigms. In this paper, two typical unsupervised neural network algorithms are used to verify the accuracy of unsupervised neural network classification through the simulation of TEP.

TEP process variables can be divided into two parts, a part has 41 measurements, XMEAS module (1–41), which includes 19 sampling process measurements and 22 continuous process measured values. The other part includes 12 manipulated variables, XMV module (1–12). This experiment also is the same as some of the author [21], with XMEAS (35) as a variable quality.

There are three main types of TEP quality failures: step failures, random failures and valve failures. During the data collection, the data were recorded every 6 min, and the competitive neural network and SOM neural network were used to simulate the data, and then the three kinds of faults were classified and identified. The classification steps are as follows:

Generate Training Set and Test Set
In order to verify the accuracy of this method, the data used for training are obtained from offline and online acquisition processes in industrial processes. In this paper, 22 groups of samples were selected as the training set, including 12 groups of step faults, 8 groups of random faults and 2 groups of valve faults, of which 800 data were collected in each group. Due to the large number of data, here is not an example. Three groups of samples are selected as the test set. These three groups of samples are from three types of faults.

Create a Competitive Neural Network
According to the algorithm in 2.2 above, the competitive neural network model is established.

Create SOM Neural Network
According to the algorithm in 3.2 above, SOM neural network model is established.

Simulation Test
In order to demonstrate the feasibility of this method, the network adopts the method of randomly sampling training samples in the training process, so the results of each operation will be different. The results of a run are shown in Table 1.

For a competitive neural network, most step fault samples correspond to the first neuron in the competitive layer (the only sample No. 11 corresponds to the second neuron in the competitive layer), so it can be identified that for step fault samples, the first neuron in the competitive layer is the winning neuron. Similarly, it can be concluded that, for random faults, the third neuron in the competition layer is the winning neuron (the only sample No. 16 corresponds to the second neuron in the competition layer). However, for valve fault samples, it is difficult to identify the corresponding winning neurons. According to the corresponding relationship, it can be found that the classification of No. 23 and No. 24 samples in the test set is correct, while the classification of No. 25 sample is difficult to determine.

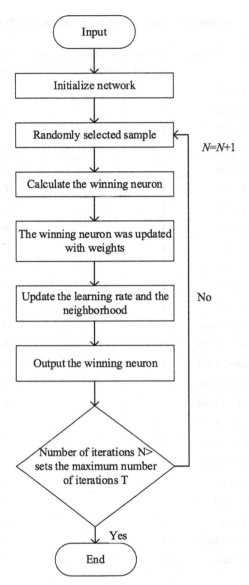

Fig. 4. SOM algorithm process

For SOM neural network, the winning neurons corresponding to step fault samples are numbered 1, 6, 7, 9, 10, 11, 13, 16. The winning neurons corresponding to random fault samples were numbered 4, 5, 8, 13, 15. The winning neurons corresponding to the valve fault samples were 3 and 14. According to the corresponding relationship, the number of neurons corresponding to step faults in the test set is 16. The number of neurons corresponding to random faults in the test set was 5. The neuron number corresponding to the valve faults in the test set was 3, and these numbers were all in the

Table 1. Comparison of prediction results between competitive and SOM neural network

Sample number	Actual category	Competitive neural network winning neuron	SOM neural network winning neuron
1	Step fault	1	1
2	Step fault	1	6
3	Step fault	1	1
4	Step fault	1	6
5	Step fault	1	9
6	Step fault	1	6
7	Step fault	1	11
8	Step fault	1	6
9	Step fault	1	7
10	Step fault	1	10
11	Step fault	2	16
12	Step fault	1	13
13	Random fault	3	15
14	Random fault	3	8
15	Random fault	3	5
16	Random fault	2	2
17	Random fault	3	8
18	Random fault	3	4
19	Random fault	3	13
20	Random fault	3	4
21	Valve fault	2	2
22	Valve fault	1	1
23	Step fault	1	16
24	Random fault	3	5
25	Valve fault	2	3

neuron number set corresponding to the winning neuron in the training set. Therefore, the accuracy of discrimination could be determined to be 100%.

If the predicted winning neuron number of a fault sample in the test set is 12, it is difficult to determine which kind of fault sample it belongs to. This is because, during the training process, the competitive layer 12 neuron never won the chance to win, and has been in a state of inhibition, which is known as the "dead" neuron. This can also be intuitively observed from the statistics of the number of times each neuron in the competitive layer becomes the winning neuron in Fig. 5. But since it's very unlikely to happen, let's ignore it for now.

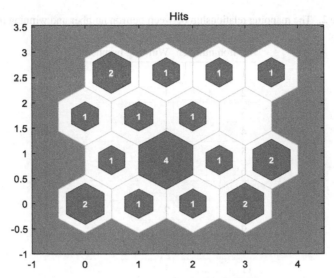

Fig. 5. Statistical map of the winning neurons

In Fig. 5, the winning neurons are numbered from left to right and from bottom to top, with the number of neurons gradually increasing, that is, the neurons in the lower-left corner are numbered 1 and the neurons in the upper right corner are numbered 16. The mapping relationship between neuron number and winning times is shown in Table 2 and the node diagram of the SOM neural network is shown in Fig. 6.

Figure 7 shows the distance distribution diagram between each neuron in the competitive layer and its neighboring neurons. The color of the filling area between adjacent neurons indicates the distance between two neurons. The darker the color (close to black), the farther apart the neurons are. As can be seen from the figure, neuron No. 15 and neuron No. 16 are darker in color. As can be seen from Table 1, neuron No. 16 belongs to the step fault sample, while neuron No. 15 belongs to the random fault sample, which indicates that the distance between winning neurons corresponding to different categories is relatively far. Similarly, it can be observed that the distance between the winning neurons corresponding to the same category is relatively close, such as No. 6, No. 7, No. 10, and No. 11, which all correspond to the step fault samples.

It can be clearly seen from the above results that, compared with the competitive neural network, the SOM neural network has a better classification effect even if there is a "dead" neuron problem. In terms of learning algorithm, it simulates the biological nervous system and uses the excitability, coordination, inhibition, and competition between neurons to carry out the dynamic principle of information processing, to guide the learning and work of the network. At the same time, SOM neural network gradually reduces its neighborhood scope and "repels" neighboring neurons. This combination of "collaboration" and "competition" makes its performance superior.

Table 2. The mapping relationship between neuron number and winning times

Neuron number	Winning number
1	2
2	1
3	1
4	2
5	1
6	4
7	1
8	2
9	1
10	1
11	1
12	0
13	2
14	1
15	1
16	1

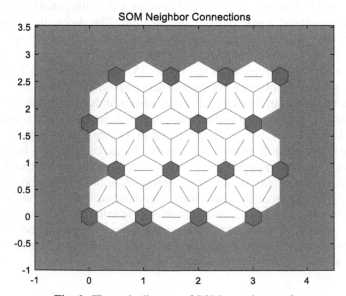

Fig. 6. The node diagram of SOM neural network

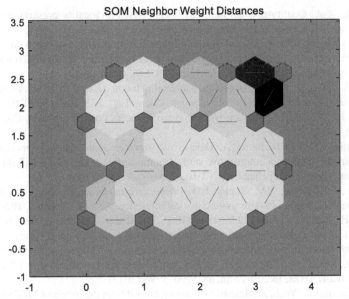

Fig. 7. Distance distribution of adjacent neurons

5 Conclusion

In this paper, the competitive neural network algorithm and SOM neural network algorithm are applied to TEP for the first time. Based on the input sample, the method can automatically adjust the weight and threshold without obtaining the output value, so it has good performance and strong generalization ability. Compared with the competitive neural network, although only one neuron wins each time in the SOM neural network, the corresponding weights of the neurons and their adjacent ranges in the SOM neural network are modified at the same time to adjust in a direction more conducive to their winning. By the simulation of TEP, the results show that SOM neural network has strong operability, fast training speed, and can obtain better classification results.

Acknowledgments. This works is partly supported by the Natural Science Foundation of Liaoning, China under Grant 2019MS008, Education Committee Project of Liaoning, China under Grant LJ2019003.

References

1. Ragab, A., El-Koujok, M., Yacout, S., et al.: Fault detection and diagnosis in the Tennessee Eastman process using interpretable knowledge discovery. In: Reliability and Maintainability Symposium RAMS2017. IEEE (2017)
2. Luo, K., Li, S., Deng, R., Zhong, W., Cai, H.: Multivariate statistical Kernel PCA for nonlinear process fault diagnosis in military barracks. Int. J. Hybrid Inf. Technol. **9**(1), 195–206 (2016). https://doi.org/10.14257/ijhit.2016.9.1.17

3. Wang, J., et al.: Quality-relevant fault monitoring based on locality-preserving partial least-squares statistical models. Ind. Eng. Chem. Res. **56**(24), 7009–7020 (2017)
4. Zhou, J., Wei, G., Zhang, S., et al.: Quality-relevant fault monitoring based on locally linear embedding enhanced partial least squares statistical models. In: 2017 IEEE 6th Data Driven Control and Learning Systems Conference (DDCLS). IEEE (2017)
5. Zheng, S., Zhao, J.: A new unsupervised data mining method based on the stacked autoencoder for chemical process fault diagnosis. Comput. Chem. Eng. **135**, 106755 (2020)
6. Downs, J.J., Vogel, E.F.: A plant-wide industrial process control problem. Comput. Chem. Eng. **17**(3), 245–255 (1993)
7. Vidal-Puig, S., Vitale, R., Ferrer, A.: Data-driven supervised fault diagnosis methods based on latent variable models: a comparative study. Chemom. Intell. Lab. Syst. **187**, 41–52 (2019). https://doi.org/10.1016/j.chemolab.2019.02.006
8. Montero, F.P., Seguí, Y.V., Zuppa, L.A.: Wind turbine fault detection through principal component analysis and multivariate statistical inference. Adv. Sci. Technol. **101**(1), 3 (2016)
9. Park, J.H., Jun, C.Y., Jeong, J.Y., et al.: Real-time quadrotor actuator fault detection and isolation using multivariate statistical analysis techniques with sensor measurements. In: 2020 20th International Conference on Control, Automation and Systems (ICCAS) (2020)
10. Long, W., Xin, Y., et al.: A new convolutional neural network-based data-driven fault diagnosis method. IEEE Trans. Ind. Electron. **65**, 5990–5998 (2017)
11. Zhou, L.J., Jie, Z., Yuan, Z.: Research on analog circuit fault diagnosis of MFCS based on BP neural network information fusion technology. In: International Conference on Mechatronic Sciences. IEEE (2014)
12. Zheng, Y., Qian, Z.C., Yang, X.I.: Fault diagnosis of coal mine hoist bearing based on wavelet neural network. Coal Mine Mach. **42**(03), 177–179 (2021)
13. Tang, Q., Xie, C.J., Zeng, M.M., Shi, Z.: Research on fault diagnosis of power battery based on fuzzy neural network. Chin. J. Power Sources **44**(12), 1779–1783 (2020)
14. Yu, B., Yang, K.Y.: Research on rotating machinery fault diagnosis based on labview and BP neural network. Ind. Instrum. Autom. **6**, 32–34 (2020)
15. Ouhibi, R., Bouslama, S., Laabidi, K.: Faults classification of asynchronous machine based on the probabilistic neural network (PNN). In: 2016 4th International Conference on Control Engineering & Information Technology (CEIT). IEEE (2017)
16. Zhang, K., Guoyong, L.I., Han, F.: Diagnosis model of elevator fault based on fault tree analysis and improved PSO-PNN network. J. Saf. Sci. Technol. **13**(9), 175–179 (2017)
17. Yang, X., Chen, W., Li, A., Yang, C., Xie, Z., Dong, H.: BA-PNN-based methods for power transformer fault diagnosis. Adv. Eng. Inf. **39**, 178–185 (2019). https://doi.org/10.1016/j.aei.2019.01.001
18. Khan, W., Ansell, D., Kuru, K., et al.: Flight guardian: autonomous flight safety improvement by monitoring aircraft cockpit instruments. J. Aerosp. Inf. Syst. **15**(4), 1–12 (2018)
19. Khan, W., Kuru, K.: An intelligent system for spoken term detection that uses belief combination. IEEE Intell. Syst. **32**(1), 70–79 (2017)
20. He, Y.S., Huang, Y., Xu, Z.M., et al.: Motor bearing fault identification based on wavelet singular entropy and sofm neural network. J. Vibr. Shock **2017**(10) (2017)
21. Sun, C., Hou, J.: An improved principal component regression for quality-related process monitoring of industrial control systems. IEEE Access **5**, 21723–21730 (2017)

Constraint Interpretable Double Parallel Neural Network and Its Applications in the Petroleum Industry

Yunqi Jiang[1], Huaqing Zhang[2], Jian Wang[3(✉)], Kai Zhang[1], and Nikhil R. Pal[4]

[1] Department of Reservoir Engineering, China University of Petroleum, Qingdao 266580, Shandong, China
[2] College of Control Science and Engineering, China University of Petroleum, Qingdao 266580, China
[3] College of Science, China University of Petroleum, Qingdao 266580, China
wangjiann1@upc.edu.cn
[4] Electronics and Communication Sciences Unit, The Centre for Artificial Intelligence and Machine Learning, Indian Statistical Institute, Kolkata 700108, India

Abstract. In this article, a constraint interpretable double parallel neural network (CIDPNN) has been proposed to characterize the response relationships between inputs and outputs. The shortcut connecting synapses of the network are utilized to measure the association strength quantitatively, inferring the information flow during the learning process in an intuitive manner. To guarantee the physical significance of the model parameters, the weight matrices are constraint by the square function operator in the non-negative interval. Meanwhile, the sparsity of model parameters has been fully improved by the constraint. Hence, the proposed network can retrieval the critical geophysical parameters sparsely and robustly, through the injection and production signals from the reservoir recovery history. Finally, a synthetic reservoir experiment is elaborated to demonstrate the effectiveness of the proposed method.

Keywords: Interpretable neural network · Connectivity analysis · Reservoir engineering

1 Introduction

Waterflood is one critical recovery method in the petroleum industry, by injecting water to supply the underground energy, and driving the residual oil flowing to the production wells. Affected by the strong heterogeneity of geophysical properties, (i.e., the permeability represents the capability of fluid flow in the porous media), the injected water is easy to enter the high-permeation channels but difficult to sweep the oil at the low-permeation areas. Thus, there would be a cycle of "high water injection and high water production", causing a low oil recovery rate and consuming a lot of water resources. To overcome these problems, the interwell connectivity analysis has to be implemented in

© Springer Nature Switzerland AG 2021
D.-S. Huang et al. (Eds.): ICIC 2021, LNCS 12836, pp. 415–423, 2021.
https://doi.org/10.1007/978-3-030-84522-3_34

the waterflooding process, aiming to provide a quantitative evaluation of the connecting strength between an injector and a producer, thereby guiding reservoir management.

The commercial simulators can make a precise simulation of the waterflood, while these simulators are developed on the grid computation, requiring huge geological properties and costing a lot of time. To characterize the interwell connectivity effectively, a variety of simplified data-driven models [1–9] have been proposed. These models are established on the dynamic observed data from the oil fields, i.e., the water injection rates (WIR) and liquid production rates (LPR), which are often available during the production history. Generally, the simplified models can be grouped into two classes: the physics-based models and the machine learning (ML) based models.

The physical models [1–7] are usually derived from the physical laws (i.e., the material balance equation) of the waterflooding process, inferring the interwell connectivity by solving partial differential equations (PDEs). These models can generate comparable connectivity characterization results with commercial simulators under certain assumptions. For instance, the capacitance resistance models [1–3] are developed on the assumption that the flow is stable (the production rate is a linear function of the difference between the average underground pressure and the bottom hole pressure of the producer). However, the subsurface flow is very complex in actuality, which means that the reservoir engineers have to choose different models in different conditions.

Different with physical models, ML-based models [8, 9] simulate the waterflood by learning the nonlinear mapping relationships from the injection signals to the production signals, and there is not any physical knowledge considered for the model development. However, these ML models are "black box" models, whose parameters are not interpretable. Even these models can generate a great history matching accuracy for the injection and production data, the reliability of connectivity analysis results by these approaches is doubted.

In the case of a neural network, the connection between two neurons would be strengthened, if these two neural cells are connected by an excitatory synapse [10]. Strikingly, there are some high-level similarities between the activities of neural cells and the behaviors of injection-production wells in a hydrocarbon field. If there is a high connection channel between an injector and a producer, the signal from that injector would have a strong influence on the producer. Thus, the contribution of every feature to the outputs should be reflected through their corresponding weights. Based on this idea, we construct a modular neural network with the specific architecture to infer the information-flow, or the strength of the energy. To guarantee the physical significance of the connectivity analysis results, original weights are constrained to be non-negative by a square function. Furthermore, the sparsity feature of weights is significantly improved, conducive to characterize the dominant connection channels among the complex interwell connecting relationships.

The remaining parts of this paper are arranged as follows. The structures and workflow of CIDPNN are introduced and explained in Sect. 2. Next, Sect. 3 exhibits the results of CIDPNN in one numerical experiment. Finally, we give some conclusions of this work in Sect. 4.

2   Constraint Interpretable Double Parallel Neural Network

Figure 1 shows the structure of a neural network with a skip connection. In ResNet [11], the shortcuts are essential to achieve the function of identity mapping, and the results of related experiments show that identity mapping is an adequate and economical way for solving the vanishing gradient problem. In addition, the structure with shortcut connections is quite easy to understand. Through these shortcut structures, the information can transfer across layers, thus avoiding the information loss caused by the activation functions. Therefore, we aim to capture the direct response relationships from the input variables to the output variables via the shortcut connections.

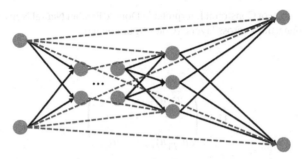

Fig. 1. The architecture of the skip connection in the neural network. The blue circles represent the nodes of the network; the red dashed lines are the shortcut connections across layers; and the black lines are the connections between two adjacent layers. (Color figure online)

To achieve high classification accuracy, and reduce learning and recall times of multilayer perceptrons (MLPs), Parallel, Self-Organizing, Hierarchical Neural Networks (PSHNN) [12] were proposed in 1990. Besides, Wang et al. [13] demonstrated that the Double Parallel Feedforward Neural Network (DPFNN) is faster in convergence and stronger in generalization, compared with common MLPs. To interpret the relationship between inputs and outputs and to demonstrate the learning mechanism, an algorithm is proposed along with specific structures appropriate for our problem. All weights are restricted to be non-negative, to discover the well connectivity by analyzing the contribution of different inputs to outputs via the weights. In this case, the contribution value obviously cannot be negative in the realistic case. In this part, we give a deep sight into the structures of CIDPNN, as shown in Fig. 2, introducing the feedforward and backpropagation process and explaining the function of the skip connection.

Because in the petroleum simulation experiments, the weights are regarded as the characterization of the fluid flow in the reservoir, and in the physical sense, fluid flow is non-negative, we model the weights using the square operation so that effectively all weights of CIDPNN are non-negative. Based on this assumption, CIDPNN is capable of meaningful representation of the information transportation process. This makes the CIDPNN more interpretable than the conventional networks with unrestricted weights, particularly for our application. We note here that instead of using the square function, we can use other functions, such as the exponential function, to realize non-negative weights.

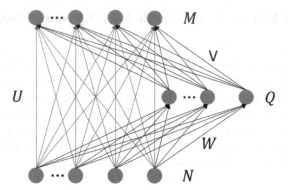

Fig. 2. The architecture of Constraint Interpretable Double Parallel Neural Network. The red lines are the shortcut connections across layers. (Color figure onine)

Let

$$\mathbf{U} = \begin{pmatrix} u_{11}^2 u_{12}^2 & \cdots & u_{1N}^2 \\ \vdots & \ddots & \vdots \\ u_{M1}^2 u_{M2}^2 & \cdots & u_{MN}^2 \end{pmatrix},$$

$$\mathbf{W} = \begin{pmatrix} w_{11}^2 w_{12}^2 & \cdots & w_{1N}^2 \\ \vdots & \ddots & \vdots \\ w_{M1}^2 w_{M2}^2 & \cdots & w_{MN}^2 \end{pmatrix},$$

$$\mathbf{V} = \begin{pmatrix} v_{11}^2 v_{12}^2 & \cdots & v_{1N}^2 \\ \vdots & \ddots & \vdots \\ v_{M1}^2 v_{M2}^2 & \cdots & v_{MN}^2 \end{pmatrix}.$$

All the elements in $\mathbf{U}$, $\mathbf{W}$ and $\mathbf{V}$ take the square form to guarantee the non-negative value. Therefore, the network output can be calculated in the following form:

$$\mathbf{Y} = g(\mathbf{V}f(\mathbf{WX}) + \mathbf{UX}), \tag{1}$$

where $\mathbf{X}$ is the input matrix, $\mathbf{Y}$ is the output matrix, g and f are activation functions. The mean square error (MSE) is used as the loss function:

$$E = \frac{1}{2J}\left(\mathbf{Y} - \hat{\mathbf{Y}}\right)^2, \tag{2}$$

where $\hat{\mathbf{Y}}$, is the ideal output, and J denotes the number of training samples.

The batch learning is used in CIDPNN, and the partial derivatives of the loss function in respect of $\mathbf{W}$, $\mathbf{V}$, and $\mathbf{U}$ are given separately by:

$$\Delta\mathbf{W} = \frac{\partial E}{\partial \mathbf{W}} = \frac{1}{J}\mathbf{V}^T\left(\mathbf{Y} - \hat{\mathbf{Y}}\right)^T\left(\mathbf{Y} - \hat{\mathbf{Y}}\right) \odot g'(\mathbf{V}f(\mathbf{WX}) + \mathbf{UX}) \odot f'(\mathbf{WX})\mathbf{X} \odot \mathbf{W},$$
$$\tag{3}$$

$$\Delta \mathbf{V} = \frac{\partial E}{\partial \mathbf{V}} = \frac{1}{J}\left(\mathbf{Y} - \hat{\mathbf{Y}}\right)^{T}\left(\mathbf{Y} - \hat{\mathbf{Y}}\right) \odot g'(\mathbf{V}f(\mathbf{WX}) + \mathbf{UX}) \odot f(\mathbf{WX})^{T} \odot \mathbf{V}, \quad (4)$$

$$\Delta \mathbf{U} = \frac{\partial E}{\partial \mathbf{U}} = \frac{1}{J}\left(\mathbf{Y} - \hat{\mathbf{Y}}\right)^{T}\left(\mathbf{Y} - \hat{\mathbf{Y}}\right) \odot g'(\mathbf{V}f(\mathbf{WX}) + \mathbf{UX})\mathbf{X} \odot \mathbf{W}, \quad (5)$$

where $\odot$ denotes Hadamard product; $\Delta \mathbf{W}$, $\Delta \mathbf{V}$ and $\Delta \mathbf{U}$ are the gradients of $\mathbf{W}$, $\mathbf{V}$ and $\mathbf{U}$, respectively.

Then, based on the backpropagation algorithm, $\mathbf{W}$, $\mathbf{V}$ and $\mathbf{U}$ can be upgraded by the following equations:

$$\mathbf{W}^{t+1} = \mathbf{W}^{t} - \eta \Delta \mathbf{W}^{t}, \quad (6)$$

$$\mathbf{V}^{t+1} = \mathbf{V}^{t} - \eta \Delta \mathbf{V}^{t}, \quad (7)$$

$$\mathbf{U}^{t+1} = \mathbf{U}^{t} - \eta \Delta \mathbf{U}^{t}, \quad (8)$$

where t means the t-th iteration step and η is the learning rate. The pseudocode of CIDPNN training procedure is illustrated in Table 1.

Table 1. Pseudocode of CIDPNN training procedure

Algorithm Constraint Interpretable Double Parallel Neural Network (CIDPNN) training procedure	
Input: X	
Output: Y	
1	/ *** **start CIDPNN training** *** /
2	**Initialization of weights:** Initialize $\mathbf{W}$, $\mathbf{V}$ and $\mathbf{U}$ randomly from 0 to 1, respectively
	While the max iteration number is not met
3	/ *** **feedforward calculation** *** /
3-1	Calculate the output of CIDPNN, using Eq. 1
3-2	Evaluate the loss of CIDPNN, using Eq. 2
3-3	/ *** **backpropagation** *** /
	Optimize $\mathbf{W}$, $\mathbf{V}$ and $\mathbf{U}$ using their gradients to the loss function, with Eq. 3 to Eq. 7
	End While
	/ *** **end CIDPNN training** *** /

CIDPNN can be separated into two modules: a nonlinear mapping module and a linear mapping module. Through the activation function in the hidden layers, the input can be transformed into a nonlinear space, giving the neural network a strong fitting capacity. Meanwhile, the linear module (shortcut connection) plays as a linear combiner in CIDPNN, transmitting the linear input information to the output. Furthermore, the weight matrix of the shortcut connections is expected to behave like good indicators to measure the contribution from the input to the output, which would be demonstrated later in detail by a numerical experiment in Sect. 3.

3 Case Study

The streak case is reconstructed from [1], built by Eclipse 2011, a professional reservoir simulator. WIR and LPR are used as inputs and outputs, respectively. There are 9 wells in this synthetic experiment, including 5 injection wells and 4 production wells, named I1, I2, I3, I4, I5, and P1, P2, P3, P4, respectively. As shown in Fig. 3, the permeability of the matrix is 5 md, except for two high-permeability streaks. One streak is 1000 md, between I1 and P1, and the other is 500 md, between I3 and P4. The fluids are supposed to flow through the areas with high permeability. The learning rate, the numbers of hidden nodes and iterations are set to be 0.1, 20, 1000, respectively, and the reported results are the average of 10 repetitions of the experiment. The interwell connectivity is characterized through the skip connection weight matrix $\mathbf{U}$ directly.

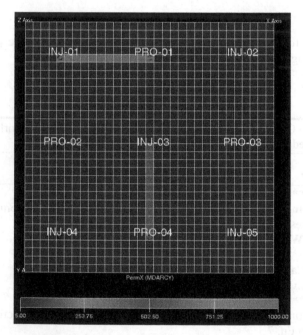

Fig. 3. The permeability of the streak case.

It must be noted that the difference of the permeability between two dominant connecting well pairs, (I1–P1 and I3–P4), and the other weak connecting pairs are significant, and thus the connectivity results are expected to demonstrate this discrepancy. Figure 4 illustrates the interwell connectivity characterization obtained by CIDPNN, where two high connecting injector-producer pairs, I1–P1 and I3–P4, are inferred visually by the heatmap. In detail, the connectivity values of I1–P1 and I3–P4 are 1 and 0.4544, respectively, and the other weak connecting well pairs are assigned with values, at least an order of magnitude lower than that of I3–P4. Thus, the permeability distribution of the streak case can get an intuitive reflection via the shortcut matrix of CIDPNN.

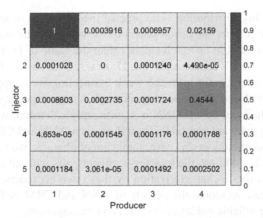

Fig. 4. The heatmap for the streak case using CIDPNN method.

As shown in Fig. 5, in the skip connection matrix, the different weights vary quite differently during learning. Especially for the first and the eighteenth weight in Fig. 5, they denote the skip connection weights that connect I1 to P1 and I3 to P4, respectively. The connectivity values of all injector-producer pairs in the streak case obtained by ANN, CRM and CIDPNN are demonstrated in Table 2. The results of ANN and CRM methods are taken from [9].

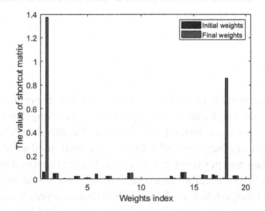

Fig. 5. The changes of shortcut matrix for the streak case using CIDPNN method. The blue bars are the initial weights of the shortcut connections; the orange bars represent the final weights of the shortcut connections after training. (Color figure online)

In CRM and ANN methods, the connectivity values of both I1–P1 and I3–P4 are 1.0. However, the permeability between I1 and P1 is 1000 md, two times higher than that between I3 and P4. Thus, these two methods cannot accurately reflect the relative strength of these high permeability streaks. Contrast with these two methods, the CIDPNN approach can acquire a more accurate characterization of two high connectivity channels, with 1.0 for I1–P1 and 0.5 for I3–P4, providing an exact inversion of the actual permeability distribution. The precise measurement of the low connecting wells is also very helpful for the development of the oil field. The values of these weak connecting well pairs should be much smaller than the value of I3–P4, since their permeability is only 5 md, 1% of that of I3–P4. However, as shown in Table 2, both ANN model and CRM model cannot infer the weak connecting wells appropriately, where the values marked with red are higher than normal. As our expectation, the sparsity of the connectivity values got by CIDPNN is significantly enhanced, where the values of all weak connecting pairs are around 0. Therefore, even ANN and CRM could infer the dominant connecting streaks in this case, they cannot characterize the internal relative differences exactly. In addition, the injector-producer pairs without high flow channels are possible to get a rough characterization by these two approaches. From these two aspects, CIDPNN can overcome the defects of ANN and CRM methods in the streak case, thus providing reliable guidance for reservoir management.

Table 2. Comparison with the current methods on the streak case.

	ANN				CRM				CIDPNN			
	P1	P2	P3	P4	P1	P2	P3	P4	P1	P2	P3	P4
I1	**1.0**	0.2	0.3	0.1	**1.0**	0.0	0.0	0.0	**1.0**	0.0	0.0	0.0
I2	0.1	0.0	0.0	0.1	0.5	0.0	0.1	0.3	0.0	0.0	0.0	0.0
I3	0.0	0.2	0.3	**1.0**	0.1	0.0	0.0	**1.0**	0.0	0.0	0.0	**0.5**
I4	0.3	0.6	0.2	0.5	0.1	0.2	0.0	0.7	0.0	0.0	0.0	0.0
I5	0.3	0.4	0.3	0.4	0.1	0.0	0.1	0.8	0.0	0.0	0.0	0.0

4 Conclusions

In this paper, an interpretable network has been proposed to evaluate the contributions from inputs to outputs through a shortcut weight matrix, avoiding the complex calculation of indexes to identify the information path. To ensure the physical meaning of the model parameters, the weights of the proposed network are constrained within 0 to 1 by a square function. The simulation experiment has revealed that the proposed approach enables to provide a sparser and more accurate characterization of the actual interwell connectivity condition. The easy interpretation of the CIDPNN reveals a very promising future of the network in oil reservoir engineering. In the future, we would like to do further research on the interpretability of neural networks with more general forms.

References

1. Yousef, A.A., et al.: A capacitance model to infer interwell connectivity from production and injection rate fluctuations. SPE Reservoir Eval. Eng. **9**(06), 630–646 (2006)
2. Sayarpour, M., et al.: The use of capacitance–resistance models for rapid estimation of waterflood performance and optimization. J. Petrol. Sci. Eng. **69**(3–4), 227–238 (2009)
3. de Holanda, R.W., et al.: A state-of-the-art literature review on capacitance resistance models for reservoir characterization and performance forecasting. Energies **11**(12), 3368 (2018)

4. Zhao, H., et al.: A physics-based data-driven numerical model for reservoir history matching and prediction with a field application. SPE J. **21**(06), 2175–2194 (2016)
5. Guo, Z., Reynolds, A.C., Zhao, H.: A physics-based data-driven model for history matching, prediction, and characterization of waterflooding performance. SPE J. **23**(02), 367–395 (2017)
6. Guo, Z., Reynolds, A.C.: INSIM-FT-3D: a three-dimensional data-driven model for history matching and waterflooding optimization. In: SPE Reservoir Simulation Conference, Galveston, Texas, USA, Paper number, SPE-193841-MS. Society of Petroleum Engineers (2019)
7. Zhao, H., et al.: Flow-path tracking strategy in a data-driven interwell numerical simulation model for waterflooding history matching and performance prediction with infill wells. SPE J. **25**(02), 1007–1025 (2020)
8. Panda, M.N., Chopra, A.K.: An integrated approach to estimate well interactions. In: SPE India Oil and Gas Conference and Exhibition, New Delhi, India, Paper number SPE-39563-MS. Society of Petroleum Engineers (1998)
9. Artun, E.: Erratum to: Characterizing interwell connectivity in waterflooded reservoirs using data-driven and reduced-physics models: a comparative study. Neural Comput. Appl. **28**(7), 1905–1906 (2017)
10. Rosen, L.D., Weil, M.M.: The Organization of Behavior: A Neuropsychological Theory. Psychology Press (2005)
11. He, K., et al.: Deep residual learning for image recognition. In: 2016 IEEE Conference on Computer Vision and Pattern Recognition, Las Vegas, NV, USA, pp. 770–778. IEEE (2016)
12. Ersoy, O.K., Deng, S.: Parallel, self-organizing, hierarchical neutral networks with continuous inputs and outputs. In: Proceedings of the 24th Annual Hawaii International Conference on System Sciences, Kauai, HI, USA, pp. 486–492 IEEE (1991)
13. Wang, J., et al.: Convergence of gradient method for double parallel feedforward neural network. Int. J. Numer. Anal. Model. **8**(3), 484–495 (2011)

Alcoholism Detection via 5-Layer Customized Convolution Neural Network

Lijia Deng[(✉)]

School of Informatics, The University of Leicester, University Road, Leicester LE1 7RH, UK
ld232@leicester.ac.uk

Abstract. Alcoholism is a serious disease, which can cause a variety of mental damage, physical damage and social damage. It is one of the top three global public health problems. Current detection of alcoholism is determined by long-term observation. We hope to detect alcoholism based on brain MRI scans to improve the detection efficiency. Existing alcoholism detection methods via the traditional medical image analysis are complex. Based on the advantages of fewer parameters, fast training and good migration of convolution neural network, a 5-layer customized convolution neural network is proposed to detect alcoholism. We performed multiple rounds of experiments to evaluate the performance of our method. The experimental results show that our method outperforms six state-of-the-art methods on all metrics tested.

Keywords: Alcoholism · Alcohol use disorder · MRI · CNN

1 Introduction

Alcohol is an addictive neurotropic substance. Alcohol addiction caused by long-term excessive drinking will lead to alcoholism. It will cause a variety of physical and mental disorders, even irreversible pathological damage [1]. According to WHO, 208 million people worldwide have alcohol use disorders (AUD), while there are approximately 3 million alcohol-related deaths each year. It has important practical implications for the diagnosis and treatment of patients with alcoholism. According to the definition of alcoholism in the fourth edition of the Diagnostic and Statistical Manual of Mental Disorders (DSM-IV), it requires maintaining three of the seven phenomena of alcoholism for at least one year to be diagnosed. Studies have shown that alcohol abuse damages brain neurons and, as a result, patients with long-term alcoholism have a smaller volume of white matter and gray matter compared with age-matched controls. Therefore, viewing brain images to diagnose alcoholism is a relatively efficient practice. At present, the identification of brain atrophy is done manually by radiologists. While, in the early stage of alcoholism, the phenomenon of brain atrophy is not obvious, so it is difficult to detect brain atrophy manually. An advanced artificial intelligence method via magnetic resonance imaging (MRI) to monitor the patient's brain and provide early diagnosis is needed.

© Springer Nature Switzerland AG 2021
D.-S. Huang et al. (Eds.): ICIC 2021, LNCS 12836, pp. 424–434, 2021.
https://doi.org/10.1007/978-3-030-84522-3_35

In the past few decades, computer vision based techniques have played a great advantage in image detection. Computer vision based techniques can discover minute detail changes in brain MRI scans more accurately, and greatly save time and effort in alcoholism diagnosis. In 2012, Monnig [2] proposed the use of computer vision techniques to detect white matter atrophy. Alweshah and Abdullah [3] proposed the method of hybrid firefly algorithm (FA) and probabilistic neural network (PNN) to detect brain changes. Hou [4] used predator-prey adaptive-inertia chaotic PSO (PACPSO) for alcoholism identification. Nayak, et al. [5] proposed a random forest based algorithm classification method via brain MRI scans. Zhang *et al.* proposed the method of Hu moment variant (HMI) based on support vector machine (SVM) to detect pathological changes in the brain [6]. Zhang et al. [7] proposed the method of using Pseudo Zernike Motion (PZM) to detect brain changes. Wang et al. [8] proposed a method based on data enhancement to detect alcoholism based on MRI scans. Muhammad et al. proposed an assay for alcoholism based on wavelet Renyi entropy and three-segment encoded Jaya algorithm (TSEJaya) [9]. Qian [10] used cat swarm optimization (CSO) for alcoholism detection. Chen [11] used wavelet energy entropy (WEE) to extract features from the alcoholism patient's brain image. Although these methods can be used to detect alcoholism, the neural networks used are complex, and their mobility on MRI scans is not good enough.

To solve the above problems, we propose a new MRI scans based alcoholism identification method via convolutional neural network (CNN). In recent years, CNN has been widely used in the field of medical image processing [12–15]. Compared with the traditional neural network, CNN can effectively reduce the number of network parameters, reduce the complexity of the network, improve the migration ability of the model, and enhance the robustness of the network.

The rest of this paper is organized as follows: Sect. 2 describes the dataset of this study. Section 3 describes the method of our customized convolutional neural network. Section 4 describes the design of the experiment and discuss the experimental results.

2 Dataset

With the development of technology, we can intuitively detect brain atrophy through brain imaging. Existing brain imaging techniques mainly include three types: Computed Tomography (CT), Positron Emission Tomography (PET), and Magnetic Resonance Imaging (MRI). Both CT and PET cause radiation damage to the patient during imaging. MRI technology is different from other imaging technologies. This method does not produce artifacts in CT, and does not need to inject contrast agents which means there is no radiation injury and no adverse effect on humans. Moreover, the information provided by MRI scanning is much larger than that produced by other imaging technologies in medical imaging. Therefore, the use of MRI has great advantages for the early diagnosis of alcoholism.

In this study, we used the Nanjing MRI dataset. This dataset collected brain MRI scans from a total of 235 samples over three consecutive years. These samples comprised 114 samples with long-term chronic alcoholic (58 males and 56 females) and 121 nonalcoholic control samples (59 males and 62 females). These volunteers were all tested by the Alcohol Use Disorders Identification Test (AUDIT) to ensure that the

426   L. Deng

data were authentic and reliable [16]. The obtained MRI data were subjected to FMRIB
software library (FSL) V5.0 software to extract images of brain sections [17, 18]. We
selected slices with $Z = 80$ (8 mm) at MNI 152 coordinates from the brain image after
normalizing to a standard MNI template. The 80th slice most clearly shows two features
of alcoholic patients: "the shrunk gray matter" and "the enlarged ventricle" [19–21]. The
selected MRI slice images are shown in Fig. 1.

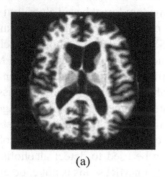 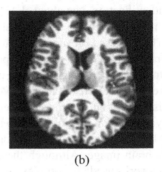

(a)   (b)

Fig. 1. MRI slice image examples: (a) an alcoholic brain and (b) a nonalcoholic brain [9].

3 Methodology

Convolutional neural network (CNN) is a kind of feed-forward network. Its feature is
that each layer of neurons only responds to the neurons in the local range of the previous
layer. Compared with the traditional neural network, the parameter sharing mechanism
of CNN can effectively reduce the number of parameters and reduce the complexity of
the network. In addition, the sparsity of connections in CNN can effectively improve the
migration ability of the model and enhance the network robustness. Nowadays, CNN
has many successful applications in the field of computer vision [22–25]. These studies
inspired us to apply CNN to the detection of alcoholism. Next, we will describe the
customized convolution neural network.

3.1 Convolution

The process of doing scalar product for images and filters is called convolution. This is
the most important step of convolutional neural networks. During the convolution, each
element of the filter will multiply the appropriate pixel in the input image. The product
of it will sum together as one new element value of the output matrix:

$$y = \sum_i w_i x_i + b \qquad (1)$$

where y is the output matrix value of convolutional layer, x_i is the pixel value of the input
image, w_i is the weight of the filter, b is the bias of the filter and i is the channel of the
image.

To prevent image shrinkage during convolution, researchers usually padded the input image to maintain the size of the image. Moreover, padding images also improves the utilization of pixels at the edges of the image. Padding will add a circle of new pixels around the input image, usually with a value of 0.

$$d_{out} = \frac{d_{in} - d_{kernel} + 1}{stride} \tag{2}$$

where, d_{out} is the size of output matrix, d_{in} is the size of input image, d_{kernel} is the size of convolutional filter and *stride* is the step size of convolutional layer.

3.2 Pooling

In convolutional neural networks, a pooling layer is often added between adjacent convolutional layers. The pooling layer can compress the matrix from the convolutional layer to reduce parameters and computation [26] while retaining the main features of the input matrix [27]. The use of pooling layer avoids the involvement of more redundant information that could prevent overfitting and improve model generalization [28]. Commonly used pooling layers are max-pooling and average-pooling. Max-pooling segments the whole input matrix into several independent submatrices with equal size. The largest element within each submatrix was taken to constitute the output which has the same planar structure as the original input [29]. As shown in Fig. 2, this is a max-pooling with a 2×2 filter, which can find out the biggest value from the input matrix.

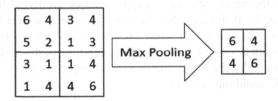

Fig. 2. The process of max-pooling

3.3 Batch Normalization

We added Batch Normalization (BN) to whiten the activation values of each hidden layer in the neural network which shows in Fig. 3. The use of BN pulls the distribution of the input value of any neuron in each layer back to the standard normal distribution with the mean value of 0 and the variance of 1 [30]. The equation of BN is shown in Eq. (6). The input activation $\hat{x}_k$ of T layer corresponding to the k neuron is converted by subtracting the mean $E[x_k]$ of activation x_k obtained from instances in mini batch and dividing by the variance $Var[x_k]$. This is similar to dropout's regularized expression that prevents overfitting [31]. In addition, the use of BN could make neural network training less demanding for initialization, and it allows the use of larger learning rates to speed up network convergence and improve the training speed [32].

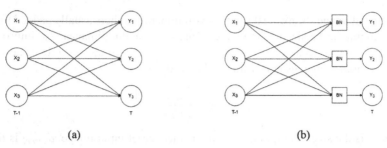

(a)   (b)

Fig. 3. Adding Batch Normalization before hidden layer. (a) Two layers of the neural network; (b) Neural networks using Batch Normalization

$$\hat{x}_k = \frac{x_k - E[x_k]}{\sqrt{Var[x_k]}} \tag{3}$$

3.4 Rectified Linear Unit

The activation layer will provide a nonlinear function to the neural network. In real world, some problems cannot easily use linear modules to solve [33]. At this time, the nonlinear factor should be added to the modules to solve this problem. Rectified Linear Units (ReLU) is a very popular activation function in deep learning. The equation of ReLU function is shown in Eq. (4). Compared with other activation function like Sigmoid function, the calculation of ReLU function is simpler [34], which saves a lot of computation on the neural network. As Fig. 4 shows, the ReLU function is closer to the activation model of biology than the Sigmoid function. ReLU layer leaves a subset of neurons with an output of 0, which creates sparsity of the network and reduces parameter interdependencies [35]. This alleviates overfitting problems in neural network training [36]. Additionally, the ReLU function does not present the problem of gradient disappearance as Sigmoid function does.

$$f(x) = max(0, x) \tag{4}$$

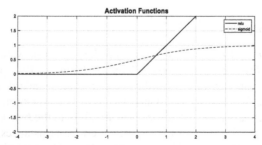

Fig. 4. The image of activation functions: Real lines represent RuLU functions, dashed lines represent Sigmoid functions.

3.5 Structure of Customized CNN

The structure of our customized CNN and the number of parameters required for individual layers are shown in Table 1. Like Fig. 5 shows, the network detects the input MRI scan with a size of 176×176 and determines whether the input image has brain changes caused by alcoholism. The network contains 3 sets of convolutional layers and 2 fully connected layers (FCL). Each set of convolutional layers contains one convolutional layer, one BN layer and one activation layer. The network is simple in structure and has a small number of parameters, which can save a lot of computation and training time.

Table 1. The structure of customized CNN

Layer	Size	Parameters
Input	176×176	
Conv layer 1	$44 \times 44 \times 32$	32 3×3 stride = 2, pooling size = 2
Conv layer 2	$22 \times 22 \times 64$	64 3×3 stride = 2
Conv layer 3	$11 \times 11 \times 128$	128 3×3 stride = 2
FCL	1000	15488×1000
FCL	2	2×1000

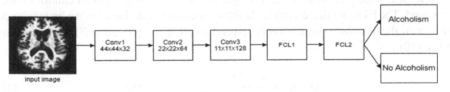

Fig. 5. The flow of MRI image detection

3.6 Measures

This study used multiple metrics to evaluate the performance of neural networks which include: Sensitivity, Specificity, Precision, Accuracy, F1 score, Fowlkes Mallows Index (FMI) and Matthews Correlation Coefficient (MCC). The higher the value of these metrics, the better the performance of the method.

Sensitivity (Sen) measures the proportion of samples that are correctly identified as positive which is also called true positive rate. The equation of Sensitivity is shown in Eq. (5), in which true positive (TP) is the sample that has the disease and the test is positive; False negative (FN) is the sample that has the disease but the test is negative.

$$Sensitivity = \frac{TP}{TP + FN} \times 100\% \tag{5}$$

Specificity (Spc) (also known as the true negative rate) is the proportion of samples that are correctly identified as negative. The equation of Specificity is shown in Eq. (6), in which true negative (TN) is the sample does not have the disease and the test is negative; false positive (FP) is the sample does not have the disease but the test is positive.

$$Specificity = \frac{TN}{TN + FP} \times 100\% \qquad (6)$$

Precision (Pre) is the proportion of samples that are correctly identified as positive overall samples identified as positive. The equation of Precision is shown in Eq. (7). The Fowlkes Mallows Index (FMI) is the geometric mean of the Recall and Precision. Equation (8) shows the equation of FMI.

$$Precision = \frac{TP}{TP + FP} \times 100\% \qquad (7)$$

$$FMI = \frac{TP}{\sqrt{(TP + FP)(TP + FN)}} \times 100\% \qquad (8)$$

Accuracy (Acc) is the proportion of samples correctly identified as positive over the total sample. The equation of Accuracy is shown in Eq. (9).

$$Accuracy = \frac{TP}{TP + FP + TN + FN} \times 100\% \qquad (9)$$

Whether it is used precision or recall alone, the accuracy of the model cannot be well measured. The F1 score that combines both two metrics would have been a better choice. F1 score is the harmonic mean of precision and recall. The equation of F1 score is shown in Eq. (10).

$$F1score = \frac{2 \times Pre \times \frac{TP}{TP+FN}}{Pre + \frac{TP}{TP+FN}} \times 100\% \qquad (10)$$

When two classes have very different size, other metrics such as accuracy may not correctly evaluate the performance of the model. At this point, we need to use Matthews Correlation Coefficient (MCC). The equation of F1 score is shown in Eq. (11).

$$MCC = \frac{TP \times TN - FP \times FN}{\sqrt{(TP + FP)(TP + FN)(TN + FP)(TN + FN)}} \times 100\% \qquad (11)$$

4 Experiment Result and Discussions

To examine our method, ten rounds of experiments were carried out on the Nanjing MRI dataset. The round of the experiment is marked as 'Run 1', 'Run 2', and so on. The Sensitivity, Specificity, Precision, Accuracy, F1 score, MCC and FMI criteria of the experiment were obtained by analyzing the experimental results. The larger the values of these metrics, the better the accuracy and the better the performance of the model. In addition, we also compared several other computer vision based alcoholism detection methods which are mentioned before with our method.

4.1 Statistical Analysis

We performed multiple rounds of repeated experiments to examine the accuracy and stability of our method. The experimental results are recorded in Table 2. From this table, experimental results of the first round gave the highest scores on the Sensitivity, Accuracy, F1 score, MCC, and FMI. While the experimental results of the fifth and seventh rounds gave the highest scores on Specificity and Precision. MSD showed little fluctuation in our method, which illustrates that our method enables stable detection of alcoholism.

Table 2. Result of proposed 5-layer customized convolution neural network method. The underlined data is the best experimental result.

Run	Sen	Spc	Prc	Acc	F1	MCC	FMI
1	98.25	96.69	96.55	97.45	97.39	94.91	97.39
2	96.49	94.21	94.02	95.32	95.24	90.67	95.25
3	95.61	95.04	94.78	95.32	95.20	90.64	95.20
4	96.49	95.87	95.65	96.17	96.07	92.34	96.07
5	93.86	97.52	97.27	95.74	95.54	91.53	95.55
6	95.61	95.87	95.61	95.74	95.61	91.48	95.61
7	93.86	97.52	97.27	95.74	95.54	91.53	95.55
8	95.61	95.87	95.61	95.74	95.61	91.48	95.61
9	95.61	94.21	93.97	94.89	94.78	89.80	94.79
10	97.37	95.04	94.87	96.17	96.10	92.37	96.11
MSD	95.88 ± 1.37	95.79 ± 1.20	95.56 ± 1.20	95.83 ± 0.69	95.71 ± 0.71	91.67 ± 1.38	95.71 ± 0.71

4.2 Comparison with State-of-the-Art Approaches

In order to better reflect the advantages of our method, we compared our method with other six advanced MRI scans based methods. Seven criteria of the sensitivity, specificity, precision, accuracy, F1 score, MCC and FMI were compared. Comparative results are reported in Table 3. The underlined data in this table is the best result. This table shows that our method has the best performance in all seven metrics.

Table 3. Comparison with State-of-the-art Approaches. The underlined data is the best experimental result.

Run	Sen	Spc	Prc	Acc	F1	MCC	FMI
PACPSO [4]	88.86	89.42	88.78	89.15	88.82	78.28	88.82
HMI-SVM [6]	84.56	85.70	84.78	85.15	84.67	70.27	84.67
PZM [7]	86.23	87.02	86.23	86.64	86.23	73.25	86.23
TSEJaya [9]	93.60	93.72	93.35	93.66	93.47	87.31	93.47
CSO [10]	91.84	92.40	91.92	92.13	91.88	84.24	91.88
WEE [11]	91.54	93.66	93.45	92.61	92.48	85.26	92.47
Ours	95.88 ± 1.37	95.79 ± 1.20	95.56 ± 1.20	95.83 ± 0.69	95.71 ± 0.71	91.67 ± 1.38	95.71 ± 0.71

The above experiments could show that our method has a good effect in alcoholism detection based on brain MRI scans. On the one hand, our method has a significant improvement in detection accuracy compared with other methods; on the other hand, our method has no significant fluctuations in detection accuracy which shows good stability. The above experimental validation illustrates the ability of our method to replace part of the work done on manual detection of brain changes in alcoholism. Compared with manual reading brain MRI scans, our method can effectively reduce the workload of radiologists and improve their work efficiency.

5 Conclusion

In this paper, a new MRI scans based alcoholism detection method is proposed. Different from previous methods, this new method is based on the 5-layer customized convolutional neural network. Experimental results show that our method has better detection efficacy than other methods on MRI scans based alcoholism detection, indicating that our method can effectively assist physician diagnosis. Meanwhile, due to the excellent mobility of the CNN model, this study could contribute not only to alcoholism detection but also to other MRI scans based diagnostic difficulties. This study is the first to apply convolutional neural networks to the field of alcoholism detection, and we hope this study will shed light on the future researchers.

References

1. Maurage, P., et al.: Is the P300 deficit in alcoholism associated with early visual impairments (P100, N170)? An oddball paradigm. Clin. Neurophysiol. **118**(3), 633–644 (2007). https://doi.org/10.1016/j.clinph.2006.11.007
2. Monnig, M.: Observed power and projected sample sizes to detect white matter atrophy in neuroimaging of alcohol use disorders: 1046. Alcohol. Clin. Exp. Res. **36** (2012)
3. Alweshah, M., Abdullah, S.: Hybridizing firefly algorithms with a probabilistic neural network for solving classification problems. Appl. Soft Comput. **35**, 513–524 (2015). https://doi.org/10.1016/j.asoc.2015.06.018

4. Hou, X.-X.: Alcoholism detection by medical robots based on Hu moment invariants and predator-prey adaptive-inertia chaotic particle swarm optimization. Comput. Electr. Eng. **63**, 126–138 (2017)
5. Nayak, D.R., Dash, R., Majhi, B.: Brain MR image classification using two-dimensional discrete wavelet transform and AdaBoost with random forests. Neurocomputing **177**, 188–197 (2016)
6. Zhang, Y., Yang, J., Wang, S., Dong, Z., Phillips, P.: Pathological brain detection in MRI scanning via Hu moment invariants and machine learning. J. Exp. Theor. Artif. Intell. **29**(2), 299–312 (2017)
7. Zhang, Y.-D., Jiang, Y., Zhu, W., Lu, S., Zhao, G.: Exploring a smart pathological brain detection method on pseudo Zernike moment. Multimedia Tools Appl. **77**(17), 22589–22604 (2017)
8. Wang, S.-H., Lv, Y.-D., Sui, Y., Liu, S., Wang, S.-J., Zhang, Y.-D.: Alcoholism detection by data augmentation and convolutional neural network with stochastic pooling. J. Med. Syst. **42**(1), 1–11 (2017)
9. Wang, S.-H., Muhammad, K., Lv, Y., Sui, Y., Han, L., Zhang, Y.-D.: Identification of alcoholism based on wavelet Renyi entropy and three-segment encoded Jaya algorithm. Complexity **2018**, 3198184 (2018)
10. Zhang, Y.-D., Sui, Y., Sun, J., Zhao, G., Qian, P.: Cat swarm optimization applied to alcohol use disorder identification. Multimedia Tools Appl. **77**(17), 22875–22896 (2018)
11. Chen, X.: Alcoholism detection by wavelet energy entropy and linear regression classifier. Comput. Model. Eng. Sci. **127**, 325–343 (2021)
12. Kang, C., et al.: A heuristic neural network structure relying on fuzzy logic for images scoring. IEEE Trans. Fuzzy Syst. **29**, 34–45 (2020)
13. Yu, X., Kang, C., Guttery, D.S., Kadry, S., Chen, Y., Zhang, Y.D.: ResNet-SCDA-50 for Breast Abnormality Classification. IEEE/ACM Trans. Comput. Biol. Bioinf. **18**(1), 94–102 (2021)
14. Xiaowei, L., Xiaowei, X., Wenwen, Y., Ye, T., Xiaodong, W.: Skin diseases classification method based on SE-Inception-v4 convolutional neural network. In: Proceedings of the SPIE, vol. 11720 (2021)
15. Yao, X., Wang, X., Wang, S.-H., Zhang, Y.-.D: A comprehensive survey on convolutional neural network in medical image analysis. Multimedia Tools Appl., 1–45 (2020). https://doi.org/10.1007/s11042-020-09634-7
16. Barik, A., Rai, R.K., Chowdhury, A.: Alcohol Use-related problems among a rural Indian population of West Bengal: an application of the alcohol use disorders identification test (AUDIT). Alcohol Alcohol. **51**(2), 215–223 (2016)
17. Smith, S.M., et al.: Advances in functional and structural MR image analysis and implementation as FSL. Neuroimage **23**, S208–S219 (2004). https://doi.org/10.1016/j.neuroimage.2004.07.051
18. Woolrich, M.W., et al.: Bayesian analysis of neuroimaging data in FSL. Neuroimage **45**(1), S173–S186 (2009)
19. Hatchard, T., Mioduszewski, O., Fall, C., Byron-Alhassan, A., Fried, P., Smith, A.M.: Neural impact of low-level alcohol use on response inhibition: an fMRI investigation in young adults. Behav. Brain Res. **329**, 12–19 (2017)
20. Bach, P., et al.: Association of the alcohol dehydrogenase gene polymorphism rs1789891 with gray matter brain volume, alcohol consumption, alcohol craving and relapse risk: ADH gene effects in alcoholism. Addict. Biol. **24**(1), 110–120 (2019)
21. Schmidt, T., Roser, P., Ze, O., Juckel, G., Suchan, B., Thoma, P.: Cortical thickness and trait empathy in patients and people at high risk for alcohol use disorders. Psychopharmacology **234**(23–24), 3521–3533 (2017)

22. Jiang, X., Zhang, Y.-D.: Chinese sign language fingerspelling via six-layer convolutional neural network with leaky rectified linear units for therapy and rehabilitation. J. Med. Imaging Health Inf. **9**(9), 2031–2090 (2019)

23. Deng, L., Wang, S.H., Zhang, Y.D.: Fully optimized convolutional neural network based on small-scale crowd. In: 2020 IEEE International Symposium on Circuits and Systems (ISCAS), pp. 1–5 (2020)

24. Hong, J., Wang, S.-H., Cheng, H., Liu, J.: Classification of cerebral microbleeds based on fully-optimized convolutional neural network. Multimedia Tools Appl. **79**(21–22), 15151–15169 (2018). https://doi.org/10.1007/s11042-018-6862-z

25. Yao, X., Wang, X., Karaca, Y., Xie, J., Wang, S.: Glomerulus classification via an improved GoogLeNet. IEEE Access **8**, 176916–176923 (2020)

26. Bizopoulos, P., Koutsouris, D.: Sparsely activated networks (in English). IEEE Trans. Neural Netw. Learn. Syst. **32**(3), 1304–1313 (2021)

27. Hebbar, R., et al.: Deep multiple instance learning for foreground speech localization in ambient audio from wearable devices. EURASIP J. Audio Speech Music Process. **2021**(1), 1–8 (2021). https://doi.org/10.1186/s13636-020-00194-0

28. Guttery, D.S.: Improved breast cancer classification through combining graph convolutional network and convolutional neural network. Inf. Process. Manage. **58**, 102439 (2021)

29. Wanda, P., Jie, H.J.: DeepFriend: finding abnormal nodes in online social networks using dynamic deep learning. Soc. Netw. Anal. Min. **11**(1), 1–12 (2021)

30. Rani, B.M.S., Rajeev Ratna, V., Prasanna Srinivasan, V., Thenmalar, S., Kanimozhi, R.: Disease prediction based retinal segmentation using bi-directional ConvLSTMU-Net. J. Ambient Intell. Humanized Comput., 1–10 (2021). https://doi.org/10.1007/s12652-021-03017-y

31. Cheng, X.: PSSPNN: patchshuffle stochastic pooling neural network for an explainable diagnosis of COVID-19 with multiple-way data augmentation. Comput. Math. Meth. Med. **2021** (2021). Art. no. 6633755

32. Yeh, R.A., Hu, Y.T., Schwing, A.G.: Chirality Nets for human pose regression. In: Advances in Neural Information Processing Systems 32, La Jolla, vol. 32, pp. 35–42. Neural Information Processing Systems (NIPS) (2019)

33. Oh, S., et al.: Energy-efficient Mott activation neuron for full-hardware implementation of neural networks. Nat. Nanotechnol. **16**, 680–687 (2021). https://doi.org/10.1038/s41565-021-00874-8

34. Zhang, Y.D., Nayak, D.R., Zhang, X., et al.: Diagnosis of secondary pulmonary tuberculosis by an eight-layer improved convolutional neural network with stochastic pooling and hyperparameter optimization. J. Ambient Intell. Human. Comput. (2020). https://doi.org/10.1007/s12652-020-02612-9

35. Wu, X.: Diagnosis of COVID-19 by wavelet Renyi entropy and three-segment biogeography-based optimization. Int. J. Comput. Intell. Syst. **13**(1), 1332–1344 (2020)

36. Mittal, V., Gangodkar, D., Pant, B.: Deep graph-long short-term memory: a deep learning based approach for text classification. Wirel. Pers. Commun. **119**(3), 2287–2301 (2021). https://doi.org/10.1007/s11277-021-08331-4

A Heterogeneous 1D Convolutional Architecture for Urban Photovoltaic Estimation

Alvaro Valderrama[1]($\boxtimes$) (ID), Carlos Valle[2] (ID), Marcelo Ibarra[3], and Hector Allende[1] (ID)

[1] Universidad Técnica Federico Santa María, Valparaíso, Chile
alvaro.valderrama.13@sansano.usm.cl
[2] Universidad de Playa Ancha, Valparaíso, Chile
[3] Universidad de Chile, Santiago, Chile

Abstract. Global energy transition to renewable sources is among the substantial challenges facing humanity. In this context, the precise estimation of the renewable potential of given areas is valuable to decision-makers. This is particularly difficult for the urban case. In Chile, valuable data for solving this problem is available, however, standard machine learning algorithms struggle with their variable-length input. We take advantage of the ability of 1-D Convolutional Neural Network to fuse feature extraction and learning over a heterogeneous representation of the data. In the present manuscript, we propose an architecture to estimate the PV potential of Chilean cities and extract the relevant features over heterogeneous representations of available data. To this end, we describe and examine the performance of said architecture over the available data. We also extract its intermediate convolutional features and use them as inputs of other machine learning algorithms to compare performances. The network outperforms all other tested machine learning algorithms, while the intermediate learned convolutional representations improve the results of all non-linear algorithms explored.

Keywords: Convolutional neural networks · Heterogeneous variable-length data · Representation learning · Urban photovoltaic estimation · Renewable energy transition

1 Introduction

The growing concern over global warming consequences has pushed several countries into long-term commitments of climate change mitigation, such as the Paris Accord, which Chile has signed and ratified [19]. These commitments imply profound changes in long-term energy planning policies. Therefore, the precise and large-scale estimation of potential renewable energy output is a valuable input for decision-makers.

This estimation faces diverse challenges, the principals being: (i) different technologies (*e.g.* wind, photovoltaic, concentrated solar power, etc.); (ii) different applicability areas (open country, urban landscape); (iii) kind of potential being estimated (theoretical, technical, economic, etc.). In Chile, the theoretical potential of solar energy (total radiation) is well described [16], nonetheless, to the best of our knowledge, there is no precise and large-scale estimation for the photovoltaic (PV) potential of urban areas.

© Springer Nature Switzerland AG 2021
D.-S. Huang et al. (Eds.): ICIC 2021, LNCS 12836, pp. 435–449, 2021.
https://doi.org/10.1007/978-3-030-84522-3_36

That is not a coincidence, considering the particularities of the urban landscape and the scarcity of adequate data.

To tackle this problem, the Ministry of Energy has provided 196 examples of urban PV potentials, which combined with variable-length registries obtained from the Tax Administration (*cf.* [17]) could allow learning the problem at hand. Machine Learning (ML) models repeatedly achieve state-of-the-art performance in several real-world applications and some proposals [2, 14] use ML algorithms for estimating urban potentials, particularly when not all required data is available for other procedures (such as 3D modeling of cities).

A general ML problem, whose examples are drawn from an unknown distribution $\mathcal{P}$ over $\mathcal{X} \times \mathcal{Y}$, can be stated as the following optimization problem:

$$h_{\mathcal{H}}^{tr} := \underset{h \in \mathcal{H}}{\text{argmin}} \sum_{(x,y) \in (X_{tr}, Y_{tr})} \ell(h(x), y),$$

where $\mathcal{H} \subset \{\mathcal{X} \rightarrow \mathcal{Y}\}$ is the hypothesis space defined by the structure and hyperparameter of our networks, X_{tr} and Y_{tr} are the train set examples and ℓ is the loss function (usually a distance). However, the aim in ML is not to learn the training examples, rather to accurately predict future data. Therefore, comparisons between different algorithms or models (also, hypothesis in this context) must be performed over a validation set, not presented to the algorithm during training. Therefore, the real optimization we perform is in practice:

$$\underset{\mathcal{H}}{\min} \sum_{(x,y) \in (X_{val}, Y_{val})} \ell\left(h_{\mathcal{H}}^{tr}(x), y\right),$$

where X_{val} and Y_{val} are our validation sets, not used during the training, that is not involved in the choosing of $h_{\mathcal{H}}^{tr}$. Changes to the structure of the networks or the data representation effectively modify the hypothesis space $\mathcal{H}$.

In our particular case, the data examples are of variable-length: each example corresponds to a collection of 3 descriptors (number of floors, built surface, and fiscal valuation) of the buildings in a particular block. Different blocks have a different number of buildings, as shown in Fig. 1. This leads us to use 1D convolutional neural networks (1D-CNN) and representing each data example as a padded sequence. However, preliminary experimentation shows that a heterogeneous representation (when different positions in the sequence correspond to different attributes rather than different buildings) improves performance over homogeneous representations (when different positions in the sequence correspond to different instances of the same attribute).

The present manuscript proposes a heterogeneous 1D Convolutional Neural Network architecture trained over heterogeneous representations of the available variable-length data. The main contributions of this paper are:

- We propose and describe a network able to model the urban PV potential of Chilean cities from the few available examples.
- We show that the network can simultaneously learn the targets and useful convolutional data representations, which can be extracted from its intermediate layers.

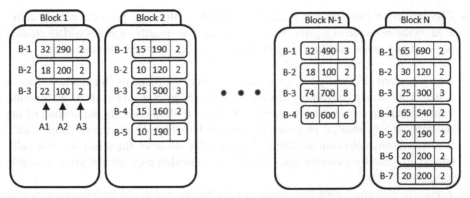

Fig. 1. Schematic representation of available data. Each block corresponds to a single example, composed of a variable number of buildings (enumerated B-1, B-2, etc.), described by 3 numerical attributes (A1, A2, and A3). The original data can be modeled as variable-length matrices, where the column number is fixed as 3 and the row number corresponds to the number of building in the given block.

- We propose a heterogeneous representation, which is beneficial to this problem given the particularities of the urban PV estimation problem.
- We provide a comparison of the proposed network and five ML algorithms, over summarized representations of the input data, a principal component analysis decomposition, and learned representations given by the CNN architecture.

The remainder of the manuscript is organized as follows. Firstly, Sect. 2 presents a conceptualization of the general PV estimation problem and its different components. Then, Sect. 3 describes a few related proposals for the urban PV estimation problem, followed by a theoretical description of the particularities of 1D-CNN in Sect. 4. Afterward, Sect. 5 presents our current proposed architecture, the used data representation and its interaction with the convolution operation. Section 6 presents our experimental results and finally, Sect. 7 presents our conclusions and discussion.

2   PV Estimation Problem

There is an interdisciplinary consensus on the urgency of the decarbonization of the global energy matrix to lessen the direst consequences of climate change [8]. Recent technological developments are enabling this transition, while the urban landscape is expected to play a key role in the coming decades [20], both from energy consumption and generation standpoints. In this context, distributed urban PV generation can contribute significantly to the decarbonization effort.

Therefore, precise and large-scale estimation of the PV generation potential in urban areas becomes an important input for decision-makers, to enable planning and guidance of the renewable transition [1]. As mentioned before, this estimation comes with different sets of challenges, and following the framework used by several authors [1, 10, 21] the estimation process can be broken up into five distinct, hierarchical successive steps:

- *Theoretical or Physical Potential*: Corresponds to the total physical energy delivered by the resource in a given area. This estimation usually involves widely studied physical processes, satellite imagery, and *in situ* measurements as validation.
- *Geographic or Urban Potential*: It consists of the energy received over areas where the energy can effectively be captured, considering the plane inclination and shadow losses. Thus, in the geographic case topographic shadows must be taken into account, while the urban case must estimate the available rooftop area, the inclination of the rooftops, and the shadow projected by adjacent buildings. While the geographic case is solved through physical modeling of topographic shadows, the urban case is usually solved in a case-by-case manner, with *ad hoc*. methodologies tailored given available local data.
- *Technical Potential*: Potential amount of electricity that could be harnessed by real PV systems installed over the available area, considering losses and inefficiencies. These are calculated using widely studied physical equations.
- *Economic Potential*: This takes into account the economic feasibility of real PV projects. It considers technology costs, operation and maintenance, and local energy costs. It is usually estimated using the Levelized Cost of Energy (LCOE).
- *Market Potential*: This last potential takes into account market frictions, local restrictions, and social reluctance to the technology. It also must consider local regulations and policies, to evaluate the real feasible potential.

This sequential framework allows researchers to focus on one (or several potentials), rather than dealing simultaneously with all different domains. By abstracting or assuming previous potentials, researchers can focus their efforts on a particular domain, allowing for better accuracies.

In the current manuscript, we deal with the urban estimation problem, which can be tackled from different domains. While physical modeling of the three-dimensional nature of cities is ideal, this is seldom possible at a large-scale given the unevenness of data availability and the prohibitive cost of LIDAR drone flights. In this context, ML appears as an interesting set of tools, as it allows the learning of hard-to-model relationships between different data sources, which can later be generalized to other areas where data is missing.

The following section presents some proposals that tackle the urban PV estimation problem.

3 Related Work

One particular proposal for the Chilean case corresponds to [3], where Campos *et al.* present a detailed estimation of the potential in Concepción city, using a simulated 3D city model. They accomplish this by gathering footprint, height, and roof characteristics from building permits. Using this simulation, the authors estimate the available area for PV installations and shadowing effects. The authors also evaluate the Levelized Cost of Energy (LCOE) as an economic restriction for the total potential. After obtaining this detailed result over the city of Concepción, the authors extrapolate their findings on an aggregate level, by assuming several constant ratios, such as the available rooftop area per inhabitant or fraction of usable rooftop area. The authors make this large-scale estimation at a resolution of cities, that is, they propose a single potential PV output for a whole city. The authors are unable to compare their estimation to any other previous results.

Another similar proposal [15] estimates the urban rooftops PV potential from publicly available geodata (building footprints, open street map) and satellite measurements of the German city of Freiburg. The authors use a standard deterministic model for the theoretical estimation; several image processing and GIS algorithms to detect rooftops and randomization of their tilts from measured values to estimate urban potential. They also include an ML algorithm, particularly using a modified version of Alexnet [13] to identify existing rooftop installations and obstacles. Finally, the authors calculate the estimated LCOE and match it to the local price of energy to obtain an economic potential. The authors evaluate the accuracy of their GIS processings, by measuring the precision of the roof azimuth estimation against a preexisting 3D model, obtaining a 70% accuracy. On the other hand, the authors evaluate the accuracy of the ML model by using manually classified images, obtaining a 90% accuracy.

Assouline *et al.* [2] evaluate the theoretical, urban, and technical potential over Switzerland's rooftops. The authors report the monthly average PV output estimated over the Swiss territory, with a spatial resolution of 200 m. To this end, they use several different data sources, including government building databases (whether from building descriptors or LiDAR 3D models), among many other varied data sources. The authors use the random forest model to learn the task of predicting several variables: climate variables, Global Horizontal Radiation, the available area over rooftops, etc. The authors train an instance of the random forest algorithm for each variable and use the predicted values to calculate the potential. They can compare their results to a previous Swiss study, finding a normalized root mean square error between both results of 26%.

From the presented proposals, we see that different methodologies can vary drastically given the availability of data of the concerned areas.

4 1D Convolutional Neural Networks

Originally inspired by the animal visual cortex and its neuronal organization [7], CNNs are a family of ML algorithms well suited for pattern recognition. The architecture and convolutional operation were inspired from observations of the neuron pathways of the animal cortex: different pathways activate when confronted with different visual patterns (e.g., horizontal or vertical lines). The CNN operation allows the network to recognize a pattern in an image regardless of its position in the image. While originally proposed for image recognition (i.e. 2-dimensional data), CNN can be easily extended to other dimensions.

The main operation of a CNN is convolution. In the general case, it operates over an arbitrary input sequence $\xi : \{1, \ldots, \alpha\} \rightarrow \mathbb{R}^\beta$ of length $\alpha \in \mathbb{N}$ and $\beta \in \mathbb{N}$ dimensions. The convolution is defined by a kernel $g : \{1, \ldots, \gamma\} \rightarrow \mathbb{R}^\beta$ (transformed as a padded sequence, that is for values outside the definition of the kernel, it outputs zeros) whose dimension corresponds to the number of dimensions of the input (i.e. β) and with $\gamma \in \mathbb{N}$ it's length, fixed as a hyperparameter and smaller than the input's length α. Then, the convolution $\xi * g$ is defined as follows:

$$(\xi * g)(i) = \sum_{k=1}^{\beta} \left(\sum_{p=1}^{\alpha} \xi^{(k)}(p)g^{(k)}(i-p+1) \right), \tag{1}$$

where $\xi^{(k)}$ denotes the kth dimension of ξ, and $i \in \{1, \ldots, \alpha\}$. Notice that while the inner sum iterates over α different values, in practice only γ values of $\xi^{(k)}$ contribute to the output for each k, since for all $(i - p + 1) \notin \{1, \ldots, \gamma\}$, $g^{(k)}(i - p + 1) = 0$.

Inside a CNN, the output sequence y (or convolutional feature in this context) from a convolutional filter defined by a kernel g operating over our input ξ is transformed by an activation function, and can be computed following the formula:

$$y(i) = \sigma(b + (\xi * g)(i)),$$

where $b \in \mathbb{R}$ corresponds to a bias (a trainable parameter) and $\sigma : \mathbb{R} \to \mathbb{R}$ is the activation function (non-linear function, chosen as a non trainable hyperparameter) [18].

A single layer of a CNN, is composed of several neurons, defined each by a filter and bias. Then, if a particular layer is composed of N neurons, with N associated filters g_j and biases b_j, the output y also has N dimensions $y^{(j)}$, all for $j \in \{1, \ldots, N\}$. These are defined as follows:

$$y^{(j)}(i) = \sigma \left(b_j + \left(\xi * g_j \right)(i) \right).$$

In other words, each neuron of a layer has a particular filter and bias. The weights in all filters and all biases are the trainable parameters of the layer, which are usually optimized using the backpropagation algorithm [6].

A fully convolutional neural network is composed of several convolutional layers, each connected to the previous one. Thus, the convolutional features extracted from a layer, are propagated to the following layer. As we advance through the layers, each layer's output is the next layer's input. This architecture allows CNN to successively filter attribute over attribute, achieving increasingly filtered and refined representations of the data, allowing for the extraction of learned representation from intermediate layers [22]. Thus, different neurons in each layer can specialize in different patterns while different layers in the network operate over different feature spaces, of progressive sophistication.

1D-CNN are particularly efficient at fusing the learning of the underlying phenomena with the learning of meaningful representations of the involved data [12]. We take advantage of this property of 1D-CNN not only to learn the problem at hand but also to use the optimized network to create fixed length learned convolutional representations useful to other ML algorithms.

5 Our Proposed Approach

We now describe our current proposal, a 1D-CNN architecture over heterogeneous representation of data for the problem of the estimation of potential photovoltaic generation in urban settings. While usually, input data is homogeneous, that is to say, that all values of a given sequence correspond to the same attributes, however, in our case data proved more easily learned over heterogeneous representations, that is where different positions of the sequence correspond to different attributes.

A schematic representation of the proposed architecture is presented in Fig. 3. Our architecture is composed of a convolutional section with 5 convolutional layers, skip connections, and batch normalization layers, followed by a dense section composed of 3 dense layers.

Our examples X_{tr} were zero-padded, that is 0 values were inserted into each example up to the largest example. Formally if μ_n corresponds to the number of buildings in example $x_n \in X_{tr}$, we define $L = max_n\{\mu_n\}$, and $J \in \mathbb{N}$ corresponds to the number of different descriptors available for the buildings, we obtain our padded examples $\bar{x}_n$ by the following formula:

$$\bar{x}_n^{(j)}(i) = \begin{cases} x_n(i,j) \ \textit{if} \ 1 \leq j \leq \mu\_n \\ 0 \qquad \textit{if} \ j > \mu_n, \end{cases} \tag{2}$$

for $i \in \{1, \ldots, L\}, j \in \{1, \ldots, J\}$ and remembering $x_n \in \mathbb{R}^{J \times \mu_n}$, or in other words a matrix with μ_n columns corresponding to the different buildings and J rows corresponding to the different descriptors. As defined, $\bar{x}$ corresponds to the homogeneous representation of the data, a sequence of buildings of length L, each described by J different dimensions. However, the input of our network is the heterogeneous sequence representation of our padded examples (cf. Eq. (2)), which is defined by:

$$\tilde{x}_n^{(i)}(j) := \bar{x}_n^{(j)}(i), \ \textit{for} \ j \in \{1, \ldots, J\}, i \in \{1, \ldots, L\}. \tag{3}$$

Here, the super index (i) corresponds to the different dimensions of the sequence and indexes the different buildings, while j corresponds to the position in the heterogeneous sequence and indexes the different descriptors. Thus, we obtain a padded set of heterogeneous examples $\{\tilde{x}_n | x_n \in X_{tr}\} \subset \mathbb{R}^{L \times J}$ of fixed length, in other words, L-dimensional sequences of length J: the dimensional and temporal roles are inverted between homogeneous and heterogeneous representations.

The particular available data from the Tax Administration contains three relevant numerical values. Namely the total built area, the number of floors, and the fiscal value of each building. For a particular data example in the homogeneous representation $\bar{x}_n$, the convolution operation defined in Eq. (1) becomes:

$$(\bar{x}_n * g)(i) = \sum_{k=1}^{J} \left(\sum_{p=1}^{L} \bar{x}_n^{(k)}(p)g^{(k)}(i - p + 1) \right), \tag{4}$$

while in the proposed heterogeneous representation $\tilde{x}_n$, the convolution becomes:

$$(\tilde{x}_n * g)(i) = \sum_{k=1}^{L} \left(\sum_{p=1}^{J} \tilde{x}_n^{(k)}(p)g^{(k)}(i - p + 1) \right). \tag{5}$$

This operation is depicted in Fig. 2 for a single kernel of length 1. Notice that in Eq. (4), for each $i \in \{1, \ldots, L\}$ (that is, for each position in the output sequence), γ consecutive buildings participate in the output for each computation. On the other hand, in Eq. (5) all buildings participate in each position computation, while only γ dimensions contribute to the output. Therefore, we see that in a heterogeneous representation of the available data, the convolutional operation combines all buildings present in the block simultaneously, while the homogeneous representation only combines a subset of buildings adjacent in the sequence.

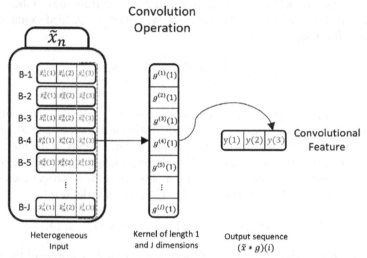

Fig. 2. Schematic representation of the convolutional operation of a single neuron over the described heterogeneous data representation. The mathematical definition of the depicted convolution is presented in Eq. (5).

Given the lack of geospatial data regarding building placement inside each block, the induced order of the homogeneous sequence representation is arbitrary. Moreover, the simultaneous computation of all buildings in the proposed heterogeneous convolutional operation allows for the whole block to be characterized rather than the characteristics of individual buildings. This could be advantageous in the urban PV estimation problem: rather than only characterizing individual buildings, the urban estimation has to deal with shadow projections between adjacent buildings, therefore the differences of buildings could be as relevant as the characteristics of each building. These reasons lead us to prefer the heterogeneous representation for our proposed network.

A graphical representation of such an operation for the proposed heterogeneous representation is presented in Fig. 2, where our transformed data $\tilde{x}_n$ is filtered through a convolutional neuron. Here the length of the input sequence corresponds to 3, while the kernel has a length 1. Both the kernel and the input homogeneous sequence have J different dimensions. The output at the right corresponds to the convolutional feature, a sequence of length 3. Each position of the output sequence is computed from a position of the input sequence, which in the homogeneous representation corresponds to a particular descriptor of the buildings. In the Figure, the color of each convolutional feature computed corresponds to that of the corresponding input data attribute. This filter of length 1 is applied over each of the 3 columns, simultaneously combining all rows into a single convolutional feature.

Moreover, we prefer kernels of size 1 in all but the last convolutional layer thus, for each position in the output, only one position of the input sequence (in heterogeneous representation that is only one descriptor) is considered, as depicted in Fig. 2. However, for the final convolution, we use a size 3 kernel. Given that only 3 building descriptors are available, the input sequences have a length of 3, and therefore a kernel of size 3 processes all values of the sequence simultaneously. This final layer is intended to combine all learned attributes into a simpler sequence which is later processed by the dense section.

We also propose to use the outputs of the labeled layers (Rep. 1 and Rep. 2, *cf*. Fig. 3) as learned representations. Rep. 1 corresponds to learned features that will be used by the rest of the convolutional section in addition to the input sequence, while Rep. 2 corresponds to the convolutional output received by the last convolutional layer. The two learned representations correspond to a different depth in the network and therefore have undergone a different number of sequential convolutional filters.

Following standard practices, we include Batch Normalization layers [9] and Skip Connections [5]. Batch normalization layers help to accelerate and stabilize the learning process, while skip connections allow for deeper architectures to be effectively trained, thus taking advantage of the increased expressive power of deeper networks [12]. We also implement a final dense section: several successful architectures use a final dense section [13, 14], which allows the network to predict the target value from learned convolutional attributes. The architecture is trained using the Mean Absolute Error (MAE) loss. If h is the hypothesis being evaluated over the train set X_{tr}, Y_{tr} and we denote by $|X_{tr}|$ the number of training examples, then the MAE loss is defined by:

$$MAE = \frac{1}{|X_{tr}|} \sum_{(x,y)\in(X_{tr},Y_{tr})} |h(x) - y|. \tag{6}$$

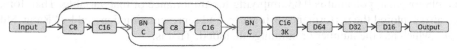

Fig. 3. Schematic representation of the proposed 1D CNN architecture. Where appropriate, activation functions correspond to hyperbolic tangent, except for the Output layer with linear activation. Layers labeled CN correspond to a 1D convolution with N neurons, BN C to a Batch Normalization layer followed by a concatenation operation. DN represents a dense layer with N neurons. 3K represents a kernel of length 3, otherwise, the convolutional layers have a kernel of length 1. Curved arrows indicate skip connections. The feature map represented by the red arrow corresponds to Rep. 1 and the green one to Rep. 2.

6 Experimental Results

A small number of 1D-CNN architectures were implemented and compared, over both heterogeneous and homogeneous representations of data, obtaining the best results for the proposed network over heterogeneous representation. Moreover, no homogeneous architecture achieved performances better than 50% MPE.

To compare the performance of 1D-CNN with other ML algorithms, we used summarized representations of data to overcome the variable-length of examples. In standard ML algorithms, increasing the input dimensions can greatly increase the dimensionality of the parameter space, leading to overfitting [4]. Therefore, instead of using zero-padding, we constructed simple, summarized representations, by using statistical operators over each of the attributes of the block's buildings. That is, if $S : \mathcal{P}(\mathcal{X}) \to \mathbb{R}$ is our statistic (where $\mathcal{P}(\mathcal{X})$ denotes the power set of $\mathcal{X}$), we extract the following values of a particular block $x_n \in X_{tr}$:

$$\Xi_n^S(j) := S(\{ x_n^{(j)}(i) | for\ i \in \{1, \ldots, \mu_n\}\}),$$

for $j \in \{1, 2, 3\}$. Thus, for a particular j, we calculate the statistic S over the set of all building attributes associated to the dimension j of the block x_n.

To allow for different features of the blocks to be summarized, we combined different statistics, particularly the mean, the range (that is the maximum minus the minimum), and the standard deviation. By joining all obtained values we have our first summarized representation which we denote by "RangeStd N/N'", defined as follows:

$$R_n = \{ \left. \Xi_n^S(j) \right| for\ j \in \{1, 2, 3\}\ and\ S \in Stats\}$$

where $Stats = \{$ "Mean", "Range", "Standard Deviation"$\}$.

While this representation is of fixed length and could give an overview of the differences and distribution of building characteristics inside a given block, a big drawback is the loss of all information regarding the number of buildings. A block with 2 buildings is represented in the same structure that a block of 20, and could even give rise to the same numerical values. However, we know that larger blocks tend to have larger photovoltaic potential, given that usually more buildings equal more available rooftop space. The observed correlation in the train set between the number of buildings in a block and the photovoltaic potential is 0.63, implying the importance of that variable. Therefore, we also explored the performance using a second representation denoted by "RangeStd N/I", which has the number of buildings included. That is:

$$R_n^* = R_n \cup \{\mu_n\}.$$

We also performed tests over learned representations from the trained 1D-CNN architecture. For this, we transformed the input data by using truncated versions of the fully trained architecture. Figure 3 identifies the layers where each truncation was performed, by the labels "Rep. 1" and "Rep. 2". These feature maps correspond to sequences of length 3 and 16 dimensions. We then obtained the transformed data by feeding the input data through the network from the "Input" layer up to the layer corresponding to "Rep. 1" or "Rep. 2" correspondingly. This allows us to obtain the intermediate feature maps of the corresponding layers, taking advantage of the 1D-CNN combination of feature extraction and task learning [12]. A third learned representation "Rep. 1 + 2" is evaluated, by concatenating both representations "Rep. 1" and "Rep. 2", equivalent to the input of the last batch normalization layer of our network (*cf*. Fig. 3) without the "Input" raw data. Therefore Rep. 1 + 2 is a sequence of length 6 and 16 dimensions.

To assess the representation learning performance of the proposed network, we also evaluate the Principal Component Analysis (PCA) representation. For this, we transform the input data using the PCA algorithm. We performed a small hyperparameter optimization of the explained variance. We explored values between 0.8 and 0.98 with a step of 0.02, evaluating the quality of the representation by the best performance of 5 evaluated ML algorithms (see below). We obtained an optimal value of 0.82, corresponding to a 7 components representation.

We performed our experiments with a train, validation, and test set composed of 75%, 15%, and 10% of the 196 available examples respectively. The target corresponds to the estimated PV potential of each block and therefore corresponds to a numerical value. Also, we performed 5 different runs to assess the performance and stability of the learned representations. We present the results of our experiments as a Mean Percentage Error (MPE), in percentage, that is:

$$MPE = \sum_{x,y} \frac{|h(x) - y|}{y},$$

where h corresponds to the hypothesis being evaluated.

We trained the architecture using Adam [11] with a 0.0001 learning rate and other hyperparameters by default. Early Stopping was also applied, using patience of 100 epochs.

The proposed architecture presents a performance of 27.46%, the best MPE of all experiments carried out over this dataset. This result is encouraging since we are approaching the reported values by Assouline *et al.*, while only using a reduced amount of training examples. We expect further improvements once more data is available. Moreover, compared to other ML algorithms, deep neural networks, including CNN, exhibit good performance despite their high expressive power: CNN has perfect expressivity, that is, given enough training, these networks can achieve zero error over a finite training set [23]. However, in neural networks, this does not necessarily lead to overfitting. Particularly, 1D-CNN has been observed to have superior performance in applications with limited amounts of labeled data [12], as is our case.

We evaluate 5 different algorithms using both summarized data transformations based on statistical operators and the three learned convolutional representations described in Sect. 4. The 5 evaluated ML algorithms are:

- Linear Regression (LinReg): with intercept and least-squares optimization,
- Regression Tree (RegTree): mean absolute error criteria, best split in each node, minimum sample for split of 2, minimum of 1 samples for leaf node, minimum impurity decrease of 0, maximum features to consider for best split equal to the number of total features, no cost-complexity pruning, and a maximum depth of 5,

- Random Forest (RF): 100 stump trees, with mean absolute error as criteria, minimum sample for split of 2, 1 for leaf nodes, minimum impurity decrease of 0, maximum features to consider for best split equal to the number of total features, no cost-complexity pruning, and bootstrap samples,
- Support Vector Machine (SVM): radial basis function kernel with scaled gamma coefficient (equal to the inverse of the variance of the examples times the number of features), 0.001 tolerance, C value of 1 and epsilon of 0.1,
- Dense Artificial Neural Network (ANN): 3 hidden layers of 20, 16, and 16 neurons respectively all with hyperbolic tangent activation, with bias, Glorot initialization and zero initialization for biases, optimized by stochastic gradient descent and mean absolute error loss. The last layer has linear activation.

The artificial neural network was implemented using the *keras* library in Python 3. The other algorithms were implemented using the *sk-learn* library in Python 3. Table 1 presents results for the battery of ML algorithm over all different considered representations, while Table 2 presents the best results of each ML algorithm and representation combination, compared to the proposed arquitechture.

Table 1. Results of comparative experiments over different ML algorithms and representations. MPE values averaged over 5-folds, in percentage.

Representation	LinReg	RF	RegTree	SVM	ANN
RangeStd N/N	269.64±97.68	164.78±52.82	238.79±128.14	183.91±55.5	54.84±11.49
RangeStd N/I	298.51±105.48	151.97±53.47	261.9±116.93	214.19±34.72	40.47±17.67
PCA	418.72 ±348.17	174.44 ±56.9	275.23 ±70.68	205.37 ±43.54	55.7 ±6.56
Rep. 1	282.45 ±84.72	142.43 ±50.22	169.05 ±84.61	154.64 ±42.85	52.47 ±15.13
Rep. 2	258.37 ±89.39	137.69 ±45.18	193.07 ±80.07	157.14 ±35.41	51.08 ±17.84
Rep. 1+2	310.36 ±82.89	139.71 ±46.54	198.51 ±70.97	161.33 ±40.85	**33.65** ±3.73

Table 2. Results of experiments over different ML algorithms the best representation. MPE values averaged over 5-folds, in percentage.

Algorithm	Representation	Performance
LinReg	Rep. 2	258.37 ± 89.39
RF	Rep. 2	137.69 ±45.18
RegTree	Rep.1	169.05 ± 84.61
SVM	Rep. 1	154.64 ± 42.85
ANN	Rep. 1+2	33.65 ± 3.73
Our Proposal		**27.46**±8.57

If we focus only on performance, from Table 2 we see that the best combination of models and representations is the ANN with both learned representations. This is intuitive given the structure of the network which learns said representations. However, the ANN outperforms all other algorithms regardless of the representation chosen, even when considered the arbitrary summarized representations. We conclude that ANN is in general better suited than other ML algorithms for this problem.

Moreover, focusing on the different representations, all algorithms except the linear regression achieve better results over the learned representations. Our interpretation of this is that the learned representations do not transform the problem into a linear one. Indeed, the problem seems to be a highly non-linear one: if we focus on the linear regression, using both learned representations an MPE error of 310.36% is obtained while using the same representation an ANN presents 33.65% MPE error. This is not only due to the learned representations, since over the summarized representations including building numbers (RangeStd N/I), the linear regression's performance is still close to 300%, while the ANN reaches 40.47% MPE over the same representation.

Finally, we find the difference of the performance of the ANN over the learned representations separately when compared to both concatenated interesting. While achieving 33.65% over the concatenation, the ANN is only able to achieve 51.08% over single representations. This reinforces our understanding that propagating intermediate representations inside the network's architecture is useful since different level representations bring complementary information of the problem to the table, even if both are learned in unison (and indeed Rep. 2 is calculated from Rep. 1 and inputs directly).

7 Conclusions

The present manuscript proposes a heterogeneous 1D convolutional neural network architecture, showing it can learn the relationship between descriptors of buildings provided by the Tax Administration and the PV potential in urban landscapes at a spatial level of city blocks. Also, learned representations are extracted from the trained 1D-CNN, which are compared to summarized representations over 5 different standard ML algorithms.

Even given the scarce amount of available training data, we were able to find competitive results with our heterogeneous 1D-CNN architecture. Moreover, the learned convolutional features allowed significant improvements in all non-linear ML algorithms when compared to summarized representations. Thus, the heterogeneous 1D-CNN architecture is not only able to model the underlying relationship between the scarce variable-length data but in the process also learns valuable fixed-length representations which can later be used to train different models.

Once more data is available, an interesting line of research would be to systematically explore different network configurations and the performance of the learned representations. A large-scale hyperparameter optimization would allow to further improve the current performances of the network, which could then be used to predict the real-world photovoltaic potential of urban rooftops over all areas with available internal revenue data. Moreover, more sophisticated representation learning techniques such as autoencoders could be optimized and evaluated.

References

1. Assouline, D., Mohajeri, N., Scartezzini, J.-L.: Estimation of large-scale solar rooftop pv potential for smart grid integration: a methodological review. In: Amini, M.H., Boroojeni, K.G., Iyengar, S.S., Pardalos, P.M., Blaabjerg, F., Madni, A.M. (eds.) Sustainable Interdependent Networks. SSDC, vol. 145, pp. 173–219. Springer, Cham (2018). https://doi.org/10.1007/978-3-319-74412-4_11
2. Assouline, D., Mohajeri, N., Scartezzini, J.-L.: Large-scale rooftop solar photovoltaic technical potential estimation using Random Forests. Appl. Energy **217**, 189–211 (2018)
3. Campos, P., Troncoso, L., Lund, P.D., Cuevas, C., Fissore, A., Garcia, R.: Potential of distributed photovoltaics in urban Chile. Sol. Energy **135**, 43–49 (2016)
4. Hawkins, D.M.: The problem of overfitting. J. Chem. Inf. Comput. Sci. **44**(1), 1–12 (2004). https://doi.org/10.1021/ci0342472
5. He, K., Zhang, X., Ren, S., Sun, J.: Deep residual learning for image recognition. In: Proceedings of the IEEE Computer Society Conference on Computer Vision and Pattern Recognition, December 2016, pp. 770–778 (2016)
6. Hecht-Nielsen, R.: Theory of the backpropagation neural network, based on "nonindent" by Robert Hecht-Nielsen, which appeared in Proceedings of the International Joint Conference on Neural Networks, June 1989, vol. 1, pp. 593–611. Academic Press, Inc. (1992)
7. Hubel, D.H., Wiesel, T.N.: Receptive fields and functional architecture of monkey striate cortex. J. Physiol. **195**(1), 215–243 (1968)
8. International Energy Agency: Energy Technology Perspectives 2016: towards sustainable urban energy systems (2016)
9. Ioffe, S., Szegedy, C.: Batch normalization: accelerating deep network training by reducing internal covariate shift. In: 32nd International Conference on Machine Learning, ICML 2015, vol. 1, pp. 448–456 (2015)
10. Izquierdo, S., Rodrigues, M., Fueyo, N.: A method for estimating the geographical distribution of the available roof surface area for large-scale photovoltaic energypotential evaluations. Sol. Energy **82**(10), 929–939 (2008)
11. Kingma, D.P., Ba, J.L.: Adam: a method for stochastic optimization. In: 3rd International Conference on Learning Representations, ICLR 2015 - Conference Track Proceedings, pp. 1–15 (2015)
12. Kiranyaz, S., Avci, O., Abdeljaber, O., Ince, T., Gabbouj, M., Inman, D.J.: 1D convolutional neural networks and applications: a survey. Mech. Syst. Sig. Process. **151**, 107398 (2021). https://doi.org/10.1016/j.ymssp.2020.107398
13. Krizhevsky, A.: Imagenet classification with deep convolutional neural networks alex. In: Handbook of Approximation Algorithms and Metaheuristics, pp. 1–1432 (2007)
14. Lecun, Y., Bottou, L., Bengio, Y., Ha, P.: LeNet. In: Proceedings of the IEEE, November, pp. 1–46 (1998)
15. Mainzer, K., Killinger, S., McKenna, R., Fichtner, W.: Assessment of rooftop photovoltaic potentials at the urban level using publicly available geodata and image recognition techniques. Sol. Energy **155**, 561–573 (2017)
16. Molina, A., Falvey, M., Rondanelli, R.: A solar radiation database for Chile. Sci. Rep. **7**(1), 1–11 (2017)
17. Servicio de Impuestos Internos: Avalúos y contribuciones de bienes raíces. https://www4.sii.cl/mapasui/internet/#/contenido/index.html. Accessed Jan 2021
18. Sharma, S.: Activation functions in neural networks - towards data science, vol. 6 (2017)
19. Simsek, Y., Lorca, Á., Urmee, T., Bahri, P.A., Escobar, R.: Review and assessment of energy policy developments in Chile. Energy Policy **127**, 87–101 (2019). https://doi.org/10.1016/j.enpol.2018.11.058

20. US EIA: International Energy Outlook 2013 - DOE/EIA-0484 (2013). Outlook 2013, p. 312 (2013)
21. Wiginton, L.K., Nguyen, H.T., Pearce, J.M.: Quantifying rooftop solar photovoltaic potential for regional renewable energy policy. Comput. Environ. Urban Syst. **34**(4), 345–357 (2010)
22. Zeiler, M.D.: Visualizing and understanding convolutional networks. Anal. Chem. Res. **12**, 40–46 (2017). https://doi.org/10.1016/j.ancr.2017.02.001
23. Zhang, C., Recht, B., Bengio, S., Hardt, M., Vinyals, O.: Understanding deep learning requires rethinking generalization. In: 5th International Conference on Learning Representations, ICLR 2017 - Conference Track Proceedings (2017)

Adversarial Attacks and Defenses in Deep Learning: A Survey

Chengyu Wang, Jia Wang$^{(\boxtimes)}$, and Qiuzhen Lin

Shenzhen University, Shenzhen 518060, China
jia.wang@szu.edu.cn

Abstract. In recent years, researches on adversarial attacks and defense mechanisms have obtained much attention. It's observed that adversarial examples crafted with small perturbations would mislead the deep neural network (DNN) model to output wrong prediction results. These small perturbations are imperceptible to humans. The existence of adversarial examples poses great threat to the robustness of DNN-based models. It is necessary to study the principle behind it and develop their countermeasures. This paper provides a survey to evaluate the existing attacking methods and defense techniques. Meanwhile, we discuss the relationship between the defense schemes and the robustness of the model in detail.

Keywords: Deep neural network · Adversarial perturbation · Adversarial attack · Adversarial defense · Robustness

1 Introduction

Deep Neural Networks have been widely adopted in applications such as computer vision, speech recognition and natural language processing tasks. However, recent studies found that the DNN has a fatal flaw: just a small amount of perturbations would fail the model, while these perturbations are typically in visible or visible but inconspicuous [42]. Such perturbed inputs are called adversarial samples. Moreover, the adversarial perturbations obtained in one model may also be effective for other models [31]. These adversarial samples have created great obstacles to the application of deep learning technology to the real world.

Adversarial attacks have two basic design golds: one is to mislead the targeted model to output error results, and the other is to make the perturbation in conspicuous, which is usually constrained by the l_p norm. In the past few years, many attacking methods have been proposed, such as L-BFGS [42], FGSM [13], PGD [23], MIT [11], JSMA [34], C&W attack [6] and Deepfool [32]. Researchers also have made their effort in improving these methods in terms of enhancing attack intensity and reducing the computational cost. The transferability of adversarial perturbations was also analyzed in [31], which inspired the black-box adversarial attacks.

As to the defense methods against adversarial attacks, one intuitive way is to destroy the perturbations and obtain clean images from adversarial samples, so that it does not

© Springer Nature Switzerland AG 2021
D.-S. Huang et al. (Eds.): ICIC 2021, LNCS 12836, pp. 450–461, 2021.
https://doi.org/10.1007/978-3-030-84522-3_37

need to modify the network. Another way is to enhance the robustness of the network to adversarial samples. Many researchers also demonstrate that using multiple different models to work simultaneously could also mitigate the abovementioned problem.

The rest of this paper is organized as follows. Section 2 introduces the attacking methods; Sect. 3 summarizes the defense methods. In Sect. 4 we give some discussions in this field, and Sect. 5 concludes this paper.

2 Adversarial Attacks

Since adversarial examples had been observed by [42] in deep learning, several effective attacking methods have been proposed to generate adversarial samples, with the knowledge of the model parameters. Such attacking methods are called white-box attack. The majority of the existing attacking methods belong to this category. Later, it was found that adversarial perturbations are transferable among images and models [31]. Gray-box attacks that only know the structure of the model and black-box attacks without any model information have been designed. More recently, the universal adversarial perturbations have also been discovered. In this section, we will give an introduction to existing attacking methods.

2.1 L-BFGS

[42] firstly found that natural images, by adding some special perturbations which are inconspicuous for human, can lead DNN-based classifiers to make wrong output. The authors proposed the L-BFGS algorithm to craft the adversarial examples as following:

$$\min_{x'} \|x - x'\|_p \quad s.t. \quad f(x') \neq y'. \tag{1}$$

y' is the incorrect label of benign image x, and $\|x - x'\|_p$ is the l_p norm of the adversarial perturbations. For the convenience of optimization, the loss function was designed as:

$$\min_{x'} \|cx - x'\|_p + J(\theta, x', y'), \tag{2}$$

where $J(\theta, x', y')$ is the loss function of the classifier for the adversarial sample and the incorrect label.

2.2 Attack on Gradient

Fast Gradient Sign Method (FGSM). [42] presented an effective method to generate adversarial samples. But the calculation of L-BFGS [42] is still complicated. For the purpose of generating adversarial samples more efficiently, [13] proposed Fast Gradient Sign Method (FGSM). The main idea of this method is to find the direction of the largest gradient of the DNN and adversarial perturbations in the same direction. Specifically, the adversarial samples of an untargeted attack is generated as:

$$x' = x + \epsilon \cdot sign(\nabla J(\theta, x', y)), \tag{3}$$

and the adversarial sample of a targeted attack is generated as:

$$x' = x - \epsilon \cdot sign(\nabla J(\theta, x', y')), \tag{4}$$

where y is the correct label of the benign image x, and y' is the incorrect label.

Projective Gradient Descent (PGD). FGSM [13] is a one-step method with low computational cost. Later researchers made some improvements on this method. One of the improvement methods is to decompose FGSM into multiple iterations. [23] realized it and proposed Basic Iterative attacking Method (BIM). The iteration step is as follows:

$$x'_{t+1} = x_t + \alpha \cdot sign(\nabla J(\theta, x', y)). \tag{5}$$

The constraint of perturbations $\epsilon = \alpha T$ is obtained by T iterations, and α is the maximum perturbation in each step. PGD [30] is a special BIM without α. Instead, PGD constrains x' after each iteration, by making $\|x - x'\|$ back to the range of $\epsilon \sim l_\infty$.

Momentum Iterative Attack (MIT). By applying momentum optimizer to FGSM, [11] proposed the momentum iterative FGSM (MI-FGSM). The iteration step was changed to:

$$x'_{t+1} = x_t + \alpha \cdot sign(g_{t+1}). \tag{6}$$

And g_t is updated by momentum as:

$$g_{t+1} = \mu \cdot g_t + \frac{\nabla J_t(\theta, x', y)}{\|\nabla J_t(\theta, x', y)\|_1}. \tag{7}$$

[52] further transformed the image in each iteration with the probability p. Then, the DI2-FGSM and M-DI2-FGSM, based on I-FGSM and MI-FGSM, were proposed in this way. Experimental results demonstrated that the new attacking model proposed by [52] has a high attack success rate on both white-box and black-box models.

Jacobian-Based Saliency Map Attack (JSMA). The direction of the adversarial samples' generation mentioned above is the gradient direction of the loss functions. But in [34], the direction of the adversarial samples' generation is the gradient direction of the predicted value on the target category label.

$$\nabla F(X) = \frac{\partial F(X)}{\partial X} = \left[\frac{\partial F_i(X)}{\partial X_i} \right]. \tag{8}$$

The attack crafted by this way was named as Jacobian-based Saliency Map Attack (JSMA).

2.3 Carlini and Wagner Attack (C&W Attack)

[6] proposed an attacking method based on optimization. For effective distillation defense models, the authors proposed a new optimization method as follows:

$$\min D(x, x + \delta) + c \cdot f(x + \delta) \quad s.t. \quad x + \delta \in [0, 1]^n \tag{9}$$

$$\delta_i = \frac{1}{2}(\tanh(\omega_i) + 1) - x_i$$

For this equation, $f(x + \delta) \leq 0$ holds when the prediction is targeted attack label. There were three attacks schemes named CW_0, CW_2 and CW_∞ limited by l_0, l_2 and l_∞ norms. C&W attacks can achieve 100% attack success rate on trained DNNs for MINST or CIFAR-10 without defense.

2.4 Deepfool

[32] tried to generate the adversarial samples using the smallest perturbations. The proposed method named Deepfool determines the decision boundary of the current points from all non-real classes in the high-dimensional space, and then make the nearest boundary as the attack label. For similar attacking performance, the perturbations generated by Deepfool are much smaller than that of FGSM [13].

2.5 GAN-Based Attacks

[48] firstly used GAN to generate adversarial samples, and called this method advGAN. The method is to use a generator to craft adversarial perturbations, a discriminator to constraint the perturbations by the l_p norm, and a classifier to fit the incorrect labels. [18] added random noise to the generator, resulting in more powerful adversarial attacks. [58] provided a black-box attacking method based on GAN, named DaST, to use the generated data fit the input data.

2.6 Universal Adversarial Perturbations

It was found by [31] that the adversarial perturbations trained on a single network may be effective for other networks. Furthermore, authors designed algorithms to generate these universal perturbations. Most images can be misclassified by adding such perturbations directly.

2.7 Transfer-Based Attacks

The transferability of perturbations provides a feasible solution for black-box attacks. [7] used the gradient of the surrogate model as a prior, and then applied them to the targeted black-box model. [56] used the adversarial perturbations, which are trained in the previous epoch, as additional inputs in the next epoch. It can both speed up the generation of adversarial samples and improve the attack power. [46] showed that prioritizing the key features of the attack concerned by various models can improve the transferability of the adversarial perturbations. In light of this discovery, [46] used model attention to search for adversarial samples with high transferability.

3   Adversarial Defense

In this section, we will survey on the latest adversarial defense schemes, mainly including 1) reduce or eliminate the adversarial attack; 2) enhance the robustness of network; 3) detect only and other defense methods. The first category aims to reduce or eliminate the impact of adversarial samples using randomization, denoising and other methods. These schemes do not need to retrain the network, and can be flexibly applied to various basic networks. The results demonstrate that its performance against black-box attacks is better than that against white-box attacks. The second kind of schemes aim to strengthen the model to obtain better generalization ability. The third kind schemes are mainly focus on detecting adversarial samples.

3.1   Reduce or Eliminate the Adversarial Attack

Randomization. In the study of DNN, we observe that DNN is usually robust when we add randomization to the image. But for the adversarial samples, especially those generated by the gradient, perturbations are fragile under randomization. Based on this observation, it's easy to think of using randomization to mitigate adversarial attacks. Several randomization-based defense methods are introduced as follows.

Random Input Transformation. [50] transformed the input image randomly before feeding it into CNNs. According to this paper, there are two kinds of transitions. The first one is to randomly resize the image, and the second one is to randomly zero padding around the image. [14] used more transformation methods, such as bit-depth, JPEG compression and image quilting. Experimental results demonstrated that these transformation methods have significant effects on gray-box attacks and black-box attacks, but their performance against the white-box attacks could be improved, especially for the EoT method [3]. Inspired by this, [35] considered that only a single transformation model is not reliable enough, and the exposure of transformation models may bring risks. Thus, they randomly selected some transformations and applied them in a random order. A fly in the ointment is that it may lead to the decline of classification accuracy. In order to alleviate this faultiness, [21] indicated that the distribution of SoftMax after image transformation contains the right prediction. They trained a distribution classifier on SoftMax outputs to make the final predictions, leading to an increase in accuracy.

Random Nosing. [53] added Gaussian noises into the input data. It both enhanced the generalization ability of the model and mitigated the adversarial attacks. [27] designed a new defense method named Random Self-Ensemble (RSE). RSE inserts Gaussian noise layers into the DNN before every convolution layer. The different random noise constitutes the ensemble, and stable result is obtained by combining them. [25] added random noises to the pixels of adversarial examples before inputting them into the classifier to eliminate the effects of adversarial perturbations.

Random in DNNs Structure. [51] proposed a new CNN structure. The output of the feature maps in this method are randomly masked. Thus, it forces filters to learn complete information from partial features, like dropout. As a result, it can enhance the ability of CNN and resist the adversarial attacks effectively.

Denoising. In order to eliminate the influence of adversarial attacks, removing adversarial perturbations is a straightforward way. These schemes are mainly divided into two categories. One is to directly denoise the input images or feature maps, the other is to generate clean images, which are close to the original images.

Feature Denoising. [12] showed that small perturbations in the pixel space may lead to very substantial "noise" in the feature maps. They added denoising blocks in intermediate layers of convolutional networks to improve adversarial robustness. This denoising block wraps the denoising operation with a 1×1 convolution and an identity skip connection. They compared four denoising operations (non-local means, bilateral filtering, mean filtering, and median filtering). Non-local means have the best performance. It shows that, by end-to-end adversarial training, denoising blocks work well on white-box attacks (e.g., PGD [11]) and black-box attacks.

Compression-Based. Researchers attempted to use compression methods to mitigate the adversarial examples. [10, 12] removed adversarial perturbations from the inputs by using standard JPEG compression. [29] pointed out that standard JPEG compression cannot effectively remove the adversarial perturbations, and may reduce classification accuracy on the benign images. Thus, they reconstructed the JPEG compression framework. The proposed method was to insert a new pair of quantization/dequantization processes on standard JPEG decompression after the original dequantization stage. [19] proposed a method similar to autoencoder method to purify adversarial perturbations. They designed a network named ComCNN to compress image and remove the redundant information. After adding the Gaussian noise, they used the networks named RecCNN to reconstruct the clean images.

Generate Clean Images. [40] proposed a network layer structure called sparse transformation layer (STL) to generate clean images. By stratifying convolutional sparse coding, authors converted images (including adversarial images and benign images) from a natural image space to a low-dimensional quasi-natural space, and then rebuild the image from the quasi-natural space. [43] proposed the Probabilistic Adversarial Robustness (PAR). It transforms the adversarial images to benign images by learning a probabilistic model, and the benign images sampled on clean space are close to the original images. In their work, they used the PixelCNN as the probabilistic model. It was proved that the lower bound of the loss function of PAR can be achieved. Thus, the model can sample images from clean space.

It's a straightforward way to use a GAN to get the clean image distribution.[39] proposed a GAN named APE-GAN to obtain benign images. APE-GAN is trained by the adversarial inputs and its corresponding clean inputs. The APE-GAN performs well on attacks that have been used during training, but weakly on the adaptive white-box CW_2 attack [6]. [37] proposed the Defense-GAN to learn the clean image distribution. Defense-GAN is trained by image adding random noise instead of adversarial examples. When the adversary example is input, an approximate sample satisfying the clean sample distribution, which is close to the adversary sample, is generated by Defense-GAN. Then the approximate sample, instead of original image, is feed into the classifier for classification. The biggest problem of GAN-Based methods is their unstable training.

3.2 Enhance the Robustness of Network

Adversarial Training. The basic idea is that, by sending the adversarial samples into the model as extra inputs, the model can learn the distribution of the adversarial samples. Adversarial training is one of the most effective defenses against adversarial attacks. The most important thing about this method is how to get adversarial samples. Adversarial samples can be generated by specific attack methods such as L-BFGS, FGSM, PGD. Then the model is trained with both benign and generated adversarial samples [13, 30, 42]. Researchers had made some improvements on these methods. [22] added random noise before generating FGSM attack samples. The adversarial training methods against definite attack have two major disadvantages. First, it is difficult to deal with the complex classification task because of the large computational cost for generating adversarial samples. Second, although the defense against the targeted attack is effective, it does not perform well against black-box attack and white-box attack of other attacking methods. Fortunately, adversarial samples can transfer among images and models [31]. Based on this discovery, adversarial samples generated on other static pre-training simple models can be used to expand the inputs of the local model [44, 47]. [56] proposed a method named Adversarial Training with Transferable Adversarial samples (ATTA), which use adversarial samples generated from the previous batch to generate the adversarial samples in the next batch. These abovementioned methods could both reduce the computational cost and improve robust against black-box attack.

[54] proposed an adversarial training method by using feature scattering. It takes the variation of the feature level as adversarial perturbations into account instead of the label layer. As an unsupervised training method, it does not leak the label. At the same time, it considered the whole batch samples, which make the model obtain better generalization ability. There're also some other schemes in this category, such as [45] generated both perturbed images and perturbed labels for adversarial training, [38] trained on universal adversarial samples.

Optimize Network Structure. [17] divided the training into two steps: the first one is the normal training stage and the other one is regularization training stage using the Frobenius norm of the Jacobian of the network. It was proved theoretically and experimentally that the proposed network improved its robustness without deceasing its accuracy. [16] proposed a method named Parametric Noise Injection (PNI) by adding trainable Gaussian noise to the activation or weight of each layer. [5] proved that the susceptibilities of features in the same layers of CNN have disparities. Thus, they designed feature regenerate units to regenerate the most 50% susceptible features to resist adversarial perturbations. [15] designed robust neural architecture search framework based on one-shot NAS. The concrete method was to design a super-net, and randomly change the architecture parameters α ($\alpha \in [0,1]$) in the adversarial training against PGD. The target networks are sampled based on the super-net performance. The networks searched out by this way were called Robnet family, which had less parameters. The experimental results demonstrated that densely connected pattern can benefit the network robustness and adding convolution operations in direct edges is more effective to improve the model's robustness. The robustness of robnet is about 5% higher than that of the basic network (such as Resnet and Densenet) under white-box and black-box attacks.

Consistency-Based. Adversarial perturbations is imperceptible for its slight change on clean image. [2] thought that adversarial perturbations only appear in the lower bit planes. They proposed the Bit Plane Feature Consistency (BPFC)that used the higher bit planes to limit the range of predicted results, and used the lower bit planes only to refine the prediction. Unfortunately, for clean images, the accuracy rate has decreased. [41] proposed Peernet that aggregates similar pixels from k nearest images for each pixel. In this way, it reduced the sensitivity to adversarial perturbations.

Bayesian Model-Based. Uncertainty of bayesian neural networks (BNN) is increasing with attacking strength. [36] proposed that adversarial samples can be detected by quantizing this observation. [28] firstly used the BNN in adversarial training. [28] assumed that all weights in the network are random, and used the normal techniques in BNN for training. In the experiment, authors observed that BNN itself had no defense ability, but when combined with adversarial training, its robustness against the attack was significantly improved. [26] proposed the utilization of deep bayesian classifier to model the distribution of the input, and used the logit values of generative classifier to detect adversarial samples by rejecting inconsistent inputs and low confidence predictions.

3.3 Ensemble Learning

Ensemble learning for adversarial defense can be summarized as: training a model ensemble, so that each model of this ensemble can achieve the same classification function, but each model should be distinguished as much as possible for better generalization ability. [1] proposed a method that determined specialists on different subsets of classes by confusion matrices, and did the classification by using voting mechanism. [4] used multiple instances of a base model. The models were trained to minimize the cross-entry loss while reducing the consistency of their classification scores. In the prediction, rank voting mechanism was used to detect adversarial samples. It works well on FGSM attack, but was not so well for iteration attack (26.4% accuracy on classification and 27.1% on detection). [33] proposed the adaptive diversity promoting (ADP) training method. After training by the ADP method, the non-maximal predictions of each network would tend to be mutually orthogonal, and lead to different feature distributions and decision domains. The above methods cannot optimize the generalization ability of the ensemble well, which leads to low resistance rate against white-box attacks. [20] presented the joint gradient phase and magnitude regularization (GPMR) scheme for improving the robustness of the ensemble against adversarial perturbations. [9] defined the gradient diversity promoting loss to increase the angle between the gradients on different models, and defined the gradient magnitude regularization loss to regularize the joint interactions of the members.

3.4 Detect Only

Some researchers focus on detecting adversarial examples. [24] used Gaussian discriminant analysis (GDA) to predict the feature distribution, and then use Mahalanobis distance to calculate the confidence of adversarial samples. The authors used the output of all DNN layers, instead of only the last layer. This method can simultaneously detect out-of-distribution samples and adversarial samples. [8] proposed that the performance of DNN can be represented by functional graph. Different test samples brought different changes to the functional graph. And thus this could be used to determine whether the network works correctly during testing and reliably identify adversarial attacks. [57] studied the output distribution of hidden neurons in DNN classifier. Then they proposed that the hidden states of DNN are quite different between adversarial samples and benign samples, which was used to reject adversarial inputs.

4 Discussions

4.1 Attack and Defense

For practical applications of the deep learning models, the robustness of the models must be evaluated and improved, especially in the areas with high safety requirements (e.g., autonomous driving and medical diagnosis). It is necessary to study advanced attack schemes for improving the model defense performance. Generally speaking, if the opponent's attack/defense method is known, then the corresponding defense/attacking method is easy to develop. Therefore, many researchers focus on the generalization ability of attack or defense. However, in terms of defense methods, adversarial training is still one of the best defense ways, although it requires large computational cost.

4.2 Defense and Model Performance

Typically, improving the robustness against adversarial attacks will reduce the correct rate of benign samples. However, some researchers have found that the performance of some models can be improved by adversarial examples. [49] found that using adversarial samples skillfully can improve the accuracy of image classifier. [55] proposed that using the attack algorithm with early end iteration for adversarial training can improve the robustness without reducing the accuracy of the model. Defense schemes against adversarial attacks inspired new approaches on improving model capability.

5 Conclusions

This paper mainly introduces the typical attacks and defense schemes, and makes some classification on their ideological connotation. Although deep neural networks have great achievements on various tasks, the appearance of adversarial samples is worth further studying. Existing defense methods have made great progress in improving the adversarial robustness of the model, but they still can't have a good effect on all attacks. The research in this field needs further development.

References

1. Abbasi, M., Gagné, C.: Robustness to adversarial examples through an ensemble of specialists. arXiv preprint arXiv:1702.06856 (2017)
2. Addepalli, S., Baburaj, A., Sriramanan, G., Babu, R.V.: Towards achieving adversarial robustness by enforcing feature consistency across bit planes. In: Proceedings of the IEEE/CVF Conference on Computer Vision and Pattern Recognition, pp.1020–1029 (2020)
3. Athalye, A., Engstrom, L., Ilyas, A., Kwok, K.: Synthesizing robust adversarial examples. In: International conference on machine learning, pp. 284–293. PMLR (2018)
4. Bagnall, A., Bunescu, R., Stewart, G.: Training ensembles to detect adversarial examples. arXiv preprint arXiv:1712.04006 (2017)
5. Borkar, T., Heide, F., Karam, L.: Defending against universal attacks through selective feature regeneration. In: Proceedings of the IEEE/CVF Conference on Computer Vision and Pattern Recognition, pp. 709–719 (2020)
6. Carlini, N., Wagner, D.: Towards evaluating the robustness of neural networks. In: 2017 IEEE Symposium on Security and Privacy (SP), pp. 39–57. IEEE (2017)
7. Cheng, S., Dong, Y., Pang, T., Su, H., Zhu, J.: Improving black-box adversarial attacks with a transfer-based prior. arXiv preprint arXiv:1906.06919 (2019)
8. Corneanu, C.A., Madadi, M., Escalera, S., Martinez, A.M.: What does it mean to learn in deep networks? And, how does one detect adversarial attacks? In: Proceedings of the IEEE/CVF Conference on Computer Vision and Pattern Recognition, pp. 4757–4766 (2019)
9. Dabouei, A., Soleymani, S., Taherkhani, F., Dawson, J., Nasrabadi, N.M.: Exploiting joint robustness to adversarial perturbations. In: Proceedings of the IEEE/CVF Conference on Computer Vision and Pattern Recognition, pp. 1122–1131 (2020)
10. Das, N., et al.: Keeping the bad guys out: protecting and vaccinating deep learning with jpeg compression. arXiv preprint arXiv:1705.02900 (2017)
11. Dong, Y., et al.: Boosting adversarial attacks with momentum. In: Proceedings of the IEEE Conference on Computer Vision and Pattern Recognition, pp. 9185–9193 (2018)
12. Dziugaite, G.K., Ghahramani, Z., Roy, D.M.: A study of the effect of jpg compression on adversarial images. arXiv preprint arXiv:1608.00853 (2016)
13. Goodfellow, I.J., Shlens, J., Szegedy, C.: Explaining and harnessing adversarial examples. arXiv preprint arXiv:1412.6572 (2014)
14. Guo, C., Rana, M., Cisse, M., Van Der Maaten, L.: Countering adversarial images using input transformations. arXiv preprint arXiv:1711.00117 (2017)
15. Guo, M., Yang, Y., Xu, R., Liu, Z., Lin, D.: When NAS meets robustness: In search of robust architectures against adversarial attacks. In: Proceedings of the IEEE/CVF Conference on Computer Vision and Pattern Recognition, pp. 631–640 (2020)
16. He, Z., Rakin, A.S., Fan, D.: Parametric noise injection: trainable randomness to improve deep neural network robustness against adversarial attack. In: Proceedings of the IEEE/CVF Conference on Computer Vision and Pattern Recognition, pp. 588–597 (2019)
17. Jakubovitz, D., Giryes, R.: Improving DNN robustness to adversarial attacks using Jacobian regularization. In: Ferrari, V., Hebert, M., Sminchisescu, C., Weiss, Y. (eds.) ECCV 2018. LNCS, vol. 11216, pp. 525–541. Springer, Cham (2018). https://doi.org/10.1007/978-3-030-01258-8_32
18. Jang, Y., Zhao, T., Hong, S., Lee, H.: Adversarial defense via learning to generate diverse attacks. In: Proceedings of the IEEE/CVF International Conference on Computer Vision, pp. 2740–2749 (2019)
19. Jia, X., Wei, X., Cao, X., Foroosh, H.: Comdefend: An efficient image compression model to defend adversarial examples. In: Proceedings of the IEEE/CVF Conference on Computer Vision and Pattern Recognition, pp. 6084–6092 (2019)

20. Kariyappa, S., Qureshi, M.K.: Improving adversarial robustness of ensembles with diversity training. arXiv preprint arXiv:1901.09981 (2019)
21. Kou, C., Lee, H.K., Chang, E.C., Ng, T.K.: Enhancing transformation-based defenses against adversarial attacks with a distribution classifier. In: International Conference on Learning Representations (2019)
22. Kurakin, A., Goodfellow, I., Bengio, S.: Adversarial machine learning at scale. arXiv preprint arXiv:1611.01236 (2016)
23. Kurakin, A., Goodfellow, I., Bengio, S., et al.: Adversarial examples in the physical world (2016)
24. Lee, K., Lee, K., Lee, H., Shin, J.: A simple unified framework for detecting out-of-distribution samples and adversarial attacks. arXiv preprint arXiv:1807.03888 (2018)
25. Li, B., Chen, C., Wang, W., Carin, L.: Certified adversarial robustness with additive noise. arXiv preprint arXiv:1809.03113 (2018)
26. Li, Y., Bradshaw, J., Sharma, Y.: Are generative classifiers more robust to adversarial attacks? In: International Conference on Machine Learning, pp. 3804–3814. PMLR (2019)
27. Liu, X., Cheng, M., Zhang, H., Hsieh, C.-J.: Towards robust neural networks via random self-ensemble. In: Ferrari, V., Hebert, M., Sminchisescu, C., Weiss, Y. (eds.) ECCV 2018. LNCS, vol. 11211, pp. 381–397. Springer, Cham (2018). https://doi.org/10.1007/978-3-030-01234-2_23
28. Liu, X., Li, Y., Wu, C., Hsieh, C.J.: Adv-BNN: improved adversarial defense through robust Bayesian neural network. arXiv preprint arXiv:1810.01279 (2018)
29. Liu, Z., et al.: Feature distillation: DNN-oriented JPEG compression against adversarial examples. In: 2019 IEEE/CVF Conference on Computer Vision and Pattern Recognition (CVPR), pp. 860–868. IEEE (2019)
30. Madry, A., Makelov, A., Schmidt, L., Tsipras, D., Vladu, A.: Towards deep learning models resistant to adversarial attacks. arXiv preprint arXiv:1706.06083 (2017)
31. Moosavi-Dezfooli, S.M., Fawzi, A., Fawzi, O., Frossard, P.: Universal adversarial perturbations. In: Proceedings of the IEEE Conference on Computer Vision and Pattern Recognition, pp. 1765–1773 (2017)
32. Moosavi-Dezfooli, S.M., Fawzi, A., Frossard, P.: DeepFool: a simple and accurate method to fool deep neural networks. In: Proceedings of the IEEE Conference on Computer Vision and Pattern Recognition, pp. 2574–2582 (2016)
33. Pang, T., Xu, K., Du, C., Chen, N., Zhu, J.: Improving adversarial robustness via promoting ensemble diversity. In: International Conference on Machine Learning, pp. 4970–4979. PMLR (2019)
34. Papernot, N., McDaniel, P., Jha, S., Fredrikson, M., Celik, Z.B., Swami, A.: The limitations of deep learning in adversarial settings. In: 2016 IEEE European Symposium on Security and Privacy (EuroS&P), pp. 372–387. IEEE (2016)
35. Raff, E., Sylvester, J., Forsyth, S., McLean, M.: Barrage of random transforms for adversarially robust defense. In: Proceedings of the IEEE/CVF Conference on Computer Vision and Pattern Recognition, pp. 6528–6537 (2019)
36. Rawat, A., Wistuba, M., Nicolae, M.I.: Adversarial phenomenon in the eyes of Bayesian deep learning. arXiv preprint arXiv:1711.08244 (2017)
37. Samangouei, P., Kabkab, M., Chellappa, R.: Defense-GAN: protecting classifiers against adversarial attacks using generative models. arXiv preprintarXiv:1805.06605 (2018)
38. Shafahi, A., Najibi, M., Xu, Z., Dickerson, J., Davis, L.S., Goldstein, T.: Universal adversarial training. In: Proceedings of the AAAI Conference on Artificial Intelligence, vol. 34, pp. 5636–5643 (2020)
39. Shen, S., Jin, G., Gao, K., Zhang, Y.: APE-GAN: adversarial perturbation elimination with GAN. arXiv preprint arXiv:1707.05474 (2017)

40. Sun, B., Tsai, N., Liu, F., Yu, R., Su, H.: Adversarial defense by stratified convolutional sparse coding. In: Proceedings of the IEEE/CVF Conference on Computer Vision and Pattern Recognition, pp. 11447–11456 (2019)
41. Svoboda, J., Masci, J., Monti, F., Bronstein, M.M., Guibas, L.: PeerNets: ex-ploiting peer wisdom against adversarial attacks. arXiv preprint arXiv:1806.00088 (2018)
42. Szegedy, C., et al.: Intriguing properties of neural networks. arXiv preprint arXiv:1312.6199 (2013)
43. Theagarajan, R., Chen, M., Bhanu, B., Zhang, J.: ShieldNets: defending against adversarial attacks using probabilistic adversarial robustness. In: Proceedings of the IEEE/CVF Conference on Computer Vision and Pattern Recognition, pp. 6988–6996 (2019)
44. Tramèr, F., Kurakin, A., Papernot, N., Goodfellow, I., Boneh, D., Mc-Daniel, P.: Ensemble adversarial training: attacks and defenses. arXiv preprintarXiv:1705.07204 (2017)
45. Wang, J., Zhang, H.: Bilateral adversarial training: towards fast training of more robust models against adversarial attacks. In: Proceedings of the IEEE/CVF International Conference on Computer Vision, pp. 6629–6638 (2019)
46. Wu, W., et al.: Boosting the transferability of adversarial samples via attention. In: Proceedings of the IEEE/CVF Conference on Computer Vision and Pattern Recognition, pp. 1161–1170 (2020)
47. Xiao, C., Zheng, C.: One man's trash is another man's treasure: Resisting adversarial examples by adversarial examples. In: Proceedings of the IEEE/CVF Conference on Computer Vision and Pattern Recognition, pp. 412–421 (2020)
48. Xiao, C., Li, B., Zhu, J.Y., He, W., Liu, M., Song, D.: Generating adversarial examples with adversarial networks. arXiv preprint arXiv:1801.02610 (2018)
49. Xie, C., Tan, M., Gong, B., Wang, J., Yuille, A.L., Le, Q.V.: Adversarial examples improve image recognition. In: Proceedings of the IEEE/CVF Conference on Computer Vision and Pattern Recognition, pp. 819–828 (2020)
50. Xie, C., Wang, J., Zhang, Z., Ren, Z., Yuille, A.: Mitigating adversarial effects through randomization. arXiv preprint arXiv:1711.01991 (2017)
51. Xie, C., Wu, Y., Maaten, L, Yuille, A.L., He, K.: Feature denoising for improving adversarial robustness. In: Proceedings of the IEEE/CVF Conference on Computer Vision and Pattern Recognition, pp. 501–509 (2019)
52. Xie, C., et al.: Improving transferability of adversarial examples with input diversity. In: Proceedings of the IEEE/CVF Conference on Computer Vision and Pattern Recognition, pp. 2730–2739 (2019)
53. Zantedeschi, V., Nicolae, M.I., Rawat, A.: Efficient defenses against adversarial attacks. In: Proceedings of the 10th ACM Workshop on Artificial Intelligence and Security, pp. 39–49 (2017)
54. Zhang, H., Wang, J.: Defense against adversarial attacks using feature scattering-based adversarial training. arXiv preprint arXiv:1907.10764 (2019)
55. Zhang, J., et al.: Attacks which do not kill training make adversarial learning stronger. In: International Conference on Machine Learning, pp. 11278–11287. PMLR (2020)
56. Zheng, H., Zhang, Z., Gu, J., Lee, H., Prakash, A.: Efficient adversarial training with transferable adversarial examples. In: Proceedings of the IEEE/CVF Conference on Computer Vision and Pattern Recognition, pp. 1181–1190 (2020)
57. Zheng, Z., Hong, P.: Robust detection of adversarial attacks by modeling the intrinsic properties of deep neural networks. In: Proceedings of the 32nd International Conference on Neural Information Processing Systems, pp. 7924–7933 (2018)
58. Zhou, M., Wu, J., Liu, Y., Liu, S., Zhu, C.: DaST: data-free substitute training for adversarial attacks. In: Proceedings of the IEEE/CVF Conference on Computer Vision and Pattern Recognition, pp. 234–243 (2020)

Pattern Recognition

Fine-Grained Recognition of Crop Pests Based on Capsule Network with Attention Mechanism

Xianfeng Wang, Xuqi Wang, Wenzhun Huang[✉], and Shanwen Zhang

School of Information Engineering, Xijing University, Xi'an 710123, China
huangwenzhun@xijing.edu.cn

Abstract. Crop pest detection and recognition in the field is one of the crucial components in pest management involving detection, localization in addition to classification and recognition which is much more difficult than generic object detection because of the apparent differences among pest species with various shapes, colours and sizes. A crop pest identification method is proposed based on capsule network with attention mechanism (CNetAM). In CAN, capsule network is used to improve classification performance of the traditional convolutional neural network (CNN) and an attention module is added to reduce the noise influence and speedup the network training. The experimental results on a pest image dataset demonstrated that the proposed method is effective and feasible in classifying various types of insects in field crops, and can be implemented in the agriculture sector for crop protection.

Keywords: Crop pest detection · Capsule Networks (CNet) · Attention mechanism · CNet with attention mechanism (CNetAM)

1 Introduction

The growth of the most field crops such as rice, wheat, maize, soybean, sugarcane and other crops are often affected by the various pests. Crop pests seriously affect crop production and quality. Accurate detection and identification of pests is the premise of pest control. The detection and identification of all types of crop insects correctly is a difficult and challenging task due to the similar appearance and complex background in the earlier stage of crop growth. The images of crop pests collected in the natural environment are often affected by illumination, insect morphology, image size and shooting Angle, etc., which bring great difficulties to pest detection and recognition. With the continuous development of image processing and computer technology, many crop insect pests detection and identification methods have been presented [1, 2], which can be roughly divided into two categories: feature extraction based methods and deep learning based methods.

Feature extraction and the classifier design are the crucial parts for image recognition of insect pests on agriculture field crops. Nanni et al. [3] utilized three different saliency methods as image preprocessing and created three different images for every saliency method, and then tested the approach on both a small dataset and the large IP102 dataset.

© Springer Nature Switzerland AG 2021
D.-S. Huang et al. (Eds.): ICIC 2021, LNCS 12836, pp. 465–474, 2021.
https://doi.org/10.1007/978-3-030-84522-3_38

Deng et al. [4] used Saliency Using Natural statistics model (SUN) to generate saliency maps and detect region of interest (ROI) in a pest image, and utilized Scale Invariant Feature Transform (SIFT) to increase the invariance to rotational changes. Wang et al. [5] designed an automatic identification system to identify insect specimen images at the order level according to the methods of image processing, pattern recognition and the theory of taxonomy. They did tests on eight- and nine-orders with different features and compared the advantages and disadvantages of their system and provided some advice for future research on insect image recognition.

Convolutional neural networks (CNN) can perform automatic feature extraction and learn complex high-level features in image classification applications. Due to the ability to learn data-dependent features automatically from the data, many CNN and its variant models have been applied to pest identification task. Thenmozhi and Reddy [6] proposed a convolutional neural network (CNN) model to classify insect species on three publicly available insect datasets. The proposed model was evaluated and compared with pre-trained deep learning architectures such as AlexNet, ResNet, GoogLeNet and VGGNet for insect classification. The experiment results validated that CNN can comprehensively extract multifaceted insect features. Fuentes et al. [7] proposed a tomato disease and insect pest detection method based on robust deep learning, which detected tomato disease and insect pest images with different resolutions, and proposed a local and global class annotation and data expansion method to improve the detection accuracy, which could effectively identify 9 different types of diseases and pests. Ayan et al. [8] implemented crop pest classification with a genetic algorithm-based weighted ensemble of deep CNNs. They modified and re-trained seven different pre-trained CNN models (VGG-16, VGG-19, ResNet-50, Inception-V3, Xception, MobileNet, SqueezeNet) by using appropriate transfer learning and fine-tuning strategies on publicly available D0 dataset with 40 classes. Xia et al. [9] proposed an improved CNN model for multi-classification of crop insects. In the model, a a region proposal network is adopted rather than a traditional selective search technique to generate a smaller number of proposal windows. It is important for improving prediction accuracy and accelerating computations. Li et al. [10] proposed a deep learning-based pipeline for localization and counting of agricultural pests in images by self-learning saliency feature maps by integrating a CNN of ZF (Zeiler and Fergus model) and a region proposal network (RPN) with Non-Maximum Suppression (NMS) to remove overlapping detections. Liu et al. [11] proposed a pipeline for the visual localization and classification of agricultural pest insects by computing a saliency map and applying deep CNN model, and explored different architectures by shrinking depth and width, and found effective sizes that can act as alternatives for practical applications. The model is optimized the critical parameters, including size, number and convolutional stride of local receptive fields, dropout ratio and the final loss function. Liu et al. [12] proposed a region-based end-to-end approach named PestNet for large-scale multi-class pest detection and classification based on deep learning. PestNet consists of three major parts, i.e., module channel-spatial attention (CSA), region proposal network (RPN) and position-sensitive score map (PSSM). To achieve pest identification with the complex farmland background, Cheng et al. [13] proposed a pest identification method based on deep residual learning. Compared to the traditional support vector machine (SVM) and BPNNs, the proposed method is noticeably improved in the complex farmland background.

Because the within-class and different-class pests in fields are various and irregular with a wide variety of shapes, poses and colors and complex backgrounds, as shown in Fig. 1, it is difficult to extract the robust and invariant classification features from the pest images, while the modeling ability and classification performance of the CNN and its improved models for geometric deformation mainly comes from the expansion of data sets, the deepening of network layers and the artificial design of the model, but it does not fundamentally solve the deformation problem of field pests [14].

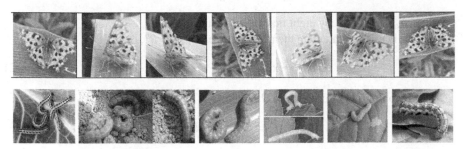

(C) Some kinds of crop pests with various poses, colours, shapes, scales and rotations

Fig. 1. Crop pest images collected in nature fields

In 2017, Hinton proposed Capsule Network (CNet) [15, 16], which replaces the neural nodes in the traditional deep neural network with the neuron vector, and trains the new neural network with the dynamic routing protocol instead of the maximum pooling method in the deep CNN. Unlike the maximum pooling in CNN, CNet does not discard accurate location information between entities within the region, retains semantic information and spatial relationships between various features in the text classification, while working to detect features and its various variants. At present, the effectiveness of CNet has been confirmed on the handwritten MNIST data set, and the training speed is fast and the accuracy is high. In recent years, attention mechanism has been widely applied in various types of deep learning tasks, such as natural language processing, image recognition and speech recognition [17, 18]. The attention mechanism in CNN comes from the biological system of human beings. Inspired by CNet and attention mechanism, a pest recognition method is proposed based on CNet with attention mechanism (CNetAM). Its contributions are given as follows.

(1) CNet is used to extract the invariant features from the various pest images.
(2) Attention mechanism is used to capture rich contextual relationships for better feature extraction and improving network training.
(3) A lot of experiments are implemented to validate the proposed method on the pest image dataset.

The rest of the paper is organized as follows. Section 2 introduces attention mechanism and CNets. Section 3 presents an improved CNet with attention mechanism in detail. The experiments and results are presented in Sect. 4. Section 5 concludes and recommends future works.

2 Related Works

2.1 Attention Mechanism

Attention mechanisms have been successfully applied to deep learning [19, 20]. Suppose the input matrix $H = [h_1, h_2, \ldots, h_n]$, h_i is the i-th vector, n is the length of a sentence. The output weight component $\alpha_i \in R^n$ of H is obtained by the tanh function,

$$\alpha_i = \tanh(W_\alpha H + b) \tag{1}$$

where $W_\alpha \in R^{1 \times n}$ is a weight matrix, and b is a bias.

Softmax is used to calculate the attention weight $\beta_i \in R^n$,

$$\beta_i = \exp(\alpha_i) / \sum_{j=1}^{n} \exp(\alpha_j) \tag{2}$$

Then the output feature vector is obtained by

$$\gamma = \sum_{i=1}^{n} \beta_i H \tag{3}$$

2.2 Capsule Network (CNet)

CapNet consists of encoder and decoder. The encoder consists of convolution layer, PrimaryCaps (main capsule layer), and digitalcaps (digital capsule layer). The input data is an image of 28×28 pixels. After the convolution operation, the features extracted by the convolutional layer are transformed into vector capsules by the main capsule layer. Then, the final results are output after the calculation of digital caps (digital capsule layer) and PrimaryCaps (main capsule layer) mapping by the dynamic routing algorithm. In the classical capsule network model, the convolution layer selects the convolution kernel with step size of 1, depth of 32 and size of 9×9. ReLU was selected as the activation function. In the second PrimaryCaps layer (main capsule layer), 8 groups of convolution kernels with step size of 2, depth of 32 and size of 9×9 were selected, and 8 times of convolution operations were carried out on the feature images output by the convolution layer to obtain $6 \times 6 \times 8 \times 32$ feature vectors, and 1152 capsules were obtained by pating the feature vectors. Each capsule consists of an 8-dimensional vector. The third layer digitCaps (layer of digital capsules) outputs tensors of 16×10. The encoder structure is shown in Fig. 2. The decoder is composed of three fully connected layers, and its structure is shown in Fig. 3. The decoder reconstructs the image from the output of the final capsule, and reconstructs the image of 28×28 pixels by accepting the 10×16 dimensional vector output of DigitCaps (digital capsule layer).

CNet classifies the input features by adopting dynamic routing algorithm instead of the pooling layer of traditional convolutional neural network. The more similar features, the stronger such features will be, which is equivalent to a feature selection process [19]. The main idea of the dynamic routing algorithm is that all the sub-capsule outputs can predict the instantiation parameters of the parent capsule through the alternating matrix. And when the bottom capsule prediction is the same, the parent capsule is activated and outputs the eigenvector. Dynamic routing algorithm is mainly composed of routing selection and vector calculation.

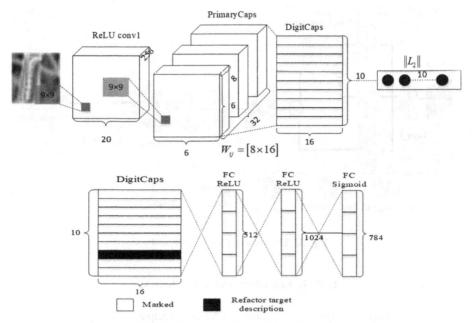

Fig. 2. Structures of encoder and decoder

3   Capsule Network with Attention Mechanism

The classical CNet model is not effective in large scale image data, which limits the application range of the model. Shared this research model by introducing the local dynamic routing algorithm instead of classical capsule network model of dynamic routing algorithm, are Shared the same capsule through window control weighting method, namely child capsule only through the window of the pre-determined route to the parent capsule, greatly reduce the network model calculation, so that the model can deal with more pixels in the image. In order to solve the over-fitting problem caused by too long training time in the neural network model, the two-stage model training method is used to reduce the over-fitting problem. In the second stage of model training, the model parameters are initialized by less training and then the model training of all the data is carried out. In the Section, a CNet with attention mechanism (CNetAM) model is constructed. Its architecture and corresponding parameters are shown in Fig. 3 and Table 1, respectively.

Input: image to be trained.

Conv 1. Passed to the first convolution layer for operation, the convolution kernel with 7×7 steps of 2 is used to extract features.

Attention Block. The output features of Conv1 are convolved into the dimensionality reduction 1×1 convolution kernel, $f(x)$, $g(x)$ and $h(x)$, respectively in the attention module and output after convolution. Then, the output of $f(x)$ and, $g(x)$ is multiplied by matrix. After the similarity output matrix is obtained, mask calculation of global Attention is carried out with the passing output.

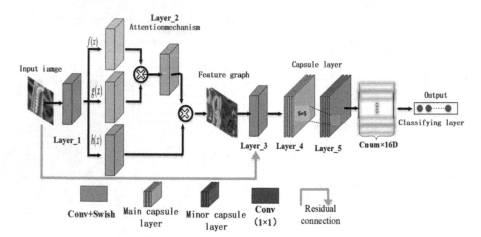

Fig. 3. The architecture of CNetAM

Table 1. Related parameters of CNetAM

Layer	Type	Size	Channel	Output
Layer_1	Conv	7 × 7	128	117 × 117 × 128
Layer_2	Attention	–	–	–
Layer_3	Conv	3 × 3	256	115 × 115 × 256
Layer_4	Main CNet	5 × 5		56 × 56 × 32 × 16D
Layer_5	Minor CNet	5 × 5	–	26 × 26 × 32 × 16D
Layer_6	Conv	1 × 1	156	26 × 26 × 156
Layer_7	Deconv	5 × 5	128	32 × 32 × 128

Conv 2. The output features from the self-attention module are transferred to the second convolution layer, and the 3 × 3 convolution kernel is used to further improve the feature description ability of the model.

Main capsule layer. Vectorized the input characteristic diagram of Conv2, in which the capsule layer is divided into the first capsule layer and the second capsule layer. Each layer is composed of 32 capsules, and each capsule is composed of 16 5 × 5 convolution nuclei.

Classification capsule layer: The classification capsule layer is composed of Cnum × 16D dimension vector, which is used to predict the type of input image. Routing between capsule layers is realized by local dynamic routing algorithm.

Update parameter. Calculate the loss function $L_{loss}(Vector_{out}, R_r)$, and then update the model parameters according to the back propagation algorithm.

4   Experiments and Results

In this Section, eight common crop pests, such as mucilworms, corn bores, moths, caterpillars, ladybugs, aphids, cotton bollworms and flying cicadas, were studied. The images were collected from the experimental base in Baoji City, Shaanxi Province. In different periods of time in natural field environment, nearly 2000 images of pests were collected by using image acquisition devices such as smart phones, cameras and the Internet of Things. About 250 images of each pest were collected. At the same time, some network images are used to supplement the data set to ensure the integrity of the data set. To improve the training efficiency of the subsequent network model, Photoshop was used to cut the images into JPG color images with size of 256 × 256 pixels. The original pest image examples are shown in Fig. 4.

Fig. 4. Original pest images

To verify the effectiveness of the proposed algorithm for crop pest detection in the field, the experimental results are compared with other four CNN models (ICNN, VGG16, ResNet and CNet) based on the augmented pest image database. All experiments use TensorFlow as the deep learning framework, using Python3.7 programming development language, operating environment of the system is Windows10 64Bit, hardware development environment is Intel Xeon E5-2643v3 @3.40 GHz CPU, 64 GB memory, graphics card NVIDIA Quadro M4000 GPU. (ReLU) is used as the activation function to ensure the nonlinear ability of the model. The initial learning rate is set as 0.01, the momentum factor is set as 0.9, and the Batch size is set as 128. After 1200 iterations of the model, the learning rate is set as 0.001. Figure 5 shows the feature graphs by CNetAM. Table 2 gives the recognition results of crop pests by five methods.

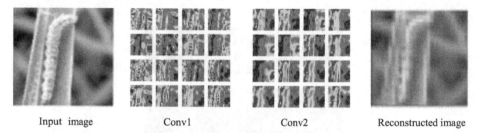

| Input image | Conv1 | Conv2 | Reconstructed image |

Fig. 5. The feature graphs by CNetAM

Table 2. The recognition results of crop pests by five methods

Method	Accuracy rate %	Recall %
ICNN	85.28	76.85
VGG16	82.49	72.62
ResNet	92.24	80.63
CNet	75.41	68.30
CNetAM	89.52	81.22

From Table 2, it is found that CNetAM achieves the highest recognition accuracy and recall rate in the same condition.

5 Conclusions

Aiming at the difficult problem of field crop pest detection and identification, a new crop pest identification method based on attention capsule network was proposed. By introducing the attention mechanism module into the capsule network, the global dependence among pixels is realized to reduce the influence of noise on the model and increase the accuracy of model recognition. The recognition experiment of rice pest image under complex background was carried out, and compared with ICNN, CAPS network, VGG16 and RESNET. The results show that the method proposed in this paper can better accomplish the identification task of rice pests in different forms and different backgrounds, and has higher identification accuracy and recall rate. The experimental results show that the method proposed in this study can meet the requirements of crop pest recognition in complex background.

References

1. Zhang, H.T., Hu, Y.X., Zhang, H.Y.: Extraction and classifier design for image recognition of insect pests on field crops. Adv. Mater. Res. **756–759**, 4063–4067 (2013)
2. Deng, L., Wang, Z., Wang, C., et al.: Application of agricultural insect pest detection and control map based on image processing analysis. J. Intel. Fuzzy Syst. **38**, 379–389 (2020)
3. Nanni, L., Maguolo, G., Pancino, F.: Insect pest image detection and recognition based on bio-inspired methods. Ecol. Inf. **57**, 101089 (2020)
4. Deng, L., Wang, Y., Han, Z., et al.: Research on insect pest image detection and recognition based on bio-inspired methods. Biosys. Eng. **169**, 139–148 (2018)
5. Wang, J., Lin, C., Ji, L., et al.: A new automatic identification system of insect images at the order level. Knowl. Based Syst. **33**(3), 102–110 (2012)
6. Thenmozhi, K., Srinivasulu Reddy, U.: Crop pest classification based on deep convolutional neural network and transfer learning. Comput. Electron. Agric. **164**, 104906 (2019). https://doi.org/10.1016/j.compag.2019.104906
7. Fuentes, A., Yoon, S., Kim, S.C., et al.: A robust deep-learning-based detector for real-time tomato plant diseases and pests recognition. Sensors **17**(9), 2022 (2017)
8. Ayan, E., Erbay, H., Varn, F.: Crop pest classification with a genetic algorithm-based weighted ensemble of deep convolutional neural networks. Comput. Electron. Agric. **179**(4), 105809 (2020)
9. Xia, D., Chen, P., Wang, B., et al.: Pest detection and classification based on an improved convolutional neural network. Sensors **18**, 4169 (2018). https://doi.org/10.3390/s18124169
10. Li, W., Chen, P., Wang, B., Xie, C.: Automatic localization and count of agricultural crop pests based on an improved deep learning pipeline. Sci. Rep. **9**(1), 7024 (2019). https://doi.org/10.1038/s41598-019-43171-0
11. Liu, Z., Gao, J., Yang, G., et al.: Localization and classification of paddy field pests using a saliency map and deep convolutional neural network. Sci. Rep. **6**, 20410 (2016)
12. Liu, L., Wang, R., Xie, C., et al.: PestNet: an end-to-end deep learning approach for large-scale multi-class pest detection and classification. IEEE Access **7**, 45301–45312 (2019)
13. Cheng, X., Zhang, Y., Chen, Y., et al.: Pest identification via deep residual learning in complex background. Comput. Electron. Agric. **141**, 351–356 (2017)
14. Wang, Q., Zheng, Y., Yang, G., et al.: Multiscale rotation-invariant convolutional neural networks for lung texture classification. IEEE J. Biomed. Health Inform. **22**(1), 184–195 (2018)
15. Patrick, M.K., dekoya, A.F., Mighty, A.A., Edward, B.Y.: Capsule networks – a survey J. King Saud Univ. Comput. Inf. Sci. (2019). https://doi.org/10.1016/j.jksuci.2019.09.014
16. Yujia, W., Li, J., Jia, W., et al.: Siamese capsule networks with global and local features for text classification. Neurocomputing **390**, 88–98 (2020)
17. Lorenzo, P.R., Tulczyjew, L., Marcinkiewicz, M., et al.: Hyperspectral band selection using attention-based convolutional neural networks. IEEE Access **8**, 42384–42403 (2020)
18. Chen, J., Wen, S., Wang, Z.: Crowd counting with crowd attention convolutional neural network. Neurocomputing **382**, 210–220 (2020)

19. Zhang, J., Wang, P., Gao, R.X.: Attention mechanism-incorporated deep learning for AM part quality prediction. Procedia CIRP **93**, 96–101 (2020)
20. Zhou, T., Canu, S., Su, R.: Automatic COVID CT segmentation using Ukmet integrated spatial and channel attention mechanism. Int. J. Imaging Syst. Technol. **31**(1), 16–27 (2021)

Small Object Recognition Based on the Generative Adversarial Network and Multi-instance Learning

Lin Zhiyong[✉]

Xiamen Road and Bridge Information Co. Ltd., Floor 18, Building A06,
Software Park Phase III, Xiamen 361000, Fujian, China

Abstract. In recent years, object recognition has experienced impressive progress. Despite these improvements, there is still a significant gap in the performance between the detection of small and large objects. We analyze that the limitation of existing algorithms for small target detection, such as: (1) the high computational overhead of image resolution increase and (2) the non-semantic data augmentation of small-object- copy-based strategy, leading a worse result in mAP. So, we figure out that the limited number of semantic training samples is a key impediment for this task due to the high cost of collecting and labelling nature images. In this paper, we propose a simple but effective framework for small object recognition. With an improved generative model, we propose a multiply instance learning detector based on CNN, which jointly learns from the labeled nature datasets and unlabeled generated images. Our method shows a state-of-the-art performance for small objects, obtained by Mask R-CNN, on MS COCO.

Keywords: Small object detection · Generative adversarial net · Multiply instance learning

1 Introduction

In recent years, artificial intelligence (AI) has developed very rapidly [1–4]. Substantial progress has been made in many applications, including speech recognition [3], image recognition [2], machine translation [5], autonomous driving [6–10], etc. In the military field, the development of AI is even more remarkable. In particular, the rapid development of deep learning (DL) has broken through many traditional methods in various fields. With the development of recognition technique in battlefield, the detection and resolution have become important directions for target, especially concealed and small target on the ground [30]. Meanwhile, it requires the device consuming as less power and holding as smaller size as possible. Under this background, applying a lightweight detector on intelligent mobile embedded devices will become the next hot spot of DL.

Aiming at the common problems of contactless human-computer interaction, underground object detection and multi-modal imaging signal real-time processing, this project conducts a real-time multi-object recognition with a lightweight deep neural

© Springer Nature Switzerland AG 2021
D.-S. Huang et al. (Eds.): ICIC 2021, LNCS 12836, pp. 475–483, 2021.
https://doi.org/10.1007/978-3-030-84522-3_39

network (DNN). Specifically, we mainly focus the problem of multitarget detection, especially small target detection. But, existing algorithms for small target detection has some limitations, for example: (1) the high computational overhead of image resolution increase [29] and (2) the non-semantic data augmentation of small-object-copy-based strategy [28].

So, we figure out that the limited number of nature training samples is a key impediment for this task due to the high cost of collecting and labelling images. Meanwhile, due to the high computational complexity and high occupation of memory and hard disk of DNN, we focus on the offline semantic data augmentation rather than stacking more convolutional layers, which can be embedded in mobile phones and devices.

In order to solve the difficulty of detecting small objects, and labeling samples in images taken from drones, we propose a new framework: small object recognition based on the generative adversarial network (GAN) [11–14] and Convolutional Neural Network (CNN), to solve the small-sample learning problems. Meanwhile, we explore the joint learning problem between GAN-based sample probability density estimation and virtual sample generation, breaking through the limitation of prior sample distribution in traditional GAN and slowly sampling of Markov chain [15, 16]. In addition, we propose a strongly generalized and practical CNN detector based on multiply instance learning (MIL) [20–23] for small object recognition.

In sum, our research is corresponding to the following three aspects: (1) se-mantic synthetic sample generation based on generative adversarial nets (GAN); (2) object recognition based on multiply instance learning; (3) acceleration of DNN based on hardware platforms. And our key contributions are listed as follows:

- We propose a new module which based on the GAN and MIL to solve the small-sample learning problems.
- We explore the joint learning problem between nature and synthetic images which is generated by GAN, breaking through the limitation of prior sample distribution in traditional GAN and sampling of slow Markov chain.
- We propose a strongly generalized, practical and lightweight model for small object recognition, which shows a state-of-the-art performance for small objects, obtained by Mask R-CNN [25, 26], on MS COCO.

2   Small Object Recognition Based on GAN and CNN

2.1   Framework

Images with small targets generated from GAN will solve the few-shot learning problem through joint learning between nature and synthetic samples, while small object recognition algorithms based on CNN will be explored.

The framework of our work is shown in Fig. 1, consists of two modules: a HIS data augmentation module based on GAN as well as a small robust object recognition module based on multiple instance learning (MIL). The first module outputs additional training samples by learning the generator's distribution over the datasets. The second module achieves robust small object recognition by using a CNN-based MIL network.

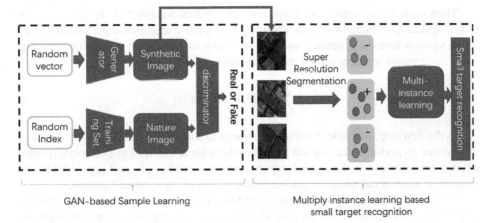

<table>
<tr><td>GAN-based Sample Learning</td><td>Multiply instance learning based small target recognition</td></tr>
</table>

Fig. 1. The framework of small object recognition based on GAN and CNN.

2.2 Data Augmentation Based on GAN

The limited number of training samples is a key impediment for small object recognition due to the high cost of collecting and labelling images. So, we propose to use data augmentation to improve the performance of system that have very few data. Specifically, this module learns the distribution over training datasets based on GAN, inputting real images and outputting synthetic images with diverse target distributions.

In our proposed adversarial nets, we train generator and discriminator alternatively. From a random noise z, the generative model tries to produce fake images $G(z; \theta_g)$. Meanwhile, a discriminative model that learns to determine whether a sample is from the real and synthetic sets, and outputs the confidence of classification $G(z; \theta_d)$. Ideally, the discriminator D estimates whether the input data is a real or fake image, generator G tries to fool the discriminator D as far as possible. Competition in this game drives them to improve performance. Based on the above description of the training process, we formulate the loss function as:

$$Loss = \frac{1}{m} \sum_{i=1}^{m} \left[log D(x^i) + log(1 - D(G(z^i))) \right] \qquad (1)$$

Where xi is the real image sampled from training set and zi is a random vector. The training process of GAN is equivalent to optimizing the loss function in a minimax way:

$$\min_{G} \max_{D} Loss \qquad (2)$$

In the offline training stage, to ensure the convergence of the target function, we don't use the traditional optimization method to update parameters θ_d and θ_q the detail ed training algorithm described as below:

Train the discriminator firstly: in the process of offline training, a set of random vectors $z = z_1, z_2, \ldots, z_m$ and images $x = x_1, x_2, \ldots, x_m$ are inputted, and then we fix θ_q to update θ_d by SGD. The random gradient information of generator is shown below:

$$\nabla_{\theta_d} \frac{1}{m} \sum_{i=1}^{m} \left[log D(x^i) + log(1 - D(G(z^i))) \right] \qquad (3)$$

Then train the generator: we input a set of random vectors $z = z_1, z_2, \ldots, z_m$ to GAN, update the generator parameters θ_q by fixing the discriminator parameters θ_d by SGD, to make sure the generated image is as similar as possible to the real images. The generator random gradient information is shown below:

$$\nabla_{\theta_d} \frac{1}{m} \sum_{i=1}^{m} log\left(1 - D\left(G\left(z^i\right)\right)\right) \tag{4}$$

In the first step, multiple training adjustment of the discriminator can theoretically ensure that its performance is good enough, otherwise it will lead to poor performance against the generator. In the second step, because the real image is not generated by the generator, we only calculate the gradient information of the synthetic images.

It is difficult for the optimization target functions to converge with each other. In response to this question, we propose the following key training points through our study:

1. First, we replace all pooling layers in discriminator D with the strided convolutions layer, and replace all pooling layers in generator G with the fractional strided convolutions layer;
2. Second, we apply the patch normalization layer in the discriminator and generator to normalize it;
3. Unlike traditional networks, there is no full connected layer in GAN;
4. Most of the generator network uses ReLU as the activation function, where the output layer uses tanh;
5. Finally, the Leaky ReLU is applied as all activation functions in the discriminator network.

The GAN's structure is shown in Fig. 2, which is similar to an encoder of traditional autoencoder networks. For example, if the input random vector's dimension is $100 \times 200 \times 1 \times 1$, which represent the batch size, channel number, height and width respectively. Then a 4×4 convolution is applied, resulting in a feature map in $100 \times 1024 \times 4 \times 4$. Then to enable the generator's output feature to cover image shape, the channel number becomes smaller with a higher image resolution, and eventually outputs 100 synthetic image of $3 \times 64 \times 64$.

2.3   Small Object Recognition Based on MIL

Traditional supervised learning uses examples with annotated information as training sets, but multiply instance learning uses bag as the basic unit instead of the images, which contains several instances. During training, the image area is marked as a negative bag when there is no small target, and as a positive bag when it contains at least one small target. Train multi-instance classifiers with hyper-pixel segmentation. Under supervise, the multi-instance classifier will learn to correctly label instances based on bags. The final classifier categorizes the instances in all bags into corresponding space.

The objective function of multi-instance learning is shown as below, which aims at both optimal markup and hyperplane segmentation. To compute the mixed integer

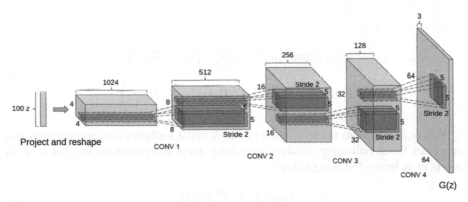

Fig. 2. The network structure of the generator.

planning problem with brute-force search algorithm, the computation complexity will grow exponentially. In the case of few training data, this problem can't be effectively solved by existing algorithms.

$$
\begin{aligned}
&mi - svm \quad \min_{y_i} \quad &&\min_{w,b,\xi} \tfrac{1}{2}\|w\|^2 + C\sum_i \xi_i \\
&s.t. \; \forall i : y_i f(x_i) \geq \quad &&1 - \xi_i, \xi_i \geq 0, y_i \in -1, 1 \\
&\sum_{i \in I} \tfrac{y_i + 1}{2} \geq 1, \forall I \; s.t. \; Y_I = 1; \; y_i = -1, \forall I \; s.t. \; Y_I = -1
\end{aligned}
\tag{5}
$$

Our method requires few training data in small. Get n images $I_1, I_2 \cdots I_n$, that represent the sift features corresponding to the p pixels in image I_i. The point marker information of image I_i is represented by $p_i = P_1, P_2 \cdots P_{C(i)}$, where $C(i)$ is the number of small targets in Image I_i. Based on this, the density values of the image area can be calculated as below:

$$
D_i(p) = \sum_{p \in p_i} N\left(p; P, \sigma^2 1_{2x2}\right), \forall p \in I_i
\tag{6}
$$

where p is the pixel in the image, $N\left(p; P, \sigma^{21}_{2x2}\right)$ is the density value of the p-pixels in the two-dimensional normal distribution.

Linear regression from image pixel characteristics to density values is learned. For a new inputting image, the pixel feature is regressed to a density values to achieve a small target estimate, the density estimation is calculated as below:

$$
D_i'(p) = w^T x_i^p, \forall p \in I_i
\tag{7}
$$

where w is the weight vector of the linear transformation.

The optimization objective function minimizes the distance between the true density value and the estimated density value. During the offline training optimization process, we update the weight vector of the area with the largest error first through the 2D max subarray algorithm:

$$w^* = \underset{w}{argmin}\|w\|^2 + \lambda \sum_{i=1}^{N} F\left(D_i(\cdot), D'_i(\cdot)\right)$$

$$F(D_1, D_2) = \underset{B \in I_i}{max}\left(\sum_{p \in B} D_1(p) - \sum_{p \in B} D_2(p)\right)$$

(8)

Where B represents the area in the image I_i and λ balances the regularization and error item. When estimating density, an inputting image can predict the number of small targets in an image following below:

$$Density_i = \sum_{p \in I_i} D'_i(p)$$

(9)

Combining multi-instance CNN with density estimation methods, our algorithm process listed as follows.

Algorithm 1 Framework of recognition based on multiply instance learning.

Input:
    The image set $I = \{i_1, i_2, \cdots, i_n\}$;
    The predict picture x;
Output:
    The sort window confidence score;
    The output top-m ranked confidence windows
1: According to the optimization target function of Eq.8, calculate the weight parameter w;
2: Initialize each instance label within the j bag $y_i = L_j$, i.e. initialize the instance label with the bag label.
3: **repeat**
4:    Train CNN based on a given example;
5:    Calculate the confidence of each instance in the positive bag $f(i)$;
6:    According to the instance label within the positive bag $y_i = sgn(f(i))$;
7:    **for** each positive package B_i **do**
8:        **if** $i \in I$ and $(1 + y_i)/2 == 0$ **then**
9:            $i^* = \underset{i \in I}{argmax} f(i)$;
10:        $y_{i^*} = 1$;
11:      **end if**
12:    **end for**
13: **until** The instance label does not change;
14: Use sliding window to calculate small target confidence $f(i)$ of image x.

---

3 Experimental Results and Analysis

Dataset. Our experiments are conducted on the challenging MS COCO [27] benchmark using the standard metrics for object detection. All models are trained on the COCO train2017 split (118k images) and evaluated with val2017 (5k images).

Training Details. ResNet-50 [2] is used as the backbone unless otherwise specified. The optimizer is SGD with weight decay 0.0001. The mini-batch is 16 images and all models are trained with 8 GPUs. Default training schedule is 136 epochs and the initial learning rate is set to 2.5×10^{-5}, divided by 10 at epoch 27 and 33, respectively. The backbone is initialized with the pre-trained weights in ImageNet [1].

Augmentation. Our method aims to generate semantic image containing multiply small objects. In above analyze we point out that the data augmentation with copy-paste strategy is out of semantic and will lead a context confuse for model. For example, birds fly in the air, but not fly on the ground or under- ground. Based on this, we propose a GAN-based small target augmentation to generate semantic and realistic image, where small target has a reasonable distribution.

Table 1 presents the Detection results of our method on MS COCO. We compare our method with the Mask R-CNN and small object augmentation with copy-paste strategy. Our method can be used with any detectors. The performance of ouR method surpasses the state-of-art copy-past-based method by 2%. And other data augmentation such as random erasing, Mixup, Augmix, etc., can further improve the performance.

Table 1. Augmentation experiments. The best performance, in terms of both small objects and overall, is achieved when the original images with small objects and generated images for training.

	Detection AP			
	small	medium	large	all
baseline	0.167	0.329	0.393	0.303
aug	0.161	0.328	0.4	0.302
original + aug	0.179	0.329	0.386	0.304
ours	0.186	0.332	0.41	0.317
original + ours	**0.201**	**0.341**	**0.416**	**0.325**

4   Conclusion

We propose a simple but effective framework for small object recognition. With an improved generative model, breaking through the preset defects and limited Markov chain sampling in vanilla model, we propose a MIL-based CNN detector, which jointly learns from the labeled nature datasets and unlabeled generated images. Our methods showed state-of-the-art performance for small objects, obtained by Mask R-CNN, on MS COCO.

Acknowledgments. This work is supported by Xiamen Major Science and Technology Projects (No. 3502Z20201017).

References

1. Krizhevsky, A, Sutskever, I., Hinton, G.E.: ImageNet classification with deep convolutional neural networks. In: Advances in Neural Information Processing Systems, pp. 1097–1105 (2012)
2. He, K., Zhang, X., Ren, S., et al.: Deep residual learning for image recognition. In: Proceedings of the IEEE Conference on Computer Vision and Pattern Recognition, pp. 770–778 (2016)
3. Deng, L., Hinton, G., Kingsbury, B.: New types of deep neural network learning for speech recognition and related applications: an overview. In: IEEE International Conference on Acoustics, Speech and Signal Processing, pp. 8599–8603 (2013)
4. LeCun, Y., Bengio, Y., Hinton, G.: Deep learning. Nature **521**(7553), 436–444 (2015)
5. Vaswani, A., et al.: Tensor2Tensor for neural machine translation. arXiv preprint arXiv:1803. 07416 (2018)
6. Caesar, H., Bankiti, V., Lang, A.H., et al.: nuScenes: a multimodal dataset for autonomous driving. In: Proceedings of the IEEE/CVF Conference on Computer Vision and Pattern Recognition, pp. 11621–11631 (2020)
7. Agarwal, S., Awan, A., Roth, D.: Learning to detect objects in images via a sparse, part-based representation. IEEE Trans. Pattern Anal. Mach. Intell. **26**(11), 1475–1490 (2004)
8. Weber, M., Welling, M., Perona, P.: Unsupervised learning of models for recognition. In: Vernon, D. (ed.) ECCV 2000. LNCS, vol. 1842, pp. 18–32. Springer, Heidelberg (2000). https://doi.org/10.1007/3-540-45054-8_2
9. Fu, K., Gong, C., Gu, I.Y.H., et al.: Spectral salient object detection. In: IEEE International Conference on Multimedia and Expo, pp. 1–6 (2014)
10. He, K., Zhang, X., Ren, S., et al.: Spatial pyramid pooling in deep convolutional networks for visual recognition. IEEE Trans. Pattern Anal. Mach. Intell. **37**(9), 1904–1916 (2015)
11. Girshick, R.: Fast R-CNN. In: Proceedings of the IEEE International Conference on Computer Vision, pp. 1440–1448 (2015)
12. Uijlings, J.R.R., Van De Sande, K.E.A., Gevers, T., et al.: Selective search for object recognition. Int. J. Comput. Vis. **104**(2), 154–171 (2013)
13. Cheng, M.M., Zhang, Z., Lin, W.Y., et al.: BING: binarized normed gradients for objectness estimation at 300fps. In: Proceedings of the IEEE Conference on Computer Vision and Pattern Recognition, pp. 3286–3293 (2014)
14. Goodfellow, I., Pouget-Abadie, J., Mirza, M., et al.: Generative adversarial nets. In: Advances in Neural Information Processing Systems, pp. 2672–2680 (2014)
15. Ledig, C., Theis, L., Huszar, F., et al.: Photo-realistic single image super-resolution using a generative adversarial network. arXiv preprint arXiv:1609.04802 (2016)
16. Zhu, J.-Y., Krähenbühl, P., Shechtman, E., Efros, A.A.: Generative visual manipulation on the natural image manifold. In: Leibe, B., Matas, J., Sebe, N., Welling, M. (eds.) ECCV 2016. LNCS, vol. 9909, pp. 597–613. Springer, Cham (2016). https://doi.org/10.1007/978-3-319-46454-1_36
17. Brock, A., Lim, T., Ritchie, J.M., et al.: Neural photo editing with introspective adversarial networks. arXiv preprint arXiv:1609.07093 (2016)
18. Isola, P., Zhu, J.Y., Zhou, T., et al.: Image-to-image translation with conditional adversarial networks. arXiv preprint arXiv:1611.07004 (2016)
19. Li, J., Monroe, W., Shi, T., et al.: Adversarial learning for neural dialogue generation. arXiv preprint arXiv:1701.06547 (2017)
20. Reed, S., Akata, Z., Yan, X., et al.: Generative adversarial text to image synthesis. In: Proceedings of the 33rd International Conference on Machine Learning (2016)
21. Ho, J., Ermon, S.: Generative adversarial imitation learning. In: Advances in Neural Information Processing Systems, pp. 4565–4573 (2016)

22. Mirza, M., Osindero, S.: Conditional generative adversarial nets. arXiv preprint arXiv:1411. 1784 (2014)
23. Maron, O., Lozano-Pérez, T.: A framework for multiple-instance learning. In: Advances in Neural Information Processing Systems, pp. 570–576 (1998)
24. Babenko, B., Yang, M H., Belongie, S.: Robust object tracking with online multiple instance learning. IEEE Trans. Pattern Anal. Mach. Intell. **33**(8), 1619–1632 (2010)
25. Wu, J., Yu, Y., Huang, C., et al.: Deep multiple instance learning for image classification and auto-annotation. In: Proceedings of the IEEE Conference on Computer Vision and Pattern Recognition, pp. 3460–3469 (2015)
26. He, K., Gkioxari, G., Dollár, P., et al.: Mask R-CNN. In: Proceedings of the IEEE International Conference on Computer Vision, pp. 2961–2969 (2017)
27. Lin, T.-Y., et al.: Microsoft COCO: common objects in context. In: Fleet, D., Pajdla, T., Schiele, B., Tuytelaars, T. (eds.) ECCV 2014. LNCS, vol. 8693, pp. 740–755. Springer, Cham (2014). https://doi.org/10.1007/978-3-319-10602-1_48

Deep Learning Based Semantic Page Segmentation of Document Images in Chinese and English

Yajun Zou and Jinwen Ma[✉]

Department of Information and Computational Sciences, School of Mathematical Sciences and LMAM, Peking University, Beijing 100871, China
jwma@math.pku.edu.cn

Abstract. Semantic page segmentation of document images is a basic task for document layout analysis which is key to document reconstruction and digitalization. Previous work usually considers only a few semantic types in a page (e.g., text and non-text) and performs mainly on English document images and it is still challenging to make the finer semantic segmentation on Chinese and English document pages. In this paper, we propose a deep learning based method for semantic page segmentation in Chinese and English documents such that a document page can be decomposed into regions of four semantic types such as text, table, figure and formula. Specifically, a deep semantic segmentation neural network is designed to achieve the pixel-wise segmentation where each pixel of an input document page image is labeled as background or one of the four categories above. Then we can obtain the accurate locations of regions in different types by implementing the Connected Component Analysis algorithm on the prediction mask. Moreover, a Non-Intersecting Region Segmentation Algorithm is further designed to generate a series of regions which do not overlap each other, and thus improve the segmentation results and avoid possible location conflicts in the page reconstruction. For the training of the neural network, we manually annotate a dataset whose documents are from Chinese and English language sources and contain various layouts. The experimental results on our collected dataset demonstrate the superiority of our proposed method over the other existing methods. In addition, we utilize transfer learning on public POD dataset and obtain the promising results in comparison with the state-of-the-art methods.

Keywords: Semantic page segmentation · Document layout analysis · Document reconstruction · Deep learning

1 Introduction

With the rapid development of the Internet and digital equipment, there are a huge number of text documents in electronic version (e.g. camera captured page images) generated every day. So, document reconstruction and digitalization become particularly important. Actually, we need to convert those document pages into editable and searchable forms so that they can be further utilized in information extraction and retrieval.

© Springer Nature Switzerland AG 2021
D.-S. Huang et al. (Eds.): ICIC 2021, LNCS 12836, pp. 484–498, 2021.
https://doi.org/10.1007/978-3-030-84522-3_40

For the reconstruction and digitalization of an input text document image, it is effective to firstly segment the regions in different semantic types and then recognize the contents of the segmented regions by the type-related recognition systems. For example, text can be recognized by the OCR system. As a result, the page is reconstructed by assembling the recognized contents of the regions according to their location. Therefore, page segmentation is a crucial step in the document reconstruction workflow. Generally, page segmentation aims at segmenting a page into a set of homogenous regions which can be categorized into several semantic types, like tables and figures. As well known, there are text documents with various styles and layouts. For instance, a document page can be single-column or multi-column. And documents with different languages may contain different texture features with respect to the text types. In addition, there is a high similarity between different semantic types, e.g. table and figure. The grid chart has the same structure of intersecting horizontal and vertical ruling lines as the table. Moreover, regions of a specific type vary greatly in aspect ratios and scales among them. Therefore, it is rather challenging to make the page segmentation in multi-language document images effective and robust.

Most of the conventional document segmentation methods [1–3] consist of unsupervised segmentation and supervised classification. They usually make an assumption on page layouts and segment a page into a number of regions by certain heuristic rules for multiple cases. Then for region classification, they extract a group of hand-craft features and then employ machine learning algorithms to classify a segmented region into different types. In this way, they have high experience dependency that can't fit in diverse documents. Nevertheless, some of the deep learning based methods [3] adopt an end-to-end trainable convolutional network to automatically extract features for the better robustness. Besides, some of the deep learning based methods [4, 5] formulate this problem as a typical object detection in natural images. These methods take the document image as an input and then output the bounding boxes of objects with corresponding labels. Moreover, there are some methods based on deep semantic segmentation network where each pixel is classified into one semantic type [6–9]. The pixel level understanding is more precise than the bounding box level one. However, they usually consider only a few semantic types. For instance, most of them only distinguish text from non-text in a page or assume that no formula regions exist in a document. This is not sufficient for document reconstruction. Moreover, their experiments are typically performed on English documents.

In this paper, we propose a deep learning based method to achieve better semantic page segmentation. For the goal of document reconstruction, four semantic types are taken into consideration, i.e. text, table, figure, formula. For an input text document image, we firstly use a semantic segmentation neural network to classify each pixel as either background or one of the four categories above. Our network leverages context features and local features of a document image to get more precise segmentation results. Then, by implementing the Connected Component Analysis (CCA) algorithm on the prediction mask, we obtain the accurate locations of regions in different types. We further develop a simple Non-Intersecting Region Segmentation Algorithm (NIRSA) to improve the segmentation result and facilitate the future page regeneration task. Furthermore, to address the issue of lacking annotated training data, we manually annotate a dataset consisting of Chinese and English documents that contain various styles and layouts. And we perform transfer leaning and domain adaption during our training procedures. Finally, we conduct the experiments on our collected dataset and public POD (Page Object Detection) dataset to demonstrate the effectiveness of our proposed method.

The rest of the paper is organized as follows. We firstly review the related work in Sect. 2. Our proposed method is then presented in Sect. 3. In Sect. 4, we summarize the experiment results and comparisons on a collected dataset and several public datasets. We finally make a brief conclusion in the last section.

2 Related Work

In recent years, there have been many methods for semantic page segmentation in document images. Most of the conventional methods [1–3] have two stages, i.e., unsupervised segmentation and supervised classification. The unsupervised segmentation stage is usually based on bottom-up or top-down structure. The bottom-up structure [1] starts to segment characters or lines and gradually groups them into homogenous regions. While the top-down structure [2] operates directly on the entire document and recursively segments the resulting regions. The greatest shortcoming of these methods is to decide a large amount of parameters by experience, which leads to poor robustness. During the classification procedure, hand-craft features of the segmented regions are firstly extracted and then fed into a classifier to determine the semantic labels.

Nowadays, the CNN based networks [3] are utilized to complete automatic feature extraction with better generalization ability. Currently, some methods formulate page segmentation as a typical object detection problem. They usually focus on a specific type, e.g. table region segmentation. DeepDeSRT [10] model adjusts the convolution kernel of the backbone in Faster R-CNN to detect the table regions. Prasad et al. [11] propose the Cascade Mask Region-based CNN High-Resolution Network that solves both problems of table detection and structure recognition simultaneously.

In addition, both PubLayNet [5] and GOD [4] use Faster R-CNN [12] and Mask R-CNN [13] to detect regions in different types. PubLayNet [5] considers five semantic types that can be applied to most documents. But it's not suitable for some statistical reports because formula type is excluded. GOD [4] only detects regions of three semantic types, leaving the text type out. However, text is the most common semantic type in documents. Cross-domain DOD model [14] is built on top of the Feature Pyramid Networks [15], which mainly addresses the domain shift problem that arises in the absence of labeled data.

There are also some methods based on semantic segmentation models. Yang et al. [9] first introduce semantic segmentation to page segmentation. But an additional tool is adopted to specify the segmentation boundary. Lee et al. [7] propose trainable multiplication layers (TMLs) and incorporate them into U-Net architecture [16] to gain better performance. But they only perform binary segmentation that only pixels in text type are identified. And they only complete pixel-wise segmentation task. DeepLayout [8] doesn't distinguish text type from background. They choose the DeepLab v2 structure [17] to segment these pixels that belong to table, figure and formula types. As a result, text regions can't be segmented during the subsequent post-processing procedure. He et al. [6] train FCN [18] to segment three types of document elements: text, table, and figure. They use multi-scale training strategy to capture multi-scale information. Also, they add a contour detection branch to improve the results of semantic segmentation. However, they only segment table regions by an additional verification net without results of regions in other semantic types.

3   Methodology

As is shown in Fig. 1, our proposed method mainly consists of two parts. A semantic segmentation network begins to classify each pixel to a certain type. Then a series of regions that do not overlap each other are generated through Connected Component Analysis and Non-Intersecting Region Segmentation Algorithm. We introduce these two main steps.

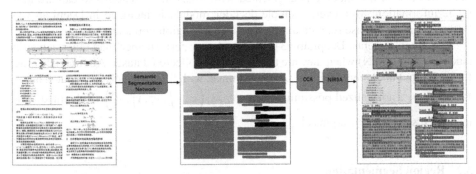

Fig. 1. Illustration of our method. The results of both pixel-wise and region-wise segmentation are shown in the rightmost image, where each region is represented by a bounding box, corresponding label and confidence score (white: background, red: text, green: table, blue: figure, black: formula) (Color figure online).

3.1   Semantic Segmentation Network

In our framework, a deep semantic segmentation network is firstly utilized to assign a semantic label to each pixel. There are five categories including the background label. As shown by the observations in [6], unlike general semantic segmentation in natural images, a large receptive field is required in the semantic segmentation network for document images to guarantee sufficient context information. For example, the text block in a table can't be recognized as part of the table without a large context. However, there is an inherent conflict between context information and spatial information in the segmentation network. The acquisition of a large context weakens the details for region boundary prediction. So we alleviate this problem by aggregating multi-scale information as in [6]. An image pyramid model based on FCN is adopted in [6], where several images with different scales are all taken as input. Since it's obvious that the image pyramid model is time-consuming, we adopt several improved networks based on Skip Connection or Atrous Spatial Pyramid Pooling (ASPP) to achieve multi-scale information fusion, e.g. U-Net [16], FPN [15], DeepLab series networks [17, 19, 20].

U-Net [16] is based on a typical encoder-decoder structure. The skip connections between low layers in encoder phase and high layers in decoder phase promote the fusion of low-level and high-level features. The low-level features contain abundant spatial information while the context information is included in the high-level features. In fact, features in different layers can be regarded as the corresponding features at different scales. FPN [15] shares the similar core idea, but the difference is that the prediction layer is added to every feature map during decoder process so as to enhance the supervision information at different scales. Instead of typical convolution, atrous convolution is used to enlarge receptive field and attain spatial information at the same time, which doesn't increase the number of parameters. DeepLab series networks use the Atrous Spatial Pyramid Pooling (ASPP) module to capture multi-scale information by concatenating these feature maps output from atrous convolution layers with different rates. Besides, ASPP module is augmented with image-level features to capture long range information. Moreover, a simple decoder module is included in DeepLab v3+ [20] to get more precise segmentation especially for region boundary.

In our experiments, DeepLab v3+ achieves better performance than other networks (such as U-Net, FPN) when they are trained on our collected dataset. At inference time, for each input document image, a prediction mask with five channels is output. That is, for pixel p_j, a normalized possibility vector $v_j = \left(v_j^0, v_j^1, v_j^2, v_j^3, v_j^4 \right)$ is obtained. And its label l_j satisfies $l_j = \underset{k}{\mathrm{argmax}} \left\{ v_j^k \right\}$.

3.2   Region Segmentation

CCA. To restore the definite region in different types, we extract the connected components of each category from the prediction mask respectively. Then each connected component with its corresponding label is regarded as a candidate region. And we take the rectangular bounding boxes of connected components to specify the boundary of regions. For the bounding box b_i with label c_i, its confidence score s_i is defined as follows.

$$s_i = \frac{1}{N_i} \sum\nolimits_{p_j \in b_i} v_j^{c_i} \tag{1}$$

Here, N_i is the number of pixels in b_i.

NIRSA. As is shown in Fig. 3, there are some intersecting bounding boxes after CCA due to the error from semantic segmentation (unclear boundary). For image (a), text and figure regions are confined into one box. For image (c), two text boxes overlap each other and one of them contains incomplete word. Thus, a Non-Intersecting Region Segmentation Algorithm is proposed to obtain more precise page decomposition results. And it can also eliminate the position conflict that may appear in the document reconstruction workflow.

Our proposed algorithm is similar to the non-maximum suppression algorithm (NMS) in object detection task. Firstly, we sort the candidate bounding boxes by their corresponding confidence sore. The pipeline of our proposed algorithm is to generate the bounding box one by one on an empty page. As is shown in Fig. 2, b_1 that has the highest confidence score is first generated on the page. At the same time, we mark all pixels of its corresponding region on the page with a non-empty flag (blue fill). Each pixel of the page is marked with an empty flag at the beginning. Then for the next selected box, we consider three possible cases:

1. If the non-empty pixels' portion of its corresponding region on the page is below a certain percentage, we drop the box directly (b_2)
2. If pixels of its corresponding region on the page are all marked empty flag, we generate the box directly on the page (b_3).
3. Otherwise, we use several small boxes to approximate the empty part of its corresponding region (b_4).

We accomplish the case 3 by exploiting the local information. There are two operations performed: splitting and merging. The foreground area of the empty part is firstly identified by simple threshold method. Next, for each row, we perform horizontal run length smoothing algorithm (RLSA). Thus the empty area is split into a series of connected components (black lines inside b_4) with a height of 1. It should be noticed that for text region, we add an additional vertical RLSA to extract text lines. This prevent character from being destroyed. It's obvious that these connected components are non-intersecting. Then we scan the connected components from top to bottom and left to right and merge them into several boxes. Suppose we currently have two merged boxes mb_1, mb_2, for the connected component cc_k, there is a new box generated because it can't be merged to mb_1 or mb_2. In Fig. 3, we show the effectiveness of our proposed NIRSA.

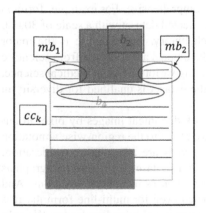

Algorithm 1 Non-Intersecting Region Segmentation Algorithm (NIRSA)

Input:
1. Bounding box set $B = \{b_1, b_2, \cdots, b_n\}$ where $s_1 > s_2 > \cdots > s_n$;
2. Page $P[:]$;
3. Pixel p;
4. Threshold th;

Output:
1. Non-intersecting bounding box set O;

1: Initialize $P[:] = 0, O = \{\}$;
2: $\forall p \in b_1$, set $P[p] = 1$;
3: delete b_1 from set B;
4: for b_i in B do
5:     if $\forall p \in b_i, P[p] = 0$ then
6:         add b_i into set O;
7:     else
8:         $emp = \frac{NUM_{(P(p)=0, p \in b_i)}}{NUM_{(p \in b_i)}}$
9:         if $emp < th$ then
10:            drop b_i;
11:        else
12:            split the empty part into multiple connected components $\{cc_k\}$;
13:            merge the connected components to several boxes $\{mb_j\}$;
14:            for each mb_j do
15:                add mb_j into O;
16:                $\forall p \in mb_j$, set $P[p] = 1$;

Fig. 2. The description of Non-Intersecting Region Segmentation Algorithm procedure.

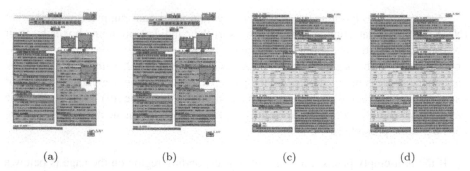

(a)   (b)   (c)   (d)

Fig. 3. Examples of the results of region segmentation before and after NIRSA. (a) and (c) represent the results after CCA. (b) and (d) represent the results after CCA and NIRSA.

4 Experiments

4.1 Datasets

POD [21]. This is a competition dataset consisting of 2,000 English document images, about 800 of which are used for testing and the rest of which are for training. These document images are extracted from scientific papers that have simple white background and vary in layout styles. There are regions with three semantic types: table, figure, formula. And Each region is presented as a rectangular bounding box and a corresponding label. Meanwhile, each subfigure is labelled as a separate region. Each formula line of multi-line formula is labelled in the similar way. In our experiments, we use the POD extend training dataset as in [8], which contains about 10,000 training images and all text line regions are appended in the ground truth.

Collected Dataset. To our knowledge, current dataset contains only document images in English, and most documents are from a certain range of fields thus are weak in diversity. Moreover, the semantic types are not comprehensive. For example, formula type is usually absent. Therefore, we annotate a collected dataset with a scale of 30,000 document images which are from a large search library. And Chinese is the major language of these documents. In addition, these documents are collected from scientific papers, magazines and statistical yearbooks, involving various fields of medical science, literature, education, natural science, etc. Thus the dataset is qualified for diversity in document layouts and contents.

As for the ground truth format, we should mark document images by pixel in our framework. However, it's not cost-effective. So we resort to the region-wise annotation in the same way as POD dataset. Four semantic types are taken into consideration: text, figure, table and formula. Different from the POD dataset, we regard paragraph as a region unit for text type. And the titles are also taken as independent regions. And for formula type, an entire formula region is labeled even for multi-line formula. It is worth noting that there is no overlap between the annotated bounding boxes. We split the intersecting area and label it with several rectangular boxes when encountering the inevitable intersection of rectangular bounding boxes, e.g. when a picture is surrounded by text. In this way, the ground truth mask for pixel-wise segmentation can be acquired, i.e. the ground truth label of pixels in the bounding boxes are set to be the same as the label of the bounding box. Examples in our collected dataset are shown in Fig. 5.

4.2 Metrics

We use the IOU metric to evaluate the segmentation results that output from semantic segmentation networks. Based on the confusion matrix of pixel classification, the IOU metric computes the intersection over union on each category. And the mean IOU (mIOU) over all classes is also calculated to measure the overall performance.

While for region segmentation task, we refer to the POD competition evaluation tool [21] where F1 score and average precision (AP) metric are both adopted. Specifically, given a IoU threshold α, a segmented region b_i is regarded as a true positive if it satisfies:

$$\text{IoU}(b_i, gt_j) = \frac{b_i \cap gt_j}{b_i \cup gt_j} > \alpha \tag{2}$$

Here, gt_j is the corresponding ground truth region. Then we can compute the precision and recall over regions. F1 score and AP are both comprehensive evaluation metric based on precision and recall. mAP and Avg. F1 calculate the mean AP and F1 over all semantic types, respectively.

In addition, we propose a page-wise evaluation method which is inspired by [22]. In order to facilitate future recognition, the segmented regions must be complete and pure. Therefore, a segmented box b_i is allowable when it contains a complete ground truth region and does not have overlap with the remaining ground truth regions. On the one hand, all b_i should be allowable. On the other hand, all gt_j should be segmented. In this case, the page segmentation is regarded as exactly correct. We count the percentage of the exact segmentation. Besides, similar to [22], we add an additional allowable case (Merging Text) when a segmented text box b_i contains multiple complete text ground truth boxes. That means multiple paragraph can be merged into one bounding box.

4.3 Experimental Results on Our Collected Dataset

We randomly select 1,000 images for testing, and the rest images are for training (about 29,000 images). We take DeepLab v3+ as our semantic segmentation network. During the training procedure, the input image is firstly random rescaled within a scale range from 0.5 to 1.5. Then an image patch with a size of 768×768 is randomly cropped from the rescaled images. The batch size is set to 4. Besides, we take the Stochastic Gradient Descent algorithm as our optimizer. And the initial learning rate is set to 0.001, which is decreased by a factor of 0.5 after 10 epochs. Moreover, we use transfer learning to accelerate the training process and enhance performance, especially in areas where there is a lack of adequate annotated data. So each network is pre-trained on ImageNet dataset and fine-tuned on our collected training set. At the inference time, we pad each input image to confirm that its width and height are both divisible by 32. The padding image is then fed into the trained network and further get the final region-wise segmentation results.

Comparison with other Methods. As we mentioned in Sect. 2, current related state-of-the-art methods [4, 5] are mainly based on Faster R-CNN and Mask R-CNN. As for parameter setting, we follow the values in [4] which achieves the best results in POD up to now. Our code inherits from Detectron2 as with [4, 5]. As with the official POD evaluation tool, the IoU threshold is set to 0.8. The region segmentation results are shown in Table 1. Among them, our results are bolded. "Ours" represents the results after semantic segmentation and CCA while "Ours + NIRSA" represents the results that are performed by CCA and NIRSA. Both of them achieve significant improvement over other methods, especially in text and formula category. "Ours + NIRSA" boosts the average F1 for 4.4% and mAP for 2.4% compared with the Mask R-CNN. This is probably because the aspect ratios of regions in these two categories vary a lot and usually extreme, which increase the difficulty of the object detection task. Among all categories, the overall performance on table category is the best while the segmentation of formula category is the hardest. This may be caused by the class-imbalance problem on the document page. For most pages, text blocks take the main part while formula blocks usually only appear in specific documents. In addition, we also label Chinese formulas which is composed of operators and Chinese characters. This makes it difficult to distinguish text from formula.

Table 1. The results of region segmentation on our collected dataset. The last two rows represent our results.

Methods	F1 score					AP				
	Text	Table	Figure	Formula	Avg.F1	Text	Table	Figure	Formula	mAP
Faster R-CNN	0.760	0.981	0.830	0.575	0.786	0.701	0.982	0.860	0.449	0.748
Mask R-CNN	0.774	0.975	0.828	0.624	0.800	0.718	0.980	0.860	0.531	0.771
Ours	0.837	0.991	0.823	0.692	0.836	0.796	0.996	0.781	0.629	**0.801**
Ours + NIRSA	0.841	0.997	0.825	0.714	**0.844**	0.795	0.996	0.788	0.602	0.795

The page-wise segmentation results are shown in Table 2. When merging multiple paragraphs into one region is allowable, about 658 pages out of 1000 pages are correctly segmented with the "Ours + NIRSA" method. It's meaningful to realize the automatic semantic segmentation of pages. Through analysis, semantic segmentation network has richer detail information than Faster R-CNN and Mask R-CNN. So we can restore more precise location information with our method, as Fig. 4 shows. But Faster R-CNN and Mask R-CNN are better at distinguishing different instances. This is also indicated by the differences between the results with "Merging Text" and without "Merging Text". For text regions, our method is likely to merge two paragraph into one region. When "Merging Text" is allowable, the percentage of exact segmentation increased by 30%.

Table 2. The percentage of exact segmentation (page-wise) on our collected dataset.

Methods	Merging text	
	✓	✗
Faster R-CNN	0.132	0.151
Mask R-CNN	0.154	0.172
Ours	0.340	0.602
Ours + NIRSA	**0.358**	**0.653**

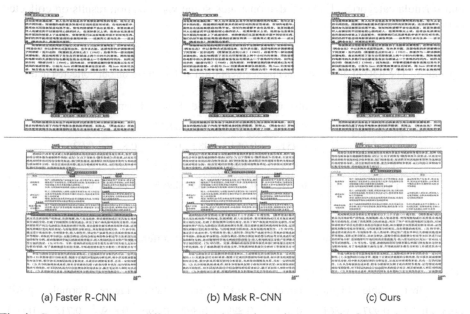

(a) Faster R-CNN   (b) Mask R-CNN   (c) Ours

Fig. 4. Comparison between different methods. Each row is an example. Our results are showed in the rightmost. First line: incomplete text region. Second line: split table region.

Moreover, our proposed NIRSA also boosts the performance. It can make up for the error of semantic segmentation network and improve the whole page segmentation.

Discussions. We adopt some mainstream semantic segmentation networks and compare their performance on our task. As is shown in Table 3, all of them can produce promising results. Based on the mean IOU metric, DeepLab v3+ achieves the best segmentation results while FPN is closely behind. DeepLab v3+ makes a significant improvement on the formula category in comparison with other architectures. Here, we contain the DeepLab v3 architecture in our contrast experiments, which lacks the decoder structure compared with the DeepLab v3+ architecture. It can be observed that the decoder structure is effective because it can raise the segmentation results especially for formula and text categories. Besides, atrous convolution in DeepLab series can increase the receptive field and retain the spatial information simultaneously, i.e., it doesn't reduce the size of the output feature map. Thus it leads to a high time consumption. To make a balance between time complexity and accuracy, we set the output stride of DeepLab v3+ to 16. The segmentation of formula categories is harder than other categories as the IOU of formula type is lower by a great margin. In fact, we try to tackle the class-imbalance problem by adopting some effective loss functions, e.g. focal loss. But it doesn't bring a significant improvement. It requires further investigation. Moreover, we explore the impact of IoU threshold on our results of the region segmentation as [4] does. A high IoU threshold requests that a true positive should have high overlap with the ground truth. As Table 4 shows, with the increase of IoU threshold, the performance of our method does not decrease sharply. In fact, our method on the table category is robust in contrast to text and formula categories. Figure 5 demonstrates some visualization results on collected dataset with our method.

Table 3. The IOU of different semantic segmentation architectures on our collected dataset.

Networks	Background	Text	Table	Figure	Formula	mIOU
U-Net	0.936	0.949	0.964	0.874	0.780	0.901
FPN	0.951	0.961	0.983	0.901	0.823	0.924
DeepLab v3	0.939	0.948	0.978	0.899	0.801	0.913
DeepLab v3+	**0.951**	**0.961**	**0.983**	**0.905**	**0.846**	**0.929**

Table 4. The results of region segmentation with different IoU threshold.

IoU	F1 score					AP				
	Text	Table	Figure	Formula	Avg.F1	Text	Table	Figure	Formula	mAP
0.5	0.926	0.997	0.861	0.811	0.899	0.892	0.996	0.839	0.715	0.860
0.6	0.913	0.997	0.847	0.789	0.887	0.876	0.996	0.811	0.692	0.844
0.7	0.894	0.997	0.837	0.760	0.872	0.854	0.996	0.801	0.659	0.827
0.8	0.841	0.997	0.825	0.714	0.844	0.795	0.996	0.788	0.602	0.795

4.4 Experimental Results on Public POD Dataset

We test the performance on POD competition dataset by adopting different training strategies to illustrate the effect of transfer learning. In Table 5, "POD" represents the results that training on POD dataset. "Collected" represents the results that training on our collected dataset. "Collected + POD" represents the results that pre-training on our collected dataset and fine-tuning on POD dataset, which achieves the best performance. The mean IOU has increased by up to 5% than the model without collected dataset

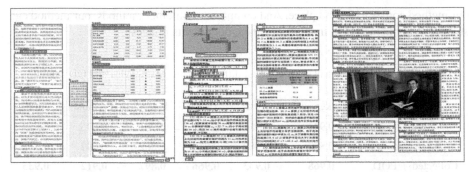

Fig. 5. Three examples of the results of our method on the collected dataset.

pre-trained. This proves that effectiveness of transfer learning. In addition, "Collected" is worse than "POD". Especially for text and formula type, there is a huge decline. This demonstrates the difference between POD data and our collected data. As we introduced above, our collected dataset has a different ground truth format from the POD dataset in these two semantic types. So domain adaption is necessary to enhance the performance.

Furthermore, in Table 6, we show the comparison between our method and other state-of-the-art methods for region segmentation. The first three lines are the results in article papers. The middle four lines are the top results of competition participants. As for formula category, our method achieves the best AP value. However, we have a poor performance at figure category compared to the top results. As Fig. 6 shows, our method tends to split a figure into several figures (b) or merge several subfigures (c). And the segmented regions in table and figure categories by our method usually contain the corresponding caption parts (a). So there is still a certain gap between our result and the top results. However, these methods are all concentrated on the segmentation of table, figure and formula regions. It's not sufficient for some real applications like page construction. The segmentation of text regions is ignored in other methods while our method can additionally achieve 0.931 F1 score and 0.911 AP for text category. Overall,

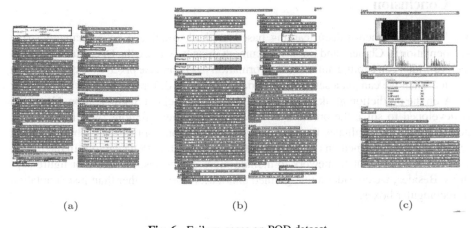

(a)   (b)   (c)

Fig. 6. Failure cases on POD dataset.

our method comes to a good place. And compared with other methods, our method can still get a promising result in multi-language documents images.

Table 5. The pixel segmentation results of different training strategies on POD dataset.

Training	Background	Text	Table	Figure	Formula	mIOU
POD	0.945	0.916	0.899	0.761	0.875	0.879
Collected	0.852	0.687	0.819	0.862	0.642	0.772
Collected + POD	**0.968**	**0.933**	**0.966**	**0.897**	**0.912**	**0.935**

Table 6. Comparison of our proposed method with the state-of-the-art methods. It should be noted that our method can additionally achieve 0.931 F1 score and 0.911 AP for text type.

Methods	F1 score				AP			
	Formula	Table	Figure	Avg.F1	Formula	Table	Figure	mAP
Li [23]	**0.932**	0.959	**0.917**	**0.936**	0.863	0.923	**0.854**	0.880
GOD [4]	0.919	**0.968**	0.912	0.933	0.869	**0.974**	0.818	**0.887**
DeepLayout [8]	0.716	0.911	0.776	0.801	0.506	0.893	0.672	0.690
NLPR-PAL [21]	0.902	0.951	0.898	0.917	0.816	0.911	0.805	0.844
icstpku [21]	0.841	0.763	0.708	0.770	0.815	0.697	0.597	0.703
FastDetector [21]	0.636	0.896	0.616	0.717	0.427	0.884	0.365	0.559
Vislnt [21]	0.241	0.826	0.643	0.570	0.117	0.795	0.565	0.492
Ours	0.923	0.914	0.812	0.883	**0.910**	0.944	0.731	0.862

5   Conclusion

We have established a deep learning based method to achieve better semantic page segmentation in Chinese and English document images. We use the DeepLab v3+ architecture which can capture multi-scale information to get precise pixel-wise classification results. Then we can get a series of candidate regions in different categories by making CCA on the prediction mask. And a Non-Intersecting Region Segmentation Algorithm is developed to solve the problem of intersection between regions, which boosts the performance and facilitates the document reconstruction applications. The promising results are both obtained on our collected dataset and public POD dataset. In the future, we plan to extend our framework to include more categories, like table or figure captions. Besides, we consider a fine-grained annotation format rather than just annotation of rectangular boxes.

Acknowledgment. This work was supported by the Natural Science Foundation of China under the grant 62071171.

References

1. Cesarini, F., Lastri, M., Marinai, S., Soda, G.: Encoding of modified XY trees for document classification. In: Proceedings of 6th International Conference on Document Analysis and Recognition, pp. 1131–1136. IEEE (2001)
2. Chen, K., Yin, F., Liu, C.L.: Hybrid page segmentation with efficient whitespace rectangles extraction and grouping. In: 12th International Conference on Document Analysis and Recognition, pp. 958–962. IEEE (2013)
3. Yi, X., Gao, L., Liao, Y., Zhang, X., Liu, R., Jiang, Z.: CNN based page object detection in document images. In: 14th IAPR International Conference on Document Analysis and Recognition, vol. 1, pp. 230–235. IEEE (2017)
4. Saha, R., Mondal, A., Jawahar, C.V.: Graphical object detection in document images. In: 15th International Conference on Document Analysis and Recognition, pp. 51–58. IEEE (2019)
5. Zhong, X., Tang, J., Yepes, A.J.: PubLayNet: largest dataset ever for document layout analysis. In: 15th International Conference on Document Analysis and Recognition, pp. 1015–1022. IEEE (2019)
6. He, D., Cohen, S., Price, B., Kifer, D., Giles, C.L.: Multi-scale multi-task FCN for semantic page segmentation and table detection. In: 14th IAPR International Conference on Document Analysis and Recognition, vol. 1, pp. 254–261. IEEE (2017)
7. Lee, J., Hayashi, H., Ohyama, W., Uchida, S.: Page segmentation using a convolutional neural network with trainable co-occurrence features. In: 15th International Conference on Document Analysis and Recognition, pp. 1023–1028. IEEE (2019)
8. Li, Y., Zou, Y., Ma, J.: DeepLayout: a semantic segmentation approach to page layout analysis. In: De-Shuang Huang, M., Gromiha, M., Han, K., Hussain, A. (eds.) Intelligent Computing Methodologies, pp. 266–277. Springer, Cham (2018). https://doi.org/10.1007/978-3-319-95957-3_30
9. Yang, X., Yumer, E., Asente, P., Kraley, M., Kifer, D., Lee Giles, C.: Learning to extract semantic structure from documents using multimodal fully convolutional neural networks. In: Proceedings of the IEEE Conference on Computer Vision and Pattern Recognition, pp. 5315–5324. IEEE (2017)
10. Schreiber, S., Agne, S., Wolf, I., Dengel, A., Ahmed, S.: DeepDeSRT: deep learning for detection and structure recognition of tables in document images. In: 14th IAPR International Conference on Document Analysis and Recognition, vol. 1, pp. 1162–1167. IEEE (2017)
11. Prasad, D., Gadpal, A., Kapadni, K., Visave, M., Sultanpure, K.: CascadeTabNet: an approach for end to end table detection and structure recognition from image-based documents. In: Proceedings of the IEEE/CVF Conference on Computer Vision and Pattern Recognition Workshops, pp. 572–573. IEEE (2020)
12. Ren, S., He, K., Girshick, R., Sun, J.: Faster R-CNN: towards real-time object detection with region proposal networks. IEEE Trans. Pattern Anal. Mach. Intell. **39**(6), 1137–1149 (2017)
13. He, K., Gkioxari, G., Dollár, P., Girshick, R.: Mask R-CNN. In: Proceedings of the IEEE International Conference on Computer Vision, pp. 2961–2969. IEEE (2017)
14. Li, K., et al.: Cross-domain document object detection: benchmark suite and method. In: Proceedings of the IEEE/CVF Conference on Computer Vision and Pattern Recognition, pp. 12915–12924. IEEE (2020)
15. Lin, T.Y., Dollár, P., Girshick, R., He, K., Hariharan, B., Belongie, S.: Feature pyramid networks for object detection. In: Proceedings of the IEEE Conference on Computer Vision and Pattern Recognition, pp. 2117–2125. IEEE (2017)

16. Ronneberger, O., Fischer, P., Brox, T.: U-Net: convolutional networks for biomedical image segmentation. In: Navab, N., Hornegger, J., Wells, W.M., Frangi, A.F. (eds.) Medical Image Computing and Computer-Assisted Intervention – MICCAI 2015: 18th International Conference, Munich, Germany, October 5-9, 2015, Proceedings, Part III, pp. 234–241. Springer, Cham (2015). https://doi.org/10.1007/978-3-319-24574-4_28

17. Chen, L.C., Papandreou, G., Kokkinos, I., Murphy, K., Yuille, A.L.: DeepLab: semantic image segmentation with deep convolutional nets, atrous convolution, and fully connected CRFs. IEEE Trans. Pattern Anal. Mach. Intell. **40**(4), 834–848 (2017)

18. Long, J., Shelhamer, E., Darrell, T.: Fully convolutional networks for semantic segmentation. In: Proceedings of the IEEE Conference on Computer Vision and Pattern Recognition, pp. 3431–3440. IEEE (2015)

19. Chen, L.C., Papandreou, G., Schroff, F., Adam, H.: Rethinking atrous convolution for semantic image segmentation. arXiv preprint arXiv:1706.05587.

20. Chen, L.-C., Zhu, Y., Papandreou, G., Schroff, F., Adam, H.: Encoder-decoder with atrous separable convolution for semantic image segmentation. In: Ferrari, V., Hebert, M., Sminchisescu, C., Weiss, Y. (eds.) ECCV 2018. LNCS, vol. 11211, pp. 833–851. Springer, Cham (2018). https://doi.org/10.1007/978-3-030-01234-2_49

21. Gao, L., Yi, X., Jiang, Z., Hao, L., Tang, Z.: ICDAR2017 competition on page object detection. In: 14th IAPR International Conference on Document Analysis and Recognition, vol. 1, pp. 1417–1422. IEEE (2017)

22. Antonacopoulos, A., Bridson, D.: Performance analysis framework for layout analysis methods. In: 9th International Conference on Document Analysis and Recognition, vol. 2, pp. 1258–1262. IEEE (2007)

23. Li, X.H., Yin, F., Liu, C.L.: Page object detection from pdf document images by deep structured prediction and supervised clustering. In: 24th International Conference on Pattern Recognition, pp. 3627–3632. IEEE (2018)

Non-central Student-t Mixture of Student-t Processes for Robust Regression and Prediction

Xiaoyan Li and Jinwen Ma[✉]

Department of Information and Computational Sciences, School of Mathematical Sciences
and LMAM, Peking University, Beijing, China
jwma@math.pku.edu.cn

Abstract. The Mixture of Gaussian Processes (MGP) is a general regression model for the data from a general stochastic process. However, there are two drawbacks on the parameter learning of the MGP model. First, it is sensitive to outliers. Actually, when the data are disturbed by heavy noise, its regression results are affected greatly by the noise, which makes it difficult to reflect the overall characteristics of the data. Second, the kernels of Gaussian processes in the MGP model do not have a simple parametric form to represent uncertain intuition by nonparametric prior over the covariance matrix. In order to overcome these problems, we propose the non-central student-t Mixture of student-t Processes (tMtP) model for robust regression and prediction. Specifically, the student-t process takes the invers Wishart distribution as its conjugate prior for the covariance matrix. The learning of the mixture parameters in the tMtP model can be implemented under the general framework of the hard-cut EM algorithm while the learning of the hyperparameters in each student-t process is implemented by maximizing the log-likelihood function of the square exponential kernel in output region. It is demonstrated by the experimental results on synthetic data sets that the tMtP model is effective for robust regression. Moreover, the tMtP model also obtains good prediction performance on a coal production data set.

Keywords: Student-t process · Mixture of student-t processes · Robust regression · Outlier · Covariance matrix

1 Introduction

Gaussian process is a powerful machine learning model for time series regression and classification. In fact, it has been widely applied in many fields of information processing and data mining [1–3]. However, it is sensitive to the data with outliers as shown in Fig. 1. Actually, Gaussian process regression is affected greatly on the data set contaminated with heavy noise. Moreover, a parameterized covariance kernel usually determines the properties of likely function under a Gaussian process. And a fully Bayesian nonparametric treatment of regression always place a nonparametric prior over the Gaussian process covariance kernel to represent uncertain intuition but the Gaussian process kernel does not have a simple parametric form [4]. It is natural to consider more general

© Springer Nature Switzerland AG 2021
D.-S. Huang et al. (Eds.): ICIC 2021, LNCS 12836, pp. 499–511, 2021.
https://doi.org/10.1007/978-3-030-84522-3_41

elliptical processes like Gaussian processes with kernel functions [5]. A student-t distribution is known to enhance the robustness and has heavy tails. Recently multivariant t distributions on covariance matrices have been used to model the relation data [6]. Base on multivariant t distribution, many equal forms of Student-t processes are introduced with different covariance priors such as inverse Wishart distribution [4] and Gamma distribution [7].

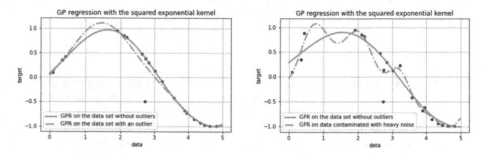

Fig. 1. The GP regression model is sensitive to output data with outliers.

Tresp [8] proposed the mixture of Gaussian processes for instance to deal with multimodal data set. By implementing a Dirichlet process prior over the Gaussian processes, we can allow the mixture to automatically determine the number of components for the given dataset [9]. Moreover, the mixture model can release the computational complexity of single Gaussian processes [10]. Yu [7] proposed the multi-task learning with student-t processes to improve the mixture model robustness. However, it takes degrees of freedom in student-t processes as the hyperparameters and sets them manually. Moreover, he supposed that the input data in each component are subject to a Gaussian distribution in the input region. As well known, Gaussian distribution will be disturbed by outliers and its tail is not heavy enough to use. Zhang et al. [11] also proposed the multi-task generalized t processes and used the mixture of Gaussian distribution model to classify the data in the input region. However, the Gaussian mixture model is not robust to outliers, which may lead to misclassifying the data in the input region.

In this paper, we propose a non-central student-t Mixture of student-t Processes (tMtP) model as well as its learning algorithm under the framework of the Hardcut EM algorithm [13, 14]. Specifically, we utilize the non-central student-t mixture for the input data and the mixture of student-t processes for the output data. The non-central student-t distributions over input region make the mixture model more robust on the input data with strong overlap between input components, while the student-t processes over output region makes the mixture model more robust on the output data with outliers or heavy noise. In addition, because both the student-t distributions and student-t processes have prior distributions of covariance kernels, the tMtP model is more convenient for parameter learning.

The rest of paper is organized as follow. In Sect. 2, we introduce the related models and learning algorithms. In Sect. 3, we propose the non-central student-t mixture of student-t processes model as well as its learning algorithm. The experimental results are summarized in Sect. 4. Finally, we briefly conclude in Sect. 5.

2 Related Models and Learning Algorithms

2.1 Non-central Student-t Distribution

The Student-t distribution is well suited to deal with the heavy-tailed and leptokurtic features than the Gaussian distribution. A p dimensional random $X \sim t_p(\mu, \sigma, v)$ with a center $\mu \in R$, $\Sigma \in \Pi(n)$, and a degree $v \in (0, +\infty)$ of freedom. Given a weight τ, X has the multivariate normal distribution [12],

$$X|v, \sigma, \tau \sim \mathcal{N}_p(\mu, \frac{\sigma}{\tau}), \tag{1}$$

The prior distribution of τ is subject to a Gamma distribution,

$$\tau|\mu, \sigma, v \sim Gamma(\frac{v}{2}, \frac{v}{2}). \tag{2}$$

By integrating τ from the joint density of (X, τ), we can get the density of X,

$$p(X) = \frac{\Gamma\left(\frac{v+p}{2}|\sigma|^{\frac{1}{2}}\right)}{(\pi v)^{\frac{1}{2}}\Gamma\left(\frac{v}{2}\right)}\left[1 + \frac{\delta_X(\mu, \sigma)}{v}\right]^{-\frac{v+p}{2}}, \tag{3}$$

Where $\delta_X(\mu, \sigma) = (X - \mu)'\sigma^{-1}(X - \mu)$.

The parameter learning function to $\{\mu, \sigma, v\}$ through the ML estimation method. On the use of the latent variable $\tau = \{\tau_1, \tau_2, \ldots, \tau_N\}$ with the expectation

$$\mathbb{E}(\tau|\mu, \sigma, v) = \frac{v + p}{v + \delta_X(\mu, \sigma)}, \tag{4}$$

we get the ML estimation function of $\hat{\mu}$ and $\hat{\sigma}$ as

$$\hat{\mu} = \frac{\sum_{i=1}^{N} \tau_i X_i}{\sum_{i=1}^{N} \tau_i} \tag{5}$$

$$\hat{\sigma} = \frac{1}{N}\sum_{i=1}^{N} \tau_i (X_i - \hat{\mu})(X_i - \hat{\mu})', \tag{6}$$

and get $\hat{v}$ by solving

$$-\phi\left(\frac{v}{2}\right) + ln\left(\frac{v}{2}\right) + \frac{1}{N}\sum_{i=1}^{N} (ln(\tau_i) - \tau_i) + 1 = 0 \tag{7}$$

for v. Detail function derivation is discussed in the reference essay [12].

2.2 Gaussian Process

The Gaussian Process is a powerful stochastic process model and it is easy to learn parameters for making inferences. Suppose we have data sets $\mathcal{D} = \{(x_i, y_i)\}_{i=1}^{N}$ in which x_i and y_i are a pair of input and output variables at sampling time i. The Gaussian process can be mathematically defined by

$$(y_1, y_2, \ldots, y_n) \sim GP\left(m(X), K\left(X, X'\right) + \sigma^2 I\right), \tag{8}$$

where $m(X)$ is a mean function, $K\left(X, X'\right) = [k(x_i, x_j)]_{N \times N}$ is a covariance matrix, and σ^2 dominates the noise globally. The most commonly used covariance function is the squared exponential function, which is defined by

$$k(x_i, x_j) = l^2 \exp\left(-\frac{1}{2}f^2 ||x_i - x_j||^2\right) \tag{9}$$

We generally set $m(X) = 0$ for simplify the model.

Parameters learning of the Gaussian process can be obtained through the Maximum Likelihood Estimation (MLE) method. The predictive output of the Gaussian process regression is given by

$$y_* \mid X, y, x_* \sim \mathcal{N}\left(\hat{y}_*, \text{cov}(y_*)\right), \tag{10}$$

where

$$\hat{y}_* = \mathbb{E}[y_* | X, y, x_*] = K(X, x_*)\left[C(X, X) + \sigma^2 I\right]^{-1} y, \tag{10}$$

$$\text{cov}(y_*) = K(x_*, x_*) - K'(X, x_*)\left[K(X, X) + \sigma^2 I\right]^{-1} K(X, x_*). \tag{11}$$

Here $y = [y_1, y_2, \ldots, y_N]$ is the output vector, $K(X, X) = [k(x_i, x_j)]_{N \times N}$ and $K(X, x_*) = [k(x_i, x_*)]_{N \times 1}$ denotes the covariance relationship vector of training inputs to the test input.

2.3 Student-t Process

The student-t process is an extension to the multivariate student-t distribution. For a base kernel K_θ parameterized by θ, and a mean function $\phi : \chi \to R$, the student-t process is generated by an inverse Wishart distribution as a conjugate prior for the covariance matrix of a Gaussian likelihood,

$$\sigma \sim IWP(\nu, K_\theta)$$

$$y|\sigma \sim GP(\phi, (\nu - 2)\sigma) \tag{13}$$

We analytically marginalize of σ in the generative model of (13). Therefor the density is defined as [4]

$$p(y) = \frac{\Gamma\left(\frac{v+n}{2}\right)}{((v-2)\pi)^{\frac{n}{2}}\Gamma\left(\frac{v}{2}\right)}|K_\theta|^{-1/2} \times \left(1 + \frac{(y-\phi)^T K_\theta^{-1}(y-\phi)}{v-2}\right)^{-\frac{v+n}{2}} \tag{14}$$

Like student-t distribution, it is also a marginally distribution from a conditional distribution

$$y|r \sim \mathcal{N}_N(\phi, \frac{r(v-2)K}{\rho}) \tag{15}$$

$$y|\sigma \sim GP(\phi, (v-2)\sigma) \tag{16}$$

With $v \in R_+\backslash[0,2]$, $\phi \in R^n$, $K \in \Pi(n)$ and ρ is a scale parameter with $\rho > 0$. And we write $y \sim MVT_n(v, \phi, K)$. After the definition of multivariable student-t distribution, Shah A [4] define a Student-t process as follows.

Definition [4]. f is a student-t process on χ with parameters $v > 2$, mean function Φ : $\chi \rightarrow R$, and kernel function $k : \chi \times \chi \rightarrow R$ if any finite collection of function values have a joint multivariate student-t distribution, i.e. $(f(x_1), \ldots, f(x_N))^T \sim MVT_n(v, \phi, K)$, where $K \in \Pi(n)$ with $K_{ij} = k(x_i, x_j)$ and $\phi \in R^n$ with $\phi_i = \phi(x_i)$. We write $f \sim TP(v, \Phi, K)$. The student-t process generalizes the Gaussian process. A Gaussian process can be seen as a limiting case of a student-t process, which is proposed in [4].

Parameters learning of the student-t process can be also obtained through the Maximum Likelihood Estimation (MLE) method. The marginal likelihood function is [4]

$$\log p(y|v, K_\theta) = -\frac{n}{2}\log((v-2)\pi) - \frac{1}{2}\log(|K_\theta|) + \log\left(\frac{\Gamma\left(\frac{v+n}{2}\right)}{\Gamma\left(\frac{v}{2}\right)}\right) - \frac{v+n}{2}\log(1 - \frac{\beta}{v-2}),$$

where $\beta = (y-\phi)^T K_\theta^{-1}(y-\phi)$ and its derivate with respect to a hyperparameter is [4]

$$\frac{\partial}{\partial\theta}\log p(y|v, K_\theta) = \frac{1}{2}Tr((\frac{v+n}{v+\beta-2}\alpha\alpha^T - K_\theta^{-1})\frac{\partial K_\theta}{\partial\theta}), \tag{17}$$

where $\alpha = K_\theta^{-1}(y-\phi)$. We learn v using gradient methods and the following derivative [4]

$$\frac{\partial}{\partial v}\log p(y|v, K_\theta) = -\frac{n}{2(v-2)} + \psi\left(\frac{v+n}{2}\right) - \psi\left(\frac{v}{2}\right) - \frac{1}{2}\log\left(1 + \frac{\beta}{v-2}\right) + \frac{(v+n)\beta}{2(v-2)^2 + 2\beta(v-2)}, \tag{18}$$

where ψ is the digamma function.

The conditional distribution for a multivariate student-t is still following a multivariate student-t distribution [4]. Then

$$y_2|y_1 \sim MVT(v + n_1, \tilde{\phi}_2, \frac{v+\beta_1-2}{v+n_1-2} \times \tilde{K}_{22}), \tag{19}$$

where

$$\tilde{\phi}_2 = K_{21} K_{11}^{-1}(y_1 - \phi_1) + \phi_2, \tag{20}$$

$$\beta_1 = (y_1 - \phi_1)^T K_{11}^{-1}(y_1 - \phi_1), $$

and

$$\tilde{K}_{22} = K_{22} - K_{21} K_{11}^{-1} K_{12}. \tag{21}$$

3   The tMtP Model and Its Learning Algorithm

The mixture of student-t processes model mixes the Student-t process by the Gaussian mixture model in many of multi-task papers [7, 11]. However, the Gaussian distribution is sensitive to samples with outliers. The student-t distribution has heavy tail to robust the regression. In this section, we use the non-central student-t mixture model instead of the Gaussian mixture model as the mixing method in input region. We proposed the non-central student-t Mixture of student-t Processes (tMtP) model.

We describe the non-central student-t mixture of student-t processes model mathe-matically as follow. Let $\mathcal{D}$ be the data set and Z be the indicator set, and the indicator variable Z subject to a multinomial distribution with proportion π_i. However, samples in c th input region follow the non-central student-t distribution

$$x_i | (z_i = c) \sim t\big(\mu_c, \sigma_c, v_c^x\big) \tag{22}$$

Where μ_c, σ_c and v_c are the center, the covariance and the degree of freedom of the student-t distribution belongs to the cth component. Finally, the predictive output of the cth component follows the student-t process.

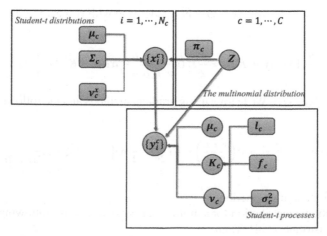

Fig. 2. The flowchart of date generation by the non-central student-t Mixture of student-t Process processes (tMtP) model. We use the squared exponential function as covariance function.

Table 1. The algorithm for learning parameters of the tMtP model.

Algorithm

Input: The set of data is $\mathcal{D} = \{(x_i, y_i)\}_{i=i}^{N}$, and the number of components C;

Indicator: The set of latent variables $Z = \{z_i\}_{i=1}^{N}$;

Output: Mixing proportions $\{\pi_c\}_{c=1}^{C}$, parameters of the input discussed $\alpha_c = \{\mu_c, \Sigma_c, v_c^x\}$, and parameters of Student-t processes discussed $\theta_c = \{l_c, f_c, \sigma_c^2, v_c\}$;

Initialization: Initialize $\{z_i\}_{i=1}^{n}$ via k-means;

Iteration:

1: **while** not converge **do**

2: **for** $c = 1, 2, ..., C$ **do**

3: Calculate the proportion of each component

$$\pi_c \leftarrow \frac{N_c}{N}$$

4: **while** not converge **do**

5: Calculate $\omega_{i.c}$, μ_c and Σ_c,

$$\omega_{i.c} \leftarrow \frac{v_c^x + 1}{v_c + \delta_{x_i}(\mu_c, \Sigma_c)}$$

$$\mu_c \leftarrow \frac{\sum_{i=1}^{n} \omega_{ic} x_i}{\sum_{i=1}^{n} \omega_{ic}}$$

$$\Sigma_c \leftarrow \frac{1}{n} \sum_{i=1}^{n} \omega_{ic}(x_i - \mu_c)^2$$

6: Update v_c^x by solving equation below

$$-\phi\left(\frac{v_c^x}{2}\right) + \ln\left(\frac{v_c^x}{2}\right) + \frac{1}{n}\sum_{i=1}^{n}(\ln(\omega_{ic}) - \omega_{ic}) + 1 + \phi\left(\frac{v_c^x + 1}{2}\right) - \ln\left(\frac{v_c^x + 1}{2}\right) = 0$$

7: **end while**

8: Obtain the Student-t process parameters θ_c by the maximizing the likelihood function

$$\log p(y_i|v_c, K_c) = -\frac{N_c}{2}\log((v_c - 2)\pi) - \frac{1}{2}\log(|K_c|)$$

$$+ \log\left(\frac{\Gamma\left(\frac{v_c + N_c}{2}\right)}{\Gamma\left(\frac{v_c}{2}\right)}\right) - \frac{v_c + N_c}{2}\log\left(1 + \frac{\beta_c}{v_c - 2}\right)$$

9: (E-step) Repartition the samples into the corresponding component based on the maximizing the posterior criterion

$$z_i \leftarrow \underset{c}{arg\,max}\,Pr(z = c|\mathcal{D}, x^*)$$

$$= \underset{c}{arg\,max}\,\pi_c t(x^*, \mu_c, \Sigma_c, v_c^x)TP(y_i|0, l_c^2 + \sigma_c^2, v_c)$$

10: **end for**

11: **end while**

And the posterior distribution of latent variable z_i is

$$p(z_i = c|x_i, y_i) = \frac{\pi_i \cdot t(x_i|\mu_c, \sigma_c, v_c^x) \cdot TP(y_i|0, l_c^2 + \sigma_c^2, v_c)}{\sum_{c=1}^{C} \pi_i \cdot \mathcal{N}(x_i|\mu_c, \sigma_c, v_c^x) \cdot TP(y_i|0, l_c^2 + \sigma_c^2, v_c)}. \tag{23}$$

We proposed an algorithm base on the un-v hard-cut EM [15] to learn parameters in the tMtP model.

4 Experimental Results

In this section, we conduct experiments on synthetic data set. Since the multi-task model is optimized for data sets with components mixing, we set three S_3 and five S_5 components over data sets. The data set S_3 with $C = 3$ student-t processes components and $C = 5$ student-t processes components. Aiming at comparing the mixture of Gaussian processes model, the non-central student-t mixture of Gaussian processes model and the non-central student-t mixture of student-t processes model, we construct the component in each data set have same proportion $\pi_c = \frac{1}{c}$, $\sigma_c = 1$, degrees $v_c^x = 2.5$ of freedom over input region, and degrees $v_c^y = 2.5$ of freedom over output region. Hyperparameters of the square of exponential kernel in S_3 are $\theta_3 = [\ln f_3, \ln l_3]$ with each component is $\ln f_3 = [3.0, 1.5, 2.0]$ and $\ln l_3 = [1.0, 2.0, 0.5]$. Hyperparameters of the square of exponential kernel in S_5 are still $\theta_5 = [\ln f_5, \ln l_5]$ with $\ln f_5 = [3.0, 1.5, 2.0, 1.5, 0.5]$ and $\ln l_5 = [1.0, 2.0, 0.5, 3.5, 2.0]$. Global noisy variances in each component in S_3 are $\sigma_3^2 = [0.05, 0.01, 0.03]$ and noisy variances in S_5 are $\sigma_5 = [0.05, 0.01, 0.03, 0.0001, 0.005]$. The implementation is based on the Sklearn toolbox [16]. All experiments are conducted on a personal computer (Inter® core™ i7-8550U CPU 1.8 GHz, 8G RAM).

4.1 Experimental Results on Synthetic Datasets.

In this section, we conduct experiments on synthetic data sets which are composed of three and five student-t processes. We generate 500 samples to get the prior distribution of model parameters. Then we give the tMtP model predicted distribution. The line in image is the mean line and the shaded area represents a $\mu \pm \sigma$ predictive interval of the predicted distribution. The same color of data belongs to the same component. In Regression results images of the tMtP model over S_3 and S_5 have illustrated in Fig. 3 and Fig. 4 separately. From regression results of the tMtP model, the model can fit multi-modal data well.

4.2 Comparison on Synthetic Datasets

To further investigate the tMtP model, we compare it with some typical models as well as their corresponding learning algorithms on synthetic data sets. We use more representative synthetic data sets S_1 and S_2 for model comparison. Compared with the previous experimental data sets, we set a higher degree of noise $v_c^y = 3.5$ in the output region in each component.

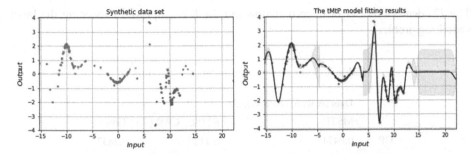

Fig. 3. Regression results of the tMtP model over $\mathcal{S}_3$ with RMSE $= 0.0125$.

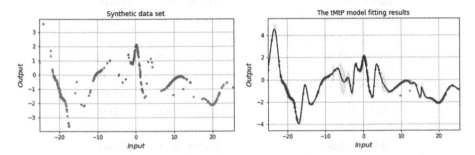

Fig. 4. Regression results of the tMtP model over $\mathcal{S}_5$ with RMSE $= 0.0175$.

We use a single Gaussian process and a single student-t process as base models. Both models use the maximum likelihood estimation to get models parameters. We use multi-task model to make comparison as well such as the MGP model and TMGP [15] model. Parameters learning of multi-task models are based on the Hard-cut EM algorithms. We conduct experiments on synthetic data sets which are composed of three and five experts.

Models performances are measured by the root of mean square of error (RMSE), which is defined as

$$RMSE = \sqrt{\frac{1}{n} \sum_{j=1}^{n} \left(\hat{y}_j - y_j\right)^2},$$

where $\hat{y}_j$ is the predicted result and $\hat{y}_j$ is the ground-truth value. Models robustness is measured by the mean of standard covariances (MSTD), which is defined as

$$MSTD = \sqrt{\frac{1}{n} \sum_{j=1}^{n} \sigma_{x_i}^2},$$

where $\sigma_{x_i}^2$ is the covariance of prediction distribution at point x_i. We calculate the classification accuracy rate (CAR) to validate the effectiveness of multi-task models and it is defined as

$$CAR = \frac{1}{n} \max_{\xi \in \Pi_c} \sum_{j=1}^{n} \mathbb{I}\big(z_i = \xi\left(\hat{z}_j\right)\big)$$

The Π_c denotes the set of C permutation and the ξ is employed to account the label switching problem.

Table 2. Comparison of different model regression

	Data set with 3 components			
	MSTD	RMSE	CAR (%)	Time (s)
GP	0.0474	0.1137	-	0.2837
TP	0.0591	0.4446	-	0.3322
MGP	0.0266	0.0549	99.6	0.3389
TMGP	0.0281	0.0566	99.6	0.4816
tMtP	**0.0140**	**0.0452**	99.6	0.5346
	Data set with 5 components			
	MSTD	RMSE	CAR(%)	Time(s)
GP	0.0836	0.2556	-	0.2348
TP	0.0363	0.1027	-	0.1206
MGP	0.0315	0.0618	99.1	0.9138
TMGP	0.0314	0.0618	99.1	0.8847
tMtP	**0.0206**	**0.0594**	99.1	0.7581

In the Table 2 shown the tMtP model performance well over both dataset S_1 and S_2. As well the best robustness of prediction is the tMtP model. Obviously, multi-task models perform better than single processes. The MGP model, the TMGP model and the tMtP model have the same classification accuracy rate (CAR). The model prediction time is greatly affected by searching the minimum value of the log marginal likelihood function, which is perturbated by random initial parameters.

Data sets S_1 and S_2 are shown in the first image in Fig. 5 and the first image in Fig. 6. The solid line is the mean, the shaded area represents a $\mu \pm \sigma$ predictive interval and points in the same color are in the same component. In Fig. 5 shows the tMtP model after hyperparameters training is better predictive mean specially outlier points than the MGP model and TMGP model. In Fig. 6 shows the mean line of the predictive results is smoothly when data points have big ups and downs.

4.3 Application to a Real Word Dataset

In this section, we apply the tMtP model to model the coal data set which recorded the coal production between January 2009 and December 2018 in a province in China. However, data from 2014 to 2015 are missing. Based on the characteristics of the data, we separate the data before 2014 and after 2015 to two components, and use a multi-task model to make predictions.

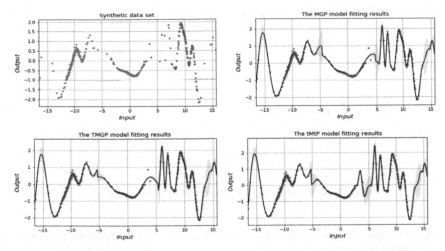

Fig. 5. The MGP model, the TMGP model and the tMtP model regression on the data set S_1.

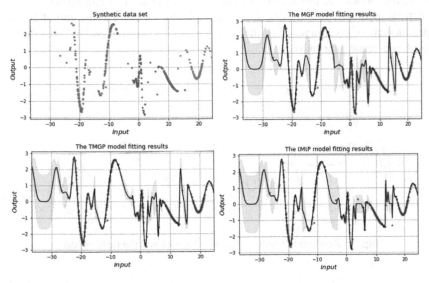

Fig. 6. The MGP model, the TMGP model and the tMtP model regression on the data set S_2.

First, we deal with time data. We take January 2009 as the first day rest of the date transforming the number of days after the first days. Then we normalize the days set as the input data set. We use the daily average coal productions as the output data, and after normalization, the data is shown in Fig. 7.

Regression results of the MGP model and the tMtP model are shown in Fig. 7. First, there are obvious discontinuities in the data set because of missing data in 2014 and 2015. Both MPG and tMtP models can automatically divide the data set into two components, and predict each component. The Fig. 8 shown that for data sets with large noise and small

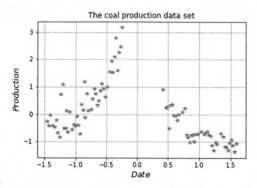

Fig. 7. The coal production data set after transformation.

number of samples in practice, the MGP model will have an over-learning phenomenon and it is difficult to reflect the overall characteristics of the data. But the tMtP model has better stability for such a data set. It can be seen from the figure that in the noisy data set, the model can predict the overall trend of the data. It can be seen that coal productions increased year by year from the beginning of 2009 to the end of 2013, but the productions of coal mines from the beginning of 2016 to the end of 2018 showed a wave-like decrease year by year.

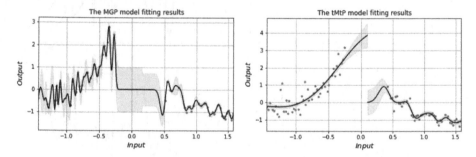

Fig. 8. Regression results of the MGP model and the tMtP model

5   Conclusion

We have proposed the non-central student-t Mixture of student-t Processes (tMtP) model and its learning algorithm. The tMtP model works well on the time series data with potential outliers and disturbed by heavy noisy. It is demonstrated by the experimental results on synthetic and coal production data sets that this tMtP model is less sensitive to outliers and more robust to noise data distributions. However, according to our experiments, there is still a problem. The squared exponential kernel is often a default choice for Gaussian process regression. However, sample functions with this covariance function are unrealistically smooth for practical optimization problems. We need to manually adjust the appropriate optimization interval to obtain better regression prediction results. Therefore, in the future, we will improve the tMtP model approach by investigating the problem.

Acknowledgement. This work is supported by the National Key R & D Program of China (2018YFC0808305).

References

1. Wang, J.M., Fleet, D.J., Hertzmann, A.: Gaussian process dynamical models. In: Proceedings of the 18th International Conference on Neural Information Processing Systems (NIPS'05), MIT Press, Cambridge, MA, USA, pp.1441–1448 (2005)
2. Wilson, A.G., Knowles, D.A., Ghahramani, Z.: Gaussian process regression networks. arXiv preprint arXiv:1110.4411 (2011)
3. Wang, J.M., Fleet, D.J., Hertzmann, A.: Gaussian process dynamical models for human motion. IEEE Trans. Pattern Anal. Mach. Intell. **30**(2), 283–298 (2007)
4. Shah, A., Wilson, A., Ghahramani, Z.: Student-t processes as alternatives to Gaussian processes. Proceedings of the Seventeenth International Conference on Artificial Intelligence and Statistics, PMLR **33**, 877–885 (2014)
5. Fang, K.W.: Symmetric multivariate and related distributions. CRC Press (2018)
6. Xu, Z., Yan, F., Qi, Y.: Sparse matrix-variate t process blockmodels. In: Proceedings of the AAAI Conference on Artificial Intelligence, vol. 25. No. 1 (2011)
7. Yu, S., Tresp, V., Yu, K.: Robust multi-task learning with t-processes. In: Proceedings of the 24th International Conference on Machine Learning, pp. 1103–1110 (2007)
8. Tresp, V.: Mixtures of Gaussian processes. In: Advances in Neural Information Processing Systems, pp. 654–660 (2001)
9. Meeds, E., Osindero, S.: An alternative infinite mixture of Gaussian process experts. In: Advances in Neural Information Processing Systems, 18, p. 883 (2006)
10. Yuan, C., Neubauer, C.: Variational mixture of Gaussian process experts. In: Advances in Neural Information Processing Systems, pp. 1897–1904 (2009)
11. Zhang, Y., Yeung, D.Y.: Multi-task learning using generalized t process. In: Proceedings of the thirteenth international conference on artificial intelligence and statistics. JMLR Workshop and Conference Proceedings, pp. 964–971 (2010).
12. Liu, C., Rubin, D.B.: ML estimation of the t distribution using EM and its extensions, ECM and ECME. In: Statistica Sinica, pp.19–39 (1995)
13. Chen, Z., Ma, J., Zhou, Y.: A precise hard-cut EM algorithm for mixtures of Gaussian processes. In: International Conference on Intelligent Computing. Springer, Cham, pp. 68–75 (2014).
14. Chen, Z., Ma, J.: The hard-cut EM algorithm for mixture of sparse Gaussian processes. In: International Conference on Intelligent Computing. Springer, Cham, pp. 13–24 (2015)
15. Li, X., Li, T., Ma, J.: The Unν-Hardcut EM algorithm for non-central student-t mixtures of gaussian processes. In: 2020 15th IEEE International Conference on Signal Processing (ICSP). IEEE, vol: 1, pp. 289–294 (2020)
16. Pedregosa, et al.: Scikit-learn: son. JMLR **12**, 2825–2830 (2011)

Multi-class Tissue Classification in Colorectal Cancer with Handcrafted and Deep Features

Nicola Altini[1(✉)], Tommaso Maria Marvulli[1], Mariapia Caputo[2], Eliseo Mattioli[3], Berardino Prencipe[1], Giacomo Donato Cascarano[1,4], Antonio Brunetti[1,4], Stefania Tommasi[2], Vitoantonio Bevilacqua[1,4], Simona De Summa[2], and Francesco Alfredo Zito[3]

[1] Department of Electrical and Information Engineering (DEI), Polytechnic University of Bari, 70126 Bari, Italy
nicola.altini@poliba.it
[2] Molecular Diagnostics and Pharmacogenetics Unit, IRCCS Istituto Tumori Giovanni Paolo II, Viale Orazio Flacco 65, Bari, Italy
[3] Pathology Department, IRCCS Istituto Tumori Giovanni Paolo II, Viale Orazio Flacco 65, Bari, Italy
[4] Apulian Bioengineering Srl, Via delle Violette 14, 70026 Modugno, BA, Italy

Abstract. Multi-class tissue classification from histological images is a complex challenge. The gold standard still relies on manual assessment by a trained pathologist, but it is a time-expensive task with issues about intra- and inter-operator variability. The rise of computational models in Digital Pathology has the potential to revolutionize the field. Historically, image classifiers relied on handcrafted feature extraction, combined with statistical classifiers, as Support Vector Machines (SVMs) or Artificial Neural Networks (ANNs). In recent years, there has been a tremendous growth in Deep Learning (DL), for all the image recognition tasks, including, of course, those concerning medical images. Thanks to DL, it is now possible to also learn the process of capturing the most relevant features from the image, easing the design of specialized classification algorithms and improving the performance. An important problem of DL is that it requires tons of training data, which is not easy to obtain in medical domain, since images have to be annotated by expert physicians. In this work, we extensively compared three classes of approaches for the multi-class tissue classification task: (1) extraction of handcrafted features with the adoption of a statistical classifier; (2) extraction of deep features using the transfer learning paradigm, then exploiting SVM or ANN classifiers; (3) fine-tuning of deep classifiers. After a cross-validation on a publicly available dataset, we validated our results on two independent test sets, obtaining an accuracy of 97% and of 77%, respectively. The second test set has been provided by the Pathology Department of IRCCS Istituto Tumori Giovanni Paolo II and has been made publicly available (http://doi.org/10.5281/zenodo.478 5131).

Keywords: Colorectal cancer · Deep learning · Handcrafted features · Histological tissue classification

---

N. Altini, T.M. Marvulli, M. Caputo, S. De Summa, F.A. Zito—Equally contributed to this paper.

© Springer Nature Switzerland AG 2021
D.-S. Huang et al. (Eds.): ICIC 2021, LNCS 12836, pp. 512–525, 2021.
https://doi.org/10.1007/978-3-030-84522-3_42

1 Introduction

Colorectal cancer (CRC) is the second cause of death for cancer with mortality ranging almost 35% over the CRC patients [1]. In the last years, new therapeutic approaches have been introduced in the clinical practice but, due to the high mortality, genomic-driven drugs are under evaluation. In particular, the advent of immunotherapy has represented a promising approach for many tumours (e.g., melanoma, non-small cell lung cancer) but results of clinical trials related to CRC have revealed that patients do not benefit from such therapeutical approaches. The chance to molecularly classify this tumour could lead to a better assessment of the regimen to be administered. Many research groups are focusing on these aspects and a multilayer approach could lead in a substantial improvement of the clinical outcomes.

The advent of different computational models allows to perform multilayer analyses including deep study of histological images. Such an approach relies on the automatic assessment of tissue types.

The classical pipeline for building an image classifier involves handcrafted feature extraction and statistical classification. A typical choice was Support Vector Machines (SVMs), or Artificial Neural Networks (ANNs), plus eventual stages of preprocessing and dimensionality reduction.

Linder et al. addressed the problem of classification between epithelium and stroma in digitized tumour tissue microarrays (TMAs) [2]. The authors exploited Local Binary Patterns (LBP), together with a contrast measure C (they referred to their union as LBP/C) as input for their classifier, an SVM. In the end, they compared LBP/C classifier with those based on Haralick texture features and Gabor filtered images, and the LBP/C classifier resulted the best model (area under the Receiver Operating Characteristic – ROC – curve was 0.995).

In the context of colorectal cancer histology, it is worth of note the multi-class texture analysis work of Kather et al. [3], which combined different features (considering the original RGB images as grey-scale ones), namely: lower-order and higher-order histogram features, Local Binary Patterns (LBP), Grey-level co-occurrence matrix (GLCM), Gabor filters and Perception-like features. As statistical classifiers, they considered: 1-nearest neighbour, linear SVM, radial-basis function SVM and decision trees. Even though they get good performances, by repeating the experiment with the same features, we noted that adopting the red channel leads to better results than using grey-scale images (data not shown). This consideration does not hold after that staining normalization techniques are applied.

Later works exploited the power of Deep Learning (DL), in particular of Convolutional Neural Networks (CNNs), for classifying histopathological images.

Kather et al. employed CNN for performing automating tissue segmentation of Hematoxylin-Eosin (HE) images from 862 whole slide images (WSIs) of The Cancer Genome Atlas (TCGA) cohort. Then, they exploit the output neuron activation in the CNN for calculating a "deep stroma score", which proved to be an independent prognostic factor for overall survival (OS) in a multivariable Cox proportional hazard model [4].

Kassani et al. proposed a Computer-Aided Diagnosis (CAD) system, composed of an ensemble of three pre-trained CNNs: VGG-19 [5], MobileNet [6] and DenseNet [7], for binary classification of HE stained histological breast cancer images [8]. They came to the conclusion that their ensemble performed better than single models and widely adopted machine learning algorithms.

Bychkov *et al.* introduced a DL-based method for directly predicting patient outcome in CRC, without intermediate tissue classification. Their model consists in extracting features from tiles with a pretrained model (VGG-16 [5]), and then applying a LSTM [9] to these features [10].

In this work, we extensively compare three classes of approaches for the multi-class tissue classification task: (1) extraction of handcrafted features with the adoption of a statistical classifier; (2) extraction of deep features using the transfer learning paradigm, then exploiting ANN or SVM classifiers; (3) fine-tuning of deep classifiers. We also proposed a feature combination methodology in which we concatenate the features of different pretrained deep models, and we investigate the effect of dimensionality reduction techniques. We identified the best feature set and classifier to perform inferences on external datasets. We investigated the explainability of the considered models, by looking at t-distributed Stochastic Neighbour Embedding (t-SNE) plots and saliency maps generated by Gradient-weighted Class Activation Mapping (Grad-CAM).

2 Materials

The effort of Kather *et al.* resulted in the development and diffusion of different datasets suitable for multi-class tissue classification [3, 4, 11, 12].

[3, 11] describe the collection of N = 5.000 histological images, with size of 150 × 150 pixels (corresponding to 74 × 74 μm).

[4, 12] introduce a dataset of N = 100.000 image patches from HE stained histological images of human colorectal cancer (CRC) and normal tissue. Images have size of 224 × 224 pixels, corresponding to 112 × 112 μm. This dataset is the designated train set in their experiments, whereas a dataset of N = 7.180 has been used as validation set. We denote the first one with **T** and the latter one with **V1**. For the train set, they provide both the original version and a normalized version exploiting the Macenko's method [13].

In order to harmonize some differences between the class names of the two collections, we considered the following classes:

- **TUM**, which represents tumour epithelium.
- **MUSC_STROMA**, which represents the union of SIMPLE_STROMA, as tumour stroma, extra-tumour stroma and smooth muscle, and COMPLEX_STROMA, as single tumour cells and/or few immune cells.
- **LYM**, which represents immune-cell conglomerates and sub-mucosal lymphoid follicles.
- **DEBRIS_MUCUS**, which represents necrosis, hemorrhage and mucus.
- **NORM**, which represents normal mucosal glands.
- **ADI**, which represents adipose tissue.
- **BACK**, which represents background.

Starting from the Dataset of [11], SIMPLE_STROMA and COMPLEX_STROMA have been merged, resulting into a MUSC_STROMA class. For the dataset of [12], DEB and MUC classes have been merged, resulting in a DEBRIS_MUCUS class, and MUS and STR classes have been merged, resulting in a MUSC_STROMA class. Of note, the merging procedure has been performed according to class definition of the **T** training dataset. At the end of the merge, our training dataset is reduced obtaining N = 77.805 images, keeping half of the images of each of the two combined classes, and maintaining the balance across classes. After the same merge, the external validation set **V1** resulted to have N = 5.988 images.

An additional dataset of N = 5.984 HE histological image patches, provided by IRCCS Istituto Tumori Giovanni Paolo II, has been used as another independent test set. The institutional Ethic Committee approved the study (Prot n. 780/CE). This dataset, hereinafter denoted with **V2**, has been made publicly available [14]. The class subdivision has been done according to the list mentioned above and classified by an expert pathologist, in order to gain the ground truth of the **V2** dataset. We made our dataset publicly available, in order to ease the development and comparison of computational techniques for CRC histological image analysis.

Some test images from both the **V1** and **V2** datasets can be seen in Fig. 1.

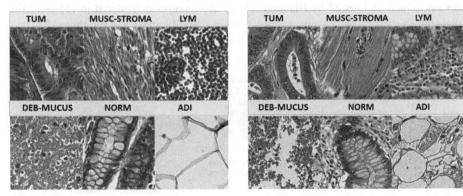

Fig. 1. Test dataset example patches for each class. Left: **V1** dataset; right: **V2** dataset. All images have been pre-processed with Macenko's method.

3 Methods

3.1 Image Features

Different features can be extracted from single channel histogram of an image. In [3], the authors only considered the grey-scale version of the image, but also other color channels may be considered. For HE images, red channel can be more informative.

According to the convention used in [3], we can consider two sets of features from the histogram: a histogram-lower, which contains mean, variance, skewness and kurtosis index, and a histogram-higher, composed of the image moments from 5th to 11th.

Another set of features used was Local Binary Patterns (LBP). An LBP operator considers the probability of occurrence of all the possible binary patterns that can arise from a neighbourhood of predefined shape and size. A neighbourhood of eight equally spaced points arranged along a circle of radius 1 pixel has been considered. The resulting histogram was reduced to the 38 rotationally-invariant Fourier features proposed by [15]; these are frequently used for histological texture analysis. To extract this set of features it is possible to use the MATLAB tool from Center for Machine Vision and Signal Analysis (CMVS) available at the link http://www.cse.oulu.fi/CMV/Downloads/LBPMatlab [16, 17].

Kather *et al.* also considered the Grey-level co-occurrence matrix (GLCM); in particular, they considered four directions (0°, 45°, 90° and 135°) and five displacement vectors (from 1 to 5 pixels). To make this texture descriptor invariant with respect to rotation, the GLCMs obtained from all four directions were averaged for each displacement vector. From each of the resulting co-occurrence matrices the following four global statistics were extracted: contrast, correlation, energy and homogeneity, as described by Haralick *et al.* in [18], thereby obtaining 20 features for each input image.

As the latest set of features, Kather *et al.* considered Perception-like features, that included features based on image perception. Tamura *et al.* in [19] showed that the human visual system discriminates texture through several specific attributes that were later on refined and tested by Bianconi *et al.*; the features considered in [3] were the following five: coarseness, contrast, directionality, line-likeness and roughness [20].

This procedure leads to the extraction of a feature vector with 74 elements.

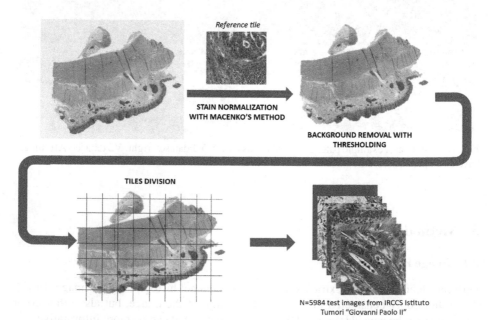

Fig. 2. Stain normalization with Macenko's method [13] and tiling, analogously to Kather *et al.* [4, 12]. This procedure has been followed to generate the test patch images for our dataset.

3.2 Stain Normalization

Stain Normalization is necessary due to the pre-analytical bias specific to different laboratories; it can lead to miscalculation of images by ANN or CNN. Techniques for handling stain color variation can be grouped into two categories: stain color augmentation, which mimics a vast assortment of realistic stain variations during training and stain color normalization, which intends to match training and test color distributions for the sake of reducing stain variation [21].

In order to normalize the images coming from different datasets, we exploited the Macenko's normalization method [13], as reported by Kather *et al.* [4, 12], allowing comparability across different datasets.

The procedure adopted for the stain normalization is depicted in Fig. 2.

3.3 Deep Learning Models

Deep Learning refers to the adoption of hierarchical models to process data, extracting representations with multiple levels of abstraction [22]. Convolutional Neural Network (CNN) have a prominent role in image recognition problems. A huge amount of literature data regarding the construction of DL-based classifiers for images [5, 23–29]. Some example of application in histological images include classification of breast biopsy HE images [30], semantic segmentation, detection and instance segmentation of glomeruli from kidney biopsies [31, 32].

An important concern about CNN is that training a network from scratch requires tons of data. One interesting possibility is that offered by transfer learning, which is a methodology for training models by using data which is more easily collected compared to the data of the problem under consideration. Refer to [33] for a comprehensive survey of the transfer learning paradigm, here we will consider models pre-trained on ImageNet as feature extractors for histological images, as done also in [10, 34–38]. The paradigm of DL-based transfer learning has led to the term Deep Transfer Learning [39]. It has been noted that, although histopathological images are different from RGB images of everyday life, they share common basic structures as edges and arcs [40]. Earlier layers of CNN capture this kind of elementary patterns, so transfer learning may be useful also for digital pathology images.

One potential drawback of deep feature extractor is the high dimensionality. Cascianelli *et al.* attempted to solve this problem by considering different technique of dimensionality reduction [38]. We investigated the combinations of deep features extracted by pretrained models, also considering different levels of compression, after having applied Principal Component Analysis (PCA). In particular, we concatenated the features coming from the ResNet18, GoogleNet and ResNet50 models, obtaining a feature set of 3584 elements. Then, different numbers of features, ranging from 128 to 3584, have been considered for training our classifiers. To ensure that deep features are relevant for the problem under consideration, we compared them to smaller sets of handcrafted features. In particular, we checked: (1) that they tend to represent similar tissue types into defined regions of the feature space, by considering a 2D scatter plot after having applied t-SNE [41] on the deep and handcrafted features; (2) that they lead to the training of an accurate model, without overfitting problems; (3) the saliency maps highlighted by Grad-CAM [42]. t-SNE can both capture the local structure of high dimensional data and reveal global structure at several scales (e.g. the presence of clusters), as image features in this case. Grad-CAM is a class-discriminative localization technique for CNN-based models to make them more transparent by producing a visual explanation.

We considered three different topologies of deep networks: ResNet18, ResNet50 [28] and GoogLeNet [25]. For each architecture, we compared the ImageNet [43] pretrained version (the network is working only as feature extractor in this case) with the fine-tuned version on our data.

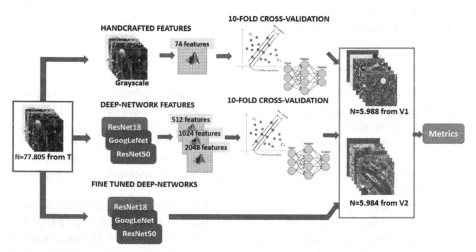

Fig. 3. Training procedure. Starting from a subset of the dataset **T**, we compared three kinds of models. 10-fold Cross-validation was performed to find the best model. **Validation procedure.** We externally validated the models found as best from internal cross-validation on two datasets: **V1** and **V2**. **T** refers to the Training set from Kather *et al.*; **V1** stands for Test set from Kather *et al.*; **V2** refers to the Test set from IRCCS Istituto Tumori Giovanni Paolo II.

4   Experimental Results

We considered three types of experiments: (1) training of ANN and SVM classifiers after handcrafted feature extraction; (2) training of ANN and SVM classifiers after deep feature extraction with models pretrained on ImageNet; (3) fine-tuning of deep classifiers. The workflow is depicted in Fig. 3. For the ANN and SVM trained after handcrafted feature extraction or pretrained deep feature extraction, we made a 10-fold cross validation (90% train, 10% test for each iteration) on the train dataset **T**, after having pre-processed it as described in Sect. 2.

Then, we exploited the best classifier for each category for testing it on the validation datasets **V1** and **V2**. Performances reported in Table 1, Table 2, Table 3 and Table 4 are assessed in terms of accuracy.

Table 1. Results of 10-fold cross-validation on the train dataset **T**. Performances for SVM and ANN are expressed as accuracy: mean ± std.

Feature set	# Features	SVM	ANN
Handcrafted Grey	74	94.64 ± 1.69%	94.59 ± 0.92%
Handcrafted Red	74	83.23 ± 0.85%	85.24 ± 0.08%
Pretrained ResNet18	512	**97.91 ± 0.14%**	97.15 ± 0.03%
Pretrained GoogleNet	1024	94.41 ± 0.23%	93.92 ± 0.05%
Pretrained ResNet50	2048	97.61 ± 0.08%	**98.04 ± 0.03%**

Table 2. Results on the **V1** dataset. Performances are reported as accuracy measure.

Feature set	# Features	Best SVM	Best ANN
Handcrafted Grey	74	85.97%	86.84%
Handcrafted Red	74	83.76%	81.66%
Pretrained ResNet18	512	**95.61%**	94.56%
Pretrained GoogleNet	1024	90.36%	90.20%
Pretrained ResNet50	2048	95.11%	95.24%
Fine-tuned ResNet18	512	97.06%	
Fine-tuned GoogleNet	1024	96.99%	
Fine-tuned ResNet50	2048	**97.26%**	

Table 3. Results on the **V2** dataset. Performances are reported as accuracy measure.

Feature set	# Features	Best SVM	Best ANN
Handcrafted Grey	74	36.48%	24.62%
Handcrafted Red	74	9.71%	19.72%
Pretrained ResNet18	512	**77.19%**	71.19%
Pretrained GoogleNet	1024	62.47%	60.80%
Pretrained ResNet50	2048	75.94%	71.59%
Fine-tuned ResNet18	512	66.34%	
Fine-tuned GoogleNet	1024	68.82%	
Fine-tuned ResNet50	2048	**72.31%**	

Table 4. Proposed methodology. Feature set is given by the concatenation of pretrained ResNet18, GoogleNet, ResNet50, considering different numbers of principal components after the PCA. Results are shown on both the **V1** and **V2** datasets. Percentages represent accuracies.

# Features	128	256	512	1024	3584
ANN on V1	94.17%	94.47%	93.33%	94.42%	95.94%
SVM on V1	95.72%	96.34%	96.37%	95.24%	95.86%
ANN on V2	58.02%	64.43%	63.15%	67.67%	73.15%
SVM on V2	64.10%	64.15%	62.85%	61.56%	76.36%

For the best classifier of each category (handcrafted features, pretrained deep features, finetuned deep model), we computed the confusion matrix to assess how errors are distributed across the different classes. Confusion matrices are reported in Tables 5, 6 and 7.

Table 5. Confusion matrix on the **V2** dataset for the best handcrafted model.

	TUM	MUSC-STROMA	LYM	DEBRIS-MUCUS	NORM	ADI	BACK
TUM	44	25	0	32	0	0	2
MUSC-STROMA	292	1782	0	702	1	4	94
LYM	4	15	0	48	0	1	4
DEBRIS-MUCUS	46	82	0	329	0	1	27
NORM	1492	386	0	462	16	0	32
ADI	0	41	0	7	0	12	1
BACK	0	0	0	0	0	0	0

TRUE CLASS (vertical label) / PREDICTED CLASS (horizontal label)

4.1 Discussion and Explainability

Looking at the confusion matrices, we observed that handcrafted features are not able to well generalize on our dataset, whilst deep features are better suited for the task. In particular, the model trained with handcrafted features is not able to recognize any LYM tissue from our **V2** dataset. For the proposed method which combines features of different deep architectures, we showed that PCA could be a useful tool for reducing dimensionality without incurring in a decrease of accuracy. Among the pretrained models on the **V1** dataset, the proposed methodology slightly outperforms the best pretrained model alone, ResNet18, using also less features. For the SVM classifiers on the **V1** dataset, using more than 256 features after PCA does not result in measurable improvements.

Table 6. Confusion matrix on the **V2** dataset for the best pre-trained deep features model.

TUM	100	0	1	0	2	0	0
MUSC-STROMA	**234**	2047	32	**377**	118	21	46
LYM	1	3	41	9	15	0	3
DEBRIS-MUCUS	27	14	14	401	12	0	17
NORM	**129**	36	57	**147**	2011	0	8
ADI	0	8	0	28	6	19	0
BACK	0	0	0	0	0	0	0
	TUM	MUSC-STROMA	LYM	DEBRIS-MUCUS	NORM	ADI	BACK

PREDICTED CLASS

Table 7. Confusion matrix on the **V2** dataset for the best deep fine-tuned model.

TUM	101	1	0	1	0	0	0
MUSC-STROMA	**370**	1856	37	**434**	41	56	81
LYM	2	13	23	13	11	0	10
DEBRIS-MUCUS	8	14	3	423	11	1	25
NORM	**226**	65	56	**133**	1890	2	16
ADI	0	1	0	26	0	34	0
BACK	0	0	0	0	0	0	0
	TUM	MUSC-STROMA	LYM	DEBRIS-MUCUS	NORM	ADI	BACK

PREDICTED CLASS

We observed that frequent misclassification errors involved NORMAL and MUSC-STROMA patches which are predicted as TUMOUR or DEBRIS-MUCUS.

In order to assess the explainability of the obtained results, we considered different techniques. First, we looked at the t-SNE embeddings, to understand if deep features, also those obtained by pre-training on ImageNet, are meaningful for the problem under consideration. Figure 4a displayed that clusters are much better defined from the **V1** dataset. It is important to highlight that they considered tiles clearly belonging to only one class, whereas we also allowed the inclusion of patches more difficult to be classified.

The presence of a sub-cluster of TUM tiles can be seen within the MUSC_STROMA cluster. As stated above, MUSC_STROMA derives from the merging of simple and complex stroma classes, the latter including also sparse tumor cells. Thus, the TUM sub-cluster and the misclassification could be explained by both the class definition and, from a biological perspective, the fact that tumor tissue invades the surrounding stroma. Moreover, it could be observed in Fig. 4b that NORM cluster includes DEBRIS_MUCUS sub-cluster. Such a result makes sense because in this case mucus containing exfoliated

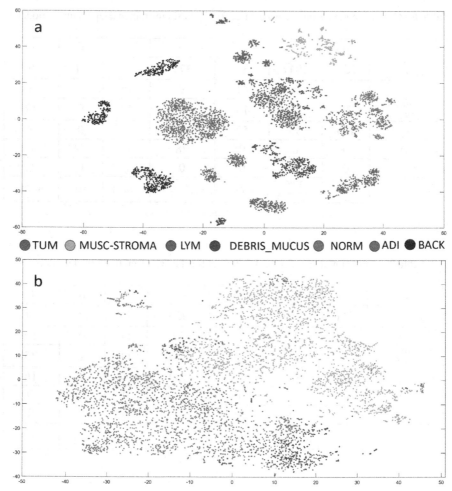

Fig. 4. t-SNE of the best classifier features; a) Fine-tuned ResNet50 features t-SNE of V1 dataset b) Pretrained ResNet18 features t-SNE of V2 dataset.

epithelial cells is mainly produced by the glands of the normal tissue component at the periphery of the tissue sample.

Then, we tried to see the activations of the fine-tuned deep models exploiting Grad-CAM method [42]. We can see from Fig. 5a-c and Fig. 5e-g the highlighted regions of sample images from **V1** and **V2** datasets. Figure 5d and Fig. 5h represent patches which have not been included into **V2** dataset since they were not clearly classifiable. In particular, Fig. 5d contains both MUSC_STROMA and TUM classes, whereas Fig. 5h contains both DEBRIS_MUCUS and NORM.

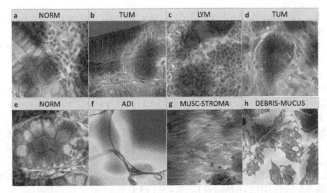

Fig. 5. Grad-CAM of Deep fine-tuned classifiers on the test set; **a, c, f,** from dataset V1 and **b, g, e** from dataset V2; **d, h** are patches not belonging to only one class. Labels are those in output of the classifier.

5   Conclusions and Future Works

In this work, three different methods have been compared for multi-class histology tissue classification in CRC. The most promising approach resulted to be to extract pretrained ResNet18 deep features from tiles combined with classification through SVM; in this way the classifier is able to generalize well on external datasets with good accuracy.

We also investigated explainability of our trained deep models observing that some misclassification issues are related to the biology of CRC. The multi-class tissue classification is a useful task in CRC histology, in particular to exploit a multi-layer approach including genomic data (mutational and transcriptional status).

The present paper could be considered a proof-of-concept because the multi-class tissue classification of digital histological images could, not only be extended to other malignancies, but also be considered as the preliminary step to explore, e.g., the relationship between the tumor, its microenvironment and genomic features.

Acknowledgments. This research was funded by Italian Apulian Region "Tecnopolo per la medicina di precisione", CUP B84I18000540002.

References

1. Siegel, R.L., et al.: Colorectal cancer statistics, 2020. CA. Cancer J. Clin. **70**, 145–164 (2020). https://doi.org/10.3322/caac.21601
2. Linder, N., et al.: Identification of tumor epithelium and stroma in tissue microarrays using texture analysis. Diagn. Pathol. **7**, 22 (2012). https://doi.org/10.1186/1746-1596-7-22
3. Kather, J.N., et al.: Multi-class texture analysis in colorectal cancer histology. Sci. Rep. **6**, 1–11 (2016). https://doi.org/10.1038/srep27988
4. Kather, J.N., et al.: Predicting survival from colorectal cancer histology slides using deep learning: a retrospective multicenter study. PLoS Med. **16**, 1–22 (2019). https://doi.org/10.1371/journal.pmed.1002730

5. Simonyan, K., Zisserman, A.: Very Deep Convolutional Networks for Large-Scale Image Recognition, 1–14 (2014)
6. Howard, A.G., et al.: Mobilenets: efficient convolutional neural networks for mobile vision applications. arXiv Prepr. arXiv:1704.04861 (2017)
7. Huang, G., Liu, Z., Van Der Maaten, L., Weinberger, K.Q.: Densely connected convolutional networks. In: 2017 IEEE Conference on Computer Vision and Pattern Recognition (CVPR), pp. 2261–2269. IEEE (2017). https://doi.org/10.1109/CVPR.2017.243
8. Kassani, S.H., Kassani, P.H., Wesolowski, M.J., Schneider, K.A., Deters, R.: Classification of histopathological biopsy images using ensemble of deep learning networks. In: CASCON 2019 Proc. - Conf. Cent. Adv. Stud. Collab. Res. - Proc. 29th Annu. Int. Conf. Comput. Sci. Softw. Eng., pp. 92–99 (2020)
9. Hochreiter, S., Schmidhuber, J.: Long Short-Term Memory. Neural Comput. **9**, 1735–1780 (1997). https://doi.org/10.1162/neco.1997.9.8.1735
10. Bychkov, D., et al.: Deep learning based tissue analysis predicts outcome in colorectal cancer. Sci. Rep. **8**, 1–11 (2018). https://doi.org/10.1038/s41598-018-21758-3
11. Kather, J.N., et al.: Collection of textures in colorectal cancer histology (2016). https://doi.org/10.5281/zenodo.53169
12. Kather, J.N., Halama, N., Marx, A.: 100,000 histological images of human colorectal cancer and healthy tissue (2018). https://doi.org/10.5281/zenodo.1214456
13. Macenko, M., et al.: A method for normalizing histology slides for quantitative analysis. In: 2009 IEEE International Symposium on Biomedical Imaging: From Nano to Macro, pp. 1107–1110 (2009). https://doi.org/10.1109/ISBI.2009.5193250
14. Altini, N., et al.: Pathologist's annotated image tiles for multi- class tissue classification in colorectal cancer (2021). https://doi.org/10.5281/zenodo.4785131
15. Ahonen, T., Matas, J., He, C., Pietikäinen, M.: Rotation invariant image description with local binary pattern histogram fourier features. In: Salberg, A.-B., Hardeberg, J.Y., Jenssen, R. (eds.) SCIA 2009. LNCS, vol. 5575, pp. 61–70. Springer, Heidelberg (2009). https://doi.org/10.1007/978-3-642-02230-2_7
16. Ojala, T., Pietikainen, M., Maenpaa, T.: Multiresolution gray-scale and rotation invariant texture classification with local binary patterns. IEEE Trans. Pattern Anal. Mach. Intell. **24**, 971–987 (2002). https://doi.org/10.1109/TPAMI.2002.1017623
17. Pietikäinen, M., Hadid, A., Zhao, G., Ahonen, T.: Computer vision using local binary patterns. Presented at the (2011). https://doi.org/10.1007/978-0-85729-748-8_14
18. Haralick, R.M., Dinstein, I., Shanmugam, K.: Textural features for image classification. IEEE Trans. Syst. Man Cybern. **SMC-3**, 610–621 (1973). https://doi.org/10.1109/TSMC.1973.4309314
19. Tamura, H., Mori, S., Yamawaki, T.: Textural features corresponding to visual perception. IEEE Trans. Syst. Man Cybern. **8**, 460–473 (1978). https://doi.org/10.1109/TSMC.1978.4309999
20. Bianconi, F., Álvarez-Larrán, A., Fernández, A.: Discrimination between tumour epithelium and stroma via perception-based features. Neurocomputing **154**, 119–126 (2015). https://doi.org/10.1016/j.neucom.2014.12.012
21. Tellez, D., et al.: Quantifying the effects of data augmentation and stain color normalization in convolutional neural networks for computational pathology. Med. Image Anal. **58**, 101544 (2019). doi:https://doi.org/10.1016/j.media.2019.101544
22. LeCun, Y., Bengio, Y., Hinton, G.: Deep learning. Nature **521**, 436–444 (2015). https://doi.org/10.1038/nature14539
23. Lecun, Y., Bottou, L., Bengio, Y., Haffner, P.: Gradient-based learning applied to document recognition. Proc. IEEE. **86**, 2278–2324 (1998). https://doi.org/10.1109/5.726791
24. Krizhevsky, A., Sutskever, I., Hinton, G.E.: 2012 AlexNet. Adv. Neural Inf. Process. Syst. (2012). https://doi.org/10.1016/j.protcy.2014.09.007

25. Zeng, G., He, Y., Yu, Z., Yang, X., Yang, R., Zhang, L.: InceptionNet/GoogLeNet - going deeper with convolutions. CVPR **91**, 2322–2330 (2016). https://doi.org/10.1002/jctb.4820
26. He, K., Girshick, R., Dollár, P.: Rethinking ImageNet Pre-training, pp. 1–10 (2018)
27. He, K., Zhang, X., Ren, S., Sun, J.: Spatial pyramid pooling in deep convolutional networks for visual recognition. IEEE Trans. Pattern Anal. Mach. Intell. **37**, 1904–1916 (2015). https://doi.org/10.1109/TPAMI.2015.2389824
28. He, K., Zhang, X., Ren, S., Sun, J.: Deep residual learning for image recognition. In: Proceedings of the IEEE Comput. Soc. Conf. Comput. Vis. Pattern Recognit. 2016-Decem, pp. 770–778 (2016). https://doi.org/10.1109/CVPR.2016.90.
29. Zeiler, M.D., Fergus, R.: Visualizing and understanding convolutional networks. In: Fleet, D., Pajdla, T., Schiele, B., Tuytelaars, T. (eds.) ECCV 2014. LNCS, vol. 8689, pp. 818–833. Springer, Cham (2014). https://doi.org/10.1007/978-3-319-10590-1_53
30. Araujo, T., et al.: Classification of breast cancer histology images using convolutional neural networks. PLoS ONE **12**, 1–14 (2017). https://doi.org/10.1371/journal.pone.0177544
31. Altini, N., et al.: Semantic segmentation framework for glomeruli detection and classification in kidney histological sections. Electronics. **9**, 503 (2020). https://doi.org/10.3390/electronics9030503
32. Altini, N., et al.: A deep learning instance segmentation approach for global glomerulosclerosis assessment in donor kidney biopsies. Electronics **9**, 1768 (2020). https://doi.org/10.3390/electronics9111768
33. Weiss, K., Khoshgoftaar, T.M., Wang, D.: A survey of transfer learning. J. Big Data **3**(1), 1–40 (2016). https://doi.org/10.1186/s40537-016-0043-6
34. Campanella, G., et al.: Clinical-grade computational pathology using weakly supervised deep learning on whole slide images. Nat. Med. **25**, 1301–1309 (2019). https://doi.org/10.1038/s41591-019-0508-1.
35. Schmauch, B., et al.: A deep learning model to predict RNA-Seq expression of tumours from whole slide images. Nat. Commun. **11**, 3877 (2020). https://doi.org/10.1038/s41467-020-17678-4
36. Fu, J., Singhrao, K., Cao, M., Yu, V., Santhanam, A.P., Yang, Y., Guo, M., Raldow, A.C., Ruan, D., Lewis, J.H.: Generation of abdominal synthetic CTs from 0.35T MR images using generative adversarial networks for MR-only liver radiotherapy. Biomed. Phys. Eng. Express. **6** (2020). https://doi.org/10.1088/2057-1976/ab6e1f.
37. Levy-jurgenson, A.: Spatial Transcriptomics Inferred from Pathology Whole-Slide Images Links Tumor Heterogeneity to Survival in Breast and Lung Cancer. 1–16 (2020).
38. Cascianelli, S., et al.: Dimensionality reduction strategies for CNN-based classification of histopathological images. In: De Pietro, G., Gallo, L., Howlett, R.J., Jain, L.C. (eds.) KES-IIMSS 2017. SIST, vol. 76, pp. 21–30. Springer, Cham (2018). https://doi.org/10.1007/978-3-319-59480-4_3
39. Tan, C., Sun, F., Kong, T., Zhang, W., Yang, C., Liu, C.: A survey on deep transfer learning. In: Kůrková, V., Manolopoulos, Y., Hammer, B., Iliadis, L., Maglogiannis, I. (eds.) ICANN 2018. LNCS, vol. 11141, pp. 270–279. Springer, Cham (2018). https://doi.org/10.1007/978-3-030-01424-7_27
40. Komura, D., Ishikawa, S.: Machine learning methods for histopathological image analysis. Comput. Struct. Biotechnol. J. **16**, 34–42 (2018). https://doi.org/10.1016/j.csbj.2018.01.001
41. der Maaten, L., Hinton, G.: Visualizing data using t-SNE. J. Mach. Learn. Res. **9**, (2008)
42. Selvaraju, R.R., Cogswell, M., Das, A., Vedantam, R., Parikh, D., Batra, D.: Grad-CAM: visual explanations from deep networks via gradient-based localization. Int. J. Comput. Vision **128**(2), 336–359 (2019). https://doi.org/10.1007/s11263-019-01228-7
43. Deng, J., Li, K., Do, M., Su, H., Fei-Fei, L.: Construction and analysis of a large scale image ontology. Presented at the (2009)

Plant Leaf Recognition Network Based on Fine-Grained Visual Classification

Wenhui Liu[1]([✉]), Changan Yuan[2], Xiao Qin[3], and Hongjie Wu[4]

[1] Institute of Machine Learning and Systems Biology, School of Electronics and Information Engineering, Tongji University, Shanghai, China
[2] Guangxi Academy of Science, Nanning 530007, China
[3] School of Computer and Information Engineering, Nanning Normal University, Nanning 530299, China
[4] School of Electronic and Information Engineering, Suzhou University of Science and Technology, Suzhou 215009, China

Abstract. Plant classification and recognition research is the basic research work of botany research and agricultural production. It is of great significance to identify and distinguish plant species and explore the relationship between plants. In recent years, most of the research methods focus on feature extraction and feature engineering related aspects. In this paper, a plant leaf recognition method based on fine-grained image classification is proposed, which can better find the regional block information of different species of plant leaves. In this study, the hierarchical and progressive training strategy is adopted, the method of cutting and generating jigsaw is used to force the model to find information of different granularity levels. The experiment proves that the model trained by the fine-grained classification method can better solve the problems of large intra-class spacing and small inter-class spacing of plant slices.

Keywords: Plant leaf recognition · Progressive training fine-grained visual classification

1 Introduction

The purpose of this study is to recognize and classify plant leaves. The research on plant classification and recognition is the basic research work of botany research and agricultural production, which is of great significance. Compared with other parts of plants, the shape and structure of plant leaves are generally more stable and suitable for two-dimensional image processing. At the same time, the digital image of the blade can be easily collected. Its color, shape, texture and other characteristics can be used as a classification basis. Therefore, the classification of plant species based on plant leaf image is the most direct, simple and effective method, which is also the trend of future digital plant research. From the perspective of research field, plant leaf images have the characteristics of small intra-class spacing and large inter-class spacing. The pictures of plant leaves of different plant species are very similar, and the pictures of the same plant leaf have great intra-class differences due to the difference in posture, background and shooting angle.

Fine-grained image classification (FGVC), also known as subcategory image classification, is a hot research topic in computer vision, pattern recognition and other fields in recent years [1, 4, 5, 8]. Its purpose is to perform a more detailed sub-category division of images belonging to the same basic category. The main goal of fine-grained classification is to find a distinguishable area block in the image that can distinguish the two categories of images, and can better represent the characteristics of these distinguishing regional blocks. Since deep convolutional network can learn very robust image feature representation, most of the methods for fine-grained image classification are based on deep convolutional network. These methods can be roughly divided into four directions: fine-tuning method based on conventional image classification network, method based on fine-grained feature learning, method based on detection and alignment of target blocks, and method based on visual attention mechanism. So far, the most effective solutions have relied on extracting fine-grained feature representations in local discriminant regions, or aggregating these local discriminant features for computation by either explicitly detecting semantic portions or implicitly by salient localization [2, 7].

Based on the idea of FGVC, this paper designs a hierarchical progressive training method. The training network is designed into different levels, from shallow to deep, the shallow network is responsible for the extraction of finer grained feature information, and the deep network is responsible for the extraction of coarser grained feature information. We adapt a jigsaw generation method to form different levels of granularity, allow the network to focus on different scales of features as per prior work. We not only accumulate the granularity information at different levels, but also transfer the granularity characteristic information obtained at the shallow level to the deep network structure. With the help of the previous network, the deep network can identify the more coarse-grained confidence. This method of training the network in a hierarchical and progressive way can effectively combine the characteristic information of different granularity.

2 DataSet and Methods

We will introduce in detail the new fine-grained classification method in with figures, and explain how we train the plant leaf dataset through a progressive training method. We will introduce the loss function we used, and discuss the advantages of the hierarchical training method in the plant leaf dataset.

2.1 DataSet

We use ICL plant leaf dataset in this paper. The ICL plant leaf data set was organized by the Institute of machine learning and systems biology, Tongji University, and was established in cooperation with Hefei botanical garden. The ICL dataset covers 220 plant species, and each plant category contains 26–1078 pictures, a total of 16,851 plant pictures as shown in Fig. 1:

each plant category contains 26–1078 pictures

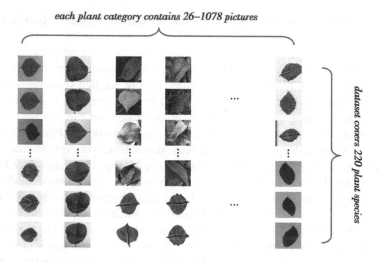

dataset covers 220 plant species

a total of 16,851 plant pictures

Fig. 1. The ICL dataset

Due to the different instruments used in the shooting, the size of each image is different, data preprocessing is required for the dataset before the experiment. For each image, the longest side of each image is scaled to 256 pixels in length, and the other side is scaled in the same proportion according to the original proportion of the image, finally obtaining a standard data set with 256 * 256 pixels in size (256 pixels are selected to facilitate the operation of sliced and jigsaw of each image, so the image pixel should be in $2^n * 2^n$ format).

We need data augmentation because there are few samples in ICL dataset. In this paper, the method of rotation, flipping and cutting is adopted. Each image will be flipped horizontally, vertically, mirror-image and rotated 180°. In addition, each image is also randomly selected from the upper left corner, upper right corner, lower left corner, lower right corner and middle of the original image to intercept images of 256 * 256 size. After the interception is completed, the final data set is 10 times the size of the original data set.

2.2 Method

The experimental method in this paper adopts hierarchical and progressive training base on Fine-Grained Visual Classification (FGVC), and divides the whole training network into N levels for training. The low-level network is trained first, and then the high-level network is trained [1]. The advantage of using progressive network training is that the representation ability at the lower level of the network is very limited, which forces the lower stage of the network to learn the discriminating local features. Compared with multi-granularity training when the whole network is trained, progressive training enables the network to gradually learn the feature information from the local to the global.

The model can use different network models as the backbone, Let's take vgg16 as an example, We will split vgg16 into five convolution blocks for training, each convolutional block consists of 2 convolutional layers and a maximum pooling layer, we will freeze the $i + 1_{th}$ to n_{th} convolution blocks when we training the $i + 1_{th}$ convolution block. Only train the output of one of the convolutional blocks at each step, for each training, the output will be directly sent to the classifier, which is composed of 3 fully connected layers, and iteratively update the obtained loss value, so that the current convolution block can learn the feature information of the granularity of this layer (Fig. 2).

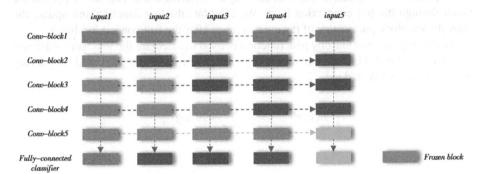

Fig. 2. Progressive training method

In each iteration training of the deeper network, the iteration parameters of the previously trained network are gradually reduced by simulated annealing method. Finally, each convolution block is connected in series to obtain a network model capable of feeling different granularity. It should be clear that the reason why the previously trained convolution blocks are constantly updated here is that for input data of different granularity, even though the shallow network may have been updated in the previous step, it still helps the model work together.

In this paper, the training data are processed in progressive training, the re-enhanced data were used for multi-batch training, the data adopted for the shallow network training is the recombination data obtained after the sliced jigsaw puzzle. The calculation method for the number of slices is 2^{n-i}, where i represents the i_{th} layer network and n represents the total number of global network layers. It means that when the network level is shallower, the resulting slice puzzle size is smaller, which forces the shallow network to be more focused on learning the local features in the data. As the training network level deepens, this incremental property allows the model to locate the different information from the local details to the global structure, and these features are gradually sent to higher stages, rather than learning all the simultaneous granularity [14, 15] (Fig. 3).

2.3 Loss Function

We adopt the progressive training method of training the shallow network first and then the deep network. In this way, rather than training the entire network directly, this incremental nature allows the model to localize differentiated information from local

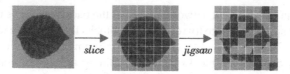

Fig. 3. Data slicing and jigsaw

details to global structures as these features are progressively carried to higher stages. In the process of progressive training, each layer of network will eventually predict the result through the full connection layer. We calculate the loss function and update the convolution block parameters of the current layer after the full connection block.

In this paper, cross entropy loss function is used to calculate the loss, for each layer of network output. The loss between the ground truth y and the predicted probability distribution is calculated as:

$$L(y, y^{truth}) = -\sum_{j=1}^{T} y_j \times \log y_j$$

3 Architecture

In this paper, the training of the input data is after slicing reassemble the 256 * 256 pixels image data, is aimed at different stages for different granularity of data input, each original plant leaf image was cut into 2^k pieces by slicing method in the experimental process, where k represented the hyperparameter. The segmented slice data will be shuffled and recombined, so that for each training process, more attention will be paid to the finer-grained information during training [12, 14].

For the slice recombination operation, it is necessary to ensure that the slice size of the deep network is larger than that of the shallow network, because in this way the deep network can capture more macroscopic information in a wider field of view.

The simulated annealing method will be adopted to gradually reduce the updating step size of the parameters of the shallow network after the training of the shallow network is completed. The purpose is to combine the progressive training method and enable the deep network to make use of the fine-grained information obtained from the shallow layer to better help the network to identify the region blocks with large differentiated information within and between classes.

The network structure in this paper is shown in the Fig. 4. Here, 4 convolution blocks are taken as an example. Each convolution block contains 2 convolution layers and a maximum pooling layer. The entire training process is divided into four steps, respectively corresponding to four different grained slicing puzzles data, corresponding to each of these steps will training level of convolution, will eventually get the results of the input to the three full connection layer and a softmax function of classifier, the classification forecast the output of the cross entropy loss function is used to calculate the loss value, update the current levels of network parameters. The maximum granularity

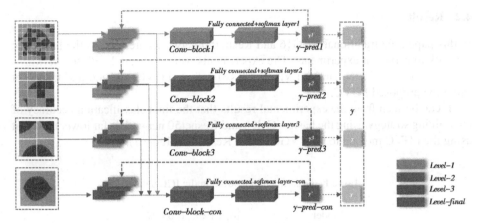

Fig. 4. The architecture of our method

of the complete image data is entered in the level-final, the parameters of the previous trained shallow network will also affect the training of the next layer.

With the help of all previously trained network layers, the last layer of network can perceive the information features of different granularity, and the training model finally obtained can perceive the information results after different granularity.

4   Experiment

In this section, we will evaluate the performance of the proposed approach on the ICL dataset and compare the performance of some other models in general. The experimental details will be introduced in Sect. 4.1. Section 4.2 will introduce the different classification results obtained by different models, and we will discuss the method proposed in this paper according to the results.

4.1   Implementation Details

We perform all experiments using PyTorch, the training phase of the experiment were completed on two Nvidia Tesla K80 graphics cards, and the memory of each graphics card was 24G.

For the proposed method, VGG16 and Resnet50 are used as the backbone network. The input image size is 256 * 256. In order to avoid gradient explosion, we use gradient clamping. Adam was used to minimize the loss function and the hyperparameters were set to beta1 = 0.9, beta2 = 0.999, epsilon = 1e−8.

Using the progressive training method, we need to initialize the learning speed of the newly added convolutional layer and full connection layer to be inconsistent with the already trained learning rate, and set the learning rate of the newly added network layer to 0.003. The learning rate for the already trained layers is reduced by following simulated annealing during training. For all the above models, we trained them with 300 epochs and set the batch size to 16.

4.2 Result

In this paper, the traditional VGG16 and Resnet50 networks are used as the backbone network to carry out experiments. The experiments were compared with the networks using the proposed Fine-Grained Visual Classification (FGVC) method and those not using the proposed FGVC.

It can be seen from the experimental results that the final classification accuracy of the training strategy using the basic VGG16 and Resnet50 network is far lower than that using the FGVC method to train VGG16 and Resnet50 (Table 1).

Table 1. Experimental results on the ICL dataset

Model	Top1 (%)	Top3 (%)
VGG16	86.44	89.23
ResNet50	87.78	90.03
VGG16 + FGVC	89.03	92.39
ResNet50 + FGVC	90.33	91.84

The traditional VGG16 and Resnet50 can achieve better results when classifying the data with large distance between classes. For the characteristics of small inter-class spacing and large intra-class spacing in plant leaf images, the training method based on progressive training based on FGVC can make the network model pay more attention to feature information of different granularity. Compared with the multi-granularity training when the whole network is trained, the progressive training based on FGVC makes the network gradually learn the feature information from the local to the global, and achieves a good classification effect.

5 Conclusion

We propose a new fine-grained classification method for processing plant leaf datasets with small inter-class spacing and large intra-class spacing. Through the progressive training and data slice jigsaw based on FGVC, the model can fuse the characteristic information of different granularity obtained from different layers of network. Experimental results show that the progressive training method enables the network to gradually learn the feature information from local to global, compared with the multi-granularity training when the whole network training is carried out simultaneously. This also proves that the fine-grained image classification method can effectively solve the problem that the feature distance in plant leaves is small and the feature distance of different categories is large.

Acknowledgements. This work was supported by the grant of National Key R&D Program of China, No. 2018AAA0100100; in part by supported by National Natural Science Foundation of China, Nos. 61861146002, 61772370, 61732012, 61932008, 61772357, 62073231, and 62002266;

in part by the Scientific & Technological Base and Talent Special Program of the Guangxi Zhuang Autonomous Region, GuiKe AD18126015; and in part by "BAGUI Scholar" Program of Guangxi Province of China.

References

1. Du, R., Chang, D., Bhunia, A.K., et al.: Fine-grained visual classification via progressive multi-granularity training of jigsaw patches (2020)
2. Technicolor, T., Related, S., Technicolor, T., et al.: ImageNet classification with deep convolutional neural networks [50]
3. Wang, X.F., Huang, D.S., Xu, H.: An efficient local Chan-Vese model for image segmentation. Pattern Recogn. **43**(3), 603–618 (2010)
4. Berg, T., Belhumeur, P.N.: POOF: part-based one-vs.-one features for fine-grained categorization, face verification, and attribute estimation. In: Computer Vision and Pattern Recognition. IEEE (2013)
5. Luo, W., Yang, X., Mo, X., et al.: Cross-X learning for fine-grained visual categorization (2019)
6. Wu, Y., Zhang, K., Wu, D., et al.: Person re-identification by multi-scale feature representation learning with random batch feature mask. IEEE Trans. Cogn. Dev. Syst. (2020)
7. Li, A.X., Zhang, K.X., Wang, L.W.: Zero-shot fine-grained classification by deep feature learning with semantics. Int. J. Autom. Comput. (2019)
8. Huang, G., Liu, Z., et al.: Densely connected convolutional networks. In: Proceedings of the IEEE Conference on Computer Vision and Pattern Recognition, vol. 1. no. 2 (2017)
9. Chang, D., Ding, Y., Xie, J., et al.: The devil is in the channels: mutual-channel loss for fine-grained image classification. IEEE Trans. Image Process. (99), 1 (2020)
10. Li, B., Huang, D.S.: Locally linear discriminant embedding: an efficient method for face recognition. Pattern Recogn. **41**(12), 3813–3821 (2008)
11. Huang, D.S.: Systematic theory of neural networks for pattern recognition. Publishing House of Electronic Industry of China (1996)
12. Huang, D.S., Du, J.-X.: A constructive hybrid structure optimization methodology for radial basis probabilistic neural networks. IEEE Trans. Neural Netw. **19**(12), 2099–2115 (2008)
13. Wei, C., Xie, L., Ren, X., et al.: Iterative reorganization with weak spatial constraints: solving arbitrary jigsaw puzzles for unsupervised representation learning. In: IEEE
14. Hinton, G.E., Salakhutdinov, R.R.: Reducing the dimensionality of data with neural network. Science **313**(5786), 504–507 (2006)
15. Won, Y., Gader, P.D., Coffield, P.C.: Morphological shared-weight networks with applications to automatic target recognition. IEEE Trans. Neural Netw. **8**(5), 1195–1203 (1997)
16. Son, K., Hays, J., Cooper, D.B.: Solving square jigsaw puzzles with loop constraints. In: Fleet, D., Pajdla, T., Schiele, B., Tuytelaars, T. (eds.) ECCV 2014. LNCS, vol. 8694, pp. 32–46. Springer, Cham (2014). https://doi.org/10.1007/978-3-319-10599-4_3
17. Bai, X., Yang, M., Huang, T., Dou, Z., Yu, R., Xu, Y.: Deep-person: learning discriminative deep features for person re-identification. arXiv: Comput. Vis. Pattern Recogn. (2017)
18. Serre, T., Riesenhuber, M., Louie, J., Poggio, T.: On the role of object-specific features for real world object recognition in biological vision. In: Bülthoff, H.H., Wallraven, C., Lee, S.-W., Poggio, T.A. (eds.) BMCV 2002. LNCS, vol. 2525, pp. 387–397. Springer, Heidelberg (2002). https://doi.org/10.1007/3-540-36181-2_39
19. Huang, D.S.: Radial basis probabilistic neural networks: model and application. Int. J. Pattern Recognit. Artif. Intell. **13**(7), 1083–1101 (1999)

20. Wang, X.-F., Huang, D.S.: A novel density-based clustering framework by using level set method. IEEE Trans. Knowl. Data Eng. **21**(11), 1515–1531 (2009)
21. Shang, L., Huang, D.S., Du, J.-X., Zheng, C.-H.: Palmprint recognition using fast ICA algorithm and radial basis probabilistic neural network. Neurocomputing **69**(13–15), 1782–1786 (2006)
22. Xiao, T., Li, S., Wang, B., Lin, L., Wang, X.: Joint detection and identification feature learning for person search. Comput. Vis. Pattern Recogn, 3376–3385 (2017)
23. Wu, D., Zheng, S., Yuan, C., Huang, D.: A deep model with combined losses for person re-identification. Cogn. Syst. Res. (2018)
24. Zhao, Z.-Q., Huang, D.S., Sun, B.-Y.: Human face recognition based on multiple features using neural networks committee. Pattern Recogn. Lett. **25**(12), 1351–1358 (2004)
25. Yin, C., et al.: Kernel pooling for convolutional neural networks. In: IEEE Conference on Computer Vision & Pattern Recognition IEEE (2017)
26. He, K., et al.: Deep residual learning for image recognition. In: Proceedings of the IEEE Conference on Computer Vision and Pattern Recognition (2016)
27. Huang, D.S., Zhao, W.-B.: Determining the centers of radial basis probabilistic neural networks by recursive orthogonal least square algorithms. Appl. Math. Comput. **162**(1), 461–473 (2005)
28. Huang, D.S.: Application of generalized radial basis function networks to recognition of radar targets. Int. J. Pattern Recognit. Artif. Intell. **13**(6), 945–962 (1999)
29. Simonyan, K., Zisserman, A.: Very deep convolutional networks for large-scale image recognition. arXiv preprint arXiv:1409.1556 (2014)
30. Huang, D.S., Ma, S.D.: Linear and nonlinear feedforward neural network classifiers: a comprehensive understanding. J. Intell. Syst. **9**(1), 1–38 (1999)

Anomaly Detection Based on Video Prediction and Latent Space Constraints

Shuanggen Fan$^{(\boxtimes)}$ ⓘ and Yanxiang Chen ⓘ

Hefei University of Technology, Hefei, China

Abstract. Anomaly detection has wide applicability in areas such as security, deception detection, and video surveillance. For video, anomaly detection usually refers to the discovery of events outside the expected behavior. Most existing anomaly detection methods are implemented using reconstruction errors, but they cannot ensure that abnormal events have sufficient reconstruction errors. To avoid these problems, this article proposes a new anomaly detection model based on video prediction that adopts dual autoencoder structure, including a discriminator and generator. The use of the same structure in the discriminator and generator can effectively prevent instability in generative adversarial network training. Use optical flow to extract the time information of the video, and use latent space constraints, gradient constraints, and intensity constraints to improve the quality of the generated prediction frames. In addition, group normalization was employed for further optimization of the overall model performance. Excellent experimental results were achieved using the Chinese University of Hong Kong, University of California—San Diego, Shanghai-Tech, and UCF_Crime100 datasets.

Keywords: Anomaly detection · Deep learning · Generative adversarial network · Prediction frame

1 Introduction

Although significant progress has been made in computer vision, this technology usually relies heavily on large-scale labeled datasets, limiting its applicability in certain fields. As anomalies are widely distributed and not fixed and the amount of abnormal data is much smaller than that of normal data, it is difficult to collect existing abnormal data and classify them as a traditional means of learning normal data distributions. To solve these problems, one-class [1] theory has been proposed. one-class uses only one type of data during training, learn distribution of such data, and does not consider the actual category of the data. A generative model based on this logic emerged, and only normal data were used during training.

Based on the characteristics of generative adversarial network (GAN) [2], which can learn data distributions, GAN models have become new tools for anomaly detection. GANomaly [3] combined a GAN with anomaly detection for the first time and could effectively improve the performance of anomaly detection models. In the detection of anomalies in video data, many studies have focused on reconstruction-based methods

© Springer Nature Switzerland AG 2021
D.-S. Huang et al. (Eds.): ICIC 2021, LNCS 12836, pp. 535–546, 2021.
https://doi.org/10.1007/978-3-030-84522-3_44

[4–6], which typically use single video frames for training, but didn't use the associated information between the video frames. The motion between video frames can be captured by rapidly changing the parameters of each frame over time, which is called optical flow [7, 8]. Liu et al. [9] were the first to introduce optical flow into video anomaly detection. Other researchers have noted the potential of learning features from videos [10–12]. In addition, abnormal events in videos are usually accompanied by blurred video frames, color distortion, and sudden changes in optical flow. Therefore, video prediction based on previous frames is a more reasonable anomaly detection method than reconstruction. However, the imbalance between the generator and discriminator in the above model can easily lead to training instability, which limits the ability of GANs to generate prediction frames.

This paper proposes a GAN-based dual autoencoder model framework that uses conventional gradients, intensity [13], and optical flow [8, 14] to limit the appearance and temporal information of the predicted frame. Unlike the previous anomaly model based on video prediction, this framework uses latent space constraints and group normalization. Inspired by VAEGAN [15], constraining the latent variables of normal data to a Gaussian distribution helps to generate reasonable and clear data. In addition, due to this characteristic and the fact that most video frames are large and the batch size is small, we used the group normalization method proposed by Wu and He [16] to replace the traditional batch normalization method to optimize the anomaly detection performance.

The proposed model makes the following contributions:

1. It can make the generation of predicted frames more stable than is possible using the existing models by employing a dual autoencoder framework.
2. Construct latent space constraints to improve the generation effect of predicted frames.
3. It employs group normalization instead of traditional batch normalization to achieve optimal performance and exhibits better performance than existing models when applied to multiple public datasets.

2   Related Work

2.1   Traditional Anomaly Detection Methods

Traditional anomaly detection methods are based on Euclidean distance and use clustering or the nearest field for evaluation. For example, support vector machines [17] and one-class neural networks [1] learn the boundaries of normal data as the main starting point. k-means clustering [18] can also be used to establish the boundary of normal data. However, these methods often do not perform well when processing high-dimensional and complex data.

2.2   Reconstruction-Based Methods

Reconstruction-based methods are widely used in image anomaly detection and have often been adapted for video anomaly detection. Jolliffe [19] used principal component analysis (PCA) to learn a representation of the normal model, but the limited features resulted in insufficient performance. Kim and Grauman [7] used probabilistic PCA to model the optical flow features. Subsequently, Zhao et al. [20] proposed a three-dimensional convolution to fit a normal frame. Mahadevan et al. [21] combined a dynamic texture mixture model with a Gaussian model. Further, Chong and Chong [10] and Luo et al. [22] proposed deep learning methods to use space and time information simultaneously. In addition, recurrent neural networks (RNNs) and long short-term memory variants have been widely used for sequential data modeling. Lu et al. [5] and Zhao et al. [23] reformulated the traditional sparse coding method and superimposed the RNN for anomaly detection. Based on the later fusion strategy to combine motion and appearance features to process the video abnormal event detection framework Unmasking [24], but it is very difficult to find suitable fusion motion and appearance features. Morais et al. [25] proposed a combination of orbit prediction and anomaly reconstruction. For video anomaly detection, these reconstruction-based methods cannot guarantee that the difference in reconstruction of abnormal events is large enough, and with the improvement of fraudulent techniques, normal data and abnormal data are more similar, which makes reconstruction-based methods encounter bottlenecks.

2.3 Video Prediction-Based Methods

Predictive learning has attracted increasing attention in video anomaly detection. Video prediction requires comparing the prediction frame $I^{'}$ and real frame I to identify differences. When the difference is small, the real frame is assumed to be normal; otherwise, it is assumed to be anomalous. Lotter et al. [26] designed a predictive neural network in which each layer makes local predictions and forwards the deviations of these predictions to subsequent networks. Their approach captures the key aspects of self-centered and object motions in a video and learns the representation of the estimated steering angle. Thus, the authors demonstrated the feasibility of utilizing predictive learning as a framework. Mathieu et al. [13] proposed a network with adversarial training to generate more natural predictions of future video frames, which they applied to rain prediction. Use video-level weak tags to learn anomalous Real-world model [27], but it shows poor performance in the face of unknown and novel anomalous data. Ravanbakhsh et al. [28] used a GAN to learn the normal distribution, and Liu et al. [9] developed a generative model-based U-net for video prediction, which introduced time information to ensure the consistency of normal data. Our model is constrained not only at the input level, but also at the latent space level. In addition, in conventional GAN architecture, the generator and discriminator compete with each other, it will produce the imbalance of capability between two subnetworks, leading to an unstable training process. We use dual autoencoders for better balance.

3 Proposed Model

The objective of this study was to train the predictor model to predict the normal frames accurately. During training, the first few frames are used as model inputs for prediction, so that the predictor can predict future frames of normal video data. During the test, event X was fed into the predictor. If the generated frame is consistent with the expected performance, the event is likely to be normal; otherwise, it may be abnormal.

For video, we divided the video dataset into training and test sets. The training set $M = \{m_1, ..., m_\alpha\}$ contained α normal events. The test set $N = \{n_1, ..., n_\beta, ..., n_{\beta+\gamma}\}$ contained β normal and γ abnormal events. Each video event X contains t frames $X = \{x_1, ..., x_t\}$, $X \in M \cup N$, and x_i represents the i-th frame of video X. $X_T = \{x_1, ..., x_{t-1}\}$ is the input to the predictive model, and x_t is a real frame of video X.

3.1 Network Architecture

The entire network model uses a GAN as the basic architecture, as shown in Fig. 1. Use optical flow to capture time information between video frames, the generator produces a prediction frame that needs to be trained with the discriminator as the opponent. To use time information effectively in the video. Calculate the optical flow $f(x_{t-1}, x_t')$ between prediction frame x_t' and previous frame x_{t-1} as well as the optical flow $f(x_{t-1}, x_t)$ between real frame x_t and previous frame x_{t-1} to obtain the corresponding optical flow loss. In addition, (x_t', x_t) is the discriminator input. A GAN was adopted as the overall network, the generator produces a prediction frame that needs to be trained with the discriminator as an adversary. To use the time information in the video effectively, the optical flows between prediction frame x_t' and previous frame x_{t-1} and between real frame x_t and previous frame x_{t-1} are calculated as f_t' and f_t, respectively, to obtain the optical flow loss. The discriminator is autoencoder. x_t and x_t' are the discriminator inputs, and the output is x_t'' or x_t'''.

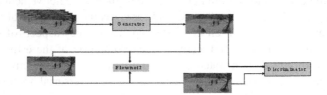

Fig. 1. The flow chart of the proposed model

Generator Network
Figure 2 shows the generator, including the encoder G_E and decoder G_D. The deep convolutional network is used to build the model framework and to standardize the dimensions of the original input video frame as 256×256. $X_T = \{x_1, ..., x_{t-1}\}$ was used as the input into G_E. Only normal data is used for training, in the latent space, inspired by VAEGAN [15], we use latent space constraints to make the latent variables of normal data tend to Gaussian distribution and improve the effect of normal data generation. The output of G_E is the mean μ and the variance σ^2, and then randomly sampled from the Gaussian distribution to obtain the latent variable of the previous $t - 1$ frame as the input of the decoder G_D. The decoder G_D corresponding to the encoder G_E uses skip connection and other methods to obtain the detailed features of the encoder G_E, which is very helpful for generating the final prediction frame. Finally, to stabilize the overall GAN training, the proposed model uses the activation function of the last output layer

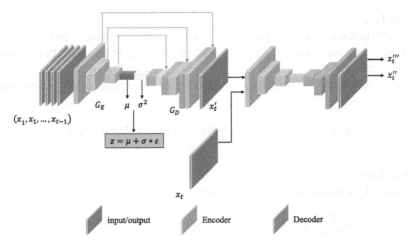

$(x_1, x_1, ..., x_{t-1})$

G_E μ σ^2 G_D x_t'

$z = \mu + \sigma * \varepsilon$

x_t''

x_t'''

x_t

▮ input/output   ▮ Encoder   ▮ Decoder

Fig. 2. Detailed structure diagram of the proposed model

of the decoder. In addition, use intensity constraints and gradient constraints to further improve the quality of the prediction frame generation in appearance.

Discriminator Network

The instability of most GANs is caused by an imbalance between the generator and discriminator. To ensure that the model is trained, the discriminator and generator have the same structure to obtain a better balance during training. The discriminator aims to achieve a discrimination effect on x_t and x_t'. The final training goal of the model is $x_t = G_D(G_E(x_1, ..., x_{t-1}))$.

3.2 Loss Function

The GAN in the model developed in this study obtains $X_T = \{x_1, ..., x_{t-1}\}$ and its latent variables z through the encoder G_E, generates the video prediction frames x_t' from the latent variables through the decoder G_D, and sends the generated prediction frames x_t' and real video frames x_t to the discriminator for discrimination. To realize the minimax game $\underset{G\ D}{minmax} V(D, G)$ between the generator and discriminator to achieve Nash equilibrium, the following definition is used:

$$V(D, G) = E_{x_t \sim p_M}\left[\log D(x_t)\right] - E_{X_T \sim p_M}[\log(D(G(X_T)))] \tag{1}$$

Traditional discriminator leads to an imbalance between the capabilities of the generator and discriminator network, which can easily lead to instability in model training and hinder the ability of the GAN to generate predicted frames, Inspired by Vu [29], the discriminator is the same as the generator, both using autoencoder.

Discriminator Loss

The objective of the discriminator is to reconstruct the real video frame x_t instead of reconstructing the generated prediction frame x_t', as shown below:

$$l_D = E_{x_t \sim p_{x_t}}||x_t - D(x_t)||_2 - ||G(x_t) - D(G(x_t))||_2 \tag{2}$$

Intensity Loss

The strength loss reduces the distance between the prediction frame x_t' and real frame x_t. The intensity loss increases the similarity of all pixels in the RGB space. In addition, the generator hopes that the produced prediction frame can fool the discriminator as much as possible, which can be optimized for learning the input data and is represented as follows:

$$l_{con} = E_{x_t \sim p_{x_t}} ||x_t - G(x_t)||_2 + ||G(x_t) - D(G(x_t))||_2 \tag{3}$$

Gradient Loss

The gradient loss is used to sharpen the generated image:

$$l_{gdl} = \sum_{i,j} |||x_t'(i,j) - x_t'(i-1,j)| - |x_t(i,j) - x_t(i-1,j)|||_1$$
$$+ |||x_t'(i,j) - x_t'(i,j-1)| - |x_t(i,j) - x_t(i,j-1)|||_1 \tag{4}$$

where $x_t(i,j)$ represents the spatial position (i,j) in the video frame x_t.

Optical Flow Loss

To use time information, we used FlowNet [30] to estimate the optical flow. The optical flow loss is given by

$$l_{fl} = E_{x_{t-1},x_t \sim p_M x_t' \sim p_{x_t'}} ||f(x_{t-1}, x_t) - f(x_{t-1}, x_t')||_1 \tag{5}$$

where f represents FlowNet [30].

Latent Space Loss

To obtain more obvious features of the video frames, the latent space of the normal and anomalous video frames can be increased. We imposed certain constraints on the latent space such that its distribution tended to a normal distribution. μ and σ^2 represent the mean and variance, respectively. KL (Kullback-Leibler) distance calculates the difference between two distributions, which is defined as follows:

$$l_{kl} = KL(N(\mu, \sigma^2)||N(0,1)) = \frac{1}{2}(-lg\sigma^2 + \mu^2 + \sigma^2 - 1) \tag{6}$$

Where $z = \mu + \sigma * \varepsilon$; ε follows a standard normal distribution with a mean of 0 and variance of 1.

When D is being trained, l_D is used; when G is being trained, only normal samples are available. Thus, the generator learns only the distributions of the prediction frames and input features. The above losses can be used to derive the total loss function of the proposed generator:

$$l_G = \lambda_{con} * l_{con} + \lambda_{gdl} * l_{gdl} + \lambda_{fl} * l_{fl} + \lambda_{kl} * l_{kl} \tag{7}$$

Among these losses, λ_{gdl}, λ_{con}, λ_{fl}, λ_{kl}, and λ_z are the weight parameters. In the actual experiment, the following values were set: $\lambda_{gdl} = 1$, $\lambda_{con} = 1$, $\lambda_{fl} = 2$, and $\lambda_{kl} = 0.4$.

3.3 Choice of Anomaly Score

In the testing phase, the anomaly score is usually selected based on the difference between the predicted and actual frames. The most commonly used method is to calculate the Euclidean distance between the predicted and actual frames, as shown in Eq. (8):

$$MSE(X) = \frac{1}{mn} \sum_{i=0}^{m-1} \sum_{j=0}^{n-1} [x_t(i,j) - x_t'(i,j)]^2 \tag{8}$$

where m, n represents the size of video frame x_t, $x_t(i,j)$ represents the spatial position of video frame x_t, and X represents the corresponding video event. However, a paper study by Mathieu et al. [12] indicated that the PSNR (Peak Signal to Noise Ratio) is a better image quality evaluation index and that its final evaluation result is often more accurate than that achievable by human vision. Therefore, we chose the PSNR indicator for abnormality assessment. The anomaly score of the test video event was calculated as follows:

$$PSNR(X) = 10\log_{10} \frac{max[x_t']^2}{MSE(X)} \tag{9}$$

Among $max[x_t']^2$ is the maximum possible pixel value of generated prediction video frame x_t'. From Eqs. (8) and (9), it can be seen that the closer the generated prediction frame is to the original real frame, the lower the MSE and the higher its PSNR, which means the higher the quality of the generation. We can obtain the final set of abnormal scores $PSNR = \{PSNR(X), X \in N\}$. Subsequently, we can normalize $PSNR(X)$ in each test video to obtain Eq. (10):

$$A(X) = \frac{PSNR(X) - min(PSNR)}{max(PSNR) - min(PSNR)} \tag{10}$$

Finally, an appropriate threshold τ is selected according to the abnormal score set to distinguish between abnormal and normal events, which can be expressed as:

$$y(X) = \begin{cases} 0 \; A(X) > \tau \\ 1 \; A(X) \le \tau \end{cases} \tag{11}$$

When $y = 0$ and 1, representative event X is a normal and abnormal event, respectively.

3.4 Datasets

To evaluate the effectiveness of the proposed model for different datasets, we applied it to the Chinese University of Hong Kong (CUHK), University of California—San Diego (UCSD), ShanghaiTech, and UCF_Crime100 datasets.

In the CUHK dataset, the size of the people in the videos may change because of the shooting location and angle. This dataset contains 16 training videos and 21 test videos with 47 anomalies, including people throwing objects, wandering, and running.

The UCSD dataset consists of pedestrian videos captured by cameras at two observation points. The anomalous conditions were related to passing vehicles, such as bicycles and cars. The dataset was divided into two subsets, denoted as ped1 and ped2. ped1 included 34 training videos and 36 test videos, whereas ped2 contained 16 training videos and 12 test videos.

ShanghaiTech is a new large-scale dataset taken from the streets of Shanghai. It contains 330 training videos and 107 tests, including 130 anomalies. There are 13 scenarios and various types of anomalies.

UCF_Crime100 contains 1900 actual surveillance videos, including 950 normal videos and 950 anomalous videos. In the experiment, we selected the area under the receiver operating characteristic (ROC) curve (AUC) to evaluate the results.

3.5 Ablation Experiment

The effects of certain constraints on the final experimental results must be considered. Thus, some of the constraints used in this study were examined separately: skip connection, loss of latent space, intensity constraint, and gradient constraint. We also noticed that the training batch was small, so we chose the group normalization method and performed ablation experiments on the UCSD dataset. Table 1 presents the results. ✓ indicates that the corresponding constraint was selected, and a blank space means that it was not selected.

Table 1. Constraint selection based on the AUC results for the UCSD datasets.

	Condition selection					
Skip connection		✓	✓	✓	✓	✓
Latent space	✓		✓	✓	✓	✓
Intensity constraint	✓	✓		✓	✓	✓
Gradient constraint	✓	✓	✓		✓	✓
Group normalization	✓	✓	✓	✓		✓
ped2 AUC	0.7764	0.8750	0.8424	0.9486	0.9352	**0.9549**
ped1 AUC	0.7263	0.7898	0.7639	0.8097	0.8134	**0.8421**

For the group normalization constraint, ✓ indicates that the group normalization method was adopted, and a blank space indicates that the batch normalization methodwas adopted.

The effect of each constraint on the generation of the predicted frames was evaluated. Skip connections significantly improved the quality of the generated prediction frames, and the loss of latent space helped artificially amplify the differences between the normal and anomalous data. Thus, these constraints were feasible. Group normalization also improved the effectiveness of the proposed model, which indicates that it is more conducive to model optimization than batch normalization for small batches. The experimental results of UCSD ped1 and UCSD ped2 demonstrate the effectiveness of the various constraints considered in this study.

3.6 Comparison of Experimental Results

Table 2 compares the performance of the proposed method with those of the existing mainstream methods, where proposed method exhibits better performance than the existing methods. However, the experimental results show that the accuracy decreases as the complexity of the dataset increases.

Table 2. Anomaly scores of each model with different datasets.

	CUHK	UCSD ped1	UCSD ped2	ShanghaiTech	UCF_Crime100
Stacked RNN [11]	0.8171	/	0.9220	/	/
New baseline [9]	0.8085	0.8213	0.9148	0.7187	/
Unmasking [24]	0.8060	0.6840	0.8220	/	/
Real world [27]	/	/	/	/	0.7541
Event detection [28]	/	0.7030	0.9350	/	/
Proposed method	**0.8210**	**0.8421**	**0.9549**	**0.7543**	**0.7816**

For practical applications, an appropriate threshold must be selected to distinguish between normal and anomalous data. Therefore, the test results were visualized. The experimental results were all obtained at the same learning rate as Ubuntu16.04, tensorflow-gpu1.4, Python 3.5, and cuda9.0, using an NVIDIA GeForce GTX 1080 series.

3.7 Experimental Results Visualization and Analysis

Prediction Model
In the UCSD dataset, normal events are ordinary pedestrians, and bicycles or other vehicles are anomalies. The first row in Fig. 3 shows the generation of predicted frames under normal events, and the second row in Fig. 3 shows the generation of predicted frames under abnormal events. Obviously, it can be found that for normal events, the difference between the predicted frame and the fact frame is small; but in abnormal situations, compared with the fact frame, the predicted frame will be obviously blurred in the abnormal area, as shown by the red box in the figure.

Frequency Histogram and Kernel Density Estimation
We used the UCSD ped2 dataset as an example to obtain the normalized PSNR (i.e., anomaly score) in intervals of 0.01 in the range [0, 1]. The normal and anomaly scores were counted in each interval to obtain the frequency histogram and the corresponding kernel-density estimation shown in Fig. 4. The intersection between the kernel density estimations of the normal and anomalous data can be used as the threshold. From Eqs. (8) and (9), the smaller the difference between the generated predicted frame and the real frame, the higher the PSNR, that is, the higher the anomaly score. If the anomaly score exceeds this threshold, an event can be considered a normal; otherwise, it is anomaly.

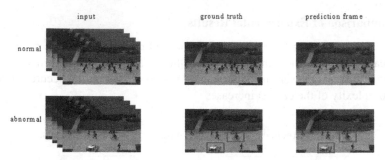

Fig. 3. Generation of prediction frames in UCSD data set

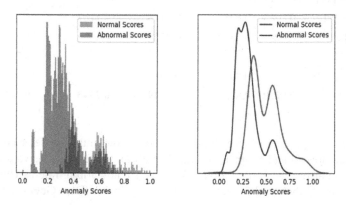

Fig. 4. Frequency histogram and kernel density estimation.

4   Conclusions

We developed a model based on two autoencoders to ensure the stable generation of predicted frames and used implicit space constraints to limit the distribution of normal data, thereby improving the generation effect of predicted frames. In addition, we used group normalization and other methods to optimize the anomaly detection model further. This model will have broad applications in elevator design, bank self-service systems, traffic, and other monitoring fields. The experiments showed that the proposed model performs better than the existing methods in video anomaly detection. Despite its numerous advantages, the proposed method has some limitations. Owing to the continuous improvement of fraud technology, abnormal and normal data have become increasingly similar. In future work, we will not only pay attention to the successful prediction of normal events, but also make some punitive losses in response to abnormal events, attempting to solve the problem of excessive similarity between abnormal and normal data.

References

1. Chalapathy, R., Menon, A.K., Chawla, S.: Anomaly detection using one-class neural networks. arXiv preprint arXiv:1802.06360 (2018)
2. Goodfellow, I., et al.: Generative adversarial nets. In: 27th Conference and Workshop on Neural Information Processing Systems, pp. 2672–2680. Curran Associates, NY (2014)
3. Akçay, S., Atapour-Abarghouei, A., Breckon, T.P.: GANomaly: Semi-supervised anomaly detection via adversarial training. In: Asian Conference on Computer Vision, pp. 622–637. Springer, Switzerland (2018)
4. Hong, R., Hu, Z., Wang, R., Wang, M.: Multi-view object retrieval via multi-scale topic models. IEEE Trans. Image Process. **25**(12), 5814–5827 (2016)
5. Lu, C., Shi, J., Jia, J.: Abnormal event detection at 150 fps in Matlab. In: Proceedings of the IEEE International Conference on Computer Vision, pp. 2720–2727. IEEE, NJ (2013)
6. Hong, R., Li, L., Cai, J., Tao, D., Wang, M., Tian, Q.: Coherent semantic-visual indexing for large-scale image retrieval in the cloud. IEEE Trans. Image Process. **26**(9), 4128–4138 (2017)
7. Kim, J., Grauman, K.: Observe locally, infer globally: a space-time MRF for detecting abnormal activities with incremental updates. In: 2009 IEEE Conference on Computer Vision and Pattern Recognition. IEEE, NJ (2009)
8. Adam, A., Rivlin, E., Shimshoni, I., Reinitz, D.: Robust real-time unusual event detection using multiple fixed location monitors. IEEE Trans. Pattern Anal. Mach. Intell. **30**(3), 555–560 (2008)
9. Liu, W., Luo, W., Lian, D., Gao, S.: Future frame prediction for anomaly detection–a new baseline. In: Proceedings of the IEEE Conference on Computer Vision and Pattern Recognition, pp. 6536–6545. IEEE, NJ (2018)
10. Chong, Y.S., Tay, Y.H.: Anomaly event detection in videos using spatiotemporal autoencoder. arXiv preprint arXiv:1701.01546 (2017)
11. Luo, W., Liu, W., Gao, S.: A revisit of sparse coding based anomaly detection in stacked RNN framework. In: Proceedings of the IEEE International Conference on Computer Vision, pp. 341–349. IEEE, NJ (2017)
12. Xu, D., Yan, Y., Ricci, E., Sebe, N.: Detecting anomalous events in videos by learning deep representations of appearance and motion. Comput. Vis. Image Underst. **156**, 117–127 (2017)
13. Mathieu, M., Couprie, C., LeCun, Y.: Deep multi-scale video prediction beyond mean square error. arXiv preprint arXiv:1511.05440 (2015)
14. Dosovitskiy, A., et al.: FlowNet: Learning optical flow with convolutional networks. In: Proceedings of the IEEE International Conference on Computer Vision, pp. 2758–2766. IEEE, NJ (2015)
15. Yu, X., Zhang, X., Cao, Y., Xia, M.: VAEGAN: a collaborative filtering framework based on adversarial variational autoencoders. In: International Joint Conference on Artificial Intelligence, pp. 4206–4212. IEEE, NJ (2019)
16. Wu, Y., He, K.: Group normalization. In: Ferrari, V., Hebert, M., Sminchisescu, C., Weiss, Y. (eds.) ECCV 2018. LNCS, vol. 11217, pp. 3–19. Springer, Cham (2018). https://doi.org/10.1007/978-3-030-01261-8_1
17. Scholkopf, B., Smola, A.J.: Learning with Kernels: Support Vector Machines, Regularization, Optimization, and Beyond. MIT Press, Cambridge (2001)
18. Zimek, A., Schubert, E., Kriegel, H.P.: A survey on unsupervised outlier detection in high-dimensional numerical data. Stat. Anal. Data Mining ASA Data Sci. J. **5**(5), 363–387 (2012)
19. Jolliffe, I.: Principal component analysis. In: International Encyclopedia of Statistical Science, pp. 1094–1096. Springer, Berlin (2011)

20. Zhao, Y., Deng, B., Shen, C., Liu, Y., Lu, H., Hua, X.S.: Spatio-temporal autoencoder for video anomaly detection. In: Proceedings of the 25th ACM International Conference on Multimedia, pp. 1933–1941. ACM, New York (2017)

21. Mahadevan, V., Li, W., Bhalodia, V., Vasconcelos, N.: Anomaly detection in crowded scenes. In: 2010 IEEE Computer Society Conference on Computer Vision and Pattern Recognition, pp. 1975–1981. IEEE, NJ (2010)

22. Luo, W., Liu, W., Gao, S.: Remembering history with convolutional lstm for anomaly detection. In: 2017 IEEE International Conference on Multimedia and Expo (ICME), pp. 439–444. IEEE, NJ (2017)

23. Zhao, B., Li, F., Xing, E.P.: Online detection of unusual events in videos via dynamic sparse coding. In: 2011 IEEE Conference on Computer Vision and Pattern Recognition, pp. 3313–3320. IEEE, NJ (2011)

24. Ionescu, R.T., Smeureanu, S., Alexe, B., Popescu, M.: Unmasking the abnormal events in video. In: Proceedings of the IEEE International Conference on Computer Vision, pp. 2895–2903. IEEE, NJ (2017)

25. Morais, R., Le, V., Tran, T., Saha, B., Mansour, M., Venkatesh, S.: Learning regularity in skeleton trajectories for anomaly detection in videos. In: 2019 IEEE Conference on Computer Vision and Pattern Recognition, pp. 11996–12004. IEEE, NJ (2019)

26. Lotter, W., Kreiman, G., Cox, D.: Deep predictive coding networks for video prediction and unsupervised learning. arXiv preprint arXiv:1605.08104 (2016)

27. Sultani, W., Chen, C., Shah, M.: Real-world anomaly detection in surveillance videos. In: 2018 IEEE Conference on Computer Vision and Pattern Recognition, pp. 6479–6488. IEEE, NJ (2018)

28. Ravanbakhsh, M., Nabi, M., Sangineto, E., Marcenaro, L., Regazzoni, C., Sebe, N.: Anomaly event detection in videos using generative adversarial nets. In: 2017 IEEE International Conference on Image Processing, pp. 1577–1581. IEEE, NJ (2017)

29. Vu, H.S., Ueta, D., Hashimoto, K., Maeno, K., Pranata, S., Shen, S.M.: Anomaly detection with adversarial dual autoencoders. arXiv preprint arXiv:1902.06924 (2019)

30. Ilg, E., Mayer, N., Saikia, T., Keuper, M., Dosovitskiy, A., and Brox, T.: FlowNet 2.0: Evolution of optical flow estimation with deep networks. In: 2017 IEEE conference on computer vision and pattern recognition, pp. 2462–2470. IEEE, NJ (2017)

A Robust Distance Regularized Potential Function for Level Set Image Segmentation

Le Zou[1,2,3], Qian-Jing Huang[4], Zhi-Ze Wu[5], Liang-Tu Song[2,3],
and Xiao-Feng Wang[1(✉)]

[1] Anhui Provincial Engineering Laboratory of Big Data Technology Application for Urban Infrastructure, School of Artificial Intelligence and Big Data, Hefei University, Hefei 230601, Anhui, China
xfwang@hfuu.edu.cn
[2] Hefei Institutes of Physical Science, Chinese Academy of Sciences, P. O. Box 1130, Hefei 230031, Anhui, China
[3] University of Science and Technology of China, Hefei 230026, Anhui, China
[4] College of Bioengineering Food and Environmental Sciences, Hefei University, Hefei 230601, Anhui, China
[5] Institute of Applied Optimization, School of Artificial Intelligence and Big Data, Hefei University, Hefei 230601, Anhui, China

Abstract. The level set is a classical image segmentation method, but during the evolution of the level set, it can produce evolutionary problems such as local spikes and deep valleys, or overly flat regions, making the iterative process of final segmentation unstable and segmentation results inaccurate. In order to ensure the stability and validity of the level set evolution during the evolution process, the level set function must be periodically initialized so that the level set is always kept as a signed distance function. We construct a new distance regularization potential function based on logarithmic and power function and give a specific analysis. During the evolution process, the level set function always approximates the signed distance function, which is stable and efficient for level set image segmentation. Experimental analyses are conducted to compare the segmentation performance of various distance regularization potential functions when combining with the classical Chan Vese model.

Keywords: Image segmentation · Level set · Double well potential function · Distance regularization energy term · Diffusion rate

1 Introduction

The applications of computer vision can be found in almost all areas of human activities. Computer vision is an interdisciplinary scientific field of artificial intelligence(AI), which deals with how computers can be made to gain high-level understanding from digital images or videos. Image segmentation is a fundamental and key problem in image processing and computer vision [1]. As a preprocessing stage, the goal of segmentation is to divide an image into different regions according to a certain consistency [2]. Over the

© Springer Nature Switzerland AG 2021
D.-S. Huang et al. (Eds.): ICIC 2021, LNCS 12836, pp. 547–556, 2021.
https://doi.org/10.1007/978-3-030-84522-3_45

past several decades, thousands of image segmentation methods have attracted scholars' attentions [3]. In recent years, active contour models (ACMs) have become the most promising frameworks and effective methods for traditional image segmentation. The level set methods (LSMs) are one kind of the most popularly and important active contour image segmentation methods over the last two decades [2].

For deriving better segment performance, it is necessary to periodically re-initialize the level set ϕ in the process of level set function evolution, so that the level set can be kept as a signed distance function (SDF) $|\nabla\phi| = 1$ near the zero-level set [4]. However, this process is complicated and faces the problem of how and when to initialize it. In order to avoid this complicated process, many researchers have done a lot of research and improvement works. There are three mainstream methods to avoid reinitialization, adding double-well potential function [5], using the Gaussian function to do a single convolution with the evolutionary function after each iteration [6], adding the diffusion energy term [7]. In studying the level set evolution equation, Li et al. [4] found that the properties of the SDF can be maintained when the gradient modulus of the evolutionary curve is equal to 1. Therefore, Li et al. [5] constructed a double-well potential function energy penalty term such that the gradient modulus of the level set is equal to 1, thus preserving the SDF properties of the level set near the zero-level set. This distance regularization term makes the corresponding computation time much shorter, but when $|\nabla\phi| < 0.5$, it makes the $|\nabla\phi|$ drop down to 0, it does not solve the problem of ϕ drastic shock when $|\nabla\phi| \to 0$. This case contradicts our expectation that the zero-level set function is always approximated as an SDF.

Many scholars constructed some new distance regularization energy terms for obtaining better segment performance. Li et al. [8] constructed a backward and forward diffusion of the distance regularized term based on the logarithmic and polynomial functions, the diffusion is forward for the steep shape region of the level set function, which keeps decreasing the gradient magnitude until it approaches 1. Otherwise, the diffusion becomes backward and increases the gradient magnitude back to 1. Sun et al. [9] reconstructed a new distance regularization term to maintain the level set function as an SDF. Wang et al. [10] presented a double-well potential penalty energy term based on a polynomial function, which can maintain the signed distance property of the level set near the zero-level set. The distance penalty energy term $P(s)$ in the paper [10] is a six-order polynomial, Zou et al. [11] used the four-order polynomial to construct distance regularized terms and avoid high order instability of polynomial functions. Cai et al. [12] constructed a distance regularization energy terms using cubic polynomials. Yu et al. [13] constructed a new six-order polynomial distance regularization term. Sun et al. [14] constructed a V-potential well function for distance regularization based on polynomial and logarithmic functions. Inspired by the idea of Li et al. [5], Wang et al. [15] constructed a new double-well potential function by logarithmic and polynomial functions to effectively solve the initialization problem of the level set. Weng et al. [16] constructed a distance regularization term based on trigonometric, polynomial and exponential functions. All these distance regularization energy terms can avoid the time-consuming reinitialization step, but they have different effectiveness. In view of the most mainstream research to avoid reinitialization methods are all improvements on the double-well potential function of Li et al. [5]. However, not all regularized terms are good at maintaining the stability of the evolutionary process and the signed distance function property of the level set function. Based on the analysis of the various types of distance regularization potential functions, we present a new distance regularization term based on the logarithmic and power function, and compare it with five typical distance regularized terms.

The remainder of this paper is organized as follows. Section 2 provides classical distance regularization energy term and discusses the analysis of potential function energy term and its diffusion rate function. We develop a novel distance regularization term based on the logarithmic and power functions, and give theoretical proof and specific analysis in Sect. 3. Experimental analyses are provided in Sect. 4. Section 5 concludes the paper.

2   Classical Distance Regularization Term

Li et al. showed [11] that level set ϕ which satisfies $|\nabla \phi| = 1$ requires the signed distance function plus a constant. Li et al. expressed the deviation between the level set function and the signed distance function by means of the following energy functional.

$$R_p(\phi) = \int_\Omega p_{Li1}(|\nabla \phi x(x)|)dx \tag{1}$$

where $p_{Li1}(s)$ is the potential function, we use $p_{Li1}(s)$ to denote the regularized term constructed by Li

$$p_{Li1}(s) = \frac{1}{2}(s-1)^2 \tag{2}$$

From Eq. (2), it can be seen that the potential function $p_{Li1}(s)$ obtains the minimum value at $|\nabla \phi| = 1$. Therefore, when minimizing the energy functional, the level set function approximately the signed distance function, and the segmentation result is stable.

Using the variational method and the steepest descent method, we obtain the gradient flow equation corresponding to the energy functional (1).

$$\frac{\partial \phi}{\partial t} = -\frac{\partial R_p(\phi)}{\partial \phi} = div(dp_{Li1}(|\nabla \phi(x)|)\nabla \phi(x)) \tag{3}$$

where

$$dp_{Li1} = \frac{p'_{Li1}(s)}{s} = 1 - \frac{1}{s} \tag{4}$$

In Eq. (3), div is the divergence operator and $dp_{Li1}(s)$ is the diffusion rate function of the level set in the evolution process.

From the gradient flow equation and the diffusion rate function, we can see that when $|\nabla \phi| > 1$, $dp_{Li1} > 0$, then the diffusion is carried forward, and then the value $|\nabla \phi|$ is reduced; when $|\nabla \phi| < 1$, $dp_{Li1} < 0$, then the diffusion is carried backward, and then the value $|\nabla \phi|$ is increased; when $|\nabla \phi| = 1$, $dp_{Li1} = 0$, then the diffusion is stopped. In the region where the level set function is relatively flat, i.e. $|\nabla \phi| \to 0$, $dp_{Li1} \to -\infty$, the diffusion will diffuse backward at a great speed, $|\nabla \phi|$ will change drastically and make the level set function produce spikes or deep valleys in the flat region, and then the level set segmentation result is unstable.

3   A Novel Logarithmic and Power Function Based Distance Regularization Potential Function

To prevent violent oscillations as well as too flat regions, we propose a novel distance regularization energy terms based on logarithmic and power functions,

$$p_{New}(s) = 2s + 4Ln(s^{0.5} + 1) + \frac{s^2}{2} - 4s^{0.5} - \frac{4s^{1.5}}{3} + c \tag{5}$$

where $c = \frac{17}{6} - 4Ln2$. The diffusion ratio equation for the gradient flow function is obtained through the potential function as shown in Eq. (6).

$$dp_{New}(s) = \frac{p'_{New}(s)}{s} = 1 - \frac{2}{s^{0.5} + 1} \tag{6}$$

Theorem 1: Let $\Omega \subset \Re^2$ be the two-dimensional image domain, $\phi : \Omega \subset R^2 \to R$ is the level set function, if $|\nabla\phi| = 1$, the energy functional defined by the potential function $p_{New}(s)$ is minimized.

Proof: The derivative of the potential function $p_{New}(s)$ is

$$p'_{New}(s) = s - \frac{2s}{s^{0.5} + 1} = s\left(\frac{s^{0.5} - 1}{s^{0.5} + 1}\right) \tag{7}$$

Let $p'_{New}(s) > 0$, we can get the monotonically increasing interval of the potential function $p_{New}(s)$ is $s \in (1, +\infty)$. Let $p'_{New}(s) < 0$, we can get the monotonically decreasing interval of the potential function $p_{New}(s)$ is $s \in (0, 1)$, so the function $p_{New}(s)$ obtain its minimum value at $s = 1$, that is, the potential function $p_{New}(s)$ obtain the minimum value when $|\nabla\phi| = 1$.

Theorem 1 illustrates that when $|\nabla\phi| = 1$, the energy functional is minimized. It shows that optimizing the energy functional defined by Eq. (5) can correct the deviation of the level set function from the signed distance function, thus ensuring the stability of the level set evolution.

Theorem 2: For $s \in (0, +\infty)$, diffusion rate function $dp_{New}(s)$ satisfies $|dp_{New}(s)| < 1$.

Proof: The derivative of the potential function $dp_{New}(s)$ is

$$dp'_{New}(s) = \frac{d}{ds}(1 - \frac{2}{s^{0.5} + 1}) = \frac{1}{(s^{0.5} + 1)^2 s^{0.5}} \tag{8}$$

Because $dp'_{New}(s)$ is a decreasing function, $\lim_{s \to \infty} dp'_{New}(s) = \lim_{s \to \infty} (\frac{1}{(s^{0.5}+1)^2 s^{0.5}}) = 0$, so for arbitrary $s \in (0, +\infty)$, $dp'_{New}(s) > 0$ is always true, so $dp_{New}(s)$ is an increasing function in its domain.

Because

$$\lim_{s \to \infty} dp_{New}(s) = \lim_{s \to \infty} (1 - \frac{2}{s^{0.5} + 1}) = 1 \tag{9}$$

$$\lim_{s \to 0} dp_{New}(s) = \lim_{s \to 0} (1 - \frac{2}{s^{0.5}+1}) = -1 \qquad (10)$$

So we have $|dp_{New}(s)| < 1$.

Theorem 2 illustrates that the proposed distance regularization term is stable, which avoids drastically changes in speed and ensures the stability of the level set evolution.

We use the gradient flow equation of the energy functional to analyze the evolution of the proposed distance regularization term.

If $|\nabla\phi| > 1$, $dp_{New}(s) > 0$, the effect of the energy penalty term is positive diffusion, the level set function $\nabla\phi$ remains smooth, and the level set will diffuse forward and decrease $|\nabla\phi|$ to 1;

If $0 \le |\nabla\phi| < 1$, $dp_{New}(s) < 0$, the effect of the energy penalty term in this case is equivalent to reverse diffusion, and it will slowly diffuse backward and increase $|\nabla\phi|$ to 1.

If $|\nabla\phi| = 1$, $dp_{New}(s) = 0$, the gradient flow equation of the energy functional stops diffusion. That is, it is determined that the level set function always approximates the signed distance function.

In order to better analyze and compare the specific effects of different potential functions and diffusion rate functions in the evolution of level sets, we choose four potential functions and five diffusion rate functions for comparison. We choose the distance regularization term Li et al. [4] and the distance regularized potential function by Zou et al. [11], the double well potential function Li et al. [5], the logarithmic and polynomial-based distance regularized potential function Li et al. [8], and the diffusion rate function by Xie [17]. For the convenience, we abbreviate them as $p_{Li1}(s)$, $p_{Zou}(s)$, $p_{Li2}(s)$ and $p_{Li3}(s)$, the corresponding diffusion ratio functions are $dp_{Li1}(s)$, $dp_{Zou}(s)$, $dp_{Li2}(s)$ and $dp_{Li3}(s)$ respectively. The diffusion rate function by Xie is abbreviated as $dp_{Xie}(s)$.

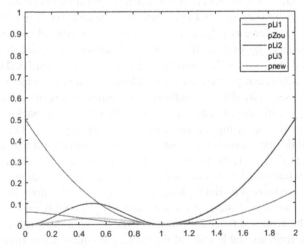

Fig. 1. Picture of the five different distance regularized potential functions

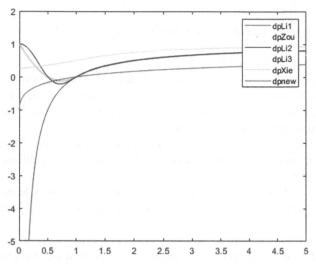

Fig. 2. Picture of the six diffusion rate functions

Figure 1 depicts the trend of the potential functions of the five different methods, we can see from Fig. 1 that the selected five distance potential functions take the minimum value at $s = 1$. $p_{Zou}(s)$ and $p_{Li2}(s)$ have two minimal value points, $s = 0, s = 1$ respectively. Figure 2 shows the trend of the six diffusion rate function, and it is clear that the main difference between the six models is at $|\nabla \phi| < 1$. When $|\nabla \phi| < 1$, $dp_{Li1}(s)$ decreases rapidly, then the corresponding diffusion ratio function will diffuse backward at a very fast speed, which will cause the level set function to change dramatically during the evolution. The diffusion rate dp_{Li2}, dp_{Zou} are negative when $1/2 < |\nabla \phi| < 1$, the diffusion rate $|\nabla \phi|$ will increase backward to 1. The diffusion rate dp_{Li2}, dp_{Zou} are positive when $|\nabla \phi| < 0.5$, the level set will diffuse forward and decrease $|\nabla \phi|$ to zero, which is not the desired situation in the evolution of the level set. When $|\nabla \phi| < 1$, dp_{Xie} still is positive, there have not $dp_{Xie}(s) = 0$. In other words, when the zero-level set reaches the boundary, the level set function is still not satisfy signed distance function (SDF) property, that is $|\nabla \phi| \neq 1$, it will continue to evolute forward and reduce $|\nabla \phi|$ to zero, and eventually causing edge leakage problem, which is not what we expect in the evolution of level set. Only $dp_{Li3}(s)$ and $dp_{New}(s)$ diffuse at a relatively slow speed and do not produce a drastic evolutionary situation. When $|\nabla \phi| < 1$, backward diffusion is performed, gradually increasing the value of $|\nabla \phi|$ until $|\nabla \phi| = 1$. This evolution will neither carry out drastic changes with fast speed changes nor form too flat regions. We can see that when $|\nabla \phi| < 1$, the proposed diffusion rate function $dp_{New}(s) \geq -1$, the equation diffuses backward, this process increases $|\nabla \phi|$ to 1, avoiding the case of $|\nabla \phi|$ dropping to zero, so there is no flat region. When $|\nabla \phi| = 1$, the diffusion ratio $dp_{New}(s)$ is zero, the evolutionary equation stops diffusing. This property ensures that the final steady state of the proposed potential function together with the level set model energy functional. Therefore, the stable state of the evolution equation determined by the proposed distance regularization potential function based on the logarithmic and polynomial functions is $|\nabla \phi| = 1$, i.e., the level set function always remains as a signed distance function. In addition, the proposed diffusion rate function is a continuous function, it is not a piecewise function, this accelerates the evolution to some extent. Therefore, the proposed distance regularization potential function can satisfy the characteristics of the zero-level set always remains as a signed distance function. We can see from the potential function and the diffusion rate function that the proposed distance regularization term in this paper is slightly better than Li3, with a more stable variation, always with a slow diffusion rate $|\nabla\phi| < 1$.

As described in many level set methods, the distance regularization term we construct in this paper can be applied to regional and edge-based level set model to improve the image segmentation performance. In this paper, we apply it to the classical Chan Vese model, the total energy functional is

$$E(c_1, c_2, \phi) = \lambda_1 \int_\Omega |I(x) - c_1|^2 H(\phi(x))dx + \lambda_2 \int_\Omega |I(x) - c_1|^2 (1 - H(\phi(x)))dx$$

$$+ \mu \int_\Omega \delta(\phi(x))|\nabla\phi(x)|dx$$

$$+ \nu \int_\Omega (2\phi(x) + 4Ln(\phi(x)^{0.5} + 1) + \frac{\phi(x)^2}{2} - 4\phi(x)^{0.5} - \frac{4\phi(x)^{1.5}}{3} + c)dx$$

$$(11)$$

where $H(z)$ and $\delta(z)$ denote the Heaviside function and Dirac function, respectively.

$$H(\phi) = \frac{1}{2}\left[1 + \frac{2}{\pi}\arctan\left(\frac{\phi}{\varepsilon}\right)\right] \tag{12}$$

$$\delta(\phi) = \frac{1}{\pi} \cdot \frac{\varepsilon}{\varepsilon^2 + \phi^2} \tag{13}$$

By minimizing the energy functional through the difference and the gradient descent methods, we can obtain

$$\frac{\partial\phi}{\partial t} = \delta(\phi)[-\lambda_1(I - c_1)^2 + \lambda_2(I - c_2)^2$$

$$+ \mu div(\frac{\nabla\phi}{|\nabla\phi|}) + div((1 - \frac{2}{|\nabla\phi|^{0.5}+1})|\nabla\phi|] \tag{14}$$

where c_1 and c_2 are the means of evolution in the classical Chan Vese (CV) model, which are calculated by the following equation,

$$c_1(\phi) = \frac{\int_\Omega I(x)H(\phi(x))dx}{\int_\Omega H(\phi(x))dx} \tag{15}$$

$$c_2(\phi) = \frac{\int_\Omega I(x)(1 - H(\phi(x)))dx}{\int_\Omega (1 - H(\phi(x)))dx} \tag{16}$$

4 Experimental Results

In order to better compare the six types of diffusion rate functions in Sect. 3, we carry out experiments based on the classical region-based Chan Vese level set image segmentation model with six different diffusion rate functions, and carry out experiments on two types of images: noisy and inhomogeneity. The experimental platform operating system is windows 10, the program is written based on matlab2018b. The parameters of the model are as follows: $\varepsilon = 1$, $\lambda_1 = \lambda_2 = 1$, $\nu = 0$, $\Delta t = 0.1$. $r = 20$.

Figure 3 and Fig. 4 show the segmentation results and the corresponding final level set function when the image reaches steady state under six different distance regularization terms. Figure 3 and Fig. 4 show the segmentation effect of the CV model with six different distance regularization term on the noisy image and the intensity inhomogeneity image. Figure 3 shows the segmentation of the noisy image, and there are tiny difference between all the segmentation results and the corresponding level set function. The experimental results show that the Chan Vese model with six different distance regularization term can segment noisy images and intensity inhomogeneity images, indicating that the distance regularization term can make the level set function approximate the signed distance function, which is robust to the target segmentation of intensity inhomogeneity images. The segmentation results of (a), (c), (e), (g), (i), and (k) in Fig. 4 are almost the same, but the changes of the corresponding level set functions of (b), (d), and (f) are oscillating, and the level set changes of (h), (j) and (I) are smoother, indicating that the distance

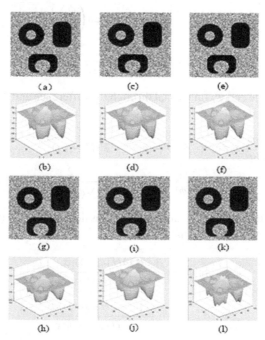

Fig. 3. The final CV model segmentation results on noise image with the six diffusion rate function and the corresponding level set functions.

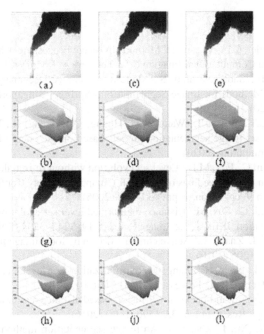

Fig. 4. The final CV model segmentation results on inhomogeneity image with the six diffusion rate function and the corresponding level set functions.

regularization term approximates the signed distance function better. The number of iterations and time for the proposed distance regularization term to reach the steady state are smaller than other types of distance regularization terms by combining the CV model.

5 Conclusion

In order to avoid more complex re-initialization in the evolution of the level set, we propose a novel distance regularization potential function based on the logarithmic and power functions. The proposed distance regularization potential function and its diffusion rate function can guarantee the signed distance function property of the level set function, making the level set functions calculate accurately and ensure the validity and stability of the level set evolution. Experiments are presented to compare the novel distance regularization term with five other distance regularized terms to demonstrate the effectiveness of the proposed regularization term based on Chan Vese model for image segmentation.

Acknowledgements. The authors would like to express their thanks to the referees for their valuable suggestions. This work was supported in part by the grant of the National Natural Science Foundation of China, Nos. 61672204 and 61806068, in part by the grant of Anhui Provincial Natural Science Foundation, Nos. 1908085MF184, 1908085QF285, in part by the Key Research Plan of Anhui Province, No. 201904d07020002.

References

1. Vese, L.A., Guyader, C.L.: Variational Methods in Image Processing, Chapman & Hall/CRC Mathematical and Computational Imaging Sciences Series. Taylor & Francis (2015)
2. Vinoth Kumar, B., Sabareeswaran, S., Madumitha, G.: A Decennary Survey on Artificial Intelligence Methods for Image Segmentation, (Springer Singapore, Singapore, 2020), pp. 291–311 (2020)
3. Zou, L., Song, L.T., Weise, T., Wang, X.F., Huang, Q.J., Deng, R., Wu, Z.Z.: A survey on regional level set image segmentation models based on the energy functional similarity measure. Neurocomputing (2020)
4. Li, C., Xu, C., Gui, C., Fox, M.D.: Level set evolution without re-initialization: A new variational formulation. In: Proceedings of the IEEE Computer Society Conference on Computer Vision and Pattern Recognition, 1, pp. 430–436 (2005)
5. Li, C., Xu, C., Gui, C., Fox, M.: Distance regularized level set evolution and its application to image segmentation. IEEE Trans. Image Process. **19**, 3243–3254 (2010)
6. Zhang, K., Song, H., Zhang, L.: Active contours driven by local image fitting energy. Pattern Recogn. **43**, 1199–1206 (2010)
7. Zhang, K., Zhang, L., Song, H., Zhang, D.: Reinitialization-free level set evolution via reaction diffusion. IEEE Trans. Image Process. **22**, 258–271 (2013)
8. Li, M., Liu, L.: Forward-and-backward diffusion-based distance regularized model for image segmentation. Appl. Res. Comput. **33**, 1596–1600 (2016)
9. Sun, L., Meng, X., Xu, J., Zhang, S.: An image segmentation method based on improved regularized level set model. Appl. Sci.-Basel. **8**, 2393 (2018)
10. Wang, X., Min, H., Zou, L., Zhang, Y., Tang, Y., Philip Chen, C.: An efficient level set method based on multi-scale image segmentation and Hermite differential operator. Neurocomputing **188**, 90–101 (2016)
11. Zou, L., et al.: Image segmentation based on local chan vese model by employing cosine fitting energy. In: Chinese Conference on Pattern Recognition 2018), pp. 466–478 (2018)
12. Cai, Q., Liu, H., Zhou, S., Sun, J., Li, J.: An adaptive-scale active contour model for inhomogeneous image segmentation and bias field estimation. Pattern Recogn. **82**, 79–93 (2018)
13. Yu, H., He, F., Pan, Y.: A scalable region-based level set method using adaptive bilateral filter for noisy image segmentation. Multimed. Tools Appl. **79**(9–10), 5743–5765 (2019). https://doi.org/10.1007/s11042-019-08493-1
14. Sun, C., Xu, Y., Bi, D., Wang, Y.: Distance regularized level set method using v-potential well function. Comput. Appl. Softw. **04**, 277–280 (2013)
15. Wang, X., Shan, J., Niu, Y., Tan, L., Zhang, S.: Enhanced distance regularization for re-initialization free level set evolution with application to image segmentation. Neurocomputing **141**, 223–235 (2014)
16. Weng, G., He, Z.: Active contour model based on adaptive sign function. J. Softw. **30**, 3892–3906 (2019)
17. Xie, X.: Active contouring based on gradient vector interaction and constrained level set diffusion. IEEE Trans. Image Process. **19**, 154–164 (2010)

Label Similarity Based Graph Network for Badminton Activity Recognition

Ya Wang[1], Guowen Pan[1], Jinwen Ma[1(✉)], Xiangchen Li[2], and Albert Zhong[3]

[1] School of Mathematical Sciences and LMAM, Peking University, Beijing, China
{wangyachn,panguowen_aais}@pku.edu.cn, jwma@math.pku.edu.cn
[2] China Institute of Sport Science, Beijing 100061, China
lixiangchen@ciss.cn
[3] Hangzhou Zhidong Sports Co. LTD, Hangzhou 310052, China

Abstract. The sensor-based human activity recognition is a key technology for modern intelligent sports. However, the complexity of sport activities and lacking of large-scale dataset give rise to the challenges on training effective deep neural networks for it. On image/video-based computer vision tasks, deep learning models can be pretrained on large-scale datasets which are semantically similar with specific tasks. However, we cannot pretrain deep learning models for sensor-based human activity recognition due to lacking public large-scale datasets. To get rid of this problem, we propose a similarity-based graph network for the sensor-based human activity recognition. Specifically, it is a Convolutional Neural Network (CNN) being enhanced with an embedded Graph Neural Network (GNN) for learning the label relationship in terms of two proposed similarity measures. The experimental results on BSS-V2 dataset demonstrate that our proposed network outperforms prior state-of-the-art work by 10.3% in accuracy and 13.3% better than backbone CNN model.

Keywords: Badminton activity recognition · Sensor-based data · Convolutional neural network · Graph neural network · Similarity

1 Introduction

Human activity recognition is very popular in recent years with wide applications. It can be divided into two main categories: video-based human activity recognition and sensor-based human activity recognition. Recognition of human activity based on sensor data is often more effective because sensor data is less computationally expensive than video data and could do better in privacy protecting than video data.

There are mainly the following two problems in sensor-based human activity recognition. Firstly, due to insufficient sample size, the model is easy to overfit. In the task of human activity recognition, it is generally difficult to obtain a large number of samples, and usually deep learning algorithms need enough sample size, otherwise it would cause the phenomenon of overfitting. Only when the sample size is large enough can the algorithm well model the specific contribution behind the data, rather than the random interfering features. Secondly, some prior works with human knowledge inserted,

© Springer Nature Switzerland AG 2021
D.-S. Huang et al. (Eds.): ICIC 2021, LNCS 12836, pp. 557–567, 2021.
https://doi.org/10.1007/978-3-030-84522-3_46

like HMC [10], rely heavily on prior knowledge. Formally, the sensor-based badminton activity recognition task is divided into two stages in HMC algorithm. In the first stage, the input sensor-based data is roughly classified into one of the major classes. Then in the second stage, the input data is further predicted for the specific category. We need to determine the main-classes of badminton activities in the first stage according to appropriate prior information. If the prior knowledge is not reasonable or suitable for the task, it will certainly affect the identification of specific categories of badminton activities in the second stage. Only if the classification of the major categories is correct and effective can the subsequent detailed sub-classifiers get satisfactory results. So prior information is very important but hard to be designed in the HMC.

In this context, we propose a novel graph-based CNN model, called Similarity Graph Based Network (SGNet) to alleviate the human design bias. It employs a GNN to enhance the extraction of CNN feature with label correlation inserted. We also propose a novel method to construct the adjacency matrix used in GNN with little human prior and extra computation but having dramatic performance bonus, which is demonstrated by sufficient experimental results.

2 Related Work

Classification task is an important research field in machine learning that has attracted a lot of attention and investment, and has been widely applied. In the context of the rapid development of deep learning, various new ideas have emerged for the solution of classification tasks.

In many literatures, various methods have been proposed to describe the relationship between labels. A novel network framework called NeXtVLAD [4, 12] in decomposed the high dimensional eigenvectors into a set of low dimensional vectors before applying NetVLAD aggregation. Similarly, a multi-layer network structure was proposed [2, 11]. In this network structure, a cyclic memory module was included, which was composed of spatial transformation layer and LSTM sub-network alternately. This kind of module had two excellent characteristics: one was its strong interpretability, which was superior to many traditional deep learning studies; the other was the ability to classify multi-label images using up and down culture. A GCN-CNN based convolution network called KSSNet [9] was proposed to solve the label relationship of multi-label image/video classification. In this work, based on the prior information on label correlation, the relationship between labels was modeled by superimposing label graphs. After that, the multi-layer convolution network was used to get the final graph, so as to get the label embedding. However, the graph used in KSSNet is composed of statistical graph and knowledge graph. The former is not available in single label classification task and the latter usually lacks exact information of certain labels, making its generalization to badminton activity recognition task unsatisfactory. To model the label relationship in single label classification, a framework called Hierarchical Multi-Classification (HMC) was proposed [10], which could well describe the relationship between different categories of badminton activities. Nevertheless, the computation overhead of HMC was several times of backbone classifier and the mapping from main-classes and sub-classes heavily depended on designed manually, making it hard to search for an optimal mapping.

Graph Neural Network (GNN) has been proved to be very effective in modeling label structure. A dual flow graph convolution network (TS-GCN) composed of classifiers and concrete examples was designed to effectively maximize the effect of knowledge graph [1]. Semi-supervised learning method was presented [7] based on the data structure and could be extended to many tasks. The method was based on the variation form of CNN which directly performed the related operations on the graphs. A new multi-label deep learning network structure ML-ZSL [3] could predict the samples of each input model, and the prediction result was more than one invisible label. The model could automatically learn a rule of information propagation according to semantic information, and then be used to model the association between visible labels and invisible labels. A dynamic neural network model, called Spatial Temporal Graph Convolutional Networks (ST-GCN), was proposed [13]. It was different from conventional GCN methods. It could automatically mine temporal and spatial rules from samples.

3 Methodology

To alleviate the dependency on large-scale training dataset and inspired by KSSNet [9], we propose a novel similarity-based graph embedded deep learning framework, Name Similarity Graph based Network (SGNet). It utilizes a CNN model with automatic feature extractor (AFEB) to extract the deep CNN feature and a graph neural network for human prior information.

In the following part, we introduce in detail how to construct a graph to embody the label relationship from two terms of similarity: label name similarity and label feature similarity.

Graph Neural Network (GNN) is one of the most powerful technology for non-euclidean tasks, especially for embedding label relationship. One of the key factors on GNN performance is the adjacency matrix among the nodes in a graph. As a typical single class activity recognition, the badminton activity recognition task lacks essential information to calculate the statistical adjacency matrix, which results in the difficulty to build an effective graph on this task. To this end, we propose a novel similarity-based adjacency matrix, which combines two kinds of similarity of labels: the name similarity and overall feature similarity. Next, we would present the construction of our proposed similarity graph.

Construction of Adjacency Matrix. Adjacency matrix is a core component of a GNN. It explains the human priors about which objects are the most essential for specific tasks and how they are correlative about each other. Generally according to information resources, there are two kinds of adjacency matrices in GNN: 1) statistical adjacency matrix which is constructed from the statistical information e.g. label co-occurrence [9]. 2) knowledge adjacency matrix which is built through knowledge graph such as ConceptNet [6]. Statistical adjacency matrix derives from specific training dataset and could be quite effective for corresponding tasks. However, the calculation of label relationship relies on the statistical information such as label co-occurrence. It is not available for single label tasks like badminton activity recognition. Knowledge graph sources from human prior knowledge about some objects. Commonly it is large enough and covers most of daily objects we meet, nevertheless it suffers from two drawbacks: 1) the weights of the edges between two nodes are determined by specific designed games or expert knowledge, both of which ignore the characteristics of certain tasks, in deed they are tradeoff values for all tasks. However, the edge weights should meet the need of specific tasks in practice. For example, the edge weight between "Washing Hand" and "Cooking" in a kitchen is intuitively more significant than that in a toilet. 2) The node set of knowledge graph could hardly contains all objects in the world. For the task of badminton activity recognition, one of the classical knowledge graphs, called ConceptNet [6], does not contain any badminton activities in Table 1. KSSNet [9] utilizes a tradeoff method to create the edge with unknown nodes. Specifically, it first splits the label phrases into individual words, then averages the weights of all edges between single words and replaces the edge weights between two primitive label phrases with the average weights of all single words.

In this part, instead of statistical and knowledge adjacency matrix, we utilize two kinds of prior information to generate the adjacency matrix for badminton activity recognition. Formally, denote the graph as $G = \{V, E, A\}$ with node (category in this task) set $V = \{v_i\}_{i=1}^K$, edge set E and adjacency matrix $A \in R^{K \times K}$.

Label Name Similarity. Assuming the phrases of all labels are $\{W^i\}_{i=1}^K$, where W^i is the phrase of the i-th label, such as "Forehand Net Shot", it is composited of several consecutive words, denoted as $W^i = \left(w_1^i, \cdots, w_{L_i}^i\right)$. L^i is the length of the i-th phrase, such as "Forehand", "Net" and "Shot" in "Forehand Net Shot". We use the Intersection over Union (IoU) of two labels v_i and v_j to construct the edge weight:

$$A_{ij} = \frac{W^i \cap W^j}{W^i \cup W^j} \tag{1}$$

"|·|" presents the cardinality of the corresponding set. Figure 1 illustrates a sample of label name similarity between "Forehand Net Shot" and "Backhand Net Shot". Equation 1 employs the IoU of two labels v_i and v_j for the similarity in terms of their names. It does not need any more expert prior or suffer from human bias, therefore it is robust for the calculation of the name-based adjacency matrix.

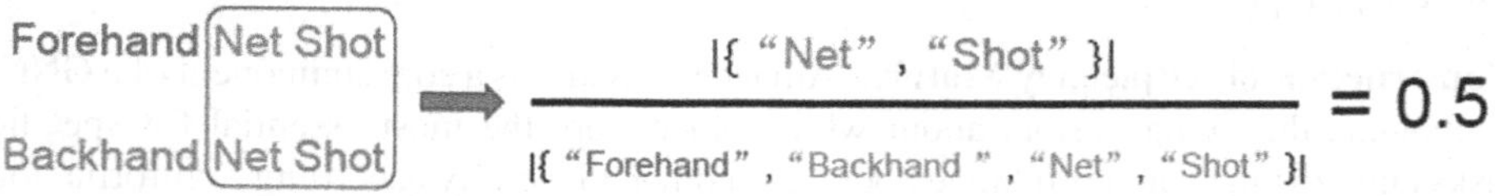

Fig. 1. Label name similarity between "Forehand Net Shot" and "Backhand Net Shot".

Feature Similarity Based Similarity. Except for the similarity in terms of label names, the feature similarity is also significant for the label relationship. Feature extraction of training samples is one of the most important procedures for calculating feature similarity of different labels. Due to the common problem "dislocation" in sensor-based badminton activity recognition, it is hard to utilize conventional machine learning methods to extract expressive features. Taking it into account, we use a CNN based shallow network AFEB-MobileNet [8] for feature extraction. Usually, features in shallow layers present shallow semantic information which are general for various tasks, such as image texture, colors and shapes. Features in deep layers are closely relative to specific tasks. Besides, they are also majorly determined by the backbone models, meaning that the feature similarity in one model is not confidently similar with that from another model. Thereby, we utilize the middle layers of backbone AFEB-MobileNet for computing feature similarity.

Assuming a well-trained AFEB-MobileNet model to be a composite function $F(X) = f_1 \circ f_2(K), X$ is the input sensor-based sample. $f_2(\cdot)$ is the composition of shallow and middle layers of backbone CNN model, $f_1(\cdot)$ is the deep and sofmax layers of backbone model. The training of AFEB-MobileNet is guided by a common cross-entropy loss:

$$l = -\frac{1}{N} \sum_{i=1}^{N} \sum_{j=1}^{K} y_{ij} \log \sigma (F(X_i)) \tag{2}$$

y_i is the ground-truth one-hot label of input sample X_i, and $y_{ij} = 1$ if X_i belongs to the j-th category, and vice versa. "$\sigma(\cdot)$" is the Softmax function, formally:

$$\sigma(z) = \left(\frac{e^{z_1}}{\sum_{j=1}^{K} e^{z_j}}, \cdots, \frac{e^{z_i}}{\sum_{j=1}^{K} e^{z_j}}, \cdots, \frac{e^{z_K}}{\sum_{j=1}^{K} e^{z_j}} \right) \tag{3}$$

where $z = (z_1, z_2, \cdots, z_K)$.

We extract the feature in middle layers, that is $f_2(X)$, as the presentative feature of input sample X for label feature similarity. Detailedly, we first employ global average pooling to transform the 2D hidden CNN feature $f_2(X)$ into a 1D feature,

$$\hat{Z} = P_C(Z) \tag{4}$$

Where $Z \in R^{C \times H \times W}$, and C, H, W are the channel, height and weight of middle hidden CNN feature. P_C is the global average pooling function among height and weight axes and the footnote denotes the shape of the output feature. Then, the mean of all training samples in each category is computed, assuming the results are $\{M_i\}_{i=1}^{K}$. After that, the Mahalanobis distance between two labels is calculated for better robustness on different dimensional magnitude and variation.

$$D_M(v_i, v_j) \sqrt{(M_i - M_j)^T \Sigma^{-1}(M_i - M_j)} \tag{5}$$

where M_j and M_j are the average of all samples in categories v_i and v_j. Σ is the covariance matrix of axes:

$$\Sigma = \begin{bmatrix} cov(a_1, a_1)cov(a_1, a_2) \dots cov(a_1, a_C) \\ cov(a_2, a_1)cov(a_2, a_2) \dots cov(a_2, a_C) \\ \dots \dots \\ cov(a_C, a_1)cov(a_C, a_2) \dots cov(a_C, a_C) \end{bmatrix} \tag{6}$$

Here $conv(\cdot, \cdot)$ is the covariance of two statistics, a_i is the output statistic of the i-th axes.

The feature similarity is then formulated as:

$$S_F\left(v_i, v_j\right) = \frac{1}{1 + D_M\left(v_i, v_j\right)} \tag{7}$$

Figure 2 presents the construction of label feature similarity between "Forehand Net Shot" and "Backhand Net Shot".

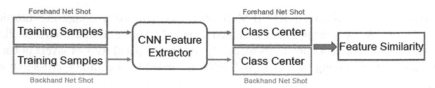

Fig. 2. Label feature similarity between "Forehand Net Shot" and "Backhand Net Shot".

Similarity Based Adjacency Matrix. The label name similarity-based graph is perceived through specific training datasets, it sources from human prior on label names, the prior is simple enough such that the human bias could be slight sufficiently. Meanwhile it would not be influenced by the lacking training samples. However, its computation is entirely from the human prior, therefore it is weak to reflect the real data distribution, which further limits its performance. Graphs based on feature similarity is a supplementary of the label name-based graph. To this end, the final graph of the proposed SGNet is constructed by superimposing label name similarity with feature similarity.

Assuming the label name-based graph is $G_N = (V, E_N, A_N)$, feature similarity-based graph is $G_S = (V, E_S, A_S)$, where A_N and A_S are adjacency matrices obtained through label name similarity and feature similarity, respectively.

Denote A'_S and A'_K as the normalized versions of A_S and A_K

$$A'_N = D_N^{-1/2} A_N D_N^{-1/2} \tag{8}$$

$$A'_S = D_S^{-1/2} A_S D_S^{-1/2} \tag{9}$$

where D_N and D_S are diagonal and $[D_N]_{ii} = \sum_j [A_N]_{ij}$, $[D_S]_{ii} = \sum_j [A_S]_{ij}$. Weighted average of A'_S and A'_K is used to superimpose the name similarity based graph into feature similarity based graph.

$$A = \lambda A'_N + (1 - \lambda) A'_S \tag{10}$$

$\lambda \in [0, 1]$ is a weight coefficient.

Meanwhile, as the elements of A'_S and A'_K are non-negative, A would have more nonzero elements than A_S and A_K, meaning that the graph constructed from A has more redundant edges than those from G_S or G_N. To suppress the redundant edges, we use a threshold $\tau \in R$ to filter the elements of A

$$[A_\tau]_{ij} = \begin{cases} 0, & \text{if } A_{ij} < \tau \\ A_{ij}, & \text{if } A_{ij} \geq \tau \end{cases} \tag{11}$$

Inspired by KSSNet [9], the performance of models drops as the number of GCN layers increases in some tasks due to the over-smoothing of deeper GCN layers. Therefore, we further use the strategy in [9], which adjusts the entries in the adjacency matrix of the superimposed graph to obtain the final adjacency matrix:

$$\widehat{A} = \eta A_\tau + (1 - \eta)I \tag{12}$$

I is the identity matrix.

At last, in order to enhance the robustness of built graph on the neighbor numbers, we re-normalize the adjacency matrix $\widehat{A}$ as

$$\tilde{A} = \widehat{D}^{-1/2}\widehat{A}\widehat{D}^{-1/2} \tag{13}$$

where $\widehat{D}$ is the diagonal matrix and $\left[\widehat{D}\right]_{ii} = \sum_j \left[\widehat{A_N}\right]_{ij}$.

With the adjacency matrix $\tilde{A}$, we construct the set of edges as

$$E = \left\{(v_i, v_j)|\tilde{A}_{ij} \neq 0, \ and \ 0 \leq i, j \leq K\right\} \tag{14}$$

and the final graph we propose is defined as $G = \left\{V, E, \tilde{A}\right\}$, which is called similarity graph.

4 Experiments

In this section, we first describe the experimental details in Sect. 4.1, then detail the comparisons and analysis against the previous state-of-the-art works on the BSS-V2 [10] benchmark and show the superiority of our proposed method in Sect. 4.2. Finally, in Sect. 4.3 we conduct a series of controlled ablation studies and more comprehensive analysis to verify the remarkable performance gain is essentially derived from the core architecture design of our work.

4.1 Experimental Setup

BSS-V2. BSS-V2 is a large-scale dataset for the badminton activity recognition. It uses a specific device to record the status of human daily activities in badminton sport. Specifically, the recording device is assembled with four units: 1) a common badminton racket as backbone, 2) a small sensor looking like a coin to record the three-dimensional acceleration vectors and spatial attitude angles (roll, pitch and yaw) of badminton activities, 3) a LED to mark the beginning and ending points of each activity, 4) a controller for the power switching. the signal frequency of the sensor is 200 Hz meaning that 200 temporal frames (called records) would be recorded every second. Totally, it has 37 major badminton activities covering almost all daily activities in badminton sport and each activity contains about 215 samples on average. The difficulty of the activity recognition in this dataset is that some badminton activities are relatively similar with each other, for example, "Forehand Net Kill" and "Backhand Net Kill", "Backhand High Clear" and "Backhand Clear". However, the limitation of lacking sufficient samples makes it hard to train an expressive deep model. Following HMC [10], we use the same random segmentation strategy where 80% of BSS-V2 is Taken as the training set, its rest as the test set.

Data Preprocessing. Per AFEB-AlexNet [8] and HMC [10], firstly we employ the low-pass ButterWorth filter to deal with sensor noise. Then, we remove the redundant temporal frames (records) based on Sliding Window and Bottom-up (SWAB) algorithm. After that, the cropped data are resized into 300 records using linear interpolation or linear compression along the temporal axis. The models without automatic feature extractor (AFEB) rely on manually extracted features. Following [8, 10], we calculate the mean value, standard deviation, skewness and kurtosis of each attribute as well as the raw attributes (three-dimensional acceleration and three-dimensional spatial attitude angles) as model inputs. LSTM based models employ sliding windows to compress the original 2D input data into 1D vectors. The width of windows is fixed to 30 records with a overlap of 15 records. As for AFEB based models, only raw attributes are taken as input. The feature embedding of feature similarity is extracted from the fourth bottleneck of MobileNetv2 [5].

Model Training. For fair comparison, we utilize the same training details with HMC [10]. In detail, The Adaptive Moment Estimation (Adam) is utilized as the optimizer with a momentum of 0.9 and a weight decay of 1e 4. The training batch size is taken as 64. All mentioned models are trained for 500 epochs in total with an initial learning rate 1e 3. The CNN based models are all pretrained on ImageNet.

Model Evaluation. Following HMC [10], the 5-folds cross evaluation is conducted for model evaluation. Speaking in detail, we randomly split the original BSS-V2 dataset into five pairs of training and test sets, then the models in Table 2 are trained and evaluated in each pair and the average accuracies are reported as the final scores.

4.2 Comparison with Baselines

We compare the performance of the proposed SGNet with commonly used models like LSTM, MobileNet, ResNet, AFEB based CNN and previous state-of-the-art model HMC. HMC follows the best setting in [10], where the mapping from Task 1 to Task2 is shown in Table 1. The experimental results are presented in Table 2.

It is shown that the proposed similarity graph architecture achieves obvious performance gain over HMC, which shares the same backbone models. Besides, in terms of efficiency, the proposed SGNet employs lightweight GCN with four layers. Demonstrated in [9], the computation overhead is so small that it keeps as efficient as backbone CNN model. While the computation of previous state-of-the-art model HMC relies on the number of subclasses, especially, it is several times of the backbone computation.

4.3 Ablations and Analysis

In this section, we perform controlled ablation studies to evaluate the core resource of the performance gain by SGNet and evaluate the influence of label graph to the recognition accuracy.

Table 1. The optimal mapping from Task1 and Task2 in HMC [10].

Task 1	M1	M2	M3
Task 2	Forehand High Serve	Forehand Net Lift	Forehand High Clear
	Backhand High Clear	Backhand Net Lift	Overhead High Clear
	Backhand Clear	Forehand Intercept	Forehand Clear
	hand Serve	Backhand Intercept	Overhead Clear
	Backhand Serve	Forehand Intercept Drive	Forehand Smash
	Forehand Net Shot	Backhand Intercept Drive	Overhead Smash
	Backhand Net Shot	Forehand Intercept Straight	Midfield Forehand Smash
	Forehand Hook Diagonal	Backhand Intercept Straight	Midfield Backhand Smash
	Backhand Hook Diagonal	Forehand Intercept Diagonal	Forehand Drop Shot
	Hand Net	Backhand Intercept Diagonal	Overhead Drop Shot
	Hand Net	-	Forehand Full Strike
	Forehand Net Kill	-	Overhead Full Strike
	Backhand Net Kill	-	-
	Forehand Return	-	-
	Backhand Return	-	-

Label Graph of SGNet. We implement four counterparts of SGNet, all of which share the same backbones AFEB-ResNet50 with labels embedded through four layers GCN. The difference is that each of them has its own adjacency matrix. Table 3 summarizes the experimental results of SGNet (Knowledge Graph), SGNet (Name Similarity Graph, SGNet (Feature Similarity Graph) and SGNet (Similarity Graph). The proposed Similarity Graph is demonstrated to outperform the others by large margins. Additionally, it is also shown that the adjacency matrix makes the key influence on the model performance. Particularly, Knowledge Graph based version even drops the accuracy by 0.4% compared with backbone AFEB-MobileNet. The reason is probably the lacking information of entire badminton activities in ConceptNet. Both Name Similarity Graph and Feature Similarity Graph show obvious performance gains over backbone models, and superimposing label name similarity and feature similarity could further achieve better accuracy.

Table 2. Quantitative results of baselines and SGNet. We report the average accuracies of five test datasets with 5-folds cross validation.

Method	Backbone	Pretrain	Accuracy
LSTM	LSTM	-	68.4%
MobileNet	mobilenetv2	ImageNet	80.0%
ResNet50	ResNet50	ImageNet	79.5%
ResNet101	ResNet101	ImageNet	79.4%
AFEB-MobileNet	mobilenetv2	ImageNet	82.2%
AFEB-ResNet50	ResNet50	ImageNet	82.0%
AFEB-ResNet101	ResNet101	ImageNet	81.7%
HMC-MobileNet	AFEB-MobileNetv2, AFEB-MobileNetv2	ImageNet	83.9%
HMC-ResNet50	AFEB-ResNet50, AFEB-ResNet50	ImageNet	83.6%
HMC-ResNet101	AFEB-ResNet101, AFEB-ResNet101	ImageNet	82.4%
SGNet	AFEB-MobileNetv2	ImageNet	89.2%
SGNet	AFEB-ResNet50	ImageNet	91.4%
SGNet	AFEB-ResNet101	ImageNet	**92.7%**

Table 3. Performance comparisons of different label graphs on BSS-V2 dataset. The "Knowledge graph" is built from ConceptNet [6] and "Similarity Graph" is the composition of "Name Similarity Graph" and "Feature Similarity Graph".

Method	Backbone	Pretrain	Accuracy
Baseline	ResNet50	ImageNet	79.5%
SGNet (Knowledge Graph)	AFEB-ResNet50	ImageNet	81.6%
SGNet (Name Similarity Graph)	AFEB-ResNet50	ImageNet	89.2%
SGNet (Feature Similarity Graph)	AFEB-ResNet50	ImageNet	90.5%
SGNet (Similarity Graph)	AFEB-ResNet50	ImageNet	**91.4%**

5 Conclusion

The conflict between lacking large-scale training datasets and numerous parameters in deep learning models makes the sensor based human activity recognition very difficult. To this end, we have proposed a novel similarity graph based architecture SGNet to enhance the performance of backbone CNN models. The adjacency matrix of the proposed SGNet is composed of two complementary priors: label name similarity and feature similarity. The former embeds the overall relationship among labels by label names. It is robust for various data noise. The latter is based on the training dataset and could better reflect the real-world sample feature similarity of specific labels. Injecting these two priors, our SGNet surpasses the backbone CNN models through extra human prior embedded. It is demonstrated by the experimental results on BSS-V2 dataset that our proposed SGNet surpasses the previous state-of-the-art model by a considerable margin.

Acknowledgment. This work was supported by the Joint Laboratory of Intelligent Sports of China Institute of Sport Science (CISS).

References

1. Gao, J., Zhang, T., Xu, C.: I know the relationships: Zero-shot action recognition via two-stream graph convolutional networks and knowledge graphs. In: AAAI (2019)
2. Kumar, V., Pujari, A.K., Padmanabhan, V., Kagita, V.R.: Group preserving label embedding for multi-label classification. Pattern Recognition (2019)
3. Lee, C.W., Fang, W., Yeh, C.K., Wang, Y.: Multi-label zero-shot learning with structured knowledge graphs. In: 2018 IEEE/CVF Conference on Computer Vision and Pattern Recognition (2018)
4. Lin, R., Xiao, J., Fan, J.: NeXtVLAD: an efficient neural network to aggregate frame-level features for large-scale video classification. In: Leal-Taixé, L., Roth, S. (eds.) ECCV 2018. LNCS, vol. 11132, pp. 206–218. Springer, Cham (2019). https://doi.org/10.1007/978-3-030-11018-5_19
5. Sandler, M., Howard, A., Zhu, M., Zhmoginov, A., Chen, L.C.: Mobilenetv2: Inverted residuals and linear bottlenecks. In: 2018 IEEE/CVF Conference on Computer Vision and Pattern Recognition (CVPR) (2018)
6. Speer, R., Chin, J., Havasi, C.: Conceptnet 5.5: an open multilingual graph of general knowledge (2016)
7. Sun, J., Zheng, W., Zhang, Q., Xu, Z.: Graph neural network encoding for community detection in attribute networks. In: IEEE Trans. Cybern. (2021)
8. Wang, Y., Fang, W., Ma, J., Li, X., Zhong, A.: Automatic badminton action recognition using CNN with adaptive feature extraction on sensor data. In: International Conference on Intelligent Computing (2019)
9. Wang, Y., He, D., Li, F., Long, X., Zhou, Z., Ma, J., Wen, S.: Multi-label classification with label graph superimposing. In: Proceedings of the AAAI Conference on Artificial Intelligence, vol. 34, pp. 12265–12272 (2020)
10. Wang, Y., Ma, J., Li, X., Zhong, A.: Hierarchical multi-classification for sensor-based badminton activity recognition. In: 2020 15th IEEE International Conference on Signal Processing (ICSP). vol. 1, pp. 371–375. IEEE (2020)
11. Wang, Z., Chen, T., Li, G., Xu, R., Lin, L.: Multi-label image recognition by recurrently discovering attentional regions. In: Proceedings of the IEEE International Conference on Computer Vision, pp. 464–472 (2017)
12. Xia, Y., Chen, K., Yang, Y.: Multi-label classification with weighted classifier selection and stacked ensemble. Inf. Sci. **557**, 421–442 (2020)
13. Yan, S., Xiong, Y., Lin, D.: Spatial temporal graph convolutional networks for skeleton-based action recognition. In: Thirty-Second AAAI Conference on Artificial Intelligence (2018)

MITT: Musical Instrument Timbre Transfer Based on the Multichannel Attention-Guided Mechanism

Huayuan Chen$^{(\boxtimes)}$ (iD) and Yanxiang Chen (iD)

Hefei University of Technology, Hefei, China

Abstract. Research on neural style transfer and domain translation has clearly demonstrated the ability of deep learning algorithms to manipulate images based on their artistic style. The idea of image translation has been applied to the task of music-style transfer and to the timbre transfer of musical instrument recordings; however, the results have not been ideal. Generally, the task of instrument timbre transfer depends on the ability to extract a separated manipulable instrument timbre feature. However, as the distinction between a musical note and its timbre is often not sufficiently clear, generated samples by current timbre transfer models usually contain irrelevant waveforms. Here, we propose a method of timbre transfer, for musical instrument sounds, capable of converting one instrument sound to another while preserving note information (duration, pitch, rhythm, etc.). The multichannel attention-guided mechanism is used to enable timbre transfer between spectrograms, enhancing the ability of the model guidance generator to capture the most distinguishable components (harmonic components) in the process. The proposed model uses a Markov discriminator to optimize the generator, enabling it to accurately learn a spectrogram's higher-order feature. Experimental results demonstrate that the proposed instrument timbre transfer model effectively captures the harmonic components in the target domain and produces explicit high-frequency details.

Keywords: Musical instrument timbre transfer · Spectrogram · Timbre · Generative adversarial network

1 Introduction

Driven by musical style transfer and image translation, musical instrument timbre transfer is increasingly attracting researcher attention. Musical instrument timbre transfer is the conversion of one instrument sound to another, while preserving note information, such as duration, pitch, and rhythm. The timbre is a perceptual characteristic, and the difference in timbre appears as a difference in harmonic distribution on the spectrogram. Although the idea of image style transfer can directly be applied to the raw audio wave's time-frequency representation, the effectiveness of doing so depends on whether independent and controllable instrument timbre features can be extracted correctly. Recently, researchers have proposed several models for implementing musical

© Springer Nature Switzerland AG 2021
D.-S. Huang et al. (Eds.): ICIC 2021, LNCS 12836, pp. 568–581, 2021.
https://doi.org/10.1007/978-3-030-84522-3_47

instrument timbre and music style transfer learning. For example, Mor et al. [1] directed raw audio wave modeling and using a shared WaveNet [2] encoder, and independent decoders, for multiple instruments to implement timbre transfer. However, WaveNet is a conventional, autoregressive model. Audio samples generated by it are required to be generated sequentially, which makes the speed of this method unsuitable for real-time applications. Timbretron [3] proposed a method that uses constant-Q transform [4] (CQT) spectrograms as the raw audio wave representation and CycleGAN [5] as the timbre transfer model. However, this model does not address the problem of musical note and timbral entanglement, which causes generated samples to contain irrelevant waveforms. Bitton et al. [6] proposed the embedding of instrument category and pitch labels into their model to achieve a many-to-many timbre transfer among instruments. Although their approach solved the one-to-one problem of timbre transfer, they did not propose a clear solution to the quality problem of spectrogram generation.

The use of spectrograms as transfer media is a common method for processing raw audio wave. However, many of the current proposed models have difficulties in extracting musical instrument timbre features and distinguishing between timbre and note information. These difficulties are attributed to the very subtle differences in the spectrograms of different instruments, which makes it difficult for the generator to perceive the difference in harmonics between the source and target domains, so that unwanted parts are likely to be generated in the samples. In 2020, Jain et al. introduced a single-channel attention-guided mechanism into the generator in ATT [7], which makes the generator can capture the most distinguishable part between the source and the target domain, realizing the separation of music note information and timbre information. ATT is the first attempt to use the attention-guided mechanism in the timbre transfer of musical instruments task. However, generative space for it is limited due to the single attention mask and the single content mask generated by the generator. In this study, we propose a musical instrument timbre transfer model based on the multichannel attention-guided mechanism [8]. Applying the multichannel attention-guided mechanism enhances the model guide generator's ability to capture the most discriminating components (harmonic components) of a spectrogram. Accordingly, the generator reduces the generation of unwanted factors. Additionally, the proposed method uses a Markov discriminator [9] to optimize the generator so that it can capture a higher-order spectrogram feature. Experimental results demonstrate that the proposed timbre transfer method effectively captures distinctions between timbre and note information, generates high-quality spectrograms, and successfully achieves timbre transfer learning.

This study's contribution is summarized as follow:

1. We propose the use of the multichannel attention-guided mechanism to guide timbre transfer. The use of the multichannel attention-guided mechanism enhances the ability of the model guidance generator to capture the most distinguishable components (i.e., harmonic components) of spectrograms.
2. To optimize the generator structure for fusing local and overall spectrogram features, the discriminator uses a superimposed convolutional layer to output an $N \times N$ matrix in which each patch is judged separately.
3. The proposed method applies the idea of image translation to musical instrument timbre transfer. CQT spectrograms are used as the medium for musical instrument

spectrogram-to-spectrogram transfer, and the converted spectrograms are classified with an accuracy of up to 96.97%.

2 Related Work

2.1 Generative Adversarial Networks (GANs)

Generative Adversarial Nets (GAN) [10] is one of the most significant model in deep learning. Since 2014, when Goodfellow proposed the GAN, its powerful generation capabilities have achieved significant results in field of image and audio generation [11, 12], image translation [13], and audio conversion [14].

2.2 Time-Frequency Representation

In many cases, time-frequency representation is an effective way to represent raw audio wave. By using filters with different center frequencies the signal can be separated (e.g. short-time Fourier transform) and the information carried by the signal can be made more clear. Short-time Fourier transform (STFT) is a commonly used signal processing method, however, its linear frequency bin spacing is known to be inadequate to some degree for analyzing and processing music signals [15]. To address this problem, we compute the constant-Q transform (CQT) [4] of a time-domain signal. CQT has the same frequency distribution as a twelve-tone scale and has a high frequency resolution at low frequencies and a high time resolution at high frequencies. The CQT of the finite-length sequence $x(n)$ is expressed as:

$$x^{cq}(k) = \frac{1}{N_k} \sum_{n=0}^{N_k-1} x(n) w_{N_k}(n) e^{-j\frac{2\pi Q}{N_k}n} \qquad (1)$$

In formula (1), N_k is the window length corresponding to the calculation of the CQ transformation of the k-th frequency f_k, w_{N_k} is the window function of length N_k, Q is the constant factor in the CQ transformation, and k is the sequence CQ spectrum Frequency subscript.

By calculating the CQT spectrum of a music signal, the amplitude value at each note frequency can be directly obtained, which is useful in recovering the fine timing of rhythms.

2.3 Image-To-Image Translation

Deep learning and GAN development has led to the rapid advancement of image translation. Examples of successful classical models include the pix2pix [9], CycleGAN [5], and UNIT [16]. These models achieved translation from source-to target-domain images, but usually only to a single target domain. With the development of StarGAN [17], MUNIT [18] and UGATIT [19], translation from sources to multiple target domains was further realized. However, these models often generated unwanted factors and could not focus on the most distinguishable components. Therefore, Tang et al. [8, 20] proposed using attention-guided mechanism to capture the most discriminative component of the image target domain and minimize variation in the background. The model's effectiveness is demonstrated experimentally. This paper draws on the idea of image translation and introduces the multichannel attention-guided mechanism into the timbre transfer of musical instruments to achieve more accurate modeling of the timbre of the target spectrogram.

3 Method

3.1 Spectrogram-To-Spectrogram Timbre Transfer

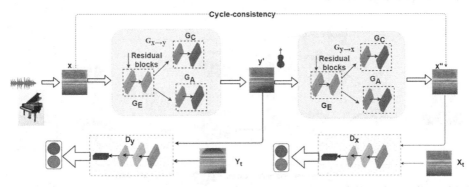

Fig. 1. MITT: musical instrument timbre transfer based on the multichannel attention-guided mechanism.

The proposed timbre transfer method aims at realizing spectrogram-to-spectrogram timbre transfer by using CQT spectrograms to represent raw audio data, and the diagram of this model is shown in Fig. 1. Inspired by the dual-learning mechanism, the proposed model uses two generators ($G_{x \to y}$, $G_{y \to x}$) and two discriminators (D_x, D_y) to form a ring network. $G_{x \to y}$, $G_{y \to x}$ represent timbre transfer from instruments X to Y and from Y to X respectively. In order to improve the ability of distinguishing timbre and music note information, the multichannel attention-guided mechanism is applied to the generator to achieve context-aware timbre transfer for instrument. Each generator consists of a parameter-sharing encoder G_E, an attention mask network G_A and a content mask network G_C. The discriminators are used to distinguish fake data from real data through punishing its corresponding generator when it produces unsatisfactory results. Otherwise, the proposed model uses Markov discriminators to optimize generators to accurately learn a spectrogram feature at higher order level. We show one mapping in this figure, i.e., $x \to G_{x \to y}(x) \to G_{y \to x}(G_{x \to y}(x)) \approx x$. We also have other mapping, i.e., $y \to G_{y \to x}(y) \to G_{x \to y}(G_{y \to x}(y)) \approx y$. The proposed model is constrained by the cycle-consistency loss and trained in an end-to-end fashion. It is also an unsupervised timbre transfer model with essentially no pairwise matching datasets and no ground truth for comparison. X_t and Y_t and represent a random sample of spectrogram in the X domain and Y domain respectively.

3.2 Multichannel Attention-Guided Mechanism

The musical instrument timbre transfer task is designed to achieve context-aware instrument timbre transfer by capturing the relationship between instrument timbre and notes, and automatically discerning which parts should be changed, and which should be left behind during the spectrum translation process. Generally, conventional instrument timbre transfer methods conduct the transference of semantic information with a primary focus on the differences between the source and target domains and cannot preserve musical note information. To achieve perceptual timbre transfer, the generator must focus on specific components (harmonics) of the spectrogram and retain those components that need not be changed. Figure 2 illustrates Framework of the proposed multichannel attention-guided generation.

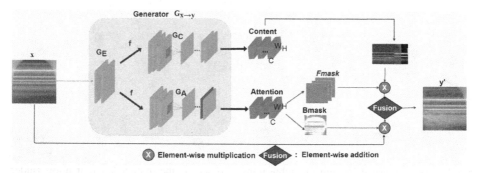

Fig. 2. Framework of the proposed multichannel attention-guided generation. We show the process of spectrogram x through generator $G_{x \to y}$ timbre transfer to generate y' in this figure.

The input spectrogram of generator $G_{x \to y}$ is $x \in R^{H \times W \times C}$, where H is the height of spectrogram, where W is the width spectrogram, where C is the number of channels. Generator $G_{x \to y}$ consists of three parts, G_E, G_A, G_C. G_E is the encoder network, G_C is the content mask network, and G_A is the attention mask network. Firstly, spectrogram x is encoded by G_E. And G_E outputs feature f. Then, the content masks $C_y \in R^{H \times W \times C}$ and attention masks $A_y \in \{0, \ldots, 1\}^{H \times W}$ can be obtained by sending f to the content mask network G_C and the attention mask network G_A respectively. Because G_C and G_A have their own parameters, they do not affect each other. Network G_A outputs $n-1$ foreground attention masks $\left\{A_y^t\right\}_{t=1}^{n-1}$ and a background attention mask A_y^b; at the same G_C outputs n-1 content masks $\{C_y^t\}_{t=1}^{n-1}$. This way, the network can simultaneously learn the novel foreground spectrogram while retaining the input spectrogram's background. The output result $G_{x \to y}(x)$ is obtained by fusing $\left\{A_y^t\right\}_{t=1}^{n-1}$, A_y^b, $\{C_y^t\}_{t=1}^{n-1}$, and the input spectrogram x as:

$$G_{x \to y}(x) = \sum_{t=1}^{n-1} (C_y^t * A_y^t) + x * A_y^b \qquad (2)$$

In this manner, the background content $x * A_y^b$ of the input spectrogram x is retained while a novel foreground content $\sum_{t=1}^{n-1}(C_y^t * A_y^t)$ is generated, which can then be combined to obtain the output spectrogram $G_{x \to y}(x)$. The formulation of generator $G_{y \to x}(y)$ and input spectrogram y can be expressed as $G_{y \to x}(y) = \sum_{t=1}^{n-1}(C_x^t * A_x^t) + y * A_x^b$, where n attention masks $\{A_x^t\}_{t=1}^{n-1}, A_x^b$ are also produced by a channel-wise Softmax activation function for normalization.

3.3 Markov Discriminator Guidance of Higher-Order Spectrogram Feature Capture

Generally, images are locally correlated, that is, neighboring pixels are more likely to have the same color and brightness. In the spectrogram, there are more harmonic correlations. The correlation is reflected in the entire frequency axis. Therefore, how to effectively learn harmonic correlation of the spectrogram is very important. The proposed model uses L1 loss to capture the lower-order spectrogram information; however, L1 has difficulties in capturing higher-order information. Accordingly, the performance of the discriminator is measured by the accurate capture of a higher-order spectrogram feature. To obtain the feature details of a higher-order spectrogram, the proposed model uses a Markov discriminator that penalizes on the spectrum patch scale. By judging individual patches rather than overall spectrum, the generator can be optimized to better focus on detail generation. In effect, the Markov discriminator models the spectrogram as a Markov random field in which each feature is described in terms of local spectrogram blocks around each pixel value. The Markov discriminator output can be understood as a measure of similarity between texture styles. Figure 3 shows the Markov discriminator's structure used here.

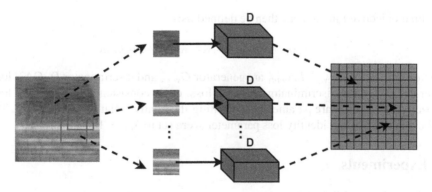

Fig. 3. Markov discriminator diagram.

3.4 Loss Function

Adversarial loss: To stabilize the model training, we use the least squares loss instead of the CrossEntropy loss. Therefore, the objective function of discriminator D_x is expressed as:

$$\min_{D_x} V(G_{y\to x}, D_x) = \frac{1}{2}E_{x\sim p_{data(x)}}\left[(D_x(x) - 1)^2\right] + \frac{1}{2}E_{y\sim p_{data(y)}}[D_x(G_{y\to x}(y))^2] \quad (3)$$

The objective function of generator $G_{y\to x}$ is expressed as:

$$\min_{G_{y\to x}} V(G_{y\to x}) = \frac{1}{2}E_{y\sim p_{data(y)}}[D_x(G_{y\to x}(y)) - 1)^2] \quad (4)$$

Generator $G_{y\to x}$ and discriminator D_x are updated alternately, and L_{GAN}^x is $G_{y\to x}$, D_x GAN loss. Similarly, L_{GAN}^y is generator $G_{x\to y}$ and discriminator D_y GAN loss.

Cyclic consistency loss: Theoretically, an output result x'' obtained after implementing $G_{y\to x}(G_{x\to y}(x))$ should be approximately equal to the original input x, and the output result y'' obtained after implementing $G_{x\to y}(G_{y\to x}(y))$ should be approximately equal to the original input y. To strengthen the overall consistency, cyclic consistency can be applied to reduce the possible set of mappings that the networks can learn and force $G_{x\to y}$ and $G_{y\to x}$ to perform opposite conversions. The cyclic consistency loss is defined as

$$L_{cycle} = E_{x\sim p_{data(x)}}\left[\|G_{y\to x}(G_{x\to y}(x)) - x\|_1\right]$$

$$+E_{y\sim p_{data(y)}}\left[\|G_{x\to y}(G_{y\to x}(y)) - y\|_1\right] \quad (5)$$

Identity loss: Identity loss can assist the generator in preserving music content, which, empirically, yields better audio quality. The identity loss $L_{identity}$ is defined as:

$$L_{identity} = E_{y\sim p_{data(y)}}\left[\|y - G_{x\to y}(y)\|_1\right] + E_{x\sim p_{data(x)}}\left[\|x - G_{y\to x}(x)\|_1\right] \quad (6)$$

The final objective function can then be defined as:

$$L = L_{GAN}^x + L_{GAN}^y + \lambda_{cyc}L_{cycle} + \lambda_{id}L_{identity} \quad (7)$$

Where $L_{GAN}^x, L_{GAN}^y, L_{cycle}, L_{identity}$ are generator $G_{y\to x}$ and discriminator D_x GAN loss, generator $G_{x\to y}$ and discriminator D_y GAN loss, cycle-consistency loss, identity loss, respectively. $\lambda_{cyc}, \lambda_{id}$ are parameters controlling the relative relation of each term. The cyclic consistency and identity loss parameter were set to $\lambda_{cyc}= 10$ and $\lambda_{id}= 0.7$.

4 Experiments

4.1 Experimental Setup

We trained the musical instrument timbre transfer model on the Nsynth dataset [21], which contains many annotated notes collected from instruments with various pitches and rates. To validate the effectiveness of our approach, we pre-processed the raw audio data properly. Specifically, we select 9300 samples for each instrument and extracted the corresponding acoustic components. In addition, we remove the silent part and then compute the constant-Q transform (CQT) for the 1s raw audio signals in time-domain.

For generating spectrograms, we used 512 samples between successive frames and FFT window of length 2048. The resolution of each spectrogram was [84*84*3]. Figure 4 shows parts of the CQT spectrogram sample. We collect the CQT spectrum into a new dataset and divide it into training, testing, and validation sets with a ratio of 7:1:2. Based on this dataset, the two-way timbre transfer experiments can be carried out using an Adam optimizer with an initialized learning rate of 0.0002, and the other parameters are: $\alpha = 0.5$, $\beta = 0.999$. We train the proposed model for 200 epochs with a batch size of 32.

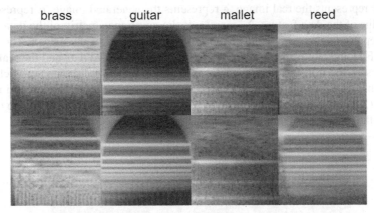

Fig. 4. Real sample of CQT spectrogram. For a spectrogram, the horizontal axis represents time and the vertical axis represents logarithmic frequency.

4.2 Evaluation Mechanism

In this paper, we propose a timbre transfer method for musical instruments in which raw audio wave are expressed as CQT spectrogram. To evaluate the reconstruction performance, several evaluation metrics are used to measure the difference between input spectrogram x and reconstructed output spectrogram x''. It includes the root-mean-square error (RMSE), the peak signal-to-noise ratio (PSNR) and the Fréchet inception distance (FID) metrics.

The root mean square error is defined as:

$$RMSE = \sqrt{\frac{1}{m}\sum_{i=1}^{m}\left(x_i - x_i''\right)^2} \tag{8}$$

For this formula, x_i represent the real sample on the verification set, and x_i'' is the reconstructed sample. RMSE calculates the root mean square error of x_i and x_i''. The smaller the value of it, the better results can get.

The peak signal-to-noise ratio is defined as:

$$PSNR = 10 * log_{10}\left(\frac{MAX_I^2}{MSE}\right) \tag{9}$$

MSE represents the mean square error between the generated image and the real image, where MAX_I^2 is the maximum pixel value of the image. Peak signal-to-noise ratio is often used as a measurement method of signal reconstruction quality in image compression and other fields. Therefore, we use it to evaluate the quality of data reconstruction. Larger PSNR value indicates a higher generation quality.

The Fréchet inception distance is defined as:

$$FID = \|\mu_r - \mu_g\|_2^2 + Tr(\Sigma_r + \Sigma_g - 2(\Sigma_r \Sigma_g)^{1/2}) \tag{10}$$

In FID, r represents the real image, g represents the generated image, μ represents the mean value of the data feature, $\sum$ represents the covariance matrix of the data feature, and T_r represents the sum of the elements on the diagonal of the matrix. The smaller value of FID, the closer the distribution between the generated data and the real data.

To assess the transfer quality, each spectrogram on the validation set is classified after timbre transfer. The transfer effect also was evaluated using the kernel inception distance (KID) and the Fréchet inception distance metrics, which measure the distance between the real spectrogram domain and the generated spectrogram domain.

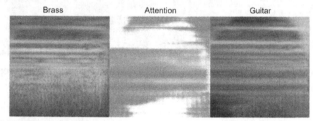

(a) Timbre transfer from brass spectrogram x to guitar spectrogram y'.

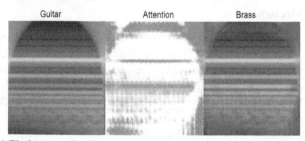

(b) Timbre transfer from guitar spectrogram y to brass spectrogram x'.

Fig. 5. The effect of timbre transfer based on spectrogram by MITT. For a spectrogram, the horizontal axis represents time and the vertical axis represents logarithmic frequency. The proposed model in this paper is an unsupervised timbre transfer model with essentially no pairwise matching datasets and no ground truth for comparison. Therefore, no ground truth samples are shown in this figure.

4.3 Musical Instrument Timbre Transfer Experiment

The musical instrument timbre transfer model is designed to maximize the distinction between timbre and musical note information while ensuring the quality of spectrogram generation, and then realize timbre transfer. To assess the converted spectrograms produced by the proposed method, we qualitatively and quantitatively analyzed and compared its output to that produced by recent methods. The compared model CycleGAN [5] is a classical model, and AGGAN [20] is a single-channel attention-guided model. Additionally, frame structure-dependent ablation experiments were conducted.

4.3.1 Qualitative and Quantitative Studies

Qualitative analysis
In spectrograms of music audio signals, differences in instrument timbre are primarily reflected as differences in the number and shape of harmonic distributions, which in turn can be represented as horizontal texture information on the spectrogram. From Fig. 5, we find that the harmonic distribution of the brass spectrogram generated from the guitar spectrogram through MITT timbre transfer has increased significantly.

In addition, we find that the attention mask generated by the model effectively grasps the nonlinear harmonic distribution of the spectrograms.

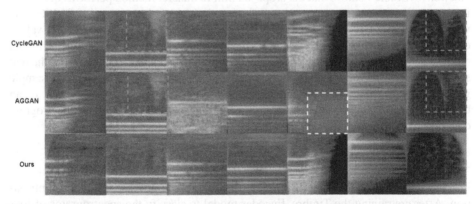

Fig. 6. Samples generated by different timbre transfer under different models. For a spectrogram, the horizontal axis represents time and the vertical axis represents logarithmic frequency.

In performing timbre transfer, it is important to generate clear texture information while reducing the creation of unwanted factor. We qualitatively compared samples generated by the transformation process of the proposed method with those produced using AGGAN and CycleGAN under the same experimental settings (see Fig. 6). Figure 6 shows the converted CQT spectrograms produced by different models, with time and frequency indicated on the horizontal and vertical axes. A comparison of the generated samples reveals that AGGAN and CycleGAN generate gray unidentified texture information in the high-frequency parts of their respective CQT spectrograms (The texture of the spectrum as shown in the red box in the figure). Additionally, several artifacts appear in the spectrogram following AGGAN timbre transfer (The texture of the spectrum as shown in the white box in the figure). Conversely, the proposed model's spectrogram generates a clearer texture and higher spectrogram quality.

Quantitative Analysis:

Table 1. Results of the quantitative evaluation of model transfer and generation effects.

Instruments	Reconstruction			Transfer
	RMSE ↓	PSNR ↑	FID ↓	FID ↓
brass ↔ guitar	0.0121	30.7510	47.15	22.29
guitar ↔ reed	0.0115	30.6799	47.89	19.87
brass ↔ reed	0.0079	32.4662	20.67	16.72
brass ↔ mallet	0.1025	27.3228	50.04	22.08

To assess the quality of spectrograms generated by timbre transfer, we calculate the average RMSE, PSNR, and FID between target domain samples and the samples generated by the MITT timbre transfer in the validation set to analyze the reconstruction and transfer. Table 1 shows that the proposed timbre transfer method performs quite well in reconstructing spectrograms and transferring timbre. We found that the RMSE, FID values of the MITT reconstructed generated samples were small and the PSNR values of MITT reconstructed generated samples were large. So that we inferred that the model reconstructed generated spectrograms had clear texture and small distance from the real sample data distribution. In addition, we found that the distance FID between the sample generated by the timbre transfer and the target domain sample is also very small. It can be inferred that the converted spectrogram is similar to the target domain spectrogram data distribution.

4.3.2 Comparison with the State-of-the-Art

Further, we evaluated the effect of timbre transfer and the proposed model's contribution. After pretraining the Timbre_Encoder model to classify the spectrograms of four instruments, we performed timbre transfer on the validation set and then used the Timbre_Encoder to classify the generated samples. Figure 7(a) shows a comparison between the results produced by the proposed model, CyleGAN, and AGGAN. Among the three assessed models, MITT produced the best results in terms of timbre-transformed spectrogram classification, with an average classification accuracy of 96.97%. In order to visualize the classification effect, we extracted the classified features for PCA dimensionality reduction visualization, and the result is shown in Fig. 7(b). Figure 7(b) shows the data distributions of the brass, mallet, guitar, and reed results on the validation set (T0, T1, T2, and T3) and the sample distributions generated by the timbre transformation model (F0, F1, F2, and F3).

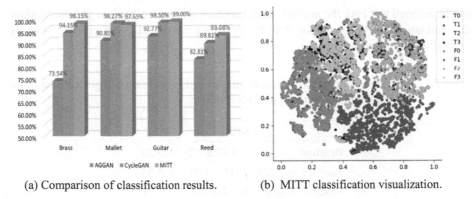

(a) Comparison of classification results.   (b) MITT classification visualization.

Fig. 7. Post-timbre transfer spectrum classification diagram.

In order to verify the validity and advancement of the proposed model, in the comparative experimental study, we conducted a comparison of the FID distances of different models, and the comparison results are shown in Table 2. We found that relative to CycleGAN, AGGAN, the distance FID values between spectrograms generated by the MITT model and the target domain spectrograms are the smallest. In other words, the samples generated by the MITT model are closer to the target domain spectrogram data distribution.

Table 2. Comparison of the FID values produced by different models for different instruments.

FID	brass ↔ guitar	guitar ↔ reed	brass ↔ reed	brass ↔ mallet
CycleGAN	38.51	36.30	38.51	47.40
AGGAN	55.28	56.51	16.70	80.82
MITT	**22.29**	**19.87**	**16.72**	**22.08**

4.3.3 Ablation Study

To quantitatively assess the effects of applying the multichannel attention-guided mechanism and Markov discriminator on MITT, we performed ablation experiments on the proposed model. To perform ablation experiments on the brass → guitar timbre transfer we removed each component sequentially while maintaining the other conditions unchanged. Among them, Baseline* is the basic model, which is composed of CyleGAN and pixel discriminator. The experimental results are shown in Table 3. It can be seen from the table that when only the multichannel attention-guided mechanism is removed, the KID increases to about 10, and the FID increases to about 38.51, that is, the quality of the generation is degraded. When only the Markov discriminator is removed, the KID value increases to about 83, and the FID value increases to about 116, which indicates that the quality of the spectrogram generation is rapidly degraded. Therefore, it can be inferred that both the multichannel attention-guided mechanism and the Markov discriminator model played a key role in improving the model's performance in terms of instrument timbre transfer.

Table 3. Ablation analysis results of the effects of the multichannel attention-guided mechanism (MAG) and Markov discriminator (MK) on timbre transfer.

	KID ↓	FID ↓
Baseline*	226.77 ±3.20	183.20
Baseline + MAG	83.15 ±1.73	115.53
Baseline + MK	10.64 ± 0.47	38.51
MITT	**5.03** ±0.46	**22.29**

5   Summary

This paper mainly introduces the combination of generative adversarial networks and musical instrument timbre transfer tasks under the current booming development of deep learning and image translation, and the realization of timbre transfer based on the idea of image translation. We propose a musical instrument timbre transfer system that uses the multichannel attention-guided mechanism to learn the timbre differences between instruments and achieve the transfer of a given instrument's sound to an exception sound while retaining its basic note information. Guided by the multichannel attention-guided mechanism, the model generator effectively captures the harmonic distributions in the spectrograms produced by individual instruments. The Markov discriminator is used to optimize the generator to accurately capture the higher-order spectrogram feature, improving the quality of spectrogram generation. Experimental results reveal that the musical instrument timbre transfer system proposed is effective at achieving timbre transfer between instruments and significantly improves the quality of converted spectrograms, achieving a classification accuracy of 96.97%. The method proposed here can effectively capture the distinctions between timbre and musical note information, generating high-quality spectrograms to successfully achieve timbre transfer learning. Based on the existing work, we will follow up on the timbre transfer of singing, and the many-to-many instrument timbre transfer for research and discussion.

References

1. Mor, N., Wolf, L., Polyak, A., et al.: A universal music translation network. arXiv preprint arXiv:1805.07848 (2016)
2. Oord, A., Dieleman, S., Zen, H., et al.: Wavenet: a generative model for raw audio. arXiv preprint arXiv:1609.03499 (2016)
3. Huang, S., Li, Q., Anil, C., et al.: Timbretron: a wavenet (cyclegan (cqt (audio))) pipeline for musical timbre transfer. arXiv preprint arXiv:1811.09620 (2016)

4. Brown, J.C.: Calculation of a constant Q spectral transform. J. Acoustical Soc. Am. **89**(1), 425–434 (1991)
5. Zhu, J.Y., Park, T., Isola, P., et al.: Unpaired image-to-image translation using cycle-consistent adversarial networks. In: Proceedings of the IEEE International Conference on Computer Vision, pp. 2223–2232. IEEE, Italy (2017)
6. Bitton, A., Esling, P., Chemla-Romeu-Santos, A.: Modulated Variational auto-Encoders for many-to-many musical timbre transfer. arXiv preprint arXiv:1810.00222 (2018)
7. Jain, D.K., Kumar, A., Cai, L., et al.: ATT: Attention-based Timbre Transfer. In: 2020 International Joint Conference on Neural Networks, pp. 1–6. IEEE, UK (2020)
8. Tang, H., Liu, H., Xu, D., et al.: Attentiongan: unpaired image-to-image translation using attention-guided generative adversarial networks. arXiv preprint arXiv:1911.11897 (2019)
9. Isola, P., Zhu, J.Y., Zhou, T., et al.: Image-to-image translation with conditional adversarial networks. In: Proceedings of the IEEE Conference on Computer Vision and Pattern Recognition, pp. 1125–1134. IEEE, Italy (2017).
10. Goodfellow, I.J., Pouget-Abadie, J., Mirza, M., et al.: Generative adversarial networks. arXiv preprint arXiv:1406.2661 (2014)
11. Reed, S., Akata, Z., Yan, X., et al.: Generative adversarial text to image synthesis. In: International Conference on Machine Learning, pp. 1060–1069 (2016)
12. Yamamoto, R., Song, E., Kim, J.M.: Parallel WaveGAN: a fast waveform generation model based on generative adversarial networks with multi-resolution spectrogram. In: ICASSP 2020–2020 IEEE International Conference on Acoustics. Speech and Signal Processing, pp. 6199–6203. IEEE, USA (2020)
13. Yi, Z., Zhang, H., Tan, P., et al.: Dualgan: Unsupervised dual learning for image-to-image translation In: Proceedings of the IEEE International Conference on Computer Vision, pp. 2849–2857. IEEE, USA (2017)
14. Pasini, M.: Melgan-vc: voice conversion and audio style transfer on arbitrarily long samples using spectrograms. arXiv preprint arXiv:1910.03713 (2019)
15. Schörkhuber, C., Klapuri, A., Sontacchi, A.: Pitch shifting of audio signals using the constant-q transform. In: Proceedings of the DAFx Conference (2012)
16. Liu, M.Y., Breuel, T., Kautz, J.: Unsupervised image-to-image translation networks. arXiv preprint arXiv:1703.00848 (2017)
17. Choi, Y., Choi, M., Kim, M., et al.: Stargan: unified generative adversarial networks for multi-domain image-to-image translation. In: Proceedings of the IEEE Conference on Computer Vision and Pattern Recognition, pp. 8789–8797. IEEE, USA (2018)
18. Huang, X., Liu, M.Y., Belongie, S., Kautz, J.: Multimodal Unsupervised Image-to-Image Translation. In: Ferrari, V., Hebert, M., Sminchisescu, C., Weiss, Y. (eds.) European Conference on Computer Vision, ECCV 2018. Lecture Notes in Computer Science, vol. 11207. Springer, Cham (2018). https://doi.org/10.1007/978-3-030-01219-9_11
19. Kim, J., Kim, M., Kang, H., et al.: U-GAT-IT: unsupervised generative attentional networks with adaptive layer-instance normalization for image-to-image translation. arXiv preprint arXiv:1907.10830 (2019)
20. Tang, H., Xu, D., Sebe, N., Yan, Y.: Attention-guided generative adversarial networks for unsupervised image-to-image translation. In: 2019 International Joint Conference on Neural Networks, pp. 1–8. IEEE, Hungary (2019).
21. Engel, J., Agrawal, K.K., Chen, S., et al.: Gansynth: adversarial neural audio synthesis. arXiv preprint arXiv:1902.08710 (2019)

Classification of Benign-Malignant Pulmonary Nodules Based on Multi-view Improved Dense Network

Li-Hua Shen[1,2], Xin-Hao Wang[1,2], Min-Xiang Gao[1,2], and Bo Li[1,2(✉)]

[1] College of Computer Science and Technology, Wuhan University of Sciences and Technology,
Wuhan 430070, China
libo@wust.edu.cn

[2] Hubei Province Key Laboratory of Intelligent Information Processing and Real-Time
Industrial System, Wuhan University of Sciences and Technology, Wuhan 430070, China

Abstract. Lung cancer is one of the most common cancers in the world, and the detection and classification of benign-malignant lung nodules are critical during the diagnosis and treatment for lung cancer. In this paper, a multi-view improved dense convolutional network is proposed for the classification of benign-malignant pulmonary nodules, where more information of input multi-scale features can be extracted from 2D views of nine different directions. The improved dense block and other layers are linked by shortcuts, which optimizes the feature extraction. The proposed network model is trained in the LIDC-IDRI dataset, and the results show that the average classification accuracy and AUC are 86.52% and 97.23% respectively, which means that the network model has significantly improved the performance of benign-malignant pulmonary nodules classification.

Keywords: Lung cancer · Dense convolution · Multi-scale features · Shortcut links

1 Introduction

Lung cancer is one of the most common cancers, which results in the highest fatal rate in the world [1]. The reason for the low survival rate is that the clinical symptoms of lung cancer usually appear in the advanced stage of lung cancer [2], so conducting an early diagnosis generally improves survival rate and provides the best chance for successful treatment. Pulmonary nodules are one of the most important early symptoms of lung cancer, which appear as irregular small structures with different sizes in CT images [3]. Pulmonary nodules can be categorized into benign and malignant: benign nodules usually have no clinical symptoms and no risk of spreading; malignant nodules can spread quickly and endanger the life of patient. If the pulmonary nodules can be found and the benign and malignant types of pulmonary nodules can be identified accurately in time through computerized tomography (CT) [4], it will greatly reduce the misdiagnosis rate of lung cancer.

© Springer Nature Switzerland AG 2021
D.-S. Huang et al. (Eds.): ICIC 2021, LNCS 12836, pp. 582–593, 2021.
https://doi.org/10.1007/978-3-030-84522-3_48

Currently, computer-aided diagnosis (CAD) system has greatly improved its identification efficiency and has become an important auxiliary means to ease the work of doctors, and it also assists radiologists in the rapid diagnosis of pulmonary nodules [5]. Generally, diagnosis of pulmonary nodules has two steps: candidate area of pulmonary nodule detection and pulmonary nodule benign-malignant classification [6]. In the process of candidate pulmonary nodule area detection, it is required to find all suspicious nodules as much as possible, which inevitably contains a large number of benign pulmonary nodules in results. It is well-known from multiple research data that less than 5% of the diagnosed pulmonary nodules are finally confirmed as malignant nodules. Therefore, benign-malignant pulmonary nodule classification is vital to the automatic diagnosis system of pulmonary nodules and is also the research target of this article. Recently, deep learning has made extensive and in-depth developments in medical image applications, and the advancement of medical image has also benefited from this trend. Various nodule classifiers of nodule characteristics have been proposed, such as Histogram of Oriented Gradient (HOG) [7], Local Binary Pattern (LBP) [8], Scale Invariant Feature Transform (SIFT) [9] and other local feature classifiers. A large number of methods have also been applied in medical image processing, such as Convolutional Neural Networks (CNN), Deep Generative Models (DGM), Generative Adversarial Networks (GAN), supervised learning, unsupervised learning, etc. Ronneberger et al. [10] proposed a Fully Convolution Network (FCN)U-Net for biological image segmentation, which has a good performance in fusing network hierarchical and local semantic features. Hua et al. [11] respectively applied Deep Neural Network (DCNN) and Deep Belief Network (DBN) to the benign-malignant pulmonary nodule classification, which revealed that deep learning is good enough for such tasks. ELBAZ et al. [12] used a two-step labeling method to accurately observe the volume changes between the corresponding nodules, which can effectively identify benign-malignant pulmonary nodules, but it takes a lot of time to observe and track the changes of nodules. Pulmonary nodules are quite different in morphological scale. Benign nodules are generally smooth and sharp. A few of benign nodules have visible notches and long burrs, which are diversified in shape and diffusely distributed, while malignant nodules with short burrs around them, are eccentrically distributed. Although current convolutional networks have unique advantages in image recognition, their layers has gradually increased as performance required. For example, VGG network has reached 19 layers. With the increasing of the number of network layers, some issues are beginning to emerge, such as model overfitting, gradient disappearance, gradient explosion, and insignificant extracted feature effects. The above problems lead to the insignificant improvements in the classification of benign-malignant pulmonary nodules in various models.

In order to solve the problems, a multi-view improved dense convolutional network is proposed in this paper to deal with the benign-malignant pulmonary nodule classification. The study revealed that when 3D model is trained, the process is extremely complicated and requires much more training data. Currently, insufficient datasets are available, and there are obviously fewer malignant pulmonary nodules in training samples. Nevertheless, when pulmonary nodules are predicted, high robustness to malignant nodules is required. We find that different sections of pulmonary nodules show different subtle features, so we cut the 3D model of each nodule in a 2D view from various symmetrical angles to obtain 9 cross-sectional views, as a result, more subtle features for each nodule can be obtained. From the previous introduction of pulmonary nodules, it is known that fixed-size filters will cause information loss and insignificant extraction effects during features extraction for pulmonary nodules of different sizes and shapes. Therefore, three parallel improved dense network structures with the same structure but different convolution kernel sizes are proposed in this paper, which have better performance in pulmonary nodules feature capture, and then the pulmonary nodule types of each 2D cross-sectional view are predicted by Softmax classifier, finally the results of each 2D view are weighted and fused to derive the final prediction result. This design innovatively uses the 2D view of the 3Dpulmonary nodules, so that the model can learn 3D features, and secondly, through its 2D view, more subtle features can be extracted, which improves the utilization of pulmonary nodule features and also enhances the power of nodules processing and the generalization ability of the network. The experimental results show that the performance of the model proposed in this paper for the benign-malignant pulmonary nodules classification has improved significantly.

2 Dataset

The dataset used in the experiment is based on the public database LIDC-IDRI, which is composed of chest medical images and independent annotation opinions of four experienced radiologists. The LIDC-IDRI dataset is mainly used for early cancer detection in high-risk populations and was initiated by the National Cancer Institute. This dataset contains a total of 1018 CT scan examples, and the diagnosis and the labeling need to be carried out in two stages. In the first stage, each radiologist diagnoses and reviews the CT images independently, marks the location of the lesion and implements three classifications: $\geq$ 3mm nodules, $\leq$ 3mm nodules, and $\geq$ 3mm non-nodules. In the second stage, each radiologist reviews the anonymous annotations of the other three radiologists independently to give a final diagnosis. According to records, only $\geq$ 3mm nodules have malignant annotations, 1375 benign nodules and 930 malignant nodules were selected in our experiment, and all the uncertain 1386 pulmonary nodules were excluded. The number of nodules with different malignant tumor ratings in the dataset is shown in Fig. 1 below.

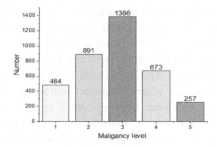

Fig. 1. The number of nodules with different malignant tumor ratings in the dataset

Pulmonary nodules are various in morphology, which also show different impacts on the final classification performance of the network. Figure 2 below shows four different types of pulmonary nodules.

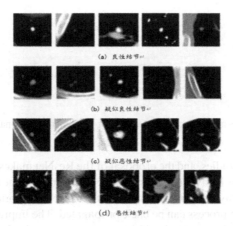

Fig. 2. Pictures of four different types of pulmonary nodules

3 Method

The multi-view improved dense convolutional network based benign-malignant pulmonary nodules classification method proposed in this paper has the following contributions. First, the 3D view of pulmonary nodule is cut in multi-direction to obtain more detailed 2D views with its features, and then taking into account the various sizes and shapes of pulmonary nodules, a multi-layer parallel network structure based on DenseNet is designed with different sizes of convolution kernels in each layer. In order to better share the information of each layer, layers are linked by shortcuts to achieve the purpose of cross-information sharing. Each view is processed by the residual network we designed, and the derived results are weighted and fused, and the Softmax classifier is used for the pulmonary nodules classification. Figure 3 shows the classification process of benign-malignant pulmonary nodules.

3.1 View Extraction

In consideration of reducing the search space and improving the accuracy of the network model, we performed lung parenchyma segmentation before extracting the view. The size of the extracted cube is $64 \times 64 \times 64$. each cube contains pulmonary nodule, and the pulmonary nodule is located in the center. Then Each cube is cut in multiple directions to extract the 2D view of 9 directions we need.

3.2 Multi-layer CrossDenseNet

In the field of computer vision, convolutional neural network is currently the most popular network model in research. Many improvements have been made to convolutional neural

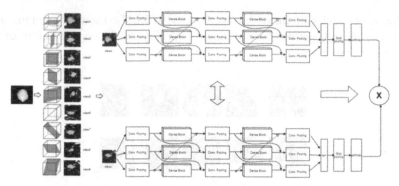

Fig. 3. Classification process of benign-malignant pulmonary nodules

networks in previous studies, and the proposal of the ResNet makes convolutional neural networks flourish. The ResNet can train a deep convolutional neural network, and the deeper and previous layers can share information through shortcut links, so that the gradient in the training process can be back-propagated. The improved DenseNet which uses denser shortcut links to reproduce features in this paper is also based on the idea of the ResNet, and it can achieve better performance than the ResNet with fewer parameters and computational costs. Considering that the size and shape of different malignant pulmonary nodules are various, if a large filter is used to extract features for small size pulmonary nodules, after multiple down-samplings, the acquired feature map will only contain fewer features information, and there will even be extreme situations such as no feature information, thereby reducing the classification accuracy. The same is true for pulmonary nodules with large size. When a smaller filter is used to extract features, due to the small acceptance field, the overall information of the nodule cannot be captured, resulting in low classification accuracy. Due to the uncertainty of the size of pulmonary nodules, it is obvious that the above-mentioned problems cannot be solved if a filter of a single size is used. Therefore, a multi-layer structure to extract features of different scales is designed in this paper, and the filters of different sizes in each layer are used to extract feature information of different scales. Each view is processed by 3 convolutional layers, 4 pooling layers, and 2 DenseNetwork layers in the model we designed. The input of each time is nine 2D views with a size of 64×64, and each view is processed by the multi-layer cross DenseNet, and the prediction results of 2D views are weighted and fused to obtain the final classification result. The multi-layer cross DenseNet has three layers with the same structure but different convolution kernel sizes, and the sizes of the convolution kernels which are used to extract the input multi-scale features are 3×3, 5×5, and 7×7 respectively. The size of the max-pooling kernel is the same as convolution kernel of each layer. Through the first convolution part, 64 feature maps with a size of 64×64 are obtained. Through the max-pooling part, 64 feature maps with a size of 32×32 are obtained. Through the dense block, 64 feature maps with a size of 32×32 are obtained. Through the second convolution part, 128 feature maps with a size of 32×32 are obtained. Throughthe max-poolingpart, 128 feature maps with a size of 16×16 are obtained.Through the dense block and the last convolution max-pooling part, 256 feature maps with a size of 8×8 are obtained. The three-layer output is combined through the cascade layer, and then the prediction classification result is obtained through the global average pooling layer and Softmax, and finally the prediction results of all the 9 2D views are weighted and fused to derive the final classification result.

3.3 Dense Residual Block

When the network performs convolution operations, the extracted features are not complete feature maps, so the correlation between the feature maps is uncertain, which affects the performance of the network to a considerable extent. Therefore, the dense residual block we designed has 3 layers, the sizes of the convolution kernels are 1×1 and 3×3. Taking into account the computational cost, no shortcut links are set between the convolution kernels inside a kernels group. Inside a layer, the results of the previous kernels group are linked with all subsequent groups. Between layers, the upper layer also links the residuals with the next layer, and the residuals of the last layer are linked with the first layer. Figure 4 shows the details of the dense residual block.

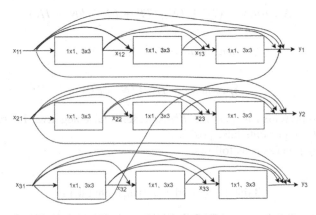

Fig. 4. Dense residual block

Assuming that the input of each layer is x_{ij}, where i represents the layer, j represents the input position of the layer, and the output is defined as y_i, then the output of each layer is as formula (1)

$$y_i = H_i([x_{i1}, x_{i2}, \ldots, x_{ij-1}, x_{i+11}])$$ (1)

where H_i is a non-linear transformation, and $[x_{i1}, x_{i2}, \ldots, x_{ij-1}, x_{i+11}]$ is the agreed input sequence for each layer.

3.4 Fusion Algorithm

In order to get the final results, the prediction results obtained from nine 2D views need to be fused. The final classification results of pulmonary nodules are shown in formula (2):

$$P = \sum_{v=0}^{v} \omega_v p_v$$ (2)

where p_v represents the prediction result of the v th view, and ω_v represents the weight of the v th view. Due to the imbalance of the dataset, the model evaluation index AUC is used to determine the weight value. In the training process, each view is trained separately without fusion, to obtain the appropriate weight value of each view, and then the training result of each view is fused with the weight value to get the final prediction result.

3.5 Training

The experiments were carried out in the python 3.7 environment and the experimental parameters are as follows. The training epoch and batch size are 300 and 200, respectively. The dropout is set to 0.5 to avoid overfitting. The initial learning rate is 1×10^{-3}, and it is set to 1×10^{-4} after half an epoch, and the learning rate in the last quarter of the epoch is 5×10^{-5}. Cross entropy is used as the loss function in training. Accuracy, Sensitivity, Specificity and AUC are used to evaluate the performance of model, where Accuracy, Sensitivity and Specificity are defined as formulas (3), (4) and (5) respectively.

$$Accuracy = (TP + TN)/(TP + FP + FN + TN) \tag{3}$$

$$Sensitivity = TP/(TP + FN) \tag{4}$$

$$Specificity = TN/(TN + FP) \tag{5}$$

where TP represents true positive, FN represents false negative, TN represents true negative, and FP represents false positive.

The training process is as follows.

Table 1. Network model algorithm steps

Input:2D views obtained from LIDC-IDRI dataset
Output:P, predicted probability of nodule types
Algorithm:
Step 1 The 9 2D views are used to train the model.
Step 2 Train to get the weight valueω_v of the view$v$.
Step 3 Calculate the AUC_v and predicted probability p_v of view v.
Step 4 Calculate $W_v = AUC_v/ sum(AUC_v)_{v=1}^9$
Step 5 Calculate the classification probability P of pulmonary nodules by formula (2)
Step 6 Return P.

4 Experiment

Several experiments are carried out to test the network structure and network performance in this paper, then the optimal network structure and corresponding parameters are obtained through comparative experiments, so that the best classification result of pulmonary nodules types can be achieved. A comparison with other sophisticated networks is also discussed in this paper to illustrate the significant improvements of our network model.

4.1 Exploration of Network Structural Hierarchy

For the exploration of the multi-layer parallel structure in the network, we adopt the network with different layers of parallel structure to compare the performance of the three-layer parallel structure network proposed in this paper. All networks have same structures and parameters except the number of parallel layers. Comparative experiments are carried out for 1 to 5 layers of parallel structure. The size of the three-layer parallel convolution kernel proposed in this paper is 3×3, 5×5, and 7×7 respectively. In the comparative experiment, for the sizes of single-layer convolution kernel is one of above three sizes, for the size of double-layer convolution kernel are two of above three sizes; for the four-layer convolution kernel, an extra size of 9×9 is included; for the five-layer convolution kernel, an extra size of 11×11 is included. The accuracies of parallel structure network with various number of layers are shown in Table 2 below. For the single-layer structure, the network with a filter size of 7 has the best performance, and the network with a filter size of 3 gets the worst. For the two-layer parallel structure, the kernel size combination 3_7 has the best performance, and the combination 3_5 gets the worst. It is worth noting that the three-layer parallel structure has the best performance among all the control groups. The accuracy of the network from the single-layer to the three-layer increases successively, reaching the highest at the three-layer structure, and the accuracy of the network from four-layer to five-layer is gradually decreasing. As the matter of fact, increasing the number of network layers after the three-layer structure not only reduces the performance, but also increases the complexity and computational cost of the model. Compared with single-layer and two-layer parallel structure, the three-layer structure uses more feature information in multi-scale to make the network have better generalization. Therefore, the three-layer parallel structure network with filter sizes of 3×3, 5×5, and 7×7 designed in this paper has the best performance.

Table 2. Accuracies of parallel structure network with various number of layers

Layer	Kernel size	Accuracy (%)
1	3/5/7	80.75/83.65/84.25
2	3_5/3_7/5_7	83.75/84.75/84.65
3	3_5_7	86.52
4	3_5_7_9	84.78
5	3_5_7_9_11	84.29

4.2 Exploration of the Number Of Dense Block

In order to get the best model accuracy of the multi-layer Dense Block while avoiding additional redundancy, we conduct comparative experiments to determine the number of dense blocks. The experimental evaluation indicators include Flops, Params, model memory and experimental error rate. The experiment is based on a three-layer dense network, and the number of dense block convolution kernel group is set to three. The experimental results are shown in Fig. 5 below. In the comparative experiment, when the number of dense blocks in the network was gradually increased from 1 to 6, the Flops, Params, and memory usage increased significantly, but the experimental error rate did not continuously and effectively decrease. It is obvious that an appropriate number of dense blocks has a significant impact on balancing model overhead and result accuracy. It can be known from the experiment that the overhead and the classification accuracy of the model are optimal when the number of dense blocks is between 1 and 3. In detail, when the number of dense blocks increased from 1 to 2, the experimental error rate decreased from 11.1% to 10.6%, and memory usage increased slightly. When the number of dense blocks increased from 2 to 3, the experimental error rate increased by 0.2% and memory usage has also risen. Therefore, the multi-layer improved dense network model with 2 dense blocks used in this paper can get both high accuracy and relatively low model memory usage.

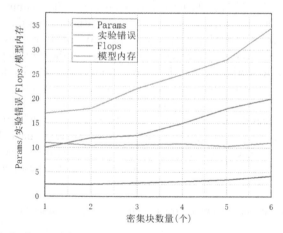

Fig. 5. Indicator values of model with various number of dense blocks

4.3 Network Evaluation

The 10-fold cross validation is used to evaluate the stability and efficiency of the network model proposed in this paper. The dataset is divided into 10 groups, 20% of the data in each group is used as the test set, and the rest are used as the training set. The average classification accuracy of our network model reaches 86.52%, and the stability of the network is evaluated by calculating the standard deviation of the 10-fold cross-validation results, then $86.52\% \pm 1.12 \times 10^{-2}$ is obtained as the final result. Compared with the ResNet network and its variants, the performance of the network we designed has improved significantly. When extracting features, the transmission of features has been significantly strengthened, and the feature information of pulmonary nodules can be fully utilized to make the network have better generalization. Relatively speaking, our network reduces the usage of parameters to a certain extent, making the network more efficient in training. The highest average accuracy of ResNet and its variants for benign-malignant pulmonary nodules classification is 85.88%, and our network model improves on this by about 0.64%.Even compared with the DenseNet network and its variants, the network model we designed is also outstanding in all aspects of evaluation indicators. It can be seen that the network model proposed in this paper is currently the best in the classification of pulmonary nodules. In addition, we also test the classification accuracy of our model for each type of pulmonary nodules, and the results are shown in Table 3 below. Our network model has a classification accuracy of 87.12% for malignant pulmonary nodules, which is 0.6% higher than the average classification accuracy. It shows that our network has the best classification performance for malignant pulmonary nodules. The network we designed also has an impressive performance in the classification of benign pulmonary nodules, with an accuracy of 85.92%. As mentioned in Sect. 1, malignant and benign pulmonary nodules are quite different in morphological scale, indeed malignant pulmonary nodules get more features for extraction, so malignant pulmonary nodules have a better classification performance than benign pulmonary nodules. In summary, the network model proposed in this paper makes significant improvements for the benign-malignant pulmonary nodules classification and has the best performance.

Table 3. Accuracy of specific pulmonary nodules type

Pulmonary nodules type	Accuracy (%)
Benign	85.92
Malignant	87.12
Average	86.52

We compared the method proposed for the LIDC-IDRI dataset in the last 5 years with our pulmonary nodules classifier, and the results are shown in Table 4 below. To ensure the fairness of the results, all methods have been evaluated in exactly the same experimental conditions. All methods are tested by 10-fold cross-validation on the LIDC-IDRI dataset, and sensitivity, specificity, and AUC are used as evaluation indices. Sensitivity indicates the proportion of malignant nodules that are correctly detected. Specificity indicates the proportion of benign nodules that are correctly detected. AUC describes the classification performance of benign-malignant pulmonary nodules, which can be calculated by summing the area under the ROC curve. The sensitivity and AUC of our network model are 92.31% and 97.23% respectively, which are the highest among the 5 methods, and the specificity of our network model is 91.85%. For the method proposed by Xie et al. [13], they obtain nine 2D views from the pulmonary nodules cube, and design a knowledge-based collaboration model for each view, and train a ResNet-50 network for features extraction. Their model achieves an AUC of 94.04% and a sensitivity of 89.67%

on LIDC-IDRI dataset. In comparison, our network model has great improvements in both AUC and sensitivity. For the multi-view convolutional network proposed by DEY et al. [14], the factors of multi-view input are mainly considered to strengthen the extraction and utilization of features by the network. In comparison, what we consider is to strengthen the multi-scale features extraction and utilization of the input network. Experiments prove that our network has stronger generalization and higher accuracy.

Table 4. Evaluation results of network models in 3 indices

Model	Sensitivity (%)	Specificity (%)	AUC (%)
3D CNNS [15]	90.37	88.27	90.48
3D DenseNet [16]	85.71	93.10	89.40
MoDenseNet [14]	90.40	90.47	95.48
Multi-view CNN [17]	89.90	85.26	94.85
Ours	92.31	91.85	97.23

5   Conclusion

The benign-malignant pulmonary nodules classification method proposed in this paper is based on a multi-view improved dense convolutional network, using nine2D views of pulmonary nodules in different directions to learn the 3D features. The improved dense convolutional network is used to extract more multi-scale features, so that it reduces the number of parameters and network over-fitting degree while optimizing feature extraction and utilization. Our network model has significantly improved the performance of benign-malignant pulmonary nodules classifier, but there are still some researches can be done on network structure, loss function and even clinical applications.

References

1. Ferlay, J., et al.: Cancer incidence and mortality worldwide: sources, methods and major patterns in GLOBOCAN 2012. Int. J. Cancer **136**(5), E359–E386 (2015)
2. Bach, P.B., Silvestri, G.A., Hanger, M., Jett, J.R.: Screening for lung cancer: ACCP evidence-based clinical practice guide-lines. Chest **132**, 69S-77S (2007)
3. Larici, A.R., Farchione, A., Franchi, P., Ciliberto, M., Cicchetti, G., Calandriello, L., del Ciello, A., Bonomo, L.: Lung nodules: Size still matters. Eur. Respir. Rev. **26**, 170025 (2017)
4. Siegel, R.L., Miller, K.D., Jemal, A.: Cancer statistics, 2018, CA. Cancer J. Clin. **68**(1), 7–30 (2018)
5. Zhao, X., Liu, L., Qi, S., Teng, Y., Li, J., Qian, W.: Agile convolutional neural network for pulmonary nodule classification using CT images. Int. J. Comput. Assist. Radiol. Surg. **13**(4), 585–595 (2018). https://doi.org/10.1007/s11548-017-1696-0
6. Filho, P.P.R., Sarmento, R.M., Holanda, G.B., de AlencarLima, D.: 'New approach to detect and classify stroke in skull CT images via analysis of brain tissue densities.' Comput. Methods Programs Biomed. **148**, 27–43 (2017)

7. Chen, S., et al.: 'Automatic scoring of multiple semantic attributes with multi-task feature leverage: a study on pulmonary nodules in CT images.' IEEE Trans. Med. Imag. **36**(3), 802–814 (2017)
8. Srensen, L., Shaker, S.B., de Bruijne, M.: 'Quantitative analysis of pulmonary emphysema using local binary patterns.' IEEE Trans. Med. Imag. **29**(2), 559–569 (2010)
9. Zhang, F., et al.: A ranking-based lung nodule image classification method using unlabeled image knowledge. In: Proceedings of the ISBI, pp. 1356–1359, April 2014
10. Ronneberger, O., Fischer, P., Brox, T.: U-Net: convolutional networks for biomedical image segmentation. In: Navab, N., Hornegger, J., Wells, W.M., Frangi, A.F. (eds.) MICCAI 2015. LNCS, vol. 9351, pp. 234–241. Springer, Cham (2015). https://doi.org/10.1007/978-3-319-24574-4_28
11. Hua, K.-L., Hsu, C.H., Hidayati, S.C., Cheng, W., Chen, Y.: Computer-aided classification of lung nodules on computed tomography images via deep learning technique. Onco-Targets Terapy **8**, 2015–2022 (2015)
12. EI-Baza, S., Gimel'farbg, et al.: Elastic phantoms generated by microfluidics technology: validation of an imaged-based approach for accurate measurement of the growth rate of lung nodules. Biotechnol. J. **6**(2), 195–203 (2011)
13. Xie, Y., et al.: Knowledge-based collaborative deep learning for benign-malignant lung nodule classification on chest CT. IEEE Trans. Med. Imag. **38**(4), 991–1004 (2019)
14. Dey, R., Lu, Z., Hong, Y.: Diagnostic classification of lung nodules using 3D neural networks. In: Proceedings of the IEEE 15th International Symposium on Biomedical Imaging (ISBI 2018), Washington, DC, USA, 4–7 April 2018, pp. 774–778 (2018)
15. Yan, X., et al.: Classification of lung nodule malignancy risk on computed tomography images using convolutional neural network: a comparison between 2D and 3D strategies. In: Chen, C.-S., Lu, J., Ma, K.-K. (eds.) ACCV 2016. LNCS, vol. 10118, pp. 91–101. Springer, Cham (2017). https://doi.org/10.1007/978-3-319-54526-4_7
16. Dey, R., Lu, Z., Hong, Y.: Diagnostic classification of lung nodules using 3D neural networks. In: Proceedings of the ISBI, pp. 774–778, April 2018
17. El-Regaily, S.A., Salem, M.A.M., Aziz, M.H.A., Roushdy, M.I.: Multi-view Convolutional Neural Network for lung nodule false positive reduction. Expert Syst. Appl. 2019, 113017. https://doi.org/10.1016/j.eswa.2019.113017

Deep Convolution Neural Network Based Research on Recognition of Mine Vehicle Head and Tail

Junqiang Li[1], Chao Wang[1,2]($\boxtimes$), Lin Cui[1], Zhiwei Zhang[1], Wenquan Tian[1],
Zhenggao Pan[1], Wanli Zhang[1], Xiaoying Yang[1], and Guolong Chen[3]

[1] School of Informatics and Engineering,
Suzhou University, Suzhou 234000, People's Republic of China
[2] Institute of Machine Learning and Systems Biology, School of Electronics and Information
Engineering, Tongji University, Shanghai 201804, People's Republic of China
[3] School of Computer Science and Information Engineering,
Bengbu University, Bengbu 233000, People's Republic of China

Abstract. The mine environmental monitoring system captures the photos of the head and the tail of the vehicle, and sometimes the system can not accurately distinct whether it is the head or the tail of the vehicle. When there are two trucks in the view of the surveillance camera, the captured image contains the head of one truck and the tail of another truck. What needs to be recognized is the head license plate number or the tail license plate number. However, because the system cannot distinguish the head and tail of the truck, it will cause more false alarms. In order to solve this problem, this paper proposes an end-to-end feature extraction and recognition model based on deep convolution neural network (Deep CNN). The Deep CNN model contains five stage CNN layer and each layer contains different kernel size to extract the features. The data set is provided by Huaibei Siyuan Technology Co., Ltd., which includes normal capture, escape and false alarm images of the trucks. The final prediction rate is 85% on the testing set, which occupied twenty percent of the whole image set. The prediction rate of our model has been higher than the prediction rate base on right-out-left-in principle, which is used in the mine environmental monitoring system. Finally, our model will be applied in the mine environmental monitoring system.

Keywords: Head and tail recognition · Deep learning · Convolution neural network · Mine environment monitoring system

1 Introduction

China is rich in mineral resources, many of which play an important role in the economy and provide important support for the sustainable and healthy development of local national economy. With the continuous expansion of the mining scale of mineral resources, the effective supervision of mine resources development has become an increasingly important issue. In recent years, the state has issued a series of laws, regulations and policies on the management of mineral resources [1], which has promoted

D.-S. Huang et al. (Eds.): ICIC 2021, LNCS 12836, pp. 594–605, 2021.
https://doi.org/10.1007/978-3-030-84522-3_49

the healthy development of mining industry, and also promoted the improvement of the level of grass-roots Mining Administration. However, it is difficult to put an end to violations of laws and regulations in the development of mineral resources [2]. There are still a series of problems, such as unlicensed mining, illegal mining, destructive mining, mining not according to the approved development and utilization plan, low resource utilization rate, concealed production, environmental pollution, safety and geological hazards. It is very difficult to rely on the personnel supervision, and the cost of human, material and financial resources is also very large, and the effect is not ideal. How to achieve accurate supervision has become an urgent problem to be solved.

In recent years, vehicle identification technology is widely used in mine supervision, traffic control and other fields, and has been widely valued and concerned by researchers [3]. Therefore, the fine identification of the license plate, head and tail of the transport vehicle and the shape of the minerals truck load in the mining area is not only of great significance for the effective management of mine resources, but also can avoid the illegal behavior of tax evasion [4]. In the mining environment, there are many kinds of vehicles in and out. In addition to transport vehicles, there are trucks, SUVs, buses and other types of vehicles. In addition, there are also waste resources such as ballast. Therefore, it is of great significance for the construction of mine comprehensive supervision platform to effectively identify the transport vehicles and the goods they trucker and eliminate the interference of other types of vehicles.

When the mine environmental supervision system captures the photos of the head and tail of the truck, it is impossible to give the impact of tax evasion caused by the problem of the head and tail of the truck distinction. The purpose of this paper is to solve the problem that the head and tail of the mine transportation vehicle could not be identified accurately. The comparison between the head and tail images is shown in Fig. 1.

Fig. 1. Vehicles' head and tail.

The research method of mining vehicle head and tail recognition based on deep convolution neural network proposed in this paper has the following contributions: First, the head and tail pictures are sorted, and the data that is difficult for people to distinguish between the head and the tail is deleted. Second, for the head and the tail have different size, the model uses different convolution kernel size, Third, The structure of the convolution tail is changed from the 1000 classes to the 2 classes that is the truck's head or tail, and then the Softmax activation function is used, and finally the Adam optimizer is used to update the parameters. Figure 2 shows the structure of proposed Deep CNN model.

The organizational structure of this paper is as follows: in the introduction part, it introduces the problems in mining management and the shortcomings of the previous solutions, and puts forward our research on vehicle head and tail recognition based on deep learning. The related work parts introduces the previous work of vehicle recognition, including vehicle license plate recognition, vehicle logo recognition and vehicle head and tail recognition, two patents and one article are introduced as the reference of our work. The third part method part, mainly introduces the principle of deep convolution neural network, the basic network framework, the network hierarchy and loss function settings used in this paper. In the experiment parts, it mainly introduces the database, the parameters set in this experiment and the comparison of the experimental results under the corresponding settings. The last part is the summary and future work. The final part is the part of acknowledgement, which mainly lists the items that support the work.

2 Related Work

The visual recognition of vehicles mainly includes license plate recognition, logo recognition, head and tail recognition, etc. License plate recognition is an early research area in vehicle recognition. The main processes of license plate recognition include image preprocessing, license plate location, character segmentation and character recognition. The preprocessing mainly includes graying, contrast enhancement, binarization, filtering, smoothing and correction of the vehicle image taken by the camera. The typical methods in license plate location include color feature based [6–8], edge feature based [9–11], morphological feature based [12, 13], support vector machine based [14], clustering method, neural network based, genetic algorithm based, hybrid feature based. The problem of character segmentation in license plate location is that the representative methods are Unicom segmentation, projection segmentation, and static boundary method. The last step is character recognition. The common methods are template matching, support vector machine, neural network. License plate recognition is a relatively mature recognition technology internationally recognized that the recognition rate of all vehicle models is over 85–95%, and the recognition time is within 200 ms.

Vehicle logo recognition is a more refined requirement for license plate recognition. This paper proposes a fast identification method for vehicle signs, which is mainly based on the filtering processing based on the characteristics of energy concentration in vertical direction of the position of the vehicle sign, and the accurate position of the vehicle sign by template matching in the rough positioning rectangle frame. According to the characteristics of high energy and concentration of the vehicle sign in the vertical direction, a new method of positioning the vehicle sign with high speed and robustness is proposed. Firstly, image filtering is truckried out by energy enhancement and adaptive morphological filtering, and then the candidate areas of the vehicle mark are segmented by adaptive threshold, Then, the vehicle logo is accurately positioned according to the characteristics of the vehicle and its relationship with the vehicle. In this paper, a fast identification method of vehicle signs based on edge histogram is proposed. The main technology is to identify the head signs by using correlation method and edge histogram based on template matching positioning.

There are not a lot of literature on vehicle head and tail identification. At present, the research on the recognition of the head and tail of large-scale transport vehicles is involved in a patent method of feature recognition of the residue truck based on convolution neural network. The patent mainly uses the deep convolution neural network to identify the head and tail of the residue truck to judge separately, If it is the tail information, it will be input into the trained traffic violation model to judge whether it violates the traffic law. The main basis is whether the tail cover of the muck truck is covered. If it is illegal to judge through the algorithm, the vehicle warning information will be notified to the supervisor. In the patent video-based accurate identification method for high-speed mobile vehicle logo, the pre-processing operation involves the positioning of the vehicle's head and tail, in which the image noise removal and contrast enhancement are mainly applied to the head, and homomorphic filtering technology is required for the tail image. In addition, vehicle logo recognition mainly uses the idea of relatively concentrated edge texture. This paper proposes that the vehicle is automatically identified by image segmentation and clustering technology in the camera assisted driving. The paper shows that this method has good recognition effect in foggy and rainy days. To sum up, it can be found in the current research that vehicle head and tail recognition is a processing link in the case that vehicle head and tail need to be distinguished in license plate recognition, and the research goal of this paper is to clearly distinguish the head and tail of the vehicle, let the surveillance camera capture the picture, and distinguish the head and tail of the vehicle through our algorithm, so as to solve the problem of background analysis. For this reason, our method is based on the deep convolution neural network (CNN) to extract features, while using the cross entropy loss function (binary entropy loss function) To measure the effect of classification.

3 Method

The method of mining vehicle head and tail recognition based on deep convolution neural network proposed in this paper, which has the following procedures: firstly, the pictures of the head and tail are sorted out, and the data that people can hardly distinguish the head and tail are filtered out. Secondly, considering the size and shape of the head and tail, the method contains total of 5 stages, each stage contains one convolution layer and one pooling layer in the experiment. Thirdly, two full connection layers and a prediction layer are used to process the convolution tail, and the softmax loss function is used to classify the head and tail of vehicles. Finally, the parameters are updated by Adam optimizer. Figure 2 shows the training process of the model.

3.1 Image Processing

In order to reduce the search space and improve the accuracy of the network model, we normalized the input pictures before training. The size of all the input image is resized as 50×50, and each image contains the head or tail of the vehicle and its located in the center. Then, all the images are grayed before input in our designed deep CNN model.

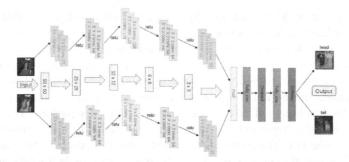

Fig. 2. The structure of the Deep CNN model.

3.2 Deep CNN

In this paper, considering the different sizes and shapes of the head and tail of the vehicle, if the large filter is used to extract the features of the head and tail of the small size vehicle, after multiple down sampling, the obtained feature map will only contain less feature information, even in extreme cases, reducing the classification accuracy. The same is true for the head and tail of large size vehicles. When using smaller filters to extract features, the overall information of the head and tail of vehicles could not be captured due to the smaller receiving area, resulting in low classification accuracy. Due to the uncertainty size of the head and tail of the vehicle, while a single size filter is used, the above problem will not be solved. Therefore, this paper designs a deep CNN to extract features of different scales. Each batch contains 32 pictures and with the picture size is 50×50. The final classification result is obtained by weighting and fusing the prediction results of 2D images. Deep CNN has three layers with the same structure but different convolution kernel sizes, and the convolution kernel sizes are 3×3, 5×5 and 7×7. These convolution kernels are used to extract input features. The size of Max pooling kernel is the same as that of convolution kernel in each layer. Finally, the output of the three layers is combined, and then the global average pooling layer and softmax are used to obtain the prediction classification results, and finally the prediction is made.

3.3 Algorithm Equations

The formula for calculating the output feature map size is as follows:

$$out_{width} = \frac{Input_{width} + 2 * Padding - Kernel_{width}}{Stride_{width}} + 1 \tag{1}$$

$$out_{length} = \frac{Input_{length} + 2 * Padding - Kernel_{length}}{Stride_{length}} + 1 \tag{2}$$

In Eqs. (1) and (2), out length and out width are the length and width of the output feature map respectively; Input length and Input width are the length and width of the input feature map; Padding is the number of edge filling pixels; Kernel length and Kernel width are convolution The size of the core; Stride length and Stride width are the step size.

After the convolution operation of the convolution layer, the result needs to be processed with a non-linear activation function. The entire process expression is as follows:

$$S(i,j) = \sigma((I * K)(i,j)) = \sigma(\sum_m \sum_n I(i+m, j+n)K(m,n)) \qquad (3)$$

In Eq. (3), I represents the two-dimensional image tensor data; K represents the convolution kernel; m and n are the length and width of the convolution kernel respectively, usually 3×3 or 7×7; i and j are the values of horizontal and vertical coordinates respectively, and "*" represents multiplication; σ is the activation function, in the convolution neural network The ReLU activation function is commonly used, which can make the network sparse, facilitate the extraction of network features, and reduce the over-fitting phenomenon in the network training process. The calculation formula is formula (4), and the schematic diagram is shown in Fig. 3.

$$\sigma(x) = max(0, x) = \begin{cases} x, x > 0 \\ 0, x \leq 0 \end{cases} \qquad (4)$$

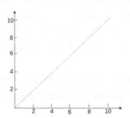

Fig. 3. ReLU activation function.

The output layer of the convolution neural network is the Softmax layer, which is connected to the last fully connected layer in Fig. 2. Softmax regression is a multi-classification algorithm, which belongs to supervised learning, which converts multiple outputs into probabilistic forms through normalization operations. For an n classification problem, given the input x belongs to an original metric h(x, yi) of the i-th category (yi), which the original metric is the input samples original class metric, the calculated probability is shown in Eq. (5).

$$P(x|y) = \frac{e^{h(x,y_i)}}{\sum_j^n e^{h(x,y_j)}} \qquad (5)$$

The loss function of Softmax regression is shown in Eq. (6). The optimization of the Loss function is achieved by iterative solution of the gradient descent method.

$$J\theta = -\frac{1}{m}\left[\sum_{i=1}^m \sum_j^k 1\{y^{(i)} = j\} \log \frac{e^{h(x,y_i)}}{\sum_{j=1}^k e^{h(x,y_j)}} \right] \qquad (6)$$

And the θ in Eq. (6) means the parameter of the model.

3.4 Algorithm Flow

CNN back propagation algorithm flow.

Input: M image samples, layer number L of CNN model and types of all hidden layers. For convolution layer, define the size k of convolution kernel, dimension F of convolution kernel matrix, filling size P and step s. For pooling layer, the size of pooling area K and pooling standard (max or average) are defined. For fully connected layer, the activation function (except output layer) and the number of neurons in each layer are defined. Gradient iteration parameter iteration step α, Max and threshold of stopping iteration ϵ。

Process:

    1) Initialize W, B of each hidden layer and output layer

The value of is a random value.

    2）for iter to 1 to MAX：

2-1) for i =1 to m：

a) Set CNN input A1 to the tensor corresponding to Xi

b) For L = 2 to L-1

b-1) if the current layer is fully connected: then ai,l=σ(zi,l)=σ(Wlai,l−1+bl)

b-2) if the current convolution layer is: then there is ai,l=σ(zi,l)=σ(Wl∗ai,l−1+bl)

b-3) if the current pooling layer is: ai,l=pool(ai,l−1), where pool refers to the process of reducing the input tensor according to the size of pooling area K and pooling criteria.

    c) For output layer L: ai,L=softmax(zi,L)=softmax(WLai,L−1+bL)

    c) The loss function is used to calculate the loss of the output layer δi,L

    d) For L = L-1 to 2

d-1) if the current layer is fully connected: δi,l=(Wl+1)Tδi,l+1 ⊙σ′(zi,l)

d-2) if the current layer is convolution layer: δi,l=δi,l+1∗rot180(Wl+1)⊙σ′(zi,l)

d-3) if it is pool layer at present: δi,l=upsample(δi,l+1)⊙σ′(zi,l)

    2-2) for L = 2 to L, update the WL, BL of layer 1 according to the following two situations:

    2-2-1) if the current layer is fully connected: Wl=Wl−α∑_{i=1}^{m} δ^{i,l}(ai,l−1)T , bl=bl−α∑_{i=1}^{m} δ^{i,l}

    2-2-2) if the current convolution layer is a convolution layer, for each convolution kernel there is: Wl=Wl−α∑_{i=i}^{m} a^{i,l−1}∗δi,l, bl=bl−α∑_{i=1}^{m} ∑_{u,v}(δ^{i,l})_{u,v}

    2-3) if all the changes of W and B are less than the stop iteration threshold ϵ, Jump out of the iteration cycle and go to step 3.

    3) The linear relation coefficient matrix W and bias vector B of each hidden layer and output layer are output.

Output: W, B of hidden layer and output layer of CNN model.

---

4 Experiment

This article tested the network structure and network performance through several experiments, and determined the best network structure and corresponding methods. These parameters are obtained through comparative experiments, so that the best truck head and truck tail classification results can be obtained. This article also discusses the comparison with other complex networks such as VGG16 and ResNet34 to illustrate the significant improvement of our network model.

4.1 Data Source and Composition

The data set used in the experiment is provided by the database of Huaibei Siyuan Technology Co., Ltd., who is responsible for regularly providing experimental data for our project. The data set is mainly based on image types, including 20,000 examples of the head and tail of the truck, and the division and preprocessing of the data set need to be in two stages. In the first stage, we randomly selected 16,000 pictures for training and another 4,000 pictures for testing. The rate of training set and test set is 4:1, and we manually mark the pictures in the training set as the head of the truck and the tail of the truck. In the second stage, we need to filter out non-jpg format pictures, unify the picture size and the grayscale of the picture, and finally scramble the picture. Figure 4 below shows the distribution of the number of training sets and test sets in the data set.

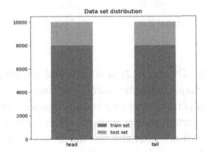

Fig. 4. The distribution of training sets and testing sets.

4.2 Parameters Setting

The vehicle head and tail algorithm designed in this paper is implemented based on the structure of deep convolution neural networks. The training of deep convolution neural networks requires a lot of calculations. In order to speed up the network training speed and parameter tuning and optimization process, this paper first compares the configuration The high workstation, namely the PC, completes the network training, and then transplants it to the server platform. The specific experimental operating environment is shown in Table 1:

The hyper parameter settings during the network training process are shown in Table 2:

A larger learning rate will accelerate network learning in the early stage of training, making it easier for the model to converge to the optimal solution. However, there will be large fluctuations in the later stage, and even the value of the loss function may fluctuate

Table 1. Network model training environment.

CPU: i7-7700HQ	OS: Window10
GPU: Nvidia Geforce 1050	Deep Learning Framework: TensorFlow1.8
Memory: 16G	Cuda: 9.1
Python: 3.6	Cudnn: 7.4
OpenCV: 4.4	TFlearn: 0.3
Compiler: vs 2017	GeForce Graphics Grive: 457.51

Table 2. Network training parameters.

Category	Set value
batch_size	32
learning_rate_begin	0.001
img_size	50
epoch_n	40
max_iter	20000

around the minimum value. The fluctuation is large and it is always difficult to reach the optimum. Therefore, this chapter uses the learning rate decay method for training. At the beginning of training, a larger learning rate is used to speed up the convergence of the network. As the number of training increases, the learning rate will gradually decrease to ensure that the model will not fluctuate too much in the later stage of training, thus getting closer optimal solution. Specifically, the learning rate is attenuated by manual setting, that is, when the number of training steps reaches the set value, the learning rate is multiplied by the attenuation coefficient to obtain the new learning rate. The learning rate set in the pre-training stage and the fine-tuning training stage is 0.01, 0.005, 0.001.

After setting up the training environment on the PC side and setting the parameters, perform network training. The accuracy change curve during the training process is shown in Fig. 5. The figure on the left is the accuracy change curve in the pre-training stage. When the number of iterations reaches 30, the training accuracy curve basically converges to about 0.95. It can be seen that the network model has room for further convergence. The figure on the right is the loss change curve in the fine-tuning training stage. Since the fine-tuning training is performed on the basis of the model obtained by the pre-training, the loss value at the beginning of the training is also small. It can be seen from the figure that the error jumps during the training process. In the end, the error of the entire model converges to about 0.1, which basically meets the convergence requirements of the network.

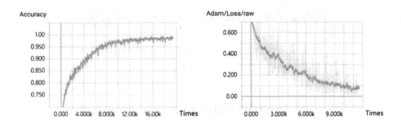

(Left) Accuracy curve of training process.   (Right) Loss curve of training process.

Fig. 5. CNN Accuracy rate and error loss change curve during training.

4.3 Comparison of Results

In order to explore the performance of the network, we use different patches and epochs to train the network to obtain the best performance of the network. From the parameter comparison Table 3, it can be seen that when the values of patch and epoch are too large or too small, the accuracy of prediction is about 0.7, so when we fine-tune these two parameters, we find that the accuracy of prediction is gradually increasing. According to Table 3, we finally receive the conclusion that when the patch is 32 and the epoch is 30, the performance of the network is the best 85.26%. The patch and epoch parameter fine-tuning comparison is shown in Table 3:

Table 3. Comparison of patch and epoch parameter.

Patch	Epoch	Val_accuracy
16	5/10/30	70.23/76.35/79.12
32	10/30/40	82.43/85.26/84.35
64	15/30/50	79.65/82.56/80.82
128	30/50/80	70.32/75.65/72.36

4.4 Compare with VGG16 and ResNet34

In Fig. 6, the figure on the left is the curve of accuracy in the pre training stage, and the curve of training accuracy basically converges to about 0.77. The figure on the right is the loss curve in the training stage. It can be seen from the figure that with the increase of training times, the value of loss function is gradually decreasing, and the error of the whole model finally converges to about 0.45.

In Fig. 7, the left figure is the curve of accuracy change in the pre training stage, and the accuracy curve of training is almost converging to about 0.81. The right figure is the curve of loss change in the training stage. It can be seen from the figure that with the increase of training times, the value of loss function is gradually decreasing, and the error of the whole model converges to about 0.40.

Table 4. Comparison with VGG16 and ResNet34.

Methods	Val_accuracy
VGG16	77.36
ResNet34	81.03
DeepCNN (ours)	**85.26**

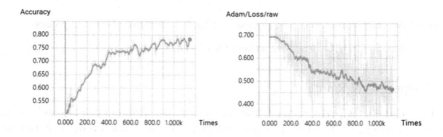

(Left) Accuracy curve of training process.   (Right) Loss curve of training process.

Fig. 6. VGG16 Accuracy rate and error loss changing curve during training.

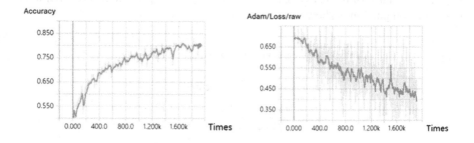

(Left) Accuracy curve of training process.   (Right) Loss curve of training process.

Fig. 7. ResNet34 Accuracy rate and error loss changing curve during training.

5 Summary and Future Work

In this paper, we proposed a method of mine vehicle head and tail recognition based on deep convolution neural network is proposed. The accuracy rate in the training set is more than 0.95, and the accuracy rate in the test set is about 0.85, which can better classify the head and tail. In the following work, we want to further improve the existing network based on the idea of attention module and knowledge distillation, hoping it can have better performance.

Acknowledgement. This work was supported by The Key Research and Technology Development Projects of Anhui Province (No. 202004a0502043; No. 202004a06020045; No. 202004b11020023) and supported by the Open Project of Suzhou University Scientific Research platform Grant No. 2017ykf12 and supported by the Natural Science Foundation of Anhui Province No. 1908085QF283 and supported by the Doctoral Start-up Research Foundation No. 2019jb08 and supported by Overseas Visiting and Study Program for Outstanding Youth No. gxgwfx2020063. This work was also supported in part by the Key Natural Science Project of Anhui Provincial Education Department under Grant KJ2019A0668 and supported by Anhui province's key R&D projects include Dabie Mountain and other old revolutionary base areas, Northern Anhui and poverty-stricken counties in 2019 No.201904f06020051 and supported by Natural science research project of Anhui Provincial Education Department No. kj20200a0733.

References

1. Xi, Y., Li, H.: Prospect of the construction of China's mineral resources legal system. In: Annual Meeting of China Association of Geology and mineral economics. China Association of Geology and mineral economics (2012)
2. Cai, G.: On the shortcomings and Countermeasures of current mineral resources management . Modern Commerce **000**(008), 115–116 (2011)
3. Wang, J.: Design and implementation of mine vehicle monitoring system based on Android. Huaibei Normal University
4. Li, L.: Business tax planning in commercial mineral resources exploration project. China land and resources economy (2012)
5. Zhang, S.: Algorithm review of license plate recognition system. Electron. Technol. Softw. Eng. **04**, 128–130 (2021)
6. Zhang, L., Sun, J., Yinxiaoyu. License plate location method based on HSV color space. Microcomputer Inf. **2008**(7), 247–248 (2008)
7. Chang, Q., Gao, M.: License plate location research based on HSV color space and mathematical morphology. J. Graph. **34**(4), 159–162 (2013)
8. Zhikun, T.H.: A comprehensive method of license plate location based on HSV color space . Comput. Appl. Chem. **28**(7), 903–906 (2011)
9. He, G.: Analysis and comparison of several classic image edge detection operators. Comput. Optical Disk Softw. Appl. **17**(9), 182–183 (2014)
10. Liang, J.: An improved algorithm based on Sobel image edge detection. Softw. Introduction **13**(12), 78–82 (2014)
11. Luo, F., Chen, S., Wang, M., et al.: An algorithm for locating license plate based on edge characteristics. J. Huazhong Univ. Sci. Technol. (NATURAL SCIENCE EDITION) **S1**, 108–110 (2004)
12. Yang, L.: The study of license plate location based on mathematical morphology. Inf. Commun. **158**(2), 64–66 (2016)
13. Luyaqin, W., Ling, C.: License plate location method based on mathematical morphology . Comput. Eng. **31**(3), 224–226 (2005)

Compact Finite-State Super Transducers for Grapheme-to-Phoneme Conversion in Highly Inflected Languages

Žiga Golob[1]([✉]), Boštjan Vesnicer[1], Mario Žganec[1], Vitomir Štruc[2], Simon Dobrišek[2], and Jerneja Žganec Gros[1]

[1] Alpineon Research and Development Ltd, Ljubljana, Slovenia
{ziga.golob,bostjan.vesnicer,mario.zganec,
jerneja.gros}@alpineon.si
[2] Faculty of Electrical Engineering, University of Ljubljana, Ljubljana, Slovenia
{vitomir.struc,simon.dobrisek}@fe.uni-lj.si

Abstract. Finite-state transducers are suitable for compact representation of pronunciation dictionaries, which are an important component of speech synthesis systems. In this paper, we first revise and analyse several properties of finite state transducers regarding their size minimization, which can be achieved by their determinization and minimisation. In scope of a novel experiment, we demonstrate that for highly inflected languages, their minimum size starts to decrease when the number of words in the presented pronunciation dictionary reaches a certain threshold. This phenomenon motivated us to introduce a new type of finite-state transducers, called finite-state super transducers, which allow the representation of pronunciation dictionaries with a smaller number of states and transitions using the existing determinization and minimization algorithms. A finite-state super transducer can accept and convert words that are not in the original represented pronunciation dictionary. The resulting phonetic transcriptions of these words may be incorrect, but we demonstrate on new data that the error rates are comparable to the performance of the state-of-the-art grapheme-to-phoneme conversion methods.

Keywords: Speech synthesis · Pronunciation dictionaries · Finite-state super transducers · Automatic grapheme-to-phoneme conversion

1 Introduction

Consistent and accurate pronunciation of words is crucial for the success of many speech technologies. To achieve the highest possible accuracy, the process of converting graphemic representation of words into their phonemic representation usually consists of two steps. First, a lookup into the available pronunciation dictionary is performed for a given word, and this typically provides its most accurate grapheme-to-phoneme conversion. However, many words may not be included in the available dictionary, as languages are constantly evolving, and new words are emerging at a high rate. The phonetic

© Springer Nature Switzerland AG 2021
D.-S. Huang et al. (Eds.): ICIC 2021, LNCS 12836, pp. 606–616, 2021.
https://doi.org/10.1007/978-3-030-84522-3_50

transcription for such non-dictionary words needs to be determined by not necessarily accurate automatic grapheme-to-phoneme conversion methods, such as rule-based [15, 16] or machine learning [17, 18] methods.

Both pronunciation dictionary lookup and in particular machine learning methods can be extremely memory consuming, especially for highly inflected languages, where pronunciation dictionaries often contain more than a million words. In some systems with limited memory resources, e.g. multilingual speech engines for embedded systems, the use of large pronunciation lexicon and direct lookup methods is not appropriate. To overcome this limitation, memory-efficient representations of pronunciation dictionaries are needed.

The most efficient methods for compact representations of pronunciation dictionaries include numbered automata [1, 2], tries [3] and finite-state transducers (FSTs) [4, 5, 6]. In this paper, a new type of finite-state transducers, called finite-state *super* transducers, is presented and discussed. The concept of finite-state *super* transducers (FSSTs) was first introduced in [7] and [11]. Finite-state *super* transducers enable more compact representations of pronunciation dictionaries, and also grapheme-to-phoneme conversions of words that are not contained in the original dictionary [7]. We prove this with novel experiments on extended language resources.

2 Language Resources

Pronunciation dictionaries of three languages from three different language groups were used to experiment with different representations of pronunciation dictionaries. For the Slavic language group, the Slovenian SI-PRON pronunciation dictionary [8] with 1,239,410 lexical entries was used. It has been augmented with the initial version of the OptiLEX pronunciation dictionary [19] comprising 57.000 lexical entries, whereby all the duplicates had been eliminated. For the Germanic language group, we used the freely available CMU-US pronunciation dictionary for North American English consisting of 133,720 lexical entries. For the Romance language group, the Italian pronunciation dictionary FST-IT from the Festival TTS toolkit [14] with 402,962 lexical entries was used. Although the size of the three dictionaries seems to be very different, the number of lemmas is comparable, as the Slovenian and the Italian language are more inflected than the English language.

3 FST Representations of Pronunciation Dictionaries

An FST can be built to accept all the words from a given dictionary and to output their corresponding phonetic transcriptions. To ensure fast dictionary lookups and small size, the FST can be converted to a minimal deterministic FST (MDFST) using efficient determinization and minimization algorithms [9]. The resulting MDFST has the smallest number of states and transitions among all equivalent FSTs [10]. Figure 1 depicts an MDFST for a simple example pronunciation dictionary containing 9 two-letter words *aa, ab, ac, ba, bb, bc, ca, cb,* and *cc*. The simple example words are represented by all possible pairs of letters from an alphabet A, containing three letters (graphemes), $A = \{a, b, c\}$. For the sake of simplicity, the corresponding phonetic symbols (allophones) are the same as the graphemes from the alphabet A, and all the phonetic transcriptions of the nine words are identical to their graphemic representations.

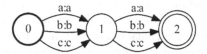

Fig. 1. MDFST representing a simple example dictionary with 9 two-letter lexical entries.

As it can be noticed from Fig. 1, the nine lexical entries can be represented with an MDFST containing three states and six transitions. We can now consider another case where the cc:cc entry is removed from the example dictionary. To represent such a dictionary, a more complex MDFST with more states and transitions is needed, as shown in Fig. 2.

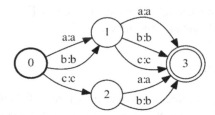

Fig. 2. MDFST representing the dictionary from Fig. 1 without the lexical entry cc:cc.

From the above examples, it can be deduced that in some cases pronunciation dictionaries containing more words can be represented by smaller and simpler MDFSTs.

In order to test the dependency between the size of the MDFST and the size of the corresponding pronunciation dictionaries, 11 sub-dictionaries of different sizes were created for each of the three pronunciation dictionaries introduced in Sect. 2. The lexical entries of the sub-dictionaries were randomly selected from the original dictionaries. An MDFST was then constructed for each sub-dictionary. Figure 3 shows the number of states for all the resulting MDFSTs in relation to the number of states of the MDFST representing the original entire dictionaries.

The MDFSTs with the highest number of states were expected to correspond to the original dictionaries containing all the lexical entries. Interestingly, this is not the case for all the three languages (Fig. 3), which confirms the initial observations reported in [7]. For Slovene and Italian, which are both heavily inflected, the size of the corresponding MDFST is up to 160% larger for smaller dictionaries, and it reaches the maximum number of states for dictionaries covering approximately 60% of the lexical entries from the original dictionary. Similar results for a highly inflected Portuguese language dictionary were reported in [1].

On the other hand, the size of the MDFST representing the English language dictionary is almost linearly dependent on the size of the dictionary. This phenomenon seems to strongly correlate with the number of inflected word forms. A possible explanation could be that inflection rules in the highly inflected languages follow a similar pattern with few exceptions, so inflected forms can often be represented by a similar FST substructure for many different lemmas. If an inflected form is missing for a particular lemma, the remaining forms have to be represented by a different FST substructure, which may result in a larger final MDFST size.

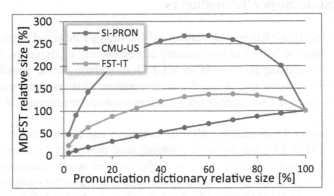

Fig. 3. Number of states of the created MDFSTs representing dictionaries for three different languages. For each language, 11 sub-dictionaries with different sizes were used, whereby their lexical entries were randomly chosen from the original dictionaries.

To further test this hypothesis, we repeated the experiment with the augmented SI-PRON pronunciation dictionary, where the sub-dictionaries were created by randomly selecting only lemmas, and then all the corresponding inflected forms were added to the created dictionary from the original dictionary. The results are presented in Fig. 4 and confirm the results initially reported in [7].

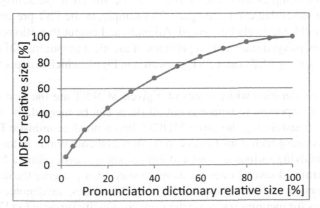

Fig. 4. The number of MDFST states representing sub-dictionaries for Slovenian with different sizes, where first only lemmas were chosen randomly from the original dictionary, and later all the corresponding inflected forms were added.

As it can be seen from Fig. 4, the size of the MDFSTs representing the augmented SI-PRON sub-dictionaries for Slovenian, which were created as described, increases monotonically. It can be thus concluded that the missing inflected word forms significantly increase the complexity of the MDFST representing the dictionary.

4 Finite-State Super Transducers

In the previous section it has been demonstrated that missing inflected forms are actually increasing the size of MDFSTs representing corresponding dictionaries. Therefore, it is beneficial for reducing the size of an MDFST to include all the possible inflected forms for all the lemmas that are in the corresponding dictionary represented by the MDFST, even if they are very rare or not needed in a particular pronunciation dictionary implementation.

In the endeavor for minimizing a MDFST dictionary representation, another question arises. Can an MDFST representing a pronunciation dictionary be even smaller if it accepts also other words that are not part of the original source dictionary? We have found that this can be achieved by defining several specific rules for merging FST states. By analogy to *supersets*, we refer to the resulting FST as a *finite-state super transducer* (FSST), as initially proposed in [7].

It should be noted that when new words are represented by such an extended FST, information about which words or lexical entries belong to the original dictionary and which have been added as the result of such an extension is lost.

4.1 Construction of FSSTs

The main idea behind the construction of FSSTs is to find new relevant words or strings that would allow additional merging of states. Instead of searching for such words, which is a very complex task, the problem can be solved by searching for relevant non/equivalent states that can be merged. For example, in the FST presented in Fig. 2, state 1 and 2 could intuitively be merged. All input and output transitions of both states are part of a new merged state, only repetitions of identical output transitions need to be omitted. The resulting FST is an FSST, presented in Fig. 1, which accepts the additional word *cc*.

We need to determine which states of a given MDFST are the best candidates to be merged. It is important to note that not all states can be merged, as certain merges could result in transforming the given MDFST into a nondeterministic FST, which is not desirable because such transducers can be slow and ambiguous in translating input words. After studying various options and limitations in transforming a MDFST into a FSST, we determined several rules for merging states that preserve the determinism of the obtained transducer. Two states, which satisfy these rules, are denoted as *mergeable* states. The rules for merging states are the same as initially proposed in [7]:

- Mergeable states do not have output transitions with the same input symbols and different output symbols.

- Mergeable states do not have output transitions with the same input and output symbols and different target states.
- If one of the two mergeable states is final, both states do not have output transitions with input symbol that is an empty character ε.

The above rules are stricter than necessary to preserve the determinism of the final FSST and are also very easy to be verified as well. The goal of this paper is not to aim at building the smallest possible FSST, but to demonstrate a potential of FSSTs.

To build an FSST from an MDFST, we defined an algorithm that searches for mergeable states by checking all possible combinations of two states of the input MDFST. Each time a pair of mergeable states is found, they are merged into a single state. Since merging affects the other states that have already been verified against the rules mentioned above, several iterations are normally needed until no new mergeable states have been found.

In our experiments, which we repeated on the extended language resources described in Sect. 2, we observed that the final size of an FSST depends on the order of state merging. The implementation size or memory footprint of an FST mostly depends on the number of transitions and less on the number of states [11]. We experimentally proved the findings in [7] that the smallest final number of transitions is obtained if only the states with the highest number of identical transitions are merged in the first iterations, since the repetitions of all the identical transitions can be immediately removed from the transducer. In order to lower the number of the additional words that are accepted by the resulting FSST, those mergeable states whose merging did not decrease the number of transitions are not merged.

4.2 Experimental Results

The experiments were conducted using the methodology proposed in [7]. Initially, two MDFSTs were built for each available pronunciation dictionary, using the open-source toolkit OpenFST [12]. For the second type of MDFSTs, denoted as MDFST-2, the output strings of transitions were constrained to the length 1 (in contrast to the first type denoted as MDFST-1, which does not have this restriction). The second type of MDFSTs normally has more states and transitions than the first type. However, it exhibits a simpler implementation structure, which results in a smaller implementation size.

Table 1. A comparison of the number of states and transitions between MDFSTs and FSSTs obtained with the first type of FSTs.

	MDFST-1	MDFST-1	FSST	FSST
Dictionary	States	Transitions	States	Transitions
ALP-SI	69.524	239.852	56.120	210.665
CMU-US	76.035	184.232	60.477	164.235
FST-IT	57.323	172.210	44.563	148.657

FSSTs were then built from all the MDFSTs using the rules presented in Sect. 4.1. The results are presented in Table 1 and Table 2.

Table 2. A comparison of the number of states and transitions between MDFSTs and FSSTs obtained with the second type of FSTs.

	MDFST-2	MDFST-2	FSST	FSST
Dictionary	States	Transitions	States	Transitions
ALP-SI	224.123	533.561	170.832	426.225
CMU-US	185.053	307.027	153.746	269.677
FST-IT	157.379	318.805	124.717	260.447

5 Non-dictionary Words

A system for converting graphemes to phonemes usually consists of two parts. First, it checks whether the input word is contained in the considered pronunciation dictionary. If not, a phonetic transcription of the input word is determined using appropriate statistical or machine learning methods. On the other hand, if an FSST is used to represent a pronunciation dictionary and it accepts the input word, then it is not possible to determine whether the output phonemic transcription is correct, since FSST may also accept words that are not part of the original dictionary, and these non-dictionary words may have incorrect phonemic transcriptions.

In order to evaluate this error, the augmented Slovenian pronunciation dictionary described in Sect. 2 was divided into a training and a test set. The training set contained about 90% of the lexical entries from the original dictionary, which were randomly selected. The remaining entries represented the test set. An MDFST-2 and an FSST were then created from the training set. The results for the number of states and transitions are presented in Table 3.

Table 3: MDFST and FSST sizes for the training set.

		MDFST-2	FSST	Reduction
1 output symbol	*States*	246.262	186.476	24.3%
	Transitions	556.723	441.234	20.7%

The results in Table 3 show that the reduction of the number of FSST states and transition is much higher when not all inflected forms are contained in the dictionary. The number of states and transitions of the obtained FSST is even lower than that of the MDFST representing all the inflected forms.

When the FSST was built from the training set, words from the test set, which represented non-dictionary words, were used as its input. Accepted and rejected words were then enumerated and counted, and the correctness of the phonetic transcriptions were checked on the output. The results of the experiment are shown in Fig. 5.

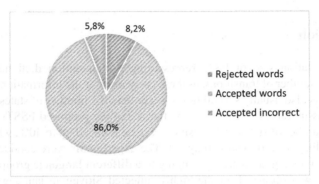

Fig. 5. Results for non-dictionary words used as the input to the FSST.

Interestingly, similarly to [7], only 8.24% of non-dictionary words were not accepted by the FSST and for only 5.83% of the accepted words the output phonetic transcription was incorrect. For the Slovenian language, one of the commonly used grapheme-to-phoneme conversion system that is based on machine learning methods correctly determine phonetic transcription for up to 83% of the words [13].

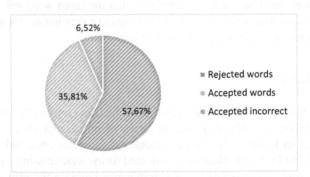

Fig. 6. Results for the non-dictionary test words as input to the FSST; only lemmas were randomly chosen.

Nevertheless, it should be noted that since test entries were chosen randomly from the original dictionary, the inflected forms that belong to the same lemma could be present both in the training as well as in the test dictionary. In this way, both dictionaries were partly similar, even though the word forms were different. Therefore, we repeated the experiment by first choosing randomly lemmas and then all the corresponding inflected forms were added to the training dictionary.

Figure 6 shows the final results. In this experiment only 42.33% of the words were accepted by the FSST. Among the accepted words the phonetic transcriptions were still correct for 84,59% of the words, which is similar to the accuracy of state-of-the-art methods for grapheme-to-phoneme transcription for Slovenian, which in any case need to be used to process the words rejected by the FSST.

6 Discussion

The implementation size of FSTs representing pronunciation dictionaries depends mainly on the number of transitions, as these carry most of the information. The number of transitions is also usually several times larger than the number of states.

The results in Tables 1, 2 and 3 show that using the proposed FSSTs, their size in terms of the number of states and transitions can be reduced up to 40%, which confirms the initial findings of a previous study [7]. The experiments were conducted language resources in three languages each belonging to a different language group. The highest reduction rate was observed for the highly inflected Slovenian language when some inflected forms were missing in the original dictionary. As can be seen in Fig. 3, non-included or missing inflected forms can significantly increase the size of an MDFST. If missing inflected forms are not added to the dictionary, an FSST can always be built from an MDFST. In this way, most of the inflected forms are automatically added to the pronunciations dictionary represented by an FSST.

The proportion of the accepted non-dictionary words with correct phonetic transcriptions is surprisingly high for FSSTs. For rejected words the phonetic transcription can always be determined using other grapheme-to-phoneme approaches. This is particularly important when there is a high probability that the input word will be rejected by an FSST. This happens mostly when a word belongs to a new lemma for which inflected forms are not represented in the pronunciation dictionary.

7 Conclusions

The presented results, obtained on an augmented data set confirm that by using FSSTs as initially introduced in [7], their size in terms of number of states and transitions can be reduced up to 40%. The highest reduction rate was observed for the highly inflected Slovenian language when some inflected forms were missing from the original dictionary. Missing inflected forms can significantly increase the size of an MDFST. When missing inflected forms are not added to the dictionary, an FSST can always be built from an MDFST, and in this way, most of the missing inflected forms are automatically represented by an FSST.

FSSTs can be thus used as very compact and memory efficient computational models for grapheme-to-allophone transcriptions. Compared to MDFSTs, their implementation size can be reduced by up to 20%. The reduction seems to be much higher for highly inflected languages, when not all the inflected forms are originally contained in the represented dictionary.

An important finding of this study is that a large proportion of the out-of-dictionary words that are accepted by an FSST are also correctly converted to their phonetic transcriptions. For the words rejected by an FSST, the phonetic transcription can always be determined using other grapheme-to-phoneme conversion methods. In our experiments, the phonetic transcriptions of the accepted non-dictionary words were correctly determined for up to 86% of the words. This is particularly useful in systems with limited memory resources, where no other or only rather basic grapheme-to-phoneme transcription methods can be used.

Acknowledgements. The work presented in this paper was partially supported by the applied research project L7-9406 OptiLEX, and by the national research program P2-0250(C) Metrology and Biometric Systems.

References

1. Lucchesi, C.L., Kowaltowski, T.: Applications of finite automata representing large vocabularies. Softw. Practice Exp. **23**(1), pp. 15–30. Wiley, New Jersey (1993)
2. Domaratzki, M., Salomaa, K. (eds.): CIAA 2010. LNCS, vol. 6482. Springer, Heidelberg (2011). https://doi.org/10.1007/978-3-642-18098-9
3. Ristov, S.: LZ trie and dictionary compression. Softw. Practice Exp. **35**(5), 445–465 (2005)
4. Mehryar, M.: Compact representation by finite-state transducers. In: 32nd Meeting of the Association for Computational Linguistics (ACL 94), Las Cruces (1994)
5. Kiraz, G.A.: Compressed storage of sparse finite-state transducers. In: Boldt, O., Jürgensen, H. (eds.) WIA 1999. LNCS, vol. 2214, pp. 109–121. Springer, Heidelberg (2001). https://doi.org/10.1007/3-540-45526-4_11
6. Golob, Ž., Žganec-Gros, J., Žganec, M., Vesnicer, B.: FST-Based Pronunciation Lexicon Compression for Speech Engines. International Journal of advanced robotic systems, vol. 9 (2011)
7. Golob, Ž, Žganec Gros, J., Štruc, V., Mihelič, F., Dobrišek, S.: A composition algorithm of compact finite-state super transducers for grapheme-to-phoneme conversion. In: Sojka, P., Horák, A., Kopeček, I., Pala, K. (eds.) TSD 2016. LNCS (LNAI), vol. 9924, pp. 375–382. Springer, Cham (2016). https://doi.org/10.1007/978-3-319-45510-5_43
8. Žganec-Gros, J., Cvetko-Orešnik, V., Jakopin, P.: SI-PRON pronunciation Lexicon: a new language resourcxe for Slovenian. Informatica **30**, 447–452 (2006)
9. Mohri, M.: Minimization algorithms for sequential transducers. Theoret. Comput. Sci. **234**(1–2), 177–201 (2000)
10. Mohri, M.: Finite-state transducers in language and speech processing. Comput. Linguist. **23**(2), 269–311 (1997)
11. Golob, Ž.: Reducing Redundancy of Finite-State Transducers in Automatic Speech Synthesis for Embedded Systems. Ph.D. thesis, University of Ljubljana, Faculty of Electrical Engineering, Tržaška 25, SI-1000 Ljubljana, Slovenia (2014)
12. Allauzen, C., Riley, M., Schalkwyk, J., Skut, W., Mohri, M.: OpenFst: a general and efficient weighted finite-state transducer library. In: Holub, J., Žďárek, J. (eds.) CIAA 2007. LNCS, vol. 4783, pp. 11–23. Springer, Heidelberg (2007). https://doi.org/10.1007/978-3-540-763 36-9_3
13. šef, T., škrjanc, M., Gams, M.: Automatic lexical stress assignment of unknown words for highly inflected slovenian language. In: Sojka, P., Kopeček, I., Pala, K. (eds.) TSD 2002. LNCS (LNAI), vol. 2448, pp. 165–172. Springer, Heidelberg (2002). https://doi.org/10.1007/3-540-46154-X_23

14. Black, A., Taylor, P., Caley, R.: The Festival Speech Synthesis System: System Documentation (2.4.0). Technical report, Human Communication Research Centre (2014)
15. Hahn, S., Vozila, P., Bisani, M.: Comparison of grapheme-to-phoneme methods on large pronunciation dictionaries and lvcsr tasks. In: Interspeech, pp. 2538–2541, Portland, OR, USA (2012)
16. Jiampojamarn, S., Kondrak, G.: Letter-phoneme alignment: an exploration. In: Proceedings of the 48th Annual Meeting of the Association for Computational Linguistics, ACL 2010, pp. 780–788. Association for Computational Linguistics, Stroudsburg, PA, USA (2010)
17. Lehnen, P., Allauzen, A., Lavergne, T., Yvon, F., Hahn, S., Ney, H.: Structure learning in hidden conditional random fields for grapheme-to-phoneme conversion. In: Interspeech, Lyon, France, pp. 2326–2330 (2013)
18. Yolchuyeva, S., Németh, G., Gyires-Tóth. B.: Transformer based Grapheme-to-Phoneme Conversion. In: Interspeech, Graz, Austria, pp. 2095–2099 (2019)
19. Žganec-Gros J., Golob, Ž., Dobrišek, S.: Učinkovita predstavitev slovarskih jezikovnih virov pri govornih tehnologijah. In: Pejović et al. (eds.) Proceedings of the Human-Computer Interaction in Information Society Conference, Information Society – IS 2020, vol. H, Ljubljana, Slovenia, pp. 45–48 (2020)

Swarm Intelligence and Optimization

An Improved SMA Algorithm for Solving Global Optimization Problems

Heng-wei Guo, Hong-yan Sang$^{(\boxtimes)}$, Jun-qing Li, Yu-yan Han, Biao Zhang, and Lei-lei Meng

Liaocheng University, Liaocheng 252000, China
{sanghongyan,lijunqing,hanyuyan,zhangbiao,mengleilei}@lcu-cs.com

Abstract. In this paper, an improved slime mould optimization algorithm (iSMA) is proposed to solve global optimization problems. The iSMA adopts a dominant-non-dominant swarm partitioning strategy and introduces a Gauss-Cauchy hybrid mutation mechanism, which helps iSMA find a balance in exploration and exploitation by using the different perturbation capabilities of the Gauss mutation operator and the Cauchy mutation operator. Finally, the 13 benchmark test functions are selected for qualitative and quantitative analysis of iSMA and compared with several state-of-the-art algorithms proposed in recent years. The computational results show that the iSMA algorithm is an effective optimization algorithm for solving global optimization problems.

Keywords: Global optimization · Swarm partitioning strategies · Slime mould optimization algorithm · Mutation operators

1 Introduction

Optimization is the process of searching for the best values for the variables of a specific problem to minimize or maximize the objective function. Optimization problems exist widely in various research fields. Many optimization problems in the real world become more and more complex, so it is more challenging to solve these problems with optimality. Mathematical optimization methods and meta-heuristics are usually used to solve optimization problems, but the former relies heavily on gradient-based information of functions to find the best solution. With the increase of computational scale, mathematical optimization methods tend to fall into local optima and fail to find a satisfactory solution within a reasonable time [1]. On the contrary, meta-heuristics can obtain better optimization results because of its simple structure, strong applicability and strong ability to avoid local optima, so it has been widely used in various optimization fields [2]. Recently, many nature-inspired meta-heuristic algorithms have been proposed, which can be broadly classified into three groups: evolution-based algorithms, population-based algorithms and physics-based algorithms [3].

The evolution-based meta-heuristic algorithms are mainly inspired by the concept of evolution in nature, where the candidate solutions are iteratively improved until the termination condition is satisfied, so that the whole evolution process will increase the possibility of obtaining better results near the global optima, such as evolutionary strategies

© Springer Nature Switzerland AG 2021
D.-S. Huang et al. (Eds.): ICIC 2021, LNCS 12836, pp. 619–634, 2021.
https://doi.org/10.1007/978-3-030-84522-3_51

(ES) [4], biogeography-based optimization (BBO) [5], etc. The physics-based algorithms are imitating physics principles, such as the black hole algorithm (BHA) [6] which simulates the phenomenon of black holes, the lightning search algorithm (LSA) [7] inspired by the natural phenomenon of lightning. The population-based algorithms, which mainly mimic the social behavior of flocks of organisms in nature, consider multiple candidate solutions and have greater exploratory power, being able to use information from multiple candidate solutions to explore towards more promising regions. In recent years, more popular population-based meta-heuristics such as fruit fly optimization algorithm (FOA) [8] inspired by fruit fly foraging behavior, grasshopper optimization algorithm (GOA) [9] developed according to the social behavior of grasshoppers, and barnacle mating optimization algorithm (BMO) [10] mimicking the natural mating behavior of barnacles.

In 2020, the slime mould algorithm (SMA), a novel meta-heuristic algorithm proposed by Li et al., simulates the behavior of slime mould in foraging [11]. The SMA has been applied in several fields with its adaptive weighting method and excellent optimization capability, such as using the SMA to process photovoltaic cell technical parameters [12], feature selection [13] and image segmentation problem [14].

The SMA, as a meta-heuristic, is essentially a stochastic optimization method, finding a reasonable balance between exploration and exploitation has been a direction of improvement for meta-heuristics. Therefore, in order to further enhance the search capability of SMA for global optimization problems, this paper proposes an improved SMA algorithm (iSMA), which introduces a dominant-non-dominant swarm partitioning strategy and a Gauss-Cauchy hybrid mutation mechanism to make full use of the information of individuals and improve the search efficiency and convergence accuracy of SMA. Under the effect of the swarm partitioning strategy and the hybrid mutation mechanism, individuals identify and move away from the local optima and gradually converge to the global optima. A large number of computational simulations and comparisons have verified the effectiveness and superiority of iSMA in solving 13 benchmark optimization problems.

The rest of the paper is organized as follows. Section 2 introduces the standard SMA algorithm. Section 3 details the improved SMA, which includes the initialization method, the dominant-non-dominant swarm partitioning strategy and the Gauss-Cauchy hybrid mutation mechanism. The 13 benchmark test functions and the analysis of experimental results are listed in Sect. 4. Finally, Sect. 5 gives a summary and outlook.

2   The Standard SMA Algorithm

Recently, Li et al. proposed a new meta-heuristic algorithm called the slime mold algorithm (SMA), which mimics the behavior of slime moulds in foraging, and the mathematical model of SMA can be summarized as the search for food phase, the wrap food phase, and the oscillation phase [11].

In the search for food stage, the slime moulds use the smell information to approach the food. According to the specific behavior of the slime mould, it can be formulated as follows,

$$X(t+1) = \begin{cases} V_c \cdot X(t), r \geq p \\ X_b(t) + V_b(W \cdot X_A(t) - X_B(t)), r < p \end{cases} \tag{1}$$

Where $V_b \in [-a, a]$, V_c decreases linearly from 1 to 0, t is the number of iterations, c $X_b(t)$ denotes the individual with the best fitness value, $X(t)$ denotes the position of the slime mould at the current iteration number t, X_A and X_B are two randomly selected individuals from population, and W represents the weight of the slime mould. The parameter p in the Eq. (1) is calculated according to the Eq. (2),

$$p = tanh|f(i) - DF| \tag{2}$$

Where $i \in 1, 2, 3...n$, $f(i)$ denotes the current individual fitness value and the parameter DF is the global optimal fitness value so far, V_b, V_c and W are calculated as follows,

$$V_b = [-a, a] \tag{3}$$

$$a = arctanh(-(\frac{t}{t_{max}}) + 1) \tag{4}$$

$$V_c = 1 - \frac{t}{t_{max}} \tag{5}$$

$$W(SmellIndex_i) = \begin{cases} 1 + r * \log\left(\frac{bF - f(i)}{bF - wF} + 1\right), condition \\ 1 - r * \log\left(\frac{bF - f(i)}{bF - wF} + 1\right), others \end{cases} \tag{6}$$

$$SmellIndex = sort(f) \tag{7}$$

Where r is a random number between 0 and 1, bF and wF are the best and worst fitness values in the current iteration respectively, $SmellIndex$ denotes the index of the sorted fitness values, and condition refers to $f(i)$ ranks of the top half of the population.

In the wrapped food phase, Eq. (8) is used to simulate the contraction pattern in the venous structure of the slime mould, Ub and Lb are the upper and lower bounds of the problem, z is used to determine the probability of SMA searching near the current best food source. The parameters $rand()$ and r are random numbers between 0 and 1.

$$X(t+1) = \begin{cases} rand() \cdot (Ub - Lb) + Lb, rand() < z \\ \begin{cases} X_b(t) + V_b \cdot (W \cdot X_A(t) - X_B(t)), r < p \\ V_c \cdot X(t), r \geq p \end{cases}, rand() \geq z \end{cases} \tag{8}$$

During the oscillation phase, the propagating waves generated by the bio-oscillator are used to alter the cytoplasmic flow in the vein, and make the slime mould move to a better position. In the oscillation phase, V_b, V_c, W are used to simulate the venous width variation. V_b oscillates randomly between $[-a,a]$ and gradually approaches 0. V_c oscillates between $[0,1]$ and eventually decreases to 0 *as the number of iteration increases*. Finally, the pseudo code for SMA is shown in Algorithm 1.

Algorithm 1: The Slime mould algorithm

Input: the size of population N, Maximum iteration t_{max}
Output: the best solution X_b
Initialization, $X_i(i = 1,2,3...N)$
While($t<t_{max}$)
  Calculate the fitness value $f(i)$ for X_i
  Update the best fitness value and X_b
  Calculate the weight W // Eq.(4)
    for $i=1 : N$
      Update V_b, V_c, p
      Update positions X_i //Eq.(6)
    end for
    $t = t+1$
end while

---

3   The Improved Slime Mould Algorithm (iSMA)

3.1   Initialization

A good initial position may lead to faster convergence of the slime mould to a better solution. In order to obtain good population positions, one possible method is the random method, where the initial position of the population is generated randomly. Although this method ensures the diversity of the population, it affects the algorithm's search performance as the randomly generated individuals may be far from the best solution. Another method is to improve the initial position of the population by checking the opposite solution of each randomly generated solution, i.e. the opposition-based learning (OBL) [15]. In the process of evaluating individual X_i, the OBL guesses that the opposite solution of X_i leads to a better solution X_i', which further reduces the distance between X_i and the best solution. By doing so, not only can the quality of the solution be improved, but also the efficiency of the solution can be enhanced. A simple example is if X_i is -5 and the optimal solution is 10, the opposite solution X_i' is 5 and the distance between X_i and the optimal solution is 15, while the distance between X_i' and the optimal solution is only 5. The opposite solution X_i' is closer to the optimal solution.

In the iSMA, we use the OBL method to generate N/2 individuals and select the best individual. To make full use of this best solution, select it as the N/2 individuals of the initial population, while the remaining N/2 individuals are generated by the random method. This initialization method not only ensures the quality of the initial population, but also takes into account the diversity of the population to a certain extent.

3.2   The Dominant-Non-dominant Swarm Partitioning Strategy

In meta-heuristics, when the population is updated, it cannot make full use of the information of the whole population, resulting in an individual being over-exploited, which makes the algorithm converge prematurely. It pays average attention to all individuals, and does not pay enough attention to the dominant individuals, so that the algorithm cannot converge with high accuracy. Therefore, when the algorithm is improved, it is necessary to adopt appropriate strategies to treat individuals differently in order to make rational use of the whole population and improve the optimization performance of the algorithm.

Therefore, we propose a partitioning strategy to classify individuals, namely the dominant-non-dominant swarm partitioning strategy, which divides the population into dominant and non-dominant swarm, it is dynamically adjusting according to the change of individual fitness values during each iteration (the number of individuals included in dominant and non-dominant swarms during evolution is non-fixed). During evolution, we record the fitness values bF and wF for the best and worst individuals, and then calculated the distance of each individual from bF and wF according to Eq. (7)(8). If Eq. (9) is satisfied, individual i will be classified to the dominant swarm, otherwise, individual i will be classified to the non-dominant swarm.

$$db_i = |f(i) - bF| \qquad (9)$$

$$dw_i = |f(i) - wF| \qquad (10)$$

$$db_i \leq e^{-\frac{t}{t_{max}}} \cdot dw_i \qquad (11)$$

Where db_i denotes the distance of the ith individual from the best fitness value, dw_i denotes the distance of the ith individual from the worst fitness value, t is the number of current iterations and t_{max} is the maximum number of iterations.

3.3 The Gauss-Cauchy Hybrid Mutation Mechanism

In addition to the problem mentioned in the previous section, meta-heuristics tend to fall into local optima, especially in the multimodal optimization problem, where there are usually multiple consecutive local optima. If the algorithm does not have a good jump ability, there is a great probability of stagnation at local optima.

Considering the mutation operation has two main advantages: on the one hand, it can speed up the convergence of the algorithm to the optimal solution when it performs local exploitation; on the other hand, when the diversity of the population decreases and faces the risk of falling into the local optima, the mutation operation can improve the diversity of the population and guide the individual away from the current search area [16].

Therefore, we propose the Gauss-Cauchy hybrid mutation mechanism utilizes two mutation operations: Gauss mutation and Cauchy mutataion. The former is to apply a random term conforming to the standard Gauss distribution to the individual [16], whose $\mu = 0, \sigma = 1$, the later applies the random term which conforms to the Cauchy distribution to the individual. The Cauchy distribution is a special distribution function where neither μ nor σ exists. We can see from Fig. 1 that the random number generated by Gauss mutation is more likely to be closer to the origin, while the random number generated by Cauchy mutation is more likely to be far away from the origin. There-fore, Gauss mutation has weaker perturbation and stronger local exploitation ability, which is suitable for deeper exploitation of the region near the better value [17]; while Cauchy mutation has stronger perturbation ability than Gauss mutation, and has a greater probability of jumping out of the local optima, so it is suitable for global search [18].

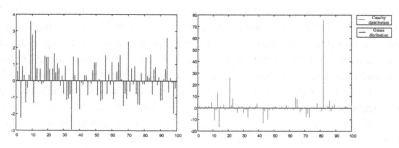

Fig. 1. Gauss and Cauchy function distributions

In order to make full use of the features of both Gauss mutation operator and Cauchy mutation operator, the Gauss-Cauchy hybrid mutation mechanism is introduced in iSMA to perform Gauss mutation operation on the dominant swarm to go deeper into local exploitation, and to perform Cauchy mutation operation with a wider range of perturbation on the non-dominant swarm to avoid the interference of local optima. The hybrid mutation mechanism is applied as follows.

Standard SMA uses random number to determine individual positions randomly in the search space, which to some extent reduces the quality of solutions and reduces the efficiency of algorithm optimization. Moreover, the iSMA is improved by using the Gauss-Cauchy hybrid mutation mechanism to guide individuals through exploitation and exploration on the basis of the dominant-non-dominant swarm. When the global best fitness value was not updated four times, performing the Gauss mutation operation around the optimal value (X_b) at the current iteration for the dominant swarm and performing the Cauchy mutation operation for individuals in the non-dominant swarm. The hybrid mutation mechanism makes full use of the exploration capabilities of the different mutation operators and greatly increases the probability of the algorithm jumping out of the local optima. The pseudo code for the Gauss-Cauchy hybrid mutation mechanism is shown in Algorithm 2.

Finally, it is worth mentioning that the completely random selection of individuals A and B in standard SMA has a high degree of uncertainty and affects the performance of the algorithm. Therefore, the iSMA abandons method of randomly selecting individuals A and B in the wrapped food stage and instead adopts a method of learning from the best individuals, in which individual A is randomly selected from the best N/2 individuals, while individual B is randomly selected from the whole population. The pseudo code for the iSMA is shown in Algorithm 3.

Algorithm 2: The Gauss-Cauchy hybrid mutation mechanism

Input:X_i
Output: updated X_i
for i = 1 : N
  if $X_i \in dominantSwarm$ // Whether X_i belongs to the dominant
   $X_i = X_b + X_b * Gauss(0,1)$
  else
   $X_i = X_i + X_i * Cauchy(0,1)$
  end if
end for

Algorithm 3: The improved SMA

Input: the size of population N, Maximum iteration t_{max}

Output: the best solution X_b
Initialization, $X_i(i = 1,2,3 \dots N)$ //The OBL method and random method
While t<t_{max}
  Calculate the fitness value f(i) for X_i
  [SmellOrder,SmellIndex] = sort(f)
  Update the best fitness value X_b,calculate W
  Population partitioning // Eq. (7-9)
  if noImp>4 // Global optimum 4 times not updated
   Algorithm 2//Gauss-Cauchy hybrid mutation mechanism
  else
   for i = 1 : N
    Update v_b, v_c, p
    if rand() < p
     $X_i = X_b(t) + v_b \cdot (W \cdot X_A(t) - X_B(t))$//select X_A randomly from
     //half of the SmellIndex, select X_B randomly from the whole
     //population.
    else
     $X_i = v_c \cdot X_i$
    end if
   end for
  end if
  t=t+1
end while

---

4 Simulation Experiment and Analysis

In this section we describe the benchmark function problems and analyze the iS-MA qualitatively and quantitatively through experiments. All experiments are run on an Inter® i7-6700HQ, 16 GB RAM computer with windows 10 professional operating system and the programming software is MATLAB R2016a.

4.1 Benchmark Funtions

For verifying the performance of the iSMA, 13 benchmark functions with dimension equals to 30 are selected for testing, the specific functions are shown in Table 1, where functions $f_1 - f_7$ are unimodal functions, $f_8 - f_{13}$ are multimodal functions. dim represents the dimension of the function, Range represents the upper and lower bound of the function, and f_{min} represents the minimum value of the function. The images of several typical functions are shown in Fig. 2.

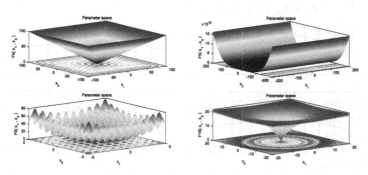

Fig. 2. Typical functions images

4.2   Performance Analysis and Algorithm Comparison

In this section, in order to solve these test function, the relevant parameters are set to: the number of agents $N = 30$, dimension set to $dim = 30$, $t_{max} = 500$. To visualise the search behavior and superiority of iSMA, qualitative and quantitative analyses were carried out using different performance metrics.

Additionally, we compared the iSMA algorithm with global optimization algorithms proposed in recent years, including SMA [11], BMO [10], MFO [3], and SFFO [15]. We obtained the control parameters of the algorithms from the corresponding literature, $z = 0.2$ for SMA, pl $= 7$ for BMO, and $P_0 = \frac{5}{n}$, $\lambda_{max} = \frac{Ub-Lb}{2}$, $\lambda_{min} = 10^{-7}$ for SFFO.

Qualitative Analysis and Discussion

In this section, in order to observe the behavior and performance of the iSMA qualitatively, we take the unimodal function f_4 and the multimodal function f_{10} as examples to draw the convergence curve, the search history curve (only the first two dimensions, the red dot indicates the global optimal point), and the trajectory curve of the first slime mould of the iSMA.

As the search history curve (Fig. 3(a) and Fig. 4(a)) shows that the iSMA tends to search extensively for promising areas in space and use the best areas for local exploitation. The dominant-non-dominant swarm partitioning strategy guides individuals to move closer to promising areas for local exploitation. In addition, the trajectory curve in Fig. 3(b) and Fig. 4(b) show that the iSMA maintains a large range of variation at the beginning of the iterations, and with the iteration progress, the fluctuations gradually decrease and converge to an optimal state, which is the result of the action of the Gauss-Cauchy hybrid mutation mechanism that helps the iSMA to strike a balance between exploration and exploitation. Finally, the convergence curves (Fig. 3(c) and Fig. 4(c))

Table 1. Benchmark test functions

Function	Range	f_{min}
$f_1(x) = \sum_{i=1}^{n} x_i^2$	$[-100,100]$	0
$f_2(x) = \sum_{i=1}^{n} |x_i| + \prod_{i=1}^{n} |x_i|$	$[-10,10]$	0
$f_3(x) = \sum_{i=1}^{i} \left(\sum_{j-1}^{i} x_j \right)^2$	$[-100,100]$	0
$f_4(x) = max\{|x_i|, 1 \leq i \leq n\}$	$[-100,100]$	0
$f_5(x) = \sum_{i=1}^{n-1} [100(x_{i+1} - x_i^2)^2 + (x_i - 1)^2]$	$[-30,30]$	0
$f_6(x) = \sum_{i=1}^{n} (x_i + 0.5)^2$	$[-100,100]$	0
$f_7(x) = \sum_{i=1}^{n} ix_i^4 + random[0,1)$	$[-1.28,1.28]$	0
$f_8(x) = \sum_{i=1}^{n} -x_i sin\left(\sqrt{|x_i|}\right)$	$[-500,500]$	$-418.98 * dim$
$f_9(x) = \sum_{i=1}^{n} \left[x_i^2 - 10\cos(2\pi x_i + 10) \right]$	$[-5.12,5.12]$	0

(continued)

Table 1. (*continued*)

Function	Range	f_{min}
$f_{10}(x) = -20exp(-0.2\sqrt{\frac{1}{n}\sum_{i=1}^{n}x_i^2}) - exp(\frac{1}{n}\sum_{i=1}^{n}\cos(2\pi x_i)) + 20 + e$	$[-32,32]$	0
$f_{11}(x) = \frac{1}{4000}\sum_{i=1}^{n}x_i^2 - \prod_{i=1}^{n}\cos(\frac{x_i}{\sqrt{i}}) + 1$	$[-600,600]$	0
$f_{12}(x) = \frac{\pi}{n}\left\{10\sin(\pi x_i) + \sum_{i=1}^{n-1}(y_i-1)^2[1+\sin^2(\pi y_{i+1})] + (y_n-1)^2\right\} + \sum_{i=1}^{n}u(x_i, 10, 100, 4)$	$[-50,50]$	0
$y_i = 1 + \frac{x_i+1}{4}$		
$u(x_i, \alpha, \kappa, m) = \begin{cases} \kappa(x_i - \alpha)^m, x_i > \alpha \\ 0, -\alpha < x_i < \alpha \\ \kappa(-x_i - \alpha)^m, x_i < -\alpha \end{cases}$		
$f_{13}(x) = 0.1\left\{\sin^2(3\pi x_1) + \sum_{i=1}^{n}(x_i-1)^2\left[1+\sin^2(3\pi x_i+1)\right] + (x_n-1)^2\left[1+\sin^2(2\pi x_n)\right]\right\} + \sum_{i=1}^{n}u(x_i, 5, 100, 4)$	$[-50,50]$	0

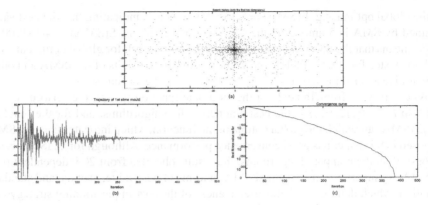

Fig. 3. Curves of f_4

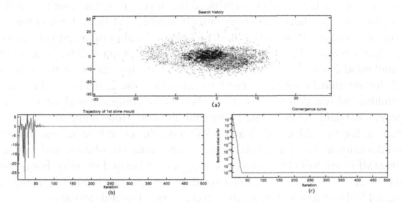

Fig. 4. Curves of f_9

clearly show the descent behavior of both the unimodal and multimodal functions, which proves that the iSMA ideally improves the convergence accuracy during the search.

The results discussed above demonstrate that the iSMA can solve function optimization problems, but the qualitative results cannot determine how much better the iSMA is compared to other algorithms, so next we compare iSMA with the state-of-the-art meta-heuristics of recent years.

Quantitative Analysis and Algorithm Comparison

To better compare the performance of the algorithms, each benchmark function is run 20 times independently to generate statistical results. We use the best value, the average, the standard deviation, the excellence rate and success rate as quantitative indicators to evaluate the performance of the algorithm. The best value reflects the convergence ability of the algorithm, while the average and standard deviation reflect the overall performance and stability of the algorithm. The excellence rate (defined as E) is the number of times the best value performs best among all compared algorithms, while the success rate (defined as S) is the number of times the algorithm successfully searches for the global optima, e.g. E(SMA) = 5/13*100 = 38.5%, indicating that the best value obtained by SMA is 5 among all the comparison algorithms, S(SMA) = 4/13*100 = 30.7% means that the SMA algorithm can successfully search for global optima in 4 out of all the tested functions. Table 2 shows the optimization results of the iSMA and other comparative algorithms when solving the 13 benchmark test functions.

As can be seen from Table 2, for the unimodal function $f_1 - f_7$, the performance of iSMA in $f_1, f_2, f_3, f_4, f_5, f_7$ is the best among the five algorithms, and for the function $f_1 - f_4$ iSMA can search the global optima of the function, which indicates that the iSMA proposed in this paper has good convergence performance. Although for f_6, iSMA is not the best of all the compared algorithms, the results obtained from 20 independent runs of iSMA are superior to the standard SMA in terms of best value, average and standard deviation, which demonstrates the effectiveness of the swarm partitioning strategy and the hybrid mutation mechanism in this paper.

For the multimodal function $f_8 - f_{13}$, due to the existence of multiple local optima of these functions, it is very likely to mislead the direction of the algorithm to find the global optima, so that the algorithm eventually stagnates at these local optima. If the algorithm does not have the ability to identify these local optima and stay away from them in time, the algorithm's ability to find the global optima will be greatly reduced, so the multimodal function is very suitable for testing the ability of the algorithm to research for the global optima. As can be seen from Table 2, the best values of iSMA in the multimodal function $f_8 - f_{11}$ are the best in the comparison algorithms, and the average and standard deviation are better than the other four comparison algorithms. In particular, for f_{10}, SMA, iSMA and BMO have the same best values, average and standard deviation and they all converge to the same optimal solution, and they are also better than MFO and SFFO in terms of convergence effect and stability. For f_{11} and f_{13}, although the iSMA is not the best of all algorithms (second only to SFFO), it outperforms the standard SMA in terms of best value, average and standard deviation.

Besides, the excellent rate and success rate of the iSMA are E(iSMA) = 76.9% and S(iSMA) = 53.8% respectively, which are much higher than the comparison algorithm, E(BMO) = 38.5%, S(BMO) = 30.7%, E(MFO) = 0%, S(MFO) = 0%, E(SFFO) = 23.1%, S(SFFO) = 0%, respectively, which also proves that the iSMA proposed in this paper has good convergence ability and jump ability when facing the local optima.

Finally, in order to further observe the performance of the iSMA, we take two functions f_4 and f_{10} as examples to draw the convergence curve of the algorithm. As shown in Fig. 5(a), for the unimodal function f_4, the convergence curves of iSMA, SMA and BMO all decrease significantly with the increase of the number of iterations, indicating that the convergence effect of them significantly better than SFFO and MFO. Although the BMO shows slightly better performance than the iSMA in the early iterations, iSMA starts to outperform BMO at 350 iterations and converges to the global minimum at 450 iterations, while BMO eventually fails to converge to the global minimum. Note: the global minimum of f_4 is 0. The Fig. 5(a) does not show the final convergence of iSMA to the global minimum due to the axis scale. As can be seen from Fig. 5(b), for the multimodal function f_{10}, the iSMA, SMA and BMO eventually converge to the same local minimum value, but iSMA's convergence speed is obviously faster than other algorithms. Therefore, on the whole, the proposed iSMA algorithm has the best performance.

In conclusion, through qualitative and quantitative analysis, the results demonstrate the superior performance of the iSMA algorithm in solving continuous function optimization compared to other algorithms. The iSMA has great advantages in terms of solution quality, avoidance of local optima and convergence.

Table 2. Comparison with 5 algorithms on 30-dimensional functions (minimum best are in bold)

		SMA	iSMA	BMO	MFO	SFFO
f_1	Best	**0.00E + 00**	**0.00E + 00**	**0.00E + 00**	7.20E−01	9.53E−13
	Average	0.00E + 00	0.00E + 00	0.00E + 00	5.06E + 02	1.70E−12
	Std	0.00E + 00	0.00E + 00	0.00E + 00	2.23E + 03	5.30E−13
f_2	Best	2.28E−281	**0.00E + 00**	1.41E−309	1.01E + 01	3.17E−06
	Average	1.10E−161	0.00E + 00	2.02E−283	3.36E + 01	4.29E−06
	Std	4.92E−161	0.00E + 00	0.00E + 00	2.03E + 01	6.51E−07
f_3	Best	**0.00E + 00**	**0.00E + 00**	**0.00E + 00**	2.88E + 03	8.03E + 03
	Average	0.00E + 00	0.00E + 00	0.00E + 00	2.07E + 04	1.39E + 04
	Std	0.00E + 00	0.00E + 00	0.00E + 00	1.09E + 04	4.51E + 03
f_4	Best	3.72E−308	**0.00E + 00**	6.68E−307	4.57E + 01	2.97E + 01
	Average	1.80E−160	0.00E + 00	3.29E−283	6.99E + 01	5.05E + 01
	Std	8.05E−160	0.00E + 00	0.00E + 00	1.09E + 01	1.73E + 01
f_5	Best	2.81E + 01	**5.97E−03**	2.72E + 01	2.65E + 02	2.31E + 01
	Average	6.45E + 00	2.87E + 01	2.77E + 01	1.57E + 04	5.78E + 02
	Std	1.01E + 01	2.87E−01	2.78E−01	3.23E + 04	7.78E + 02
f_6	Best	4.66E + 00	1.33E−03	8.94E−02	5.34E−01	**1.02E−12**
	Average	6.09E + 00	5.34E−03	4.00E−01	1.98E + 03	1.95E−12
	Std	6.74E−01	3.14E−03	2.72E−01	4.06E + 03	7.35E−13
f_7	Best	3.82E−05	**8.53E−07**	1.93E−06	1.50E−01	1.40E−01
	Average	2.12E−04	5.66E−05	4.47E−05	1.73E + 00	2.58E−01
	Std	1.87E−04	7.01E−05	5.87E−05	2.67E + 00	9.71E−02
f_8	Best	−1.29E + 04	**−1.26E + 04**	−9.34E + 03	−1.05E + 04	−1.01E + 04
	Average	−5.87E + 04	−1.26E + 04	−7.54E + 03	−8.82E + 03	−9.38E + 03
	Std	6.18E + 04	4.26E−01	8.25E + 02	0.24E + 02	4.58E + 02
f_9	Best	**0.00E + 00**	**0.00E + 00**	**0.00E + 00**	9.65E + 01	2.98E + 01
	Average	4.57E + 01	0.00E + 00	0.00E + 00	1.75E + 02	5.01E + 01
	Std	9.61E + 01	0.00E + 00	0.00E + 00	3.60E + 01	1.07E + 01

(*continued*)

Table 2. (*continued*)

		SMA	iSMA	BMO	MFO	SFFO
f_{10}	Best	**8.88E−16**	**8.88E−16**	**8.88E−16**	1.77E + 00	7.24E−07
	Average	8.88E−16	8.88E−16	8.88E−16	1.37E + 01	1.29E−01
	Std	1.01E−31	1.01E−31	1.01E−31	7.71E + 00	4.13E−01
f_{11}	Best	**0.00E + 00**	**0.00E + 00**	**0.00E + 00**	5.97E−01	1.13E−13
	Average	2.42E−01	0.00E + 00	0.00E + 00	9.92E + 00	6.53E−03
	Std	5.02E−01	0.00E + 00	0.00E + 00	2.75E + 01	6.10E−03
f_{12}	Best	6.28E−01	6.25E−06	2.69E−03	2.44E + 00	**6.90E−15**
	Average	8.51E−01	4.91E−03	1.80E−02	2.30E + 03	3.08E−01
	Std	1.49E−01	6.60E−03	1.26E−02	7.51E + 03	5.28E−01
f_{13}	Best	2.71E + 00	2.48E−04	5.01E−01	3.04E + 00	**1.21E−13**
	Average	2.84E + 00	9.72E−03	2.85E + 00	4.90E + 02	5.80E−03
	Std	7.07E−02	1.36E−02	5.52E−01	1.95E + 03	1.45E−02

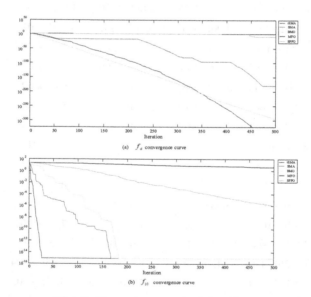

(a) f_4 convergence curve

(b) f_{10} convergence curve

Fig. 5. Convergence graph of algorithms

5   Conclusions

The global optimization problem is a very important research topic in academia, and the iSMA proposed in this paper provides a new and effective way to solve such problems. The iSMA incorporates a dominant-non-dominant swarm partitioning strategy, which partition the slime mould into dominant and non-dominant swarms according to certain constraints. This strategy changes the disadvantage of fuzzy positioning for all individuals in the traditional population-based meta-heuristics, and makes full use of the characteristics of different individuals, which is conducive to the iSMA balance global and local search. Furthermore, the Gauss-Cauchy hybrid mutation mechanism performs different operation on different swarm based on swarm partitioning strategy, which provides a guarantee for iSMA to identify and keep away from the local optima. This mutation operator based on swarm partitioning strategy makes full use of the information of individuals of different quality, which benefits the iSMA convergence to the global optima. Qualitative and quantitative analysis shows that the proposed iSMA has strong optimization ability. Future research could consider adding adaptive mechanisms to SMA and using the improved SMA for practical engineering optimization problems.

Acknowledgements. This work was supported by the National Science Foundation of China under Grant (61803192,61773192), Shandong province colleges and universities youth innovation talent introduction and education program, Special fund plan for local science and technology development lead by central authority, Research fund project of Liaocheng university under Grant NO. 318011922.

References

1. Li, Y., Zhao, Y.R., Liu, J.S.: Dimension by dimension dynamic sine cosine algorithm for global optimization problems. Appl. Soft Comput. J. **98**, 106933 (2021)
2. Jin, Q.B., Xu, Z.H., Cai, W.: An improved whale optimization algorithm with random evolution and special reinforcement dual-operation strategy collaboration. Symmetry. **13**(2), 238–238 (2010)
3. Mirjalili, S.: Moth-flame optimization algorithm: a novel nature-inspired heuristic paradigm. Knowl.-Based Syst. **89**, 228–249 (2015)
4. Hansen, N., Müller, S.D., Koumoutsakos, P.: Reducing the time complexity of the derandomized evolution strategy with covariance matrix adaptation (CMA-ES). Evol. Comput. **11**(1), 1–18 (2014)
5. Simon, D.: Biogeography-based optimization. IEEE Trans. Evol. Comput. **12**(6), 702–713 (2009)
6. Hatamlou, A.: Black hole: a new heuristic optimization approach for data clustering. Inf. Sci. **222**(3), 175–184 (2013)
7. Shareef, H., Ibrahim, A.A., Mutlag, A.H.: Lightning search algorithm. Appl. Soft Comput. **36**, 315–333 (2015)
8. Pan, W.T.: A new fruit fly optimization algorithm: taking the financial distress model as an example. Knowl.-Based Syst. **26**(2), 69–74 (2012)
9. Saremi S., Mirjalili S., Lewis A.: Grasshopper optimization algorithm: theory and application. Adv. Eng. Softw. **105**(MAR), 30–47 (2017)
10. Sulaiman, M.H., Mustaffa, Z., Saari, M.M.: Barnacles Mating Optimizer: a new bio-inspired algorithm for solving engineering optimization problems. Eng. Appl. Artif. Intell. **87**, 103330 (2020)
11. Li, S., Chen, H.L., Wang, M.J.: Slime mould algorithm: a new method for stochastic optimization. Fut. Gener. Comput. Syst. **111**, 300–323 (2020)
12. Mostafa, M., Rezk, H., Aly, M.: A new strategy based on slime mould algorithm to extract the optimal model parameters of solar PV panel. Sustain. Energy Technol. Assess. **42**, 100849 (2020)

13. Abdel-Basset, M., Mohamed, R., Chakrabortty, R.K.: An efficient binary slime mould algorithm integrated with a novel attacking-feeding strategy for feature selection. Comput. Ind. Eng. **153**, 107078 (2021)

14. Abdel-Basset, M., Chang, V., Mohamed, R.: HSMA_WOA: a hybrid novel Slime mould algorithm with whale optimization algorithm for tackling the image segmentation problem of chest X-ray images. Appl. Soft Comput. **95**, 106642 (2020)

15. Sang, H.Y., Pan, Q.K., Duan, P.Y.: Self-adaptive fruit fly optimizer for global optimization. Nat. Comput. **18**, 785–813 (2019)

16. Xu, J., Xu, F.: Adaptive mutation frog leaping algorithm based on function rate of change. Softw. Guide **17**(09), 77–80 (2018)

17. Song, X.X., Jian, L., Song, Y.Q.: A chunk updating LS-SVMs based on block Gaussian elimination method. Appl. Soft Comput. **51** (2017)

18. Zhou, H., Wang, J., Gu, X.: Research on LSSVM model for gauss-cauchy mutation operator optimization. Comput. Dig. Eng. **48**(01), 19–24 (2020)

An Improved Chicken Swarm Optimization Algorithm with Fireworks Factor

Baofeng Zheng[1], Xiuxi Wei[2($\boxtimes$)], and Huajuan Huang[2]

[1] College of Electronic Information, Guangxi University for Nationalities,
Nanning 530006, China
[2] College of Artificial Intelligence, Guangxi University for Nationalities,
Nanning 530006, China

Abstract. Considering the problem that the original chicken swarm optimization algorithm ethnic diversity and easy to fall into local optimum problems, an improved chicken swarm optimization algorithm based on fireworks (FW-ICSO) algorithm was proposed. In this algorithm, roulette was introduced to die out some chickens, produce new chicken's methods is implemented in fireworks algorithm, and adding an inertia factor to balance search capability. Finally, FW-ICSO is tested with twenty benchmark function and compared with other similar algorithms to confirm their effectiveness in terms of the accuracy and convergence rate of the results.

Keywords: Chicken swarm optimization (CSO) algorithm · Roulette algorithm · Fireworks optimization (FWA) algorithm · Function optimization

1 Introduction

Swarm intelligence optimization algorithms are novel type of computational method by simulating the swarm behavior or predatory behavior of natural creatures. At present, researchers have proposed a variety of swarm intelligence optimization algorithms, such as particle swarm optimization (PSO) [1], Genetic algorithm (GA) [2], firefly algorithm (FA) [3].

The chicken swarm optimization (CSO) algorithm was originally presented by Meng et al. [4] at the Fifth International Conference on swarm intelligence (ICSI) in 2014. The algorithm simulates the regular pattern of foraging and the hierarchy order in the chicken swarm, the proposed algorithm has become the focus of a growing number of scholars, they are extensively used in applications. Such as, Dinghui Wu et al. [5] have used the Markov chain to analyze the convergence of clustering algorithm and verified the global convergence of CSO; Deb S et al. [6] summarized the research progress of chicken swarm olony optimization algorithm;Hafez A I et al. [7] proposed an innovative approach for feature selection based on chicken swarm optimization; Cui D [8] proposed projection pursuit model for evaluation of flood and drought disasters based on chicken swarm optimization algorithm; Osamy W et al. [9] proposed the Chicken Swarm Optimization based Clustering Algorithm to improve energy efficiency in Wireless Sensor Network.

© Springer Nature Switzerland AG 2021
D.-S. Huang et al. (Eds.): ICIC 2021, LNCS 12836, pp. 635–651, 2021.
https://doi.org/10.1007/978-3-030-84522-3_52

Although the CSO algorithm has fast convergence rate and high optimum accuracy, there is easy to trap in the local optimum. Therefore, the optimization of the CSO algorithm becomes the focus of the researcher. Such as, Dinghui Wu et al. [10] proposed a method, which added the part of chicken learning cock, and introduced inertia weight and learning factor; Wang J et al. [11] proposed the improved Chicken Swarm Optimization algorithm to solve the interruptible load scheduling scheme; Bin Li et al. [12] proposed algorithm is a combination of the grey wolf optimizer (GWO) and the particle swarm optimization (PSO) algorithm with CSO; N. Bharanidharan et al. [13] proposed a method, which Improved chicken swarm optimization to classify dementia MRI images using a novel controlled randomness optimization algorithm.

In this paper, an improved chicken swarm optimization algorithm based on fireworks (FW-ICSO) algorithm was developed. Firstly, the roulette algorithm is used to select and eliminate individuals. Secondly, combines the advantage of the firework algorithm (FWA) [14] to generate new particles and improve the population diversity. Then, adding inertia factor to balance search capability. Finally, FW-ICSO is tested with a set of test functions and compared with other algorithms. The experimental data show that FW-ICSO has obvious advantages.

The rest of the paper is organized as follows: Sect. 2 introduces the idea of the chicken swarm optimization algorithm and update formula. The improvements of the FW-ICSO algorithm are introduced in Sect. 3. In Sect. 4, simulation results of FW-ICSO are presented. Finally, conclusions and expectation are stated in Sect. 5.

2   CSO Algorithm Introduction

The CSO algorithm was proposed by observing chicken swarm behavior. The best quality food is taken as the target point, through the continued transport of position information among chickens of different grades in each subgroup, and comparison with their best position, the next direction of each chicken is determined and the food finally funds. The central idea of CSO is as follows:

1) The chicken swarm is divided into multiple subgroups, which are composed of roosters, hens, and chicks.
2) Each subgroup consists of only one rooster and several hens and chicks, roosters have the best fitness and act as the leader in a subgroup. The individuals with the worst fitness values will be defined as chicks and the rest are hens.
3) The hierarchy and mother-child relationships are updated every few times.
4) The roosters lead its subgroup foraging, hens always looking for food follow the roosters, each chick follows their mothers to search for food.

In the algorithm, the entire swarm population was set to N, then N_R, N_M, N_H, N_C were the number of roosters, mother hens, hens, and chicks. x denotes the position of each chicken, then the position update of the rooster can be expressed as:

$$x_{i,j}(t + 1) = x_{i,j}(t) \times \left[1 + Randn\left(0, \sigma^2\right) \right] \tag{1}$$

$$\sigma^2 = \begin{cases} 1, & \text{if } f_i \leq f_k \\ \exp(\frac{f_k - f_i}{|f_i| + \varepsilon}), & \text{otherwise} \\ k \in [1, N_R], & k \neq i \end{cases} \tag{2}$$

In formula (1), $x_{i,j}(t)$ denote the position of the i th rooster on the j th dimension in the t th iteration($j \in [1, d]$, d is the dimension of the search space); $Randn(0, \sigma^2)$ is a Gaussian distribution with mean 0 and standard deviation σ^2. Formula (1) simulates the rooster's position moves.

In formula (2), k is a rooster's index, which is randomly selected from the roosters' group $(k \neq i)$, ε is the smallest constant in the computer. Formula (2) simulates the competitive relationship between different roosters.

The position update of the hens (include mother hens) can be expressed as:

$$x_{i,j}(t+1) = x_{i,j}(t) + C_1 \times \text{Rand} \times (x_{r_1,j}(t) - x_{i,j}(t)) + C_2 \times \\ \text{Rand} \times (x_{r_2,j}(t) - x_{i,j}(t)) \tag{3}$$

$$C_1 = \exp(\frac{f_i - f_{r_1}}{abs(f_i) + \varepsilon}) \tag{4}$$

$$C_2 = \exp(f_{r_2} - f_i) \tag{5}$$

In the formula (3), $Rand$ represents a random number between [0,1]; r_1 is the rooster in the same subgroup, r_2 is chicken(rooster or hen), which randomly calculated in the entire swarm($r_1 \neq r_2$), and the fitness of r_2 is better than the fitness of i.

The position update of the chicks can be expressed as:

$$x_{i,j}^{t+1} = x_{i,j}^t + FL(x_{m,j}^t - x_{i,j}^t) \tag{6}$$

In the formula (6), m is the index of the mother hen of chicks i; FL is the adjustment parameter of chick following its mother, $FL \in [0, 2]$.

The algorithm CSO is shown in Algorithm 1.

Algorithm 1. CSO

Initialization *pop, M, G, Dim, rPercent, hPercent, mPercent, t;*
Initialization *x*, save *best_fitness, best_location;*
While $t \leq M$ **do**
    If $t\%G$ **do**
        Sort fitness;
        Role assignment;
        subgroup division;
    End
    For each chicken x **do**
        If x is a rooster **do**
            Update x location using formula (1) ~ (2);

        End
        If x is a hen **do**
            Update x location using formula (3) ~ (5);
        End
        If x is a chick **do**
            Update x location using formula (6);
        End
        Update fitness based on x ;
        Save *best_fitness, best_location;*
    End
End

---

3 FW-ICSO Algorithm Introduction

Although convergence accuracy and rate of the traditional CSO algorithm is quite remarkable, the ethnic diversity is relatively single, and the chicks easily fall into local optimum, which leads to the depression of algorithm efficiency. Therefore, this paper proposed the FW-ICSO algorithm, additional the elimination mechanism, and use the FWA algorithm to generate new individuals, which is conducive to jumping out of the local optimum; additional the inertia factor, and balancing the searchability.

3.1 Elimination Mechanism

In this paper, using the roulette algorithm, after the G cycle, the poorly fitness part of chickens will be eliminated. Roulette algorithm: Link the probability of each individual being selected to its fitness, with better fitness comes a lower probability of being selected, and with worse fitness comes a greater probability of being selected.

Firstly, calculate the probability of each individual being selected:

$$P(x_i) = \frac{f(x_i)}{\sum_{j=1}^{N} f(x_j)} \tag{7}$$

Secondly, calculate the cumulative probability:

$$Q(x_i) = \sum_{k=1}^{i} P(x_k) \tag{8}$$

If the individual's fitness is poor, the corresponding selection probability will be greater. After the selection probability is converted to the cumulative probability, the corresponding line segment will be longer, and the probability of being selected will be greater.

Finally, generate a random number *judge*, *judge* $\in$ [*rPercent*, 1]. If $Q(x_i) > $ *judge*, eliminate the individual.

3.2 Creates New Individuals

In order to create new individuals, this paper introduces the FWA [12] algorithm. Assuming that n individuals are eliminated in part 3.1, then set the first n individuals with better fitness as the center. Calculate the explosive strength, explosive amplitude and displacement, then create new individuals, and select individuals with good fitness to join the algorithm.

(a) Explosive strength
   This parameter is used to determine the number of new individuals. Good fitness individuals will create more new individuals, and poor fitness individuals will create fewer new individuals.

   In formula (9): S_i represents the number of new individuals produced by the i th individual; m is the constant used to control the number of new individuals of the maximum explosive amplitude, Y_{max} is the worst fitness.

$$S_i = m \frac{Y_{max} - f(x_i) + \varepsilon}{\sum_{i=1}^{N}(Y_{max} - f(x_i)) + \varepsilon} \tag{9}$$

(b) Explosive amplitude
   The explosive amplitude is used to limit the range of individuals to create new individuals. Good fitness solutions can be very close to the global solution, so the smaller the explosive amplitude, on the contrary, the larger the explosive amplitude.

$$A_i = \hat{A} \frac{f(x_i) - Y_{min} + \varepsilon}{\sum_{i=1}^{N}(f(x_i) - Y_{min}) + \varepsilon} \tag{10}$$

In formula (10): A_i represents the explosive amplitude of the i th individual creating new individuals; $\hat{A}$ is the constant of the maximum explosive amplitude; Y_{min} is the best fitness.

(c) Displacement operation

  With the explosive strength and explosive amplitude, *and according to the* Displacement operation, S_i new individuals can be created.

$$\Delta x_i^k = x_i^k + rand(0, A_i) \tag{11}$$

In formula (11): x_i^k represents the location of the ith individual; Δx_i^k represents the location of the new individual;$rand(0, A_i)$ is the random displacement distance.

3.3 Inertial Factor

In the test without adding inertial factor ω, although the FW-ICSO algorithm can achieve better results than other algorithms, its stability is not strong, so the inertial factor ω is added to solve this problem.

The updated roosters' position formula is:

$$x_{i,j}(t+1) = \omega \times x_{i,j}(t) \times \left[1 + N\left(0, \sigma^2\right)\right] \tag{12}$$

The updated hens' position formula is:

$$x_{i,j}(t+1) = \omega \times x_{i,j}(t) + C_1 \times \text{Rand} \times \left(x_{r_1,j}(t) - x_{i,j}(t)\right) + C_2 \times \\ \text{Rand} \times \left(x_{r_2,j}(t) - x_{i,j}(t)\right) \tag{13}$$

The updated chicks' position formula is:

$$x_{i,j}^{t+1} = \omega \times x_{i,j}^t + F\left(x_{m,j}^t - x_{i,j}^t\right) \tag{14}$$

3.4 The Flow of the FW-CSO Algorithm

The algorithm FW-ICSO is shown in Algorithm 2.

Algorithm 2. FW-ICSO

Initialization *pop, M, G, Dim, rPercent, hPercent, mPercent, t;*
Initialization *x,* save *best_fitness, best_location*;
While ($t \leq M$)
    If $t\%G$ **then**
        If $t \neq 1$ **do**
            Select individual elimination using formula (7) ~ (8);
            Creates new individuals using formula (9) ~ (11);
        End
        Select individuals with good fitness and save to x ;
        Sort fitness;
        Role assignment;
        subgroup division;
    End
    For each chicken x **do**
        If x is a rooster **do**
            Update x location using formula (12);
        End
        If x is a hen **do**
            Update x location using formula (13);
        End
        If x is a chick **do**
            Update x location using formula (14);
        End
    End
    Update fitness based on x ;
    Save *best_fitness, best_location*;
End

---

4 Experimental Comparison and Analysis

4.1 Experimental Parameter Settings

Fifteen benchmark functions in Table 1 are applied to compare FW-ICSO, CSO, ISCSO[15], PSO, BOA [16]. Set the $D = 50$; The search bounds are $[-100,100]$; The total particle number of each algorithm to 100; The maximum number of iterations is 1000; The algorithms run 50 times independently for each function. The parameters for algorithms are listed in Table 2.

Table 1. Fifteen benchmark functions

ID	Function name	ID	Function name
F1	Bent Cigar	F9	Discus
F2	Sum of different power	F10	Ackley's
F3	Zakharov	F11	Powell
F4	Rosenbrock's	F12	Griewank's
F5	Rastrigin's	F13	Katsuura
F6	Rotated Rastrigin's	F14	HappyCat
F7	Levy	F15	HGBat
F8	High conditioned elliptic		

Table 2. The main parameter settings of the five algorithms

Algorithm	Parameters
FW-ICSO	$N_R = 0.15 * NN_H = 0.7 * NN_C = N - N_R - N_H N_M = 0.5 * N_H$ $G = 10, F \in (0, 2), m = 50, \hat{A} = 40, \omega = 0.8$
CSO	$N_R = 0.15 * NN_H = 0.7 * NN_C = N - N_R - N_H N_M = 0.5 * N_H$ $G = 10, F \in (0, 2)$
ISCSO	$N_R = 0.15 * NN_H = 0.7 * NN_C = N - N_R - N_H N_M = 0.5 * N_H$ $G = 10, F \in (0, 2)$
PSO	$c1 = c2 = 1.5, \omega = 0.8$
BOA	$p = 0.8, \alpha = 0.1, c = 0.01$

4.2 Experiment Analysis

As can be seen from Figs. 1, 2, 3, 4, 5, 6, 7, 8, 9, 10, 11, 12, 13, 14, 15, 16, 17, 18, 19, 20, 21, 22, 23, 24, 25, 26, 27, 28, 29 and 30 the FW-ICSO has a considerable convergence speed. In Fig. 1, 15, and 17, the convergence speed of the FW-ICSO algorithm is much faster, the PSO and BOA has fallen into the local optimum. CSO and ISCSO converge at the same speed, but compared with FW-ICSO, the speed is much slower, and it falls into the local optimum. Therefore, FW-ICSO can avoid falling into the local optimal solution.

In Fig. 3, 7, 9, 11, and 23, FW-ICSO has a considerable speed of convergence. Although other algorithms perform well, they are still slower than FW-ICSO in terms of speed. Therefore, the convergence speed of the FW-ICSO algorithm is excellent. It can be seen from their variance graphs, the variance of FW-ICSO is small and stable. Therefore, FW-ICSO is not only fast, but also very stable.

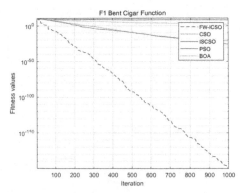

Fig. 1. Comparison of algorithm convergence in F1 function

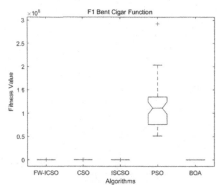

Fig. 2. Comparative ANOVA tests of algorithms in F1 function

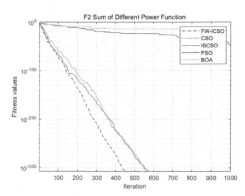

Fig. 3. Comparison of algorithm convergence in F2 function

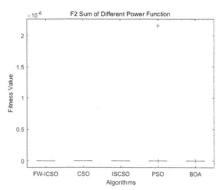

Fig. 4. Comparative ANOVA tests of algorithms in F2 function

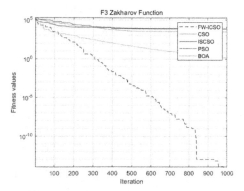

Fig. 5. Comparison of algorithm convergence in F3 function

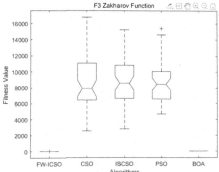

Fig. 6. Comparative ANOVA tests of algorithms in F3 function

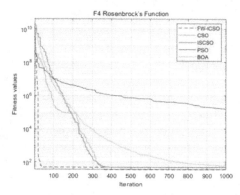

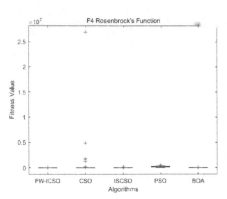

Fig. 7. Comparison of algorithm convergence in F4 function

Fig. 8. Comparative ANOVA tests of algorithms in F4 function

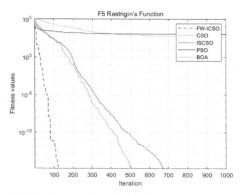

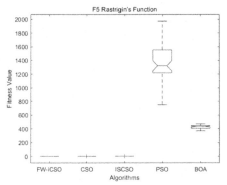

Fig. 9. Comparison of algorithm convergence in F5 function

Fig. 10. Comparative ANOVA tests of algorithms in F5 function

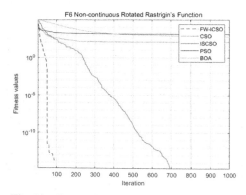

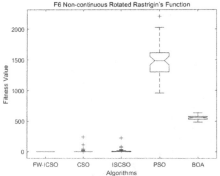

Fig. 11. Comparison of algorithm convergence in F6 function

Fig. 12. Comparative ANOVA tests of algorithms in F6 function

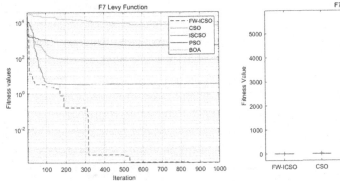

Fig. 13. Comparison of algorithm convergence in F7 function

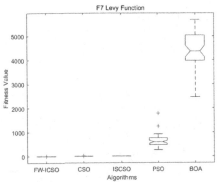

Fig. 14. Comparative ANOVA tests of algorithms in F7 function

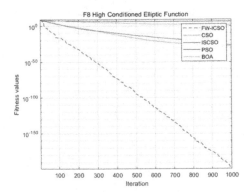

Fig. 15. Comparison of algorithm convergence in F8 function

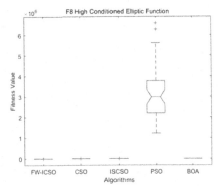

Fig. 16. Comparative ANOVA tests of algorithms in F8 function

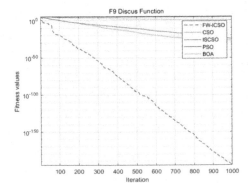

Fig. 17. Comparison of algorithm convergence in F9 function

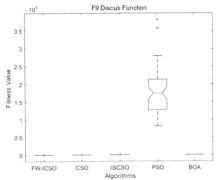

Fig. 18. Comparative ANOVA tests of algorithms in F9 function

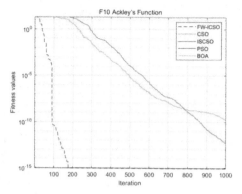

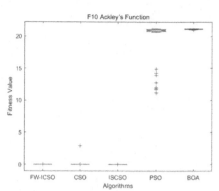

Fig. 19. Comparison of algorithm convergence in F10 function

Fig. 20. Comparative ANOVA tests of algorithms in F10 function

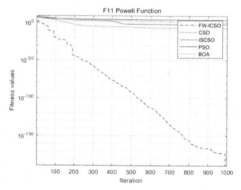

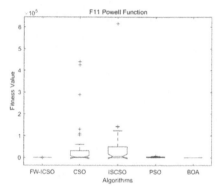

Fig. 21. Comparison of algorithm convergence in F11 function

Fig. 22. Comparative ANOVA tests of algorithms in F11 function

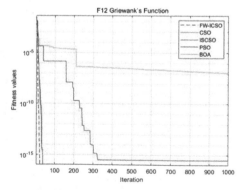

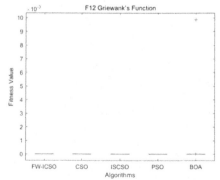

Fig. 23. Comparison of algorithm convergence in F12 function

Fig. 24. Comparative ANOVA tests of algorithms in F12 function

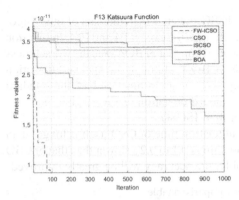

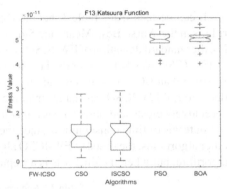

Fig. 25. Comparison of algorithm convergence in F13 function

Fig. 26. Comparative ANOVA tests of algorithms in F13 function

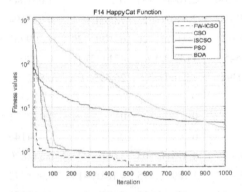

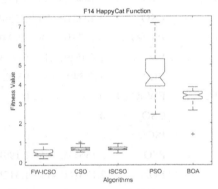

Fig. 27. Comparison of algorithm convergence in F14 function

Fig. 28. Comparative ANOVA tests of algorithms in F14 function

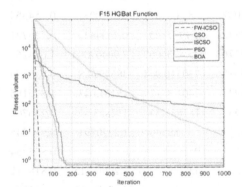

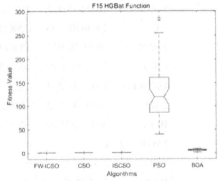

Fig. 29. Comparison of algorithm convergence in F15 function

Fig. 30. Comparative ANOVA tests of algorithms in F15 function

It can be seen from Table 3 that the FW-ICSO is better than the data of other algorithms in the Worst, Best, Mean, and Std of each function except F4 and F14. In function F4, the standard deviation of FW-ICSO was worse than that of BOA, and the worst fitness of FW-ICSO was better than the best fitness of BOA. although the standard deviation is worse than BOA, it is also within the acceptable range. In function F14, the standard deviation of FW-ICSO was larger than the standard deviation of ISCSO, but the value is relatively close, and other values are better than ISCSO.

In terms of time consumption, since the algorithm time of CSO itself is longer than other algorithms, the speed of FW-ICSO algorithm is 0.1 to 0.2 s slower than that of CSO algorithm, but when the time loss is acceptable, the accuracy will be greatly improved.

Table 3. Accuracy comparison table

Function	Algorithm	Worst	Best	Mean	Std.	Time
F1	FW-ICSO	6.1894E−186	1.13E−200	2.104E−187	0	1.051045
	CSO	3.40331E−20	6.238E−27	2.0407E−21	7.003E−21	0.957194
	ISCSO	4.06791E−19	8.924E−27	1.3122E−20	6.049E−20	1.309967
	PSO	292452978.1	51403345	114868034	47144279	0.182071
	BOA	212.5405395	95.420394	152.697557	26.59569	0.832887
F2	FW-ICSO	0	0	0	0	1.114456
	CSO	0	0	0	0	1.047371
	ISCSO	0	0	0	0	1.454144
	PSO	2.15657E−06	1.199E−72	4.3131E−08	3.05E−07	0.244820
	BOA	2.17614E−14	1.312E−23	1.1895E−15	4.188E−15	0.991472
F3	FW-ICSO	4.90599E−11	2.075E−41	2.3296E−12	7.68E−12	0.968975
	CSO	16779.68466	2552.4041	8772.98815	3746.6	0.815973
	ISCSO	15233.94168	2817.5755	8725.16405	2847.7072	1.276224
	PSO	15385.06415	4684.2087	8844.41871	2761.0037	0.154354
	BOA	1.694109795	0.5827842	1.1987125	0.2434447	0.775721
F4	FW-ICSO	48.68285915	13.8433763	45.4735067	7.94407003	1.074267
	CSO	26919621.79	46.715618	730510.944	3854974.8	0.935579
	ISCSO	107363.5283	46.806922	2401.37179	15161.043	1.376185
	PSO	481169.9612	32591.93	183831.55	112823.36	0.256701
	BOA	59.03452539	52.354011	56.1893463	1.2720791	0.956245
F5	FW-ICSO	0	0	0	0	0.895254
	CSO	5.32907E−15	0	4.9738E−16	1.296E−15	0.818512
	ISCSO	3.55271E−15	0	1.0658E−16	5.571E−16	1.341090
	PSO	1973.247062	750.83661	1365.0541	271.31223	0.272896
	BOA	469.8530387	369.29832	427.495741	24.207126	1.028247

(continued)

Table 3. (*continued*)

Function	Algorithm	Worst	Best	Mean	Std.	Time
F6	FW-ICSO	0	0	0	0	2.454195
	CSO	240.7162391	0	10.1121123	37.742581	2.361683
	ISCSO	223.6098307	0	11.7887804	35.079087	3.135543
	PSO	2205.574795	960.40339	1491.23701	298.63115	1.622343
	BOA	637.9790385	485.56995	556.661562	36.401121	3.875971
F7	FW-ICSO	0.003964923	4.788E−10	0.00055943	0.0009584	1.279563
	CSO	13.84271935	2.7653595	3.56474086	1.5273858	1.079681
	ISCSO	4.026353675	2.6930204	3.30681594	0.3092122	1.609634
	PSO	1786.141083	263.95178	626.270917	251.02204	0.390569
	BOA	5684.007969	2463.7525	4403.996	735.67077	1.363668
F8	FW-ICSO	1.208E−190	6E−211	2.744E−192	0	2.201493
	CSO	6.38208E−23	1.92E−30	2.0776E−24	9.318E−24	1.829214
	ISCSO	6.84164E−21	1.806E−29	1.383E−22	9.674E−22	2.293385
	PSO	6581731.942	1223689.8	3233924.77	1330427.8	1.028134
	BOA	47.33434907	21.019203	35.3138648	7.0328411	2.601756
F9	FW-ICSO	2.2114E−196	1.42E−212	5.33E−198	0	1.145240
	CSO	3.61323E−24	9.983E−31	1.41E−25	5.668E−25	0.962248
	ISCSO	7.48824E−24	1.185E−31	3.0069E−25	1.185E−24	1.389841
	PSO	37983.62703	8127.6735	18080.3354	6694.1586	0.191719
	BOA	3.057654745	0.5670763	1.85472434	0.4892557	0.871098
F10	FW-ICSO	4.44089E−15	8.882E−16	9.5923E−16	5.024E−16	1.158565
	CSO	2.883589661	1.643E−13	0.05767179	0.4078012	1.036261
	ISCSO	1.81687E−07	7.994E−15	3.7897E−09	2.568E−08	1.527546
	PSO	21.21403019	11.214801	19.6998291	3.0778906	0.373748
	BOA	21.24352341	21.027682	21.1774251	0.0441896	1.193409
F11	FW-ICSO	3.3649E−149	2.416E−211	6.73E−151	4.76E−150	1.266613
	CSO	381039.5695	3.8513E−19	49370.8543	85243.512	0.983339
	ISCSO	378745.4295	5.2773E−33	42672.9934	88703.772	1.418165
	PSO	6861.689017	368.528453	2175.08293	1208.4472	0.310457
	BOA	2.898785526	0.36426493	1.5823845	0.5714033	1.156489

(*continued*)

Table 3. (*continued*)

Function	Algorithm	Worst	Best	Mean	Std.	Time
F12	FW-ICSO	0	0	0	0	1.279460
	CSO	0	0	0	0	1.234044
	ISCSO	0	0	0	0	1.674844
	PSO	0	0	0	0	0.420399
	BOA	0.009864687	6.013E−12	0.0003946	0.0019527	1.299393
F13	FW-ICSO	0	0	0	0	24.39417
	CSO	2.75E−11	1.33E−12	1.11E−11	6.365E−12	24.27090
	ISCSO	2.89E−11	1.53E−13	1.03E−11	7.749E−12	25.52178
	PSO	5.61E−11	4.11E−11	5.10E−11	4.205E−12	23.32075
	BOA	5.61E−11	4.63E−11	5.08E−11	2.203E−12	47.55279
F14	FW-ICSO	0.93545686	0.1853164	0.47910787	0.1902337	1.033239
	CSO	1.003017025	0.5063184	0.67737218	0.1142708	0.868092
	ISCSO	0.945762978	0.461541	0.69000236	0.1085319	1.357010
	PSO	7.144064581	2.4240018	4.53295436	0.9515144	0.142176
	BOA	3.844123963	1.3945829	3.34472732	0.3847439	0.806925
F15	FW-ICSO	0.500000001	0.4950851	0.49980061	0.0008948	1.052220
	CSO	0.586131408	0.4261263	0.48958174	0.0265692	0.826003
	ISCSO	0.683222701	0.4419355	0.49830781	0.0404395	1.349768
	PSO	286.5622008	39.581567	132.858233	58.787859	0.128205
	BOA	8.357891343	0.8503189	5.2778622	1.9406607	0.767732

5   Conclusion and Future Work

In order to improve the CSO algorithm, this paper proposes the FW-ICSO algorithm to select individuals for elimination through the roulette algorithm, introduces the explosion strength, explosion amplitude, and displacement operations in the FWA algorithm, generates new individuals to join the algorithm, and introduces the searchability of the inertial balance algorithm. Finally, tested FW-ICSO with Fifteen benchmark functions, the results demonstrate that it is true, the convergence speed, accuracy, and robustness of the FW-ICSO algorithm are considerable.

Acknowledgments. This work is supported by the National Natural Science Foundation of China (61662005). Guangxi Natural Science Foundation (2018GXNSFAA294068) ;Research Project of Guangxi University for Nationalities (2019KJYB006); Open Fund of Guangxi Key Laboratory of Hybrid Computation and IC Design Analysis (GXIC20–05).

References

1. Kennedy, J., Eberhart, R.: Particle swarm optimization. In: IEEE Proceedings of ICNN'95-International Conference on Neural Networks, vol. 4, pp. 1942–1948 (1995)
2. Yong-Jie, M.A., Yun, W.X.: Research progress of genetic algorithm. Appl. Res. Comput. **29**(4), 1201–1206 (2012)
3. Iztok, F., et al.: A comprehensive review of firefly algorithms. Swarm Evol. Comput. **13**, 34–46 (2013)
4. Meng, X., Liu, Y., Gao, X., Zhang, H.: A new bio-inspired algorithm: chicken swarm optimization. In: Tan, Y., Shi, Y., Coello, C.A.C. (eds.) ICSI 2014. LNCS, vol. 8794, pp. 86–94. Springer, Cham (2014). https://doi.org/10.1007/978-3-319-11857-4_10
5. Wu, D., Shipeng, X., Kong, F.: Convergence analysis and improvement of the chicken swarm optimization algorithm. IEEE Access **4**, 9400–9412 (2016)
6. Deb, S., Gao, X.-Z., Tammi, K., Kalita, K., Mahanta, P.: Recent studies on chicken swarm optimization algorithm: a review (2014–2018). Artif. Intell. Rev. **53**(3), 1737–1765 (2019). https://doi.org/10.1007/s10462-019-09718-3
7. Ahmed Ibrahem, H., et al.: An innovative approach for feature selection based on chicken swarm optimization. In: 2015 7th International Conference of Soft Computing and Pattern Recognition (SoCPaR), pp. 19–24. IEEE (2015)
8. Cui, D.: Projection pursuit model for evaluation of flood and drought disasters based on chicken swarm optimization algorithm. Adv. Sci. Technol. Water Resour. **36**(02), 16–23+41 (2016)
9. Osamy, W., El-Sawy, A.A., Salim, A.: CSOCA: chicken swarm optimization based clustering algorithm for wireless sensor networks. IEEE Access **8**, 60676–60688 (2020)
10. Wu, D., et al.: Improved chicken swarm optimization. In: 2015 IEEE International Conference on Cyber Technology in Automation, Control, and Intelligent Systems (CYBER), pp. 681–686. IEEE (2015)
11. Wang, J., Zhang, F., Liu, H., et al.: A novel interruptible load scheduling model based on the improved chicken swarm optimization algorithm. CSEE J. Power Energy Syst. **7**(2), 232–240 (2020)
12. Bin, L.I., Shen, G.J., Sun, G., et al.: Improved chicken swarm optimization algorithm. J. Jilin Univ. (Eng. Technol. Ed.) **49**(4), 1339–1344 (2019)
13. Bharanidharan, N., Rajaguru, H.: Improved chicken swarm optimization to classify dementia MRI images using a novel controlled randomness optimization algorithm. Int. J. Imaging Syst. Technol. **30**(3), 605–620 (2020)
14. Tan, Y., Zhu, Y.: Fireworks algorithm for optimization. In: International Conference in Swarm Intelligence, vol. 6415, pp. 355-364. Springer, Heidelberg (2010)
15. Tong, B., et al.: Improved chicken swarm optimization based on information sharing. J. Guizhou Univ. (Nat. Sci.) **38**(01), 58–64+70 (2021)
16. Fan, Y., et al.: A self-adaption butterfly optimization algorithm for numerical optimization problems. IEEE Access **8**, 88026–88041 (2020)

An Improved Lagrangian Relaxation Algorithm for Solving the Lower Bound of Production Logistics

Nai-Kang Yu[1,2], Rong Hu[1,2(✉)], Bin Qian[1,2], and Ling Wang[3]

[1] School of Mechanical and Electrical Engineering, Kunming University of Science
and Technology, Kunming 650500, China
[2] School of Information Engineering and Automation, Kunming University of Science
and Technology, Kunming 650500, China
[3] Department of Automation, Tsinghua University, Beijing 10084, China

Abstract. In this paper, two different lower bound models are proposed for the bottleneck process and logistics distribution in production logistics. The objective is to minimize the cost, which is a typical NP hard combinatorial optimization problem. Firstly, the optimal distribution quantity is determined for the bottleneck process in production. Secondly, the distribution problem is modeled as vehicle routing problem with time window (VRPTW), and some variables and constraints are relaxed. Thirdly, in order to improve the lower bound of the model, a Lagrangian relaxation model for VRPTW is designed, and an improved subgradient algorithm is proposed. Simulation results show that the algorithm proposed in this paper is effective for the problem and can calculate a tight lower bound.

Keywords: Lower bound · Vehicle routing problem with time window · Lagrangian relaxation · Subgradient

1 Introduction

Since 2015, China has proposed to develop intelligent manufacturing vigorously, which has opened the process of transformation from manufacturing large to manufacturing power. In manufacturing industry, the two common manufacturing forms are discrete and process industry, and discrete industry accounts for most of them. The production process of discrete manufacturing industry is generally composed of parts processing, assembly, inspection and packaging, and the processing process is independent, so the product processing process is discrete, as shown in Fig. 1. Typical representatives are automobile assembly [2], electronic equipment manufacturing [3], machining operation [4] and other industries. In the discrete accomplishment industry, time is the key factor, because the artefact processing action is carefully in accordance with the BOM. If the production capacity is insufficient, such as bottleneck process, logistics shortage, backlog of work in process and so on, the production cycle of the product will be extended, resulting in the product cannot be delivered within the delivery period. In appearance of the above

© Springer Nature Switzerland AG 2021
D.-S. Huang et al. (Eds.): ICIC 2021, LNCS 12836, pp. 652–662, 2021.
https://doi.org/10.1007/978-3-030-84522-3_53

problems, we must do a good job matching with the production. Production logistics is one of the most important links. The coordination between production logistics and production line directly determines the continuity of product manufacturing and the punctuality of delivery. In the manufacturing industry, production logistics refers to the logistics activities in the internal production process of the enterprise, that is, according to the requirements of the factory facility layout and process flow, the circulation of materials and products in the warehouse and workshops, workshops and workshops, and factory stations and stations logistics activities.

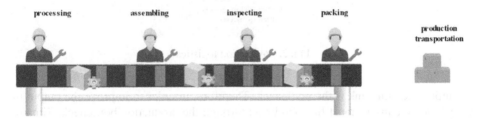

processing   assembling   inspecting   packing

production
transportation

Fig. 1. Scattered manufacturing

The optimization of production logistics is mainly divided into the upper optimization represented by strategic layout, equipment layout and facility location, and the lower optimization represented by inventory management, material distribution, vehicle scheduling, line side warehouse management and work in process (WIP) transportation cache, as shown in Fig. 2. Among them, material distribution and vehicle scheduling is an important link for enterprise logistics optimization. In the workshop production, the storage capacity of the line side warehouse is limited as an intermediate buffer area, which requires the enterprise to send the materials to the corresponding station in time according to the production speed of the production line, and ensure the smooth production process by timely replenishing the materials. For the flexible manufacturing enterprise, there are many kinds of materials, so it is more necessary to closely cooperate the processing and production links with the material distribution links.

Aiming at the production logistics process of automobile assembly workshop, this paper combines the constraint theory with vehicle routing problem. Firstly, the optimal processing lot size of the workshop is determined. Secondly, the transportation quantity is determined according to the production lot size. Then, the transportation process is modeled as a vehicle routing problem (VRP), and the effective lower bound is obtained by using the improved Lagrangian relaxation technique. Constraint theory [5] is a production management technology proposed by Dr. Goldratt, an Israeli physicist. TOC (Theory of constraints), MRP (Manufacturing Resource Planning) and JIT (just in time) are commonly used methods in production management. According to the, the capacity of each stage of the production process is different, so it is necessary to find the link that restricts the whole production efficiency in the production process. The production system of the workshop is connected by multiple production links, and the most inefficient link in the production system directly restricts the output of the whole system. As long as we find out which link in the production process hinders the maximization

Fig. 2. Production logistics

of output, then this link is the bottleneck. Therefore, in order to improve the output of the whole system, we must find and break through the production bottleneck. Simatupang et al. [6] used TOC theory to identify the bottleneck resources in the supply chain enterprises and eliminate the bottleneck resources in the process of cooperation, so as to improve the efficiency through the cooperation among enterprises. The problem model is given. Reimer and Glenn [7] combine TOC and MRP theory, use DBR as management means, and apply the above two theories to enterprise management. Cook [8] applied the traditional planned production mode, JIT production mode and TOC theory to the management of workshop production time and WIP quantity, and found that TOC theory can achieve better management effect. The material distribution problem in production logistics can be reduced to vehicle routing problem, which was proposed by Dantzig [9] in 1959, Vehicle routing problem is generally described as a series of customer points and warehouse points, arranging a certain number of fleet, under certain constraints (vehicle load, time window, etc.), vehicles serve customers in a certain order, so that the vehicle distances is the shortest, that is, the access route of each vehicle is a Hamilton cycle (loop). The goal of VRP optimization is usually travel cost or time. In practical application, VRP have different types, including VRP with capacity limitation (CVRP), VRP with time window (VRPTW), VRP with backhaul pickup (VRPBP) and so on. VRPTW is the most important derivative problem of VRP, which has been proved to be NP-hard problem [10]. At present, there are two main methods to solve VRP, one is intelligent algorithm represented by ant colony algorithm and genetic algorithm, the other is accurate algorithm represented by branch pricing and Lagrangian relaxation. Wang y et al. [11] proposed a heuristic algorithm based on improved ant colony algorithm and simulated annealing algorithm to solve the periodic vehicle routing problem with time window and service selection. Ursani Z et al. [12] used genetic algorithm and introduced local optimization framework (LOF) to solve VRPTW. For the vehicle routing problem (VRPTW-ST) with hard time window and random service time. ERICO f et al. [13] established the VRPTW-ST model by using a chance constrained program, and proposed a new set partition formula, which included a set of constraints on the minimum success probability of vehicle routing. By extending the label dimension and providing new advantage rules to adapt to the resource consumption of dynamic programming algorithm. The effectiveness of the method and model is proved by experiments. Kalle-hauge B et al. [14] aimed at vehicle routing problem with time windows, a constrained shortest path subproblem was generated by using Lagrangian relaxation of constraint set. A stable cutting plane algorithm for Lagrangian dual problem is presented. Because there are few negative cycles, the algorithm is more likely to produce constrained shortest path subproblems.

There have been relevant researches on workshop logistics, but the research combined with workshop production has not attracted the attention of relevant scholars. Compared with the traditional vehicle routing problem, the demand of customer points in workshop logistics distribution will change with the process of workshop production line. In this paper, for the vehicle routing problem with time window and lot size, the optimal delivery quantity in the bottleneck process is given firstly, and then the vehicle routing problem in the yard is planned according to the delivery quantity, and a tight lower bound is given by using Lagrangian relaxation algorithm.

2   The Vehicle Routing Problem with Time Window and Lot Size

2.1   Optimal Batch Model

Different from the traditional VRP Problem, the logistics distribution in the workshop pays more attention to the balance of logistics, in addition to considering the optimal transport volume, but also consider the time buffer. In order to quantify this factor, referring to the model of [15], the time buffer is transformed into inventory buffer. There are the following formulas for calculating the amount of transportation and processing.

$$Q_{bfp} = c_{bft}(Q_o - Bu)/c(t_{bfp} - t_{bft}) \tag{1}$$

$$Q_{bbp} = Q_o * \sqrt{(c_{bbtc} + c_{bsc} + c_{bbsc})/cQ_o(t_{bp} + t_{bbp})} \tag{2}$$

In the above formula, Q_{bfp} and Q_{bbp} are the processing batch before and after the bottleneck, that is, the transportation volume in the next stage. Q_o is the order volume, c_{bft} is the unit handling charge before the bottleneck, Bu is the buffer inventory, c is the inventory cost per unit product, t_{bfp} is the processing time per unit product before the bottleneck, t_{bft} is the unit transportation time per unit product before the bottleneck, c_{bbtc} is the unit handling charge after the bottleneck cost, c_{bsc} and c_{bbsc} bottleneck and equipment adjustment preparation cost after bottleneck, t_{bp} and t_{bbp} are the processing time of unit product after bottleneck and bottleneck.

2.2   The Construction of VRP Model

The goal of workshop logistics distribution and ordinary VRP is basically the same, that is, the shortest distance and the lowest cost. If the workshop station also has requirements for material distribution time, it is the VRPTW. We take minimization as the objective function and establish the corresponding model.

656   N.-K. Yu et al.

Table 1. Symbol description

	Symbol definition
Decision variables	$x_{ijk} = \begin{cases} 1 \text{ if customer } i \text{ is visited before customer } j \text{ by the vehicle } k \\ 0 \text{ otherwise} \end{cases}$
	$y_{ik} = \begin{cases} 1 \text{ if customer } i \text{ is visited by the vehicle } k \\ 0 \text{ otherwise} \end{cases}$
	t_i, The time the vehicle arrives at customer i
	w_i, The waiting time for the vehicle to arrive at customer i
Parameters	K, Total number of vehicles
	n, Total number of customers
	Q, Maximum vehicle capacity
	c_{ij}, Cost from customer i to customer j
	t_{ij}, Time from customer i to customer j
	d_i, Demand of customer i
	$[e_i, l_i]$, Time window of customer i
	e_i, The earliest time that customer i can be visited
	l_i, The latest time that customer i can be visited
	s_i, Service time of customer i

OVRPTW are defined in Table1, and its mixed integer programming model is described as follows:

$$\min \sum_{i=0}^{n}\sum_{j=0}^{n}\sum_{k=1}^{K} c_{ij}x_{ijk} \tag{3}$$

Subject to :

$$\sum_{k=1}^{K}\sum_{i=0}^{n} x_{ijk} = 1, \forall j = 1, 2, \ldots, n, i \neq j \tag{4}$$

$$\sum_{k=1}^{K}\sum_{j=0}^{n} x_{jik} = 1, \forall i = 1, 2, \ldots, n, i \neq j \tag{5}$$

$$\sum_{i=1}^{n} d_i y_{ik} \leq Q, \forall k = 1, 2, \ldots, K \tag{6}$$

$$\sum_{j=1}^{n} x_{0jk} = 1, \forall k = 1, 2, \ldots, K \tag{7}$$

$$\sum_{i=1}^{n} x_{i0k} = 1, \forall k = 1, 2, \ldots, K \tag{8}$$

$$\sum_{k=1}^{K}\sum_{i=0}^{n} x_{ijk}(t_i + t_{ij} + w_i + s_i) = t_j, \forall j = 1, 2, \ldots, n \tag{9}$$

$$e_i \leq t_i + w_i \leq l_i, \forall i = 1, 2, \ldots, n \tag{10}$$

$$x_{ijk} \in \{0, 1\}, \forall k = 1, 2, \ldots, K, \forall i = 1, 2, \ldots, n, \forall j = 1, 2, \ldots, n \tag{11}$$

$$y_{ik} \in \{0, 1\}, \forall i = 1, 2, \ldots, n, \forall k = 1, 2, \ldots, K \tag{12}$$

The optimization objective (3) is to minimize the distance the vehicle is travelling. Constrain (4–5) to ensure that each station has a transport vehicle to provide services. Constraint (6) is the vehicle capacity constraint. Constraints (7) and (8) indicate that all vehicles will leave the warehouse to serve the process. Constraints (9) and (10) represent the time window limit of the customer. Constraints (11) and (12) are used to indicate that the decision variable is a 0–1 integer variable. Our problem assumes that the distance between customers satisfies the triangle inequality. The above model is a mixed integer programming model, which is quite difficult to solve. However, by relaxing the values of variables in constraints (11)–(12) from 0–1 integer variables to 0–1 linear variables, we get the first lower bound model, that is, the lower bound of linear relaxation.

3 Solution Approach

3.1 Introduction of Lagrangian Relaxation Algorithm

For large-scale optimization algorithms such as vehicle routing problem, apart from heuristic search algorithm, model decomposition is another common strategy. The decomposition algorithm can calculate the optimal upper and lower bounds of the current objective value, while the Lagrangian relaxation algorithm is often used in the decomposition of discrete optimization models because of its relatively simple and effective operation, and the bounds obtained by Lagrangian relaxation are sometimes even tighter than those obtained by linear programming.

Linear relaxation is introduced in the previous section. The basic idea is to remove the integer constraints in mixed integer programming, while Lagrangian relaxation is to relax some constraints in the model, and keep integer constraints and other linear constraints. The relaxed constraint is to use Lagrangian multiplier to add constraints as penalty items to the objective function to punish the solutions that do not meet the constraints. In the Lagrangian relaxation algorithm, the subgradient method is mostly used for optimization. However, the convergence speed of this method is slow. The step size and direction of the algorithm in the iterative process will affect the performance of the algorithm. In the iteration of the sub gradient algorithm, the so-called 'zigzagging phenomenon' will appear, which means that the current value is very close to the value of the next iteration. So how to design the direction and step length of the algorithm is the key factor.

3.2 Improved Lagrangian Relaxation Model of VRPTW

In order to get a tighter lower bound, the Lagrangian relaxation is performed for the 2.2 model and the Lagrangian multiplier vector is updated by the sub gradient method. In

Sect. 2.2, some constraints are called 'complicated constraint' that couple all vehicle information together, which means that these constraints are difficult to solve. For this nonsmooth objective function, it is superior to use the subgradient algorithm to optimize the function. The complicated constraints (4), (5) in the model proposed in the previous section are combined. The Lagrangian relaxation model is obtained by adding multiplier λ, $\lambda = (\lambda_1, \lambda_2, ..., \lambda_m) \in R_+^m$ to the objective function.

$$D = \max_{\lambda \in R_+^m} z(\lambda) = \max(\min \sum_{i=0}^{n} \sum_{j=0}^{n} \sum_{k=1}^{K} c_{ij} x_{ijk} + \lambda^T (1 - \sum_{k=1}^{K} \sum_{i=0}^{n} x_{ijk}) \qquad (13)$$

Subject to (6)–(12).

The value D in (13) is a new lower bound of the problem presented in this paper.

The procedure of the subgradient algorithm is given as follows:

Step1: Let Lagrangian multiplier $\lambda = 0.1$, loop $= 1$.
Step2: Calculate the subgradient s^{loop}, If $s^{loop} = 0$ or s^{loop} less than the setting accuracy, the algorithm is terminated and the current solution is the optimal solution, otherwise go to step 3.
Step3: $\lambda^{loop+1} = \max\{\lambda^{loop} + \eta^{loop} s^{loop}, 0\}$, Where η^t denotes the step-size.
Step4: loop $=$ loop $+ 1$, go to step 2.

From the algorithm flow, it can be concluded that the subgradient algorithm is easy to implement, and each iteration can proceed in the direction of the subgradient. The difficulty is how to select the appropriate step-size η. This paper adopts the following way to update the step:

$$\eta^{loop} = \theta \frac{ub - d^{loop}}{\|\eta^{loop}\|}, 0 < \theta < 2 \qquad (14)$$

Taking the above subgradient calculation method, the algorithm has better convergence. Where ub is a feasible upper bound that can be corrected in the iteration, D^{loop} represents a lower bound of the original problem, θ is a parameter used to adjust the step size when the D^{loop} does not change in several steps.

4   Test and Comparisons

4.1   Numerical Experiment

In order to illustrate the effectiveness of the algorithm, this paper first gives a small example to illustrate the implementation process of the algorithm. The optimization problem is as follows: the left represents the original problem, and the right represents the Lagrange relaxation problem.

$$\min 2x_1 + 3x_2 \qquad\qquad D = \max z(\lambda) = \max(\min 2x_1 + 3x_2 + \lambda(4 - 5x_1 - 2x_2)$$
$$s.t.\ 5x_1 + 2x_2 \ge 4 \qquad\qquad s.t.\ 4x_1 + 4x_2 \ge 1$$
$$4x_1 + 4x_2 \ge 1 \qquad\qquad 0 \le x_1 \le 1, 0 \le x_2 \le 2, x \in \mathbb{Z}$$
$$0 \le x_1 \le 1, 0 \le x_2 \le 2, x \in \mathbb{Z}$$

The optimization process is described as follows: let $\lambda = 0.1$, the corresponding lower bound is 1.9, the solution is x $= (1, 0)$, and this solution satisfies the constraints of the original problem, so it can be brought in as the initial solution of the original problem, and the upper bound is 2. Then update the subgradient $4 - 5x_1 - 2x_2 = -1$, according to formula (14) update step $\eta = 0.1$, and the next multiplier is $\lambda = 0.0$. At this time, it is brought into the Lagrange relaxation problem again, and the lower bound is 2 and the upper bound coincides. At this time, the upper and lower bounds coincide, and the optimal target value of the original problem is 2, and the optimal solution is x $= (1, 0)$.

Through the above small examples, it can be seen that the Lagrange relaxation algorithm is easy to operate.

4.2 Optimal Batch Instance

In order to verify the effectiveness of the proposed algorithm, we take an automobile parts processing enterprise as an example. As there are many kinds of automobile assembly enterprises, this paper takes the production of the subframe needed by the ordinary car as an example to analyze. First, determine the transfer quantity required by each station. The production of subframe is mainly divided into three processes: welding (process 1), bending (process 2), polishing (process 3), and the three processes are completed by 10 stations respectively. The process 1 needs to be conducted for 40 min, process 2 needs 25 min, process 2 needs 30 min. The processing cost of each station is ¥35. The transportation cost from process 1 to process 2 is ¥9, and that of process 2 to process 3 is ¥12. The inventory cost of the work in process is ¥120.

Taking the order quantity $Q_o = 100$ as an example, the investigation shows that process 2 is the bottleneck process. From the formula given in Sect. 2.1, we can get the optimal $Q_{bfp} \approx 27, Q_{bbp} \approx 10$.

4.3 Lower Bound of VRPTW

According to the calculation formula in the previous section, we can get the optimal processing lot before and after the bottleneck, and allocate the transportation volume to the stations according to the processing lot. Among the seven stations, there are 3 stations for process 1 processing, three stations for process 4 processing and 3 stations for process 3 processing. The transfer batch is determined by the results given in the previous section. In order to simplify the calculation, we assume that the material transported is only the main material, that is, a subframe needs one main material. The transportation distance is given in the Table 2. Suppose there are 12 AGV vehicles available in the depot, the service time of each station is 10 min, and the time window [0–100] of the station is generated randomly. According to the formula in Sect. 3.2, two optimal lower bounds are calculated and compared. Our algorithm is coded in Python3.6 and tested on a PC with Intel Core 3.60 GHz and 16.0-GB memory, and the mixed linear programming problems MIP are solved by the solver GUROBI.

In Table 3, we divide the distribution of 10 stations into five groups to solve the distribution problem. The second and fourth columns are the lower bounds of linear relaxation and Lagrangian relaxation respectively, and the last column is the quantity bound. Compared with each other, where $Gap = \frac{LP - LR}{LR} * 100\%$, we can see that the bound obtained by LR is tighter than that obtained by LP.

Table 2. Distance between depot and stations

Num.	0	1	2	3	4	5	6	7	8	9	10
0	0	15	20	25	10	16	19	12	9	5	16
1	15	0	5	10	6	2	5	4	8	11	8
2	20	5	0	7	10	5	2	8	11	15	7
3	25	10	7	0	15	10	8	14	18	21	14
4	10	6	10	15	0	5	9	3	4	6	9
5	16	2	5	8	9	0	4	4	7	10	8
6	19	5	2	8	9	4	0	7	10	14	6
7	12	4	8	14	3	4	7	0	4	7	6
8	9	8	11	18	4	7	10	4	0	4	11
9	5	11	15	21	6	10	14	7	4	0	8
10	16	8	7	14	9	8	6	6	11	8	0

Table 3. Comparison of LP and LR lower bounds

LB	LP	Time(sec)	LR	Time(sec)	Gap(%)
3	44.74	1.3	41.41	1	8.0
5	45.40	9	44.53	15	1.9
7	48.15	10	48.13	25	0.04
9	49.92	12	49.03	40	1.8
10	55.36	25	55.09	61	0.5

5 Conclusions and Future Research

In this paper, for the production line with bottleneck process in production logistics, we first determine the optimal material transportation volume, then transform the distribution model into vehicle routing problem with time window, and propose two different lower bound models, including mixed integer programming model and Lagrangian relaxation model. The significance is that for a series of combinatorial optimization problems such as vehicle scheduling problem, it is very difficult to find the optimal solution because of its NP hard problem characteristics. Although some heuristic algorithms can get the optimal solution in a reasonable time, it is difficult to evaluate the quality of the solution. However, it is an effective method to get a tight lower bound to evaluate the quality of the solution through an accurate algorithm such as Lagrangian relaxation. Exact mathematical model can make up for the blindness of intelligent algorithm search mechanism. The effectiveness of the proposed lower bound model is verified by simulation experiments. The future research direction will start from improving the lower bound model, analyzing the validity of inequality and calculating the lower bound model of other combinatorial optimization problems.

Acknowledgments. This research is partially supported by the National Science Foundation of China (61963022), and National Science Foundation of China (51665025).

References

1. Zhenyun, P., Yi, G., Zhaolin, T.: Fundamentals and Applications of MES, 7. China Machine Press, Beijing (2021)
2. Zhihong, L., Hongchao, W., Chenhao, Z.: A distributed intelligent control system solution for discrete manufacturing industry. Ordnance Autom. (1), 7–10 (2017)
3. Dejun, Z., Xiucai, Z., Pengfei, C., et al.: Research on networked testing technology of electronic equipment production and testing line. Electron. Test. **435**(06), 66–68 (2020)
4. Chunting, X., Chunlin, Z., Chong, Z., et al.: Research and implementation of MES in discrete manufacturing workshop. Aviat. Manuf. Technol. **11**, 86–89 (2012)
5. Watson, K.J., Blackstone, J.H., Gardiner, S.C.: The evolution of a management philosophy: the theory of constraints. J. Oper. Manag. **25**(2), 387–402 (2007)
6. Simatupang, T.M., Wright, A.C., Sridharan, R.: Applying the theory of constraints to supply chain collaboration (2004)
7. Reimer, G.: Material requirements planning and theory of constraints: can they coexist? a case study. Prod. Invent. Manag. J. **32**(4), 48–52 (1991)
8. Cook, D.P.: Simulation comparison of traditional, JIT, and TOC manufacturing systems in a flow shop with bottlenecks (1994)
9. Dantzig, G.B., Ramser, J.H.: The truck dispatching problem. Manag. Sci. **6**(1), 80-91 (1959)
10. Savelsbergh, M.: Local search for routing problems with time windows. Ann. Oper. Res. **4**, 285–305 (1985)
11. Wang, Y., Wang, L., Chen, G., et al.: An improved ant colony optimization algorithm to the periodic vehicle routing problem with time window and service choice. Swarm Evol. Comput. **55**, 100675 (2020)
12. Ursani, Z., Essam, D., Cornforth, D., et al.: Localized genetic algorithm for vehicle routing problem with time windows. Appl. Soft Comput. **11**(8), 5375–5390 (2011)

13. Errico, F., Desaulniers, G., Gendreau, M., Rei, W., Rousseau, L.-M.: The vehicle routing problem with hard time windows and stochastic service times. EURO J. Transp. Logist. **7**(3), 223–251 (2016). https://doi.org/10.1007/s13676-016-0101-4
14. Kallehauge, B., Larsen, J., Madsen, O.: Lagrangian duality applied to the vehicle routing problem with tim windows. Comput. Oper. Res. **33**(5), 1464–1487 (2006)
15. Hongbo, W.: Research on Production Logistics Optimization of Structural Parts Manufacturing Workshop of North Communications Heavy Industry Group. Shenyang University of technology, Shenyang (2013)

Multidimensional Estimation of Distribution Algorithm for Distributed No-Wait Flow-Shop Scheduling Problem with Sequence-Independent Setup Times and Release Dates

Sen Zhang[1], Rong Hu[1,2(✉)], Bin Qian[1,2], Zi-Qi Zhang[1,2], and Ling Wang[3]

[1] School of Information Engineering and Automation, Kunming University of Science and Technology, Kunming 650500, China
[2] School of Mechanical and Electrical Engineering, Kunming University of Science and Technology, Kunming 650500, China
[3] Department of Automation, Tsinghua University, Beijing 10084, China

Abstract. This paper proposes a three-dimensional matrix Estimation of Distribution Algorithm (TDEDA) for distributed no-wait flow-shop scheduling problem (NFSSP) with sequence-independent setup times (SISTs) and release dates (RDs) to minimize the total completion time, which is a typical NP-hard combinatorial optimization problem with strong engineering background. First, a population is initialized in a hybrid way by modified NEH heuristic algorithm and the random method. Secondly, probabilistic model is developed to learn knowledge by accumulating the information of the blocks and the order of jobs from the elite individuals. Then, four search methods are developed to optimize the quality of solutions. Finally, computational results and comparisons demonstrate TDEDA obviously outperforms other considered optimization algorithms for addressing DNWFSP_SISTs_RTs.

Keywords: Estimation of distribution algorithm · Distributed no-wait flow-shop scheduling problem · Sequence-independent setup times · Release dates

1 Introduction

Flow-shop scheduling is the common scheduling problem in industrial production. It is to assume that the job can be waited in the buffer between different machines. However, in many actual productions such as steel, food manufacturing, biopharmaceuticals, and chemical processing industries [1, 2], often the job must be continuous throughout the processing process, that is, once the job is required to be processed continuously from start to end without waiting either on or between machines. This kind of scheduling problem is called the no-wait flow shop scheduling problem (NWFSP), In many real-life NWFSP, such as steel manufacturing sequence-independent setup times (SISTs) and release dates (RDs) are two very common constraints.

© Springer Nature Switzerland AG 2021
D.-S. Huang et al. (Eds.): ICIC 2021, LNCS 12836, pp. 663–672, 2021.
https://doi.org/10.1007/978-3-030-84522-3_54

In recent years, with the deepening of economic globalization, competition among enterprises has become increasingly fierce. In order to improve the production efficiency of enterprises, reduce production costs and manage risks [3], enterprise decision makers have begun to gradually transform traditional centralized single production factories into Multiple coordinated factories distributed in different areas [4]. Different from traditional single-factory production scheduling, distributed production scheduling requires decision makers to assign the job to the different factories. secondly, they need to determine a scheduling plan for each factory. Obviously, these two issues are coupled with each other and cannot be solve separately. For the distributed permutation flow-shop scheduling problem (DPFSP) Naderi and Ruiz [5] established 6 mixed integer programming models and proposed two job assignment rules, and then designed 14 heuristic algorithms to solve this problem. Gao et al. [6] designed a neighborhood effective heuristic method by inserting a set of job into a factory at a time instead of only one job. Gao et al. [7] adopted a tabu search (TS) algorithm to define the neighborhood structure by exchanging sub-sequences between factories, instead of inserting a single job into other position. Insert_Jobs, Exchange_Job and Move_Jobs are adopted to enhance local search ability. Wang et al. [8] proposed an estimation of distribution algorithm (EDA), and proposed ECF rules for decoding a feasible scheduling scheme for production. For NWFSP, a large number of intelligent algorithms have been applied to solve this problem, including differential evolution((HDE)) algorithm, discrete particle swam optimization (DPSO), genetic algorithm (GA), iterated greedy (IG) algorithm, etc. Regarding the approaches for solving the DNWFSP, an iterated cocktail greedy (ICG) was first proposed by Lin and Ying [9], which contains two self-tuning mechanisms and a cocktail destruction mechanism. Then, Komaki and MalaKooti [10] proposed a general variable neighborhood search algorithm (GVNS) to solve it. it can be seen that there is less research for DNWFSP with SISTs and RTs. Hence, this paper considers such important problem.

Estimation of distribution algorithm (EDA), first introduced in the field of evolutionary computation [11], is a prevailing population based evolutionary algorithm which explores the space of solutions by sampling an explicit probabilistic model updated according to the promising solutions found so far, Therefore, EDA has strong global search capabilities, and it has been widely used to solve production scheduling problems [12, 13], so we proposed a block-based three-dimensional EDA to solve DNWFSP with SISTs and RTs.

2   DNWFSP with SISTs and RTs

2.1   Problem Descriptions

The DNWFSP_SISTs_RT can be described as follows: There are n jobs $J = \{J_1, J_2, \cdots, J_n\}$ must be processed in f identical factory $\{F_1, F_2, \cdots, F_f\}$, each of which is a permutation flow shop with m machines $\{M_1, M_2, \cdots, M_m\}$. Each job $J_i, i = 1, 2, ..., n$ consists of a sequence of m operations $\{O_{i1}, O_{i2}, \cdots, O_{i,m}\}$, that are processed through m machines with same order. All jobs can be scheduled at time zero. At the first assignment stage, all jobs must all jobs must be assigned to a certain factory. Once a job is assigned to a certain factory, all its operations must be processed only in this factory. Then at the second stage, under the no-wait constraints, each job must be processed without interruptions between consecutive machines and each machine just can process np more than one job, it indicates that the start of a job must be delayed on the first machine when necessary. In the DNWFSP with SISTs, the setup times must be performed between the completion time of one job and the start time of another job on each machine, and the setup times depend on both the current and the immediately preceding jobs at each machine. Moreover, In the DNWFSP with RTs, if a machine is ready to process a job but the job is still not released, it must stay idle until the release time of the job satisfies the demands.

2.2 DNWFSP with SISTs

The notation used is presented below.

$\pi = \{\pi_1, \pi_2, \cdots, \pi_n\}$ A feasible schedule

$\pi^k = [\pi^k(1), \pi^k(2), \ldots, \pi^k(n^k)]$ The job processing sequence in factory k

n^k The total number of the jobs assigned to factory k

$t_{j,l}$ The processing time of job J_j on machine l $(t_{0,l}, l = 1, 2, \ldots m)$

st_j The total processing time of job J_j

$D_{\pi^k}(j - 1, j, l)$ The minimum delay on the machine l between the Completion of job J_{j-1} and J_j

$L_{\pi^k}(j - 1, j)$ The minimum delay on the first machine between the start of job J_{j-1} and J_j

$s_{j,l}$ The sequence independent setup time

r_j The release time of job J_j $(r_0 = 0, l = 1, 2, \ldots m)$

$St_{\pi^k(j)}$ The start processing time of job J_j on machine 1

$C_{\pi^k(j)}$ The Completion time of J_j job from factory k

In the DNWFSP with SISTs, $D_{\pi^k}(j - 1, j, l)$ and $L_{\pi^k}(j - 1, j)$ can be calculated as follows:

$$D_{\pi^k}(j - 1, j, l) = \begin{cases} \max\{s_{\pi^k(j),1} + t_{\pi^k(j),1} - t_{\pi^k(j-1),2}, s_{\pi^k(j),2}\} + t_{\pi^k(j),2}, l = 2 \\ \max\{D_{\pi^k}(j - 1, j, l - 1) - t_{\pi^k(j-1),l}, s_{\pi^k(j-1),l}\} + t_{\pi^k(j),l}, l = 3, 4, \ldots m \end{cases} \quad (1)$$

$$L_{\pi^k}(j - 1, j) = D_{\pi^k}(j - 1, j, m) + st_{\pi^k(j-1)} - st_{\pi^k(j)}. \quad (2)$$

2.3 DNWFSP with SISTs and RTs

In the NFSSP with SDSTs and RTs, $St_{\pi^k(j)}$ can be written as follows:

$$St_{\pi^k(j)} = \begin{cases} \max\{D_{\pi^k}(j - 1, j, l) - st_{\pi^k(j)}, r_{\pi^k(j)}\}, i = 1 \\ St_{\pi^k(j-1)} + \max\{L_{\pi^k}(j - 1, j), r_{\pi^k(j)} - St_{\pi^k(j-1)}\} i = 1, 2, \ldots, n \end{cases} \quad (3)$$

Then, the makespan $C_{\max}(\pi)$ of the sequence π is:

$$C_{\pi^k(j)} = St_{\pi^k(j)} + st_{\pi^k(j)} \tag{4}$$

$$C_{\max}\pi^k = \sum_{j=1}^{n^k} C_{\pi^k(j)} \tag{5}$$

$$C_{\max}(\pi) = \max\{C_{\max}(\pi^k)\}\, k = 1, 2, ..., f. \tag{6}$$

The aim of this paper is solving the DBPFSP with SISTs and RTs constraint and to find a feasible solution π^* with the minimum makespan (Fig. 1).

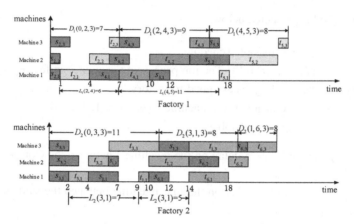

Fig. 1. Shows a small example of DNWFSP with SISTs when $n = 6$, $m = 3$ and $f = 2$.

3 TDEDA for DNWFSP with SISTs and RTs

3.1 Solution Representation

The NEH heuristic algorithm is the most common initialization method, combined with the characteristics of the DNWFSP_SISTs_RT, we design a modified NEH heuristic algorithm to initialize the population, it can be summarized as follows.

Step1: Calculate the total processing time of each job on all machine $T_j = \sum_{l=1}^{m} (t_{j,l} + s_{j,l})$ $j = 1, 2, ..., n$, and sort all jobs in descending order to T_j, denoted as $T = \{J_1, J_2, \cdots, J_n\}$.
Step2: Let $k = 1$, $\pi^k(1) = J_k$.
Step3: Let $k = k + 1$, if $k \leq f$, go to step 2, otherwise, go to step 4.
Step4: Let $k = f + 1$.
Step5: Insert the job in all possible position of all existing factories, place the job in the position of the current factory the minimum makespan.
Step6: Let $k = k + 1$, if $k \leq f$, go to step 2, otherwise, end.

3.2 Three-Dimensional Matrix Cube

For a feasible solution, the block structure refers to two consecutive job in the job permutation π. For example, two elite individuals $\pi_1 = [1, 2, 3, 4]$ and $\pi_2 = [2, 1, 3, 4]$, where $[1, 2], [2, 3], [3, 4], [2, 1], [1, 3], [3, 4]$ are block structures, both π_1 and π_2 have $[3, 4]$, so $[3, 4]$ is a similar block. A three-dimensional matrix $MC_{n \times n \times n}^{G}$ is devised to reserve the valuable information of elite individuals. The details of $MC_{n \times n \times n}^{G}$ are given as follows:

$$I_{MC_{n \times n \times n}^{G,k}(x,y,z)} = \begin{cases} 1, & y = \pi_{sbest,x}^{G,k}, z = \pi_{sbest,x+1}^{G,k} \\ 0, & else \end{cases} , x = 1, 2, ..., n-1; y, z = 1, 2, ..., n; k = 1, \dots, sps. \quad (7)$$

$$MC_{n \times n \times n}^{G}(x, y, z) = \sum_{k=1}^{sps} I_{MC_{n \times n \times n}^{G,k}(x,y,z)}, x = 1, 2, ..., n-1; y, z = 1, 2, ..., n, \quad (8)$$

$$MC_{n \times n \times n}^{G}(x, y) = [MC_{n \times n \times n}^{G}(x, y, 1), MC_{n \times n \times n}^{G}(x, y, 2), ..., MC_{n \times n \times n}^{G}(x, y, n)]_{1 \times n},$$
$$x = 1, 2, ..., n-1; y = 1, 2, ..., n. \quad (9)$$

$$MC_{n \times n \times n}^{G}(x) = \begin{bmatrix} MC_{n \times n \times n}^{G}(x1) \\ \vdots \\ MC_{n \times n \times n}^{G}(xn) \end{bmatrix}_{n \times 1} = \begin{bmatrix} MC_{n \times n \times n}^{G}(x, 1, 1) & \cdots & MC_{n \times n \times n}^{G}(x, 1, n) \\ \vdots & \ddots & \vdots \\ MC_{n \times n \times n}^{G}(x, n, 1) & \cdots & MC_{n \times n \times n}^{G}(x, n, n) \end{bmatrix}_{n \times n} . \quad (10)$$

Where, $S\_pop(G) = \{\pi_{sbest}^{G,1}, \pi_{sbest}^{G,2}, ...\pi_{sbest}^{G,sps}\}$ are the superior sub-population or elite individuals selected from $pop(G) = \{\pi_{sbest}^{G,1}, ..., \pi_{sbest}^{G,sps}, \pi_{sbest}^{G,sps+1}, ..., \pi_{sbest}^{G,sp}\}$. $\pi_{sbest}^{G,k} = [\pi_{sbest,1}^{G,k}, \pi_{sbest,2}^{G,k}, ..., \pi_{sbest,n}^{G,k}]$ is the kth individual in the superior sub-population $S\_pop(G)$. $I_{MC_{n \times n \times n}^{G,k}(x,y,z)}$ is an indicator function of the kth elite individual in the superior sub-population, $MC_{n \times n \times n}^{G}(x, y, z)$ saves the number of times that the similar block structure appears in the same position of the elite individual, $MC_{n \times n \times n}^{G}(x)$ reserve all the ordinal numbers of jobs and the corresponding blocks at the xth positions of individuals in $S\_pop(G)$. Therefore, $MC_{n \times n \times n}^{G}$ reserves the valuable information of excellent individuals at generation G, it can be learned and reserved in an intuitive and effective way.

3.3 Updating Mechanism

Probability model determines the global exploitation ability and the search direction of EDA. A three-dimensional probability matrix $P\_MC_{n \times n \times n}^{G}$ are proposed, which accumulate the valuable information from $MC_{n \times n \times n}^{G}$. therefore, $P\_MC_{n \times n \times n}^{G}$ is described as follows:

$$P\_MC_{n \times n \times n}^{G}(x) = \begin{bmatrix} P\_MC_{n \times n \times n}^{G}(x, 1, 1) & \cdots & P\_MC_{n \times n \times n}^{G}(x, 1, n) \\ \vdots & \ddots & \vdots \\ P\_MC_{n \times n \times n}^{G}(x, n, 1) & \cdots & P\_MC_{n \times n \times n}^{G}(x, n, n) \end{bmatrix}_{n \times n} \quad (11)$$

When $G = 0$, $P\_MC_{n \times n \times n}^{G}(x)$ is updated by

$$P\_M_{n \times n \times n}^{0}(x, y, z) = \begin{cases} 0, & x = 1, y, z = 1, ..., n \\ 1/n^2, & x = 2, 3, \cdots, n-1, y, z = 1, ..., n \end{cases} . \quad (12)$$

When $G = 1$, $P\_MC^G_{n \times n \times n}(x)$ is updated by

$$P\_MC^1_{n \times n \times n}(x, y, z) = \begin{cases} MC^0_{n \times n \times n}(x, y, z)/Sum\_MC^0(x), & x = 1, \ y, z = 1, ..., n, \\ (P\_MC^0_{n \times n \times n}(x, y, z) + MC^0_{n \times n \times n}(x, y, z))/ \\ (Sum\_P\_MC^0(x) + Sum\_MC^0(x)), \\ x = 2, 3, \cdots, n - 1, \ y, z = 1, ..., n. \end{cases}$$

$$(13)$$

When $G = 2$, $P\_MC^G_{n \times n \times n}(x)$ is updated by

$$P\_MC^G_{n \times n \times n}(x, y, z) = (1 - r) \times P\_MC^G_{n \times n \times n}(x, y, z) + r \times MC^G_{n \times n \times n}(x, y, z)/$$
$$Sum\_MC^G(x), \ x = 1, 2, \cdots, n - 1, \ y, z = 1, ..., n.$$

$$(14)$$

Where, $Sum\_MC^G(x)$ is the summation of all elements in $MC^G_{n \times n \times n}(x)$, $Sum\_MC^G(x) = \sum_{y=1}^{n} \sum_{z=1}^{n} MC^G_{n \times n \times n}(x, y, z)$, $Sum\_P\_MC^G(x)$ is the summation of all elements in $P\_MC^G_{n \times n \times n}(x)$, $Sum\_P\_MC^G(x) = \sum_{y=1}^{n} \sum_{z=1}^{n} P\_MC^G_{n \times n \times n}(x, y, z)$.

3.4 New Population Generation

In the process of population evolution, sample to generate offspring from $P\_MC^G_{n \times n \times n}$. Because the individual $\pi^{G,k}_{sbest}$ does not exist $[\pi^{G,k}_0, \pi^{G,k}_1]$, we design a specially sampling method to determine the position jobs, it details as follows:

Step1: $sum\_P\_MC(y) = \begin{cases} \sum_{y=1}^{sp} i/sp, & G = 1, y = 1, ..., n \\ \sum_{z=1}^{n} P\_MC^{G-1}(1, y, z), & G \neq 1, y = 1, ..., n \end{cases}$.

Step2: Randomly generate a probability p_r where $p_r \in [0, sum\_P\_MC(n))$.

Step3: If $p_r \in [0, sum\_P\_MC(1))$, then set $\pi^{G,k}_1 = 1$, otherwise, go to **step 4**.

Step4: If $p_r \in \left[\sum_{h=1}^{t} P\_MC^{G-1}_{n \times n \times n}(j - 1, \pi^{G,k}_{i-1}, h), \sum_{h=1}^{t+1} P\_MC^{G-1}_{n \times n \times n}(j - 1, \pi^{G,k}_{i-1}, h)\right)$ then set $\pi^{G,k}_1 = j$.

The job at the other position of the individual according to the following:

Step1: Randomly generate a probability p_r where $p_r \in \left[0, \sum_{h=1}^{n} P\_MC^{G-1}_{n \times n \times n}(i - 1, \pi^{G,k}_{i-1}, h)\right)$.

Step2: If $p_r \in \left[0, P\_MC^{G-1}_{n \times n \times n}(i - 1, \pi^{G,k}_{i-1}, 1)\right)$, then set $\pi^{G,k}_i = 1$, otherwise, go to step 4.

Step3: If $p_r \in \left[\sum_{h=1}^{t} P\_MC^{G-1}_{n \times n \times n}(j - 1, \pi^{G,k}_{i-1}, h), \sum_{h=1}^{t+1} P\_MC^{G-1}_{n \times n \times n}(j - 1, \pi^{G,k}_{i-1}, h)\right)$, then set $\pi^{G,k}_i = t + 1$.

3.5 Local Search

It is widely recognized that local search is helpful to intensify the exploitation ability of EDA. For DNWFSP_SISTs_RT, the completion time of the entire production process is determined by the factory with the latest makespan (denoted as the critical factory). Obviously, decreasing the makespan of the critical factories can changing the makespan of a solution. Therefore, we design local search method for our EDA. The detail of local search methods can be described as follows.

1) Swap move within factory: swap the Randomly select two distinct job $\pi^k(i)$ and $\pi^k(k)$ from the critical factory f_k and then swap them.
2) Insert move within factory: Randomly select a job $\pi^k(i)$ from the critical factory f_k, and removing $\pi^k(i)$,and then re-inserting it into all the possible positions and choose the job permutation vector with minimum completion time.
3) Swap move between factories: Randomly select a job $\pi^k(i)$ from the critical factory f_k. Randomly select a factory $F_j(j \neq k)$ from the other $f - 1$ factories. Swap the job $\pi^k(i)$ with each job in factory F_j. The position leading to the best makespan for that factory is chosen.
4) Insert move between factories: Randomly select a job $\pi^k(i)$ from the critical factory f_k. Randomly select a factory $F_j(j \neq k)$ from the other $f - 1$ factories. The job $\pi^k(i)$ is took out and re-inserted into all possible positions of the sequence of factory F_j.The position leading to the best makespan for that factory is chosen.

4 Computational Result and Comparisons

To further show the effectiveness of the proposed TDEDA, we compare the TDEDA with the HIA [14], EDA [15] and IG [16]. For this problem, 16 problems size each with the following size ($n \in \{10, 30, 50, 70\}$, $m \in \{5, 20\}$,$f \in \{2, 4\}$)are randomly generated, and the related data are generated from the discrete uniform distributions as follows: $t_{j,l} \in U(1, 100)$, $st_j \in U(0, 100)$, $r_j \in U(0, 150)$,The parameters are:*popsize* $= 100$, *sps* $= 0.1 \times popsize$, $r = 0.3$. We coded all the algorithms in Delphi XE8 and conducted experiments on a computer of CPU, 2.6 GHz with 16 GB memory. To make a fair comparison, all algorithms adopted the same time, the termination conditions are the same (Fig. 2).

In order to ensure the stability and credibility of algorithm performance, all the tests independently run 30 times. The performance metrics are given as follows:

$$SD(C_i) = \sqrt{\frac{1}{20} \sum_{i=1}^{20} (C_i - C_{avg})^2}, \tag{15}$$

Where C_i is the makespan generated in the ith replication by an algorithm, the C_{avg} is the average value of C_i, $i = 1, 2, ..., 20$. In addition, to make the experimental results clearer, the best values in each row are represented by using the bold.

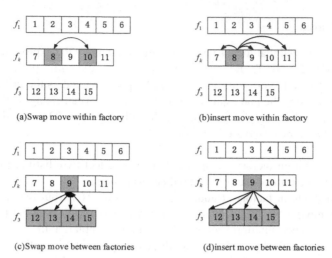

Fig. 2. Swap move within factory, insert move within factory, swap move between factories, insert move between factories

From Table 1, Only when $n = 10, m = 20, f = 2$ and $n = 10, m = 20, f = 4$ the smallest BST and AVG value appears in the EDA and HIA, and other optimal results appear in TDEDA. Therefore, we can draw the conclusion than TDEDA can achieve better performance than HIA, EDA and IG in the different scale problems. The reason can be attributed to two points (1) Three-dimensional matrix and the updating mechanism are helpful to explore the search space. (2) With the local search, it improves the good solutions by enhancing the exploitation capability.

Table 1. Comparison of AVE, BST and SD of HIA, EDA and IG

Instance	HIA			EDA			IG			TDEDA		
n,m,f	AVG	BST	SD	AVG	BST	SD	AVG	BST	SD	AVG	BST	SD
10,5,2	805	**793**	5.98	854	820	19.19	802	796	18.54	**796**	795	**0.79**
10,5,4	536	533	1.66	548	529	13.14	535	533	1.67	**533**	**533**	0
10,20,2	1783	**1759**	12.16	1829	1793	19.81	1778	1769	5.7	**1768**	1768	0
10,20,4	**1395**	**1385**	5.88	1408	1393	10.46	1403	1397	3.44	1396	1396	0
30,5,2	2155	2130	17.71	2399	2335	51.53	2119	2092	16.26	**1960**	1948	8.53
30,5,4	1237	1188	27.32	1488	1449	24.44	1202	1167	20.3	**1143**	1131	7.36
30,20,2	3701	3550	105.23	4226	4120	55.18	3715	3557	94.5	**3382**	3349	20.23
30,20,4	2344	2279	38.52	2645	2596	24.58	2325	2247	43.17	**2210**	2196	8.03
50,5,2	3624	3573	30.03	4264	4168	56.95	3419	3372	30	**3202**	3178	14.09
50,5,4	2045	1985	37.06	2536	2475	40.51	1989	1923	34.97	**1825**	1813	6.69
50,20,2	5719	5637	49.27	6479	6361	61.32	5414	5342	52.34	**5117**	5089	17.42
50,20,4	3376	3333	30.59	3873	3794	58.42	3224	3185	24.13	**3044**	3028	8.01

(continued)

Table 1. (*continued*)

Instance	HIA			EDA			IG			TDEDA		
n,m,f	AVG	BST	SD	AVG	BST	SD	AVG	BST	SD	AVG	BST	SD
70,5,2	5005	4939	54.78	6101	5981	73.23	4711	4631	56.5	**4327**	**4299**	**19.98**
70,5,4	2830	2747	58.68	3727	3624	64.29	2755	2663	48.09	**2519**	**2510**	**6.05**
70,20,2	7825	7717	65.89	9057	8859	129.66	7542	7427	69.15	**6903**	**6871**	**16.12**
70,20,4	4492	4424	39.65	5419	5295	70.07	4292	4227	48.1	**4001**	**3981**	**13.18**

5   Conclusions and Future Research

This paper presented a three-dimensional matrix Estimation of Distribution Algorithm (TDEDA) for distributed no-wait flow-shop scheduling problem (NFSSP) with sequence-independent setup times (SISTs) and release dates (RDs). First, a population initialization method based on heuristics and randomness is designed to ensure the quality of the initial solution; Secondly, three-dimensional matrix and the updating mechanism are utilized to accurately record the corresponding blocks with their exact positions or ordinal numbers in the excellent individuals; then, a local search based on the characteristics of the problem is designed to improve local optimization of the high-quality regions found by the global search, Finally, the effectiveness of TDEDA have been verified by simulation experiments and algorithm comparisons on standard test problems of different scales. For future research, the proposed algorithm can be extended to study for other kinds of scheduling, such as low-carbon scheduling.

Acknowledgements. This research is partially supported by the National Science Foundation of China (61963022) and National Science Foundation of China (51665025).

References

1. Allahverdi, A.: A survey of scheduling problems with no-wait in process. Eur. J. Oper. Res. **255**, 665–686 (2016)
2. Qian, B., Wang, L., Huang, D.X., Wang, W.L., Wang, X.: An effective hybrid DE-based algorithm for multi-objective flow shop scheduling with limited buffers. Comput. Oper. Res. **36**(1), 209–233 (2009)
3. Kahn, K.B., Castellion, G., Griffin, A.: The PDMA Handbook of New Product Development, 2 ed, pp. 65–79. John Wiley & Sons, Hoboken (2005)
4. Wang, L., Shen, W.: Process planning and scheduling for distributed manufacturing. Springer **47**(4), 1151–1152 (2007)
5. Naderi, B., Ruiz, R.: The distributed permutation flow shop scheduling problem. Comput. Oper. Res. **37**(4), 754–768 (2010)
6. Gao, J., Chen, R.: An NEH-based heuristic algorithm for distributed permutation flow shop scheduling problems. Sci. Res. Essays **6**(14), 3094–3100 (2011)

7. Gao, J., Chen, R., Deng, W.: An efficient tabu search algorithm for the distributed permutation flow-shop scheduling problem. Int. J. Prod. Res. **51**, 1–11 (2013)
8. Wang, S.Y., Wang, L., Min, L.: An effective estimation of distribution algorithm for solving the distributed permutation flow-shop scheduling problem. Int. J. Prod. Econ. **145**, 387–396 (2013)
9. Lin, S.W., Ying, K.C.: Minimizing makespan for solving the distributed no-wait flow shop scheduling problem. Comput. Ind. Eng. **99**, 202–209 (2016)
10. Komaki, M., Malakooti, B.: General variable neighborhood search algorithm to minimize makespan of the distributed no-wait flow shop scheduling problem. Prod. Eng. Res. Devel. **11**(3), 315–329 (2017). https://doi.org/10.1007/s11740-017-0716-9
11. Mühlenbein, H., Paaß, G.: From recombination of genes to the estimation of distributions I. Binary parameters. In: Voigt, Hans-Michael., Ebeling, Werner, Rechenberg, Ingo, Schwefel, Hans-Paul. (eds.) PPSN 1996. LNCS, vol. 1141, pp. 178–187. Springer, Heidelberg (1996). https://doi.org/10.1007/3-540-61723-X_982
12. Jarboui, B., Eddaly, M., Siarry, P.: An estimation of distribution algorithm for minimizing the total flowtime in permutation flow shop scheduling problems. Comput. Oper. Res. **36**(9), 2638–2646 (2009)
13. Pan, Q.K., Ruiz, R.: An estimation of distribution algorithm for lot-streaming flow shop problems with setup times. Omega **40**(2), 166–180 (2012)
14. Xu, Y., Wang, L., Wang, S.: An effective hybrid immune algorithm for solving the distributed permutation flow-shop scheduling problem. Eng. Optim. **46**(9), 1269–1283 (2014)
15. Wang, S.Y., Wang, L., Liu, M.: An effective estimation of distribution algorithm for solving the distributed permutation flow-shop scheduling problem. Int. J. Prod. Econ. **145**(1), 387–396 (2013)
16. Ruiz, R., Stützle, T.: An Iterated Greedy heuristic for the sequence dependent setup times flow shop problem with makespan and weighted tardiness objectives. Eur. J. Oper. Res. **187**(3), 1143–1159 (2008)

Hybrid Whale Optimization Algorithm for Solving Green Open Vehicle Routing Problem with Time Windows

Wen Jiang[1], Rong Hu[1($\boxtimes$)], Bin Qian[1], Nai-Kang Yu[1], and Bo Liu[2]

[1] School of Information Engineering and Automation, Kunming University of Science and Technology, Kunming 650500, China

[2] Academy of Mathematics and Systems Sciences, Chinese Academy of Sciences, Beijing 10084, China

Abstract. Based on the perspective of energy efficiency and minimizing fuel consumption as the optimization goal, this paper establishes a Green Open Vehicle Routing Problem (GOVRP) model, and designs a Hybrid Whale Optimization Algorithm (HWOA). To solve. HWOA integrates the specific operations in the mechanism on the premise that the individual update mechanism in the standard WOA remains unchanged, so that it can quickly perform the search operation of the solution space. A local search operation based on 4 kinds of neighborhood operations is designed to perform a more detailed search in the high-quality solution domain. Through the simulation experiments and algorithm comparison of several test cases with different customer sizes, it is verified that HWOA can effectively solve GOVRP.

Keywords: Open vehicle routing problem · Green · Whale optimization algorithm · Variable domain search

1 Introduction

The traditional vehicle routing problem (VRP) was first proposed by Dantzig and Ramser [1]. The VRP satisfying vehicle load, volume and customer service requirements, etc. The vehicle routing is reasonably arranged, so that the total mileage or total cost is the lowest. With the rapid development of sharing economy, modern logistics enterprises have developed rapidly, and the demand for third-party logistics distribution has increased significantly. Today's environmental protection requirements are becoming more and more stringent [2]. In this context, the study of the Green Open Vehicle Routing Problem (GOVRP) has very important practical significance. In addition, in terms of computational complexity, VRP is an NP-hard problem, and VRP is belonged to GOVRP, so GOVRP is also an NP-hard problem. Therefore, studying the modeling of GOVRP and its solving algorithm has important theoretical value and practical significance. Compared with traditional logistics scheduling, the research of green logistics scheduling mainly achieve the purpose of energy saving and emission reduction [3, 4].

© Springer Nature Switzerland AG 2021
D.-S. Huang et al. (Eds.): ICIC 2021, LNCS 12836, pp. 673–683, 2021.
https://doi.org/10.1007/978-3-030-84522-3_55

Therefore, the study of green logistics scheduling problem has more practical significance. Aiming at the open vehicle routing problem with time windows (GOVRPTW), Niu et al. [5] established the mathematical model of GOVRPTW based on the integrated emission model with cost minimization as the optimization objective, and designed a hybrid tabu search algorithm to solve it. Araghi et al. [6] aimed at the green stochastic open location routing problem (GSOLRP), constructed the GSOLRP model with the objective of minimizing CO_2 emissions, and designed a hybrid algorithm combining imperialist competition algorithm and variable neighborhood search to solve it. Li et al. [7] studied the vehicle routing problem with time window with the goal of minimizing total fuel consumption, and established a mathematical programming model based on fuel consumption, than, proposed a random variable neighborhood tabu search algorithm to solve it. Soto et al. [8] address the Multi-Depot Open Vehicle Routing Problem (MDOVRP), A multi neighborhood search method based on tabu search strategy is proposed to solve the problem, which proves the effectiveness of the strategy and determines a unified view of ejection chain, so that it can deal with multiple cells simply. Mirhassani et al. [9] presents a real-value version of particle swarm optimization (PSO) for solving the open vehicle routing problem (OVRP) that is a well-known combinatorial optimization problem. Gu et al. [10] Aiming at the vehicle routing problem under multi-center collaborative distribution, a total cost minimization model is established. The model built meets the characteristics of multi-center, multi-demand point and semi-open. Considering the complexity of the problem, a three-stage algorithm based on ant colony algorithm and K-mediods is designed to solve the problem. Fan et al. [11] Aiming at the logistics distribution of fresh products, a semi-open multi-distribution center joint distribution model is proposed. Considering the timeliness requirements of fresh product transportation, the corresponding time window and penalty cost are designed, and the optimization model with the minimum sum of vehicle transportation cost, dispatch cost, time penalty cost and fresh food loss cost is constructed, and the ant is designed. The group algorithm solves it.

The ranking model can be established for GOVRP, in which the service order and constraints of vehicles are implicit, and the corresponding algorithm is intelligent algorithm [12–15]. Using certain mechanisms that simulate the natural world to guide the algorithm search and the intelligent algorithm does not require in-depth analysis of the mathematical analytical formula of the problem model. Algorithms can often obtain satisfactory solutions or approximate optimal solutions to problems in a short time. In recent years, intelligent algorithms have been widely studied and applied due to their simplicity and superiority.

Whale optimization algorithm (WOA) is a new group intelligent optimization algorithm proposed by seyedali et al. [16]. WOA originates from the simulation of hunting behavior of humpback whales in nature, and realizes the purpose of searching for high-quality solution through the process of searching, encircling and attacking. WOA has been widely used in many fields [17–20] due to its advantages of simple operation, few parameters, easy understanding and strong optimization performance. Literature research shows that the current application of WOA mainly focuses on solving continuous optimization problems. For discrete optimization problems, real number solutions need to be discretized. There is no research on using WOA to solve vehicle routing problems. For GOVRP, due to its complex model and huge solution space, this paper designs a hybrid whale optimization algorithm (HWOA) to solve. In this paper, the GOVRP model with the lowest fuel consumption as the optimization objective is established, and HWOA is designed to solve it. The remaining parts are arranged as follows: Sect. 2 briefly introduces GOVRP and its mathematical model, Sect. 3 introduces the design of the algorithm for solving GOVRP, Sect. 4 analyzes the results of simulation experiments, Sect. 5 summarizes and summarizes the full text.

2 GOVRP

2.1 Problem Description

GOVRP can be described as: in the directed network graph $G = (V, E)$, $V = \{0, 1, 2, \cdots, n\}$ Represents the vertex set, vertex 0 represents the central warehouse, $V' = V \setminus \{0\}$ Represents customer set. $E = \{(i, j)|i, j \in V, i \neq j\}$ Represents the edge set. As shown in Fig. 1, In the central warehouse, there are homogeneous vehicles to provide transportation and distribution services, Q_k is the maximum load of each vehicle. q_i is The demand of customer i. GOVRP meets the following constraints:

1. The vehicle departs from the unified distribution center warehouse and does not need to return to the central warehouse after the service all customers;
2. Every customer must be served and only served once;
3. The total demand of each vehicle service for all customers does not exceed the maximum load capacity of the vehicle;
4. The number of vehicles used does not exceed the maximum number of vehicles in the central warehouse.

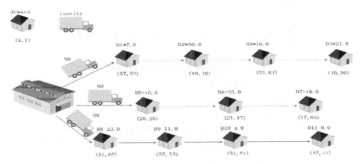

$\pi \sim 0, 1, 2, 3, 4, 0, 5, 6, 7, 0, 8, 9, 10, 11$
vehicle1:0, 1, 2, 3, 4
vehicle2:0, 5, 6, 7
vehicle3:0, 8, 9, 10, 11

Fig. 1. Vehicle routing

2.2 Energy Consumption Model

This paper considers the impact of vehicle speed, load, road conditions, etc. and gives a vehicle unit fuel travel distance model [30]. The driving distance of the vehicle from the customer i to the customer j per unit of fuel F_{ij} is shown in formula (1):

$$F_{ij} = a_0 + a_1 \cdot v P_{ij} Y_{ij} \tag{1}$$

The Y_{ij} is the gradient coefficient between customer i and customer j. In this article, it is assumed that the road inclination angle is 0, so set Y_{ij} to 1, and v is 60 km/h. P_{ij} is the load factor, as shown in formula (2):

$$P_{ij} = b_0 + \frac{b_1 \cdot L_{ij}}{b_0} + b_1 \cdot L_v \tag{2}$$

Where L_{ij} represents the load capacity of the vehicle from customer i to customer j, L_v is the average load capacity of the vehicle under long-term operation, and other fuel consumption parameters are constant.

2.3 Mathematical Model of GOVRP

The mathematical model of 2L-GOVRP as follows:

$$\min f = \sum_{k=1}^{K} \sum_{i=0}^{N} \sum_{j=0}^{N} \frac{d_{ij} x_{ijk}}{F_{ij}} \tag{3}$$

$$\sum_{i=1}^{N} q_i y_{ik} \le Q_k \, k \in \{0, 1, \ldots, K\} \tag{4}$$

$$\sum_{k=1}^{K} y_{ik} = 1 \, \forall i \in V' \tag{5}$$

$$\sum_{i=1}^{N} x_{ijk} = y_{ik} \, \forall j \in V', k \in \{0, 1, \ldots, K\} \tag{6}$$

$$\sum_{j=1}^{N} x_{ijk} = y_{ik} \, \forall i \in V', k \in \{0, 1, \ldots, K\} \tag{7}$$

$$x_{ijk} = \begin{cases} 1 & \text{vehicle k from customer i to j} \\ 0 & \text{other} \end{cases} \tag{8}$$
$$i, j \in v', k \in \{0, 1, \ldots, K\}$$

$$y_{ik} = \begin{cases} 1 & \text{vehicle k service customer j} \\ 0 & \text{other} \end{cases} \tag{9}$$
$$i \in v', k \in \{0, 1, \ldots, K\}$$

In the above model, formula (3) gives the objective function of GOVRP (total fuel consumption); formula (4) guarantees that the load of each vehicle does not exceed its maximum load; formula (5) guarantees that each customer Can be served; Eqs. (6) and (7) ensure that each customer is served only once; Eqs. (8) and (9) represent the value range of decision variables.

3   Algorithm Design

This section will introduce the population initialization, individual location update and local search of HWOA, and finally the overall flow of the algorithm is given.

3.1   Population Initialization

Set the size of the whale population is *popsize*, search space is n-dimensional. $\pi_i^{gen} = [\pi_i^{gen}[1], \pi_i^{gen}[2], \ldots, \pi_i^{gen}[g], \ldots, \pi_i^{gen}[S]](\pi_i^{gen}[g] \in V, n + 1 \leq S \leq 2n)$ is the first individual in the first generation population (a distribution plan). $P_i^{gen} = [P_i^{gen}[1], P_i^{gen}[2], \ldots, P_i^{gen}[g], \ldots P_i^{gen}[n]](P_i^{gen}[g] \in V')$ is the discrete sorting based on customers in the corresponding individual (the sequence not included the central warehouse) As shown in Fig. 1, three vehicles in the central warehouse are used to serve 11 customers (n = 11), the routing of vehicle 1 is *route*[1] = 0, 1, 2, 3, 4. The routing means that vehicle 1 starts from the central warehouse and serves customer 1, customer 2, customer 3 and customer 4; Similarly, the routing of vehicle 2 is *route*[2] = 0, 5, 6, 7 the routing served by vehicle 3 is *route*[3] = 0, 8, 9, 10, 11, which get a complete service routing $\pi = 0, 1, 2, 3, 4, 0, 5, 6, 70, 8, 9, 10, 11$(Feasible solution). $X_i^{gen} = [X_i^{gen}[1], X_i^{gen}[2], \ldots, X_i^{gen}[g], \ldots X_i^{gen}[n]](X_i^{gen}[g] \in [a, b])$ Represents the real number sequence corresponding to the customer's discrete sort P_i^{gen}.

$$X_j^{gen} = \begin{cases} rand\left(\overline{X}, X_i^{gen}\right) X_i^n > \overline{X} \\ rand\left(X_i^{gen}, \overline{X}\right) X_i^n \leq \overline{X} \end{cases} \quad \overline{X} = \frac{a+b}{2} \tag{10}$$

The quasi-opposition strategy refers to finding the reverse solution $X_j^{gen} = [X_j^{gen}[1], X_j^{gen}[2], \ldots, X_j^{gen}[g], \ldots, X_j^{gen}[n]]$ $(X_j^{gen}[g] \in [a, b])$ of the opposite of the individual X_i^{gen}(quasi-opposition solution) in the real number field [a,b] through the calculation method of formula (10). Therefore, when a population is randomly generated, its quasi-oppositional solution (another population) can be obtained. calculating the two populations fitness value and select the first popsize individual as the initial population. However, the position vector of the individual is a sequence of real numbers in the continuous domain, which cannot be directly used for the vehicle routing problem in the discrete solution space. In this paper, the LOV(Largest-order-value) rule is used to successfully transform the individual real number position information (real number sequence) X_i^{gen} into a discrete customer sequence P_i^{gen}. The specific steps of the LOV rule are as follows:

Step1: The position information $X_i^{gen} = [X_i^{gen}[1], X_i^{gen}[2], \ldots, X_i^{gen}[g], \ldots X_i^{gen}[n]]$ is arranged in descending order as $X_i^{gen\prime}$, and then the middle sequence of the corresponding sequence number is φ_i^{gen};

Step2: The Discrete customer sequence is calculated by $P_i^{gen}[\varphi_i^{gen}[g]] = g$.

By calling the LOV rules, the individual location information can be discretized into the customer's service sequence, realizing the transformation from a real number vector to a discrete vector. If the customer service order changes, the individual location information can also be adjusted by RLOV (Reverse largest-order-value) rule.

3.2 The HWOA for GOVRP

Seyedali et al. [16] designed the Whale Optimization Algorithm (WOA) by learn-ing the foraging behavior of whale colony. WOA mainly includes three parts: surround search mechanism, spiral search mechanism and random search mechanism. Although WOA can quickly search for high-quality solution regions, it has the disadvantages of low solution accuracy, slow convergence speed and easy to fall into local optimality. In order to overcome these problems, this paper designs HWOA that adaptively adjusts the search strategy and integrates the search mechanism in the whale algorithm. HWOA consists of two parts: the early stage and the later stage. $A = 2 \cdot \left(1 - \frac{gen}{T_{\max}}\right) \cos\left(\frac{\pi}{2} \cdot \frac{gen}{T_{\max}}\right)$, $T_{\max}$ is the maximum number of iterations, and gen is the current number of iterations. In the early search process ($A > 1$), it is mainly to search for high-quality solutions in the huge population solution space; in the later search process ($A \leq 1$), the detailed search is mainly carried out around the optimal solution area obtained in the early period. This search strategy can effectively balance the relationship between global search and local search of HWOA. In addition, in order to prevent the algorithm from falling into the local optimum, in the early search process, the individual chooses the update method of the early search according to the update factor Q and the threshold w_q, as shown in Eqs. (12) and (13). The definition of the update factor is as shown in Eq. (11):

$$Q = \frac{f(\pi_i^{gen}) - f(\pi^{gen*})}{f_{\max}(\pi^{gen}) - f(\pi^{gen*})} \tag{11}$$

Where $f(\pi^{gen*})$ is the fitness value of the best individual in the current population, $f_{\max}(\pi^{gen})$ is the fitness value of the worst individual in the population, $f(\pi_i^{gen})$ is the fitness value of the current individual i. In the later search, if the fitness value of the updated individual π_i^{gen+1} is better than the optimal individual, the new individual π^{gen*} will be taken as the optimal solution. The mathematical model of HWOA is as Eqs. (12)–(14):

$$X_i^{gen+1} = X^{gen'} - A \cdot \left| C \cdot X^{gen'} - X_i^{gen} \right| \quad Q \geq w\_q \tag{12}$$

$$X_i^{gen+1} = X^{gen'} + A \cdot r_1(b-a) \quad Q < w\_q \tag{13}$$

$$X_i^{gen+1} = X_i^{gen} + A\left(X^{gen*} - X_i^{gen}\right) \tag{14}$$

In the early stage, HWOA can search for high-quality solutions on a global scale to ensure the diversity of solutions. In the later stage, the enveloping in-depth search is carried out for the high-quality solution areas obtained in the early stage to further improve the search ability of the algorithm.

3.3 Local Search

In this paper, four variable neighborhood structures (Insert operation, Swap operation, Inverse operation, and Interchange operation) are used to construct a variable neighbor-hood search (Variable Neighborhood Search, VNS) operation, which can carry out local search more carefully and achieve the purpose of updating the current optimal solution.

Insert operation: Randomly select two non-adjacent customer k and l in the route *i* of the vehicle service, and insert the customer k into the customer l position.

Swap operation: Customer k and l are randomly selected for exchange in the route of the vehicle service.

Inverse operation: In the route *route*[*i*] served by the vehicle *i*, several customers who are continuously served are randomly selected and their order is reversed (The start and end positions are k and l).

Interchange operation: Selecting customer k and customer l in route *route*[*i*] of vehicle *i* service and route *route*[*j*] of vehicle *j* service respectively for exchange.

According to the local search operation, the search operation process between the vehicles, after performing an operation, if a better solution that meets the load and packing constraints is generated, the generated solution will replace the current optimal solution, otherwise. The solution is not accepted. The variable neighborhood local search execution steps are as follows:

Step1: Choose the best individual π_i and convert it to the permutation π_0 according to the LOV rule.

Step2: Random perform a neighborhood operation on π_0 to get π .

Step3: Set loop =1 max-loop = 40
  n = 1, max-n=20
    While n < max-n do
      Random select m-k , m-k $\in$ {1,2,3}
      Randomly select k and l, where k $\neq$ l
      If m-k =1, π'=insert(π,k,l)
      If m-k =2, π'=swap(π,k,l)
      If m-k =3, π'=inverse(π,k,l)
      If π' p π ,then $\pi = \pi'$ else n = n+1
    End while
  π'=interchange(π,k,l)
  If π' p π ,then $\pi = \pi'$
  loop =loop+1
  Until loop = max-loop
  If π p π_0, then $\pi_0 = \pi$;

Step4: Convert π_0 back to π_i according to the RLOV rule.

4   Computational Results

In this paper, 18 different customer sizes proposed by Iori [2] are used for experimental comparative analysis. The scale of customers in the data ranges from 15 to 100. The algorithm programming tool uses Python3.7, the operating system is windows 10, the computer memory is 16G, the CPU is Inter(R)Core(TM) i5–8500, and the main frequency is 3.00 GHz.

4.1 Settings of Parameters

For GOVRP, the parameters in the fuel consumption model are mainly obtained from the transportation and energy departments [31–33], which is set as 74.342% of the maximum vehicle load. $a_0 = 2.8196$, $a_1 = 0.0658$, $b_0 = 9.701$, $b_1 = 0.0000749$. The maximum vehicle load Q is different in different cases, so L_v needs to be adjusted accordingly (Table 1).

Table 1. Flow chart fuel consumption model parameter values

	Parameter value	Data sources
a_0	2.8196	U.S.Department of
a_1	0.0658	energy
b_0	9.701	Department of
b_1	0.00007491	transport

The key parameters in HWOA are the size of the population pop, the threshold for selecting search mechanism w_q, and the number of local searches max $\_L$. Select three level values for these three parameters respectively, and establish the L9 orthogonal experiment table. The specific values are shown in Table 2 below. In order to explore the influence of parameters on the algorithm, this paper uses a number of examples of 20 customers 2L _cvrp0403 to carry out the L9 (3^4) scale parameter orthogonal test verification. Each parameter combination is run independently for 20 times, and the average value (AVG)of the obtained fuel consumption is compared. this paper conducts a parameter test on the algorithm. Based on the analysis of experimental results, the parameters of HWOA are set to $pop = 80$, $w\_q = 1/2$, max $\_L = 40$.

4.2 Computational Results

This paper compares HWOA with recent algorithms SA [21] and HEDA [22] to verify the effectiveness of HWOA, and Take the minimum fuel consumption as the optimization objective, 20 independent tests were carried out with 18 different customer scale test cases. The best value (BST) and average value (AVG) are selected as performance indexes and bold the best results. The experimental results are shown in Table 2. It can be seen from Table 2 that HWOA is significantly better than the other comparison algorithms in most of the calculation examples, which verifies the effectiveness of the proposed HWOA in solving GOVRP. TSOA is significantly better than the other four comparison algorithms in most cases, which verifies the effectiveness of TSOA in solving GOVRPTW. In fact, as the size of customers increases, the feasible solution space of GOVRPTW is gradually increasing, and the importance of boxing becomes more prominent. The above-mentioned comparison algorithms have achieved better solutions when solving the corresponding problems, but because they ignore the "installation" The importance of "box" leads to a better solution in terms of routing, but the solution has to be abandoned due to the failure of box packing, which leads to weaker optimization ability of the overall algorithm.

Table 2. Comparison of HWOA, SA, and HEDA

Inst.	Size	HWOA		SA		EDA	
		Best	AVG	Best	AVG	Best	AVG
2l_cvrp103	15	**86.23**	**98.99**	88.67	101.30	92.29	99.56
2l_cvrp403	20	**126.00**	**134.39**	**126.00**	**134.39**	129.09	136.63
2l_cvrp603	21	133.85	138.64	132.11	141.19	**128.64**	**138.52**
2l_cvrp703	22	133.76	**142.81**	**127.54**	145.57	135.09	145.14
2l_cvrp903	25	**158.85**	**168.85**	160.36	171.10	161.35	171.62
2l_cvrp1003	29	176.05	**187.78**	179.50	196.41	**174.30**	192.28
2l_cvrp1203	30	**189.85**	**200.85**	191.63	200.87	195.18	203.19
2l_cvrp1403	32	**185.80**	**208.28**	191.31	208.94	187.46	211.63
2l_cvrp1603	35	**223.88**	**234.28**	**223.88**	**234.28**	229.28	236.55
2l_cvrp1703	40	**251.41**	**266.14**	253.25	270.63	260.02	270.85
2l_cvrp1803	44	274.19	**293.59**	**272.65**	300.57	280.54	298.20
2l_cvrp1903	50	313.53	**332.22**	**298.92**	333.52	321.15	333.06
2l_cvrp2003	71	**389.66**	**416.80**	400.69	425.91	406.56	417.71
2l_cvrp2403	75	484.58	**505.29**	483.07	505.40	**482.67**	507.28
2l_cvrp2703	100	**641.41**	**669.76**	655.37	680.01	653.85	670.61
2l_cvrp2803	120	775.84	809.06	787.02	812.69	**772.34**	**796.81**
2l_cvrp3103	199	**1307.22**	**1336.49**	1337.51	1357.39	1321.52	1344.84
2l_cvrp3503	252	1648.35	**1687.69**	1657.48	1704.79	**1624.47**	1692.08

5 Conclusions and Future Research

In this paper, the GOVRP model is established with the optimization goal of minimizing fuel consumption, and designed HWOA to solve it.it can be effective by adjusting and updating the search mechanism of HWOA and adding adaptive adjustment search strategies. Balance the early search and late search of the algorithm; construct a variable domain local search based on four domain operations, perform a more detailed search on the high-quality solution area found by the global search, and enhance the overall search ability of the algorithm.

This study provides a reference for food logistics, cold chain logistics and other logistics distribution enterprises to develop vehicle routing distribution scheme under the requirements of energy reduction and efficiency improvement. Future work will focus on multi-objective GOVRP with fuzzy requirements, refine the nature of the problem, and design efficient algorithm to solve it. So that the research is more in line with the actual distribution of logistics transportation.

Acknowledgements. This research is partially supported by the National Science Foundation of China (51665025), National Natural Science Fund for Distinguished Young Scholars of China (61525304), and the Applied Basic Research Foundation of Yunnan Province (2015FB136).

References

1. Dantzig, G.B., Ramser, J.H.: The truck dispatching problem. Manag. Sci. **1**(6), 80–81 (1959)
2. Sbihi, A., Eglese, R.: Combinatorial optimization and green logistics. 4OR Q. J. Oper. Res. **2**(5), 97–116 (2007)
3. Nosrati, M., Arshadi Khamseh, A.: Distance discount in the green vehicle routing problem offered by external carriers. SN Appl. Sci. **2**(8), 1–14 (2020). https://doi.org/10.1007/s42452-020-03245-5
4. Moghdani, R., Salimifard, K., Demir, E., et al.: The green vehicle routing problem: a systematic literature review. J. Clean. Prod. **279**(123691), 1–19 (2020)
5. Niu, Y., Yang, Z., Chen, P., et al.: Optimizing the green open vehicle routing problem with time windows by minimizing comprehensive routing cost. J. Clean. Prod. **171**(2), 962–971 (2018)
6. Araghi, M.E.T., Tavakkoli-Moghaddam, R., Jolai, F., et al.: A green multi-facilities open location-routing problem with planar facility locations and uncertain customer. J. Clean. Prod. **282**(124343), 1–21 (2020)
7. Li, J., Fu, P.: Model and simulation for vehicle routing problem with time windows based on energy consumption. J. Syst. Simul. **25**(06), 1147–1154 (2013)
8. Soto, M., Sevaux, M., Rossi, A., et al.: Multiple neighborhood search, tabu search and ejection chains for the multi-depot open vehicle routing problem. Comput. Ind. Eng. **107**(MAY), 211–222 (2017)
9. Mirhassani, S.A., Abolghasemi, N.: A particle swarm optimization algorithm for open vehicle routing problem. Expert Syst. Appl. **38**(9), 11547–11551 (2011)
10. Gu, Y., Yuan, Y.Y., Zhang, L., Duan, J.J.: Multi-center and semi-open vehicle routing problem with time windows. China Mech. Eng. **31**(14), 1733–1740 (2020)
11. Fan, H.M., Yang, X., Li, D., Li, Y., Liu, P.C., Wu, J.X.: Semi-open vehicle routing problem based on multi-center joint distribution of fresh products. Comput. Integr. Manuf. Syst. **25**(01), 256–266 (2019)
12. Ibn Faiz, T., Vogiatzis, C., Noor-E-Alam, M.: A column generation algorithm for vehicle scheduling and routing problems. Comput. Ind. Eng. **130**(1), 222–236 (2019)
13. Xia, Y., Fu, Z.: Improved tabu search algorithm for the open vehicle routing problem with soft time windows and satisfaction rate. Clust. Comput. **1**(2), 1–9 (2018)
14. Lahyani, R., Gouguenheim, A.L., Coelho, L.C.: A hybrid adaptive large neighbourhood search for multi-depot open vehicle routing problems. Int. J. Prod. Res. **57**(22), 1–14 (2019)
15. Qian, B., Wang, L., Wang, X., Huang, X.: An effective hybrid DE-based algorithm for flow shop scheduling with limited buffers. Int. J. Prod. Res. **36**(1), 209–233 (2009)
16. Mirjalili, S., Lewis, A.: The whale optimization algorithm. Adv. Eng. Softw. **95**(1), 51–67 (2016)
17. Cai, Y., Du, P.: Path planning of unmanned ground vehicle based on balanced whale optimization algorithm. Control Decis. **1**, 1–9 (2021)
18. Yue, X., Peng, X., Lin, L.: Short-term wind power forecasting based on whales optimization algorithm and support vector machine. Electric Power Syst. Autom. **32**(02), 146–150 (2020)
19. Zheng, W., Li, Z., Jia, H., Gao, C.: Research on prediction model of steelmaking end point based on LWOA and LSSVM. Acta Electronica Sinica. **47**(03), 700–706 (2019)

20. Pankaja, K., Suma, V.: Plant leaf recognition and classification based on the whale optimization algorithm (WOA) and random forest (RF). J. Inst. Eng. (India) Ser. B **101**(5), 597–607 (2020). https://doi.org/10.1007/s40031-020-00470-9
21. Wei, L., Zhang, Z., Zhang, D., et al.: A simulated annealing algorithm for the capacitated vehicle routing problem with two-dimensional loading constraints. Eur. J. Oper. Res. **265**(3), 843–859 (2018)
22. Perez-Rodriguez, R., Hernandez-Aguirre, A.: A hybrid estimation of distribution algorithm for the vehicle routing problem with time windows. Comput. Ind. Eng. (S0360–8352) **130**(1), 75–96 (2019)

Hybrid Grey Wolf Optimizer for Vehicle Routing Problem with Multiple Time Windows

Nan Li[1], Rong Hu[1(✉)], Bin Qian[1], Nai-Kang Yu[2], and Ling Wang[3]

[1] School of Information Engineering and Automation, Kunming University of Science and Technology, Kunming 650500, China
[2] School of Mechanical and Electronic Engineering, Kunming University of Science and Technology, Kunming 650500, China
[3] Department of Automation, Tsinghua University, Beijing 10084, China

Abstract. Aiming at the vehicle routing problem with multiple time windows (VRPMTW), a hybrid grey wolf optimizer (HGWO) is designed to minimize the comprehensive distribution cost. Firstly, on the premise of retaining the standard grey wolf optimizer (GWO) individual update mechanism, the large order value (LOV) rule based on random key coding is used to implement the global search of the GWO in the discrete problem solution space. Secondly, a new nonlinear decreasing distance control parameter is used to better balance the ability of local search and global search. Finally, a variable neighborhood local search algorithm with five different forms of neighborhood operators is designed for local search in and between vehicles, which further enhances the local search ability of the algorithm. Through simulation experiments and algorithm comparisons on different scale problems, it is verified that HGWO can effectively solve VRPMTW.

Keywords: Vehicle routing problem · Multiple time windows · Gray wolf optimizer · Variable neighborhood local search

1 Introduction

Vehicle routing problem with multiple time windows (VRPMTW) widely exists in industrial gas distribution [1], long-distance transportation [2], team orientation [3] and other problems. VRPMTW is an extension of vehicle routing problem (VRP), which was first proposed by Favaretto et al. [4] in 2007. Then, some scholars extended the VRPMTW model and algorithm. For example, Belhaiza et al. [5] proposed a hybrid variable neighborhood tabu search heuristic algorithm to solve VRPMTW with the objective of minimizing the total duration. Beheshti et al. [6] proposed a cooperative evolutionary multi-objective quantum genetic algorithm to solve the multi-objective vehicle routing problem with multiple priority time windows and the optimization objectives of minimizing travel cost and maximizing customer satisfaction. Belhaiza et al. [7] established a multi-objective VRPMTW model with the objective of minimizing the total travel cost, maximizing the minimum customer utility and maximizing the minimum driver utility from the perspective of game theory, and proposed a pareto non dominated method

© Springer Nature Switzerland AG 2021
D.-S. Huang et al. (Eds.): ICIC 2021, LNCS 12836, pp. 684–693, 2021.
https://doi.org/10.1007/978-3-030-84522-3_56

to solve the model. Hoogeboom et al. [8] proposed an adaptive variable neighborhood search algorithm and an accurate polynomial time algorithm to recalculate the path duration based on the forward and backward start time intervals for VRPMTW with variable starting time. Bogue et al. [9] proposed a column generation algorithm and post optimization heuristic algorithm based on variable neighborhood search for VRPMTW. According to the above literature research, the research on VRPMTW is still limited. In addition, in terms of computational complexity, VRP belongs to NP-hard problem [10], and VRP can be reduced to VRPMTW, so VRPMTW also belongs to NP-hard problem. Therefore, the research on VRPMTW modeling and its solution algorithm has important theoretical value and practical significance.

Grey wolf optimizer (GWO) [11] is a swarm intelligent optimization algorithm proposed by Australian scholar Mirjalili in 2014 by simulating the social class and hunting mechanism of grey wolves. GWO has a wide range of applications in real life because of its advantages of few control parameters, strong global optimization ability, fast convergence and easy implementation. However, like other swarm intelligence algorithms, GWO has some disadvantages, such as weak local development ability, and difficult to jump out of local optimization when optimizing complex problems. In order to improve the optimization performance of GWO, researchers mainly improve the algorithm from adjusting control parameters, updating search mechanism, introducing new operators and algorithm fusion. At present, GWO is mainly used in continuous optimization problems [12, 13], and less used in discrete optimization problems [14].

This paper studies the vehicle routing problem with multiple time windows, and proposes a hybrid gray wolf optimizer to solve it. The rest of this paper is arranged as follows. In Sect. 2, the model of VRPMTW is briefly introduced. In Sect. 3, HGWO is proposed and described in detail. In Sect. 4, simulation results and comparison are given. Finally, In Sect. 5, we end this paper and draw some conclusions and future work.

2   Problem Description

VRPMTW can be described as a set of customer points in a complete digraph $G = (V, E)$. $V = \{0\} \cup V_0$, $V_0 = \{1, 2, 3 \cdots, n\}$ is the collection of customer points. 0 represents the distribution center. $E = \{(i, j) | i, j \in V\}$ is the set of edges. Each customer has a set of non overlapping time windows $W_i = \left\{ \left[e_i^1, l_i^1\right], \left[e_i^2, l_i^2\right], \cdots, \left[e_i^{T_i}, l_i^{T_i}\right] \right\}$. Let $K = \{1, 2, \cdots m\}$ be the available vehicle set of the distribution center. Q is the maximum load of the vehicle. Q_{k0} is the load of vehicle k from the distribution center. q_i is the demand of customer i. $[e_0, l_0]$ is the opening time window of the distribution center. d_{ij} is the distance from customer i to customer j. t_{ij} is the travel time of the vehicle from customer i to customer j. a_i^k is the time when vehicle k arrives at customer i. b_i^k is the time when vehicle k leaves customer i. w_i^k is the waiting time of vehicle k at customer i. s_i^k is the service time of vehicle k at customer i. r^k is the total number of vehicles required by the distribution scheme. c is the fixed cost of vehicle departure. c_1 is the unit driving cost of vehicle. c_2 is the waiting cost of vehicle arriving early. x_{ijk} is the decision variable, if the vehicle k from customer i to customer j is 1, otherwise it is 0. y_{ik} is the decision variable, if customer i is served by vehicle k, it is 1, otherwise it is 0.

z_{ik}^p is the decision variable, if the vehicle k serves the customer i in the p th time window, it is 1, otherwise it is 0. Starting from the full load of the distribution center, the vehicles serve the distribution center in a number of time windows specified by customers. The optimization objective is to reasonably arrange the vehicle routes to minimize the total distribution cost under the constraint conditions.

The constraints are as follows: 1. The total customer demand of each distribution route does not exceed the maximum vehicle load; 2. The vehicle can only serve within a certain time window specified by the customer. If the vehicle arrives before the specified time window, the vehicle will wait until the time window is opened; 3. Each customer has and can only be visited once.

According to the above description, the mathematical model of VRPMTW is established as follows:

$$\min \sum_{k \in K} cr^k + \sum_{i \in V} \sum_{j \in V} \sum_{k \in K} c_1 d_{ij} x_{ijk} + \sum_{k \in K} \sum_{i \in V_0} c_2 w_i^k, \tag{1}$$

$$\text{s.t.} \sum_{i \in V} x_{ihk} = \sum_{j \in V} x_{hjk}, \ \forall h \in V_0, \ \forall k \in K, \tag{2}$$

$$x_{ijk} = 0, \ \forall i = j, \ \forall i, j \in V, \ \forall k \in K, \tag{3}$$

$$\sum_{j \in V_0} x_{0jk} = \sum_{j \in V_0} x_{j0k} \le 1, \forall k \in K, \tag{4}$$

$$\sum_{i \in V} x_{ijk} = y_{jk}, \forall j \in V_0, \ \forall k \in K, \tag{5}$$

$$\sum_{j \in V} x_{ijk} = y_{ik}, \forall i \in V_0, \ \forall k \in K, \tag{6}$$

$$Q_{k0} = \sum_{i \in V} \sum_{j \in V_0} q_j x_{ijk}, \forall k \in K, \tag{7}$$

$$0 < Q_{k0} \le Q, \forall k \in K, \tag{8}$$

$$b_{0k} = e_0, \forall k \in K, \tag{9}$$

$$a_{0k} \le l_0, \forall k \in K, \tag{10}$$

$$a_j^k = a_i^k + w_i^k + s_i^k + t_{ij}, \forall i, j \in V, \forall k \in K, \tag{11}$$

$$a_i^k + w_i^k \ge \sum_{p \in W_i} e_i^p z_{ik}^p, \forall i \in V, \forall k \in K, \tag{12}$$

$$a_i^k + w_i^k \le \sum_{p \in W_i} l_i^p z_{ik}^p, \forall i \in V, \forall k \in K, \tag{13}$$

$$\sum_{p \in W_i} z_{ik}^p = 1, \forall i \in V, \forall k \in K, \tag{14}$$

$$x_{ijk} \in \{0, 1\}, \forall i, j \in V, \forall k \in K, \tag{15}$$

$$y_{jk} \in \{0, 1\}, \forall j \in V_0, \forall k \in K, \tag{16}$$

$$z_{ik}^p \in \{0, 1\}, \forall i \in V, \forall k \in K. \tag{17}$$

Objective (1) is the optimization objective function, which means to minimize the total comprehensive cost, including the departure cost of all vehicles, the driving cost of all vehicles and the waiting cost of all vehicles arriving early; Constraints (2) indicates that the vehicles arriving at each customer and leaving it are the same; Constraints (3) indicates that the same customer has no path to connect; Constraints (4) indicates that each vehicle has only one service path, and it starts from the distribution center and returns to the distribution center after distribution; Constraints (5) and (6) ensures that when the customer point is served by the vehicle, there must be a path to connect with it; Constraints (7) and (8) indicates that the load of the vehicle when it departs from the distribution center is the total demand of the customers served by the vehicle, and the load cannot exceed the maximum load of the vehicle; Constraints (9) and (10) indicates that the vehicle departure time is the earliest opening time of the distribution center, and it should return before the closing time of the distribution center; Constraints (11) shows that the time to customer j is equal to the time to customer i, plus the waiting time and service time at customer i; Constraints (12) and (13) indicates that the customer is served within the time window; Constraints (14) indicates that only one time window can be selected; Constraints (15) (16) and (17) denotes the attributes of decision variables.

3 HGWO for VRPMTW

3.1 Encoding and Decoding

The population size of gray wolf is *popsize*. $\pi_i^{gen} = \left[\pi_i^{gen}[1], \pi_i^{gen}[2], \cdots, \pi_i^{gen}[L]\right]$ $(2 + n \leq L \leq 2n + 1)$ is individual i in the *gen* generation of gray wolf population. L is the length of π_i^{gen}. $C_i^{gen} = \left[C_i^{gen}[1], C_i^{gen}[2], \cdots, C_i^{gen}[n]\right]$ is the arrangement based on customer serial number. $X_i^{gen} = \left[X_i^{gen}[1], X_i^{gen}[2], \cdots, X_i^{gen}[n]\right]$ is the real location information of individual customer corresponding to C_i^{gen}. Standard GWO is mainly used to solve continuous variable optimization problems, and most of the problems are solved by real coding. However, VRPMTW is a discrete optimization problem, so the coding method of standard GWO cannot be directly used in VRPMTW. In order to use the position update strategy in the standard GWO, HGWO performs the global search of standard GWO in the solution space of discrete problems. In this paper, we still use the real coding method, and then use the large order value (LOV) rule to transform the continuous gray wolf position vector to the discrete customer sequence. When the customer sequence changes, we can update the individual position information through the reverse large order value (RLOV) rule.

The LOV rule are as follows:

Step 1: Arrange all the elements in real sequence $X_i^{gen} = [x_{i,1}^{gen}, x_{i,2}^{gen}, \ldots, x_{i,n}^{gen}]$ in descending order.

Step 2: Through step 1, intermediate integer sequence $\varphi_i = [\varphi_1, \varphi_2, \ldots, \varphi_n]$ can be obtained.

Step 3: Through formula $c_{i,\varphi_{i,k}} = k$, the customer sequence corresponding to real number sequence X_i^{gen} can be obtained.

Table 1. Coding method based on LOV

k	1	2	3	4	5	6	7	8
$X_{i,k}^{gen}$	0.19	0.65	0.86	0.28	0.32	0.51	0.76	0.97
$\varphi_{i,k}$	8	4	2	7	6	5	3	1
$C_{i,k}$	8	3	7	2	6	5	4	1

As shown in Table 1, $x_{i,2} = 0.65$, in descending order, $\varphi_{i,2} = 4$, according to the formula $c_{i,\varphi_{i,3}=2} = k = 3$ can be obtained, other values, and so on. Can get $C_i^{gen} = [8, 3, 7, 2, 6, 5, 4, 1][8, 3, 7, 2, 6, 5, 4, 1]$. The above integer sequence obtained according to LOV rule is based on customer number sequence, which is not a complete distribution scheme. In this paper, the customer sequence obtained is allocated to each vehicle according to the allocation strategy of routing before grouping. According to the load of vehicles and the time window constraints of customers, the customers in C_i^{gen} are assigned to each vehicle from left to right in turn, and the distribution route $\pi_i^{gen} = [0, 8, 3, 7, 2, 0, 6, 5, 4, 1, 0]$ of all vehicles can be obtained, it means that there are two vehicles in the distribution scheme, vehicle 1 starts from distribution center 0 and serves customers 8, 3, 7 and 2 in turn before returning to the distribution center, vehicle 2 starts from distribution center 0 and serves customers 6, 5, 4 and 1 in turn before returning to the distribution center. When the solution corresponding to the real position information of gray wolf population changes, the real position information of gray wolf individual can be obtained by RLOV rule.

3.2 Nonlinear Decreasing Distance Control Parameter

The convergence factor used to control the hunting behavior of gray wolf in standard GWO is linearly decreasing according to the number of iterations. However, in the optimization process of high-dimensional complex problems, because the search process of the algorithm is extremely complex, the linear decreasing strategy of convergence factor is difficult to meet the actual situation of the search. In this paper, the nonlinear decreasing convergence factor proposed in reference [15] is used, as shown in Eq. (18). Figure 1 shows the dynamic curve of the convergence factor used in this paper and the convergence factor in standard GWO as the number of iterations increases. As can be seen from Fig. 1, in the early stage of the algorithm iteration, the nonlinear decreasing convergence factor formula can provide a larger convergence factor, so that the algorithm has a larger search step, improves the exploration ability of the algorithm, and avoids premature convergence of the algorithm; In the later stage of iteration, the formula of nonlinear decreasing convergence factor can provide a smaller convergence factor, which makes the algorithm concentrate on a certain region to search and accelerate the convergence speed of the algorithm.

$$a = \left(a_{initial} - a_{final}\right) \times \left(1 - \sin\left(\left(t/T\right)^2 \cdot \pi/2\right)\right) \tag{18}$$

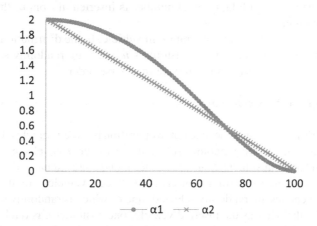

Fig. 1. Dynamic curves of different convergence factors a

When gray wolf individuals update their positions, firstly, α, β, δ and ω, are determined according to the fitness of all individuals in the gray wolf population. Secondly, the real number position of each individual in ω is close to α, β, δ and ω by individuals update formula to update the group position. Then, the real number position of gray wolf is corresponding to the customer sequence according to LOV rule. Finally, according to the strategy of routing before grouping, the corresponding solution is obtained and the fitness of each solution is calculated. After the above operation, the whole gray wolf population is updated. The updated gray wolf population is arranged according to the fitness of the solution corresponding to the real number position of the gray wolf individual, and the new optimal position is taken as the position of α, β, δ wolf. The specific principle and steps of GWO are shown in reference [11].

3.3 Variable Neighborhood Local Search

In order to enhance the search ability of the algorithm and guide the algorithm to increase the search depth in a more compact solution space, this paper designs a variable neighborhood local search strategy for two kinds of local search operators in vehicles and between vehicles, which have five kinds of local search operators. In the local search, the local search operator in the vehicle is used to search the vehicle first, and then the local search operator between vehicles is used to search the vehicle after reaching the number of searches. The five local operators are as follows:

(1) Local operator in vehicle

1. $Swap\_in(\pi, u, v)$, customer exchange operation in vehicle. In distribution scheme π, a car is randomly selected, and two customers u and v, that are not adjacent are randomly selected to exchange customers u and v.

2. $Insert\_in(\pi, u, v)$, insertion operation in vehicle. In distribution scheme π, a vehicle is randomly selected, and two customers u and v ($u < v$) are randomly selected in the vehicle. The customer v with large serial number is inserted in front of the customer u with small serial number.

3. $Reverse\_in(\pi, u, v)$, reverse operation in vehicle. In the distribution scheme π, a vehicle is randomly selected, and two customers u and v are randomly selected in the route, and the customers between u and v are in reverse order.

(2) Local operator between vehicles

4. $Swap\_out(\pi, u, v)$, customer exchange operation between vehicles. In distribution scheme π, two vehicles are randomly selected, customer u is selected in vehicle 1, customer v is selected in vehicle 2, and customer u and v are exchanged.

5. $Insert\_out(\pi, u, v)$, customer insertion between vehicles. In the distribution scheme π, two vehicles are randomly selected, one of which is randomly selected as the inserted vehicle, the other as the inserted vehicle, one customer u is randomly selected in the inserted vehicle, one customer v is randomly selected in the inserted vehicle, and the customer u is inserted in front of customer v.

The following is an example of the local search step in the vehicle to illustrate the local search of variable neighborhood.

Step 1: Input the best individual π_i^{gen}. Let $\pi_0 = \pi_i^{gen}$.

Step 2: Disturbance phase.

Step 2.1: Set loop_ 0=1.

Step 2.2: Randomly select i and j .($i \neq j$), $\pi_1 = Swap(\pi_0, i, j)$.

Step 2.3: $\pi_0 = \pi_1$, if loop_0<=5, go to step 2.2. otherwise, go to step 3.

Step 3: Development phase.

Step 3.1: Set loop_ 1=1.

Step 3.2: Randomly generated $K(1 \leq K \leq 3)$.

Step 3.3: If $K = 1$, then $\pi_2 = Swap\_in(\pi_1, u, v)$;

If $K = 2$, then $\pi_2 = Insert\_in(\pi_1, u, v)$;

If $K = 3$, then $\pi_2 = Reverse\_in(\pi_1, u, v)$;

Step 3.4: If $f(\pi_2) < f(\pi_1)$, then $\pi_1 = \pi_2$.

Step 3.5: loop_1=loop_1+1.

Step 3.6: If loop_1<=10, go to step 3.2, otherwise go to step 4.

Step 4: If $f(\pi_1) < f(\pi_i^{gen})$, then $\pi_i^{gen} = \pi_1$.

3.4 HGWO Process

Let $P(gen)$ be the *gen* generation population of the algorithm, the population size is *popsize*, and the maximum running algebra of the algorithm is *gen_max*. the specific steps of the algorithm are as follows:

Step 1: Set algorithm parameters, Randomly generate an initial population of population size *popsize*.

Step 2: Calculate the fitness value of all individuals in the population, and determine the first generation α, β, δ according to the size of the fitness value.

Step 3: According to Sect. 3.2, the individuals in ω are moved closer to α, β, δ wolves respectively to update the population position of ω. The individuals in the population are sorted by the solution after population position update according to the order of fitness from small to large, and α, β, δ is redetermined.

Step 4: According to Sect. 3.3, the variable neighborhood local search is carried out for α, β, δ in the order of first in vehicle and then between vehicles. After local search, α, β, δ is determined again according to fitness.

Step 5: If $gen \leq gen\_$ max, go to step 2, otherwise, output the optimal value.

4 Simulation Result and Comparisons

4.1 Experimental Setup

In order to verify the performance of the proposed algorithm, this paper uses 18 VRPMTW examples in RM1, RM2, CM1, CM2, RCM1 and RCM2 in the literature [5] to verify. In the distribution of customers, RM1 and RM2 have random and uniform CM1, CM2 are clustered; RCM1, RCM2 are semi-clustered. In the time window distribution, RM1, CM1, RCM1 have a narrow time window; RM2, CM2, RCM2 have a wider time window. All the algorithms and test programs in this paper are coded by Python 3.7. The operating system is Windows 10, the CPU frequency is 3.30 GHZ, and the memory is 8 GB. Each algorithm runs 20 times independently, and the running time of each algorithm is 60 s. The termination condition of the algorithm is t greater than 60 s. *BST*, *AVG* are the optimal value and average value deviation of the algorithm running 20 times.

4.2 Results and Comparison

In order to verify the effectiveness of HGWO, HGWO is compared with HAVNTS [5] and AVNS [8] for VRPMTW. All test problems were performed 20 independent experiments at the same time, and the optimal results corresponding to each problem are shown in bold. The solution results of different problems are shown in Table 2.

It can be seen from Table 2 that the test results of HGWO on most problems are significantly better than AVNS and HAVNTS, indicating that HGWO has good performance for solving VRPMTW.

Table 2. Comparison of HGWO with HAVNTS and AVNS

Problem	HAVNTS		AVNS		HGWO	
	BST	AVG	BST	AVG	BST	AVG
rm102	3542.24	3867.33	3224.00	3619.40	**3211.81**	**3328.45**
rm104	3629.50	4221.86	3579.86	4019.45	**3428.19**	**3678.42**
rm106	3873.14	4144.70	3642.38	4094.95	**3483.40**	**3704.02**
rm202	5700.39	5868.97	5681.66	5828.73	**4721.48**	**4788.60**
rm204	5850.64	6042.70	5786.79	5901.71	**4739.54**	**4829.25**
rm206	5821.80	6674.34	5668.87	6551.05	**4696.30**	**4802.80**
cm102	4449.24	4667.89	**4435.60**	**4615.74**	4487.83	4633.72
cm104	4690.57	4862.85	4703.78	4843.59	**4673.23**	**4795.30**
cm106	4400.51	4526.71	4319.84	4488.26	**4315.91**	**4389.23**
cm202	4879.89	5078.20	**4727.86**	4956.83	4829.67	**4950.86**
cm204	5354.43	5530.80	**5170.00**	5399.84	5185.49	**5375.40**
cm206	5173.63	5406.17	5082.45	5264.09	**4978.16**	**5143.38**
rcm102	3672.86	4058.99	3673.51	4057.77	**3373.34**	**3583.66**
rcm104	4024.91	4385.86	3877.04	4315.16	**3706.41**	**3868.92**
rcm106	4058.83	4244.57	4027.96	4259.84	**3756.07**	**3915.75**
rcm202	5746.98	6746.48	5488.82	6437.37	**3399.55**	**3446.38**
rcm204	6257.16	7166.08	6135.73	6981.81	**4757.33**	**4990.07**
rcm206	6163.14	7139.36	5760.20	6918.73	**4779.23**	**4931.48**

5 Conclusion and Future Work

In this paper, a hybrid gray wolf optimizer (HGWO) is proposed to solve the vehicle routing problem with multiple time windows (VRPMTW) with the objective of minimizing the comprehensive cost. In the encoding and decoding part of HGWO, a code method based on LOV rules is designed to realize the transformation from continuous position vector to discrete customer sequence. In the part of individual location updating, a nonlinear decreasing convergence factor is used to balance the ability of local search and global search. In the local search part, the variable neighborhood local search algorithm based on the vehicle and between vehicles is designed to guide the algorithm to increase the search depth in a more compact solution space. The simulation experiment and algorithm comparison verify that HGWO has good robustness and stability when solving VRPMTW. The HGWO proposed in this paper can be used as a reference for other intelligent algorithms to solve discrete scheduling problems. In the future, we will design effective algorithms to study time-dependent vehicle routing problem with multiple time windows.

Acknowledgements. This research is partially supported by National Science Foundation of China (61963022), and National Science Foundation of China (51665025).

References

1. Pesant, G., Gendreau, M., Potvin, J.Y., et al.: On the flexibility of constraint programming models: from single to multiple time windows for the traveling salesman problem. Eur. J. Oper. Res. **117**(2), 253–263 (1999)
2. Asvin, G.: The Canadian minimum duration truck driver scheduling problem. Comput. Oper. Res. **39**(10), 2359–2367 (2011)
3. Souffriau, W., Vansteenwegen, P., Berghe, G.V., et al.: The multiconstraint team orienteering problem with multiple time windows. Transport. Sci. **47**(1), 53–63 (2013)
4. Favaretto, D., Moretti, E.: Ant colony system for a VRP with multiple time windows and multiple visits. J. Interdisc. Math. **10**(2), 263–284 (2007)
5. Belhaiza, S., Hansen, P., Laporte, G.: A hybrid variable neighborhood tabu search heuristic for the vehicle routing problem with multiple time windows. Comput. Oper. Res. **52**, 269–281 (2014)
6. Beheshti, A.K., Hejazi, S.R., Alinaghian, M.: The vehicle routing problem with multiple prioritized time windows: a case study. Comput. Ind. Eng. **90**, 402–413 (2015)
7. Belhaiza, S., M'Hallah, R.: A pareto non-dominated solution approach for the vehicle routing problem with multiple time windows. In: 2016 IEEE Congress on Evolutionary Computation (CEC). IEEE (2016)
8. Hoogeboom, M., Dullaert, W., Lai, D., et al.: Efficient neighborhood evaluations for the vehicle routing problem with multiple time windows. Transport. Sci. **54**(2), 400–416 (2020)
9. Bogue, E.T., Ferreira, H.S., Noronha, T.F., et al.: A column generation and a post optimization VNS heuristic for the vehicle routing problem with multiple time windows. Optim. Lett. (6) (2020)
10. Lenstra, J.K., Kan, A.: Complexity of vehicle routing and scheduling problems. Networks **11**(2), 221–227 (2010)
11. Mirjalili, S., Mirjalili, S.M., Lewis, A.: Grey wolf optimizer. Adv. Eng. Softw. **69**, 46–61 (2014)
12. Sahoo, B.P., Panda, S.: Improved grey wolf optimization technique for fuzzy aided PID controller design for power system frequency control. Sustain. Energy. Grids. **16**, 278–299 (2018)
13. Esmaeil, H., Sobhan, M., Parham, S.: A grey wolf optimizer-based neural network coupled with response surface method for modeling the strength of siro-spun yarn in spinning mills. Appl Soft Comput. **72**, 1–13 (2018)
14. Korayem, L., Khorsid, M., Kassem, S.S.: Using grey wolf algorithm to solve the capacitated vehicle routing problem. IOP Conf. Ser. Mater. Sci. Eng. **83**(1), 012014 (2015)
15. Huang, C.C., Wei, X., Huang, D.Q., Ye, J.H.: Hybrid frog leaping gray wolf optimization algorithm for solving high dimensional complex functions. Contl. Theor. Appl. **37**(07), 1655–1666 (2020)

Spatial Prediction of Stock Opening Price Based on Improved Whale Optimized Twin Support Vector Regression

Huajuan Huang, Xiuxi Wei$^{(\boxtimes)}$, and Yongquan Zhou

College of Artificial Intelligence, Guangxi University for Nationalities, Nanning 530006, China

Abstract. It is difficult to predict the opening price of stock accurately, so it is very important to predict the change space of the opening price. Twin support vector regression (TSVR) has good generalization performance in dealing with large-scale data sets, but the selection of its parameters has great blindness and randomness, which has a great impact on its learning performance. In order to solve this problem, an improved whale optimization algorithm is used to optimize the parameters of TSVR. So we propose improved whale optimized twin support vector regression (IWOA-TSVR) in this paper. Finally, IWOA-TSVR used to predict the change space of stock opening price. The experimental results show that the method has high accuracy, good stability and easy operation.

Keywords: Stock prediction · Twin support vector regression · Shanghai stock · Whale optimization algorithm

1 Introduction

In the stock market, high risk and high return exist at the same time. Accurate forecasts of stock prices will mean high market returns. Therefore, the stock market composite index and stock price prediction has become an important focus of the academic community [1, 2]. In the traditional stock market forecasting modeling research, the most common methods are multiple regression analysis, time series analysis, index smoothing, etc. [3, 4]. However, the stock market is a complex nonlinear dynamic system, and its prediction is a problem of estimation and extrapolation of nonlinear function values. At the same time, the market behavior is influenced by many factors, which has significant nonlinear and time-varying characteristics. It is difficult to predict the accurate value by using the traditional statistical prediction technology [5, 6]. Since it is difficult to predict the accurate value of the stock index, it is particularly important to predict the change trend and change space in the last 3–5 days.

Twin support vector regression (TSVR) is a new machine learning method proposed by Peng in 2010 [7]. TSVR aims at generating two nonparallel functions such that each function determines the ε-insensitive down- or up- bounds of the unknown regressor. TSVR only need to solve a pair of smaller quadratic programming problems, instead of solving the large one in support vector regression (SVR). Furthermore, the number of

© Springer Nature Switzerland AG 2021
D.-S. Huang et al. (Eds.): ICIC 2021, LNCS 12836, pp. 694–706, 2021.
https://doi.org/10.1007/978-3-030-84522-3_57

constraints of each quadratic programming problem in TSVR is only half of the classical SVR, which makes TSVR work faster than SVR. Similar to SVR, the parameter selection of TSVR is random and blind, which seriously affects the learning performance and generalization ability of TSVR [8, 9].

Whale optimization algorithm (WOA) is a new swarm intelligence optimization algorithm which imitates the hunting. It was proposed by Mirjalili and Lewis [10] in 2016. Compared with other swarm intelligence optimization algorithms, such as particle swarm optimization (PSO) algorithm [11], artificial bee colony (ABC) [12] algorithm and genetic algorithm (GA) [13], WOA has the advantages of simple adjustment parameters and fast operation speed. However, the convergence accuracy of WOA and its ability to jump out of local optimum need to be improved. In this paper, an improved WOA algorithm is proposed to optimize the regularization parameters and kernel parameter of TSVR, which is called IWOA-TSVR. This can improve the performance and accuracy of the prediction model of stock opening price range. The specific measures are as follows. The basic WOA algorithm is improved by introducing backward learning to initialize the population position to enhance the local search ability, improve the convergence accuracy and speed up the convergence.

The paper is organized as follows. In Sect. 2, our prediction model of stock opening price range is formulated. Improved WOA strategy for model selection of TSVR is arranged in Sect. 3. Experimental results are described in Sect. 4, and concluding remarks are given in Sect. 5.

2 The Prediction Model of Stock Opening Price Range Granulated by Fuzzy Information

The spatial prediction of opening price in stock index is a kind of forecasting problem about time series. Time series is an important complex type of data. The analysis of time series is to discover and reveal the development and change law of a certain phenomenon from a large number of data, or to carve out the internal quantitative relationship and its change law between a phenomenon and other phenomena from a dynamic perspective, so as to extract the accurate information we need as much as possible, and combine these knowledge with Information is used for forecasting. Generally, the analysis of time series is based on the operation of a large time series data set, which will inevitably lead to computational difficulties. Therefore, how to simplify the representation of time series without losing the information and knowledge contained in time series becomes more and more important.

In 1979, Professor L.A. Zadeh proposed the concept of information granule, which studies a group of similar research objects as a whole or divides a whole into parts to study, and the objects put together into a whole are called information granules [14]. Information granular theory has shown its good performance in solving large-scale data, and has become an important tool in intelligent data analysis, computing science and image processing. The information granulation is introduced into the time series analysis, and the time series is divided into sub columns (information granules) for analysis, which will greatly reduce the amount of calculation.

We use the fuzzy information granulation in this paper. The fuzzy granulation of time series with fuzzy sets can be divided into two steps: granulation and fuzziness. The so-called granulation is to divide the given time series into small columns (information granules) as operation windows, while fuzzification is to fuzzify each window generated in the first step to generate fuzzy sets, that is, fuzzy particles. This method of fuzzy information granulation is called f-granulation.

In this paper, we mainly predict the change space of opening price in the next five days, so we take five trading days as the size of a window, and each window will generate a fuzzy particle. After defining the size of particles, the next step is how to blur the particles, and create a fuzzy set in each window, so that it can replace the given data to represent the relevant information.

Giving a subsequence $X = \{x_1, x_2, \cdots, x_N\}$, the task of fuzzification is to establish a fuzzy particle P in X, that is, if a reasonable description of the fuzzy concept G is determined, the fuzzy particle P will be determined:

$$P \triangleq x \text{ is } G, x \in X \tag{1}$$

Therefore, fuzzy granulation process is actually a process to determine the function A. A is a membership function of fuzzy concept G, i.e. $A = \mu_G$. Thus, the form of particles can be simply described as:

$$P = A(x), x \in X \tag{2}$$

According to the fuzzy time-series information granulation theory proposed by L.A. Zadeh, basic fuzzy particles forms contain triangular form, trapezoidal form, gaussian form and parabolic form, etc. In this paper, we choose the triangular form for fuzzy time-series information granulation. Generally, the triangular fuzzy membership function of the particle is descripted as follows:

$$A(x, a, m, b) = \begin{cases} 0 & \text{if } x \leq a \\ \frac{x-a}{m-a} & \text{if } x \in [a, m] \\ \frac{b-x}{b-m} & \text{if } x \in [m, b] \\ 0 & \text{if } x \geq b \end{cases} \tag{3}$$

If the form of particle has been determined, the next step is to discuss how to determine the parameters of the membership function, namely, how to determine the specific membership function. Generally, establishing the membership function must be based on the following basic idea of establishing particle:

1) Particles can reasonably represent the original data.
2) Particles must have certain speciality.

To meet the above two requirements and find the best balance between the two, a related function is given by the idea of the Pedryce fuzzy granulated theory as follows:

$$Q_A = \frac{M(A)}{N(A)} \tag{4}$$

Here, $M(A)$ must guarantee to meet the first requirement of the particle fuzzy granulation, meanwhile, $N(A)$ should meet the second requirement. In order to meet the requirements, we take $M(A) = \sum_{x \in X} A(x)$, $N(A) = measure(\sup p(A)) = b - a$, then the Eq. (4) becomes:

$$Q_A = \frac{\sum\limits_{x \in X} A(x)}{measure(\sup p(A))} \tag{5}$$

Thus, in order to meet the above two requirements, the value of Q_A should be as large as possible, namely, $\sum_{x \in X} A(x)$ should be required as large as possible while $measure(\sup p(A)) = b - a$ is required as small as possible. Following the above ideas, the method to determine the specific parameters of fuzzy membership function is described as follows:

1) the value of m: take the middle number of subsequence $X = \{x_1, x_2, \cdots, x_N\}$ as the value of m. So if N is odd, set $m = x_{\frac{N+1}{2}}$, otherwise, set $m = x_{\frac{N}{2}}$.

2) the value of a and b: traverse all points of the data set, and then choose the value of a who can make the function $Q(a) = \frac{\sum_{k=1, x_k < m}^{N} A(x_k)}{|m-a|}$ take the maximum value. Here, $A(x_k)$ is the membership of x_k. When $a \leq x_k \leq m$, set $A(x_k) = \frac{|x-a|}{|m-a|}$ and when $x_k \leq a$, set $A(x_k) = 0$. The value of b can be determined by a similar method.

After the above two steps, three parameters (a, m, b) of the triangular fuzzy membership function have been determined. Therefore, a triangular fuzzy information particle can be described as $P = (a, m, b)$.

3 The Prediction Model of Stock Opening Price Range Using IWOA-TSVR

3.1 Twin Support Vector Regression (TSVR)

TSVR would generate two nonparallel functions around the data points.

For the linear case, TSVR aims at finding a pair of nonparallel functions

$$f_1(x) = w_1^T x + b_1 \tag{6}$$

$$f_2(x) = w_2^T x + b_2, \tag{7}$$

such that each function determines the ε-insensitive down- or up- bounds regressor. The two functions are obtained by solving the following quadratic programming problems:

$$\min \quad \frac{1}{2}\|Y - e\varepsilon_1 - (Aw_1 + eb_1)\|^2 + C_1 e^T \xi \tag{8}$$
$$s.t. \quad Y - (Aw_1 + eb_1) \geq e\varepsilon_1 - \xi, \quad \xi \geq 0,$$

$$\min \quad \frac{1}{2}\|Y + e\varepsilon_2 - (Aw_2 + eb_2)\|^2 + C_2 e^T \eta \tag{9}$$

$$s.t. \quad (Aw_2 + eb_2) - Y \geq e\varepsilon_2 - \eta, \quad \eta \geq 0,$$

where, $C_1, C_2 > 0$; $\varepsilon_1, \varepsilon_2 > 0$ are the parameters, ξ, η are the slack vectors and e is the vector of ones of appropriate dimensions.

Introducing the Lagrangian multiplier vectors α, γ, considering the KKT conditions, the dual QPPs of (8) and (9) can be obtained as follows:

$$\max \quad -\tfrac{1}{2}\alpha^T G(G^T G)^{-1} G^T \alpha + f^T G(G^T G)^{-1} G^T \alpha - f^T \alpha \tag{10}$$

$$s.t. \quad 0 \leq \alpha \leq C_1 e$$

$$\max \quad -\tfrac{1}{2}\gamma^T G(G^T G)^{-1} G^T \gamma - h^T G(G^T G)^{-1} G^T \gamma + h^T \gamma \tag{11}$$

$$s.t. \quad 0 \leq \gamma \leq C_2 e$$

where, $G = [A \quad e], f = Y - \varepsilon_1$ and $h = Y + \varepsilon_2 e$.

After optimizing (10) and (11), we can obtain the regression function of TSVR as follows:

$$f(x) = \frac{1}{2}(f_1(x) + f_2(x)) = \frac{1}{2}(w_1 + w_2)^T x + \frac{1}{2}(b_1 + b_2) \tag{12}$$

where, $[w_1 \quad b_1]^T = (G^T G)^{-1} G^T (f - \alpha), [w_2 \quad b_2]^T = (G^T G)^{-1} G^T (h + \gamma)$.

For the nonlinear case, TSVR considers the following kernel-generated functions:

$$f_1(x) = K(x^T, A^T)w_1 + b_1, f_2(x) = K(x^T, A^T)w_2 + b_2 \tag{13}$$

Similarly, solving (9) can be obtained by dealing with the following quadratic programming problems:

$$\min \quad \frac{1}{2}\left\|Y - e\varepsilon_1 - (K(A, A^T)w_1 + eb_1)\right\|^2 + C_1 e^T \xi \tag{14}$$

$$s.t. \quad Y - (K(A, A^T)w_1 + eb_1) \geq e\varepsilon_1 - \xi, \quad \xi \geq 0$$

$$\min \quad \frac{1}{2}\left\|Y + e\varepsilon_2 - (K(A, A^T)w_2 + eb_2)\right\|^2 + C_2 e^T \eta \tag{15}$$

$$s.t. \quad (K(A, A^T)w_2 + eb_2) - Y \geq e\varepsilon_2 - \eta, \quad \eta \geq 0$$

According to the KKT conditions, the dual problems of (14) and (15) are as follows:

$$\max \quad -\tfrac{1}{2}\alpha^T H(H^T H)^{-1} H^T \alpha + f^T H(H^T H)^{-1} H^T \alpha - f^T \alpha \tag{16}$$

$$s.t. \quad 0 \leq \alpha \leq C_1 e$$

$$\max \quad -\tfrac{1}{2}\gamma^T H(H^T H)^{-1} H^T \gamma - h^T H(H^T H)^{-1} H^T \gamma + h^T \gamma \tag{17}$$

$$s.t. \quad 0 \leq \gamma \leq C_2 e$$

where, $H = [K(A, A^T) \quad e]$. After optimizing (16) and (17), we can obtain the augmented vectors for $f_1(x)$ and $f_2(x)$, which are.

$$[w_1 \quad b_1]^T = (H^T H)^{-1} H^T (f - \alpha), \quad [w_2 \quad b_2]^T = (H^T H)^{-1} H^T (h + \gamma) \tag{18}$$

Then the regression function of nonlinear TSVR is constructed as follows:

$$f(x) = \frac{1}{2}(f_1(x) + f_2(x)) = \frac{1}{2}K(x^T, A)(w_1 + w_2) + \frac{1}{2}(b_1 + b_2) \tag{19}$$

In short, compared with SVR, TSVR is comprised of a pair of smaller quadratic programming problems. In TSVR, each quadratic programming problem determines the one of up- or down-bound function by using only one group of constraints. This strategy makes TSVR obtain approximately four times faster than SVR.

Similar to SVR, the size of TSVR parameters has a considerable impact on its learning ability and generalization ability. Therefore, how to select the most suitable parameters is the key to improve the performance of TSVR. In this paper, the improved whale optimization algorithm is used to find the optimal solution of TSVR parameters.

3.2 Whale Optimization Algorithm and Its Improvement

3.2.1 The Standard Whale Optimization Algorithm

Whale optimization algorithm (WOA) is a new optimization algorithm based on the modeling of whale predation behavior. The whale looks for the location and encircles it by the smell of its prey. Assuming that the prey position reflected by the smell is the current optimal position or close to the optimal position, a certain number of virtual humpback whales are defined as the search agent. By comparing the feasible solutions of various search agents, the optimal solution is found as the next position vector of humpback whale. At the same time, other search agents update their positions. To find the optimal solution strategy.

In order to describe this strategy, the following mathematical model is adopted.

$$D_1 = \left| CX^*(t) - X(t) \right| \tag{20}$$

$$X(t+1) = X^*(t) - AD_1 \tag{21}$$

where t is the current number of iterations, $X(t)$ is the coordinate vector of the current humpback whale, $X(t+1)$ is the target coordinate vector after the next iteration and $X^*(t)$ is the coordinate vector to get the best solution so far. And if there is a better feasible solution, $X^*(t)$ should be updated immediately. A and D are coefficients given by the following equation.

$$A = 2ar_1 - a \tag{22}$$

$$C = 2r_2 \tag{23}$$

where r_1 and r_2 are random numbers between 0 and 1. a is obtained from Eq. (24).

$$a = 2 - 2t/T_{\max} \tag{24}$$

In order to find a better target location to get close to prey, humpback whales randomly use arbitrary whale coordinate vectors instead of humpback whales to generate the next iteration of whale coordinate vectors. This can achieve the purpose of deviating from the prey to avoid falling into local optimum. The description uses the following mathematical model:

$$X(t+1) = \begin{cases} X^*(t) - AD_1, & |A| < 1, \ p < 0.5 \\ X_{rand}(t) - AD_2, & |A| \geq 1, \ p < 0.5 \\ X^*(t) + D_3 e^{dl} \cos(2\pi l), & p \geq 0.5 \end{cases} \qquad (25)$$

$$\begin{aligned} D_1 &= \left| CX^*(t) - X(t) \right| \\ D_2 &= \left| CX_{rand} - X(t) \right| \\ D_3 &= \left| X^*(t) - X(t) \right| \end{aligned} \qquad (26)$$

where t represents the current number of iterations, $X^*(t)$ is the best position vector so far, $X_{rand}(t)$ is the random whale position vector, $X(t)$ represents the current whale position vector, d is a constant with a default value of 1, which is used to control the hunting path shape. l is given by the following formula:

$$l = (a_2 - 1)r_3 + 1 \qquad (27)$$

$$a_2 = -1 - t/T_{\max} \qquad (28)$$

where r_3 is a random number between 0 and 1, t is the current number of iterations, $T_{\max}$ is the maximum number of iterations.

3.3 Improved Whale Optimization Algorithm

Dynamic inertia weight is an important mechanism to balance and adjust the global survey and local mining capacity of the algorithm. A larger inertia weight means a larger search step size, which is conducive to the algorithm to carry out global search, enhance the ability to jump out of local extremum and find the global optimal solution; the smaller inertia weight is conducive to the algorithm for local mining, improving the convergence accuracy of the algorithm and speeding up the recovery Convergence speed. Based on the above analysis, this paper divides the shrinking encirclement strategy and spiral predation strategy of the whale optimization algorithm into three evolutionary stages: the early stage, the middle stage and the late stage. In the early stage of evolution, the inertia weight of the algorithm remains unchanged, that is, it still uses a large fixed value, so that the algorithm can fully search in the global space. In the middle and later stage, the nonlinear decreasing dynamic inertia weight is introduced to make the whale approach the global optimal solution, which accelerates the convergence speed of the algorithm. The weight formula introduced is as follows:

$$if \ (t < \frac{1}{3} * T_{\max})$$

$$w = 1$$

$$else$$

$$w = 1 - e^{rand*(\frac{t}{T_{\max}} - 1)}$$

$$end$$

In the above formula, while maintaining the overall decreasing trend of inertia weight from 1 to 0, it shows certain randomness. This randomness reduces the risk that the algorithm fails to search for the theoretical optimal value in the early stage, but falls into the local extreme value directly due to the monotonous decline of inertia weight in the later stage, which is conducive to the algorithm getting rid of the local optimum. When inertia weight is added, the formula of contraction encirclement strategy and spiral predation strategy is as follows:

$$X(t+1) = \begin{cases} wX^*(t) - AD_1, & |A| < 1, \ p < 0.5 \\ wX_{rand}(t) - AD_2, & |A| \geq 1, \ p < 0.5 \\ wX^*(t) + D_3 e^{dl} \cos(2\pi l), & p \geq 0.5 \end{cases} \qquad (29)$$

3.4 Fuzzy Information Granulation Based on IWOA-TSVR for Prediction Model of Stock Opening Price Range

In this paper, the improved whale optimization algorithm is used to optimize the parameters of TSVR, and a twin support vector regression machine based on the improved whale optimization is proposed. Finally, it is used to predict the change space of the stock opening index within 5 days. The algorithm flow chart of this paper is as follows:

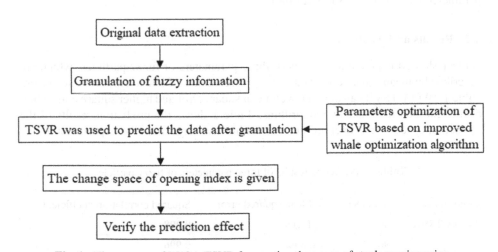

Fig. 1. The process of IWOA-TSVR forecasting the space of stock opening price

4 Experimental Results and Analysis

4.1 Experimental Data and Algorithm Parameter Setting

In order to verify the performance of IWOA-TSVR, three algorithms of IWOA-TSVR, WOA-TSVR and TSVR were used to analyze the opening data of Shanghai Stock Exchange (hereinafter referred to as Shanghai Stock Exchange) for 4946 trading days from December 19, 1990 to February 28, 2011 http://wenku.baidu.com/view/3c3b7d 878762caaedd33d4b0.html) Regression analysis was carried out and the forecast was made within 5 trading days after February 28, 2011 The change space of disk data.

This experiment is carried out in matalab7.11 environment. The related parameters are set as follows:

The population number of whale optimization algorithm is $N = 30$, $T_{\max} = 1000$, $d = 1$. Table 1 shows the prediction results of three fuzzy particle parameters by the three algorithms, Table 2 shows the actual opening index values of Shanghai Stock Exchange in the five trading days after February 28, 2011, and Table 3 shows the spatial prediction results of the three algorithms for the index change space of Shanghai Stock Exchange in the five trading days after February 28, 2011. Figure 2 shows the opening data of Shanghai stock exchange for 4946 trading days from December 19, 1990 to February 28, 2011. Figure 3 is the result of granulation of opening price data, Fig. 4 is the fitting result of parameter a by IWOA-TSVR algorithm, and Fig. 5 is the fitting error chart of parameter a by IWOA-TSVR algorithm.

4.2 Results and Analysis

Table 1 shows the prediction results of three parameters of fuzzy particles, which are described by mean square error and square correlation coefficient. As can be seen from Table 1, IWOA-TSVR can obtain lower mean square error and higher square correlation coefficient than the other two algorithms. Table 3 shows the prediction results of the prediction space of the opening number within five days after the experimental data

Table 1. Prediction results of three parameters of fuzzy particle

Algorithm	Parameter	Mean squared error	Squared correlation coefficient
IWOA-TSVR	a	**13.15**	**0.998**
	m	**8.02**	**0.996**
	b	**7.87**	**0.995**
WOA-TSVR	a	16.54	0.945
	m	13.58	0.937
	b	10.21	0.921
TSVR	a	18.02	0.885
	m	16.13	0.921
	b	13.85	0.867

deadline. The prediction result is described by a fuzzy particle. The value of particle parameter a represents the lowest opening price in these five days, the parameter m represents the median value of the opening price in five days, and parameter B describes the highest opening price in five days. From the actual opening price of these five days, the forecast range of IWOA-TSVR is effective. Moreover, compared with other algorithms, the prediction result of IWOA-TSVR is closer to the actual opening price shown in Table 2. Figure 2 shows an example of 4946 opening prices of the original data, and Fig. 3 shows an example of the fuzzy granulation of the original data. The number of particles is 990, greatly reducing the number of samples. Figure 4 shows the fitting process of parameter a by IWOA-TSVR. From the figure, we can see that the fitting effect of parameter a is very good. Figure 5 shows the fitting error results of parameter a by IWOA-TSVR, and the error of most particles is close to 0, which shows the good fitting effect of parameter a. At the same time, we also use this model to predict other indexes of Shanghai Stock Exchange, and we also obtain satisfactory results.

Table 2. The actual opening price within five days

Date	2011.3.1	2011.3.2	2011.3.3	2011.3.4	2011.3.7
Actual opening price	2906.28	2905.51	2918.73	2902.19	2952.72

Table 3. The predicted range opening price within five days

Algorithm	The forecast range of opening price of march 1 to march 4 and march 7 (described by the fuzzy particle)
IWOA-TSVR	[a,m,b] = **[2905.1,2915.3,2954.6]**
WOA-TSVR	[a,m,b] = [2803.9,2912.3,2918.6]
TSVR	[a,m,b] = [2554.3,2885.8,2907.3]

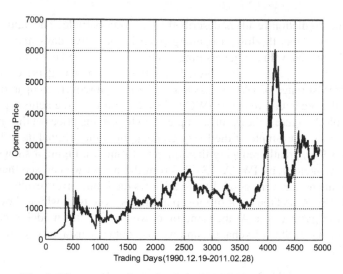

Fig. 2. The number of opening price (1990.12.19-2011.02.28)

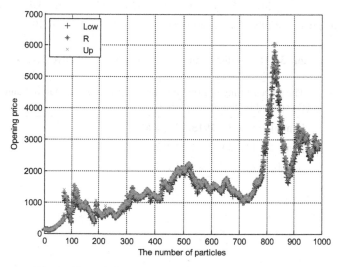

Fig. 3. Fuzzy information granulation visualization map

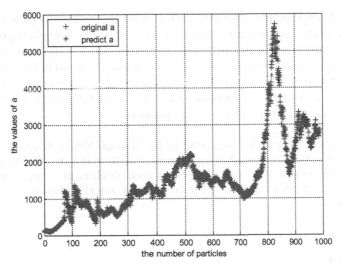

Fig. 4. Fitting result of parameter a by IWOA-TSVR

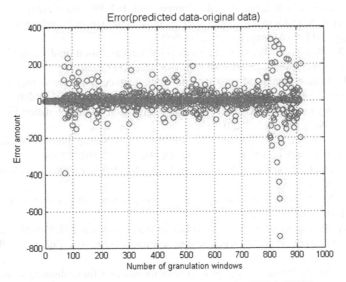

Fig. 5. Fitting error chart of parameter a by IWOA-TSVR

5 Conclusions

The stock market system is one of the most active economic systems in the economic system, and it is very difficult to predict the stock market due to the influence of many changing factors. Most of the time, we can't accurately predict the stock index. At this time, if we can predict the change space of the stock index's opening index, it is particularly important. TSVR is a regression model with good learning performance and generalization ability, but its performance is affected by the parameters. In this paper, the improved whale optimization algorithm is used to optimize its parameters, and a twin support vector regression machine learning method based on the improved whale optimization is proposed to establish the prediction model of stock index change space. The experimental results show that the prediction model proposed in this paper has high accuracy and stable performance. It is an effective tool to predict the change space of stock index. In view of the good performance of this prediction model, it will be the next step to apply it to other fields of time series prediction.

Acknowledgments. This work is supported by the National Natural Science Foundation of China (61662005). Guangxi Natural Science Foundation (2018GXNSFAA294068); Basic Ability Improvement Project for Young and Middle-aged Teachers in Colleges and Universities in Guangxi (2019KY0195); Research Project of Guangxi University for Nationalities (2019KJYB006).

References

1. Vijh, M., Chandola, D., Tikkiwal, V.A., Kumar, A.: stock closing price prediction using machine learning techniques. Procedia Comput. Sci. **167**, 599–606 (2020)
2. Das, G., et al.: Rule discovery from time series. In: 4th International Conference on Knowledge Discovery and Data Mining, pp. 16–22 (1998)
3. Zhang, X., Huang, Y., Du, J., et al.: Stock price forecast based on discrete hidden Markov model. J. Zhejiang Univ. Technol. **48**(02), 148–153+211 (2020)
4. Shen, J., Omair Shafiq, M.: Short-term stock market price trend prediction using a comprehensive deep learning system. J. Big Data **7**(1), 10696–10707 (2020)
5. Jiang, C., Maasoumi, E., Xiao, Z.: Quantile aggregation and combination for stock return prediction **39**(7), 715–743 (2020)
6. Nti, I.K., Adekoya, A.F., Weyori, B.A.: Efficient stock-Market prediction using ensemble support vector machine. Open Comput. Access **10**(1), 153–163 (2020)
7. Peng, X.: TSVR: An efficient Twin Support Vector Machine for regression. Neural Netw. **23**, 365–372 (2010)
8. Wang, Y., Xu, Y.: Scaling up twin support vector regression with safe screening rule. Inf. Sci. **465**, 174–190 (2018)
9. Hao, P.-Y.: Pairing support vector algorithm for data regression **225**, 174–187 (2017)
10. Seyedali, M., Andrew, L.: The whale optimization algorithm. Adv. Eng. Softw. **95**(5), 51–67 (2016)
11. Sun, P., Shan, R.: Predictive control with velocity observer for cushion robot based on PSO for path planning. J. Syst. Sci. Complexity **33**(4), 988–1011 (2020). https://doi.org/10.1007/s11424-020-8375-x
12. Zhou, X., Jiaxin, L., Huang, J., et al.: Enhancing artificial bee colony algorithm with multi-elite guidance. **543**, 242–258 (2021)
13. Barkaoui, M., Berger, J.: A new hybrid genetic algorithm for the collection scheduling problem for a satellite constellation. **71**(9), 1390–1410 (2020)
14. Zadeh, L.A.: Discussion from imprecise to granular probabilities. Fuzzy Sets Syst. **154**, 370–374 (2005)

Channel Assignment Algorithm Based on Discrete BFO for Wireless Monitoring Networks

Na Xia[1], Lin-Mei Luo[1(✉)], Hua-Zheng Du[1], Pei-Pei Wang[1], Yong-Tang Yu[2], and Ji-Wen Zhang[2]

[1] School of Computer Science and Information Engineering, Hefei University of Technology,
Hefei 230601, China
2019111048@mail.hfut.edu.cn
[2] China Ji-Kan Research Institute of Engineering Investigations and Design Company Ltd.,
Xi'an 710043, China

Abstract. Wireless monitoring networks employ distributed sniffers to capture the transmissions of wireless users. It can be used for wireless network status analysis, fault diagnosis, and resource management, etc. Due to the limited number of sniffers, it is a key topic to optimize sniffers' channel assignment to collect the maximum transmitted data, so as to maximize the Quality of Monitoring (QoM) of the network. In this paper, a channel assignment algorithm based on discrete Bacterial Foraging Optimization is proposed. A 2D multi-radio multi-channel (MRMC) coding is designed to represent the bacterial individual; the bacterial foraging and position updating can achieve the optimized channel assignment scheme for wireless monitoring networks. This algorithm is with low complexity and has provable convergence performance. Extensive experiments also demonstrate that the proposed algorithm is efficient and outperforms the existing algorithms.

Keywords: Wireless monitoring networks · Channel allocation · Discrete BFO

1 Introduction

With the growing applications of wireless networks (WLAN, WiFi, Mesh, etc.), it is becoming very important to estimate the network conditions and performance characteristics, so as to complement the network status analysis, fault diagnosis, and resource management [1, 2]. Wireless monitoring networks composed of a set of hardware devices called *sniffers* are used to monitor the activities in wireless networks. Sniffers capture transmissions of wireless devices or activities of interference sources in their vicinity, and store packet level or PHY layer information in trace files, which can be analyzed distributedly or at a central location. Wireless monitoring networks [3–5] has been shown to complement wire side monitoring using SNMP and basestation logs since it reveals detailed PHY (e.g., signal strength, spectrum density) and MAC behaviors (e.g., collision, retransmissions), as well as timing information (e.g., backoff time), which are essential for network analysis and diagnosis [6, 7].

© Springer Nature Switzerland AG 2021
D.-S. Huang et al. (Eds.): ICIC 2021, LNCS 12836, pp. 707–724, 2021.
https://doi.org/10.1007/978-3-030-84522-3_58

With the development of hardware technique, wireless sniffers typically are equiped with multi-radio, which means a sniffer can collect information using multi radios with different channels. Thus, it is a key topic to determine the optimal channel assignment for sniffers' radios to maximize the information collected. This is called the *multi-radio multi-channel assignment* problem in wireless monitoring networks.

The rest of the paper is organized as follows. In Sect. 2, we analyze the related works of recent years. The problem is described and formulated in Sect. 3. A channel assignment algorithm based on discrete Bacterial Foraging Optimization (DBFO) is detailed in Sect. 4. Simulations and practical network experiments are conducted in Sect. 5. Finally, we conclude the paper with some future work in Sect. 6.

2 Related Work

In recent years, wireless monitoring network has been an active research area. It is constructed and deployed among wireless network to collect and store data, carry out analysis and feedback in real-time. The relationship between wireless monitoring network and wireless network is shown in Fig. 1. Wireless network can be WLAN, WIFI, WiMAX, Mesh, Ad hoc, etc., and the wireless users in wireless network can be wireless routers, access points or mobile users.

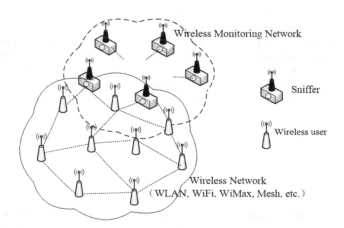

Fig. 1. Wireless monitoring network model

There has been much work done on wireless monitoring network including monitoring equipment development, network framework design, sniffers' channel selection/assignment, data analysis and fusion, etc. Among these, sniffers' channel selection/assignment is vital to improve the monitoring quality of the network effectively.

In 2010, Chhetri *et al.* [8] formulated the problem of sniffer channel selection, and proved it to be NP-hard to maximize the quality of monitoring (QoM) of the network under universal network model. Greedy and LP algorithms were also employed to try to solve the problem. Greedy always misses the optimal or approximate optimal solution.

Although LP can achieve better solution than Greedy, its complexity is too high to meet the real-time requirement in dynamic wireless networks.

In 2012, Shin *et al.* [9] proposed a Dynamic-optimal Sniffer-Channel Assignment algorithm, combining proximal optimization algorithm with dual approach to solve the problem of LP relaxation, and finally proved that the algorithm can achieve the best approximation of $1 - 1/e$ after taking the integer portion.

In 2013, we presented a Monte Carlo Enhanced PSO algorithm for optimal QoM in single-radio multi-channel wireless monitoring networks [10]. The suitable coding expression of particles and the fitness function were designed. Monte Carlo method was incorporated to revise the solution. Furthermore, in 2015, Xia *et al.* [11, 12] studied Multi-channel Quantum-Inspired Immune Clone Algorithm (MQICA). The algorithm applied full interference crossover strategy, and introduced Gaussian mutation operation to jump out of the local optima effectively. The simulations demonstrated the good convergence to the global optimal solution.

In this paper, a discrete Bacterial Foraging Optimization based algorithm is proposed to achieve the optimal channel assignment for multi-radio multi-channel (MRMC) wireless monitoring networks. The theoretical prove and extensive experiments demonstrate its effectiveness.

3 Problem Descriptions

3.1 Network Model

Consider a wireless monitoring network of m monitoring sniffers, n users, q optional channels, and each sniffer has p radios ($p \geq 2$, $p < q$). $S = \{s_1, s_2, \cdots, s_m\}$ is the set of sniffers, $U = \{u_1, u_2, \cdots, u_n\}$ is the set of users, $C = \{c_1, c_2, \cdots, c_q\}$ is the set of channels, and $R = \{r_1, r_2, \cdots, r_p\}$ is the set of radios, $S^R = \{s_1^{r_1}, s_1^{r_2}, \cdots, s_1^{r_p}, \cdots, s_m^{r_1}, s_m^{r_2}, \cdots, s_m^{r_p}\}$ is the set of radios of all sniffers, where $s_i^{r_w}$ is the $w-th$ radio of sniffer s_i. When radio $s^r \in S^R$ and its adjacent user $u \in U$ operate on the same channel, s^r can capture the data transmissions from u, i.e., u is covered or monitored by s^r.

The relationship between *multi-radio* sniffers and users can be described using an undirected bi-partite graph $G^R = (S^R, U, E)$ shown in Fig. 2. If u_j is the neighbor of s^r, there will be an edge between them, denoted by $e = (s^r, u_j) \in E$. E represents the set of all connecting edges.

If a user cannot be captured by any sniffer, it is excluded from G^R. The vertex v of G^R is a sniffer or user, $v \in S^R \cup U$. $N(v)$ denotes the neighbors of vertex v. If the vertex is a sniffer s, $N(s)$ is the set of neighbor users of s, if the vertex is a user u, $N(u)$ means the set of neighbor sniffers of u. If a sniffer is inside the communication range of another sniffer, it is called adjacent sniffer. $V(s)$ denotes the set of adjacent sniffers of s. In this paper, we assume that the communication radius of sniffer is twice as its monitoring radius.

3.2 Problem Formulation

The multi-radio sniffer channel assignment problem is all sniffers' radios $s_i^{r_w}$ ($i = 1, \cdots, m$; $w = 1, \cdots, p$) select channels from $C = \{c_1, c_2, \cdots, c_q\}$ to be able to

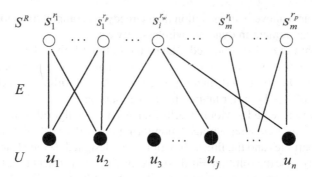

Fig. 2. Undirected bi-partite graph GR

capture the maximum data transmissions from users. Different from single-radio problem, the radios of a multi-radio sniffer have the same position, their neighbor users and sniffers are shared.

The searching space of multi-radio multi-channel assignment problem is shown in Fig. 3, where $m \times p$ columns denote m $\times$ p radios of m sniffers, and q rows denotes q optional channels. Each radio $s_i^{r_w}$ searches in the channel space $\{c_1, c_2, \cdots, c_q\}$. The shadow combination in Fig. 3 represents a channel assignment scheme for $m \times p$ radios.

The channel assignment scheme shown in Fig. 3 can be expressed using a **2-dimension binary coding** as Fig. 4. i $(i = 1, \cdots, m)$ denotes sniffer number, w $(w = 1, \cdots, p)$ denotes radio number of each sniffer, and k $(k = 1, \cdots, q)$ denotes channel number. Therefore, the coding matrix is $m \times p$ columns and q rows. Element $\theta_{k(i \times w)} = 1$ means the $i-th$ sniffer's $w-th$ radio is assigned with channel k. For example, $\theta_{2(1 \times 1)} = 1$, $\theta_{(q-1)(1 \times 2)} = 1$, $\theta_{1(1 \times p)} = 1$ means for the first sniffer, its first, second and $p - th$ radio are assigned with channel 2, $q - 1$, and 1 respectively.

$s_1^{r_1}$	$s_1^{r_2}$	$\cdots$	$s_1^{r_p}$	$s_2^{r_1}$	$s_2^{r_2}$	$\cdots$	$s_2^{r_p}$	$\cdots$	$s_m^{r_1}$	$s_m^{r_2}$	$\cdots$	$s_m^{r_p}$
c_1	c_1	$\cdots$	c_1	c_1	c_1	$\cdots$	c_1	$\cdots$	c_1	c_1	$\cdots$	c_1
c_2	c_2	$\cdots$	c_2	c_2	c_2	$\cdots$	c_2	$\cdots$	c_2	c_2	$\cdots$	c_2
$\vdots$	$\vdots$	$\vdots$	$\vdots$	$\vdots$	$\vdots$	$\vdots$	$\vdots$	$\cdots$	$\vdots$	$\vdots$	$\vdots$	$\vdots$
c_{q-1}	c_{q-1}	$\cdots$	c_{q-1}	c_{q-1}	c_{q-1}	$\cdots$	c_{q-1}	$\cdots$	c_{q-1}	c_{q-1}	$\cdots$	c_{q-1}
c_q	c_q	$\cdots$	c_q	c_q	c_q	$\cdots$	c_q	$\cdots$	c_q	c_q	$\cdots$	c_q

Fig. 3. Searching space of the channel assignment problem

Because each radio can be assigned with only one channel, i.e., in Fig. 4, each column has only one "1", the number of possible 2D coding matrix is $q^{m \times p}$, which means the number of possible channel assignment schemes is $q^{m \times p}$. If considering "sniffer should make its different radios operate on different channels to avoid collecting repeated information", the number of possible 2D coding matrix is $\left(C_q^p\right)^m$. It is a complicated combination optimization problem to search for the optimal 2D coding (channel assignment scheme).

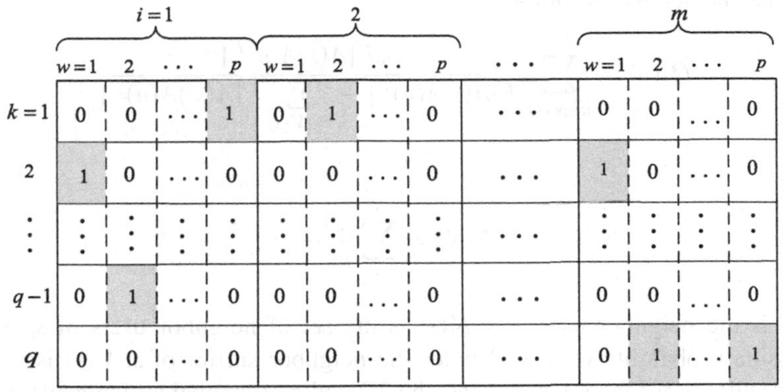

Fig. 4. A 2-dimension binary coding for a channel assignment scheme

Case 1: different radios of a sniffer can operate on the same channel.

In this condition, 2D binary coding for a sniffer will have multi "1" in a row. For example, in Fig. 4, for sniffer m, two "1" in row q, which means the sniffer's radio 2 and p are both assigned with channel q. If we accumulate the elements in the same row for each sniffer, 2D binary coding will be transferred to **2D accumulation coding** shown as Fig. 5. Each column represents the channel usage situation for a sniffer, denoted by $A(s_i) = (\sigma_{i,1}, \sigma_{i,2}, \cdots, \sigma_{i,q})$, where $\sigma_{i,k} (0 \leq \sigma_{i,k} \leq p)$ indicates the times of channel k has been assigned to sniffer i. For example, in Fig. 5, the m-th column $A(s_m) = (0, 1, \cdots, 0, 2)$, where $\sigma_{m,2} = 1$ means channel 2 has been assigned to a radio of sniffer m; $\sigma_{m,q} = 2$ means channel q has been assigned to two radios of sniffer m.

Fig. 5. A 2D accumulation coding for a channel assignment scheme (Case 1)

Although different radios operating on the same channel will capture repetitive user data, it is valuable especially for some network applications with reliable monitoring demands. It can provide multiple guarantees for tracking lost during transmission, so as to promote the fusion of information and increase the integrity of user information.

Definition 1. Quality of Monitoring of sniffer in Case 1 (**QoM-1**). When wireless monitoring networks operate on channel assignment scheme **a**, quality of monitoring of sniffer s can be defined as follows:

$$Q_s(\mathbf{a}) = \sum_{u \in N(s)} \frac{p_u F[A(s)A(u)^T]}{F[A(s)A(u)^T] + \sum_{s' \in V(s)} F[A(s')A(u)^T]} \tag{1}$$

$$F(N) = \sum_{i=1}^{N} \frac{1}{k^{i-1}} \tag{2}$$

where u is the neighbor user of s, $N(s)$ is the set of neighbor users of s, p_u is the transmission probability of u, s' denotes the neighbor sniffer of s, $V(s)$ is the set of neighbor sniffers of s, and $A(s)$ denotes the 2D coding (channel assignment) for s. The number of radios operating on the same channel of u can be computed by $A(s)A(u)^T$. Considering the radios on same channel will capture repetitive data, the contribution coefficient $F(N)$ of the radios will be computed by formula (2), and the parameter k ($k \geq 1$) can be set according to monitoring reliability requirement. It is shown that with the increasing of the radios sharing same channel, the contribution increment will decrease exponentially. E.g., a sniffer has three radios operating on same channel at same time to monitor the activity of user u, $N = 3$ can be computed by $A(s)A(u)^T$, the information contribution coefficient of the three radios is $F(3) = 1 + 1/k + 1/k^2$.

Case 2: different radios of a sniffer operate on different channels.

In this case, 2D binary coding for a sniffer will have only one "1" in a row. If the coding does not satisfy the constraint, it should be revised. For example, in Fig. 4, for sniffer m, two "1" in row q, which means the sniffer's radio 2 and p are both assigned with channel q. This coding should be revised: maintain the coding of radio 2, and modify the coding of radio p (move "1" to another bit in this column, till there is only one "1" in a row for this sniffer), shown in Fig. 6. Conduct this type of revision till there is no channel multiplex. Figure 6 is the revised coding from Fig. 4, where radio 2 and p of sniffer m are assigned with channel q and $q - 1$ respectively.

Accumulate the elements in the same row for each sniffer, 2D binary coding will also be transferred to 2D accumulation coding shown as Fig. 7. It is shown that, different from the former, the nonzero element in this coding must be "1".

Each column $A(s_i) = (\sigma_{i,1}, \sigma_{i,2}, \cdots, \sigma_{i,q})$ also represents the channel usage situation for a sniffer, where $\sigma_{i,k}$ ($\sigma_{i,k} = 1 \,/\, 0$) indicates whether channel k has been assigned to sniffer i. For example, in Fig. 7, the m-th column $A(s_m) = (0, 1, \cdots, 1, 1)$ means channel 2..., $q - 1$, q have been assigned to the radios of sniffer m.

Definition 2. Quality of Monitoring of sniffer in Case 2 (**QoM-2**). When wireless monitoring networks operate on channel assignment scheme **a**, quality of monitoring of sniffer s can be defined as follows:

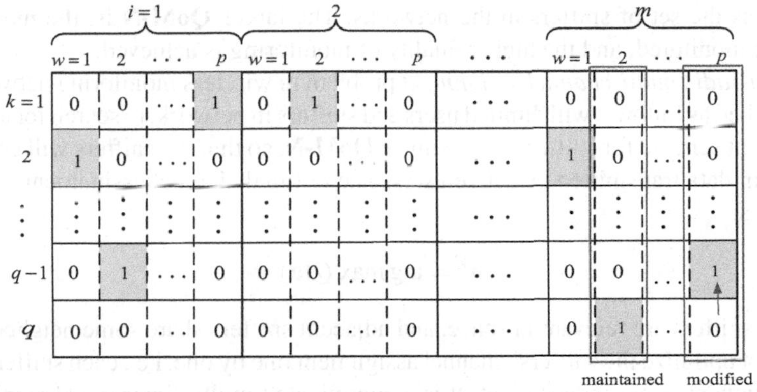

Fig. 6. A 2-dimension binary coding satisfying the constraint condition

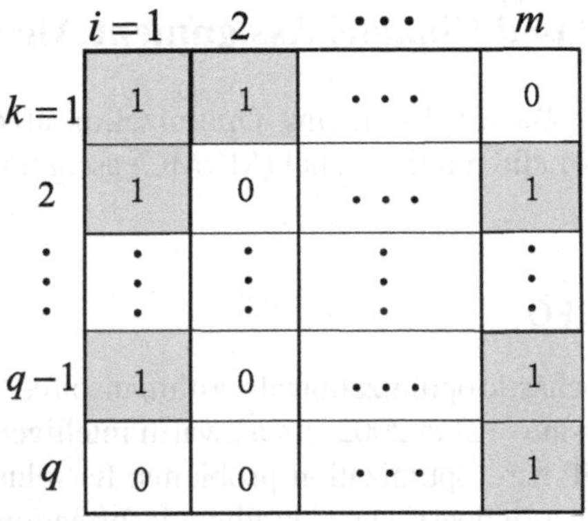

Fig. 7. A 2D accumulation coding for a channel assignment scheme (Case 2)

$$Q_s(\mathbf{a}) = \sum_{u \in N(s)} \frac{p_u A(s) A(u)^T}{1 + \sum_{s' \in V(s)} A(s') A(u)^T} \tag{3}$$

where the symbols of u, $N(s)$, $V(s)$, etc. are same as them in definition 1. Because different radios of a sniffer operate on different channels to capture different users' transmissions, there is no need to process the repetitive data using formula (2).

Definition 3. Quality of Monitoring of network (**QoM-N**). Given a channel assignment scheme **a**, the quality of monitoring of wireless monitoring networks can be defined as follows:

$$Q(\mathbf{a}) = \sum_{s \in S} Q_s(\mathbf{a}) \tag{4}$$

where S is the set of sniffers in the networks. The larger **QoM-N** is, the more active users are monitored, and the higher quality of monitoring is achieved.

Multi-radio multi-channel assignment problem in wireless monitoring networks can be described as follows: with limited users and sniffers in network, to search for a channel assignment scheme for sniffers to maximize **QoM-N**, so that the sniffers will collect the maximum data transmitted by active users. The optimal channel assignment scheme is as follows:

$$\mathbf{a}^* = \arg\max Q(\mathbf{a}) \tag{5}$$

Because sniffers are relevant in space, and adjacent sniffers share some neighbor users, we cannot optimize the sniffers' channel assignment one by one, i.e., each sniffer's channel assignment is not independent. It is a complicated multi-dimensional combination optimization problem to search for the optimal channel assignment scheme for network.

4 Discrete BFO Based Channel Assignment Algorithm

In this paper, a discrete Bacterial Foraging Optimization algorithm (DBFO) is proposed to solve the multi-radio multi-channel (MRMC) assignment problem in wireless monitoring networks.

4.1 Basic Theory of BFO

BFO algorithm is the stochastic optimization algorithm inspired by the bacterial foraging behavior studied by Passino [12] in 2002. As a swarm intelligent optimization method, it is suitable to solve NP-hard optimization problems. Its solution performance is not sensitive with initial values, it doesn't need gradient information of object function, and its searching is parallel towards the global optimization position. It has been successfully applied in several fields in recent years.

The specific process for solving BFO algorithm: initial population, fitness function value calculation, doing three basic operations of chemotaxis, reproduction and elimination-dispersal by loop, to obtain optimal or near-best optimal solution.

4.2 Process of DBFO

Assuming that there are S bacterias in the population, $\theta^i(j, k, l)$ indicates the position of bacterial i after finishing j-th chemotaxis, k-th reproduction, and l-th elimination-dispersal, for short, $\theta^i \in R^D$, where D indicates the dimension of problem. $D = m \times p \times q$, in a monitoring network of m sniffers, p radios, and q channels. $\theta^i = [\theta^i_1, \theta^i_2, \cdots, \theta^i_D]$, where θ^i_d indicates the d-th dimensional position of bacteria i. $\Delta^i = [\Delta^i_1, \Delta^i_2, \cdots, \Delta^i_D]$ indicates the directional vector of bacterial i when doing chemotaxis operation, where $\Delta^i_d \in [-1, 1]$ indicates the d-th dimensional randomly selected chemotactic direction of bacteria i. $\Phi^i = \Delta^i / \sqrt{(\Delta^i)^T \times \Delta^i}$ indicates the unit vector of Δ^i. $\Phi^i = [\phi^i_1, \phi^i_2, \cdots, \phi^i_D]$, where ϕ^i_d indicates the unit directional vector of each dimension.

Sigmoid function is used in the BFO algorithm. The followings are discrete binary value of 0-1 of each dimension.

$$\Delta\theta_d^i(j, w+1, k, l) = \Delta\theta_d^i(j, w, k, l) + \lambda^i\phi_d^i \qquad (6)$$

$$\theta_d^i(j, w+1, k, l) = \begin{cases} 1 & , \text{ if } \rho_d^i(j, w+1, k, l) < sigmoid(\Delta\theta_d^i(j, w+1, k, l)) \\ 0 & , \qquad\qquad otherwise \end{cases} \qquad (7)$$

where, w indicates swimming count, $\Delta\theta_d^i(j, k, l)$ indicates in the d-th dimension the moving distance of bacterial i after finishing j-th chemotaxis, k-th reproduction, and l-th elimination-dispersal; λ^i indicates the swimming step length; θ_d^i indicates the unit directional vector in the d-th dimension; $\theta_d^i(j+1, k, l)$ indicates the d-th dimensional position of bacterial i after finishing $(j+1)$-th chemotaxis, k-th reproduction, and l-th elimination-dispersal; $\rho_d^i(j+1, k, l)$ indicates a random sequence of bacterial i after finishing $(j+1)$-th chemotaxis, k-th reproduction, and l-th elimination-dispersal. The sequence is uniformly distributed, $\rho^i \sim U(0, 1)$, $sigmoid(x) = 1/(1 + exp(-x))$.

After executing flipping operation in the process of chemotaxis, the result needs to be verified through the QoM of fitness function. If $Q(\theta^i(j+1, k, l)) > Q(\theta^i(j, k, l))$, continue to swim a step in the direction of the chemotaxis, until the fitness function value QoM is not increasing or reaching the maximum number of swimming N_s. The count of chemotaxis updates only in flipping operation, and keep the same when swimming.

Through above, after the discretization, the BFO algorithm can be discrete from continuous amount into binary value, which contributes to solve binary discrete space searching problem. So, we can use the BFO algorithm to solve the channel coding problem in multi-radio multi-channel wireless monitoring network.

4.3 Update and Revise Channel Coding Table

Assuming that the number of sniffers is m, each sniffer has p radios, and the number of channels is q. $(m \times p) \times q$ binary numbers compose a sniffer channel coding table $A = (\sigma_{i,w,k})_{m \times p \times q}$, where $\sigma_{i,w,k}$ represents the coding of radio w in sniffer i on channel k, $\sum_{w=1}^{p} \sigma_{i,w,k} = \sigma_{i,k}$. The coding table represents the current position of bacteria. We adopt the discrete BFO algorithm updating and revising channel coding table below.

In the initial stage of the algorithm, each bacterium generates a table, the element of which is 0. That means, the initial position values of all bacteria are 0. Updating bacteria position is the same with updating channel coding table. As a channel assignment scheme, the updated coding table must follow certain principles: (1) the radio of sniffer at the same time can only be allocated on one channel, so only one element in the column vector corresponding to each sniffer radio can be encoded "1" in channel coding; (2) in order to make full use of the radio, the column vector does not allow to have all "0" states; (3) in the case of "no different radio multiplexing channel", in the convergence stage, column vector is not allowed to have more than one "1".

In the initial table, the position value is not a solution, which can't be taken into the formula QoM to solve for the first time, so it needs to flip a direction to generate a new position based on the initial one, then be used to solve the solution. Because of the random generation of each dimension position, it is difficult to guarantee once discretization fully meeting the channel assignment principle, so it needs to revise the binary coding table after each bacterium updating position.

The revising steps are as follows:

Step 1: randomly choose column elements $(\sigma_{i,w,1}, \sigma_{i,w,2}..., \sigma_{i,w,q})$ which are not running, if the number of column elements coding "1" is 1, that is $\sum_{k=1}^{q} \sigma_{i,w,k}=1$, turn to Step 4.

Step 2: if the number of column elements coding "1" is more than 1, that is $\sum_{k=1}^{q} \sigma_{i,w,k} > 1$, for each $k(\sigma_{i,w,k}=1)$, compute the active users sniffer radio monitoring on the channel: compute the active users monitored by sniffer radio on the channel: $Q_{i,w}(k) = \sum_{u \in N(s_i)} p_u \cdot A_k A(u)^T$, where A_k is a q dimensional vector, excepting for the *k-th* dimension is 1, the other elements are 0. Sort the calculated values and select the channel corresponding $\max(Q_{i,w}(k))$ and code for "1", the other elements are set "0".

Step 3: if the column elements all code for "0", that is $\sum_{k=1}^{q} \sigma_{i,w,k}= 0$, for each k ($k = 1, \cdots, q$), compute the active user's sniffer radio monitoring on the channel, select the channel corresponding $\max(Q_{i,w}(k))$ and code for "1".

Step 4: repeat step 1, until each column of the coding table only has one element of "1", that is, $\sum_{k=1}^{q} \sigma_{i,w,k}=1(i = 1, \cdots, m; \ w = 1, \cdots, p)$.

Step 5: enter the state of collecting coding. Choosing case 1, simply collect directly to make $\sigma_{i,k} = \sum_{w=1}^{p} \sigma_{i,w,k}$; choosing case 2, after collecting, judge each column whether the element coding is more than 1, if so, operate step 6.

Step 6: If the element coding is more than 1, that is $\sigma_{i,k} > 1$, firstly set the element for "1", calculate the active user's sniffer radio monitoring on the channel corresponding the coding of "0", and sort the calculated results to balance the coding of "1".

Figure 8 shows a revised process after updating encodings of 4 sniffers, 2 radios in 1 sniffer, 10 channels wireless monitoring network in channel assignment. Figure 8(a) indicates the coding after bacteria updating a position at a certain time, where identity "(1)" column indicates that violating the principle of channel assignment (1); identity "(2)" column indicates that violating the principle of channel assignment (2); identity "(3)" column indicates that violating the principle of channel assignment (3); no identity column indicates that conforming to the principle of channel assignment. Figure 8(b) indicates the coding after running revised step 2. Figure 8(c) indicates the coding after running revised step 3. Figure 8(d) indicates the collecting coding in case 1. Figure 8(d) indicates the collecting coding in case 2. Figure 8(f) indicates the coding after revision of step 6.

4.4   Theoretical Performance Analysis of DBFO

The moving distance $\Delta\theta_d^i(j, k, l)$ on each dimensional position is the probability of updated position $\theta_d^i(j, k, l)$ for 0 or 1.

If $\Delta\theta_d^i(j, k, l) = 0$, then $p(\theta_d^i(j, k, l) = 1) = 0.5$; if $\Delta\theta_d^i(j, k, l) < 0$, then $p(\theta_d^i(j, k, l) = 1) < 0.5$; if $\Delta\theta_d^i(j, k, l) > 0$, then $p(\theta_d^i(j, k, l) = 1) > 0.5$; $p(\theta_d^i(j, k, l) = 0) = 1 - p(\theta_d^i(j, k, l) = 1)$. Even if moving the same distance $\Delta\theta_d^i(j, k, l)$ on each dimensional position component, due to the randomness of $\rho_d^i(j, k, l)$, position

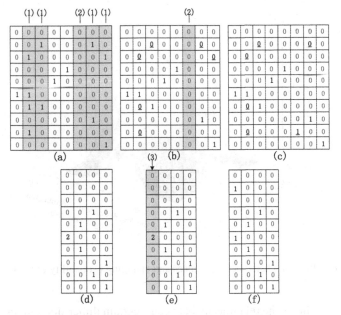

Fig. 8. Revised process after updating encodings

$\theta_d^i(j, k, l)$ may choose different values; *sigmoid* function value does not represent the changing probability on a certain dimension, only represents the probability of a certain dimension chooses 1.

Analysis of Positional Changing Probability

Literature [14] puts forward an idea of positional changing probability for the discrete DPSO, here we also analyze each dimension positional changing probability for the discrete BFO. The process of bacteria updating position is a kind of probability changing, for a single dimension is the probability of swimming distance variation as 1. So, when bacterial i doing chemotaxis j operation, the probability of a certain dimension position changing is:

$$p_{0 \leftrightarrow 1} = sigmoid(\Delta\theta_d^i(j-1, k, l)) * [1 - sigmoid(\Delta\theta_d^i(j, k, l))]$$
$$+ sigmoid(\Delta\theta_d^i(j, k, l)) * [1 - sigmoid(\Delta\theta_d^i(j-1, k, l))] \quad (8)$$

Formula shows that from a certain dimension position moving distance of doing chemotaxis j-1 and j operations, we can get the changing probability of the dimension position. Combined with sigmoid function and (8):

$$p_{0 \leftrightarrow 1} = \frac{1}{1 + \exp(-\Delta\theta_d^i(j-1, k, l))} \times [1 - \frac{1}{1 + \exp(-\Delta\theta_d^i(j, k, l))}]$$
$$+ \frac{1}{1 + \exp(-\Delta\theta_d^i(j, k, l))} \times [1 - \frac{1}{1 + \exp(-\Delta\theta_d^i(j-1, k, l))}] \quad (9)$$

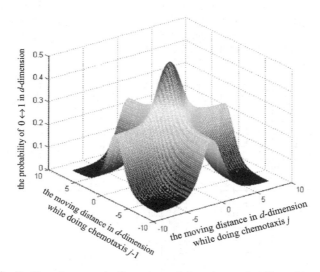

Fig. 9. Two chemotactic distances and changing probability of $0 \leftrightarrow 1$

Formula (9) shows the relationship between a certain dimension moving distance of the position component and the changing position, which is shown in Fig. 9. We can see that the changing probability of $0 \leftrightarrow 1$ is associated with the two chemotactic distances, but the movement distance is 0 according to the biggest changing probability in the two operations of chemotactic, and the maximum value is 0.5.

As a special case, when the two chemotactic operating distances are the same. The vertices are in the place of $\Delta \theta_d^i(j, k, l) = 0$ and the maximum value is 0.5, which means the bacteria does no longer move.

Discrete BFO Convergence Analysis

The bacteria D-dimensional position solution space $\theta^i = (\theta_1^i, \theta_2^i, \cdots, \theta_D^i)$ generated by DBFO algorithm is a discrete random process, and conducts a finite homogeneous Markov chain. Chemotaxis, reproduction and elimination-dispersal operations all ensure the future state of BFO is only related to the current state, and has nothing to do with other states before, which is in conformity with the definition of a Markov chain. And transition probability apparently has nothing to do with time n, only is related to the current generation of position deviation, so Markov chain can be seen as homogeneous. As a result, Θ constitutes finite homogeneous Markov chain in state space Ω^S.

The goal of reproduction and elimination-dispersal operations is to speed up the convergence, and help to jump out of local optimum. The core of DBFO algorithm is the chemotactic operation. If you can prove that in the condition of taking swimming position, chemotaxis process is convergent in probability 1, then you can prove BFO algorithm is convergent in probability 1.

5 The Experimental Results and Analysis

5.1 Simulations

In order to prove the discrete BFO algorithm in solving the problem of multi-channel assignment has a better performance, we firstly validate the convergence of BFO after discretization by the simulation experiments. Convergence experiment conducts in scale of 9 monitoring sniffers, each sniffer equipped with 2 radios, 6 available channels, and 200 users. Experimental parameters of DBFO algorithm are set in Table 1.

Table 1. Experimental parameter settings

S	N_c	N_{re}	N_{ed}	N_s	P_{ed}	$d_{attract}$	$w_{attract}$	$h_{repellant}$	$w_{repellant}$
10	50	4	2	4	0.2	0.05	0.1	0.05	0.05

Parameters in the table are obtained from a large number of experiments and good performance experience values of the discrete BFO. Convergence experiment conducts in case 1 "sniffer allows different radios select the same channel" and case 2 "sniffer forbids different radios select the same channel", shown in Fig. 10.

InitQom in the Figure represents the quality of monitoring after algorithm running the chemotaxis for the first time, different InitQoms indicate bacterial chemotaxis operates from different directions. Iterations on abscissa indicate the number of chemotactic. Case 1 and case 2 get the same quality of monitoring, and InitQom of case 2 generally is greater than case 1. case 2 is easier to reach convergence, and the complexity of the curve reduces.

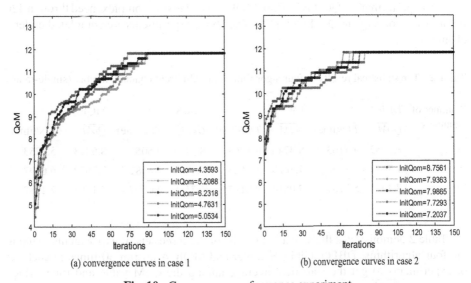

(a) convergence curves in case 1   (b) convergence curves in case 2

Fig. 10. Convergence performance experiment

We conduct the comparative experiments in network scale of 9 monitoring sniffers, single-radio, 6 channels and 200 users. Quality of Monitoring (QoM) of every algorithm is shown in Fig. 11.

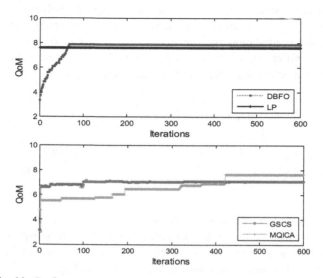

Fig. 11. Performance comparison of the four algorithms (single-radio)

As depicted in Fig. 11, after 69 iterations, the proposed DBFO algorithm converges to the extremely optimal solution (QoM = 7.8624). LP algorithm is a deterministic algorithm, only running once in the process of experiment, which achieves the QoM of 7.5629, while Gibbs Sampler is a distributed algorithm, basically achieves the centralized algorithm performance, QoM is 7.0736. MQICA is relatively complex, need through 426 iterations to converge to QoM of 7.6231. The above experiments are completed under 6 channels.

Table 2. The statistical results of four algorithms under different channel number (single-radio)

Number of channels	DBFO		LP		GSCS		MQICA	
	$\overline{QoM}$	$\overline{T}$(s/time)	$\overline{QoM}$	T(s/time)	$\overline{QoM}$	$\overline{T}$(s/time)	$\overline{QoM}$	$\overline{T}$(s/time)
3	8.5952	4.7163	8.4748	6.1964	8.3405	3.5052	8.5893	5.3253
6	7.7583	4.8456	7.5629	7.3523	7.1407	3.8481	7.7032	6.6597
9	7.2631	5.3159	7.0814	9.6197	6.7188	4.0697	7.1563	7.2987

Table 2 demonstrates the statistical results of the three sets of experiments. Among the four algorithms, DBFO, Gibbs Sampler and MQICA all run 30 times in each set of experiments to get the statistical average among the QoM value and time, which indicated of $\overline{QoM}$ and $\overline{T}$. As deterministic methods, LP only runs once. From Table 2, we can see that DBFO outperforms LP in three sets of experiments, and is evidently better than Gibbs Sampler and MQICA. Furthermore, DBFO converges fast, with shorter running time than MQICA.

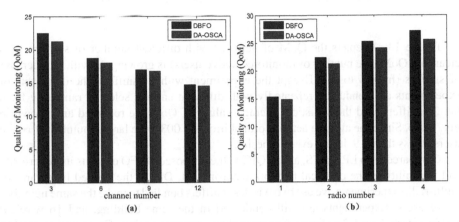

Fig. 12. Statistical comparison of DBFO and DA-OSCA in different numbers of channel and radio

Literature [13, 15] introduces multi-radio monitoring network model, and puts forward a DA-OSCA algorithm to solve channel assignment problem, which is a linear programming algorithm. This method converts multi-radio of sniffer into multiple sniffers to allocate channels, although not accurately called multi-radio channel assignment, this algorithm is similar with case 1, so we compare DBFO with DA-OSCA. To illustrate the performance of DBFO, we compare DBFO and DA-OSCA in different numbers of channel and radio. Due to DA-OSCA is a deterministic algorithm, only need to obtain one solution, and randomness exists in DBFO, so we use the mean value of 30 times of experiments. The results are shown in Fig. 12. Figure 12(a) shows the effect of channel number on QoM. With the increase of channel number, QoMs of two algorithms both reduce, but QoM of DBFO is better than that of DA-OSCA to some extent, and the gap between two algorithms reduces. Figure 12 (b) shows the effect of radio number on QoM. With the increase of radio number, QoMs of two algorithms both increase, and the gap between them relative increases.

5.2 Practical Network Experiment

In this subsection, we evaluate the proposed DBFO algorithm by practical network experiment based on campus wireless network (IEEE 802.11.b WLAN). Twenty-one WiFi sniffers are deployed in a building to collect the user information from 9:00 am to 2:00 pm (over 5 h). Totally 587 active users are monitored working on three orthogonal channels. The number of users on the three channels is 316, 113, and 158 respectively. The information transmission probabilities of these users are recorded in Table 3. It is shown that the activity probabilities of most users are less than 1%. The average activity probability is 0.0031.

Table 3. Transmission probability statistics for users

Information transmission probability	0–0.01	0.01–0.02	0.02–0.04
Numbers of users	524	28	35

Figure 13(a) depicts the QoM of network with different number of sniffers. It is clear that QoM (the number of monitored active users) is growing up with the increase of sniffers (from 5 to 21). Except the experiment with 21 sniffers, the other sets of experiments are conducted repeatedly with different sniffers selected randomly from the 21 sniffers, and the statistical average values of QoM are recorded and shown in Fig. 13(a). Since the average activity probability is 0.0031, the largest number of active users is less than 1.9 during every timeslot.

Compared with LP, GSCS, and MQICA, the proposed DBFO exhibits its superiority and feasibility in the practical network environment. Due to the limited hardware for multi-radio experiments, we use two wireless sniffers bundled to make the same neighbor active users. Experiments are still conducted in the same building, and 16 wireless sniffers treated as 8 2-radio sniffers complete user monitoring. Compared with DA-OSCA algorithm, the proposed DBFO also exhibits a better QoM shown in Fig. 13(b).

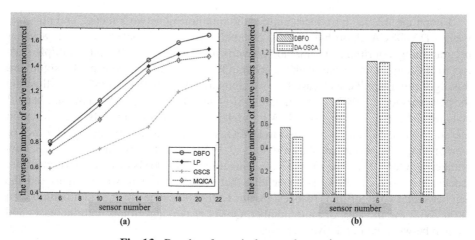

Fig. 13. Results of practical network experiment

6 Conclusion

This paper focuses on how to optimize the channel assignment to maximize the quality of monitoring in multi-channel multi-radio wireless network, converts channel assignment problem into two-dimensional channel coding problem. In the two cases of "whether different radios of the same sniffer can multiplex channel or not", we define sniffer monitoring quality and quality of monitoring in wireless monitoring network (QoM), and express them formulated, finally, adopt a discrete BFO algorithm to solve. Discrete each dimension positions of binary 0 or 1 in an iterative process, which can encode for wireless monitoring network channel assignment problem, by updating and revising of the three steps of chemotaxis, reproduction and elimination-dispersal, ultimately make the optimal channel assignment. Large numbers of experiments show that the algorithm can achieve multi-radio multi-channel assignment to maximize quality of monitoring. Compared with other algorithms, this algorithm is with low complexity, and high-quality solution, and also makes up defects of the existing literatures lacking study on multi-radio channel assignment. For the next work, we will further introduce the related theory of multi-radio channel assignment to distributed achieve, which is conducive to the applications in large-scale networks and to solving the reliable monitoring problem in wireless networks.

Acknowledgement. This work was support in part by the National Natural Science Foundation of China under Grant 61971178 and Grant 61701161; in part by Science and Technology Major Project of Anhui Province under Grant 18030901015; in part by the Technology Innovation Guidance Project (Fund) of Shaan xi Province under Grant 2020CGHJ002.

References

1. Price, R.: Fundamentals of Wireless Networking. McGraw-Hill, Irwin (2015)
2. Arora, P., Xia, N., Zheng, R.: A Gibbs sampler approach for optimal distributed monitoring of multi-channel wireless networks. In: 2011 IEEE Global Telecommunications Conference - GLOBECOM 2011, pp. 1–6 (2011). https://doi.org/10.1109/GLOCOM.2011.6133790
3. Branco, A., Sant' Anna, F., Ierusalimschy, R., Rodriguez, N., Rossetto, S.: Terra: flexibility and safety in wireless sensor networks. ACM Trans. Sensor Netw. (TOSN) 11(4), 1–27 (2015)
4. Arianpoo, N., Leung, V.C.M.: How network monitoring and reinforcement learning can improve tcp fairness in wireless multi-hop networks. EURASIP J. Wireless Commun. Netw. 2016(1), 1–15 (2016). https://doi.org/10.1186/s13638-016-0773-3
5. Arcadius Tokognon, C., Gao, B., Tian, G.Y., Yan, Y.: Structural health monitoring framework based on Internet of Things: a survey. IEEE Internet Things J. 4(3), 619–635 (2017). https://doi.org/10.1109/JIOT.2017.2664072
6. Ghanavati, S., Abawajy, J.H., Izadi, D., Alelaiwi, A.A.: Cloud-assisted IoT-based health status monitoring framework. Cluster Comput. 20(2), 1843–1853 (2017). https://doi.org/10.1007/s10586-017-0847-y
7. Bhuiyan, M.Z.A., Wu, J., Wang, G., Wang, T., Hassan, M.M.: e-sampling: event-sensitive autonomous adaptive sensing and low-cost monitoring in networked sensing systems. ACM Trans. Auton. Adapt. Syst. (TAAS) 12(1), 1–29 (2017)
8. Emde, S., Boysen, N.: Berth allocation in container terminals that service feeder ships and deep-sea vessels. J. Oper. Res. Soc. 67(4), 551–563 (2016)
9. Shin, D., Bagchi, S., Wang, C.: Distributed online channel assignment toward optimal monitoring in multi-channel wireless networks. In: 2012 Proceedings IEEE INFOCOM, pp. 2626–2630 (2012). https://doi.org/10.1109/INFCOM.2012.6195666
10. Hua-Zheng, D., Xia, N., Jiang, J.-G., Li-Na, X., Zheng, R.: A Monte Carlo enhanced PSO algorithm for optimal QoM in multi-channel wireless networks. J. Comput. Sci. Technol. 28(3), 553–563 (2013)
11. Xia, N., Xu, L., Ni, C.: Optimal QoM in multichannel wireless networks based on MQICA. Int. J. Distrib. Sensor Netw. 9(6), 120527 (2013)

12. Passino, K.M.: Biomimicry of bacterial foraging for distributed optimization and control. IEEE Control Syst. Mag. **22**(3), 52–67 (2012). https://doi.org/10.1109/MCS.2002.1004010
13. Kennedy, J., Eberhart, R.C.: A discrete binary version of the particle swarm algorithm. In: 1997 IEEE International Conference on Systems, Man, and Cybernetics. Computational Cybernetics and Simulation, vol. 5, pp. 4104–4108 (1997). https://doi.org/10.1109/ICSMC.1997.637339
14. Wu, X.L.: Continuity and convergence of set-valued function on fuzzy measure space. Southeast University Press (2019)
15. Nga Nguyen, T.T., Brun, O., Prabhu, B.J.: Joint downlink power control and channel allocation based on a partial view of future channel conditions. In: 2020 18th International Symposium on Modeling and Optimization in Mobile, Ad Hoc, and Wireless Networks (WiOPT), pp. 1–8 (2020)

A Multi-objective Particle Swarm Optimization Algorithm Embedded with Maximum Fitness Function for Dual-Resources Constrained Flexible Job Shop Scheduling

Jing Zhang[1] and Jing Jie[2(✉)]

[1] Department of Computer and Information Security, Zhejiang Police College, Hangzhou, Zhejiang, China
[2] College of Automation and Electronic Engineering, Zhejiang University of Science and Technology, Hangzhou, Zhejiang, China
jjing277@sohu.com

Abstract. This paper is concerned with a multiple-objective flexible job-shop scheduling problem with dual-resources constraints. Both time and cost-concerned objectives are taken into consideration and the corresponding mathematical model is presented. Based on Maximal fitness function, a hybrid discrete particle swarm algorithm is proposed to effectively solve the problem. The global and local search ability of the algorithm are both improved by modifying the position updating method and simulating annealing strategy with Maximal fitness function. Moreover, external archive is used to reserve better particles. Finally, the effectiveness of the proposed algorithm demonstrated by simulation examples and the results show that the obtained solutions are more uniformly distributed towards the Pareto solutions.

Keywords: Maximum fitness function · Particle swarm optimization · Dual-resource constrained scheduling problem · Multi-objective optimization

1 Introduction

In the past decades, flexible job shop scheduling problem (FJSP) has received extensive research interests due to its practical applications in modern management, supply chain, manufacturing industry [1–3]. As well known, FJSP is NP-hard and also one of the most difficult combinational optimization problems. Thus, various methods have been introduced to solve the problem which focuses on either single resource constraint or single optimization objective. However, in practical manufacturing systems, there are always other resource constraints besides machine such as worker constraint and various optimization objectives [2, 4, 5]. Moreover, the objectives may be conflict, for example, minimizing the production time requires that competent workers and machines are used more time and hence increase the cost. Thus FJSP with dual resource constraints (DRC) and multiple objectives is an interesting and practical issue that deserves investigation

© Springer Nature Switzerland AG 2021
D.-S. Huang et al. (Eds.): ICIC 2021, LNCS 12836, pp. 725–738, 2021.
https://doi.org/10.1007/978-3-030-84522-3_59

and we called it DRC-MOFJSP for short. DRC refers to machine and worker constraints here.

On the other hand, the multiple objective problem of FJSP is also a hot research topic and various hybrid intelligent algorithms were presented to solve it [6]. As for the objectives, time-based criteria are critical and always chosen as objectives in FJSP, such as flow time, makespan and tardiness time. Other objectives with respect to machine load, for instance, maximal machine workload and the total workload, are also considered [2]. In [7], the total production cost was also pointed out to be one of the main concerns and it contains the cost of raw material inventory before started and in progress and cost of tardiness and losing goodwill. Thus we choose the production cost and production time to be the optimization objectives in this paper.

It is well known that discrete particle swarm optimization (DPSO) is an effective tool in the combinational optimization area [8] and Maximin Fitness Function (MFF) is also very useful in multiple objective optimization [9, 10]. However, they have not been applied for DRC-MOFJSP in the existing literature to the best of the authors' knowledge. Thus we aim to propose a hybrid DPSO (MOPSO) based on the MFF to solve the DRC-MOFJSP in this paper. A three dimension coding scheme is used to represent a particle due to the DRC feature. A mixed initialization way is employed to improve the quality of the population. Then DSPO is presented as the global search algorithm, in which MFF is used and an improved position updating method is introduced. This is the main novelty of the proposed algorithm and the searching efficiency is improved. As for the local search one, a new simulating annealing search (NSA) strategy is designed. A modified reservation strategy is used to maintain the external archive (EA) based on MFF. Finally, simulation results are given to the results show that the obtained solutions are more uniformly distributed towards the Pareto solutions.

2 Problem Description

2.1 Literature Review

The development of research on FJSP can be found in [1–3, 11, 12] and the references therein. The considered situations varied from the most standard one to complicated ones such as dynamic scheduling with uncertainty and multiple objective scheduling. Fruitful results were obtained and here we focus on reviewing the literature on the DRC and multiple objective issues.

For the DRC issue, two kinds of methods were commonly used for such issue. One method is first to find the operation sequences by intelligent algorithms and then allocate the machines and workers by dispatching rules [4, 13]. The main advantages of this method are less computation time and the reduction of complexity. Though local optimal solutions can always be found, the global one is difficult to be obtained in this way. Another is to determine the operation sequence, machine and worker simultaneously by hybrid intelligent algorithms [14, 16, 17]. An improved variable neighbourhood search was proposed in [16] the advantage of which is the ability to explore different neighbourhoods of the current solution and thus is easy for implementation. In [17], a knowledge-guided fruit fly optimization algorithm was proposed for DRC-FJSP with a novel encoding scheme.

Survey of multiple objective optimization problems can be found in [6], which reviewed existing algorithms of many fields including FJSP. In [2], the objectives were chosen to be makespan, total machine load and maximal machine load, and a Pareto method was presented to balance the load of machines and the makespan. A memetic algorithm was presented in [15] and a novel local search mechanism together with a hierarchical strategy was introduced to improve the search ability. A set of Pareto solutions were obtained and the method was shown to be effective by comparing benchmark instances. A hybrid artificial bee colony algorithm and a simulated annealing algorithm were proposed in [18] and [19] respectively to solve MOFJSP and improved results were obtained. In [20], a discrete firefly algorithm combined with local search method was presented to enhance the searching accuracy and information sharing among fireflies.

The aforementioned results considered either DRC constraint or multiple objective issue of FJSP. However, few research results considered DRC-MOFJSP, especially considering the production cost as objective. In [7], a novel production cost objective model was presented for the scheduling of job shop and performance analysis was carried out with a rolling time horizon approach. In [21–23], the cost objective and the corresponding job scheduling problem was modified by taking DRC and flexibility into consideration and hybrid genetic algorithms were presented to solve it. In [24], a inherited genetic algorithm was proposed to for DRC-MOFJSP with objectives being production time and production cost.

2.2 Mathematical Model

In this section, the mathematical model of the DRC-MOFJSP is described. The model is based on [21, 23] and some notations are introduced first.

Notations: Let p, q, h and g be the index of job, operation and machine and worker respectively. O_{pq} is q-th operation of job p. M_{pq} and W_h are the set of candidate machines for O_{pq} and workers for machine h.

Parameters: Denote N_n, N_m, N_g and N_p as the total number of jobs, machines, workers and operations of job p. t_{pqh} is the processing time of operation O_{pq} on machine h. α_p, β_h, γ_g are the cost of material, machine and workers, respectively. η is the holding cost rate. U is a large number.

Decision Variables: S_{pq} is the starting time of operation O_{pq}. K_{hg} is the starting time of machine h processed by worker g. C_{pq} and C_p are the completing time of operation O_{pq} and job p. C_{max} is the maximum completing time of all jobs. E_p and T_p are the penalty tardiness and advance of job p, respectively. and

$$\sigma_{pqh} = \begin{cases} 1, & \text{if machine } h \in M_{pq} \text{ is} \\ \text{selected for the operation } O_{pq}, \\ 0, & \text{otherwise} \end{cases} \quad \varepsilon_{hg} = \begin{cases} 1, & \text{if worker } g \in W_h \text{ is} \\ \text{selected for the machine } h \\ 0, & \text{otherwise} \end{cases}$$

$$\chi_{hg-h'g} = \begin{cases} 1, & \text{if worker } g \text{ is performed on} \\ \text{machine } h \text{ before } h' \\ 0, & \text{otherwise} \end{cases} \quad \varsigma_{pqh-p'q'h} = \begin{cases} 1, & \text{if } O_{pq} \text{ is performed on} \\ \text{machine } h \text{ before } O_{p'q'} \\ 0, & \text{otherwise} \end{cases}$$

Some standard assumptions are made before we formulate the DRC-MOFJSP. First, the initial time for all jobs, machines and workers are zero. Second, each machine can process only one operation and every worker can operate only one machine at one time. Finally, any operation cannot be interrupted until its completion once it is started.

Under the assumptions and notations, the mathematical model for the problem is defined as follows:

$$\begin{cases} F_1 = \min C_{\max} = \min\{\max_{p=1}^{n}\{C_p\}\} \\ F_2 = \min\{F_{21} + F_{22} + F_{23}\} \end{cases} \tag{1}$$

where

$$F_{21} = \eta \sum_{p=1}^{N_n} \alpha_p S_{p(n_p+1)}$$

$$F_{22} = \eta \sum_{p=1}^{N_n} \sum_{q=1}^{N_p} (S_{p(n_p+1)} - S_{pq})(\beta_h + \gamma_g)t_{pqh}\sigma_{pqh}\varepsilon_{hg}$$

$$F_{23} = \sum_{p=1}^{N_n} \sum_{q=1}^{N_p} (\beta_h + \gamma_g)t_{pqh} + \sum_{p=1}^{N_n} (\phi_p E_p + \varphi_p T_p)$$

s.t.

$$S_{p(q+1)} \geq S_{pq} + t_{pqh}\sigma_{pqh}\varepsilon_{hg} \tag{2}$$

$$S_{p'q'} + (1 - \varsigma_{pqh-p'q'h})U \geq S_{pq} + t_{pqh} \tag{3}$$

$$K_{h'g} + (1 - \chi_{hg-h'g})U \geq K_{hg} \tag{4}$$

$$K_{hg} + (1 - \sigma_{pqh})(1 - \varepsilon_{hg})U \leq S_{pq} \tag{5}$$

$$S_{p(q+1)} + (1 - \varsigma_{p(q+1)h-p'q'h})U \geq C_{pq} \tag{6}$$

$$\sum_{h=1}^{M_{pq}} \sigma_{pqh} = 1, \sum_{g=1}^{W_h} \varepsilon_{hg} = 1 \tag{7}$$

$$C_p \leq C_{\max}, S_{pq}, t_{pqh} \geq 0 \tag{8}$$

$$\sigma_{phg}, \varsigma_{pqh-p'q'h} \in \{0, 1\}, \varepsilon_{hg}, \chi_{hg-h'g} \in \{0, 1\} \tag{9}$$

where $p, p' \in J, h \in M_{pq}, g \in W_h, q, q' = 1, 2, \ldots, n_p$, F_1 and F_2 are the production time and production cost, respectively. F_{21} and F_{22} are the cost of inventory of raw material and the incremental cost of machines and workers in progressing. F_{23} is the sum of processing cost of machines and workers and penalty of tardiness and advance of jobs. Constraints (2)–(9) are standard restrictions for FJSP and implications of them can be referred to existing literature.

3 Design of MOPSO

For the DRC-MOFJSP, we propose a MFF-based MOPSO to solve it. Firstly, a three-dimension coding scheme is used to represent a particle. Secondly, various methods are employed in a mixed way for initialization of the particles. Then an improved DPSO and a NSA mechanism are designed as the global and local search strategies, respectively. Finally, a modified reservation strategy is used to maintain the EA. Since MFF plays an important role in the design of the MOPSO, some fundamental concepts and features of MFF are introduced here first.

3.1 Maximin Fitness Function

MFF was used in [9] to determine whether a solution is non-dominated or not and hence was employed to solve multiple-objective optimization problems. For an optimization problem with size N and number of objectives M, the MFF of a solution X_i is defined by

$$F_{mf}(X_i) = \max_{j=1,2,...,N, j \neq i} \left\{ \min_{k=1,2,...,M} \left\{ F_k(X_i) - F_k(X_j) \right\} \right\} \tag{10}$$

Clearly, $F_{mf}(X_i) < 0$ and $F_{mf}(X_i) > 0$ represent that X_i is a non-dominated and dominated solution for the problem, respectively. If $F_{mf}(X_i) = 0$, then X_i can be a weakly-dominated, dominated or non- dominated with duplicated solutions, as point B in (a), points A and C in (b) and points B and D in (c) of Fig. 1. We can see that $F_{mf}(X_i) = 0$ may cause some critical issues such as elimination of non-dominated solutions, and thus need to be addressed.

Firstly, since dominated solutions have effects on the MFF of non-dominated ones, they have to be eliminated while evaluating the performance of non-dominated solutions by MFFs. Thus, if X_i is either weakly-dominated or dominated when $F_{mf}(X_i) = 0$, then X_i should be eliminated. Secondly, for the case of non-dominated with duplicated solutions, if we check them by the standard of $F_{mf}(X_i) = 0$, then the non-dominated solutions will be eliminated. We propose a method called elimination of common solutions (ECS) to avoid this situation. Find out the solutions with the same MFF, then randomly choose one and use it to calculate MFF again. Finally we set the MFF value to the other solutions with the same MFF. For example, we use ECS to handle (c) of Fig. 1. If we choose point B to calculate the MFF and set the value to point D, then we have $F_{mf}(B) = F_{mf}(D) = -2$. Otherwise, $F_{mf}(B) = F_{mf}(D) = 0$ leads to the loss of two non-dominated solutions.

Another issue is that the objectives may not be in the same order of magnitude. This will make the MFF tend to prefer the objective with smaller magnitude order. Thus a standardization procedure is employed here to avoid this deviation. The procedure is as follows:

$$std\_F_k(X_i) = \frac{F_k(X_i) - \min_{j=1,2,...N} F_k(X_j)}{\max_{j=1,2,...N} F_k(X_j) - \min_{j=1,2,...N} F_k(X_j)} \tag{11}$$

where $std\_F_k(X_i)$ is the standardized one of $F_k(X_i)$. Considering the calculation load, we only standardize the particles in the EA.

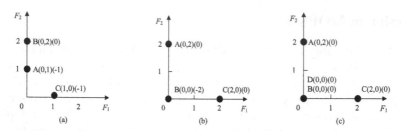

Fig. 1. Three cases of $F_{mf} = 0$

Finally, it should be pointed out that for the particles with the same MFF, if the distances among them are too small, then they cannot be identified which one is better or not. Thus a modified reservation strategy with truncation will be presented to reserve more sparse particles in EA. The crowding distance of a particle is denoted as follows:

$$D_{dis}(i) = \sum_{k=1}^{M} |std\_F_k(i+1) - std\_F_k(i-1)| \tag{12}$$

Clearly, we can see that larger crowding distance means the particle is sparser. If several particles have the same MFF and only part of them can be reserved, then the particles with smaller crowding distance should be reserved for further optimization.

3.2 Initialization and Maintenance of EA

The EA is used to store the global best particles and the length is denoted by NG. EA is very important for the MOPSO since the global best solution used in DPSO is chosen from EA. The initialized particles with $F_{mf}(X_i) < 0$ are included for the initialization of EA. As for the maintenance of EA, a modified reservation strategy with truncation is used to make the particles sparser. The benefit of this strategy is that better particles can be reserved for with the ones same MFF.

Assuming that there are j particles have the same MFF and i particles have smaller MFF than them, where $i < $ NG. If $i + j < = $ NG, then the j particles are all reserved, otherwise calculate the crowding distance and eliminate the NG particles with larger crowding distance. It should be pointed out that this strategy consumes less computation since only if $i + j > $ NG is satisfied.

3.3 Coding Scheme

A three-dimension coding scheme [24] is employed to represent a particle. Specifically, each particle has three vectors which are denoted by OS, MA and WA, respectively, where OS stands for the operation sequence, MA and WA represent for the machine and worker allocations, respectively. For OS, the *i-th* appearance of *j* means the operation O_{ij}. Each element of MA and WA refers to the index of the corresponding machine and worker, respectively. The advantage of the encoding method is the avoidance of invalid solutions. For example, the second column of the particle in Table 1 means the operation O_{13} is operated on machine 1 and by worker 2. The whole particle stands for a valid production process and objectives can be calculated accordingly.

Table 1. Code of a particle

OS	1	3	3	1	2	3	2
MA	2	1	2	2	3	3	3
WA	1	2	1	1	1	1	1

3.4 Initialization of Population

The quality of population initialization has effects on the convergence speed of the algorithm. In this paper, we choose a mixed way for the initialization. Random generation and the most number of operations remaining (MNOR) rule are used for the initialization of OS [4]. For MA and WA, three rules are used in a mixed way, which are random generating rule, a modified Approach by Localization (AL) and Earliest Finish Time (EFT) rule [2, 25]. The modified AL is to rearrange the processing time according to initialized OS, then allocate operations with minimal workload. The EFT rule is taking into account the processing time and the earliest started time of resources.

3.5 Global Search Mechanism

An improved DPSO is presented as a global search mechanism for the MOPSO. The position updating equation is given as follows:

$$X_i^{k+1} = c_2 \otimes f_3\{[c_1 \otimes f_2(w \otimes f_1(X_i^k), pB_i^k)], gB^k\} \qquad (13)$$

where x_i^k is the k-th generation of particle i, w is the inertia weight, c_1 and $c_2 \in [0,1]$ are the acceleration coefficients. pB_i^k and gB^k are the best personal of i-th particles and the best global particle at k-th generation, respectively. pB_i^k and gB^k are obtained based on MFF and will be used in the next updating step of particles. f_1, f_2 and f_3 are operators of updating. Specifically, the updating consists three part. The first one is

$$E_i^k = w \otimes f_1(X_i^k) = \begin{cases} f_1(X_i^k); & r < w \\ X_i^k; & \text{Otherwise} \end{cases} \qquad (14)$$

where $r \in (0,1)$, and if $r < w$, then f_1 is operated for x_i^k. For OS, exchange two different operations randomly while keep machine and worker allocation unchanged. For MA and WA, we randomly choose an operation and replace the machine and worker from the corresponding candidate set, respectively. The second part is adjustment with be best personal pB_i^k which is formulated as

$$F_i^k = c_1 \otimes f_2(E_i^k, pB_i) = \begin{cases} f_2(E_i^k, pB_i^k); & r < c_1 \\ E_i^k; & \text{Otherwise} \end{cases} \qquad (15)$$

If $r < c_1$, then a Precedence Preserving Order based Crossover (POX) operation is carried out for OS, and two Rand-point Preservation Crossover (RPX) are operated for MA and WA. Specific procedure of POX and RPX can be referred to [26]. They can effectively avoid the occurrence of invalid solutions and hence improve the searching efficiency. f_3 is the operation between $F_i{}^k$ and gB^k

$$X_i^k = c_2 \otimes f_3(F_i^k, gB^k) = \begin{cases} f_3(F_i^k, gB^k); & r < c_2 \\ F_i^k; & \text{otherwise} \end{cases} \tag{16}$$

If $r < c_2$, then f_3 is adjusted with gB^k and the operations are similar to f_2. Now we have a new generation of particles X^{k+1} after updating. Then the personal best particles pB_i^{k+1} and global best gB^{k+1} are obtained by calculating the MFF of X^{k+1}. If $F_{mf}(X_i^{k+1}) < F_{mf}(pB_i^k)$, then $pB_i^k = X_i^{k+1}$, Otherwise pB_i^k is unchanged. Choose gB^{k+1} from the pB_i^{k+1} with the smallest MFF. If more than one particle are the smallest, then randomly choose one.

3.6 Local Search Mechanism

SA is an effective local search algorithm [27]. In SA, the choice of neighborhood significantly influences the performance. A DFF-based NSA with two neighborhood NS$_1$ and NS$_2$, which are used to change MA and WA vectors, respectively, is presented here as the local search mechanism [26]. For NS$_1$, we first calculate the load of each machine, then randomly choose one operation and allocate it to the machine with the least load. If this machine cannot process the operation, then randomly allocate it to a machine that can process it. For NS$_2$, the procedure is similar and thus omitted here. Define T_0, T_{end} and B as the initial and final temperature, and annealing rate, respectively. $pr_1, pr_2 \in (0, 1)$ are randomly generated parameters and $pl \in (0, 1)$ is a parameter to be chosen. The procedure is as follows.

Step 1. If $0 < pr_1 < pl$, then NS$_1$ is carried out. Otherwise execute NS$_2$.

Step 2. Calculate the MFF of F_1 and $\overline{F_1}$, denoted by $MF(\overline{F_1})$ and $MF(F_1)$ respectively. If $MF(\overline{F_1}) > MF(F_1)$, then replace the solution with the new obtained one.

Step 3. Otherwise, if $pr_2 > e^{-D/T_0}$, replace the solution with the new obtained one.

Stpe 4. Let $T_0 = BT_0$ and generate new $pr_1, pr_2 \in (0, 1)$ and start another loop if $T_0 < T_{end}$.

3.7 Procedure of the MOPSO Algorithm

The MOPSO is carried out by the following procedure.

Step 1. Set the parameters such as population size P, iteration times G and so on.

Step 2. Set $k = 0$. Initialize the population and then calculate the objectives and the corresponding MFFs. Initialize the EA^0 and choose the personal best pB_i^0 and the global best $G_b{}^0$ based on the MFFs.

Step 3. Let $k = k + 1$. Carrying out the global search mechanism leads to a new generation of particles X^k. Calculate the objectives and the corresponding MFF of each particle in X^k. Then obtain new pB_i^k and gB^k.

Step 4. Choose particles $r \in [1, P/2]$ in the current population, executing local search based on NSA. Update pB_i^k and gB^k.

Step 5. Update EA^k by the gB^k, then carry out the local search mechanism of particles in EA^k. Then Maintain EA^k.

Step 6. Choose the particle from EA^k with the smallest MFF as the most updated gB^k. If $k = G$, then finish and output EA^k. Otherwise go back to Step3.

4   Simulation Results

In this section, ten randomly generated instances R1-R10 in [26] are used for the parameter analysis first. Then two instances S1 in [22] and S2 in [23] are used to compare the performance of the proposed MOPSO with other existing algorithms.

4.1   Parameter Analysis

This subsection discusses the impact of parameters w, c_1 and c_2 on the performance of the proposed MOPSO. We choose $c_2 = 2c_1 = c$ for simplicity. The reason why we make such setting is that the proposed algorithm adjust once with c_1 and twice with c_2. Additionally, choosing values too closely will cause too much unnecessary computation load and get close results. Thus, the testing way is first to set c from 0.4 to 0.9 and choose six values for w which are, $w = 0.1, 0.2,..., 0.6$, respectively. While the best $w*$ is obtained, then fix it and choose $c = 0.4, 0.5,..., 0.9$ to find the best $c*$.

Denote y_i, $i = 1, 2,... 6$, and $z_j, j = 4, 5,... 9$, and $y_i = 1$ and $z_j = 1$ mean parameters $w = 0.1 \times i$ with $c = 0.4$, and $w = w*$ with $c = 0.1 \times j$ are chosen, respectively. Each instance will be run ten times with each combination of w and c.

For each instance, it is clear that a non-dominated solution set will be obtained for $y_i = 1$, $i = 1, 2,..., 6$. The set and the number of solutions are denoted by PFw_i and r_i, respectively. Then by combining PFw_i, $i = 1, 2,..., 6$, together and calculating the MFF, we obtain a set with more non-dominated solutions and denote it by $PFtruew$. Clearly, more solutions of PFw_i in $PFtruew$ means $w = 0.1*i$ is better.

We define ew_{ij}, $i = 1, 2,..., 6, j = 1, 2,..., r_i$, to represent whether the j-th solution of PFw_i belongs to $PFtruew$ or not. If yes, then $ew_{ij} = 0$, otherwise $ew_{ij} = 1$, ERw_i is defined to as follows:

$$ERw_i = \sum_{j=1}^{r_i} ew_{ij}/r_i, \quad i = 1, 2, \ldots, 9 \qquad (17)$$

Similarly, we can obtain PFc_j and $PFtruec$ with $z_j = 1, j = 4, 5,... 9$. Define $ec_{jl}, j = 4, 5,..., 9, l = 1, 2,..., s_j$, to represent whether the l-th solution of PFc_j within $PFtruec$ or not, where s_j is number of the non-dominated solution of PFc_j. Define ERc_j as

$$ERc_j = \sum_{l=1}^{s_j} ec_{jl}/s_j, \quad j = 4, 5, \ldots, 9 \qquad (18)$$

We can see that that smaller ERw_i with $w = 0.1 * i$ and smaller ERc_j means $c = 0.1 * j$ are better.

Other parameters are chosen as $P = 50$, $G = 100$, $ND = 5$ and $NG = 10$ and the results are shown in Table 3 and 4, respectively, where $SUM\text{-}ERw$ and $SUM\text{-}ERc$ are the sum of ERw_i and ERc_j, $i = 1, 2, ..., 9$, $j = 4, 5, ..., 9$, respectively.

Explanation of the elements in Table 2 and 3 are given. For example, $ERw_1 = 0$ for R2 in Table 2 means all the solutions with $w = 0.1$ and $c = 0.4$ are in the non-dominant set $PFtruew$. While $ERw_1 = 1$ for R2 with $w = 0.5$ and $c = 0.4$ indicating that no solutions are within the non-dominant set $PFtruew$. So $w = 0.1$ is better than $w = 0.4$ for R2. Values between 0 and 1 means part of the solutions are within the set $PFtruew$.

From Table2, it can be seen that $SUM\text{-}ERw$ is the smallest when $w = 0.1$. Moreover, part or all the non-dominated solutions can be obtained for all the instances except R7 and R8. For the parameter c, we can see from Table 3 that $c = 0.9$ results in the smallest $SERc$ and part or all the non-dominated solutions can be obtained for all the instances except R3 and R6. Thus, we choose $w = 0.1$ and $c = 0.9$ as the best parameters.

Table 2. Performance test result of w

Instance	w					
	0.1	0.2	0.3	0.4	0.5	0.6
R1	0.42	0.75	0.33	1	1	1
R2	0	0.67	1	1	1	1
R3	0	1	0	0.67	1	1
R4	0.70	1	1	1	1	1
R5	0.29	0.57	1	1	1	1
R6	0.88	0.60	0.20	1	1	1
R7	1	0.91	0.89	0.88	1	0.83
R8	1	1	0.33	1	0.8	0
R9	0.39	1	1	1	1	1
R10	0.86	0.57	0.40	0.86	0.8	0.80
Sum-ERw	5.54	8.07	6.15	9.41	9.6	8.63

4.2 Comparison

To test the performance of the proposed MOPSO, instances S1 and S2 will be used for comparison. The parameters are chosen as $w = 0.1$, $c = 0.9$, $T_0 = 3$, $T_{end} = 0.01$, $B = 0.9$, $P = 50$, $G = 100$, $ND = 5$ and $NG = 10$, respectively. After 20 running times, the non-dominated solutions are obtained for S1 and S2, respectively.

As shown in Table 4, the proposed MOPSO achieves less production cost than HGAI [21] and HGAII [22] with the same production time for instance S1. Moreover, we can see from Table 5 that more non-dominated solutions are obtained by the MOPSO than HGAIII [23] and DOKIGA [24], such as the ones with $F_1 = 50, 51, 70$. For the solutions with same production time, less production cost is obtained.

Table 3. Performance test result of c

Instance	c					
	0.4	0.5	0.6	0.7	0.8	0.9
R1	1	0	1	1	0.87	0.60
R2	1	1	1	0.86	0.50	0.33
R3	0.67	1	1	1	0.40	1
R4	1	1	1	0.14	1	0.80
R5	0.92	0.57	0.60	1	1	0.44
R6	1	1	0.50	1	0.40	1
R7	0.50	1	0.89	0.90	0.62	0.63
R8	0.88	0.89	0.25	0.78	0.78	0.67
R9	1	1	0.89	0.89	1	0
R10	0.67	0.44	0.90	1	1	0.11
Sum-ERc	8.64	7.90	8.03	8.57	7.57	5.58

Table 4. Comparison of MOPSO with other algorithms (S1)

HGAI [21]		HGAII [22]		HDPSO [28]		MOPSO	
F_1	F_2	F_1	F_2	F_1	F_2	F_1	F_2
17	3032.32	17	3012.32	17	2815.47	17	2815.47

Error ratio (ER), Mean ideal distance (MID) and Maximum spread (MS) metrics are also used compare our method with HGAIII and DOKIGA. ER, MID and MS [29–31], are defined as follows:

$$ER = \sum_{i=1}^{N} e_i / N \qquad (19)$$

$$MID = \frac{\sum_{i=1}^{N} \sqrt{\sum_{j=1}^{M} f_{ij}}}{N} \qquad (20)$$

$$MS = \sqrt{\sum_{j=1}^{M} (\max_{i=1:|N|} f_{ij} - \min_{i=1:|N|} f_{ij})^2} \qquad (21)$$

where N is the number of the elements in non-dominated set and M is the number of the objectives. $e_i = 0$ if vector i is a member of the set and $e_i = 1$ otherwise. f_{ij} denotes the jth objective value of ith non-dominated solution. ER indicates the percentage of solutions that are not members of the true Pareto optimal set and smaller ER is better.

Table 5. Comparison of MOPSO with other algorithms (S2)

HGAIII [23]		DOKIGA [24]		MOPSO	
F_1	F_2	F_1	F_2	F_1	F_2
–	–	–	–	51	6608.06
52	6886.46	52	6886.46	52	6412.84
53	6865.58	53	6865.58	53	6277.05
54	6865.47	54	6821.18	54	6144.09
55	6740.96	55	6740.44	55	6117.61
56	6688.27	56	6695.81	56	6094.14
57	6450.24	57	6450.24	57	6013.39
–	–	–	–	58	5912.85
–	–	59	6426.02	59	5897.86
60	6409.74	60	6409.74	60	5896.83
61	6380.62	61	6328.48	61	5856.52
–	–	62	6325.31	62	5846.08
63	6368.11	63	6310.66	63	5815.28
64	6272.67	64	6157.51	64	5770.1
65	6206.02	65	6154.03	65	5758.97
66	6147.86	66	6051.08	66	5745.3
67	6028.21	67	6028.21	67	5726.92
–	–	68	5991.83	68	5714.96
–	–	69	5987.91	69	5667.35
				70	5583.71

MID measures the closeness between Pareto solution and an ideal point and smaller MID means more close the solution to the ideal point. The ideal point is chosen to be [0,0] here. MS shows the diversity of the Pareto curve and higher value means better diversity. Computational results are shown in Table 6. It can be seen the proposed MOPSD achieves better performance for all the three metrics than HGAIII in [21] and DOKIGA in [22].

Table 6. Performance comparison of MOPSO with other algorithms (S2)

Algorithm	HGAIII [21]	DOKIGA [22]	MOPSO
ER	1	1	0
MID	81.00	80.29	77.47
MS	858.381	898.71	1024.53

5 Conclusions

In this paper, a MOPSO algorithm was proposed to solve the DRC-MOFJSP with both machine and worker constraints. The optimization objectives were considered to be production time and production cost. A three dimension coding scheme was used for the representation of particles by considering the DRC feature. Random, MNOR and AL ways were employed together to improve the quality of the initialized population. Then a modified DPSO based on MFF was presented to enhance the search ability. As for the local search one, a MFF-based NSA strategy was designed for the local exploiting ability. A MFF-based reservation strategy was used to maintain the EA. Finally, simulation results were given to show the effectiveness of the proposed MOPSO algorithm. For the future work direction, DRC-MOFJSP with other objective functions and more realistic constraints deserve investigation.

Acknowledgment. This research work was partly supported by Natural Science Foundation of Zhejiang Province (Grant No. LGF21G030001) and General Projects of Zhejiang Educational Committee (Grant No. Y201839027).

References

1. Brandimarte, P.: Routing and scheduling in a flexible job shop by tabu search. Ann. Oper. Res. **41**(3), 157–183 (1993)
2. Kacem, I., Hammadi, S., Borne, P.: Approach by localization and multiobjective evolutionary optimization for flexible job-shop scheduling problems. IEEE Trans. Syst. Man Cybern. Part C Appl. Rev. **32**(1), 1–13 (2002)
3. Xiong, L., Qian, Q., Yunfa, F.: Review of application of genetic algorithms for solving flexible job shop scheduling problems. Comput. Eng. Appl. **55**(23), 15–22 (2019)
4. Elmaraghy, H., Patelb, V., Abdallaha, I.B.: A genetic algorithm based approach for scheduling of dual-resource constrainded manufacturing systems. J. Manuf. Syst. **48**(1), 369–372 (1999)
5. Deb, K.: A fast elitist multi-objective genetic algorithm: NSGA-II. IEEE Trans. Evol. Comput. **6**(2), 182–197 (2000)
6. Lei, D.M.: Multi-objective production scheduling: a survey. Int. J. Adv. Manuf. Technol. **43**(9–10), 926–938 (2009)
7. Shafaei, R., Brunn, P.: Workshop scheduling using practical (inaccurate) data Part1: the performance of heuristic scheduling rules in a dynamic job shop environment using a rolling time horizon approach. Int. J. Prod. Res. **37**(17), 3913–3925 (1999)
8. Kennedy, J., Eberhart, R.C.: A discrete binary version of the particle swarm algorithm. In: IEEE International Conference on Computational Cybernetics and Simulation, vol. 5, pp. 4104–4108 (1997)
9. Balling, R.: The maximin fitness function; multi-objective city and regional planning. In: Fonseca, C.M., Fleming, P.J., Zitzler, E., Thiele, L., Deb, K. (eds.) EMO 2003. LNCS, vol. 2632, pp. 1–15. Springer, Heidelberg (2003). https://doi.org/10.1007/3-540-36970-8_1
10. Menchaca-Mendez, A., Coello, C.: Selection operators based on maximin fitness function for multi-objective evolutionary algorithms. In: Purshouse, R.C., Fleming, P.J., Fonseca, C.M., Greco, S., Shaw, J. (eds.) EMO 2013. LNCS, vol. 7811, pp. 215–229. Springer, Heidelberg (2013). https://doi.org/10.1007/978-3-642-37140-0_19

11. Rajabinasab, A., Mansour, S.: Dynamic flexible job shop scheduling with alternative process plans: an agent-based approach. Int. J. Adv. Manuf. Technol. **54**(9–12), 1091–1107 (2011)
12. Wang, L., Zhou, G., Xu, Y., Wang, S.Y., Liu, M.: An effective artificial bee colony algorithm for the flexible job-shop scheduling problem. Int. J. Adv. Manuf. Technol. **60**(1–4), 303–315 (2012)
13. ElMaraghy, H., Patel, V., Abdallah, I.B.: Scheduling of manufacturing systems under dual-resource constraints using genetic algorithms. J. Manuf. Syst. **19**(3), 186–201 (2000)
14. Li, J.Y., Sun, S.D., Huang, Y.: Adaptive hybrid ant colony optimization for solving dual resource constrained job shop scheduling problem. J. Softw. **6**(4), 584–594 (2011)
15. Yuan, Y., Xu, H.: Multiobjective flexible job shop scheduling using memetic algorithms. IEEE Trans. Autom. Sci. Eng. **12**(1), 336–353 (2015)
16. Lei, D.M., Guo, X.P.: Variable neighbourhood search for dual-resource constrained flexible job shop scheduling. Int. J. Prod. Res. **52**(9), 2519–2529 (2014)
17. Zheng, X.L., Wang, L.: A knowledge-guided fruit fly optimization algorithm for dual resource constrained flexible job-shop scheduling problem. Int. J. Prod. Res. **54**(18), 1–13 (2016)
18. Li, J., Xie, S., Pan, Q., Wang, S.: A hybrid artificial bee colony algorithm for flexible job shop scheduling problems. Stud. Inform. Control **24**(2), 171–180 (2015)
19. Khalife, M.A., Abbasi, B., Abadi, A.H.K.D.: A simulated annealing algorithm for multi objective flexible job shop scheduling with overlapping in operations. J. Optim. Ind. Eng. **8**(8), 1–24 (2015)
20. Karthikeyan, S., Asokan, P., Nickolas, S.: A hybrid discrete firefly algorithm for solving multi-objective flexible job shop scheduling problems. Int. J. Bio-Inspired Comput. **7**(6), 386–401 (2015)
21. Pan, Q.K.: Multi-objective scheduling optimization of job shop in intelligent manufacturing system, Ph.D. dissertation, Nanjing University of Aeronautics and Astronautics (2003)
22. Liu, X.X., Xie, L.Y., Tao, Z., Hao, C.Z.: Flexible job shop scheduling for decreasing production costs. J. Northeastern Univ. **29**(4), 561–564 (2008)
23. Liu, X.X., Cai, G.Y., Xie, L.Y.: Research on bi-objective scheduling optimization for DRC job shop. Modular Mach. Tool Autom. Manuf. Tech. **2009**(10), 107–112 (2009)
24. Li, J.Y., Sun, S.D., Huang, Y., Niu, G.G.: Double- objective inherited genetic algorithm for dual-resource constrained job shop. Control Decis. **26**(12), 1761–1767 (2011)
25. Pezzella, F., Morganti, G., Ciaschetti, G.: A genetic algorithm for the flexible job-shop scheduling problem. Comput. Oper. Res. **35**(10), 3202–3212 (2008)
26. Zhang, J., Wang, W., Xu, X.: A hybrid discrete particle swarm optimization for dual-resource constrained job shop scheduling with resource flexibility. J. Intell. Manuf. **28**(8), 1961–1972 (2017). https://doi.org/10.1007/s10845-015-1082-0
27. Kirkpatrick, S.C.D., Gelatt, J., Vecchi, M.P.: Optimization by simulated annealing. Science **220**(4598), 671–680 (1983)
28. Zhang, J., Jie, J., Wang, W., Xu, X.: A hybrid particle swarm optimization for multi-objective flexible job-shop scheduling problem with dual-resources constrained. Int. J. Comput. Sci. Math. **8**(6), 526–532 (2017)
29. Van Veldhuizen, D.A.: Multiobjective evolutionary algorithms: classifications, analyses, and new innovations. Evol. Comput. **8**(2), 125–147 (1999)
30. Masoud, A., Amir, A.N.: Solving a multi-mode bi-objective resource investment problem using meta- heuristic algorithms. Adv. Comput. Tech. Electromagn. **2015**(1), 41–58 (2015)
31. Zitzler, E., Deb, K., Thiele, L.: Comparison of multiobjective evolutionary algorithms: empirical results. Evol. Comput. **8**(2), 173–195 (2000)

An Improved Firefly Algorithm for Generalized Traveling Salesman Problem

Yu Huang, Xifan Yao[(✉)], and Junjie Jiang

School of Mechanical and Automobile Engineering, South China University of Technology,
Guangzhou 510640, China
mexfyao@scut.edu.cn

Abstract. The material transportation problem in workshops under complex environment can be idealized as the Generalized Traveling Salesman Problem (GTSP). To solve such a problem, an improved firefly algorithm is proposed. Firstly, two-layer coding is utilized to define the firefly individual, the inter-individual distance formula and position updating formula in standard firefly algorithm are improved, and the repaired method for infeasible solutions is defined. Then, in order to improve the local optimization ability of the proposed algorithm and accelerate the convergence speed, an improved Iterative Local Search (ILS) strategy and Complete 2-opt (C2opt) optimized operator are introduced. Moreover, the greedy firefly mutation strategy is used. Finally, 20 test cases are simulated with the algorithm. Experimental results indicate that the proposed algorithm has good convergence speed and problem solving accuracy.

Keywords: Firefly Algorithm · Iterative Local Search · Generalized Traveling Salesman Problem · C2opt operator

1 Introduction

As an extension of the Traveling Salesman Problem (TSP), the Generalized Traveling Salesman Problem (GTSP) is also a kind of NP-hard Problem, but its solution is more complicated than that of TSP. In the 1960s, Heny-Labordere [1], Saksena [2] and Srivastava et al. [3] proposed GTSP in their respective research fields. Nowadays, GTSP has been widely applied in different fields, including circular logistics system design, random vehicle scheduling, customer agency service, tool path, etc. [4–6].

The goal of GTSP is to find a least cost Hamiltonian loop on an undirected fully weighted graph $G = (V, E, W)$, where $V = \{v_1, v_2, \ldots, v_n\}(n \geq 3)$ is a vertex set, $E = \{e_{ij}|v_i, v_j \in V\}$ is an edge set, $W = \{w_{ij}|w_{ij} \geq 0, w_{ii} = 0, \forall i, j \in N(n)\}$W = $\{w_{ij}|w_{ij} \geq 0, w_{ii} = 0, \forall i, j \in N(n)\}$ is a weight set between v_i and v_j, and $N(n) = \{1, 2, \ldots, n\}$. The vertex set V is divided into m groups: $V_1, V_2, \ldots, V_m$, which satisfies: $m \leq n$, $|V_j| \geq 1$, $V = \bigcup_{j=1}^{m} V_j$. GTSP can be divided into two categories, as shown in Fig. 1. The first category is that the loop can only traverse each group once and can only pass through one vertex in each group, while the second category only can pass through multiple vertices in each group. Like most existing studies, this paper only studies the first type of GTSP.

© Springer Nature Switzerland AG 2021
D.-S. Huang et al. (Eds.): ICIC 2021, LNCS 12836, pp. 739–753, 2021.
https://doi.org/10.1007/978-3-030-84522-3_60

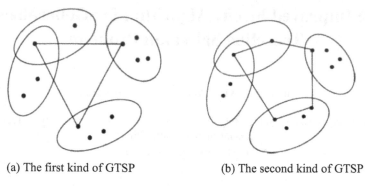

(a) The first kind of GTSP   (b) The second kind of GTSP

Fig. 1. The classification of GTSP

Dynamic programming [1], integer programming [7] and branch and bound [8] are commonly used exact approaches for solving GTSP. Some researchers also convert GTSP into TSP [9] for further solving. The above methods can show good calculation speed and precision when applied to small-scale GTSPs, but faced with difficulty to large-scale ones that are complex and time-consuming. With the rapid development of intelligent optimization algorithms, some of them have been gradually applied to solve GTSP problems, which show better computational performance in large-scale problems. For example, Wu et al. [6] proposed a generalized chromosomal genetic algorithm (GCCA) for the GTSP. Although the computational efficiency can be greatly improved, there is still a great room for improvement in the computational accuracy. Yang et al. [10] combine dynamic programming with simulated annealing algorithm to obtain better calculation results for medium-scale test cases. Pintea et al. [11] proposed a kind of enhanced-ant Colony System (RACS), and introduced a new modified criterion into the algorithm to obtain a higher solving accuracy. Krari et al. [12], on the basis of breaking through the meta-heuristic algorithm of Breakout Local Search (BLS), proposed an improved cultural gene algorithm, which provided a better solution method for GTSPs.

Firefly Algorithm (FA) was an emerging intelligent algorithm proposed by Yang [13]. The Algorithm is simple in idea and easy to implement, but it has disadvantages such as slow convergence speed and easy to fall into local optimum. The improved algorithms based on FA have been used in some TSP problems [14–16], and show good adaptability and solving performance. Since GTSP is an extension of TSP, it is a reasonable idea to apply FA extension to solve GTSP, but there are few relevant studies in the existing literature. In this paper, we propose an Improved Firefly Algorithm (IFA) in which two-layer coding is utilized to define the firefly individual, the inter-individual distance formula and position updating formula are improved, the optimization operator ILS and C2opt are introduced, and the greedy firefly mutation strategy is used to improve the local optimization ability of the algorithm and accelerate the convergence speed. Comparison results of test cases show that the proposed algorithm has the advantages of fast convergence and high precision.

2 The Standard Firefly Algorithm

In nature, fireflies use luminescent behavior to find food, find a mate, or communicate and tend to move toward brighter fireflies within the range of perception. Firefly Algorithm is a random search algorithm based on this mechanism [17]. At present, firefly algorithm has been successfully applied in numerical optimization [18], engineering technology [19], traffic signal optimization [20] and other fields.

In order to build the mathematical model of FA, the following three idealized assumptions are first given:

1. Fireflies move in a direction based only on brightness, regardless of gender.
2. The attraction between fireflies is proportional to the brightness of the fluorescence and inversely proportional to the distance between the fireflies.
3. The position and brightness of fireflies can be regarded as the solution of the objective function and the value of the objective function respectively. The updating of the position and brightness of fireflies is the process of the constant optimization of the objective function value.

Assume that m fireflies are located in the n-dimensional search space, then the position of firefly i in space can be expressed as: $x_i = (x_{i1}x_{i2}, \ldots, x_{in})$. It is defined as follows:

- The absolute brightness of firefly i is I_i:

$$I_i = f(x_i) \tag{1}$$

which represents the light intensity at its own position ($r = 0$), and $f(x_i)$ represents the fitness value of the objective function corresponding to the position of firefly i.

- The relative brightness of firefly i at the position of firefly j is I_{ij}:

$$I_{ij} = I_i e^{-\gamma r_{ij}} \tag{2}$$

where γ is the absorption coefficient of light intensity, and r_{ij} represents the distance between Firefly i and Firefly j, which is usually calculated by Euclidian distance:

$$r_{ij} = \|x_i - x_j\| = \sqrt{\sum_{k=1}^{n}(x_{ik} - x_{jk})^2} \tag{3}$$

- The attractiveness between Firefly i and Firefly j:

$$\beta_{ij} = \beta_0 e^{-\gamma r_{ij}^2} \tag{4}$$

where β_0 represents the maximum attractiveness, that is, the attractiveness at the light source ($r = 0$), and usually $\beta_0 = 1$.

- The position updating formula of Firefly i attracted by Firefly j:

$$x_i^{t+1} = x_i^t + \beta_{ij}\left(x_j^t - x_i^t\right) + \alpha\varepsilon_i \tag{5}$$

where t is the number of iterations, ε_i is the random perturbation term, and α is the perturbation term coefficient. According to Eq. (5), the position update formula consists of three parts, namely the current position information, attraction term and disturbance term.

3   The Improved Firefly Algorithm

In this section, an improved firefly algorithm, called IFA, is introduced. Firstly, two-layer coding is utilized to define the firefly individual, the inter-individual distance formula and position updating formula in standard firefly algorithm are improved, and the repaired method for infeasible solutions is defined. Then, in order to improve the local optimization ability of the proposed algorithm and accelerate the convergence speed, an improved Iterative Local Search (ILS) strategy and Complete 2-opt (C2opt) optimized operator are introduced. Finally, the greedy firefly mutation strategy is used.

3.1   The Two-Layer Coding Method

In GTSP, the vertex set V with n vertices is divided into m groups $\{V_1, V_2, \ldots, V_m\}$, so a feasible encoding can be defined by using a two-layer encoding method. The low level encodes the number of the group in the form of $(v_1, v_2, \ldots, v_i, \ldots, v_m)$; The high level code the vertex number in the form of $(a_1, a_2, \ldots, a_i, \ldots, a_m)$; where v_i represents the i th group traversed by the path, and a_i represents the ith position of the path, and for $\forall i \in [1, m]$, both satisfy $a_i \in V_{v_i}$. The formula for calculating the path length is as follows:

$$L = \sum_{i=1}^{m-1} d(a_i, a_{i+1}) + d(a_m, a_1) \tag{6}$$

where $d(a_i, a_{i+1})$ represents the distance between Vertex a_i and Vertex a_{i+1}.

For example, for a GTSP with 15 cities, being clustered into 5 groups: $V_1 = \{1, 2, 3\}$, $V_2 = \{4\}$, $V_3 = \{5, 6, 7, 8, 9, 10\}$, $V_4 = \{11, 12, 13\}$, and $V_5 = \{14, 15\}$, the low-level coding can be $(2, 3, 5, 1, 4)$, representing the city clustered order of path traversal as $V_2 \rightarrow V_3 \rightarrow V_5 \rightarrow V_1 \rightarrow V_4$, and the high-level coding can be $(4, 6, 15, 2, 11)$, representing the city order of path traversal as $4 \rightarrow 6 \rightarrow 15 \rightarrow 2 \rightarrow 11$.

3.2   The Inter-individual Distance Formula

In this study, the Hamming distance, the number of different bits corresponding to two groups of codes, is used to simply measure the distance between two firefly individuals. Assuming that high-level codes of Firefly x and Firefly y are $(x_1, x_2, \ldots, x_i, \ldots, x_m)$ and $(y_1, y_2, \ldots, y_i, \ldots, y_m)$, then the distance calculation formula is as follows:

$$d(x, y) = \sum_{i=1}^{m} x_i \oplus y_i \tag{7}$$

For example, $x = (4, 6, 15, 2, 11)$, and $y = (4, 14, 1, 9, 11)$, then the Hamming distance is 3.

3.3 The Position Update Formula

According to the formula (4)–(7), this paper further proposes an improved position update formula:

$$x_i^{t+1} = x_i^t + e^{\frac{-2.5 r_{ij}}{m}} \left(x_j^t - x_i^t \right) + \varepsilon_i \tag{8}$$

where, m is the number of city clusters, and ε_i is a sequence of random numbers with length m and magnitude between $[-0.5, 0.5]$. Formula (8) eliminates the influence of firefly coding length on the calculation of attractiveness. Since $r_{ij} \in [0, m]$, the magnitude of attractiveness will always be within the interval of $[0.082, 1]$. The encoding with the updated position is no longer an integer encoding, so rounding is used to set the encoding value to an integer.

3.4 The Repair Method for Infeasible Solutions

After updating the position of fireflies, duplicates of city cluster or city number and city cluster number lost may occur in the code of fireflies. These are all infeasible codes, so the following steps are adopted to repair them:

Step 1: Convert firefly high-level code x_i^{t+1} to low-level code x'^{t+1}_i.

Step 2: Find the lost city cluster number set, represented as $M = \{m_{V_1}, m_{V_2}, \ldots, m_{V_k}, \ldots, m_{V_K}\}$.

Step 3: The city number belonging to city cluster m_{V_k} in firefly x_j^t is recorded, represented as $S = \{s_1, s_2, \ldots, s_k, \ldots, s_K\}$.

Step 4: In the low-level code x'^{t+1}_i, the second and subsequent repeated city cluster positions are found, represent as $U = \{u_1, u_2, \ldots, u_k, \ldots, u_K\}$.

Step 5: For $k \in [1, K]$, s_k replaces the code on the u_k in the high-level code of Firefly x_i^{t+1}.

Use the example in Sect. 3.1, and assume $x_i^t = (4, 6, 15, 2, 11)$, and $x_j^t = (4, 14, 1, 9, 11)$, according to Formula (8) and rounded, $x_i^{t+1} = (4, 8, 12, 4, 11)$ can be obtained and converted to the low-level code $x'^{t+1}_i = (2, 3, 4, 2, 4)$. Due to duplicate cities and city clusters as well as city clusters lost, such x_i^{t+1} and x'^{t+1}_i are infeasible codes. Using the above method to repair, we can get $M = \{1, 5\}$, $S = \{1, 14\}$, and $U = \{4, 5\}$ from Step 2 to Step 4 respectively. Finally, according to Step 5, we replace the 4th and 5th codes in x_i^{t+1} with 1 and 14 in S respectively, so the repaired codes $x_i^{t+1} = (4, 8, 12, 1, 14)$ and $x'^{t+1}_i = (2, 3, 4, 1, 5)$ can be obtained.

3.5 The Improved Iterative Local Search

ILS is a meta-heuristic algorithm proposed by Loureno et al. [21], with good performance when used in a variety of optimization problems. The algorithm principle of ILS mainly includes four parts: initial solution construction, local search, perturbation strategy and acceptance criteria, among which the perturbation strategy has the greatest influence on algorithm performance. If the perturbation is too small, it may not be able to jump out of the local optimum. On the contrary, if the perturbation is too large, it will degenerate into a random algorithm. Therefore, the design of perturbation strategy is the key to ensure good search performance of ILS.

In this paper, the disturbance strategy and acceptance criteria are improved to form an improved ILS strategy. The specific operation steps are as follows:

Step 1: Define the parameter k_{max} ($k_{max} = \lceil m/5 \rceil$), where m is the number of city clusters, and $\lceil \rceil$ is rounded upward. Initialize $k = 1$.

Step 2: Initialize $i = 1$.

Step 3: Intercept the $[i, i + k]$ code segment in the high-level code x_{old} of the individual firefly, and then flip it to obtain the new code x_{new}.

Step 4: Calculate the changes of path length f_{old} and f_{new} corresponding to x_{old} and x_{new} respectively, and according to the acceptance criterion to decide whether to update code, we have $f_{old} = d\left(x_{old\_(i-1)}, x_{old\_i}\right) + d\left(x_{old\_(i+k)}, x_{old\_(i+k+1)}\right)$, and similarly, $f_{new} = d\left(x_{new\_(i-1)}, x_{new\_i}\right) + d\left(x_{new\_(i+k)}, x_{new\_(i+k+1)}\right)$.

Step 5: $i = i + 1$, and jump to step 3 until $i = m$.

Step 6: $k = k + 1$, and jump to step 2 until $k = k_{max}$.

The acceptance criterion is expressed as follows:

$$
x = \begin{cases} x_{new} \ f_{inew} > f_i \text{or} \frac{f_i - f_{inew}}{f_{inew}} \leq p \\ x_{old} \ \ otherwise \end{cases}
\tag{9}
$$

where p is acceptance probability. If the new code is superior to the original one, the new code is accepted or even not superior, it is also accepted with a certain probability. Otherwise, the original code is retained.

The pseudo-code for the improved ILS is shown in Fig. 2.

define parameter $k_{max} = \lceil m/5 \rceil$;
for $k = 1 : k_{max}$
  for $i = 1 : m$
    reverse coding segment $[i, i + k]$ in x_{old};
    generate x_{new};
    calculate f_{old} and f_{new};
    if $(f_{new} > f_{old}$ or $\frac{f_{old} - f_{new}}{f_{new}} \leq p)$
      $x = x_{new}$;
    else
      $x = x_{old}$;
    end if
  end for i
end for k

Fig. 2. Pseudo-code of the improved ILS

3.6 The C2opt Optimization Operator

In TSP, 2-opt [22] is a very effective local search algorithm. Usually, local optimization is carried out for 20%–30% of the cities on a path, which can eliminate cross-paths. The C2opt operator is a further optimization of the 2-opt operator, that is, the 2-opt operator is optimized for all cities on the path. Assume a path is coded as $x = (x_1, x_2, \ldots, x_i, \ldots, x_m)$. In symmetric GTSP, when a section of code (x_i, x_j) is flipped, the path length in this section of code interval will not change, and the change of path length is only caused by $d(x_{i-1}, x_i) \rightarrow d(x_{i-1}, x_j)$ and $d(x_j, x_{j+1}) \rightarrow d(x_i, x_{j+1})$. Therefore, the entire path length can be simplified to the sum of the two sections of path length in calculation. Figure 3 and Fig. 4 respectively show the effect and pseudo-code realized by using C2opt optimization operator.

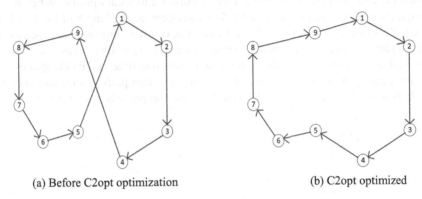

(a) Before C2opt optimization   (b) C2opt optimized

Fig. 3. Schematic diagram of optimization effect of the C2opt operator

```
for i = 1: m − 2
    for j = i + 2: m
        if j == m
            temp = 1;
        else
            temp = j + 1;
        end if
        len1 = d(xᵢ, xᵢ₊₁) + d(xⱼ, x_temp);
        len2 = d(xᵢ, xⱼ) + d(xᵢ₊₁, x_temp);
        if len2 < len1
            for k = 0: ceil(ʲ⁻ⁱ⁄₂ − 1)
                swap{xᵢ₊ₖ₊₁, xⱼ₋ₖ};
            end for k
        end if
    end for j
end for i
```

Fig. 4. Pseudo-code of the C2opt optimized operator

3.7 The Firefly Mutation

In the GTSP, a city is selected from each city cluster to be added to the path. If only the fireflies are searched in the solution space due to mutual attraction, then the search efficiency of the algorithm is very low, and the convergence speed is very slow, and it is easy to fall into the local optimum. Therefore, a greedy mutation strategy for fireflies is proposed in this study. Assuming $(x_{i+1}, x_{i+2}, ..., x_{i+K})$ is a segment of length K taken from the firefly code, and the corresponding low-level code is $(x'_{i+1}, x'_{i+2}, ..., x'_{i+K})$, and find cities $(x^{best}_{i+1}, x^{best}_{i+2}, ..., x^{best}_{i+K})$ in the K city clusters to obtain the minimum value that satisfies: $x^{best}_{i+j} \in V_{x'_{i+j}}$, and $[d(x_i, x^{best}_{i+1}) + d(x^{best}_{i+1}, x^{best}_{i+2}) + ... + d(x^{best}_{i+K-1}, x^{best}_{i+K}) + d(x^{best}_{i+K}, x_{i+K+1})]$ for $\forall j \in [1, K]$. The smaller the value of K, the more serious the greedy thought of fireflies in mutation, and the greater the probability of falling into local optimal. When $K = 1$, although the code of each position of fireflies can optimize the length of local path, the whole path is usually not optimal. If the value of K is larger, though it can increase the ability to jump out of the local optimal, it will undoubtedly greatly increase the complexity and calculation time of the algorithm, and at the same time violate the greedy idea. Therefore, after comprehensive consideration of algorithm performance and operation time, K is set to 3r. The pseudo-code of firefly mutation process is shown in Fig. 5.

```
for i = 1: m
  if i == m − 3
    t_a = m − 2; t_b = m − 1; t_c = m; t_d = 1;
  elseif i == m − 2
    t_a = m − 1; t_b = m; t_c = 1; t_d = 2;
  elseif i == m − 1
    t_a = m; t_b = 1; t_c = 2; t_d = 3;
  elseif i == m
    t_a = 1; t_b = 2; t_c = 3; t_d = 4;
  else
    t_a = i + 1; t_b = i + 2; t_c = i + 3; t_d = i + 4;
  end if
  L_min = Min{d(x_i, x_{t_a'}) + d(x_{t_a'}, x_{t_b'}) + d(x_{t_b'}, x_{t_c'}) + d(x_{t_c'}, x_{t_d'})};
  (x_{t_a} ∈ V_{x'_{t_a}}; x_{t_b} ∈ V_{x'_{t_b}}; x_{t_c'} ∈ V_{x'_{t_c}})
  [c_1, c_2, c_3] = arg{L_min};
  x_{t_a} = c_1; x_{t_b} = c_2; x_{t_c} = c_3;
end for i
```

Fig. 5. Pseudo-code of the firefly mutation process

3.8 The Improved Firefly Algorithm Flow Chart

In the proposed algorithm, the fireflies with updated position will carry out iterative local search, C2opt optimization and mutation operation in turn, so the overall flow chart is shown in Fig. 6.

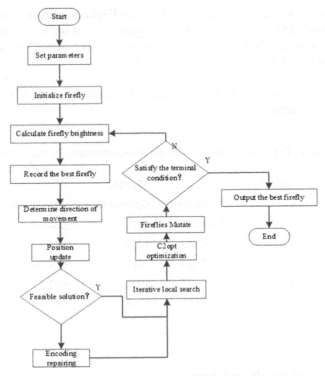

Fig. 6. The flowchart of IFA

4 Simulation Experiments and Result Analysis

In this section, three different experiments were conducted to evaluate the proposed algorithm-IFA. The acceptance probability P was set by the first experiment. The next two experiments, using different GTSP standard instances, with different length obtained from TSPLIB [23], respectively verify the excellent performance of IFA in error, relative error and other indicators, as well as the superiority of IFA compared with other algorithms.

4.1 The Setting of Acceptance Probability

In the improved iterative local search, the acceptance probability P also has a great influence on the algorithm's performance. If P is too small, the search ability of firefly will be reduced, which is not conducive to the comprehensive search of the algorithm in the whole solution space, and it is easy to fall into local optimum. On the contrary, if P is too large, it cannot make good use of the firefly algorithm optimization characteristics. For optimizing algorithm performance, we set the appropriate value of probability P based on the following experiments: select the two test instances KroA200 and gil262 from TSPLIB; according to the method proposed by Fischetti method to form the GTSP test cases 40kroA200 and 53gil262; and P, respectively, set up in five out of 0.1–0.5, the results are shown in Table 1. Other parameters are set as follows: population size 20, maximum number of iterations 20, and running times 30.

As can be seen from Table 1, when P is 0.3, both the optimal value and the average value are the minimum for the two instances. Therefore, P is set to 0.3 to conduct subsequent experiments.

Table 1. Experimental results with different Acceotance Probabilities P

P	40kroA200		53gil262	
	Optimal value	Average value	Optimal value	Average value
0.1	13429	14044.03	1026	1084.07
0.2	13491	13909.23	1039	1068.33
0.3	13417	13779.00	1017	1048.70
0.4	13577	13851.37	1025	1062.43
0.5	13818	14357.03	1088	1142.07

4.2 Experimental Results and Analysis

In order to verify the computational performance and effect of the proposed algorithm, 20 medium and small scale problems were selected from the instances of GTSP for experiments, among them, the first 12 instances are less than 200 cities, and the rest are greater than or equal to 200 cities. The results were shown in Table 2.

In the Table 3, the calculation formulas of error and relative error are as follows (Table 2):

$$error = \frac{Minimun\ value - Optimal\ solution}{Optimal\ solution} \times 100\% \qquad (10)$$

$$Relative\ error = \frac{Average\ value - Optimal\ solution}{Optimal\ solution} \times 100\% \qquad (11)$$

Table 2. Calulation results of GTSP instances

Instance	Optimal solution	Minimum value	Maximum value	Average value	Average value	Standard deviation	Error (%)	Relative error (%)
10att48	5394	5394	5394	5394	5394	0	0	0
11eil51	174	174	174	174	174	0	0	0
14st70	316	316	316	316	316	0	0	0
16pr76	64925	64925	64925	64925	64925	0	0	0
20kroA100	9711	9711	9711	9711	9711	0	0	0
20kroB100	10328	10328	10330	10328.13	10328.13	0.51	0	0
20KroC100	9554	9554	9554	9554	9554	0	0	0
21lin105	8213	8213	8213	8213	8213	0	0	0
25pr124	36605	36605	37097	36783.6	36783.6	122.14	0	0.49
30KroA150	11018	11018	11372	11120.27	11120.27	77.13	0	0.93
31pr152	51576	51576	52528	51984.6	51984.6	261.42	0	0.79
32u159	22664	22664	23186	22829.23	22829.23	140.62	0	0.73
40kroA200	13406	13417	14660	13779	13779	239.51	0.08	2.78
40kroB200	13111	13151	13963	13577.5	13577.5	205.78	0.31	3.56
53gil262	1013	1017	1127	1048.7	1048.7	22.31	0.39	3.52
60pr299	22615	22658	24364	23264.17	23264.17	475.68	0.19	2.87
64lin318	20765	20812	22602	21604.77	21604.77	443.27	0.23	4.04
84fl417	9651	9651	9996	9847.37	9847.37	83.35	0	2.03
88pr439	60099	60540	63228	61440.5	61440.5	559.9	0.73	2.23
89pcb442	21657	22581	23734	23108.67	23108.67	304.01	4.27	6.7

The Table 5 shows that those in less than 200 cities, as well as 84fl417, the proposed algorithm can obtain the optimal solutions, and the relative errors are less than 1% except 84fl417. Especially for the first 7 instances, whose relative errors are 0%, every run can reach the optimal solution, so the algorithm in dealing with small-scale cities has reliable stability. For the problems with 200 or more cities, the solutions are very close to the optimal ones, their error are all less than 0.73%, and the relative errors are not more than 4.04% except for 89pcb442. Figure 7 is the iterative evolution curve of the four instances of 14st70, 20kroA100, 32u159 and 53gil262. It can be seen that the algorithm can converge to the final value after only 4–6 iterations, so it has a fast convergence speed.

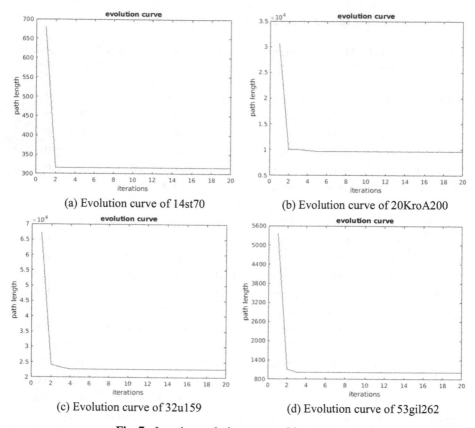

(a) Evolution curve of 14st70

(b) Evolution curve of 20KroA200

(c) Evolution curve of 32u159

(d) Evolution curve of 53gil262

Fig. 7. Iterative evolution curves of four instances

To further verify the effectiveness of the proposed algorithm (IFA), Algorithm 1 (similar to IFA, the difference only in the parameter K variation; the K is set to 1 in Algorithm 1, and set K to 3 in IFA, in G Part of Section IV) and other algorithms in the literature are compared as shown in Table 6 with the optimal results marked as bold (Table 3).

As can be seen from Table 8, compared with other algorithms, the advantage of the proposed algorithm is not obvious when the problem size is small, but it significantly performs better when the problem size is large. On the whole, the proposed IFA has better global optimization capability.

Table 3. Comparison of calculation results with IFA and other algorithms

Instance	IFA	Algorithm 1	Improved ACO [24]	SA [25]	GI^3 [26]
10att48	**5394**	**5394**	NA	NA	NA
11cil51	**174**	**174**	176	NA	**174**
14st70	316	316	**315**	NA	316
16pr76	**64925**	**64925**	65059	NA	**64925**
20kroA100	**9711**	**9711**	9758	NA	**9711**
20kroB100	**10328**	**10328**	NA	NA	**10328**
20KroC100	**9554**	**9554**	9569	NA	**9554**
21lin105	**8213**	**8213**	**8213**	NA	**8213**
25pr124	**36605**	**36605**	36903	NA	36762
30KroA150	**11018**	11040	11470	11027	**11018**
31pr152	**51576**	**51576**	51602	51584	51820
32u159	**22664**	**22664**	NA	22916	23254
40kroA200	13417	13577	NA	13454	**13406**
40kroB200	13151	13409	NA	13117	**13111**
53gil262	**1017**	1053	NA	1047	1064
60pr299	**22658**	23537	NA	23186	23119
64lin318	**20812**	21586	NA	21528	21719
84fl417	**9651**	9655	NA	10099	9932
88pr439	**60540**	61650	NA	66480	62215
89pcb442	**22581**	23385	NA	23811	22936

5 Conclusion

In this paper, an improved firefly algorithm (IFA) is proposed to solve the GTSP problem. The firefly individuals are defined in the form of two-layer coding, and the Inter-individual distance formula and position update formula are also improved. The improved iterative local search (ILS) and C2opt optimization operator are introduced to make the algorithm have better local optimization capability, and the greedy firefly mutation strategy is used to accelerate the convergence speed of the algorithm. The experimental results show that the proposed algorithm has the advantages of fast convergence and high precision, and the global optimization ability compared with the rival algorithms in the literature. In the future, the algorithm parameters need to be further studied for the optimal setting.

Acknowledgement. This work was supported by the National Natural Science Foundation of China and the Royal Society of Edinburgh (51911530245), Guangdong Basic and Applied Basic Research Foundation (2021A1515010506), China Scholarship Council (No. [2020]1509), and Zhanjiang Science and Technology Project (2020A01001, 2019A03014).

References

1. Henry-Labordere, A.L.: The record balancing problem: a dynamic programming solution of a generalized traveling salesman problem. RAIRO Oper. Res. **B2**, 43–49 (1969)
2. Saksena, J.P.: Mathematical model of scheduling clients through welfare agencies. Can. Oper. Res. Soc. J. **8**(3), 185–200 (1970)
3. Srivastava, S.S., Kumar, S., Garg, R.C., Sen, P.: Generalized traveling salesman problem through n sets of nodes. Can. Oper. Res. Soc. J. **7**(2), 97–101 (1969)
4. Laporte, G., Asef-Vaziri, A., Sriskandarajah, C.: Some applications of the generalized travelling salesman problem. J. Oper. Res. Soc. **47**(12), 1461–1467 (1996)
5. Easwaran, A.M., Pitt, J., Poslad, S.: The agent service brokering problem as a generalised travelling salesman problem. In: Conference on Autonomous Agents, pp. 414–415. ACM (1999)
6. Wu, C., Liang, Y., Lee, H., Lu, C.: Generalized chromosome genetic algorithm for generalized traveling salesman problems and its applications for machining. Phys. Rev. E Stat. Nonlinear Soft Matter Phys. **70**(1), 016701 (2004)
7. Laporte, G., Nobert, Y.: Generalized travelling salesman problem through n sets of nodes: an integer programming approach. INFOR Inf. Syst. Oper. Res. **21**(1), 61–75 (1983)
8. Fischetti, M., Toth, J.: A branch-and-cut algorithm for the symmetric generalized traveling salesman problem. Oper. Res. **45**(3), 378–394 (1997)
9. Dimitrijevic, V., Saric, Z.: An efficient transformation of the generalized traveling salesman problem into the traveling salesman problem on digraphs. Inf. Sci. **102**(1–4), 105–110 (1997)
10. Yang, W., Zhao, Y.: Improved simulated annealing algorithm for GTSP. In: International Conference on Automatic Control & Artificial Intelligence, pp. 1202–1205. IET, April 2013
11. Pintea, C.M., Pop, P.C., Chira, C.: The generalized traveling salesman problem solved with ant algorithms. Complex Adapt. Syst. Model. **5**(1), 8 (2017)
12. Krari, M.E., Ahiod, B.: A memetic algorithm based on breakout local search for the generalized travelling salesman problem. arXiv:1911.01966, October 2019
13. Yang, X.: Firefly algorithm, stochastic test functions and design optimisation . Int. J. Bio Inspired Comput. **2**(2), 78–84 (2010)
14. Zhou, Y.Q., Huang, Z.X.: Artificial glowworm swarm optimization algorithm for tsp. Control Decis. **27**(27), 1816–1821 (2012)
15. Yu, H.T., Goo, L.Q., Han, X.C.: Discrete artificial firefly algorithm for solving traveling salesman problems. Huanan Ligong Daxue Xuebao/J. South China Univ. Technol. (Nat. Sci.) **43**(1), 126–131 and 139 (2015)
16. Zhang, L.Y., Gao, Y., Fei, T.: Firefly genetic algorithm for traveling salesman problem. Comput. Eng. Des. **40**(7), 1939–1944 (2019)
17. Liu, C., Liu, L.Q., Zhang, L.N., Yang, X.S.: An improved firefly algorithm and its application in global optimization. J. Harbin Eng. Univ. **38**(004), 569–577 (2017)
18. Zhang, Q., Li, P.C.: Adaptive grouping difference firefly algorithm for continuous space optimization problems. Control and Decis. **32**(7), 1217–1222 (2017)
19. Liu, J.S., Mao, Y.L., Li, Y.: Natural selection firefly optimization algorithm with oscillation and constraint. Control Decis. **35**(10), 2363–2371 (2020)
20. Liu, C.Y., Ren, Y.Y., Bi, X.J.: Timing optimization of regional traffic signals based on improved firefly algorithm. Control Decis. **35**(12), 2829–2834 (2020)
21. Kramer, O.: Iterated local search. In: A Brief Introduction to Continuous Evolutionary Optimization. SAST, pp. 45–54. Springer, Cham (2014). https://doi.org/10.1007/978-3-319-034 22-5_5
22. Croes, G.A.: A method for solving travelling salesman problems. Oper. Res. **6**, 791–812 (1958)

23. Reinelt, G.: TSPLIB–a traveling salesman problem library. ORSA J. Comput. **3**(4), 376–384 (1991)
24. Yang, J., Shi, X., Marchese, M., Liang, Y.: An ant colony optimization method for generalized TSP problem. Prog. Nat. Sci. **18**(011), 1417–1422 (2008)
25. Tang, X., Yang, C., Zhou, X., Gui, W.: A discrete state transition algorithm for generalized traveling salesman problem. Control theory & applications. arXiv:1304.7607v1, August 2013
26. Boctor, R.: An efficient composite heuristic for the symmetric generalized traveling salesman problem. Eur. J. Oper. Res. **108**(3), 571–584 (1998)

2. Reinelt, G.: TSPLIB-a traveling salesman problem library. ORSA J. Comput. 3(4), 376–384 (1991)

Virtual Reality and Human-Computer Interaction

User Study on an Online-Training System of Activity in Daily Life for the Visually Impaired

Hotaka Takizawa[1]([✉]), Koji Kainou[1], and Mayumi Aoyagi[2]

[1] University of Tsukuba, Tsukuba 305-8573, Japan
takizawa@cs.tsukuba.ac.jp
[2] Aichi University of Education, Kariya 448-8542, Japan

Abstract. This paper describes an online system to enable a sighted trainer to train a visually impaired trainee to use various objects in a daily environment. The system consists of a robot at a trainee's house and a trainer computer at a rehabilitation facility, which are connected with each other via the Internet. The robot is equipped with a camera, a microphone, and a parametric speaker on a pan-tilt head. The parametric speaker can acoustically point to target objects to be used by the trainee. The trainer can observe the situation of the trainee by use of the camera, and teach him/her where the objects are and how to use them by use of the parametric speaker. This online-training system was proposed in our previous works, and in this paper, we report the results of user study where three blindfolded participants search for a refrigerator and a sink in a kitchen environment.

Keywords: The visually impaired · Online ADL training · Parametric speaker · User study

1 Introduction

World Health Organization reported that the number of visually impaired individuals was at least 2.2 billion worldwide [9]. There are rehabilitation facilities that provide the training of activities in daily life (ADL), such as walking, reading, writing and cooking, to visually impaired individuals. However, the number of such facilities is not sufficient yet. Visiting training can be one of the solutions for the problem, but it needs large cost to send sighted trainers to trainees' houses. Therefore, many visually impaired individuals have not been able to receive adequate ADL training.

There are several studies for visually impaired individuals. Kane et al. developed an application called Slide Rule that enabled blind people to interact with touchscreens based on gesture inputs and audio outputs [4]. Krajnc et al. developed a user interface on a touch screen which enabled a visually impaired individual to control an application on a computer [5]. Yoshikawa et al. proposed a

© Springer Nature Switzerland AG 2021
D.-S. Huang et al. (Eds.): ICIC 2021, LNCS 12836, pp. 757–762, 2021.
https://doi.org/10.1007/978-3-030-84522-3_61

haptic map system that enabled the visually impaired to recognize roads, blocks and buildings [10]. Padmavathi et al. developed a method to convert scanned braille documents to texts which can be read by a computer [6]. Seki proposed a system to train a visually impaired trainee to perceive his/her surrounding in a virtual environment, including cars, walls and so forth, by use of a three-dimensional acoustic simulator [7]. Garaj et al. proposed a network-based system that used a camera and a GPS receiver to train a visually impaired pedestrian to walk in a real road environment [1]. Ge et al. [2] proposed an image processing based shaping method of line drawings created by visually impaired people with Lensen Drawing Kit.

In our previous works [3,8], we proposed an online ADL training system, and reported the results of preliminary user study. In this paper, we conduct more practical user study and report its results.

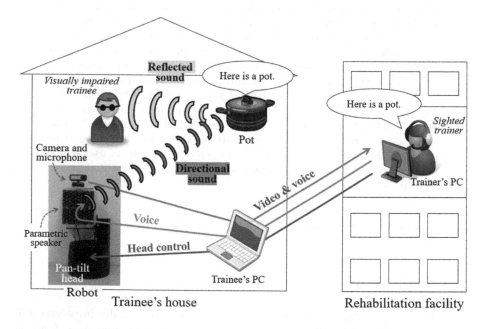

Fig. 1. Configuration of online ADL training system.

2   Configuration of Our Online ADL Training System

Figure 1 shows the configuration of our online ADL training system. A sighted trainer is in a rehabilitation facility, and a visually impaired trainee is in his/her own house. They use personal computers connected to each other via the Internet. The trainee also uses a robot equipped with a camera, a microphone and a parametric speaker on a pan-tilt head. These devices are connected with the trainee computer, and therefore the trainer can use them.

The parametric speaker consists of ultrasonic oscillators arranged on a circuit board. The speaker can generate directional sounds, which are diffusively reflected on the surfaces of objects. The visually impaired trainee can hear the reflected sounds as if they are generated from the object surfaces. By using this mechanism, the trainer can acoustically point to the objects.

The trainer can turn the parametric speaker to an arbitrary direction by controlling the pan-tilt head. The trainer can observe the situation of the trainee by use of the camera, and talk with the trainee by use of the microphone and parametric speaker.

3   Online ADL Training

Visually impaired people can receive training necessary to use several objects in daily environments as follows:

1. The trainer checks the situation of the trainee, and turns the parametric speaker toward a target object (e.g. a pot) as shown in Fig. 1.
2. The trainer inputs voices with a microphone on his/her computer. The parametric speaker irradiates directive sounds to the target object.
3. The directive sounds are reflected on the surface of the target object, and the trainee hears the reflected sounds as if they are generated from the target object. The trainee searches for the object and takes it to use.
4. By iterating the above procedure, the trainee can learn the sequence to use objects in the environment.

4   User Study

We conducted user study where three blindfolded participants in their twenties searched for a refrigerator and a sink in an experimental environment (kitchen) shown in Fig. 2. The participants were instructed by a trainer to find and touch these target objects as quickly as possible according to voice guidance from the robot. Before staring the user study, the participants visually confirmed the structure of the environment to ensure their safety.

The user study was composed of two steps. In the first step, the trainee used the parametric speaker, and in the second step, he used a conventional speaker as follows:

1. First step
   (a) The blindfolded participant was taken to one of two initial locations indicated by Ⓐ and Ⓑ, and then was randomly rotated in the range of plus or minus 30°. He was not informed of his initial location and direction.
   (b) The trainer checked the participant, and then pointed to the refrigerator or the sink by use of the *parametric* speaker.
   (c) The participant moved to the target object, and attempted to touch it.

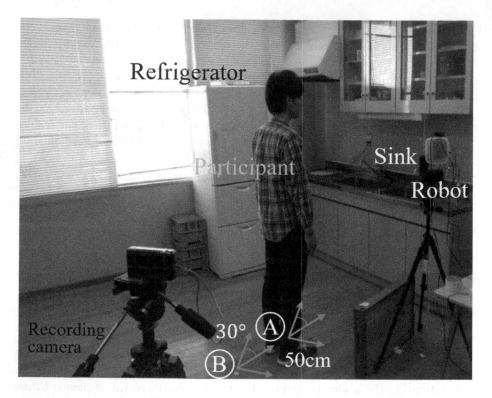

Fig. 2. Experimental environment for user study.

2. Second step
   (a) The participant was taken to the initial position and randomly rotated in the same manner.
   (b) The trainer taught the participant about the position of the target object by use of the *conventional* speaker.
   (c) The participant moved, and attempted to touch the target object.

   The above procedure was iterated 8 times for each participant. Table 1 lists the rates that the three participants successfully touched the target objects. The success rates of the parametric speaker were larger than those of the conventional speaker. Table 2 lists the mean searching time of the three participants. When the parametric speaker was used, the participants was able to find the target objects in shorter time.

5   Discussion

As mentioned above, the original system was proposed in [3, 8]. The main contribution of this paper was that the user study demonstrated that the participants were able to find the target objects more accurately and efficiently when the

Table 1. Success rates.

	Parametric speaker	Conventional speaker
Participant 1	100%	87.5%
Participant 2	87.5%	62.5%
Participant 3	100%	62.5%

Table 2. Mean searching time (second).

Parametric speaker	Conventional speaker
11.2	20.0

parametric speaker was used. The trainer and the participants were forced to have extra conversations to adjust the positions of the participants when the conventional speaker was used, whereas the positional adjustment was easier when the parametric speaker was used.

6   Conclusions

This paper proposed the online training system of ADL for the visually impaired. User study was conducted with three blindfolded participants, and its results demonstrated that the parametric speaker was effective for the online training of ADL. One of our future works is to conduct user study with actual visually impaired individuals in various and severe environments.

References

1. Garaj, V., Jirawimut, R., Ptasinski, P., Cecelja, F., Balachandran, W.: A system for remote sighted guidance of visually impaired pedestrians. Br. J. Vis. Impairment **21**(2), 55–63 (2003)
2. Ge, L., Takizawa, H., Ohya, A., Aoyagi, M., Kobayashi, M.: Line drawing aid with lensen drawing kit for the visually impaired. In: Proceedings of LifeTech 2020, pp. 264–267. IEEE (2020)
3. Kainou, K., Takizawa, H., Aoyagi, M., Ezaki, N., Mizuno, S.: A preliminary study on network-based ADL training for the visually impaired by use of a parametric-speaker robot. In: Proceedings of the 8th International Symposium on Visual Information Communication and Interaction, the Association for Computing Machinery (2015)
4. Kane, S.K., Bigham, J.P., Wobbrock, J.: Fully accessible touch screens for the blind and visually impaired. University of Washington (2009)
5. Krajnc, E., Knoll, M., Feiner, J., Traar, M.: A touch sensitive user interface approach on smartphones for visually impaired and blind persons. In: Holzinger, A., Simonic, K.-M. (eds.) USAB 2011. LNCS, vol. 7058, pp. 585–594. Springer, Heidelberg (2011). https://doi.org/10.1007/978-3-642-25364-5_41

6. Padmavathi S., Reddy, S.S., Meenakshy, D.: Conversion of braille to text in English, Hindi and Tamil languages. Int. J. Comput. Sci. Eng. Appl. **3**(3), 19–32 (2013). https://doi.org/10.5121/ijcsea.2013.3303
7. Seki, Y.: Training technology for auditory orientation by the persons with visual impairment. Synthesiology English Edition **6**(2), 66–74 (2013)
8. Takizawa, H., Kainou, K., Aoyagi, M., Ezaki, N., Mizuno, S.: A preliminary study on self ADL training for the visually impaired based on a parametric-speaker robot. In: Proceedings of the 6th International Conference on Advanced Mechatronics (ICAM 2015) (2015)
9. WHO: World health organization, media centre, visual impairment and blindness, fact sheet n° 282 (2021). http://www.who.int/mediacentre/factsheets/fs282/en/. Accessed 26 Feb 2021
10. Yoshikawa, T., Satoi, T., Koeda, M.: Virtual haptic map using haptic display technology for visually impaired. In: Parenti Castelli, V., Schiehlen, W. (eds.) ROMANSY 18 Robot Design, Dynamics and Control, pp. 375–382. Springer, Vienna (2010). https://doi.org/10.1007/978-3-7091-0277-0_44

Person Property Estimation Based on 2D LiDAR Data Using Deep Neural Network

Mahmudul Hasan[1,2]($\boxtimes$), Riku Goto[1], Junichi Hanawa[1], Hisato Fukuda[3], Yoshinori Kuno[1], and Yoshinori Kobayashi[1]

[1] Graduate School of Science and Engineering, Saitama University, Saitama, Japan
{hasan,r.goto,hanawa0801,kuno,yosinori}@hci.ics.saitama-u.ac.jp
[2] Department of Computer Science and Engineering, Comilla University, Cumilla, Bangladesh
[3] Graduate School of Science and Technology, Gunma University, Maebashi, Japan
fukuda.hisato@gunma-u.ac.jp

Abstract. Video-based estimation plays a very significant role in person identification and tracking. The emergence of new technology and increased computational capabilities make the system robust and accurate day by day. Different RGB and dense cameras are used in these applications over time. Video-based analyses are offensive, and individual identity is leaked. As an alternative to visual capturing, now LiDAR shows its credentials with utmost accuracy. Besides privacy issues but critical natural circumstances also can be addressed with LiDAR sensing. Some susceptible scenarios like heavy fog and smoke in the environment are downward performed with typical visual estimation. In this study, we figured out a way of estimating a person's property, i.e., height and age, etc., based on LiDAR data. We placed different 2D LiDARs in ankle levels and captured persons' movements. These distance data are being processed as motion history images. We used deep neural architecture for estimating the properties of a person and achieved significant accuracies. This 2D LiDAR-based estimation can be a new pathway for critical reasoning and circumstances. Furthermore, computational cost and accuracies are very influential over traditional approaches.

Keywords: Person property estimation · 2D LiDAR · Deep neural network · Height estimation · Age estimation · ResNet

1 Introduction

From the beginning of artificial intelligence, it was always challenging to accurately identify persons and their different properties. How a human think, memorizes, and makes decisions, these techniques were tried to imitate a machine. Over the period, developments of new designs, machine learning algorithms, and increased computational capabilities help us find some sustainable and reliable models for different property estimations of a person. All these developments come to a real momentum after the innovation of deep neural networks. Massive data can be processed with a brain-like model using deep learning. Our study also finds a breakthrough after using DNN models.

© Springer Nature Switzerland AG 2021
D.-S. Huang et al. (Eds.): ICIC 2021, LNCS 12836, pp. 763–773, 2021.
https://doi.org/10.1007/978-3-030-84522-3_62

Video-based surveillance and analysis took a vital role in human property estimation. Introducing new video cameras made the analysis more accurate and sophisticated day by day. Depth camera, night vision camera, RGB camera, wide-angle architectures helped us capture high-definition images and videos in different circumstances. Using machine learning algorithms and DNN architectures, it is relatively easier to process these data for any further enhancements. Above all these benefits, video cameras are not error-free and well accepted in all conditions. Some natural and privacy issues arise while using surveillance cameras in all situations. Some regulations and confidentiality policies also prevent us from using video cameras in all areas. Our study aims to focus on all these issues and find a well-fit alternative of cameras to accurately identify a person's property. Here we emphasize height and age for recognition. A very well-suited application of this property estimation is museum guided robot. By analyzing persons' walking data, it can predict his/her height and age-appropriately. Adult and a child can also be accurately identified and guided. Though walking speed and patterns vary during watching and working, our proposed system can handle these issues very well. It continuously scans the persons and can update its previous data simultaneously. Some other applications, i.e., childcare agents, elderly support robots, aircraft assistance, etc., can easily support their clients with this application.

This research has used a 2D LiDAR sensor as an alternative to a video camera for data acquisition. We considered human walking places' general situation and tried to gather their data from different analysis angles. As LiDAR sensors provide distance data of a moving object in its territory, we collected these distance data, placed them on a particular time on a frame, and created motion history images. We used these images of different persons for our network as train and test data. We further validated the model with these data that increased the efficiency of the model. For our research, we used Residual Network (ResNet) as a backbone of our network. We used twenty-nine-person data for our analysis and got a significant performance. We categorized our study in different ways to find the person's various properties with multiple considerations. Our study covered the analysis of predicting the height of a person in two and three categories. We also extended our research to indicate the age of a person with these data. Even though LiDAR-based applications have a bottleneck of accuracy, we expanded our study with different neural network models and tried to find an optimal model with various disjoint data.

In Fig. 1, we can see a block diagram of our proposed system. We collected LiDAR data by sensing the sensors continuously. For our experiments, we placed LiDAR sensors at the ankle level. We evaluated four different sensor data for our investigations and showed significant results with all the data. We processed these LiDAR data and created motion history images with varying frames per second (FPS). These images are used as input of the deep neural network (DNN) for analysis. A pre-trained ResNet is used as modeled network and analyzed for two specific person property: height and age as a research domain. Our study shows very substantial results in different conditions. We varied our datasets in various patterns and performed our experiments for fault tolerance and the model's sustainability.

Interestingly our model most accurately identified our test set data with disjoint train and validation data. This study emphasizes the use of a 2D LiDAR sensor as an alternative

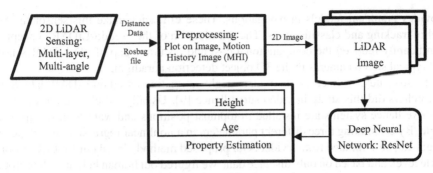

Fig. 1. Block diagram of the proposed system

to video cameras. We enhanced our research to find the persons' different properties using LiDAR sensor as cameras are less performing in some vulnerable situations.

2 Related Works

Many researchers added their contributions to recognizing a person's height and age by different techniques. Some people used camera images, some developed algorithms, and some used new technologies, i.e., artificial intelligence, machine learning even deep learning. We focused on state-of-the-art technologies to find the best performances on the topic. Here we will discuss some new approaches of person property estimation and LiDAR-based applications. Levi G. et al. [1] demonstrated a convolutional neural network-based age and gender classification. They continued their research based on facial images and automatic face recognition capabilities. This research was an introduction of deep earning on such a study. Our contributions in this research vary with this fundamental approach in using video cameras. We initiated LiDAR-based estimation rather than cameras. Some researchers tried to find a way to handle person recognition alternatively. Yamada H. et al. [2] describe a 3D LiDAR-based identification. They used LiDAR data and long short-term memory for gait recognition. A data augmentation method was used to improve the performance of the system. This approach has a bottleneck performance with multi-line LiDAR data, where we used only 2D LiDAR with improved property estimation. Some research focused on online learning [3] with the mobile robot by using 3D LiDAR data. A human classification and tracking were shown here with self-developed datasets. Benedek C. et al. [4] worked with 3D LiDAR-based gait analysis in a 4D surveillance system. Here they showed some activities of persons also.

Research on portable person behavior measurement [5] system by 3D LiDAR also tried to track the target persons. A large area of people behavior analysis was performed that was relatively tough by typical tracking mechanisms. As 3D LiDARs are not affected by lighting conditions, some researchers used this for their work in person classification [6]. But the computational cost is very high in 3D scan and data processing, which makes this system unacceptable in many applications. Despite this, 3D LiDAR scan is widely used autonomous driving [7], where deep learning models are applied on point cloud data to show a visual model. An alternative to dense depth cameras, LiDAR data provides graphical models in navigations. These efficacies make this sensor the best tool in tracking and classifications. The introduction of the pseudo-LiDAR concept [8] significantly reduced the computational cost and accuracy gap of 3D LiDAR sensors and typical stereo cameras in the 3D object detection paradigm.

In this study, we focused on human height estimation based on LiDAR data. Some researchers did this study in video surveillance [9]. Usually, all close circuit cameras in surveillance systems are installed in topmost positions and want to cover up wide-angle. By calculating three different parameters in a nonlinear regression model, person height can be estimated from a video. Our proposed method placed our LiDAR sensor at ankle level, and based on only distance data, we figured out human height with profound accuracy. Another research was done based on the image to estimate person height where camera parameters and scene geometry are unknown [10]. They tried to demonstrate that even though deep learning techniques are used, there may be a difficulty in monocular height estimation. This research has shown a bottleneck of using images for height estimation, even in deep learning. Other researchers [11] estimated persons' height from the video, taking gravity as a reference. They considered videos of motion as input for their system.

A trajectory-based clustering approach to the group, an individual people from his image [12], was performed previously. This research's main novelty was to use a person's individuality and enhance the gait analysis study. All research, as mentioned earlier, was conducted on 3D LiDAR or RGB camera-based images. Our previous approaches [13, 14] were based on a 2D LiDAR image to find an effective tracking system by adopting EDBSCAN and EOPTICS algorithms. Here, based on the LiDAR image, we find a way to estimate a person's height and age using deep learning techniques.

3 Proposed Method

3.1 Dataset Preparation

We prepared our dataset for this research. We considered 29 users in this experiment with different ages and gender. In our study, various geographical peoples also participated. Most of the people wore shoes, but few of them wore sandals. In our research person with different heights and ages were also attended. Sixteen persons were below 170 cm height, and others were above the threshold. The age limit was between 22 to 36 years old. Thirteen persons were greater or equivalent to 30 years old, and others were below 30 years. We considered four different LiDAR sensors at different altitudes and angles. Then captured all these LiDAR data through ROS (Robot Operating System) environment to a '.bag file.' Individual LiDAR data of every person was stored in a separate bag file.

All bag files are used for generating LiDAR motion history images (MHI). These images were used as an input of our proposed method. Figure 2(a) shows motion history images of LiDAR data. Here different colors (i.e., red, green, and yellow) indicating different LiDAR data captured from different layers and/or different angles. The corresponding grayscale plot of these data showing the same data. All these lines are for a specific moment of a single persons' ankle movement. We accumulated 0.5s (20frames) data from all LiDARs for generating MHI. These images are being used as an input of our system. Figure 2(b) shows the image dataset of different participants. We categorize these datasets for height estimation and age measurements. Based on height, there are two classes of data we prepared: tall and short. Sometimes we incorporated another type as the medium. For age estimation, we grouped these data as young and elder. These categories help us to accurately identify a persons' class based on their walking data.

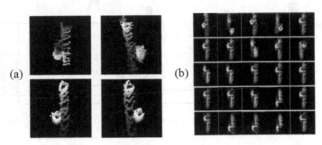

Fig. 2. (a) Motion history images (Color Image and grayscale) (b) Image dataset (Color figure online)

3.2 Person Property Estimation

For a human, the different properties can be estimated for various purposes. Here we considered height and age in this circumstance. Estimating a person's height is a vital property in other applications. Clothing, defense recruitment, safety and security agencies, rescue measurements, live programs, event management, etc., applications are closely related to human height. Some cases are restricted to disclose a person's identity. Here no video cameras are allowed. Our LiDAR-based property estimation technique is an excellent alternative to these applications.

Moreover, LiDAR data are independent of bias by light, motion, and natural calamities. We placed our LiDARs in four positions. Two LiDARs are at the same angle but at different heights, and another two sensors were 2 m apart from the first and at different heights and angles. These positionings are described here as multi-layer and multi-angle in this study. In Fig. 3, our experimental setup is shown. Four LiDAR sensors are placed in multi-layer and multi-angle positions. Persons are walking in front of these sensors, and they collect data. For training, testing, and validation, we used different LiDAR images in our experiments.

For person property estimation, we placed LiDAR sensors in ankle level height. Peoples come in front of these sensors, and it collects their data. A rosbag package is used for hardware interface with LiDAR sensor and computer. Raw data of LiDAR sensors were stored as a bag file. We considered a batch program to make a motion history image (LiDAR image) from a bag file by combining different LiDAR data. These images are used as an input of our application. We considered PyTorchLightning to train these images. With the developments of deep learning architectures, possibly deep Residual Network is the most revolutionary invention in Computer Vision and deep neural network peoples in the last couple of years. ResNet enabled us to train maximum layers, even thousands, with encouraging performances and precise constraints. We used ResNet 18 and ResNet 50 in different conditions in our application and found a significant improvement in accuracy with all disjoint datasets.

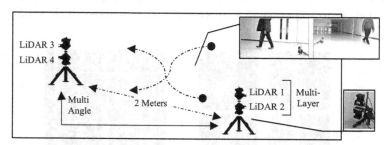

Fig. 3. Experimental setup and person walking in front of LiDAR sensors.

3.2.1 Height Estimation

In this research, we considered two properties of a person to estimate by LiDAR data. The first one is height. Accurate height estimation is essential in different geometric estimations and scientific research. Some applications are susceptible to accurate height estimation. Even though very well-established research has been done on the topic, this is still a thrust sector. In some recent studies, human height estimation was proposed based on depth and color information [15]. The human body and head were extracted from color images and predicted their height based on depth information. Mask R-CNN [16] was used for extracting data from individual frames. Here, height estimation through LiDAR data makes this invention eventually excepted to all. It convinced most of the shortcomings of traditional RGB and RBD-D-based applications efficiently. In Fig. 4, multi-angle ankle level LiDAR sensors are used for data acquisition. A motion history image based on the LiDAR sensor's distance data is used as an input of our system. We resized all the images as our application that its processing goes unique. A pre-trained ResNet18 model was used for training our model especially binary classification of LiDAR images. It requires a 224*224*3 size input image, and 71 deep layers were used for analysis. For cross-validation, we used resnet34 and resnet50 also.

A prevalent query about using the residual network is what are the benefits it. Significantly faster convergence, easy optimization, and significant improvements of precision over increased depth make ResNet well accepted to all computer vision researchers. Among all other models, ResNet18 is the best dealing model in contrast to performance. We discuss this model in detail here. In Fig. 5, a detailed explanation of ResNet18 architecture is explained. We first resize the input images gathering from the rosbag file. A very well-known image size (224*224*3) is produced from given images. There are different convolutional layers are responsible for filtering the input image. The first convolutional layer (Conv: 1) is accountable for providing low-level features, i.e., edge, gradient, color, etc.

The deeper layer provides relatively high-level features. Finally, a feature map is created by convolutional layers to predict the class probability for all gained feature

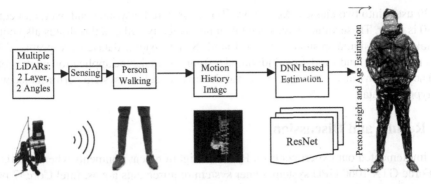

Fig. 4. DNN based height estimation through 2D LiDAR.

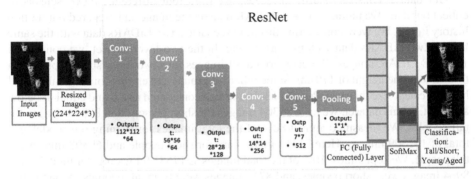

Fig. 5. ResNet structure for the application.

maps. The pooling layer lessens the spatial volume of the convolution elements. The influential details, i.e., rotational and positional invariants, are found in the pooling layer also. Fully Connected Layer (FC Layer) receives flatten output from pooling layer and acts as a feedforward network. SoftMax layer is responsible for a binary outcome. It limits the range within [0,1]. Here in our application, the result will remain the classes: tall and short or aged and young. When we changed binary output into three categories (tall, medium, and short), we changed the SoftMax layer. The promising efficacy of ResNet to incorporate the property of skipping connection to add the output of the previous layer to next ensures the proficiency from beginning to end in deep layers.

3.2.2 Age Estimation

A standalone application to estimate a person's age from LiDAR images is unique and might be interesting in different applications. A caregiver company, various public dealings organizations, transport agencies may need to know the person's age before offering any amenities to its customers. Video-based applications always suffer false identification due to facial expressions and makeup. Our proposed LiDAR-based application can handle these issues effectively. The process of age estimation is relatively like height estimation. We prepared our datasets as per the parameter of age. Here we categorize all 29 users into two classes. Age below 30 is considered as young and over, and equal to 30 is elder. These values have been taken heuristically. Other's thresholds also could be taken, and similar results could be found. Same disjoint datasets are prepared for training, testing, and validation. Individual testing sets make the application more robust and accurate. We also performed cross-validation to check the system performance with all types of data.

4 Results and Discussion

We implemented our experiments on PyTorch Lightning environments where NVIDIA GeForce GTX 1060 GPU system. Other system requirements are as, Intel Core i7 processor with 3.8 GHz clock speed, 6 GB graphics memory, 64-bit ubuntu 18.04 release, 16 GB RAM, and 1TB SSD.

Our dataset consists of different sizes. We used four different LiDAR sensors to collect our data. We plotted different LiDAR data on the image and prepared our motion history image. We were considering different scenarios: two LiDARs data with the same height, two LiDARs data with two layers, etc. In the two-layer dataset, we considered 1,76,678 LiDAR images. We categorized all images into two main subgroups with a threshold of the height of 170 cm. Some other subgroups and entries could be taken, and here we categorize this way. Persons above the size considered as tall and equal or below are short. In this dataset, we have 89304 images for short persons and 87374 images for tall peoples. We measured almost 80% of the total images for training our model as a total of 141309 images. Here 71411 images are for short people and 69898 images for tall people. We kept almost 10% of the total images as 17681 for testing our model. Here 8964 images were short peoples, and 8717 images were from tall peoples. Around 10% of disjoint images were considered as validation datasets. Here 8929 images were from short people and 8759 images from tall people. With rigorous training with the ResNet18 model, we found very impressive accuracy there. In Table 1, a confusion matrix shows the detail of the system with all disjoint datasets. This matrix is formatted from test data and significantly found its efficacy here. Among all 8717 images of testing-tall data, our system accurately identified 8658 images as tall, and only 59 were misclassified. The same scenario was found for the short dataset also. Here only 62 images were falsely identified out of 8964 images. The accuracy of 99% relieved the cumbersome of previous estimations.

Again, we considered two LiDAR sensors at the same height where the only angle is different. In this condition, we considered 226241 images as a total dataset, where 121244 images for short persons and 104997 images were from tall persons. The same 80% (181016) of total images were kept as test data. The remaining 20% (22632 and 22593) data were considered test and validation data almost equally. All data was disjoint here also, where the same persons were assessed in all three groups (train, test, and validation). We considered the same residual network ResNet18 for analysis. The right side of Table 1 shows the confusion matrix of test data. Among 10498 images of tall people, our model can accurately identify 10411 images, and for 12134 images of short persons, this can detect 12027 images ideally. The system accuracy is near about 99% in total. Both Multi-Layer and Multi-Angle show almost the same precision but combining these two will increase the overall system performance in a complex environment.

Table 1. Confusion Matrix of resnet18 based height estimation

Confusion Matrix (Test)

Tall	8658	62
Short	59	8902
	Tall	Short

Height Estimation (Multi Layer)

Confusion Matrix (Test)

Tall	10411	107
Short	87	12027
	Tall	Short

Height Estimation (Multi-Angle)

Further, we enhanced our study of height estimation for three categories. Here another class, 'medium,' was introduced. People over 175 cm are considered tall. In between or equal 170 cm to 175 cm was medium, and below 170 cm was short. In our study, we assessed a total of 216474 images of two-layer LiDAR data. 80% of the total (173180) images were considered training data, and the remaining (43294) images were equally split into the test and validation set. We segmentize all images as a tall, medium, and short with 59327, 99656, and 57491 images, respectively. The overall accuracy reduced a little of 4% in total by adding a new category. In Table 2, We placed all test data in this confusion matrix. Among 5932 images of tall persons, our system accurately identified 5676 images. Some photos were misclassified as tall, but there is no short classification here. The same scenario is in short type classification. Only 268 images were grouped as a medium but no tall. Two types of data bias the middle class. Here among 9966 images, 9523 were accurately categorized, but 330 were identified as tall, and 113 were short.

Table 2. Confusion matrix of three classes of height estimation and age estimation

Confusion Matrix (Test)

Tall	5676	330	0
Medium	256	9523	268
Short	0	113	5481
	Tall	Medium	Short

Height Estimation

Confusion Matrix (Test)

Young	11535	412
Elder	483	9217
	Young	Elder

Age Estimation

Another property of humans is the age that we estimated through 2D LiDAR data. For this purpose, we considered 216474 images as 'Age Dataset.' In this research, people over or equal to 30 years were considered elder aged and below 30 years were deemed to be young—a total of 96295 images for older people and remaining images for young people. As the previous ratio total 80% images were set for training, the remaining 20% were equally distributed as training and validation. On the right side of Table 2. we see that among 12018 images of young people in the test dataset, 11535 images could correctly identify by our model. For older people, 9217 images have correctly identified where only 412 were misclassified. The overall precision of this model is 95% approximately.

Table 3 describes the overall properties of the experiments that we have performed in this research. For our investigation, we considered different batch sizes and ten epochs for all cases. Here train, test, and validate accuracies showing system performances. On the other hand, different loss functions showing system integrity and robustness with the datasets. This research emphasizes the best use of 2D LiDAR data for person property estimation that could be an excellent alternative to RGB and RGB-D cameras.

Table 3. Overall system performance with ResNet18 network

Data	Experiments Type	Batch Size	Epoch	GPU	Model	Train Accuracy	Train Loss	Test Accuracy	Test Loss	Validation Accuracy	Validation Loss
New data 2 Layer LiDAR	Height 3 Category	24	10	Yes	Resnet18	0.963	0.0921	0.9599	0.1063	0.9593	0.1028
New Data 2 Layer LiDAR	Age 2 Category	34	10	Yes	Resnet18	0.959	0.0971	0.9575	0.1032	0.9546	0.1107
New Data 2 Layer LiDAR	Height 2 Category, Complete Random	34	10	Yes	Resnet18	0.9939	0.0173	0.9937	0.0178	0.99382	0.01776
New Data 2 Angle LiDAR	Height 2 Category, Complete Random	34	10	Yes	Resnet18	0.997	0.00976	0.9958	0.0125	0.99552	0.01187

5   Conclusion

Person property estimation is always challenging and sometimes crucial in different circumstances. Many efforts have been initiated on this topic from the beginning of computer vision. Simultaneously some applications were developed with the help of the LiDAR sensor. We concentrated on the amalgamation of these two phenomena. To find a suitable alternative of a video camera and improve the property estimation accuracy bring we out to do so. Our LiDAR-based person property estimation, especially introducing a deep residual network, gives a well-accepted benchmark here. Preparing a new dataset and finding its individual properties makes this study abundant. This dataset can be used for other 2D LiDAR-based research. In the future, we will try to enhance our study to find more properties of a person that he/she can be accurately traced. We will combine our previous tracking system with this property estimation technique to develop a LiDAR-based autonomous system. Our impending focus is on group recognition and the prediction of their behavior without compromising individual identity.

References

1. Levi, G., Hassner, T.: Age and gender classification using convolutional neural networks. In: Proceedings of the IEEE Conference on Computer Vision and Pattern Recognition Workshops (CVPR-W), pp. 34–42 (2015)
2. Yamada, H., Ahn, J., Mozos, O.M., Iwashita, Y., Kurazume, R.: Gait-based person identification using 3D LiDAR and long short-term memory deep networks. Adv. Robot. **34**(18), 1201–1211 (2020)

3. Yan, Z., Duckett, T., Bellotto, N.: Online learning for 3D LiDAR-based human detection: experimental analysis of point cloud clustering and classification methods. Auton. Robot. **44**, 147–164 (2020)

4. Benedek, C., Gálai, B., Nagy, B., Jankó, Z.: Lidar-based gait analysis and activity recognition in a 4D surveillance system. In: Proceedings of the IEEE Transactions on Circuits and Systems for Video Technology, vol. 28, no. 1, pp. 101-113 (2018)

5. Koide, K., Miura, J., Menegatti, E.: A portable three-dimensional LIDAR-based system for long-term and wide-area people behavior measurement. Int. J. Adv. Robot. Syst. **16**(2), 172988141984153 (2019)

6. Yan, Z., Duckett, T., Bellotto, N.: Online learning for human classification in 3D LiDAR-based tracking. In: Proceedings of the IEEE/RSJ International Conference on Intelligent Robots and Systems (IROS), pp. 864–871 (2017)

7. Tu, J., et al.: Physically Realizable Adversarial Examples for LiDAR Object Detection. In: Proceedings of the IEEE/CVF Conference on Computer Vision and Pattern Recognition (CVPR), pp. 13716–13725 (2020)

8. Qian, R., et al.: End-to-end pseudo-LiDAR for image-based 3D object detection. In: Proceedings of the IEEE/CVF Conference on Computer Vision and Pattern Recognition (CVPR), pp. 5881–5890 (2020)

9. Li, S., Nguyen, V.H., Ma, M., et al.: A simplified nonlinear regression method for human height estimation in video surveillance. J. Image Video Process. **32**, 1–9 (2015)

10. Günel, S., Rhodin, H., Fua, P.: What face and body shapes can tell us about height. In: Proceedings of IEEE International Conference on Computer Vision (ICCV) Workshops (2019)

11. Bieler, D., Gunel, S., Fua, P., Rhodin, H.: Gravity as a reference for estimating a person's height from video. In: Proceedings of the IEEE/CVF International Conference on Computer Vision (ICCV), pp. 8569–8577 (2019)

12. Sugimura, D., Kitani, K.M., Okabe, T., Sato, Y., Sugimoto, A.: Using individuality to track individuals: clustering individual trajectories in crowds using local appearance and frequency trait. In: ICCV (2009)

13. Hasan, M., Hanawa, J., Goto, R., Fukuda, H., Kuno, Y., Kobayashi, Y.: Tracking people using ankle-level 2D LiDAR for gait analysis. In: Ahram, T. (eds.) Advances in Artificial Intelligence, Software and Systems Engineering. AHFE 2020. Advances in Intelligent Systems and Computing, vol. 1213, Springer, Cham (2021) https://doi.org/10.1007/978-3-030-51328-3_7

14. Hasan, M., Hanawa, J., Goto, R., Fukuda, H., Kuno, Y., Kobayashi, Y.: Person tracking using ankle-level LiDAR based on enhanced DBSCAN and OPTICS. IEEJ Trans. Electr Electron. Eng. **16**(5), 778–786 (2021)

15. Lee, D., et al.: Human height estimation by color deep learning and depth 3D conversion. Appl. Sci. **10**(16), 5531 (2020)

16. He, K., Gkioxari, G., Dollár, P., Girshick, R.: Mask R-CNN. IEEE Trans. Pattern Anal. Mach. Intell. **42**(2), 386–397 (2020)

Detection of Pointing Position by Omnidirectional Camera

Yuuichiro Shiratori and Kazunori Onoguchi[(✉)]

Hirosaki University, 3, Bunkyo-cho Hirosaki, Aomori 036-8561, Japan
onoguchi@hirosaki-u.ac.jp

Abstract. In recent years, voice recognition devices such as smart speakers have been in the limelight, but intuitive communication such as instructing the target with ambiguous expressions is difficult with voice alone. In order to realize such communication, it is effective to combine gestures such as pointing with voice. Therefore, we propose a method to detect the pointing position from the image of the omnidirectional camera using a convolutional neural network. Although many methods have been proposed to detect the pointing position using a normal camera, the standing position of the person who gives instructions is limited since the observation area is small. We solve this problem by using an omnidirectional camera. First, the proposed method converts a hemisphere image taken from an omnidirectional camera to a panoramic image. Next, the bounding box surrounding the person with pointing gesture is detected in the panoramic image by the object detection network. Finally, the pointing position estimation network estimates the pointing position in the panoramic image from the image in the bounding box and its location. Since it is difficult to prepare a large number of pointing gesture images, CG images created by Unity are used for pre-training. Experiments using real images of pointing gesture shows that the proposed method is effective for pointing position detection.

Keywords: Pointing gesture · Panoramic image · Convolutional Neural Network

1 Introduction

While various smart devices such as smartphones, smart watches and smart glasses are being developed, voice recognition devices such as smart speakers are in the limelight. However, it is difficult to perform intuitive communication such as "take that" or "what is the name of it" by only voice instruction. To indicate an object in an ambiguous expression, it is effective to use pointing gestures together with voice. Therefore, this paper proposes a method to detect the pointing position from the image of the omnidirectional camera using a convolutional neural network.

Many methods using a normal camera have been proposed to detect the pointing position. However, since the observation area of a normal camera is narrow, the standing position of a person giving instructions is limited. In addition, the performance is not sufficient since most of methods use handcraft features for recognition. Therefore, it is required to realize a high performance method that can observe a wide area.

D.-S. Huang et al. (Eds.): ICIC 2021, LNCS 12836, pp. 774–785, 2021.
https://doi.org/10.1007/978-3-030-84522-3_63

In recent years, omnidirectional cameras that can easily acquire omnidirectional images have begun to spread. In addition, advances in neural networks have led to the development of methods that cut out persons and recognize their poses with high accuracy. For this reason, the proposed method uses a convolutional neural network (CNN) to detect the bounding box of the person performing the pointing gesture from the omnidirectional image and estimate the pointing position from the image in the bounding box. Since the omnidirectional camera can capture the entire room in a single image, the proposed method can detect the pointing position regardless of the standing position in the room.

We could not find a public dataset of pointing gestures taken by an omnidirectional camera and it is difficult to collect a large number of those images. Therefore, CG images created by Unity are used for pre-training.

Although only the method of estimating the pointing position in an image is described in this paper, the pointing object can be identified by combining the proposed method and the image identification network. Therefore, if the proposed method is incorporated into a voice recognition device, the usability of the device will be significantly improved.

The rest of this paper is organized as follows. Section 2 gives a brief overview of related works. Section 3 describes details of the proposed method. The experimental results are shown in Sect. 4, followed by the conclusion in Sect. 5.

2   Related Works

The pointing gesture detection method is roughly divided into a method using a depth sensor, a method using multiple cameras and a method using a single camera.

As a method using a depth sensor, Dhingra et al. [1] proposes the method in which 3D positions of the elbow and the hand are detected using the RGB-D camera and the pointing vector toward the target is estimated. Droeschel et al. [2] proposes a method that detects the positions of the face, the elbow and the hand from 3D point clouds obtained by a ToF camera and calculates the pointing direction using the Gaussian Process Regression.

Instead of depth sensors, several methods uses multiple cameras to acquire 3D information. Joic et al. [3] detect the pointing position using the dense disparity maps obtained by a stereo camera. This method detects the foreground by simple 3D background subtraction. The foreground pixels is divided into the arm extending in the direction of the target and the other body part using the Gaussian mixture model. Park et al. [4] propose the pointing gesture detection method using a stereo camera, 3D particle filters and two stage HMMs for mobile robots. Hu et al. [5] estimate hand pointing based on two orthogonal-views, the top view and the side view. The hand area in each view is detected by Haar-like features and an AdaBoost cascade detector. Feature points along the hand contour are detected and tracked by an Active Appearance Model (AAM). Combining two views of hand features, pointing direction is estimated. These methods are difficult to apply to applications that point to the entire room since they have a small observation area. Watanabe et al. [6] propose a multi-camera system to detect omnidirectional pointing gesture. This method installs eight cameras around the subject and finds which camera captures the frontal face of the subject. The cameras on both sides of this camera are used for stereo measurement. The pointing direction is estimated from the straight line connecting the face position with the hand position. Although this method can detect omnidirectional pointing gestures, many cameras must be placed around the room.

Some methods using a single camera have been proposed because they are less expensive computationally and physically. Cernekova et al. [7] detect and track the pointing hand by a snake using the gradient vector flow field. The center of the gravity of the snake is transformed directly into the canvas coordinates using linear transformation. Huang et al. [8] proposes the method to detect the pointing gesture using egocentric vision. The bounding box of the hand is detected from the image of the wearable camera by the Faster R-CNN [11]. In the bounding box, the index fingertip and index finger joint are detected by the CNN-based framework. They developed the Ego-Air-Writing system that recognizes characters drawn in the air with the index finger. This method requires the wearable camera such as the smart grass and it is difficult to detect the target other than in front of the camera. Mukherjee et al. [9] proposed the method that detects the fingertip using a standard laptop camera or a web-cam to recognize air writing. This method detects the hand region using the Faster R-CNN and extracts the binary hand mask by skin segmentation and background subtraction. The fingertip is detected by signature function called distance-weighted curvature entropy. Jaiswal et al. [10] estimates the direction of finger pointing from a single RGB image using a deep convolutional neural networks. Since these methods use a camera with a normal angle of view, they are not suitable for applications that point to a large area.

Since the omnidirectional camera can capture a large area in a single image, the proposed method is suitable for application that point to the entire room.

3   Detection of Pointing Position

3.1   Overview

Figure 1 shows the outline of the proposed method. First, the hemisphere image taken from an omnidirectional camera is converted to the panoramic image. The origin of the panoramic image is the upper left of the image. Next, the bounding box surrounding the person with pointing gesture is detected in the panoramic image by the object detection network (Faster R-CNN [11]). Finally, the pointing position estimation network estimates the pointing position in the panoramic image from the image in the bounding box and its location (x, y, w, h). (x, y) is the coordinate value of the center of the bounding box in the panoramic image and w and h are the width and height of the bounding box. The pointing position is regressed from the feature map obtained by ResNeXt [13] and the bounding box location (x, y, w, h). The proposed method can detect the pointing position regardless of the standing position in the room by inputting the location of the bounding box.

3.2   Dataset

We prepared two types of pointing gesture images, real images and CG images since it is not easy to collect a large number of real images in which both the pointing position

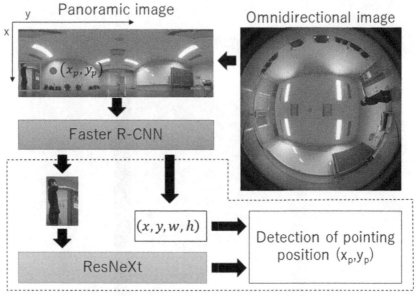

Pointing Position Estimation Network

Fig. 1. Outline of the proposed method

and the bounding box are annotated. After pre-training with CG images, the network is fine-tuned with a few real images.

Figure 2 shows the environment in which the real image are taken. An omnidirectional camera (Kodax PIXPRO SP360 4K) is placed on a table in the center of the room. Subjects perform pointing gestures anywhere around the table. The ground truth of the pointing position in the panoramic image is obtained using a laser pointer held by the subject. Panoramic images converted from hemispherical images by the polar coordinate transformation are used as training images. The size of the panoramic image is 2,048×700 and 887 real images are prepared.

CG images are created by Unity and Unity Recorder. Seven human models shown in Fig. 3 are prepared. The camera is installed at the origin of the coordinate and four walls are randomly generated at positions from 5 m to 9 m around the origin. The human model randomly points to the wall with the right hand and the scene is converted into a panoramic image with a resolution of 2,048 ×700 pixels. To reduce the difference from the real image, the area of 324 pixels at the bottom of the image is removed. Figure 4 shows the example of the CG image. The number of CG images is 77,000.

3.3 Detection of Person Area

A person area is detected from a panoramic image using Faster R-CNN [11] additionally trained by CG images. In a panoramic image, when a person stands near the left or right edge of the image, a person area may be separated into left and right as shown in Fig. 5. To avoid this problem, the panoramic image is extended to a $2,448 \times 700$ pixel image

Wall(Pointing potision)

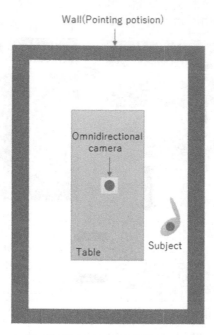

Fig. 2. Experimental environment

by copying the leftmost 200 pixels of the image to the right edge and the rightmost 200 pixels of the image to the left edge as shown in Fig. 6. The bounding box of a person is detected in this extended image. Since the same person area may be detected twice, the detection range is set to $200 \leq x \leq 2, 248$.

Fig. 3. Human model used in UNITY

Training of Faster R-CNN is performed with a batch size of 8 and training time is set to 20 epochs. The leaning rate is 1×10^{-5}. Adam [12] with $\beta_1 = 0.9$ and $\beta_2 = 0.999$ is used for optimization. First, Faster R-CNN is pre-trained using 70,000 CG images and tested with 7,000 CG images. Then, fine tuning is performed using 787 real images. Evaluation is performed using the remaining 100 real images. Data augmentation that changes brightness, contrast, saturation and hue is performed during training. Faster R-CNN is trained independently using training images with a bounding box annotated and the parameters are fixed after training.

Fig. 4. Example of CG image

Fig. 5. Separation of person area

Fig. 6. Extension of image

3.4 Estimation of Pointing Position

Figure 7 shows the outline of the pointing position estimation network. Features in the bounding box are extracted by ResNeXt [13] in which Inception module [14] is added to ResNet [15]. The bounding box image is normalized to a 224×224 pixel image by resizing and zero padding. ResNeXt inputs the normalized image and outputs 2,048 dimensional features. The location of the bounding box (x, y, w, h) is input to a fully connected layer of 64 channels and its output is combined with image features.

Finally, the pointing position (x_p, y_p) in a panoramic image is estimated from the 2,112 dimensional features.

The loss function of the pointing position estimation network is given by

$$
\text{Loss} = \frac{1}{n} \sum_{k=1}^{n} \left(\sin\left(2\pi x_p^k\right) - \sin\left(2\pi x_{gt}^k\right) \right)^2
$$
$$
+ \frac{1}{n} \sum_{k=1}^{n} \left(\cos\left(2\pi x_p^k\right) - \cos\left(2\pi x_{gt}^k\right) \right)^2
$$
$$
+ \frac{1}{n} \sum_{k=1}^{n} \left(y_p^k - y_{gt}^k \right)^2,
\tag{1}
$$

where (x_p, y_p) is the predicted value of the pointing position (x, y) and (x_{gt}, y_{gt}) is its ground truth. These values are normalized between 0 and 1.

If $\left| x_p - x_{gt} \right|$ is simply used as an error, the loss becomes extremely large when the ground truth is at the right end of the image and the predicted value is at the left end of the image, as shown by the red arrow in Fig. 8. In order to deal with such a case, the L_2 norms of $\sin(2\pi x)$ and $\cos(2\pi x)$ are used for the loss. This loss function maintains the connectivity between the left and right edges of the image because $\sin(0) = \sin(2\pi x) = 0$ and $\cos(0) = \cos(2\pi x) = 1$. When $x_{gt} = 0(0.5)$ and $x_p = 0.5(0)$, the loss of the first term of Eq. (1) becomes 0. However, a complementary relationship holds since the loss of the second term becomes large.

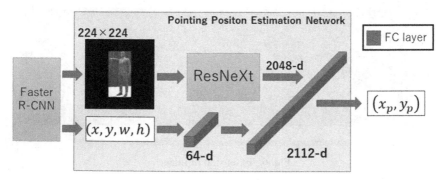

Fig. 7. Pointing position estimation network

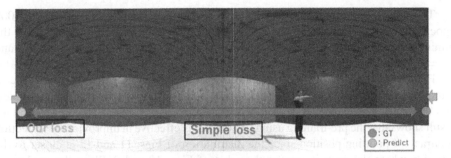

Fig. 8. Loss function

4 Experiments

Experiments were conducted in the environment shown in Fig. 2. Eight subjects perform pointing gestures anywhere around the table. The following two types of experiments were conducted.

1. The pointing position estimation network is trained using only the real image.
2. The pointing position estimation network is fine-tuned with the real image after pre-training with the CG image.

4.1 Results Using Only Real Image

The pointing position estimation network was trained and tested using only real images without pre-training with CG images. The number of training images and test images is 787 and 100.

Training is performed with a batch size of 128 and training time is set to 1,500 epochs. The initial learning rate is 1×10^{-5} and it is reduced by a factor of 0.5 if the minimum validation loss is not updated for 100 epochs. Adam [12] with $\beta_1 = 0.9$ and $\beta_2 = 0.999$ is used for optimization.

Figures 9 and 10 show a histogram representing the error distribution in the x-direction and the y-direction. The horizontal axis of Fig. 9 shows the bin that divides the range of x ($0 \sim 2,048$) into 100 sections and the vertical axis shows the number of errors within that section. The horizontal axis of Fig. 10 shows the bin that divides the range of y ($0 \sim 700$) into 70 sections and the vertical axis shows the number of errors within that section. The mean and median of horizontal error are 274 pixels and 186 pixels. The mean and median of vertical error are 61 pixels and 49 pixels. This result shows that the pointing error is considerably large when training is performed using only a small number of real images since the distribution of the error is wide.

4.2 Results Using CG Image and Real Image

The pointing position estimation network was pre-trained using 70,000 CG images created by UNITY. Then, it was fine-tuned using 787 real images. The number of test images is 100.

Training is performed with a batch size of 128 and training time is set to 1,000 epochs. The initial learning rate is 5×10^{-5} and it is reduced by a factor of 0.1 if the minimum validation loss is not updated for 50 epochs. Adam [12] with $\beta_1 = 0.9$ and $\beta_2 = 0.999$ is used for optimization.

Figures 11 and 12 show a histogram representing the error distribution in the x-direction and the y-direction. The mean and median of horizontal error are 134 pixels and 79 pixels. The mean and median of vertical error are 55 pixels and 38 pixels. This result shows that the pre-training using CG images is effective in improving the detection accuracy of pointing position since the distribution of Figs. 11 and 12 is closer to the left side where the error is smaller than that of Figs. 10 and 9. Since the median of the horizontal error is 79 pixels and the median of the vertical error is 38 pixels, a region centered on the predicted point, e.g. a rectangular region whose size is about 150×150 pixels, includes a part of the designated object. Therefore, it is considered that the designated object can be identified from an omnidirectional image.

Figure 13 shows the results of some qualitative evaluations. The ground truth of the pointing position is indicated by the light blue circle and the prediction is indicated by the yellow circle. The pointing position close to the ground truth is predicted in all images even if the subject's standing position changes in the room. Accuracy tends to decrease when the subject points to the ceiling as shown in Fig. 14. This problem is

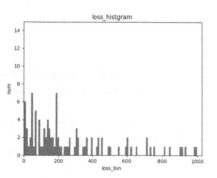

Fig. 9. Error distribution of X (Real)

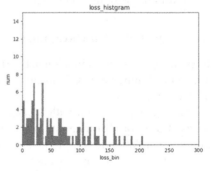

Fig. 10. Error distribution of Y (Real)

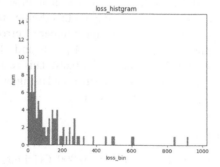

Fig. 11. Error distribution of X (CG + Real)

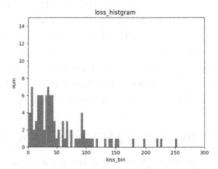

Fig. 12. Error distribution of Y (CG + Real)

Fig. 13. Example results of the proposed method

caused because the CG image dataset does not contain such images. Therefore, this problem is solved by increasing the number of CG images pointing to the ceiling.

Fig. 14. Failure example

5 Conclusions

In this paper, we proposed a method to detect the pointing position from the image of the omnidirectional camera using a convolutional neural network. The conventional method using a normal camera has a problem that the standing position of a person is limited. To solve this problem, we use an omnidirectional camera for pointing position detection. Our method detects the bounding box of the person with pointing gesture by Faster R-CNN in the panoramic image and the pointing position estimation network estimates the pointing position from the image in the bounding box and its location. Since it is difficult to prepare a large number of pointing gesture images, CG images are used for pre-training. Experimental results demonstrated that our method can detect pointing position correctly even if the subject's standing position changed in the room.

Since we aim to recognize the designated object by the image identification network, the position accuracy is evaluated by the image coordinates. Currently, since the position on the wall of the room is used for training, it can be applied only to an object existing near the wall. However, if the relationship between the image position of a point on the wall surface and its 3D position is obtained in advance, the 3D direction to the designated object can be obtained by detecting the 3D position of the fingertip using two omnidirectional cameras. This is the future work.

References

1. Dhingra, N., Valli, E., Kunz, A.: Recognition and localisation of pointing gestures using a RGB-D camera. In: arXiv:2001.03687 (2020)
2. Droeschel, D., Stuckler, J., Behnke, S.: Learning to interpret pointing gestures with a time-of-flight camera. In: Proceedings of 6th ACM/IEEE International Conference on Human-Robot Interaction (HRI) (2011)
3. Joic, J., Brumitt, B., Meyers, B., Harris, S., Haung, T.: Detection and estimation of pointing gestures in dense disparity maps. In: Proceedings of Fourth IEEE International Conference on Automatic Face and Gesture Recognition (2000).
4. Park, C., Lee, S.: Real-time 3D pointing gesture recognition for mobile robots with cascade HMM and particle filter. Image Vis. Comput. **29**(1), 51–63 (2011)
5. Hu, K., Canavan, S., Yin, L.: Hand pointing estimation for human computer interaction based on two orthogonal-views. In: Proceedings of 2010 International Conference on Pattern Recognition (2010)

6. Watanabe, H., Yasumoto, M., Yamamoto, K.: Detection and estimation of omni-directional pointing gestures using multiple cameras. In: Proceedings of IAPR Workshop on Machine Vision Applications (2000)
7. Cernekova, Z., Malerczyk, C., Nikolaidis, N.: Single camera pointing gesture recognition for interaction in edutainment applications. In: Proceedings of the 24th Spring Conference on Computer Graphics (SCCG) (2008)
8. Huang, Y., Liu, X., Zhang, X., Jin, L.: A Pointing gesture based egocentric interaction system: dataset, approach and application. In: Proceedings of 2016 IEEE Conference on Computer Vision and Pattern Recognition Workshops (2016)
9. Mukeherjee, S., Ahmed, S.A., Dogra, D.P., Kar, S., Roy, P.P.: Fingertip detection and tracking for recognition of air-writing in videos. Expert Syst. Appl. **136**, 217–229 (2019)
10. Jaiswal, S., Mishra, P., Nandi, G.C.: Deep learning based command pointing direction estimation using a single RGB camera. In: Proceedings of 5th IEEE Uttar Pradesh Section International Conference on Electrical, Electronics and Computer Engineering (UPCON) (2018)
11. Ren, S., He, K., Girshick, R., Sun, J.: Faster R-CNN: towards real-time object detection with region proposal networks. In: Proceedings of Neural Information Processing Systems (NIPS) (2015)
12. Kingma, D. P., Ba, J.: Adam: a method for stochastic optimization. In: Proceedings of Computer Vision and Pattern Recognition (CVPR) (2015).
13. Xie, S., Dollar, G. P., Tu, Z., He, K., He.: Aggregated residual transformations for deep neural networks. In: Proceedings of Computer Vision and Pattern Recognition (CVPR) (2017)
14. Szegedy, C., et al.: Going Deeper with Convolutions. In: Proceedings of Computer Vision and Pattern Recognition (CVPR) (2015)
15. He, K., Zhang, X., Ren, S., Sun, J.: Deep residual learning for image recognition. In: Proceedings of Computer Vision and Pattern Recognition (CVPR) (2016)

Optimization of Low-Speed Dual Rotor Axial Flux Generator Design Through Electromagnetic Modelling and Simulation

P. Premaratne[1(✉)], M. Q. Abdullah[1,2], I. J. Kadhim[2], B. Halloran[1], and P. J. Vial[1]

[1] School of Electrical Computer and Telecommunications Engineering,
University of Wollongong, Wollongong, NSW, Australia
prashan@uow.edu.au

[2] Electrical Engineering Technical College, Middle Technical University, Baghdad, Iraq

Abstract. Dual Rotor Axial Flux Generator (DRAFG) is a permanent magnet axial flux generator commonly used for low-speed power generation using wind power. This generator can generate useful amount of voltage even under very low revolutions per minute (RPM). Our focus has been to develop a very simple disk type DRFAG to generate 100W with RPM of no more than 100. This limit is set so that the generator could be operated from an age-old 'hit and miss' engine which is nearly noiseless consuming only a very small amount of fuel. Our design when combined with an engine is predicted to be operated for 10 h with one litre of gasoline making it very appealing for sub-Saharan countries. We accomplished a design that can generate 140 W with 100 RPM using 12 coils with a rotor dimension of 40 cm. We optimized many crucial parameters in this design including the shoulder length of coil, the gap between magnets and the coil, the coil size compared to the magnet size and so on. Cedrat Flux 12.1 was used to compute 3D Finite Element Analysis and to perform the simulation. Our design has optimized all the parameters to realize a low cost and highly efficient device.

Keywords: Dual axial flux generator · Hit and miss engine · Cedrat flux · Cogging torque

1 Introduction

Dual rotor axial flux generator (DRAFG) is a type of permanent magnet generators which consists of rare-earth magnets such as neodymium N52-type powerful magnets in rotors placed on either side of a fixed stator winding, similar to a sandwich with a very small airgap creating a very strong magnetic flux between them. Thus, this configuration induces a powerful Electro Motive Force (EMF) in the stator winding. The major advantage of this disk-type generator is that when the diameter is large consisting of many individual coils in the stator (9, 12, 16 or more coils), a voltage of around 20V can be generated with very low RPM. Hence, many designers try to develop their own DRAFG for power generation for wind turbines and tidal current-based power generation. The design can generate much more power at higher RPMs. However, there are many design

© Springer Nature Switzerland AG 2021
D.-S. Huang et al. (Eds.): ICIC 2021, LNCS 12836, pp. 786–801, 2021.
https://doi.org/10.1007/978-3-030-84522-3_64

considerations for an effective stator and rotor design that would ensure the energy is not wasted in the process as many designers have ended up with poor designs with very low efficiencies. A typical design of a DRAFG is shown in Fig. 1 (a).

The major design parameters in a dual rotor axial flux generator can be stated as the coil shape, coil thickness, number of turns in any individual coil, number of coils in the stator, the coil gap between each coil in the stator configuration, the type of connection of coils which is star or delta, the type of magnet (strength) and its shape and dimensions, number of magnets in each rotor and the gap between the magnets and finally the airgap between rotors and stator which is extremely crucial for an efficient design for given materials [1–7]. We have established simulation results that show drastic reduction in generated power when the air gap increases from even 1 mm to 2 mm.

The goal of this work is to use a typical DRAFG design which complies with both Fleming's right-hand rule and Lenz's Law to maximize the power generation and then to optimize the parameters that were stated above. We also have one additional constraint; for practical power generator with low RPM, the design should be adequate to generate around 100 W with an RPM value of around 100. We believe that this goal will lead to future designs that will lead to a very quiet power generators consuming only a very small amount of fuel compared to what is available in the market today. In 1850s in the USA, an engine called Hit-and-Miss was developed that used momentum in flywheels to maintain the speed of an engine using very small amount of fuel. These engines were very quiet as their pistons did not fire continuously as in the engines today.

Developing a generator combining such an engine with a low RPM dual rotor axial flux generator is one of our goals in the future.

Our work is based on using Cedrat Flux 12.1 to conduct simulations of the basic design with our own objectives of generating 100 W at 100 RPM to carryout finite element analysis of the magnetic flux distribution of the design. The basic rules to implement the design is the Fleming's right-hand rule and Lenz's Law. Fleming's right-hand rule provides the direction of the current flow whenever a conductor is moving inside the magnetic field. This theory applies whenever the rotor is rotating between the stator where it creates an induced current in the coils which increases the output power of the DRAF generator. Lenz's Law states that when the current is induced, it creates a magnetic field around the current flow. In order to create significant amount of electrical energy, significant amount of mechanical energy is also required in the dual rotor axial flux generator.

There are two different types of axial flux generators; a single axial flux generator and dual rotor axial flux generator. The main difference between these two types of axial flux generators is the performance, weight and the cost to build them because the DRAFG requires twice the number of permanent magnets than a single axial design [8, 9].

Our simulation of DRAFG design is implemented using Cedrat Flux 12.1 and Solidworks 2013 software. Solidworks is mainly used to model the desired shape of the design in 3-dimensional geometry and is then imported into Cedrat Flux 12.1 to simulate the mesh computation at each node. Many parameters of physics, mesh design and geometry are needed to be considered so that the simulation runs properly.

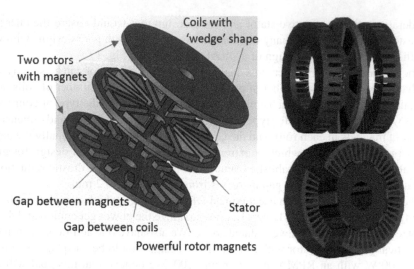

Fig. 1. Typical design of a dual rotor axial flux generator (b) top an Axial flux generator and bottom a Radial flux generator

The design has three phase axial generator that consists of 4 poles (coils) and four different alternating current circuits for each phase. The material used for each coil is copper and the permanent magnets used at each rotor were Neodymium Magnets (N52). Besides that, other parameters such as magnet specifications, coil specifications, rotations per minute, coil and magnet gaps and air gap between stator and rotors were also taken into consideration in the design process. Further explanation on the magnetic field occurrence in the dual axial flux generator will be presented next in the *Related work* section.

2 Related Work

Murphy et al. [10] describes that three-phase power is more reliable than single-phase power for generators. Three-phase power has higher starting torque and each pair is able to boost the power factor. It allows significant electrical loads control as the current distributed in each phase is lower. Besides that, the three-phase system is also capable of producing three different waves of power in sequence which ensures the power flow is constantly delivered through loads. Comparing this to single-phase system that is only able to produce single wave of power that may drop to zero during one complete cycle. Even though the zero drop is undetectable by human perception, power electrical equipment with high power demand can easily encounter problems in the long-term operation. The efficiency of three-phase comes with several drawbacks in terms of cost of installation and maintenance, but it all depends on the power needs and prioritization. Figure 1(a) shows the two types of common AC flux generators.

As shown in Fig. 2(a) an axial flux generator can comprise of either single or dual rotors with one stator. In this configuration, the magnetic rotor is sandwiched between two stators. Figure 2(b) shows a radial flux generator where the design is much more complicated as the coil-magnet separation with minimum air gap is difficult to achieve.

Pop et al. [11] evaluated the capabilities axial flux generator and the radial flux generator. The comparison made was highly focused on the electromagnetic field created during the current induction between the magnet and the conductor. The radial flux generator is a three-phase system that has configuration of 18 stator teeth that are glued to the rotor skeleton and 6 coils. Similarly, the axial flux generator is also a three-phase system that has one rotor in between of the two stators. The rotor is a 4 pole-pairs while permanent magnets used are NdFeB type which are glued to the disc rotor. The authors state that if both axial flux and radial flux have the same air gaps, the machines should produce the same amount of electromagnetic torque because the area of air gap is directly proportional to electromagnetic torque. Axial flux generator is preferred over radial flux generator because of the area of the air gap in axial flux generator can produce higher torque to weight ratio compared to conventional radial flux generator.

Nasiri-Zarandi et al. investigated the effects of dimension of air-gap and the number of stator slots available on the performance of axial flux generator [12]. The experiments they carried out were based on the Finite Element Analysis method and the performance was evaluated between simulation and the constructed prototype in order to have a reliable comparison. Three different configurations of axial flux generator were proposed in which the first design had 24 slots of stator followed by the second design that had 30 slots of stator and final design had 36 slots for stators. Each stator also had coils with the same number of turns. The experiment was performed using same RPM speeds, same dimensions and the analytical results of the Total Harmonics Distortion (THD) of the flux density and the output torque were evaluated. Their results showed that by increasing number of slots results in improving flux distribution in the machine as well as reducing the THD.

Commonly, there are two types of stators in Axial Flux Generator; coreless and cored. Coreless Axial Flux Generator mainly consists of vacuum or air such that the flux dissipation is not controlled while the cored stator is able to control the flux switching. Zhang et al. conducted an experiment on different shape of cored stator effect to the back electromotive force and the cogging torque [13]. Three different shapes of cored stator are shown in Table 1. The results shown in Table 2 suggest that the Back-EMF (voltage) is the highest when using C-core shape in order to obtain greater output torque and using E-core shape leads to better fault tolerance [13]. U-shape stator requires twice the number of permanent magnets than the E-core shape or the C-core shape. Therefore, the experiments conducted were focused more on C-core and E-core shaped stators. However, in terms of cost effectiveness, coreless-stator is preferred compared to cored-stator in order to reduce volumetric usage of metal in the machine.

Hosseini et al. have conducted an experiment on a small and low-cost axial flux coreless permanent magnet generator [14]. They used a coreless stator which was stated to be highly efficient as it could eliminate the direct flow of magnetic field in between the stator and rotors. The proposed design had two outer rotors; one stator in between the rotors and non-ferromagnetic holders to counteract the forces on magnets during the process. Materials used in the design were 12 rectangular shaped Neodymium Iron Boron (NdFeB) magnets and 6 trapezoidal shaped coils as shown in Fig. 2. The results

Table 1. a) U-core shape stator b) C-core shape stator c) E-core shape stator and their analytical results are also shown next to each shape.

U-core	C-core		E-core	
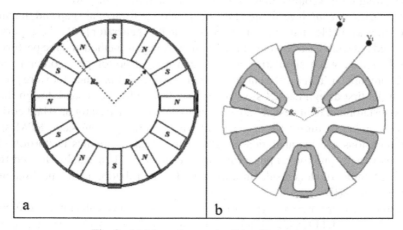		Back EMF(V) = 1.65 Cogging Torque (Nm)= 0.96 THD = 19.4		Back-EMF(V)=1.13 Cogging Torque(Nm)=1.22 THD= 10.4%

Fig. 2. (a) Magnets on rotor (b) coils on stator

they obtained show that the increase of RPM greatly increases the output voltage thus proportionally increasing the efficiency. However, when it comes to comparing the simulation and practical results, the efficiency was found to be reduced by 6.2% in practice. For this configuration, the simulation results produced Phase voltage of 56.68 with an output power of 460 W at 3000 rpm. The Hardware simulation provided 40 V per phase with 390 W at an rpm of 3000 and efficiency of 78.1% compared to 84.3% for the simulation. The authors concluded that the shape of the magnets led to poor test results.

Different shapes of perm anent magnets potentially affects the harmonic components that is also related to the back-EMF which can result in losses. Shokri *et al.* have conducted a research on how the different magnet shapes can affect the performance of axial flux generator using 3-D finite element analysis method [15]. Three different shapes of permanent magnets were used which were 'sector or wedge' like magnets, sinusoidal shaped magnets and cylindrical shaped magnet as shown in Table 2. The results analyzed in the experiments were based on the induced phase voltage that each shape of magnets could generate and the cogging torque occurrence, as shown in Fig. 3.

Based on the results, sinusoidal shaped magnets had the lowest cogging torque and the optimized induced phase voltage at an angle of 180°. Thus, the best performance can be obtained by using sinusoidal-shaped permanent magnet arrangement.

Table 2. Different shaped permanent magnet configurations

'Sector' shaped magnet	Sinusoidal shaped magnets	Cylindrical shaped magnets

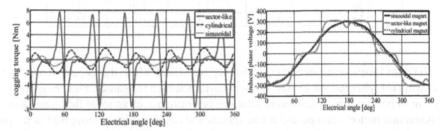

Fig. 3. Comparison between induced phase voltage of different shaped of magnets [14] on the left, and the comparison between cogging torque of different shaped of magnets [14] on the right

Cogging torque is an unwanted noise that may occur in in any machine that involves mechanical parts and permanent magnets. The cogging torque is caused by vibration of objects that usually occur when the machine is working at low speeds. Wanjiku, *et al.* have conducted an analytical quasi-3D analysis on the influence of tooth profiles and slot openings to the cogging torque of an axial flux generator [16]. In order to mitigate the cogging torque effectively, the stator and rotor need to be modified. The authors proposed analytical quasi-3D algorithm in the study of these cases. The author used two profiles for the experiments which were trapezoidal teeth and parallel teeth as shown in Table 3. Slot opening is the opening in between each magnet. The authors introduced 4 different slot openings for the experiment. The results from the experiment showed that the smaller the slot opening the lower the cogging torque and the parallel tooth profile offered lower cogging torque compared to the trapezoidal tooth profile. The results are also shown in Table 3. Woo, *et al.* [17] also conducted an analysis on cogging-torque-effect by experimenting on three different cases involving magnet pole-arc, skewing angle, back-EMF and cogging torque. Kurt *et al.* and Gor *et al.* studied the loses and efficiency of electromagnetic design of permanent magnet generator [18, 19]. Their proposed design consisted of two rotors at both sides of the stator containing 32 rare magnet discs and a stator in between with 24 coils. The machine with its 3-phase design was capable of generating 340 W at 1000 rpm with an air gap of 5 mm. The authors found that the core losses increased as the speed of rotor increased. Besides that, large amount of losses was accounted due to copper and mechanical losses.

Table 3. Cogging torque analysis on different slot opening and two different teeth profiles. Profiles are shown in the table.

Slot opening	Trapezoidal tooth		Parallel tooth	
(mm)	(Nm)	(%)	(Nm)	(%)
2.00	0.71	9.9	0.54	7.5
4.25	1.33	18.5	0.26	3.6
6.50	1.67	23.2	1.01	14
8.75	1.10	15.3	1.19	16.5

The core losses were increasing at higher resistance while the copper losses were decreasing at higher resistance due to lower current at high resistance. Hence, speed also plays an important role in delivering efficient performance of an axial flux generator.

Axial flux motors can operate at low speeds and capable of producing higher torque. Ani *et al.* have evaluated the performance of a simple core structure of a small axial flux generator for small wind turbines [20]. They explained in detail the design criteria and manufacturing process in order to develop a low-cost axial flux generator. Coreless axial flux generator was chosen because of the coreless criteria that can eliminate the cogging torque activity in the magnetic field leading to a simple design for a low-speed generator.

3 Preliminary Analysis

We can carry out a simple analysis to determine an approximate value of the EMF that could be possible with our design using fundamental theory. Using Fleming's right-hand rule and Len's law, we can clearly see that for a 'wedge' shaped coil as seen on Fig. 4 (a), when the magnets rotate, only the brown 'shoulders' of the coil (shown in Fig. 4 (b)) produce any current as the magnetic flux is only sweeping the coils at zero degrees there by not producing any current.

Magnetic Field Due to Permanent Magnets
According to Fig. 4(a), in order to analyse the emf generated by dual axial flux generator, one wedge shaped coil can be enlarged as seen on the right. Here the magnetic field strength is B, the angular velocity of clockwise rotating rotors is ω, and the area of one shoulder of the coil is A (these shoulders are the only part of the coil that produces

emf) and the number of turns in each shoulder is N, then the peak electromotive force generated without any losses for each coil is

$$E = A \times N \times 2 \times \varpi \times B.$$

The total amount of power generated from the complete generator with 12 coils at peak disregarding losses is

$$E_{Total} = 24ANB\varpi.$$

For 100RPM with 100 turns of coil, which is the target of the simulation, the angular velocity will $E_{Total} = 24ANB\varpi = 24 \times A \times 100 \times B \times 2\pi \times 100$ be.

If the length of the coil shoulder is 7 cm and the width of the section is 1 cm then the above E_{total} can be expressed in terms of magnetic flux density as $E_{Total} = 1055.5B$. The magnetic field strength B depends on the air gap between the coil and the magnets from both sides and the type of material used (Neodymium-Iron-Boron N52 in our experiment). B is a value that is difficult to calculate but can easily be modelled using the properties and the dimension of the magnets as done in the simulation.

For the dimension of the magnet shown in Fig. 5(a), the magnetic field is given by:

$$B(x) = \frac{B_r}{\pi} \left[arctan\frac{bc}{2x\sqrt{4x^2 + b^2 + c^2}} - arctan\frac{bc}{2(a+x)\sqrt{4(a+x)^2 + b^2 + c^2}} \right].$$

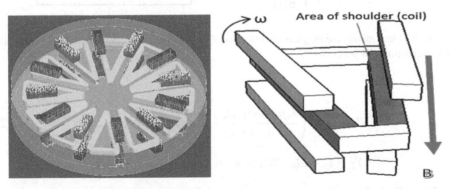

Fig. 4. (a) (left), An image captured from our DRAFG simulation (b) (right) one 'wedge' shaped coil enlarged along with the four adjoining magnets from both rotors to explain the EMF

For N52, the remanence value B_r is about 1.3 T. As shown in the diagram, the B is essentially a function of distance x from the centre of the magnet's surface as shown in Fig. 5(a). However, when two rotors, each containing 16 magnets on each plate facing each other, the above calculation does not provide any useful means to calculate the varying magnetic field and hence the total flux. In order to calculate the overall effect, the model has to be extended to what is shown in Fig. 5(b). This diagram clearly shows a section of the two-rotor magnetic plate along with the stator containing two coils. It also shows the magnetic path for four magnets and how the path is closing the loop through the air gap.

Price *et al.* [21] describes how an analytic expression for airgap magnetic flux density is determined using the coordinate system of Fig. 6(left) [21]. The circumferential and axial directions are represented by x and y coordinates respectively. Using the analytical work of both Price et al. and Brumby *et al.*, the space harmonic flux densities at position y due to magnets on the rotors 1 and 2 are found to be [21, 22],

$$B_{yn1}(x) = \left(\frac{\hat{J}_n \mu_0 \sinh(u_n l_m)}{\mu_n \sinh(u_n Y_2)} \cosh(u_n(Y_2 - y)) \right) \cos(u_n x) \tag{1}$$

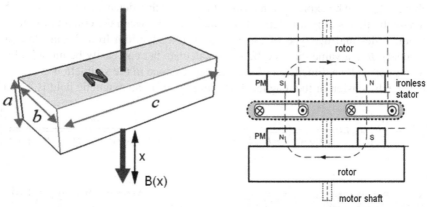

Fig. 5. (a) Determining the magnetic flux density of a typical rectangular magnet, (b Construction of a dual rotor axial generator

$$B_{yn2}(x) = \left(\frac{\hat{J}_n \mu_0 \sinh(u_n l_m)}{\mu_n \sinh(u_n Y_2)} \cosh(u_n(y)) \right) \cos(u_n x) \tag{2}$$

$$Y_2 = l_g + 2l_m \quad \text{Where } u_n = 2\pi n/\lambda; \ \lambda = 2\pi R_m/p.$$

Here, R_m is the mean core radius of the radial slice, p represents the number of pole pairs and $\hat{J}_n$ is the equivalent current sheet due to permanent magnets 1 or 2. The total flux density for each slice of the machine is determined by the superposition of Eq. (1) and Eq. (2). The magnets of each radial slice are modelled as an equivalent current sheet:

$$\hat{J}_n = \frac{4B_r}{\tau_p \mu_0 \mu_{rec}} \sin\left(\frac{n\pi \tau_m}{2\tau_p} \right) \tag{3}$$

where τ_m and τ_p are the magnet and pole pitches, respectively, for the radial slice under analysis. The terms B_r and μ_{rec} are the remanent flux density and permeability of the permanent magnets.

The total flux densities due to magnets on rotors 1 and 2 for a single radial slice are the sum of space harmonics:

$$B_{y1}(x) = \sum_{n=1} B_{ym1}(x) \tag{4}$$

$$B_{y2}(x) = \sum_{n=1} B_{ym2}(x) \tag{5}$$

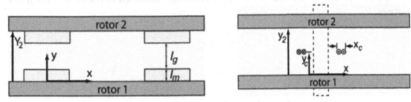

Fig. 6. (left) Coordinate system of the magnetic flux density (This is a cross sectional view of machine looking inward radially). (right) Cross sectional view showing the coordinate system for armature (looking inward radially)

Magnetic Field Due to Armature Reaction
This section discusses the magnetic field due to armature reaction which complicates the magnetic field generated by the placement of permanent magnets on both rotors. Figure 6(right) shows the coordinate system used in the expressions for the armature reaction. According to the expressions of [21, 22] current sheet K and the current sheet with peak value of $\hat{K}$, located at the mean axial position of $y_c = \frac{Y_2}{2}$ [22]. The flux density at the mean axial position using Fourier analysis is

$$B_{n-arm}(x) = \hat{K}_n \mu_0 \left(\frac{\cosh(\mu_n Y_c)}{\sinh(\mu_n Y_2)} \cosh(\mu_n Y_2 - y)) \right) \cos(\mu_n x), \tag{6}$$

where the linear current density function is

$$K(x) = \sum \hat{K}_n \sin(\mu_n x). \tag{7}$$

The peak current density is

$$\hat{K}_n = \frac{Ni}{x_c} \frac{4}{n\pi} \sin\left(\frac{n x_c p}{4 R_m} \right) \tag{8}$$

4   Simulation

Based on the well-understood theory as outlined in the *Related work*, a design of dual rotor axial flux generator has been decided. Axial flux is chosen over radial flux because the axial flux design can be easily manufactured due to its volume and round shape of rotors and stator. The proposed design of DRAFG will have one stator in between the two rotors. Each rotor will have sixteen rectangular Neodymium permanent magnets and the stator will have twelve coils. We have the desire to experiment with lower number of coils and lower number of magnets if those configurations can still generate the desired output power of 100 W at 100 RPM. The system is an AC generator with 3 phases and four pole per phase. The stator is coreless type that will have air between each coil to reduce cogging torque effect. The speed of rotor rotation for the experiment is from 100 rpm up to 400 rpm. The target output power is approximately 100 watts that any household can use for simple tasks with an estimate induced current of 4 Amperes and 25 V output.

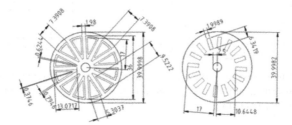

Fig. 7. 2D draft of magnet and coil placement on a 40 cm rotor and stator.

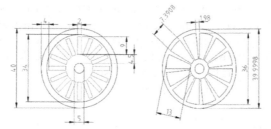

Fig. 8. 2D draft of more accurate coil and magnet positioning based on real coils and magnets in hand.

The design phase is performed in three steps. The first step is to perform dimension calculation and 2D drafting of stator and rotor for DRAFG using Qcad 3.15. It is important to attempt 2D draft before designing the 3D model of DRAFG in order to avoid miscalculating the number of permanent magnets used, find the correct angle for each position of coils and magnets, determine the size of air gap between coils and to find the best size for each coils and magnets that will fit in the proposed design. The first and second drawing are as shown in Fig. 7 and Fig. 8 with the dimensions for each drawing. This 2D draft is then used to generate 3D drawings for stator and rotor. The final step is to import each part into the simulation software and use the toolbox to perform mesh analysis and obtain the results.

Cedrat Flux 12.1 is a software that is capable of computing finite element electromagnetics for both 2D and 3D model. It is a powerful software that is also able to optimize any electromagnetic devices using physics analysis. Firstly, the application chosen is Transient Magneto Analysis to perform movement of object at y-axis. The 3D model created in Solidworks are imported into Cedrat Flux 12.1 and each of its volume is defined accordingly to its function. The volume assignment is also created in Cedrat Flux 12.1.

The pre-processing is a simulation phase to test the magnetic field and operate the DRAFG similar to hardware testing. The rotors containing the permanent magnets will rotate in y-axis and the stator containing the coils will have fixed position. Each movement of magnets will be computed to obtain any readings such as voltage induced, current induced and power output produced at each coil. The magnetic field and the flux flow also can be analyzed for each rotation of the rotor.

Simulation Results

Several dimensions are set as default value and the output will be based on the rotation speed and the external resistor value. In order to simulate the DRAFG correctly, an electrical circuit needs to be constructed and linked to each coil. Here, each coupling has a fixed copper resistance value of 0.67 Ω which is due to the 1mm diameter and the 100 turns in each copper coil. We will experiment with different external resistor values to produce different output power, induced current and voltage. There are four cases of study that has been conducted in the experiments. The output power, induced current and the voltage generated will be tabulated, analyzed and discussed thoroughly. The focus of the experiment is to find the best design that is capable of producing 100 W of output power at low rpm.

Table 4. Power generated using a rotor and stator with 35 cm diameter at 5 mm air gap configuration.

	100 RPM	200 RPM	300 RPM	400 RPM		100 RPM	200 RPM	300 RPM	400 RPM		100 RPM	200 RPM	300 RPM	400 RPM
	10Ω					1Ω					0.1Ω			
Voltage	5.4V	10.8V	15.9V	21V	Voltage	1.89V	3.66V	5.19V	6.66V	Voltage	0.67V	1.5V	2.4V	3.09V
Current	2.1A	4.11A	6A	8.19A	Current	5.79A	11.52A	17.16A	22.77A	Current	7.05A	14.1A	21A	27A
Power	11.34 W	44.4 W	95.4W	172W	Power	10.94W	42.2W	89.1W	151.65W	Power	4.73W	21.15W	50.4W	83.43W

We experimented with three different external resistor values; 10 Ω, 1 Ω and 0.1 Ω. The purpose was to find which resistance value will result in reasonable amount of voltage, low current and high output power. As can be seen from Table 4, we find that the 10 Ω is suitable for further experimentation in our simulation. Here, the output power was limited due to the air gap of 10 mm which substantially weakened the magnetic flux sweeping the stator. Usually having a larger gap between rotor and the stator will allow much more tolerance in designing a practical generator. We then moved onto reduce this gap. One of our goals in this experiment is to simulate a versatile design that is compact and uses minimum amount of resources such as copper and magnets so that a generator could be developed at the lowest cost. Keeping this in mind, we ran few simulations where the diameter of the rotors and stator were reduced to 35 cm. This essentially reduces the size of the coils as well as the size of the magnets that can be mounted on the rotors. However, the power output was negligible as shown in Table 4. This indicates that 5 cm drop in the diameter somehow resulted in dimensions that lead to cancelling of voltages leading to a very poor power output. We discontinued the 35 cm diameter and moved back to 40 cm diameter to systematically analyze the effect of reducing the airgap on the output power. Next, the simulations were attempted using multiple air gaps starting from 5 mm and lower.

The design is performing as expected and the reduction in the airgap has increased the power output. However, keeping in mind that our goal is to produce 100 W at 100 RPM, we did not rush into lower the airgap drastically as a tiny airgap demands precision workmanship. We would be content with a design that produces our power output goal with the biggest airgap.

Looking at power outputs at 3 mm and 2 mm, we realized that 100 W is attainable at 100 RPM. We wanted to see whether we could still attain 100 W if we reduce the number of magnets so that production cost for such a generator could be lowered as well. Table 5 indicates that this indeed was not the case as the total power was reduced by almost 95%. It is possible that magnet placement has resulted in cancelling of voltages generated in coils.

Table 5. Power generated using a rotor and stator with 35 cm diameter at 5 mm air gap configuration.

Speed	100 rpm 1-phase 3-phase		200 rpm 1-phase 3-phase	
Voltage	0.03 V	0.09 V	0.11 V	0.33 V
Current	0.011 A	0.033 A	0.027 A	0.081 A
Power	0.0003 W	0.0029 W	0.003 W	0.026W

The next set of experiments were conducted to decide how much power could be generated at different RPMs when the air gap was lowered. The results are shown in Table 6. In this, it is very clear that reduction in the air gap increases the generated output as expected for 40 cm diameter rotor and stator combinations. It is also evident that by increasing the RPMs of the rotors, the power can be substantially increased. According to this Table 6, highest amount of power that can be generated at 100RPM was when the air gap was only 2 mm. Next, we would observe whether the power output can be further increased at our final air gap at 1 mm. The first simulation was run with an air gap of 1 mm when 12 magnets were used. As outlined in the previous section, reducing the number of magnets will substantially lower the cost of a practical design. With this configuration, the generated power was only 5 W. This is again due to currents cancelling out when the magnets are configured in a 12-magnet configuration.

We ran our final simulation with 16 magnets and an airgap of 1 mm. The results in Table 7 show that the system is capable of generating almost 150 W at 100 RPM as desired, which would result in a practical system even with substantial loses. It is also important to look at the waveforms of 3-phase current and voltages to ensure the generator is capable of supporting a reasonable load.

Table 6. The impact of generating power at 100 rpm and 200 rpm when the air gap is reduced from 5 mm To 2 mm (Shown on the right-most column)

Speed	100 RPM 1-Phase	3-Phase	200 RPM 1-Phase	3-Phase	
Voltage	3.97V	15.88V	8.96V	35.84V	5mm
Current	1.49A	4.47A	2.98A	8.94A	
Power	5.92W	70.98W	26.07W	320.4W	
Voltage	5.01V	20.04V	9.99V	40V	4mm
Current	1.52A	4.56A	3.24A	9.72A	
Power	7.61W	91.38W	32.36W	388.8W	

Speed	100 RPM 1-Phase	3-Phase	200 RPM 1-Phase	3-Phase	
Voltage	5.35V	21.4V	11.02V	44.08V	3mm
Current	1.65A	4.95A	3.51A	10.53A	
Power	8.82W	105.93W	38.68W	464.2W	
Voltage	6V	24V	12.27V	49.08V	2mm
Current	1.8A	5.4A	3.67A	11.01A	
Power	10.4W	129.6W	45W	540W	

Table 7. Power generation at 1 mm Gap with 16 Magnets

Speed	100 rpm 1-phase	3-phase	200 rpm 1-phase	3-phase
Voltage	5.81 V	23.24 V	13.47 V	53.88 V
Current	2.12 A	6.36 A	4.21 A	12.63 A
Power	12.31 W	147.8 W	56.71 W	680.5 W

5 Conclusion

Developing an optimized DRAFG has a great impact on small scale power generation when lower RPM and low noise requirements are paramount. Our design which had a goal of generating 100 W at 100 RPM was realized with a diameter of 40 with 16 magnets and 12 coils. We have also realized that changing the diameter of the stator even by 5 cm using the same magnets and coils can change the configuration drastically reducing the power by twenty-fold due to currents cancelling each other. This points out that any DRAFG design should best be simulated first using realistic component dimensions and properties in order to avoid any inefficiencies associated with coil and magnet gap variations. We also observed that the airgap is very critical in determining the final power output and hence kept it to a minimum. This allows the magnetic flux to flow through the tiny air gap with the least reluctance generating the most amount of power.

References

1. Hwang, D.H., Lee, K.C., Kang, D.H., Kim, Y.J., Choi, K.H., Park, D.Y.: A modular-type axial-flux permanent magnet synchronous generator for gearless wind power systems. In: 30th Annual Conference of IEEE Industrial Electro. Society (IECON 2004), vol. 2, pp. 1396–1399 (2004)
2. Mahmoudi, A., Kahourzade, S., Rahim, N.A., Hew, W.P.: Design, analysis, and prototyping of an axial-flux permanent magnet motor based on genetic algorithm and finite-element analysis. IEEE Trans. Magn. **49**(4), 1479–1492 (2013)

3. Ferreira, A.P., Costa, A.F.: Direct driven axial flux permanent magnet generator for small-scale wind power applications, ICREPQ 2011, pp. 905–910 (2010)
4. Wang, R.J., Maarten, K.K., Westhuizen, K.V., Fieras, J.F.: Optimal designing of a coreless stator axial flux permanent-magnet generator. IEEE Trans. Magn. **41**(1), 55–64 (2005)
5. Lehr, M., Woog, D., Binder, A.: Design, construction and measurements of a permanent magnet axial flux machine. In: 2016 XXII International Conference on Electrical Machines (ICEM), pp. 1604–1610 (2016)
6. Ishikawa, T., Amada, S., Segawa, K., Kurita, N.: Proposal of a radial- and axial-flux permanent-magnet synchronous generator. IEEE Tran. Magn. **53**(6), 1–4 (2017)
7. Borokowski, D., Wegiel, T., Radwan-Praglowska, N.: 'Model of coreless axial flux permanent magnet generator. In: International Symposium on Electrical Machines, (SME) (2017) https://doi.org/10.1109/ISEM.2017.7993568
8. Latoufis, K.C., Messinis, G.M., Kotsampopoulos, P.C., Hatziargyriou, N.D.: Axial flux permanent magnet generator design for low cost manufacturing of small wind turbines. Wind Eng. **36**(4), 411–431 (2012)
9. Dirba, J., Levin, N., Orlova, S., Pugachov, V., Ribickis, L.: Optimization of the magnetic circuit of an axial inductor machine based on the calculation and analysis of magnetic field. In: 13th European Conference on Power Electronics and Applications, Latvia, pp.1–8 (2009)
10. Murphy, J.: Understanding AC induction, permanent magnet and servo motor technologies. http://www.leeson.com/documents/PMAC_Whitepaper.pdf Accessed 28 Sep 2016
11. Pop, A.A., Radulescu, M., Balan, H., Kanchev, H.: 'Electromagnetic torque capabilities of axial-flux and radial-flux permanent-magnet machines. In: 4th International Symposium on Electrical and Electronics Engineering, pp. 1–4 (2013)
12. Nasiri-Zarandi, R., Mirsalim, M., Ashrafi, R.: Effect of air-gap variation and the number of stator slots on performance of an axial flux hysteresis motor. In: 6th International Power Electronics Drive Systems and Technologies Conference, Tehran, Iran, pp. 609–614 (2015)
13. Zhang, W., Liang, X., Lin, M.: Analysis and comparison of axial field flux-switching permanent magnet machines with three different stator cores. IEEE Trans. Appl. Supercond. **26**(7), 1–6 (2016)
14. Hosseini, S.M., Agha-Mirsalim, M., Mirzaei, M.: Design, prototyping, and analysis of a low cost axial-flux coreless permanent-magnet generator. IEEE Trans. Magn. **44**(1), 75–80 (2008)
15. Shokri, M., Rostami, N., Behjat, V., Pyrhonen, J., Rostami, M.: Comparison of performance characteristics of axial-flux permanent-magnet synchronous machine with different magnet shapes. IEEE Trans. Magn. **51**(12), 1–6 (2015)
16. Wanjiku, J., Khan, M.A., Barendse, P.S., Pillay, P.: Influence of slot openings and tooth profile on cogging torque in axial-flux PM machines. IEEE Trans. Industr. Electron. **62**(12), 75–78 (2015)
17. Woo, D.-K., Kim, I.-W., Lim, D.-K., Ro, J.-S., Jung, H.-K.: Cogging torque optimization of axial flux permanent magnet motor. IEEE Trans. Magn. **49**(5), 2189–2192 (2013)
18. Kurt, E., Gör, H.: Electromagnetic design of a new axial flux generator. In: 6th Edition of International Conference on Electronics, Computers and Artificial Intelligence, Bucharest, pp. 39–42 (2014)
19. Gör, H., Kurt, E., Bal, G.: Analyses of losses and efficiency for a new three phase axial flux permanent magnet generator. In: 4th International Conference on Electric Power and Energy Conversion Systems, Turkey, pp. 1–6 (2015)
20. Ani, S.O., Polinder, H., Ferreira, J.A.: Low cost axial flux PM generator for small wind turbines. In: IEEE Energy Conversion Congress and Exposition, Delft, pp. 2350–2357 (2012)

21. Price, G.F., Batzel, T.D., Comanescu, M., Muller, B.A.: Design and testing of a permanent magnet axial flux wind power generator. In: Proceeding of the 2008 IAJC-IJME International Conference (2008)
22. Bumby, J.R., Martin, R., Mueller, M.A., Spooner, E., Brown, N.L., Chalmers, B.J.: Electro-magnetic design of axial-flux permanent magnet machines. IEE Proc. Electr. Power Appl. **151**(2), 151 (2004)

A Lightweight Attention Fusion Module for Multi-sensor 3-D Object Detection

Li-Hua Wen[1], Ting-Yue Xu[2], and Kang-Hyun Jo[1]([✉])

[1] Department of Electrical, Electronic and Computer Engineering,
University of Ulsan, Ulsan 44616, South Korea
acejo@ulsan.ac.kr
[2] School of Mechatronic Engineering and Automation,
Shanghai University, Shanghai 200444, China

Abstract. With the rapid development of autonomous vehicles, three-dimensional (3D) object detection has become more important, whose purpose is to perceive the size and accurate location of objects in the real world. Many kinds of LiDAR-camera-based 3D object detectors have been developed with two heavy neural networks to extract view-specific features, while a LiDAR-camera-based 3D detector runs very slow about 10 frames per second (FPS). To tackle this issue, this paper first presents an accuracy and efficiency multiple-sensor framework with an early-fusion method to exploit both LiDAR and camera data for fast 3D object detection. Moreover, we also present a lightweight attention fusion module to further improve the performance of our proposed framework. Massive experiments evaluated on the KITTI benchmark suite show that the proposed approach outperforms state-of-the-art LiDAR-camera-based methods on the three classes in 3D performance. Additionally, the proposed model runs at 23 frames per second (FPS), which is almost $2\times$ faster than state-of-the-art fusion methods for LiDAR and camera.

Keywords: Three-dimensional object detection · Multiple sensor · Early-fusion method · LiDAR and camera

1 Introduction

Recently, feature extraction [1–4] with deep learning has drawn much attention. For the RGB image, a general 2D convolutional neural network (CNN) can be used to extract its features. For the point cloud however, it is difficult to extract its features due to its irregular distribution and sparse contributions. Before the advent of highly-efficient graphics processing units (GPUs), representative studies [5–10] have converted point clouds into 2D dense images or structured voxel-grid representations and utilized 2D neural networks to extract the corresponding feature from the converted 2D image. With the development of computer technology, the authors in [11–14] directly utilized a multi-layer perceptron (MLP) to aggregate features from point clouds. Shi and Rajkumar [15] encoded the point cloud natively in a graph using the points as the graph vertices.

© Springer Nature Switzerland AG 2021
D.-S. Huang et al. (Eds.): ICIC 2021, LNCS 12836, pp. 802–815, 2021.
https://doi.org/10.1007/978-3-030-84522-3_65

To leverage the mutual advantages of point clouds and the RGB image, some researchers have attempted to fuse view-specific region of interest (ROI) features. Currently, there are two mainstream fusion methods. The first is to fuse two view-specific features and the other method is pointwise feature fusion. Chen et al. [5] and Ku et al. [6] directly fuse the ROI feature maps output with the two backbones of the point cloud and RGB image, respectively. On the other hand, Xu et al. [16] and Sindagi et al. [17] fuse pointwise features. These methods achieve better performance compared with LiDAR-based methods; however, their inference time is usually intolerable for application in real-time autonomous driving systems.

To deal with the above issues, this paper proposes a novel point-wise fusion strategy between point clouds and RGB images. The proposed method directly extracts pointwise features from the raw RGB image based on the raw point cloud first. Then, it fuses the two pointwise features and feeds them into a 3D neural network. The structure, as shown in Fig. 1, has only one backbone to extract features, making the proposed model much faster than state-of-the-art LiDAR and camera fusion methods.

The key contributions of this work are as follows:

- This paper presents an early-fusion method to exploit both LiDAR and camera data for fast multi-class 3D object detection with only one backbone, achieving a good balance between accuracy and efficiency.
- We prove that raw RGB image features benefit 3-D object detection.
- We also present a lightweight attention fusion module to further improve the performance of our proposed framework.

The presented one-stage 3D multi-class object detection framework outperforms state-of-the-art LiDAR-Camera-based methods on the KITTI benchmark [18] both in terms of the speed and accuracy.

2 Related Work

2.1 LiDAR-Based 3-D Object Detection

Recently, there have been three main 3D object detectors based on LiDAR: voxel-based detectors, point-based detectors, and graph-based detectors. Voxel-based methods [7, 8, 19, 20] first voxelize the raw point cloud over a given range and then utilize a 3D CNN or 2D CNN to extract features. Unlike VoxelNet [7], Yan et al. [19] replaced a 3D CNN by a 3D sparse convolutional network, and Lang et al. [20] directly organized point clouds in vertical columns (pillars) to generate 2D BEV images. Point-based detectors [11–14] directly deal with the raw point cloud. Charles et al. [11] pioneered the method used to deal with each point independently using their shared MLPs. Based on PointNet [11], Qi et al. [12] further introduced the metric space distances to learn local features with increasing contextual scales. Yang et al. [13] abandoned the Upsampling layers in PointNet++ to boost the inference speed. The proposed method voxelizes a point cloud using a dynamic voxelization method compared with the hard voxelization method in [7] and aims to avoid information loss during voxelization.

2.2 Multi-modal 3-D Object Detection

3-D Object detection in point clouds and RGB images is a fusion problem. As such, it is natural to extract the RGB image feature and the point cloud feature with two different backbones, respectively, which is the paradigm present in all previous works [5, 6, 9, 10, 16, 17, 21–24]. Obviously, by employing two heavy backbones, these approaches are very slow and consume a great deal of memory. In the paradigm, these methods are designed to either study how to fuse or how to improve accuracy based on state-of-the-art fusion methods, e.g., AVOD [6] changes the feature generation method in MV3D [5] from hand-crafted techniques to automation to improve the running speed of the model. According to different fusion methods, these methods can be divided into two categories: pointwise fusion [16, 17] and region of interest (ROI)-based fusion [5, 6, 21–25]. Compared with the ROI-based fusion, pointwise fusion is more flexible. Inspired by pointwise fusion, this article will explore whether it is possible to directly aggregate the point features of the raw RGB image with point cloud features.

In this paper, we first present an early-fusion method to exploit both LiDAR and camera data for fast 3D object detection with only one backbone, and it achieves a good balance between accuracy and efficiency. Thanks to the novel pointwise feature fusion module, which makes the fusion between LiDAR and camera data highly efficient.

2.3 Attention Module

Hu et al. report a Squeeze-and-Excitation block [26], which can be inserted into any 2-D networks. It is a simple attention module with a global average pooling (GAP) operation. Follow after the GAP, they adopt two fully connected (FC) layers to reduce feature dimensions. Wang et al. conduct massive experiments and show the dimension reduction after the FCs reduce the performance [27]. However, if the FCs keep the feature dimension same as the input and the computation cost must be increased. Hence, they propose a cross-channel interaction. Qin et al. proof the GAP is a special case of the feature decomposition in the frequency domain [28]. In this paper, we propose a lightweight attention fusion module.

3 Proposed Method

The proposed model, as shown in Fig. 1, takes point clouds and RGB images as inputs and predicts oriented 3D bounding boxes for cyclists, pedestrians, and cars. This model includes four main parts: (1) A point feature fusion module that extracts the point features from the RGB image and fuses the extracted features with the corresponding point cloud features, (2) a voxel feature encoder (VFE) module and a 3D backbone to process the fused pointwise features into a high-level representation, (3) a detection head that regresses and classifies the 3D bounding boxes, and (4) a loss function.

3.1 Point Feature Fusion Module

The fusion module, shown in Fig. 2, consists of three submodules: the point transform module, the voxelization of point clouds, and the pointwise fusion module. Since this

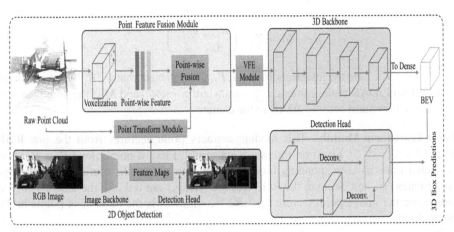

Fig. 1. The architecture of the proposed one-stage 3D object detection network for the LiDAR and camera. It mainly includes the input data, the point feature fusion module, the 3D backbone, and the detection head. The gray box and green box represent the convolutional block and feature map, respectively. (Color figure online)

module involves the input of raw data, before introducing the module, the input data is first introduced.

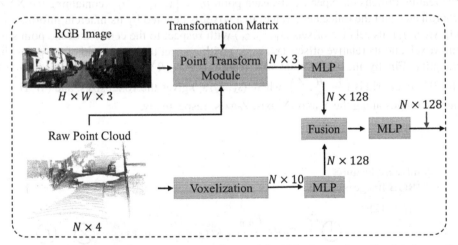

Fig. 2. Visualization of the point feature fusion module. N is the number of points in a point cloud, and MLP denotes one multiple perceptron layer.

Input Data. This model accepts point clouds and RGB images as the input. Points $\{(x, y, z) \mid x \in X, y \in Y, z \in Z\}$ and point clouds (X, Y, Z) are mapped onto the original image $(W \times H)$ plane as follows

$$(\mu, v, 1)^{\mathrm{T}} = \mathbf{M} \cdot (X, Y, Z, 1)^{\mathrm{T}}, \tag{1}$$

$$M = P_{rect} \begin{pmatrix} R_{velo}^{cam} & t_{velo}^{cam} \\ 0 & 1 \end{pmatrix}, \tag{2}$$

Where (μ, ν) are the image coordinates, P_{rect} is a project matrix, R_{velo}^{cam} is the rotation matrix from LiDAR to the camera, t_{velo}^{cam} a translation vector, and M is the homogeneous transformation matrix from LiDAR to the camera.

Point Transform Module. This module extracts point features from the raw RGB image $I \in R^{H \times W \times 3}$ based on the raw point cloud. First, a point cloud $P \in R^{N \times 3}$ is projected onto its corresponding image by Eq. 1 to obtain the corresponding image coordinates (μ_i, ν_i). Second, the RGB and the (μ_i, ν_i) are fed into the image sampler, outputting the image point feature $P_i \in R^{N \times 3}$, where N is the number of points in the point cloud.

Voxelization. Voxelization divides the point cloud into evenly spaced voxel grids and then generates a many-to-one mapping between 3D points and their corresponding voxels. Currently, there exist two voxelization methods: hard voxelization and dynamic voxelization. Compared with the former, dynamic voxelization makes the detection more stable by preserving all the raw points and voxel information. This work applies the dynamic voxelization method. Given a point cloud $P = (p_1, p_2, \cdots, p_N)$, the process assigns N points to a buffer of size $N \times F$, where N is the number of points and F denotes the feature dimension. Specifically, each point $p_i = [x_i, y_i, z_i, r_i]$ (containing the XYZ coordinates and the reflectance value) in a voxel is denoted by its inherent information (x_i, y_i, z_i, r_i), its relative offsets (x_v, y_v, z_v) with respect to the centroid of the points in the voxel, and its relative offsets (x_p, y_p, z_p) with respect to the centroid of the points in the pillar. Finally, the output point-wise feature is $P_v \in R^{N \times 10}$, and the resulting size of the 3D voxel grid is $\left(\frac{W}{s_x}, \frac{H}{s_y}, \frac{Z}{s_z} \right)$, where (s_x, s_y, s_z) gives the voxel sizes, and (W, H, D) are the ranges along the Y-axis, X-axis, Z-axis, respectively.

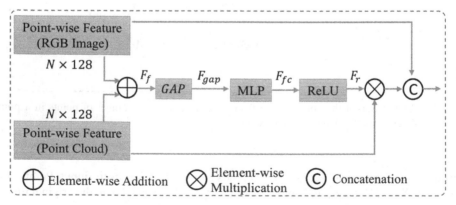

Fig. 3. The proposed lightweight attention fusion module.

Pointwise Fusion. This module fuses the pointwise features P_i and P_y. Since the dimensions of the two features are different, two multiple perceptron layers (MLP) are used, one for each feature, are used to adjust their dimensions to be the same. There are three common fusion methods for ROIs: addition, concatenation operation and the lightweight attention fusion module (LAFM), as shown in Fig. 3. Therefore, this paper will analyze which fusion method is the most suitable for the pointwise features in Table 3 in the ablation section. After the fusion operation, one FC layer is utilized to further merge the fused features and output the result as P_f.

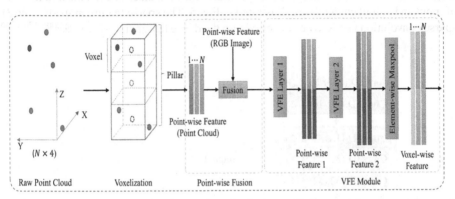

Fig. 4. The details of the voxelization, point-wise fusion, and the voxel feature encoder module (VFE). The VFE module adopts MLP to learn representative features and concatenates the representative feature to each pointwise features to generate the final voxel-wise features.

3.2 Voxel Feature Encoder Module and 3D Backbone

This section introduces the voxel feature encoder module and the 3D backbone, in that order.

Voxel Feature Encoder Module. Upon completing the pointwise fusion, the fused feature P_f is transformed through the VFE layer which is composed of a fully connected network (FCN), into a feature space, where information from the point features $f_i \in \mathbb{R}^m$ can be aggregated to encode the shape of the surface contained within the voxel [7, 8, 17], where $i \in [1, N]$ and m is the feature dimension of a point. The FCN consists of a linear layer followed by a batch normalization layer, and a ReLU layer. An elementwise max-pooling process is used to locally aggregate the transformed features and output a feature $\vec{f}$ for P_f. Finally, the max-pooled feature Ef is concatenated with each point feature fi to generate the final feature P_{vfe}. his work stacks two such VFE layers and both of the output lengths are 128. This means the shape of P_{vfe} is N × 128.

The details of the voxelization, point-wise fusion, and the voxel feature encoder module (VFE), as shown in Fig. 4.

3-D Backbone. The 3-D backbone takes the feature $\mathbf{P}_{vfe}$ and its corresponding index of 3D coordinates (X, Y, Z) as inputs. The backbone is widely used in [30], [31] and has twelve 3D sparse convolutional layers and is divided into four stages according to feature resolution, as shown in Fig. 5. The four-stage feature resolutions in the order of (W, H, D) are (1600, 1408, 41), (800, 704, 21), (400, 352, 11), and (200, 176, 2). Specifically, each stage has two kinds of 3D convolutional layers: the submanifold convolution [19] and the sparse convolution. The former does not generate new points and shares the point coordinate indices in each stage; hence, the submanifold convolution runs very fast. The latter is a sparse version of the dense 3D convolution. Usually, these two convolutions are used in conjunction to achieve the speed/accuracy balance. The details and numbers of input and output channels are illustrated in Fig. 5. The sparse feature map after the 3D sparse convolution needs to be converted into the dense feature map $\mathbf{F}_d \in \mathbb{R}^{200 \times 176 \times 256}$.

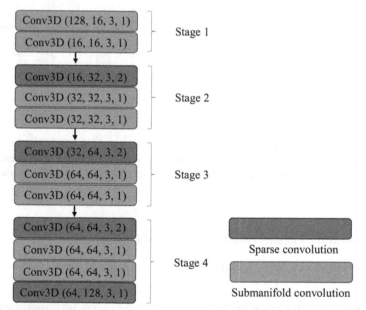

Fig. 5. The 3D backbone architecture. Conv3D (cin, cout, k, s) denotes a convolutional block, where the parameters cin, cout, k, and s represent the input-channel numbers, the output-channel numbers, the kernel size, and the stride, respectively. Each block consists of a 3D convolutional layer followed by a batch normalization layer and a ReLU layer.

Detection Head. The input data of the detection head is the dense feature map $\mathbf{F}_d$. The detection head is comprised of three convolution blocks. Block 1 has five 2D convolutional layers and outputs the feature map $\mathbf{F}_1 \in \mathbb{R}^{100 \times 88 \times 256}$. Similarly, block 2 also has five 2D convolutional layers and takes the feature map F1 as input and outputs the feature map $\mathbf{F}_2 \in \mathbb{R}^{50 \times 44 \times 256}$. Block 3 has two transpose layers and one 2D convolutional layer. $\mathbf{F}_1$ and $\mathbf{F}_2$ are transposed as the feature map $\mathbf{F}_3 \in \mathbb{R}^{100 \times 88 \times 256}$ and the feature map $\mathbf{F}_4 \in \mathbb{R}^{100 \times 88 \times 256}$, respectively. Finally, the feature maps $\mathbf{F}_3$ and $\mathbf{F}_4$ are concatenated as the feature map $F \in \mathbb{R}^{100 \times 88 \times 512}$. The feature map F is mapped to three desired learning targets: (1) a classification score map $F_{score} \in \mathbb{R}^{100 \times 88 \times 18}$, (2) a box regression map $F_{box} \in \mathbb{R}^{100 \times 88 \times 42}$, and (3) a direction regression map $F_{dir} \in \mathbb{R}^{100 \times 88 \times 12}$.

Loss Function. This work utilizes the same loss functions in PointPillars [20] and SECOND [19].

4 Experiments

4.1 Dataset

The proposed model is trained and evaluated on the KITTI dataset [18]. The KITTI object dataset possesses 7,518 testing frames and 7,481 training frames. Each frame is comprised of a point cloud, stereo RGB images (the left image and the right image), and calibration data. In this research, only a point cloud and the left image with their calibration data are used. To impartially compare the proposed approach with existing methods, the training dataset is divided into two subsets (training subset and validation subset) based on the same criteria, and the ratio of the two subsets is 1:1.

For KITTI's criteria, according to the size, truncation, and occlusion classes of objects, all objects are grouped into three difficulty classes: easy (E), moderate (M), and hard (H). Before October 8th, 2019, KITTI's object detection metric was defined as the 11-point average precision (AP) metric. Since then, the metric has been defined by 40 recall positions. Compared with the 11-point AP, the 40-point AP more properly assesses the quality of an algorithm based on the infinite approximation. Intersection-over-Union (IoU) is the generic evaluation criterion for object detection. In the evaluation of 2D, 3D, and bird's eye view (BEV) detection, the IoU is at the threshold of 0.7 for the car class and 0.5 for the pedestrian/cyclist class.

4.2 Experimental Setting

The proposed model is an end-to-end 3D detector for three classes: the car, pedestrian, and cyclist. When designing the anchors for the three classes, different classes employ different sizes (w, l, h). The sizes (1.6, 3.9, 1.56), (0.6, 0.8, 1.73), and (0.6, 1.76, 1.73) are for the car, the pedestrian, and the cyclist, respectively. Note that each anchor has two directions $\{0°, 90°\}$, which means that each location has six anchors. The detection area in the point cloud is $\{(x, y, z) | x \in [0, 70.4], y \in [-40, 40], z \in [-3, 1]\}$.

The framework is based on Pytorch and programmed by the python language. This model is trained from scratch based on Adam optimizer. The whole network is trained with a batch of size 10 and the initial learning rate is 0.003 for 80 epochs on one TITAN RTX GPU. This work also adopts the cosine annealing learning rate for the learning rate decay. The entire training time is around 12 h.

For data augmentation, this work employs the widely used augmentations found in [7, 19, 20], including global scaling [0.95, 1.05], global rotation around the Z-axis $[-45°, 45°]$, and the random flipping along the X-axis.

4.3 Results

To fairly compare the results with the existing methods, the proposed model is evaluated using the more challenging dataset: the KITTI testing dataset. In Table 1, this part only compares the proposed method with state-of-the-art methods in two aspects: BEV and 3D. Since it requires a great deal of data to compare these three performances, here, the results are simply compared based on the mean average precision (mAP). For the 3D performance, the proposed model has the best performance. For the BEV and 2D performances, the proposed method is the second-best, but the overall performance of the proposed method outperforms state-of-the-art methods when taking accuracy and speed into account.

Table 1. Performance comparison using the KITTI testing dataset. The results of cars are evaluated by the mean Average Precision with 40 recall positions. The top performance is highlighted in bold only for the mAP columns and FPS column, and the second-best is shown in blue.

Method	FPS	AP_{BEV}(IOU = 0.7)				AP_{3D}(IOU = 0.7)			
		E	M	H	mAP	E	M	H	mAP
MV3D [5]	3	86.0	76.9	68.5	77.1	71.1	62.4	55.1	62.9
F-PointNet [21]	6	88.7	84.0	75.3	82.7	81.2	70.4	62.2	71.3
AVOD [6]	13	86.8	85.4	77.7	83.3	73.6	65.8	58.4	65.9
AVOD-FPN [6]	10	88.5	83.8	77.9	83.4	81.9	71.9	66.4	73.4
ContFusion [22]	17	**94.1**	85.4	75.9	85.1	83.7	68.8	61.7	71.4
MVX-Net [17]	7	89.2	85.9	78.1	84.4	83.2	72.7	65.2	73.7
PFF3D [29]	18	89.6	85.1	80.4	85.0	81.1	72.9	67.2	73.8
Ours	**23**	90.0	**86.1**	**80.9**	**85.7**	83.9	**73.8**	**68.1**	**75.3**

4.4 Ablation Study

This section analyzes the proposed methods individually by conducting ablation experiments using the KITTI validation dataset.

Effect of the Lightweight Attention Fusion Module. This section analyzes the point feature fusion module based on the three classes in detail. In Table 2, the 'Addition' and 'Concatenation' represent the respective addition and concatenation fusion methods. The parameter 'MLP' means the multiple perceptron layer followed after the fusion operation, as shown in Fig. 2. The experimental results show that the combination of the addition operation and MLP of the proposed module is best for the three classes: the car, pedestrian, and cyclist. The data in the first row give the results of the proposed method when only taking a point cloud as input. Compared with the LiDAR-based method (the first row), the proposed method (the fourth row) achieves 0.6% gains in the 3D performance. Compared with the performance improvement of cars, the proposed model is more helpful for improving the identification of pedestrians and cyclists. Compared to other fusion operations, the proposed lightweight attention fusion module (in the fifth row) outperforms other methods in the three classes with a large margin.

Table 2. Effect of the proposed LAFM.

Method				3D Performance (IOU = 0.7)		
Addition	Concat.	MLP	LAFM	Cars	Ped.	Cyclists
				77.0	56.2	55.7
√				76.6	57.7	59.7
	√			77.0	57.4	60.5
√		√		77.6	60.2	60.0
	√	√		77.1	59.7	59.4
			√	**77.5**	**60.5**	**60.9**

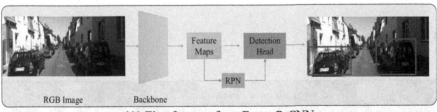

(A) Fine features from Faster R-CNN

(B) Coarse features from a backbone

(C)Raw image feature

Fig. 6. Three kinds of image features used to fuse the corresponding point cloud features.

Effect of RGB Image Features. Three kinds of image features used to fuse the corresponding point cloud features, as shown in Fig. 6. This work explores which kind of RGB image features are better for 3-D object detection. Three kinds of image features: the fine features from Faster R-CNN [30], the coarse features from a backbone, and the raw image features, as shown in Fig. 6. The ablation experiment results are shown in Table 3.

Table 3. Ablation study for RGB image features. The "Time" column denotes the training time and the "Memory" is the memory needed when the model is run for four batch sizes. The "Speed" column denotes the runtime. "Image Feature" denotes that if the model uses a full 2D image detection branch. The results of the cars are in the'Moderate' difficulty category for the BEV and 3D.

Image Feature	Backbone	Time (hour)	Memory (MB)	Speed (FPS)	BEV (%)	3-D (%)
Faster R-CNN	ResNet101	28.0	19,500	9.5	85.95	76.82
	ResNet-50	23.5	12,550	11.0	86.29	76.92
	ResNet V1–50	25.0	12,700	10.6	85.51	76.48
	VGG-11	16.0	11,900	12.0	85.96	76.44
Only Backbone	VGG-16	15.0	9,000	14.0	86.11	76.75
Raw Image	No	**11.5**	**4,200**	**23.0**	**86.67**	**77.45**

4.5 Visualization

Fig. 7. The qualitative results on 2-D RGB images. The green 3-D bounding boxes are ground truth. The red, cyan, and yellow colors denote the predictions for car, pedestrian, and cyclist classes, respectively. (Color figure online)

Figure 7 presents some qualitative results. As can be seen in the figures, each object can be detected by the proposed model and the predicted bounding boxes are well-matched with their corresponding ground truth boxes. Even in very complex scenes, the proposed model can detect objects quite well.

5 Conclusions

This paper is the first to propose a lightweight, memory-saving, and energy-saving framework for 3D object detection based on LiDAR and an RGB camera. Different from the existing frameworks, the proposed framework only employs one backbone to extract features from a point cloud and RGB image. The framework benefits from the proposed module, i.e., the point feature fusion module. The fusion module directly extracts the point features of RGB images and fuses them with the corresponding point cloud features. The experimental results using both the KITTI validation dataset and testing dataset demonstrate that the proposed method significantly improves the speed (23 FPS) of LiDAR-camera-based 3D object detection compared with other state-of-the-art approaches. Note that the proposed native model can achieve an inferring speed 23 FPS.

References

1. Wen, L., Jo, K.: Traffic sign recognition and classification with modified residual networks. In: 2017 IEEE/SICE International Symposium on System Integration (SII), Taipei, Taiwan, pp. 835-840 (2017) https://doi.org/10.1109/SII.2017.8279326
2. Vo, X.-T., Wen, L., Tran, T.-D., Jo, K.-H.: Bidirectional non-local networks for object detection. In: Nguyen, N.T., Hoang, B.H., Huynh, C.P., Hwang, D., Trawiński, B., Vossen, G. (eds.) ICCCI 2020. LNCS (LNAI), vol. 12496, pp. 491–501. Springer, Cham (2020). https://doi.org/10.1007/978-3-030-63007-2_38
3. Shahbaz, A., Jo, K.-H.: Dual camera-based supervised foreground detection for low-end video surveillance systems. IEEE Sens. J. 21(7), 9359–9366 (2021). https://doi.org/10.1109/JSEN.2021.3054940
4. Hoang, V., Hoang, V., Jo, K.: Realtime multi-person pose estimation with rcnn and depthwise separable convolution. In: 2020 RIVF International Conference on Computing and Communication Technologies (RIVF), Ho Chi Minh City, Vietnam, pp. 1–5 (2020) https://doi.org/10.1109/RIVF48685.2020.9140731
5. Chen, X., Ma, H., Wan, J., Li, B., Xia, T.: Multi-view 3D object detection network for autonomous driving. In: 2017 IEEE Conference on Computer Vision and Pattern Recognition (CVPR), Honolulu, HI, USA, pp. 6526–6534 (2017) https://doi.org/10.1109/CVPR.2017.691
6. Ku, J., Mozifian, M., Lee, J., Harakeh, A., Waslander, S.L.: Joint 3D proposal generation and object detection from view aggregation. In: 2018 IEEE/RSJ International Conference on Intelligent Robots and Systems (IROS), Madrid, Spain, pp. 1–8 (2018) https://doi.org/10.1109/IROS.2018.8594049
7. Zhou, Y., Tuzel, O., VoxelNet: End-to-end learning for point cloud based 3D object detection. In: 2018 IEEE/CVF Conference on Computer Vision and Pattern Recognition, Salt Lake City, UT, USA, pp. 4490–4499 (2018) https://doi.org/10.1109/CVPR.2018.00472
8. Wen, L., Jo, K.-H.: Fully convolutional neural networks for 3D vehicle detection based on point clouds. In: Huang, D.-S., Jo, K.-H., Huang, Z.-K. (eds.) ICIC 2019. LNCS, vol. 11644, pp. 592–601. Springer, Cham (2019). https://doi.org/10.1007/978-3-030-26969-2_56

9. Wen, L., Jo, K.-H.: LiDAR-camera-based deep dense fusion for robust 3D object detection. In: Huang, D.-S., Premaratne, P. (eds.) ICIC 2020. LNCS (LNAI), vol. 12465, pp. 133–144. Springer, Cham (2020). https://doi.org/10.1007/978-3-030-60796-8_12

10. Wen, L.-H., Jo, K.-H.: Three-attention mechanisms for one-stage 3-d object detection based on LiDAR and camera. IEEE Trans. Indus. Inf. **17**(10), 6655–6663 (2021). https://doi.org/10.1109/TII.2020.3048719

11. Charles, R.Q., Su, H., Kaichun, M., Guibas, L.J.: PointNet: deep learning on point sets for 3D classification and segmentation. In: 2017 IEEE Conference on Computer Vision and Pattern Recognition (CVPR), Honolulu, HI, USA, pp. 77–85 (2017) https://doi.org/10.1109/CVPR.2017.16

12. Qi, C., Yi, L., Su, H., Guibas, L.: PointNet++: Deep Hierarchical Feature Learning on Point Sets in a Metric Space. NIPS (2017)

13. Yang, Z., Sun, Y., Liu, S., Jia, J.: 3DSSD: point-based 3D single stage object detector. In: 2020 IEEE/CVF Conference on Computer Vision and Pattern Recognition (CVPR), Seattle, WA, USA, pp. 11037–11045 (2020) https://doi.org/10.1109/CVPR42600.2020.01105

14. Wen, L., Vo, X.-T., Jo, K.-H.: 3D SaccadeNet: a single-shot 3D object detector for LiDAR point clouds. In: 2020 20th International Conference on Control, Automation and Systems (ICCAS), Busan, Korea (South), pp. 1225–1230 (2020) https://doi.org/10.23919/ICCAS50221.2020.9268367

15. Shi, W., Rajkumar, R.: Point-GNN: graph neural network for 3D object detection in a point cloud. In: 2020 IEEE/CVF Conference on Computer Vision and Pattern Recognition (CVPR), Seattle, WA, USA, pp. 1708–1716 (2020) https://doi.org/10.1109/CVPR42600.2020.00178

16. Xu, D., Anguelov, D., Jain, A.: PointFusion: deep sensor fusion for 3D bounding box estimation. In: 2018 IEEE/CVF Conference on Computer Vision and Pattern Recognition, Salt Lake City, UT, USA, pp. 244–253 (2018) https://doi.org/10.1109/CVPR.2018.00033

17. Sindagi, V.A., Zhou, Y., Tuzel, O.: MVX-Net: multimodal VoxelNet for 3D object detection. In: 2019 International Conference on Robotics and Automation (ICRA), Montreal, QC, Canada, pp. 7276–7282 (2019) https://doi.org/10.1109/ICRA.2019.8794195

18. Geiger, A., Lenz, P., Urtasun, R.: Are we ready for autonomous driving? The KITTI vision benchmark suite. In: 2012 IEEE Conference on Computer Vision and Pattern Recognition, Providence, RI, USA, pp. 3354-3361 (2012) https://doi.org/10.1109/CVPR.2012.6248074

19. Yan, Y., Mao, Y., Li, B.: SECOND: sparsely embedded convolutional detection. Sensors **18**(10), 3337 (2018). https://doi.org/10.3390/s18103337

20. Lang, A.H., Vora, S., Caesar, H., Zhou, L., Yang, J., Beijbom, O.: PointPillars: fast encoders for object detection from point clouds. In: 2019 IEEE/CVF Conference on Computer Vision and Pattern Recognition (CVPR), Long Beach, CA, USA, pp. 12689–12697 (2019) https://doi.org/10.1109/CVPR.2019.01298

21. Qi, C.R., Liu, W., Wu, C., Su, H., Guibas, L.J.: Frustum PointNets for 3D object detection from RGB-D data. In: 2018 IEEE/CVF Conference on Computer Vision and Pattern Recognition, Salt Lake City, UT, USA, pp. 918–927 (2018) https://doi.org/10.1109/CVPR.2018.00102

22. Liang, M., Yang, B., Wang, S., Urtasun, R.: Deep continuous fusion for multi-sensor 3D object detection. In: Ferrari, V., Hebert, M., Sminchisescu, C., Weiss, Y. (eds.) ECCV 2018. LNCS, vol. 11220, pp. 663–678. Springer, Cham (2018). https://doi.org/10.1007/978-3-030-01270-0_39

23. Wang, J., Zhu, M., Sun, D., Wang, B., Gao, W., Wei, H.: MCF3D: multi-stage complementary fusion for multi-sensor 3D object detection. IEEE Access **7**, 90801–90814 (2019). https://doi.org/10.1109/ACCESS.2019.2927012

24. Liang, M., Yang, B., Chen, Y., Hu, R., Urtasun, R.: Multi-task multi-sensor fusion for 3D object detection. In: 2019 IEEE/CVF Conference on Computer Vision and Pattern Recognition (CVPR), Long Beach, CA, USA, pp. 7337–7345 (2019) https://doi.org/10.1109/CVPR.2019.00752

25. Yoo, J.H., Kim, Y., Kim, J., Choi, J.W.: 3D-CVF: generating joint camera and LiDAR features using cross-view spatial feature fusion for 3D object detection. In: Vedaldi, A., Bischof, H., Brox, T., Frahm, J.-M. (eds.) ECCV 2020. LNCS, vol. 12372, pp. 720–736. Springer, Cham (2020). https://doi.org/10.1007/978-3-030-58583-9_43

26. Hu, J., Shen, L., Sun, G.: Squeeze-and-excitation networks. In: 2018 IEEE/CVF Conference on Computer Vision and Pattern Recognition, Salt Lake City, UT, USA, pp. 7132–7141 (2018) https://doi.org/10.1109/CVPR.2018.00745

27. Wang, Q., Wu, B., Zhu, P., Li, P., Zuo, W., Hu, Q.: ECA-net: efficient channel attention for deep convolutional neural networks. In: 2020 IEEE/CVF Conference on Computer Vision and Pattern Recognition (CVPR), Seattle, WA, USA, pp. 11531–11539 (2020) https://doi.org/10.1109/CVPR42600.2020.01155

28. Qin, Z., Zhang, P., Wu, F., Li, X.: FcaNet: Frequency Channel Attention Networks, arXiv (2020)

29. Wen, L.-H., Jo, K.-H.: Fast and accurate 3D object detection for lidar-camera-based autonomous vehicles using one shared voxel-based backbone. IEEE Access 9, 22080–22089 (2021). https://doi.org/10.1109/ACCESS.2021.3055491

30. Ren, S., He, K., Girshick, R., Sun, J.: Faster R-CNN: towards real-time object detection with region proposal networks. IEEE Trans. Pattern Anal. Mach. Intell. 39(6), 1137–1149 (2017). https://doi.org/10.1109/TPAMI.2016.2577031

Regression-Aware Classification Feature for Pedestrian Detection and Tracking in Video Surveillance Systems

Xuan-Thuy Vo, Tien-Dat Tran, Duy-Linh Nguyen, and Kang-Hyun Jo$^{(\boxtimes)}$

Department of Electrical, Electronic and Computer Engineering, University of Ulsan, Ulsan
44610, Korea
{xthuy,tdat}@islab.ulsan.ac.kr, ndlinh301@mail.ulsan.ac.kr,
acejo@ulsan.ac.kr

Abstract. Pedestrian detection and tracking in video surveillance systems is a complex task in computer vision research, which has widely used in many applications such as abnormal action detection, human pose, crowded scenes, fall detection in elderly humans, social distancing detection in the Covid-19 pandemic. This task is categorized into two sub-tasks: detection, and re-identification task. Previous methods independently treat two sub-tasks, only focusing on the re-identification task without employing re-detection. Since the performance of pedestrian detection directly affects the results of tracking, leveraging the detection task is crucial for improving the re-identification task. The total inference time is computed in both the detection and re-identification process, quite far from real-time speed. This paper joins both sub-tasks in a single end-to-end network based on Convolutional Neural Networks (CNNs). Moreover, the detection includes the classification and regression task. As both tasks have a positive correlation, separately learning classification and regression hurts the overall performance. Hence, this work introduces the Regression-Aware Classification Feature (RACF) module to improve feature representation. The convolutional layer is the core component of CNNs, which extracts local features without modeling global features. Therefore, the Cross-Global Context (CGC) is proposed to form long-range dependencies for learning appearance embedding of re-identification features. The proposed model is conducted on the challenging benchmark datasets, MOT17, which surpasses the state-of-the-art online trackers.

Keywords: Pedestrian detection · Tracking and Re-identification · Video surveillance system · Convolution Neural Networks (CNNs)

1 Introduction

Nowadays, surveillance systems have been universally employed in many applications such as intelligent transportation systems, prevention of crime, military supervision systems, prisons, hospitals, industrial applications. The objective of the most surveillance system is to detect and track abnormal pedestrian activities in a video scene. The pedestrians are always walking or running on the street under a supervision camera. For

© Springer Nature Switzerland AG 2021
D.-S. Huang et al. (Eds.): ICIC 2021, LNCS 12836, pp. 816–828, 2021.
https://doi.org/10.1007/978-3-030-84522-3_66

example, the first pedestrian detection and tracking benchmark [2] is proposed, capturing real human activity on the street by CCTV. Pedestrian detection and tracking in video surveillance systems is a challenging task because of real dynamic environments such as illumination variation, crowded density scene, complicated distractor, shadows, occlusion, object deformation.

Recently, the accelerated development of deep learning, especially for Convolutional Neural Networks (CNNs), has brought a bright future in solving computer vision tasks such as pedestrian detection and tracking.

Pedestrian detection and tracking are one of the core applications of multiple object tracking for understanding visual objects in video. It includes two sub-tasks: detection and data association (re-identification). The pedestrian detection is to determine what objects are presented and where objects are located in each frame. Data association groups the same objects in different frames to output trajectories, assigning and tracking unique identification (ID) to each object across all frames. Previous methods, Sort [3], Deep-Sort [4], Poi [5] treat two sub-tasks independently. Specifically, re-ID is a secondary task in which the performance of it heavily depends on the main detection. Accordingly, leveraging the detection task is important for enhancing re-ID performance. The model complexity is calculated in both the detection and re-ID task, affecting the total inference time. Therefore, this work joins detection and re-ID task in the single end-to-end network based on the single object tracking paradigm, reducing the model complexity.

The generic detection consists of the classification and regression task. However, RetinaNet [15], BNL [14] only used classification performance for ranking detection during inference without considering regression score. There is inconsistency in object detection. PISA [16] showed that both of tasks have positive correlation. Mean that the detection has high classification quality corresponding to high regression quality, otherwise. Accordingly, this paper introduces a novel module, named Regression-Aware Classification Feature (FACF), to guide regression distribution to classification feature with ignored computational cost. During backward, the gradient is propagated from the classification branch to the regression branch.

The single object tracking follows the Siamese method learning the similarity using correlation filter of the search feature and template feature to emphasize the interest of objects. In this paper, similarity learning is employed with global feature modeling to get informative features from the input. The convolution operation is the main component of CNNs, only extracting local features. As a result, the receptive field is limited inside local neighborhoods. To overcome this problem, many convolution layers can be deeply stacked up to 50 layers or 100 layers. This strategy is not efficient, leading to high computational cost and difficulty to perform back-propagation. Inspired by BNL [14] and GCNet [19], Cross-Global Context (CGC) with an additional computational cost is proposed to model long-range dependencies, i.e., global feature, and additionally learn similarity between features of the current frame and previous frame. Moreover, CGC improves appearance embedding for learning re-ID features. In another aspect, GCNet includes context modeling utilizing the global context pooling and transformation step using two convolutional layers with channel reduction to learn channel dependencies.

Although the channel reduction strategy avoids the high computational cost, it is ineffective because channel reduction can lose important information of input. Therefore, Cross-Global Context avoids channel reduction by using lightweight 1D convolution to excite the importance of each channel without affecting the overall performance.

The proposed method is evaluated on two challenging benchmarks, that are MOT17, and MOT17Det. Compared to previous methods, the performance achieves high multiple object tracking accuracy (MOTA), ID switch, and higher order tracking accuracy (HOTA) with additional computational cost.

2 Related Works

Pedestrian Detection and Tracking. Pedestrian detection and tracking are grouped into the online method and offline method according to input frame. For the online method, the input employs the current frame and past frame, while the offline method relies on the whole frame. Most of the online methods [3–5], and offline methods utilize available object detection and only consolidate data association performance. The data association includes the Kalman filter predicting future motions and the Hungarian algorithm for tracking. Several methods such as JDE [9], Tracktor [6], and CTracker [10] introduced single end-to-end networks leveraging re-detection to improve appearance features for the re-ID step. Accordingly, this paper inherits "re-detection" method to combine detection and re-ID into one network, inspired by CTracker.

Correlation Between Classification and Regression. GA-RPN [17] presented the feature adaptation module between classification and regression branch using deformable convolution to add offset prediction into the rectangular grid sampling locations in regular convolution, thus enhancing feature representation. PISA [16] proposed the positive correlation module between classification and regression, improving the overall performance. The classification score is inserted to regression loss to re-weight prime samples, i.e., give more contribution to easy samples. However, the classification score and regressed offsets are computed independently during testing. Mean that there is inconsistent computation during training and testing. Alternatively, this work introduces a simple but effective module performing the correlation between classification and regression during training and testing without relating to the loss function.

Global Feature. GCNet [19] introduced global context module modeling long-range dependencies. This module includes the global context pooling and transformation step. The global context pooling squeezes the input tensor to vector to calculate the relationship between a query position and all positions and aggregate features of all positions by taking an averaging. The transformation step using two convolutional layers excites channel dependencies, i.e., whether certain channels are important or not. BNL [14] proposed the bidirectional non-local network by the dissecting global context module to gather and distribute features between query position and key position, which applied to object detection task.

3 The Proposed Method

This section analyzes the proposed end-to-end architecture, Cross-Global Context (CGC) module, and Regression-Aware Classification Feature (RACF) module.

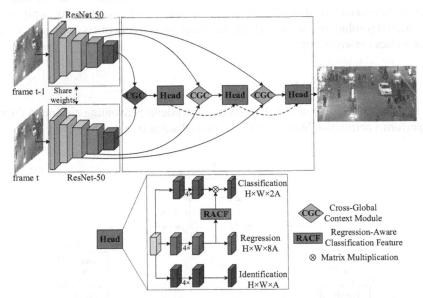

Fig. 1. The overall architecture of the proposed method. The two adjacent frames, frame t − 1, and frame t, are the input of the single end-to-end network. The backbone ResNet-50 extracts feature from two input frames. Then, the feature pyramid is constructed from stage 3, stage 4, and stage 5 of the backbone structure. CGC is the cross-global context module to model long-range dependencies and perform similarity learning between the feature at frame t − 1 and feature at frame t. Each head predicts classification score, regressed offset of paired bounding boxes of the same target, and Re-id score according to IoU score between paired bounding boxes with the same ID. 4 × denotes four convolutional layers, each convolutional layer includes 3 × 3 convolution following by group normalization and ReLU activation function. A indicates the number of anchor boxes per location.

The overall architecture is shown in Fig. 1. The input continuous video of this task is captured by CCTV, separated into discrete frames at a certain frame rate. Following the online method, the input only takes the current frame and last frame. Inspired by CTracker, two adjacent frames are used as input. The shared backbone extracting feature is ResNet-50 pre-trained on ImageNet. Similar to EFPN [13], and FPN, a feature pyramid is constructed to detect the objects with different scales, i.e., solve scale imbalance problem. For example, the large objects, medium objects, and small objects are assigned to a small feature, medium feature, and large feature, respectively. The CGC module will be discussed in Subsect. 3.1. Note that three feature maps corresponding to three heads are selected as a pyramid. Each head includes the classification, regression, and identification branch. The classification branch outputs objectness scores of each anchor box (pre-defined box) because the network only contains the pedestrian class. The regression branch predicts eight offset values corresponding to paired bounding boxes of the same target (the first four values for a target in the previous frame and last four values for a target in the current frame). The RACF module will be described in Subsect. 3.2. The identification branch predicts re-ID score learning IoU (Intersection of Union) between paired bounding boxes with the same ID. It means that the data association tracks IoU matching between paired bounding boxes of two adjacent frames without applying the Hungarian algorithm for tracking. Therefore, the proposed network is a one-shot tracker, which reduces inference time.

3.1 Cross-Global Context

The cross-global context models long-range dependencies avoiding channel reduction and performs correlation learning between two adjacent frames, shown in Fig. 2.

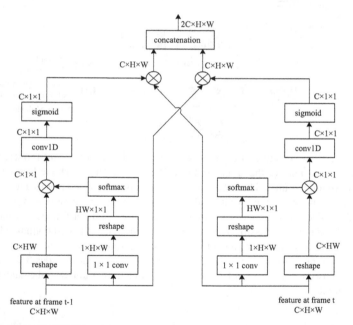

Fig. 2. The Cross-Global Context (CGC) takes two adjacent features as input. For each input, the CGC consists of the global context pooling and transformation step. The global context pooling learns the correlation between a query position and all key positions, which models long-range dependencies. The transformation step using light-weight convolution computes channel dependencies.

The input takes two adjacent features at frame $t - 1$, and frame t. For feature $\mathbf{F}^{t-1}$ with dimension $C \times H \times W$, CGC includes the global context pooling and transformation step. The global context pooling according to GCNet [19] gathers features from a query position and all key position by computing average, as follows:

$$\omega_{ij}^{t-1} = \sum_{j=1}^{H*W} \frac{\exp\left(\mathbf{W}_k \mathbf{F}_j^{t-1}\right)}{\sum_m \exp\left(\mathbf{W}_k \mathbf{F}_m^{t-1}\right)} \mathbf{F}_j^{t-1}, \tag{1}$$

where H, W, C is the height, width, and number of channels of the input feature map. ω_{ij}^{t-1} is a correlation function between the query position $\mathbf{F}_i^{t-1}$ and key position $\mathbf{F}_j^{t-1}$, in which the input tensor is squeezed to vector $C \times 1 \times 1$. $\mathbf{W}_k$ is a 1×1 convolution operation to gather feature of all positions. The exp is the exponential function. In CNNs, this function is softmax operation to output the attention map of each position, i.e., which positions contain the informative feature. Then, the matrix operation is performed between attention map and the reshaped input $\mathbf{F}_j^{t-1}$ to create a channel vector.

The transformation step learns channel dependencies by using excitation operation. In GCNet [19], the two 1×1 convolution layers with the channel reduction excite channel relationship, leading to losing information. To avoid channel reduction, CGC only utilizes lightweight 1D convolution with a kernel size of 5, learning cross-feature interaction. This computation is defined as:

$$e^{t-1} = \delta\left(\mathbf{W}_z \omega_{ij}^{t-1}\right), \tag{2}$$

where e^{t-1} is a re-scale function. δ is the sigmoid function to output the probability of each channel. $\mathbf{W}_z$ is a 1D convolution with a kernel size of 5 and padding of 2.

Similarly, the global context pooling function ω_{ij}^t and transformation function e^t of current feature at frame t are computed as Eq. 1, and Eq. 2, respectively. The rescaled features of two transformed features are crossed to learn similarity between the current and previous features, defined as:

$$\mathbf{R}^{t-1} = e^{t-1} \odot \mathbf{F}^t, \tag{3}$$

$$\mathbf{R}^t = e^t \odot \mathbf{F}^{t-1}, \tag{4}$$

where $\mathbf{R}^{t-1}$ and $\mathbf{R}^t$ are crossed features at frame t and t − 1 by using broadcast element-wise multiplication. e^{t-1} and e^t are re-scale functions computed as Eq. 2. $\mathbf{F}^t$ and $\mathbf{F}^{t-1}$ are input features at frame t and frame t-1.

Finally, two crossed features are concatenated as input of head part:

$$\mathbf{CF} = \left[\mathbf{R}^t, \mathbf{R}^{t-1}\right], \tag{5}$$

where $\mathbf{CF}$ is the concatenated feature with dimension $2C \times H \times W$.

3.2 Regression-Aware Classification Feature

The head consists of classification, regression, and identification branch, shown in Fig. 1. The detection quality depends on classification quality and regression quality. Independently learning the classification and regression branch is a straightforward way to improve detection quality. This paper introduces Regression-Aware Classification Feature (RACF) to positively correlate two branches, shown in Fig. 3. RACF module is computed consistently during training and testing. The easy samples and hard samples are measured by regression quality (IoU score). Specifically, the easy samples normally have a high IoU score corresponding to high regression quality. It means that the regression branch gives more contribution (more signals) to easy samples through the FACF module. Alternative speaking, the network focuses on easy samples. Otherwise, the hard samples have low regression quality. The regression branch gives less contribution to hard samples. During Non-maximum Suppression (NMS), the low-quality samples are filtered out.

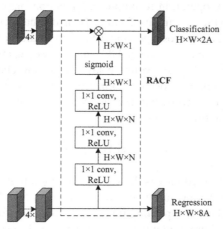

Fig. 3. The Regression-Aware Classification Feature (RACF) takes regressed offsets of paired bounding boxes as input. H, W, N is the height, width, and number of hidden channels, respectively.

To measure regression quality, the *topk* function is applied to select two maximum values from the regression distribution of each sample. Regularly, the regression distribution is Gaussian distribution. The three convolutional layers learn the correlation between regression and classification quality to improve classification feature which is aware of regression feature. Each convolutional layer includes a 1×1 convolution operation following by a ReLU activation function. Since regression quality and classification score are different ranges, the sigmoid function normalizes the regression quality between 0 and 1. Finally, the matrix multiplication is performed between learned regression quality and classification to enhance the classification feature.

During optimization, the gradient from the classification branch is propagated to the regression branch. The samples with higher classification loss will bring a larger gradient for regression quality, which means higher suppression on the regression quality.

4 Experiment Setup

The proposed method conducts the experiments on the challenging benchmark MOT17 [1]. This dataset includes 7 training videos and 7 testing videos. Pedestrian tracking is a complex task depending on classification, localization, and re-ID task. To measure whole aspects of performance, all results are evaluated by three primary metrics for tracking, detection, and ID assignment such as multiple object tracking accuracy (MOTA), and ID switch (IDF1) proposed by CLEAR MOT [11]; higher order tracking accuracy (HOTA) proposed by [12].

For implementation details, all experiments are implemented by the deep learning Pytorch framework. The parameters of the ResNet-50 [18] are pre-trained on ImageNet as initialization. The added convolutional layers in FPN, three branches in the head part, CGC module, and RACF module are initialized by selecting values from the normal distribution. The model is trained for 100 epochs with a batch size of 8. The learning rate is set to $3 \times 1e^{-5}$ for all experiments. Adam optimizer is employed to optimize objective function defined as:

$$ L = L_{reg}\left(b_i^{t-1}, g_i^{t-1}; b_i^t, g_i^t\right) + L_{cls}\left(s, \hat{s}\right) + L_{id}\left(d, \hat{d}\right), \tag{6} $$

where L_{reg} is regression loss for paired bounding boxes of the same target at frame t-1 and frame t, which uses smooth-L1 loss. L_{cls} is the classification loss using Focal loss [15] in which s, $\hat{s}$ is classification score and target label. L_{id} is defined by CTracker [10] for identification loss, utilizing Focal loss [15] to predict ID according to IoU matching.

5   Results

This section analyzes how to select the hyperparameters in the CGC, RACF module and the importance of each component in Subsect. 5.1. The results of the ablation study are measured on the sub-set of the training set since the MOT benchmark did not provide ground truth annotation for the testing set. The main results tested on the MOT17 testing set are shown in Subsect. 5.2, which are submitted to the evaluation protocol system[1].

5.1   Ablation Study

The Number of Hidden Channels in RAFA. Several implementations are conducted to select the hyperparameter N in the RAFA module. We select $N \in \{16, 32, 64, 128, 256\}$ to train the single-end-end model. The results are shown in Table 1, measured by the evaluation tool[2].

Where MOTP is Multiple Object Tracking Precision. MT, ML is Mostly Tracked Trajectories, Mostly Lost Trajectories. FP, FN is the number of False Positives and False Positive. IDS is the number of Identity Switches. #par is the additional parameter of the RAFA module to the total parameters.

The performance of the whole network is insensitive to the number of hidden channels N. The small N will output coarse features for the classification branch. Alternatively speaking, the 1×1 convolution with output channels, 16 or 32, does not satisfy to learn variables of regression distribution. The model with a large N means that RAFA can

---

Table 1. The effects of the number of hidden channels on the performance

N	MOTA↑	IDF1↑	MOTP↑	MT↑	ML↓	FP↓	FN↓	IDS↓	#par
16	74.3	65.8	85.1	259	60	1954	25927	1015	+0.4k
32	74.6	66.6	85.2	270	53	2060	25413	1032	+1.3k
64	75.7	66.3	**85.7**	270	55	**1493**	24807	998	+4.7k
128	**76.1**	**67.3**	85.6	**278**	**51**	1668	**24190**	**977**	+17.5k
256	74.8	66.2	85.2	266	57	2115	25158	1033	+67.8k

learn rich information of regression quality. Specifically, the proposed method achieves the best results at N = 128 with 76.1% of MOTA. Hence, we select N = 128 for all experiments. In another aspect, the RAFA only takes 17.5k (thousands) parameters while the total parameters of the whole network are up to million parameters. Therefore, it demonstrates the RAFA module is simple but effective.

Avoiding Channel Reduction in CGC. This experiment analyzes the effect of channel reduction on the CGC performance, which is shown in Table 2.

Table 2. The effect of channel reduction on the results

Method	MOTA↑	IDF1↑	MOTP↑	MT↑	FP↓	IDS↓	#par
CGC w/ CR	74.9	65.3	85.1	273	1836	1038	$2 \times C^2/r$
CGC w.o/ CR	75.6	65.7	85.1	283	2280	989	$k = 5$

As expected, the CGC module with channel reduction (use two 1×1 convs with a reduction ratio of r) decreases the MOTA score by 0.7% when compared CGC module without channel reduction (use 1D convolution with the kernel size of 5). Moreover, the lightweight operation only takes 5 parameters while using 2 consistent convolutions is still high computational cost. Usually, the number of channels C is 256 and r = 8, the number of parameters is 16.4k parameters in which larger than our method by 3280 times.

The Effect of Each Component. We investigate the importance of individual components on the sub-training set. The results are shown in Table 3.

The baseline is the simple version of the proposed method, which outputs 73.1% MOTA. When CGC is added to the baseline, the results gain the MOTA score of 2.4%. Similarly, the RAFA boosts the baseline performance by 1.8%. Remarkably, the full version includes the CGC and RAFA module, which increases the baseline results by a large margin, 3.0% MOTA score. It is easy to understand that the proposed model can learn finer features from global features and regression quality, effectively.

Table 3. The importance of each component

Baseline	CGC	RAFA	MOTA↑	IDF1↑	MOTP↑	MT↑	FP↓	IDS↓
✓			73.1	63.3	85.3	245	2229	2385
✓	✓		75.5	66.0	84.9	277	1747	983
✓		✓	74.9	65.3	85.1	273	1836	1038
✓	✓	✓	76.1	67.3	85.6	278	1668	977

Error Tracking Decomposition. To measure the error of the proposed method, we decompose the tracking performance into three components: detection errors, association errors, and localization errors, shown in Fig. 4.

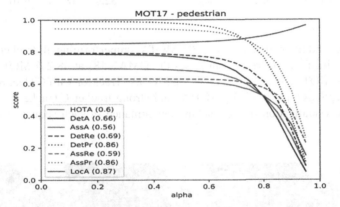

Fig. 4. HOTA and decomposed components

Where alpha indicates the performance at different localization score thresholds from 0.05 to 0.95 with step size 0.05. DetA is the detection accuracy score achieving the average score of 0.66, which decomposing to DetRe (Detection Recall) and DetPr (Detection Precision). For localization error, we compute LocA (localization accuracy score) at various alpha and average $LocA_\alpha$. Accordingly, LocA achieves a score of 0.87, high localization quality because of that the RAFA learns localization quality to guide classification features in which the detection works well at the high alpha threshold. For association errors, the AssA (association accuracy) is employed for assigning predicted IDs to ground truth trajectories, which decomposes to AssRe (association recall) and AssPr (association precision).

5.2 Comparison with State-of-the-Art Methods

This subsection shows the main results of the proposed method tested on MOT17. The performance of the whole network is compared with other methods, listed in Table 4. The bold font indicates the best results among trackers.

Table 4. Comparison with state-of-the-art online methods on MOT17 test set

Method	MOTA↑	IDF1↑	MOTP↑	MT↑	FP↓	IDS↓
DMAN [8]	48.2	55.7	75.9	19.3	26281	2194
MOTDT [7]	50.9	52.7	76.6	17.5	24069	2474
Tracktor [6]	53.5	52.3	78.0	19.5	**12201**	**2072**
Tracktor + CTDet	54.4	56.1	78.1	25.7	44109	2574
DeepSORT [4]	60.3	61.2	**79.1**	31.5	36111	2442
CTracker [10]	66.6	**57.4**	78.2	32.2	22284	5529
Ours	**67.3**	54.7	78.6	**32.9**	18771	5910

The proposed method achieves 67.3% MOTA, outperforming all trackers by a large margin. Specifically, our tracker surpasses the DMAN [8] at 48.2% MOTA, MOTDT [7] at 50.9% MOTA, Tracktor [6] at 53.5% MODA, Tracktor with CTDet [6] at 54.4% MOTA, DeepSORT [4] at 60.3% MOTA, and strong tracker CTracker [10] at 66.6% MOTA. The visualization of the tracking performance in each sequence is shown in Fig. 5.

frame 1   frame 100   frame 400

Fig. 5. The qualitative results on some sequences of MOT17 benchmark.

6 Conclusion

This paper introduced the online single end-to-end network joining detection and data association in the one-shot tracker for pedestrian detection and tracking in video surveillance systems, which reducing computation cost. Moreover, the RACF module with simple but effective learned regression distribution to guide classification feature, leveraging the positive correlation between classification and regression task. The CGC module with lightweight operation is proposed by investigating channel reduction in transform step to model long-range dependencies and similarity learning between two adjacent frames (current frame and previous frame). The performance is evaluated on the challenging benchmark MOT17, which outperformed all online trackers by a large margin.

References

1. Milan, A., Leal-Taixé, L., Reid, I., Roth, S., Schindler, K.: MOT16: A benchmark for multi-object tracking. arXiv preprint arXiv:1603.00831 (2016)
2. Dollár, P., Wojek, C., Schiele, B., Perona, P.: Pedestrian detection: a benchmark. In: 2009 IEEE Conference on Computer Vision and Pattern Recognition, pp. 304–311 (2009)
3. Bewley, A., Ge, Z., Ott, L., Ramos, F., Upcroft, B.: Simple online and realtime tracking. In 2016 IEEE International Conference on Image Processing (ICIP), pp. 3464–3468 (2016)
4. Wojke, N., Bewley, A., Paulus, D.: Simple online and realtime tracking with a deep association metric. In: 2017 IEEE International Conference on Image Processing (ICIP), pp. 3645–3649 (2017)
5. Yu, F., Li, W., Li, Q., Liu, Y., Shi, X., Yan, J.: Poi: Multiple object tracking with high performance detection and appearance feature. In: European Conference on Computer Vision, pp. 36–42 (2016)
6. Bergmann, P., Meinhardt, T., Leal-Taixe, L.: Tracking without bells and whistles. In: Proceedings of the IEEE/CVF International Conference on Computer Vision, pp. 941–951 (2019)
7. Chen, L., Ai, H., Zhuang, Z., Shang, C.: Real-time multiple people tracking with deeply learned candidate selection and person re-identification. In: 2018 IEEE International Conference on Multimedia and Expo (ICME), pp. 1–6 (2018)
8. Zhu, J., Yang, H., Liu, N., Kim, M., Zhang, W., Yang, M.H.: Online multi-object tracking with dual matching attention networks. In: Proceedings of the European Conference on Computer Vision (ECCV), pp. 366–382 (2018)
9. Wang, Z., Zheng, L., Liu, Y., Wang, S.: Towards real-time multi-object tracking. arXiv preprint arXiv:1909.12605 2(3), 4 (2019)
10. Peng, J., et al.: Chained-tracker: Chaining paired attentive regression results for end-to-end joint multiple-object detection and tracking. In: European Conference on Computer Vision, pp. 145–161 (2020)
11. Bernardin, K., Stiefelhagen, R.: Evaluating multiple object tracking performance: the clear mot metrics. EURASIP J. Image Video Processing 2008(1), 1–10 (2008). https://doi.org/10.1155/2008/246309
12. Luiten, J., et al.: HOTA: a higher order metric for evaluating multi-object tracking. Int. J. Comput. Vis. 1–31 (2020)
13. Vo, X.T., Jo, K.H.: Enhanced feature pyramid networks by feature aggregation module and refinement module. In: 2020 13th International Conference on Human System Interaction (HSI), pp. 63–67 (2020)
14. Vo, X.T., Wen, L., Tran, T.D., Jo, K.H.: Bidirectional non-local networks for object detection. In: International Conference on Computational Collective Intelligence, pp. 491–501 (2020)
15. Ross, T.Y., Dollár, G.: Focal loss for dense object detection. In Proceedings of the IEEE Conference on Computer Vision and Pattern Recognition, pp. 2980–2988 (2017)
16. Cao, Y., Chen, K., Loy, C., Lin, D.: Prime sample attention in object detection. In: Proceedings of the IEEE/CVF Conference on Computer Vision and Pattern Recognition, pp. 11583–11591 (2020)

17. Wang, J., Chen, K., Yang, S., Loy, C., Lin, D.: Region proposal by guided anchoring. In: Proceedings of the IEEE/CVF Conference on Computer Vision and Pattern Recognition, pp. 2965–2974 (2019)
18. He, K., Zhang, X., Ren, S., Sun, J.: Deep residual learning for image recognition. In Proceedings of the IEEE Conference on Computer Vision and Pattern Recognition, pp. 770–778 (2016)
19. Cao, Y., Xu, J., Lin, S., Wei, F., Hu, H.: Gcnet: non-local networks meet squeeze-excitation networks and beyond. In: Proceedings of the IEEE/CVF International Conference on Computer Vision Workshops (2019)

Efficient Face Detector Using Spatial Attention Module in Real-Time Application on an Edge Device

Muhamad Dwisnanto Putro, Duy-Linh Nguyen, and Kang-Hyun Jo$^{(\boxtimes)}$

Department of Electrical, Electronic and Computer Engineering, University of Ulsan, Ulsan 44616, South Korea
{dputro,ndlinh301}@mail.ulsan.ac.kr, acejo@ulsan.ac.kr

Abstract. The practical application requires a vision-based face detector to work in real-time. The robot application uses the face detection method as the initial process for the face analysis system. During its development, it utilizes an edge device to be used to process sensor information. Jetson Nano is a mini portable computer that is easily synchronized with sensors and actuators. However, traditional detectors can work fast on this device but have low performance for occlusion cases, multiple poses, and small faces. On the other hand, CNN-based detectors that implement deep layers are slow to run on low memory GPU devices. In this work, an efficient real-time face detector using a simple spatial attention module was developed to localize faces rapidly. The proposed architecture consists of the backbone module to efficiently extract features, the light connection module to reduce the size of the detection layer, and multi-scale detection to perform prediction of faces on various scales. As a result, the proposed detector achieves competitive performance from state-of-the-art fast detectors on several benchmark datasets. In addition, this efficient detector can run at 55 frames per second in video graphics array resolution on a Jetson Nano.

Keywords: Face detector · Efficient network · Jetson Nano · Real-time application

1 Introduction

Face detection is a popular field in computer vision for localizing facial areas in an image. This method is the initial process used for advanced vision systems such as face recognition, emotion, gender, and landmarks [1–4]. Face analysis is widely used in intelligent video technology. It encodes data from specific facial components to determine the characteristics and models of that information. Besides, face detection is a current technology required in a portable device to unlock mobile phones and log in payment accounts.

Nowadays, a security system is needed in every aspect to monitor the environment. It is used to prevent crime that occurs in public areas [5]. Even developed countries use it to find missing people. Intelligent video surveillance not only applies this monitoring

© Springer Nature Switzerland AG 2021
D.-S. Huang et al. (Eds.): ICIC 2021, LNCS 12836, pp. 829–841, 2021.
https://doi.org/10.1007/978-3-030-84522-3_67

application, but it can also explore the control system to optimize its performance. The system needs to be supported by massive process memory availability because the face detectors work with heavy compressing data. It requires expensive hardware for the detector to work optimally. Another application is for Human-robot Interaction (HRI) to implement a face detector on a robot supporting the facial recognition system's capabilities [6]. Even emotion recognition is applied to a robot service to predict the expressions of the customer. Both methods require a face detector as an initial step to localize the face area.

A face has a distinctive texture and color against the background. It is unique because every human being has a different characteristic [7]. Information from facial organs is important feature data to identify them. The eye has an ellipse shape and a characteristic color that is different from the nose. The lips contain a reddish color that is different from the eyebrows and chin. Although each organ's shape is almost the same as the other background features, the relationship between these features generates a face model strengthening the distinguisher against the background. Therefore, the extraction of faces and the relation between features is essential information to identify a face in the image.

Several works have presented conventional methods for detecting faces in an image. The Viola-Jones discovered important facial features by applying Haar-like features that were moved around using a sliding window [8]. The difference between light and dark areas is used to identify interest facial features. Combining an integral image and AdaBoost Learning generates a learning model that can work quickly to classify facial and background features. Although this detector can work in real-time, it has low accuracy in occlusion cases, multiple poses, and small faces. The Haar-like feature has limitations in capturing facial feature information that is blocked and small sizes. The rotation-invariant drops the performance of this method in a real-case application for finding parallel facial features.

The modern method introduced the Convolution Neural Network (CNN) as a robust features extractor feature [9]. It employs convolutional operations on input features and kernel shape. It adopts a neural network approach that updates the kernel weights to improve network performance and minimize error rates. This method obtains a high degree of accuracy for image classification work [10]. It can distinguish categories from images by extracting specific information.

On the other hand, CNN capabilities have been implemented to distinguish facial and background features to localize facial areas in an image. Several studies [11–13] achieved high accuracy for the complicated challenge, but practical applications prevent these methods from working in real-time. Moreover, robotic applications require a vision system to work quickly on low-cost devices. The deep backbone tends to employ huge filter layers, resulting in over a million total parameters. VGG-16 [14] and ResNet [15] are benchmark backbones that have successfully filtered out important object features, but these models also generate large parameter weights.

Additionally, MobileNet [16] and ShuffleNet [17] have been introduced as lightweight CNN backbones, but these models have stagnated in real-time work when implemented a low-cost device. Jetson Nano is an edge device generally used in IoT (Internet of Things) and robotics application [18]. The CNN method is relatively implemented in this hardware as a visual approach to sense the object and environment.

The lightweight architecture employs a few slim layers of convolution and delivers efficient computation power. Apart from preventing premature saturation, this network produces a small number of parameters. However, superficial networks do not produce high accuracy. Several methods apply an attention mechanism to improve the feature extraction performance [19]. This module summarizes the essential features and generates attention weights to update the feature input. The spatial attention module captures specific information based on the feature map's size representing valuable information from each cell element [20]. This block employs sigmoid activation to generate probability weights and produces low computation costs and parameters. It is very efficient to increase the performance of a shallow backbone implemented in a real-time detector.

Based on the above issues, this study proposes a lightweight CNN with simple spatial attention to rapidly localize facial areas. The contributions of this work are as follows:

1. A new efficient CNN architecture is used to build a face detector that is fast works in a real-time application.
2. A simple spatial attention module was introduced as a reinforcing module for shallow backbones to support the network's efficiency and effectiveness.

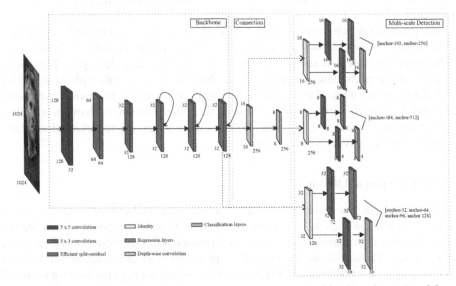

Fig. 1. The proposed architecture of face detector. It uses a combination of 1×1 and 3×3 convolution as extractor features. Multilayer detection with anchors assignment plays a role in predicting faces at various scales.

2 Proposed Architecture

In this section, the detail of our proposed architecture is explained. The proposed method applies a shallow layer of Convolutional Neural Network and combines each module to produce an efficient architecture. Instead of significantly reducing the detector performance, the main module consists of backbone, connection, and multi-scale detection layers, as shown in Fig. 1.

2.1 Backbone Module

CNN-based architectures tend to use multiple layers in the backbone to extract essential information. This layer has an impact on a large number of parameters and computations. The proposed detector employs nine layers of convolution by combining 3×3 and 1×1 filters. Specifically, the backbone module consists of a shrink layer for reducing the feature map size and a stem module that sequentially discriminates against facial and background features. The shrink module employs a 7×7 filter at the beginning of the stage to significantly reduce the feature map's size [9]. It is followed by a 3×3 convolution with a stride of 2, generating the 32×32 with 128 channel feature map. This filter effectively and efficiently captures facial features of various sizes [14]. Additionally, in order to prevent saturation and vanishing gradients, ReLU and Batch-Normalization are used after the convolution process at each layer. ReLU selects positive values and ignores negative values from feature input, while Batch-Normalization maintains the average distribution is close to 0 and the output standard deviation close to 1 for each mini-batch.

The proposed detector is designed as a low weight efficiency detector by applying a partial transfer structure at the stem module. An efficient split-residual block divides input features map into two parts and unites them at the end of the module. Figure 2 (a) shows that the split approach reduces computation at the start of the module without removing other parts' information. Half of the feature map is processed in feature extraction, sequentially applying convolution. At the same time, the attention mechanism is applied to other chunks. Finally, the efficient module concatenates the representation of the essential elements and other extraction parts.

2.2 Simple Spatial Attention

Attention mechanism increases the interest features intensity by eliminating distinctive features and reducing trivial information [20]. The feature location on each feature map is valuable information. Thus, spatial information tends to show cues of extracted facial features. The spatial attention module is proposed as an enhanced module without producing excessive computation. Figure 2 (b) shows that average pooling for each cell of the input features is applied to summarize the channel array information. The simple spatial attention is defined as follows:

$$S_{att} = \sigma(W_{c1}AVG(x_i)) \tag{1}$$

where σ is sigmoid activation to generate probability weights from a single feature map representation of the simple convolution ($c1$). This module updates each element of the input features and implements element-wise multiplication with weighted maps. Therefore, this module reduces non-facial features and enhances distinctive facial features to strengthen the discrimination process.

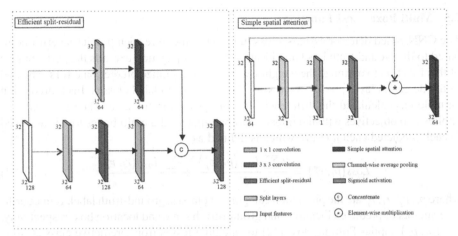

Fig. 2. The efficient split-residual as a stem module that sequentially extracts features and produces a few parameters and low-cost computation power (a). Simple spatial attention is employed at part of the input feature map to capture the representation of the valuable elements (b).

2.3 Connection Module

CNN-based detectors use the connection module to link between the prediction layers. Multilayer detection uses prediction layers with various scales and employs shrink blocks to create feature maps of different sizes. Instead of using standard convolution, it employs Depth-wise convolution with a stride of two to save the parameters. This convolution does not apply a multi convolutional filter to each input channel, but it only uses a single convolutional filter. This block emphasizes computational efficiency with fewer multiplication and addition operations on each channel. Besides, this approach also specifies that the number of relations between each kernel and input elements is equal to the number of channels. A 3×3 kernel with linear operations is used to filter specific object features. Furthermore, this is also followed by ReLU and BatchNormaliszation to maintain network performance and to avoid accuracy dropped in the training process.

2.4 Multi-scale Detection and Anchor Assignment

The proposed detector is a one-stage architecture that predicts classes and face bounding box location. It also assigns anchors of various sizes as initial bounding boxes. This feature map is generated by a 3×3 convolution, which predicts the category and regression layers. Category prediction is face and none, while regression determines the location coordinates (x, y) and fits the scale of bounding boxes, namely height (h) and width (w). The variation in face size emphasizes that the detector needs to assign predictions with various scales. A three-layered detection handles small, medium, and large faces, with feature map sizes including 8, 16, and 32, respectively. In addition, anchor assignments of various sizes are placed on each prediction layer. Anchors 32, 64, 96, and 128 were assigned to predict small faces, while 192 and 256 for medium and large faces were applied 384 and 512.

2.5 Multi Boxes Loss Function

The CNN-based detector assigns a loss function to measure each prediction error compared to the ground-truth label and location. Backpropagation exploits the performance of this function to optimize the weight neurons and minimize prediction errors. The end of the detection layer predicts regression (x, y, h, w) and label classes. Each prediction variable has calculated the difference with the ground-truth value. Then the total loss applies two objectives with the imbalanced parameter. The multi boxes loss is assigned to each predicted anchor (i-th), which is defined as

$$Loss(p_i, r_i^*) = \frac{2 \sum_i L_{cat}(p_i, p_i^*)}{N} + \frac{\sum_i L_{reg}(r_i, r_i^*)}{N}, \tag{2}$$

where p_i, p_i^*, r_i, r_i^* are the prediction category of classes, ground-truth label, four coordinate vectors of predicted location, and ground-truth scale and location box, respectively. $L_{cat}(p_i, p_i^*)$ applies Softmax-loss [12] to calculate losses from predictive class classes, while L1-smooth loss [11] is used to calculate regression losses $L_{reg}(r, r_i^*)$. It gives greater weight to the loss classification side, which tends to produce lower scores at the default training stage. Therefore, this manipulation balances the updating weight's performance in neurons to work fairly on both sides of the function.

3 Dataset and Implementation Setup

Proposed models are trained on a WIDER dataset containing 32,203 total images, with only 12,800 of the training set is used by detector as knowledge to learn the characteristics of facial features. Additionally, PASCAL face, AFW (Annotated Faces in the Wild), and FDDB (face Detection Data Set and Benchmark) are test datasets for evaluating training models. In order to enrich training data variation, random cropping, scale transformation, color distortion, and horizontal flipping are applied as augmentation methods. The end of this process produces 1024×1024 RGB as the input image size of training.

The training process divides all images dataset into 32 batches, shortening the time from the network for learning data on small partitions. Proposed models are trained through the end-to-end stage with random weights at the beginning of the epoch. The Stochastic Gradient Descent (SGD) was used to optimize the neuron weights in the backpropagation process with $5 \cdot 10^{-4}$ weight decay and 0.9 momentum. It assigns different learning rate weights for the variation in the number of epochs. The initial stage uses a 10^{-3} learning rate for 200 epochs, followed by a 10^{-4} learning rate for 100 epochs, the next 10^{-5} learning rate for 50 epochs, and the last 20 epochs at a 10^{-6} learning rate. Intersection over Union (IoU) of 0.5 is used to select predicted anchors that overlap in the evaluation process. Finally, the training, evaluation, and real-time testing processes of this detector are implemented in the PyTorch framework.

4 Experiments and Results

In this section, the proposed detector is tested for the performance of each module and evaluates on benchmark datasets. It also compares Average Precision (AP) with various competitors. Besides, the efficiency of the face detector is also tested in real-time applications on the Jetson Nano device.

4.1 Ablative Study

This ablative study comprehensively shows the strength of each proposed module, including shrink, stem with attention, connection, and multi-scale detection module. Each proposed module gradually is removed, it analyzes the accuracy and number of parameters with the same training configuration. Table 1 shows that each module increases the accuracy and number of parameters of the detector. The stem module increases the accuracy of this detector by 4.12% and adds 220K parameters. Additionally, the proposed attention module also increases the accuracy by 1.08%, but this only adds a few parameters.

Table 1. Ablative study of the proposed modules on FDDB dataset

Modules	Proposed detector				
Simple spatial attention	√				
Stem	√	√			
Connection	√	√	√		
Multi-scale detection	√	√	√	√	
Shrink	√	√	√	√	√
Number of parameter	433.363	405.700	186.448	152.656	152.656
Average precision (%)	96.46	95.38	91.26	90.50	82.71

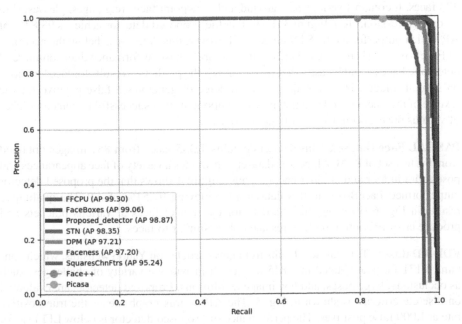

Fig. 3. Evaluation of proposed detector on AFW dataset

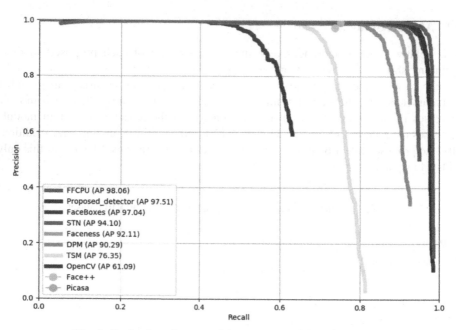

Fig. 4. Evaluation of proposed detector on PASCAL face dataset

4.2 Evaluation on Datasets

AFW Dataset. This dataset is obtained from Flickr images and has 205 images with 473 faces. It contains various background and viewpoint faces (e.g., ages, glasses, skin colors, expression, etc.). Figure 3 shows that the proposed detector achieves 98.87% of AP, which outperforms the STN detector. However, this detector is below the accuracy of FaceBoxes [11] and FFCPU [12]. It has a more robust performance than commercial detectors (Face++ and Picasa). Figure 6 (a) shows the prediction results for face locations on several images. However, the proposed detector generates a false positive for the texture of the background and the human component. It is successful in detecting faces of various poses, occluded and of various sizes.

PASCAL Face Dataset. This dataset contains 1,335 faces from 851 images obtained from the test set of PASCAL person dataset. It provides a variety of face appearances and poses with indoor and outdoor backgrounds. Figure 4 shows that the proposed detector outperformed FaceBoxes in this dataset by achieving 97.51% of AP. The qualitative results in Fig. 6 (b) show that the detector can detect faces at various brackets and produce error prediction on textures and colors similar to faces.

FDDB Dataset. This dataset obtains from news articles on Yahoo websites, which contains 5,171 faces annotated in 2,845 images. It provides a variety of challenges, such as occlusions, large poses, and low image resolutions. Proposed detectors are evaluated on discrete criteria, as shown in Fig. 5. The AP on this graph means the true-positive rate at 1,000 false positives. The performance of proposed detector is below LFFD [13], FFCPU, and FaceBoxes. The shallow layer of the detector cannot correctly discriminate

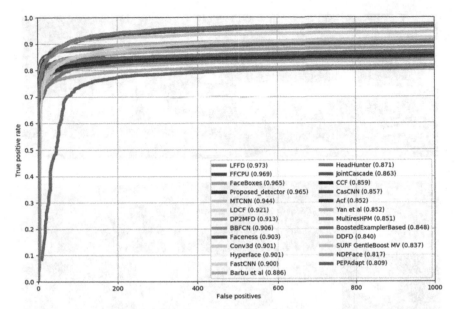

Fig. 5. Evaluation of proposed detector on FDDB dataset at 1000 false positives

facial and non-facial features, as shown in Fig. 6 (c). Hand features are predicted as facial features. However, the proposed detector produces a lower number of parameters, and it can work faster on the Jetson Nano device.

4.3 Runtime Efficiency

The CNN-based detector is useful if it can work quickly on low-cost devices. The CNN models require large and expensive GPUs to work in real-time. In general, robotics applications use a portable computer that can acquire sensor and actuator data. Jetson Nano is an edge device that is easily connected and synchronized with robotic devices. However, heavyweights detectors are slow to work in real-time on this device. The proposed detector generates 433,363 parameters which are lower than the other competitors. The LFFD detector achieved the best performance. However, this produces 2M trainable parameters, as shown in Fig. 7. Therefore, this detector works slowly in real-time applications.

Testing of the real-time application using a webcam as an input device, this data is directly processed on each detector. The speed of each detector at different video input sizes is shown in Fig. 8. The proposed detector outperformed competitors' speed by achieving 54.87 FPS at VGA resolution. It differs 13 FPS from the slower FFCPU detector. The implementation at Full HD resolution shows that the proposed detector can work in real-time by reaching a speed of 25.89 FPS, while other detectors work slowly with speeds below 20 FPS. The superficial model of the proposed detector emphasizes

(a)

(b)

(c)

Fig. 6. Visualization of result from proposed detector on AFW (a), PASCAL face (b), and FDDB datasets (c)

the computational efficiency and the number of parameters. The backbone module produces a small number of parameters, but it maintains the quality of feature extraction. Furthermore, the simple spatial attention module improves detector performance without significantly slowing down the real-time detector speed.

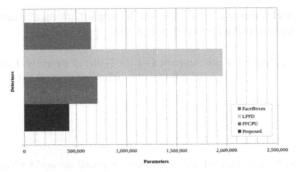

Fig. 7. Comparison of trainable parameters detector with other competitors

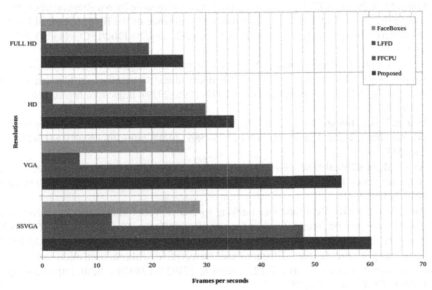

Fig. 8. Comparison of data processing speed in real-time application with other competitors at different video input sizes.

5 Conclusions

This work presented an efficient real-time face detector using a simple spatial attention module implemented on a Jetson Nano. The proposed architecture consists of three main modules, a backbone, a connection, and multi-scale detection. The backbone plays an essential role in discriminating facial and background features by applying a simple spatial attention block. This module effectively improves backbone performance without adding significant number of parameters. Proposed detectors produce trainable parameters that are lower than CNN-based fast detectors. Finally, the results showed that the proposed detector achieved competitive performance with state-of-the-art fast detectors and outperformed their speed by 55 FPS in real-time on a Jetson Nano. In the future, the quality of model training can be improved by implementing and exploring IoU loss and Focal Loss without reducing speed in real-time applications.

Acknowledgement. This work was supported by the National Research Foundation of Korea (NRF) grant funded by the government (MSIT) (No. 2020R1A2C2008972l2).

References

1. Choi, J.Y., Lee, B.: Ensemble of deep convolutional neural networks with gabor face representations for face recognition. IEEE Trans. Image Process. **29**, 3270–3281 (2020). https://doi.org/10.1109/TIP.2019.2958404
2. Putro, M.D., Nguyen, D.-L., Jo, K.-H.: A dual attention module for real-time facial expression recognition. In: IECON 2020 the 46th Annual Conference of the IEEE Industrial Electronics Society, Singapore, pp. 411–416 (2020). https://doi.org/10.1109/IECON43393.2020.925 4805
3. Zhou, Y., Ni, H., Ren, F., Kang, X.: Face and gender recognition system based on convolutional neural networks. In: 2019 IEEE International Conference on Mechatronics and Automation (ICMA), Tianjin, China, pp. 1091–1095 (2019). https://doi.org/10.1109/ICMA.2019.8816192
4. Hoang, V.-T., Huang, D.-S., Jo, K.-H.: 3-D facial landmarks detection for intelligent video systems. IEEE Trans. Industr. Inf. **17**(1), 578–586 (2021). https://doi.org/10.1109/TII.2020.2966513
5. Awais, M., et al.: Real-time surveillance through face recognition using HOG and feedforward neural networks. IEEE Access **7**, 121236–121244 (2019). https://doi.org/10.1109/ACCESS.2019.2937810
6. Putro, M.D., Jo, K.: Real-time face tracking for human-robot interaction. In: 2018 International Conference on Information and Communication Technology Robotics (ICT-ROBOT), Busan, Korea (South), pp. 1–4 (2018). https://doi.org/10.1109/ICT-ROBOT.2018.8549902
7. Li, X., Yang, Z., Wu, H.: Face detection based on receptive field enhanced multi-task cascaded convolutional neural networks. IEEE Access **8**, 174922–174930 (2020). https://doi.org/10.1109/ACCESS.2020.3023782
8. Paul, V., Michael, J.: Robust real-time face detection. Int. J. Comput. Vision **57**(2), 137–154 (2004)
9. Zeiler, M.D., Fergus, R.: Visualizing and understanding convolutional networks. In: Fleet, D., Pajdla, T., Schiele, B., Tuytelaars, T. (eds.) ECCV 2014. LNCS, vol. 8689, pp. 818–833. Springer, Cham (2014). https://doi.org/10.1007/978-3-319-10590-1_53
10. Lei, X., Pan, H., Huang, X.: A dilated CNN model for image classification. IEEE Access **7**, 124087–124095 (2019). https://doi.org/10.1109/ACCESS.2019.2927169
11. Zhang, S., Wang, X., Lei, Z., Li, S.Z.: Faceboxes: a CPU real-time and accurate unconstrained face detector. Neurocomputing **364**, 297–309 (2019). ISSN 0925-2312
12. Putro, M.D., Jo, K.-H.: Fast face-CPU: a real-time fast face detector on CPU using deep learning. In: 2020 IEEE 29th International Symposium on Industrial Electronics (ISIE), Delft, Netherlands, pp. 55–60 (2020). https://doi.org/10.1109/ISIE45063.2020.9152400
13. He, Y., Xu, D., Wu, L., Jian, M., Xiang, S., Pan, C.: LFFD: A Light and Fast Face Detector for Edge Devices (2019). arXiv:1904.10633
14. Szegedy, C., et al.: Going deeper with convolutions. In: Proceedings of the IEEE Conference on Computer Vision and Pattern Recognition, pp. 1–9 (Jun 2015). https://doi.org/10.1109/CVPR.2015.7298594

15. He, K., Zhang, X., Ren, S., Sun, J.: Deep residual learning for image recognition. In: 2016 IEEE Conference on Computer Vision and Pattern Recognition (CVPR), Las Vegas, NV, USA, pp. 770–778 (2016). https://doi.org/10.1109/CVPR.2016.90
16. Sandler, M., Howard, A., Zhu, M., Zhmoginov, A., Chen, L.: MobileNetV2: inverted residuals and linear bottlenecks. In: 2018 IEEE/CVF Conference on Computer Vision and Pattern Recognition, Salt Lake City, UT, USA, pp. 4510–4520 (2018). https://doi.org/10.1109/CVPR.2018.00474
17. Zhang, X., Zhou, X., Lin, M., Sun, J.: ShuffleNet: an extremely efficient convolutional neural network for mobile devices. In: 2018 IEEE/CVF Conference on Computer Vision and Pattern Recognition, Salt Lake City, UT, USA, pp. 6848–6856 (2018). https://doi.org/10.1109/CVPR.2018.00716
18. Süzen, A.A., Duman, B., Şen, B.: Benchmark analysis of Jetson TX2, Jetson Nano and Raspberry PI using deep-CNN. In: 2020 International Congress on Human-Computer Interaction, Optimization and Robotic Applications (HORA), Ankara, Turkey, pp. 1–5 (2020). https://doi.org/10.1109/HORA49412.2020.9152915
19. Hu, J., Shen, L., Sun, G.: Squeeze-and-excitation networks. In: 2018 IEEE/CVF Conference on Computer Vision and Pattern Recognition, Salt Lake City, UT, USA, pp. 7132–7141 (2018). https://doi.org/10.1109/CVPR.2018.00745
20. Ferrari, V., Hebert, M., Sminchisescu, C., Weiss, Y. (eds.): ECCV 2018. LNCS, vol. 11210. Springer, Cham (2018). https://doi.org/10.1007/978-3-030-01231-1

Non-tactile Thumb Tip Measurement System for Encouraging Rehabilitation After Surgery

Erika Aoki, Tadashi Matsuo[✉], and Nobutaka Shimada

Ritsumeikan University, 1-1-1 Nojihigashi, Kusatsu, Shiga 525-8577, Japan
t-matsuo@fc.ritsumei.ac.jp, nshimada@is.ritsumei.ac.jp

Abstract. The thumb is the only finger that has a high degree of freedom of articulation and can face the other four fingers, and functional training is significant after injury surgery. We propose a system to measure and visualize the range of motion of the thumb during rehabilitation in a non-contact manner using a depth sensor and a deep neural model.

Keywords: Finger tip measurement · Rehabilitation · Non-tactile

1 Introduction

When you injure your thumb, you need to do rehabilitation to recover your physical functions after treatment. When rehabilitating the thumb, it is important for the therapist to understand the degree of recovery in order to rescheduling the therapy plan. It is more important for the patient to keep motivation for rehabilitation. The thumb is the only finger that has a high degree of freedom of articulation and can face the other four fingers, and functional training is significant after injury surgery. The movable range of the thumb joint is one of difficult to measure because the thumb has complicated structure. Currently, the movable range of the thumb joint is measured by directly fitting an protractor to the joint. Figure 1 shows an example of measurement using a goniometer, which is a kind of protractors [1]. It is measured numerically but quite just roughly and difficult to visually understand how much it has become possible to move.

In order to solve the problem described above, we propose a system that measures the thumb without contact, and an application that visualizes the range of motion of the thumb tip when rehabilitating the thumb. In order to measure the thumb without contact, we propose a measurement system with a depth sensor and an estimating method thumb tip position using a deep neural model. Finally we developed a tablet application to visualize the estimated results (see Fig. 2) by creating CG trajectory mesh around the observed thumb tips.

2 Related Studies

According to Ishida [2], the movable range of joints is often measured by using an measurement method based on the experience of therapists, which is based on the "Joint

© Springer Nature Switzerland AG 2021
D.-S. Huang et al. (Eds.): ICIC 2021, LNCS 12836, pp. 842–852, 2021.
https://doi.org/10.1007/978-3-030-84522-3_68

Range of Motion Indication and Measurement Method (1995)" issued by the Japanese Orthopedic Association and the Japanese Society of Rehabilitation Medicine. He stated that the measurement method was built just based on the experience of therapists. He also stated that the inter-examiner measurement error of the movable range of joints for hip flexion and straight leg rising (SLR) had a maximum value of 20 deg (minimum 5 deg) for hip flexion and a maximum value of 30 deg (minimum 5 deg) for SLR. This means that the accuracy of measured values strongly depends on the skills of the therapist and the measured values may vary.

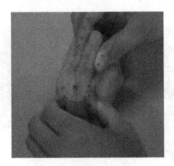   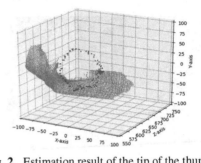

Fig. 1. Example of measurement using a goniometer [1]   **Fig. 2.** Estimation result of the tip of the thumb coordinates

As a previous study, Kaneda et al. [3] proposed a method in which a therapist measures the movable range of joints using an angle gauge on a computer by taking pictures using a 3-D scanner and using the 3-D data obtained from the images. By capturing the images on a computer, making the 3-D data translucent, and moving it around to make it easier to measure, he stated that the 3-D scanner was able to reduce the measurement error between examiners compared to the traditional method of directly fitting the angle gauge to the patient. How- ever, the problem is that the measurement time is longer due of the 3-D scanner. Miyake et al. [4] proposed a system that uses depth images of the entire hand obtained from a depth sensor to estimate thumb tip coordinates through a deep neural model. They used ResNet to estimate the thumb tip coordinates from the depth image of the hand. They also tried to visualize them in 3-D in computer display. However, it was difficult for therapists in the field to visually understand and handle thumb tip estimation coordinates because they were displayed as points. Based on this study, our solutions employing computer vision and graphics will be explained in the following sections.

3 Image Acquisition Using Depth Sensor

We employ Intel RealSense Depth Camera D435 (Realsense) as the depth sensor, which is developed and marketed by Intel Corporation. It has a depth sensor, RGB sensor, and IR projector to acquire RGB images and depth images. Realsense acquires depth information using active IR stereo method. The active IR stereo method is a method of measuring depth by projecting infrared light from an IR projector and capturing the image with a camera, and measuring the difference between the projected pattern and the captured pattern. The depth image is captured as an unsigned 16 [bit] integer, and each pixel contains depth information as an integer value from 20 to 10000 in millimeters. In order to capture images under the same conditions as in the previous study [4], the distance between the hand and Realsense was set to 60 [cm], and the left hand was placed on a table with two boxes of 10 [cm] height fixed so that the hand would appear in the center of the depth image. The booklet was placed at a distance of 80 [cm] from the Realsense in order to reduce the noise around the hand during the capture. Figure 3 shows the scene during the capture.

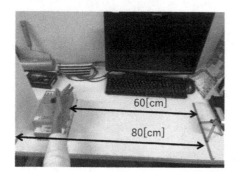 

Fig. 3. Image capturing setup

4 Thumb Tip Position Measurement by DNN Model

4.1 Extraction of Hand Areas

The depth image taken by Realsense is shown in Fig. 4. Originally, the depth information of the depth image is recorded in 16 [bits], but it is converted to 8 [bits] and visualized. Since the depth image contains not only the hand but also a table and background objects, we first perform masking on the depth image to extract the hand area. Next, in order to remove background objects and regions other than the hand, subtract 550 from the depth value to extract only the area whose depth is between 550 [mm] and 705 [mm] in the depth image, considering the capturing conditions and the movable range of the thumb, set the depth of 550 [mm] to 0, and fill the background area of the hand with 0. The size of depth image captured by Realsense is 640 × 480 [px], but it needs to be 227 × 227 [px] for the input to ResNet model, so we extracts only the hand region based on the center of the captured depth image. Figure 5 shows the depth image created by extracting the hand region through these operations.

4.2 Assigning Teacher Signals

We will use the depth image created in Sect. 4.1 and the X and Y coordinates of the tip of the thumb for machine learning, so we need to specify the correct coordinates of the tip of the thumb (hereafter referred to as the teacher signal). Click on the thumb tip

Fig. 4. Part of the depth image taken by Realsense

Fig. 5. Depth image with the stand portion removed

Fig. 6. Enlarged view of the thumb area of the screen where the teacher signal is being given

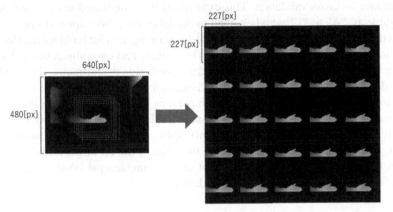

Fig. 7. Creating a dataset from depth images

in the depth image to add a teacher signal. In order to make it easier to understand the position of the tip of the finger and to click on it, the depth image of 640 × 480 [px] is doubled to 1280 × 960 [px] and visualized in 24 [bit]. In order to correct the depth value to the Z coordinate on the tip of the thumb even if the teacher signal is shifted, the median of the depth values in the range of 5 × 5 [px] centered on the X, Y coordinates clicked as the teacher signal is used as the true depth value. Figure 6 shows a magnified image of the thumb area. The clicked area is shown as a yellow cross, and a 5 × 5 [px] area around the clicked X and Y coordinates is surrounded by a red frame as a teacher signal.

4.3 Creating Data Sets

In order to improve the performance of the estimation, a large number of images are required during the learning process. Therefore, in this study, in order to prepare a large number of depth images efficiently, multiple depth images are cut out from a single depth image to create a data set. As shown in Fig. 7, from one depth image, we shift the image by 4 [px] vertically and horizontally based on the center of the image, and shift it by 2 steps vertically and horizontally. As shown in Fig. 7, 25 depth images of 227 × 227 [px] are created from one depth image by performing the above data augmentation.

After that, the dataset is divided into the created depth images in chronological order, one for validation data and two for train data, and then 1/3 of the total data is used for validation and 2/3 for training.

5   Experiments of Thumb Tip Measurement

5.1   Dataset and Prediction Model

In this section, we examine whether we can estimate the thumb tip coordinates even for unknown hand shapes. Since there are only a few subjects, we will evaluate the performance by cross-validation. This experiment was conducted with the cooperation of 10 subjects, "A" to "J", who belonged to our laboratory. We captured image data for each of them. The data of 9 subjects were used as training data for building the ResNet [5] prediction model with MSE loss function. Those of the rest one subject was not used for training but as test data to evaluate the model's prediction performance. The trials were repeated in the way of LOOCV. We also visualized the estimated thumb tip trajectory. Their finger actions were captured for each person: 1) move your thumb in a big circle (large movement), 2) move your thumb in small circles (small movements), and 3) move your thumb freely for 10 s (free movement). Table 1 shows the specifications of each subject: [size of the palm] the length from the wrist to the tip of the middle finger, and [thumb length] the length from the end of the first metacarpal bone to the tip of the thumb. There are 114750 instances in the dataset.

5.2   Estimation Results

Table 2 shows the square root of the mean square error when one subject is used as the test data and the remaining subjects are used as the training data. The average of the square root of the mean square error is 3.72 [px] (out of 227×227 [px]) and 5.98 [mm].

Figure 8 shows the relative frequency of errors for each movement of subject F. An example of the estimation results for the data of subject F is shown in Fig. 9. The white crosses in each figure represent the teacher signal and the pink crosses represent the estimated coordinates. The number in the upper left corner of each figure represents the estimated coordinates [x,y]. These figures are selected when the position of the thumb is different.

5.3   Visualization of Thumb Tip Trajectory

Figures 10, 11, 12 show the visualization of the movable range of thumb tip for the data of the three movements of subject F based on the results of Sect. 5.2 by creating a mesh. The blue points in the figure are hands, the orange points are thumb tip coordinates, and the gray mesh is a mesh that surrounds the points of the thumb tip.

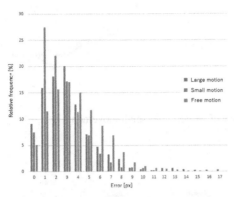

Fig. 8. Relative frequency

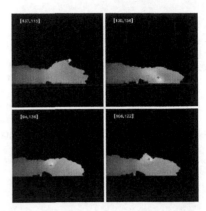

Fig. 9. Thumb tip estimation and GT for subject F

Table 1. Subject data

Subj.	Gender	Palm length (cm)	Thumb length (cm)
A	Male	18	11
B	Male	17	11
C	Female	16	9
D	Male	20	10
E	Male	19	9
F	Male	18	9
G	Male	18	8.5
H	Female	18	8
I	Female	18	10
J	Female	18	8

Table 2. SD of error [px(mm)]

Test subj.	Movement		
	Large	Small	Free
A	5.05(8.07)	5.20(8.26)	3.48(5.56)
B	2.56(4.14)	1.39(2.32)	2.21(3.43)
C	2.41(3.93)	2.25(3.68)	3.47(5.57)
D	4.36(7.04)	1.99(3.30)	3.08(4.97)
E	3.30(5.04)	3.13(4.85)	4.24(6.39)
F	6.01(8.85)	3.63(5.70)	5.08(7.97)
G	3.95(6.08)	2.13(3.42)	2.21(3.47)
H	1.84(2.92)	2.10(3.41)	1.95(3.15)
I	9.58(16.13)	9.36(16.15)	7.71(13.07)
J	2.42(3.85)	1.93(3.21)	3.44(5.54)

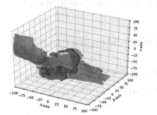
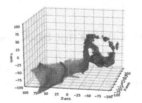
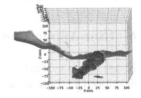

Fig. 10. Thumb tip trajectory of subject F in large movements

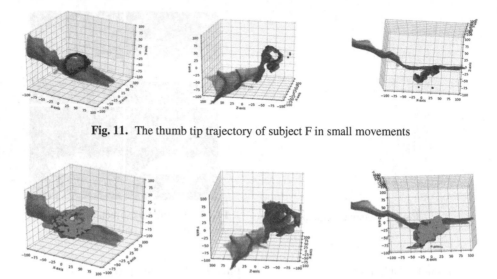

Fig. 11. The thumb tip trajectory of subject F in small movements

Fig. 12. Thumb tip trajectory of subject F in free movement

5.4 Evaluation

From Table 2, we can see that we can estimate the thumb tip coordinates, although there is small error. The reason why SD is larger for subject I is because subject I had a depth of about 3 cm deeper than the other subjects (he placed his hand about 3 cm away from the camera). In addition, Table 2 and Fig. 8 show that the larger you move your thumb, the larger the error becomes. The actual maximal error was between 32 [px] and 33 [px], here we truncated the long tail in Fig. 8, because the ResNet detector missed the thumb tip positions for the case of large movements. Although some outliers are removed, Fig. 10 shows that some outliers are also in the mesh. In some cases, the outliers are slightly different from the actual point cloud of the estimated thumb coordinates. From the above, we found that we can get some results for data of people with unknown hand and thumb shapes that are not included in the training images. If we use more thumb data and larger movements for training, we can expect to get more accurate results for unknown data. Here the movable range of thumb tip is displayed as a mesh polygon and we have some outliers. Its removal process will be explained in the following Sect. 6.2.

6 User Application for Capturing and Visualizing Measurements

Based on the model for estimating 3-D thumb tip position from depth images, we next developed an application for users to easily acquire the depth images of palm and fingers.

6.1 Depth Image Acquisition

We implemented image capturing and playback functions in the application on a tablet PC using Unity and Realsense, using the wrapper provided to enable Realsense in Unity.

It acquires a 640×480 [px], 16-bit depth image at 30 fps from Realsense, converts it to Texture2D format, and then displays it on the screen of the app. In addition, surround the 227×227 [px] area in the center of the depth image with a white frame to show the area used for machine learning. It also displays the current status (capturing or playback) in the upper left corner. The capture depth images are stored as the PNG format. To play back the captured images, select the folder that contains the PNG images to be played back on the file dialog, and then display one image per frame on the screen.

Capture Function. When you press the "Start Recording" button in the lower left corner and select the destination folder until you press the Stop Recording button in the lower left corner, a folder with the date as the folder name will be created and the PNG image will be saved in it.

Playback Function. After pressing the Play button in the lower right corner and selecting the folder containing the PNG image to be played, the PNG image will be played repeatedly until the Stop Playing button in the lower right corner is pressed.

6.2 Outlier Removal for Visualizing Thumb Tip Trajectory

For visualization of the 3-D estimated thumb tip trajectory, we convert the point cloud data into polygon mesh data for rapid rendering so that users can operate (move, rotate and scale) the displayed trajectory or information on the tablet. First we remove outlier points in the trajectory data. In order to remove outliers, we used the Local Outlier Factor (LOF) algorithm [6] using Python's scikit-learn library. LOF is an outlier detection algorithm that focuses on the density of data in space and uses the local density to describe how dense a point is with its k nearest neighbors.

6.3 Creating Mesh Visualizing Thumb Tip Trajectory

We converted the outlier-deleted point cloud created in Sect. 6.2 into voxel data. Since the point cloud was relatively sparse, we created a voxel of $9 \times 9 \times 9$ [mm] and then voted each cloud point into the voxels. Then we converted the voxel data into polygon mesh that surrounds the point group that is the movable area of the thumb tip by using marching cube method [7] as shown in Figs. 10, 11, 12. The blue dot represents the hand, the orange dot represents the tip of the thumb, and the gray object represents the mesh surrounding the point group of the tip of the thumb.

6.4 Application Displaying Thumb Tip Trajectory

Data Files. Three files are fed into the app: a text file that contains the estimated coordinates of the thumb tip, a text file that contains the contour coordinates of the hand, and an object file that contains the movable range of the thumb tip. The contour data of the hand is created by extracting the contour of the hand from a 227×227[px] 16-bit depth image of the hand using the following procedure in Python script.

1. Convert to 8-bit depth image
2. Convert to RGB image
3. Perform threshold processing
4. Perform contour extraction
5. Perform contour approximation so that it is not too heavy when displayed in the application.
6. To unify the Z-coordinates for better appearance, the median of the non-zero depth values in 7×7[px] centered on the extracted vertex is used as the depth value of the vertex of the contour, and the average of the depth values of all the vertices is taken as the Z-coordinate of all the vertices.
7. Perform coordinate system transformation to convert from image coordinate system to camera coordinate system.
8. Subtract 650 from Z and save it in a text file because you want it to appear near the origin in the app.

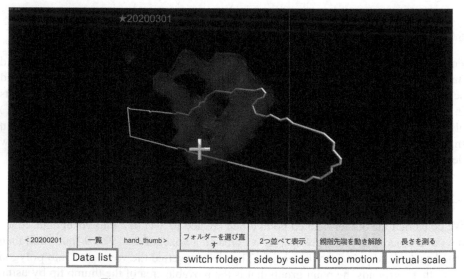

Fig. 13. Display the movement of the tip of the thumb

Displaying Thumb Tip Trajectory. The thumb tip trajectory and the con- tour of the hand are displayed on the tablet. The outline of the hand is depicted by chain of cylinders. You can freely translate, rotate and scale the displayed trajectory by drag and pinch on the tablet panel. You can also confirm the name of the currently open file at the top of the screen. In addition, the name of the currently open file is displayed at the top of the screen. If you touch the fifth button from the left at the bottom, you can change the viewing mode. Each time you touch the button, you can switch between three viewing modes: one instance view, side by side view for two instances, and overlay view stacking two instances for comparison to each other. Figure 14 shows how this works.

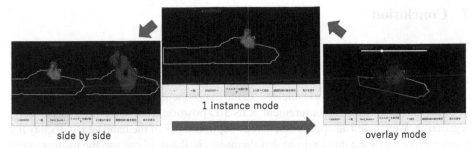

1 instance mode

side by side   overlay mode

Fig. 14. Multiple display modes

Displaying Thumb Tip Motion. When you click the "Show thumb tip movement" button (the sixth button from the left), the app displays the animation of the thumb tip motion in time order to describe the movement. Figure 13 shows how it works.

Measuring Any Length Using a Virtual Scale. The application has "virtual scale" function to measure length between any two points you want to measure. After pressing the "Virtual Scale" button, you can place the virtual Scale by swiping on the tablet screen, by stretching the scale from the starting point to the end of swiping point. When the user releases the scale, it will continue to be installed. Since the numerical length in the virtual space is automatically calculated and displayed along the scale, the measured length can be quantitatively recorded. When the user swipes it again, a new scale can be placed, allowing multiple scales to be installed. Figure 15 shows a scene in which the virtual scale is installed and measured lengths are displayed.

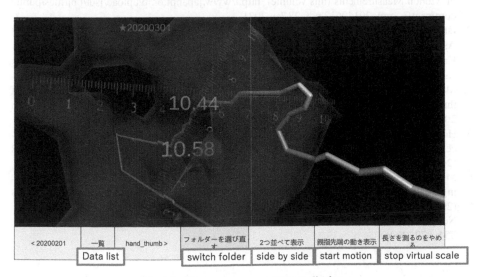

Fig. 15. Dimension measurement display

7 Conclusion

This study improved UI and UX of measuring the movable range of thumb tip by using depth images taken by depth sensors, estimating the position of the thumb tip by the DNN model, visualizing the movable range of the thumb tip in 3D space and enabling it operable on the touch screen of tablet PC. In the experiment, we showed the effectiveness of finger tip detection and measurement of its 3-D position from depth image by ResNet regression. We created an application to take depth images of the hand and to display the range of motion of the thumb tip so that therapists in the field can use the measurement system easily.

One of the future tasks is to add more training images for more accuracy. The training images created in this study were taken from the movements of nine subjects, but in order to use them in actual rehabilitation rooms, more subjects and various thumb movements in the training images are needed in order to correspond to hands with various shapes and conditions. Second one is building more sophisticated application interface. The tablet PC can be easily and intuitively operated for visualization of 3-D information, but UI and UX should be developed by listening the user's feelings and opinions. After releasing the beta version of the application, the user surveillance is the next important task.

References

1. Japan Physical Therapists Association Call for Public Comment on the Society's Version of MMT and the Society's Version of ROM Assessment Guidelines Document 3: Joint Range of Motion Measurements (this volume). http://www.japanpt.or.jp/upload/jspt/obj/files/public comment/3_rom_20140612.pdf. 28 Jan 2021
2. Ishida, K.: Guidelines for assessing range of motion of joints in the Japanese Physical Therapist Association version. Phys. Ther. **42**(8), 763–764 (2015)
3. Kaneda, et al.: A new method for measuring range of motion using a non-contact 3D scanner. J. Jpn. Soc. Phys. Ther. **36**(4), 481–487 (2017)
4. Miyake, Y., Matsuo, T., Shimada, N.: Measurement and visualization of range of motion of thumb tip for rehabilitation. In: Proceedings of the Robotics and Mechatronics Symposium 2020, 2P1-E04 (2020). (in Japanese)
5. He, K., Zhang, X., Ren, S., Sun, J.: Deep residual learning for image recognition. In: Proceedings of IEEE Conference on Computer Vision and Pattern Recognition, pp. 770–778 (2016)
6. Markus, B., et al.: LOF: identifying density-based local outliers0 In: SIGMOD 2000: Proceedings of the 2000 ACM SIGMOD International Conference on Management of Data, pp. 93–104 (2000)
7. Newman, T.S., Yi, H.: A survey of the marching cubes algorithm. Comput. Graph. **30**(5), 854–879 (2006)

Real-Time Prediction of Future 3D Pose of Person Using RGB-D Camera for Personalized Services

Yasushi Mae[✉], Akihisa Nagata, Kaori Tsunoda, Tomokazu Takahashi,
Masato Suzuki, Yasuhiko Arai, and Seiji Aoyagi

Kansai University, Suita, Osaka, Japan
mae@kansai-u.ac.jp

Abstract. This paper describes a method for prediction of 3D human pose using an RGB-D camera in real-time for personalized robot services. In input frame from RGB-D camera, 2D pose as a skeleton represented by 2D keypoints on an RGB image is estimated by OpenPose: an image-based human pose estimation technique. 3D pose represented by a set of 3D keypoints is estimated from the depth at the keypoints of 2D keypoints. Daily actions of a person, which are represented by a sequence of 3D keypoints, are measured by an RGB-D camera fixed to a daily environment. After the sequences of measured 3D keypoints are learned by a motion predictor of RNN, the future 3D keypoints are predicted by the RNN from the sequential input of 3D keypoints, which are measured in real-time by RGB-D camera. Experimental results show the predicted 3D pose and errors for standing from and siting on a chair in a room of daily environment.

Keywords: Prediction of future 3D human pose · 3D human motion estimation · RGB-D camera

1 Introduction

It is expected that service robots support and help human daily activities in the environment of everyday life. For supporting human daily activity, personalization to the user becomes important key technique in robot services. For realizing personalized robot service of supporting daily activity, it will be useful that home service robots and environmental robot systems watch a person's actions and know the required help and the best time to help for the person by predicting the personal motion.

There have been proposed many not only 2D human pose estimation methods [1–3] but also 3D human pose estimation methods [4–9]. Most of the methods learn relation between 2D human pose and 3D human pose data measured by 3D mocap. There are also many studies for prediction future human pose [10–14]. These methods also predict general motions of people by learning many persons' motion in database. In general, human pose data sets include many people, these methods will be applied to general use. However, human motions of different persons are slightly different even in the same categorized actions such as "standing" and "sitting", since each person has personal motion

© Springer Nature Switzerland AG 2021
D.-S. Huang et al. (Eds.): ICIC 2021, LNCS 12836, pp. 853–861, 2021.
https://doi.org/10.1007/978-3-030-84522-3_69

habit and features in an action. For prediction of personal motion for the personalized robot service, it is better to create and use personal motion database for learning personal motion by Deep Neural Network.

Recently, RGB-D camera become popular to use. It can be useful as a visual sensing device for robot vision and computer vision. RGB-D camera can obtain not only RGB image but also depth information of the scene in real-time like a usual camera. Advantage of using RGB-D camera is that depth information can be directly measured by RGB-D camera without using and learning general human motion database measured by 3D mocap. In [15], the methods predict a person's body future position after 0.5 s in real-time using RGB-D camera, Kinect v2. The method predicts the body position. However, it does not predict person's arm and leg motions. For personalized robot service in human daily activity, it will be useful to predict future 3D pose and motion including personal motion habit and features in actions so that service robot can start to move earlier to help the person at the best time.

In this paper, we propose a method for prediction of human motion by RGB-D camera. Human 2D pose of a person is estimated as a 2D skeleton from RGB image by OpenPose [2]: an image-based pose estimation technique. The depth information corresponding to the 2D keypoints of the skeleton is obtained from the depth measured by RGB-D camera. Then the 3D skeleton of the person is estimated. The sequence of the measured 3D skeleton of a person represents the motion of the person. The motion of the person is learned using RNN (Recurrent Neural Network), the future 3D pose represented by the 3D skeleton is predicted from the inputted sequence of the 3D skeleton of the person. The error of the prediction is evaluated based on the experimental results.

2   Proposed Method

2.1   Measurement of 3D Pose of Human Using RGB-D Camera

An RGB image and a depth image are captured at the same time by an RGB-D camera. First, 2D human pose of a person is estimated from RGB image by OpenPose [2]. OpenPose detects 25 keypoints of the 2D skeleton of a person estimated on the RGB image. Example of an estimated 2D pose for a scene "a person is sitting on a chair" is shown in Fig. 1. As an RGB-D camera for measuring 3D human pose, we use Intel RealSense L515.

The captured RGB image and depth image are aligned for corresponding the pixel position on the image. Then the depth data are obtained by corresponding aligned pixels of the 2D keypoints of skeleton and depth image. Then a 3D pose as a set of 25 3D keypoints is obtained by an RGB-D camera. Example of an estimated 3D pose for a scene "a person is sitting on a chair" is shown in Fig. 1.

Let a 3D coordinates of keypoint at time t_n is denoted by $P(t_n)$. A 3D pose $Pose(t_n)$ is represented by a set of 25 keypoints of $P(t_n)$ for representing 3D skeleton at time t_n. 3D motion is represented by a time sequence of $Pose(t_n)$. If the current measurement depth data at the 2D keypoint is not available, in this case the depth value becomes zero in RealSense L515, the current depth data is extrapolated from the recent past data. This case may happen when the reliability of the measured depth data becomes low. If the depth data is not available at the pixel corresponding to the 2D keypoint on the RGB image, the depth data at time t_n is extrapolated from the 3D coordinates of the keypoint of the previous frame at time t_{n-1}.

$$P(t_n) = P(t_{n-1}) + \frac{P(t_{n-1}) - P(t_{n-2})}{t_{n-1} - t_{n-2}}(t_n - t_{n-1}) \qquad (1)$$

Extrapolation of keypoint is illustrated in Fig. 2. In Fig. 2, extrapolated keypoint at time t_n is represented by green color.

The frame rate of the 3D pose estimation is 10 fps.

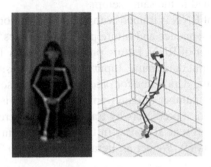

Fig. 1. Example of an estimated 2D and 3D pose for a scene "sitting on a chair".

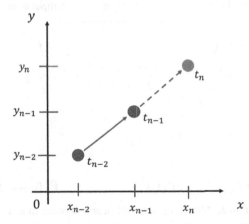

Fig. 2. Extrapolation of keypoint from the recent past keypoints. (Color figure online)

2.2   Training of 3D Motion of Human

Many 3D pose can be obtained by observing human daily activity after setting the RGB-D camera in the daily environment. We use Recurrent Neural Network (RNN) for training and prediction of 3D pose. We use Gated Recurrent Unit (GRU) for the hidden layers. The 3D pose data at current and previous m frames, totally $(m + 1)$ frames, are input, and network outputs 3D pose after T frames. Figure 3 shows a model of the RNN. 3D pose is represented by 25 keypoints. Then the number of nodes for inputting a frame data becomes 75, since a keypoint has three data (x, y, z) in a 3D space. The number of nodes of output layer is 75 also. The number of nodes of the middle layers are 4, 32, 64 from the input to output layer. In the paper, 3D pose $P(t_n), \ldots, P(t_{n+m})$ are sequentially used for input data frames to RNN, and 3D pose $P(t_{n+m+T})$ after T data frames are used as teaching data for the output layer. Mean squared error (MSE) is used as error function for evaluation of error in the output layer. Adm is used as optimizer.

In our approach, since our system is used for personal use, the RGB-D camera is set at a position in a room of everyday environment, and the measuring actions for training and prediction is performed in the same setup.

A motion predictor of RNN is trained with the following conditions. The number of input data frames m of a sequence and future prediction frames T are the conditions for training a motion predictor of RNN. Future prediction frames T represents the number of frames after the last input data frame. Input data are successive $(m + 1)$ data frames, and corresponding teaching data frame for prediction is 3D pose after T frames from the last input data frame. The number of data frames of an action is denoted by NF.

We obtain several motion predictors by training with different conditions m, T, and coordinate systems to represent 3D pose. In the experiment in Sect. 3, the number of frames of an action NF is 60. We set an iteration times N to 200 for training motion predictors of RNN.

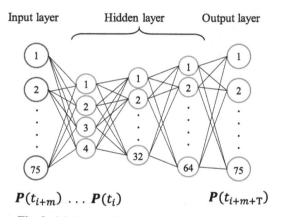

Fig. 3. Motion predictor of neural network model

2.3   Prediction of 3D Pose of Human

In future prediction of 3D pose, time sequence of a set of 25 3D keypoints of $(m + 1)$ data $Pose(t_i), \ldots, Pose(t_{i+m})$ measured in real-time by RGB-D camera are input to the input layer successively, then the network outputs 3D pose $Pose(t_{i+m+T})$ represented by a set of 25 3D keypoints at future time t_{i+m+T} after T frames from the current time t_{i+m}.

3 Experiments

3.1 Training and Prediction in Experiment

For preparing action data set for training RNN and evaluation of prediction test, a person's actions are measured and recorded. We use Intel RealSense L515 as an RGB-D camera for measuring 3D pose of a person. A device is set around 1.2 m height from the floor. A person performs 41 daily actions in a room in front of an RGB-D camera. The 41 daily actions include 18 different kinds of daily actions often occurred in everyday activity such as "standing from a chair", "sitting on a chair", "picking up a box", and "calling with mobile phone" of a person in a room. In the action data set, it includes 7 actions of "standing from a chair" and 7 actions of "sitting on a chair". We use these two actions, "standing from a chair" and "sitting on a chair", for evaluation of prediction.

For evaluation of prediction error, we obtain 12 motion predictors by training RNN with 12 combinations of input and output conditions. In the experiment, we set a number of input frames m to 3 and 5. As a future prediction, we set a number of future prediction frames T to 3, 5, 10. As coordinate systems, camera-centered coordinate system C and person-centered coordinate system P are both used for training and prediction. In the person-centered coordinate system, the origin is the coordinate of the 3D keypoint corresponding to the waist (Mid Hip) of the person. In the camera-coordinate system, the motion predictor learns the person's motion depending on the position of the person. Since the frequent actions may occur in the same position in a room, it will be suitable for the motion prediction for the actions depending on the position. On the other hand, in the person-centered coordinate system, the error of the estimated position of the person is canceled. It will be suitable for predicting motion independently of the position in a room.

Table 1 shows the list of the 12 motion predictors. In this paper, a motion predictor trained with different conditions is named as (Coordinate system) (number of input frames) (T = (future prediction frames)). For example, in a motion predictor P_3(T = 5), P represents the person-centered coordinate system, 3 represents the number of input data frames, and 5 represents the number of the future prediction frame after the last input frame.

3.2 Experimental Results

In prediction, we use the same RGB-D camera under the same condition in the training. We evaluate the error of prediction for "standing from a chair" and "sitting on a chair". The other measured action data are used for only training in the evaluation.

We predict a motion of the person for two actions of "standing from a chair" and "sitting on a chair". As examples, 6 predicted skeletons and measured skeletons are shown in Fig. 4 and Fig. 5 for the two actions, respectively. The skeletons are drawn in camera-centered coordinate system. The predicted skeleton is represented in green color. The measured skeleton is represented in orange color. These prediction results are obtained by the motion predictor P_3(T = 5). The future prediction time is about 0.5 s. In this case, a future 3D pose after 5 frames (about 0.5 s) is predicted from successive three input data frame. Next future 3D pose is predicted by inputting the next data frame

Table 1. Parameters of motion predictors

Motion predictor	Coordinate systems camera/person	Number of input frames $m + 1$	Future frames (times) T (s)	Future times s
C_3(T = 3)	Camera	3	3 (0.3 ± 0.003)	0.3 ± 0.003
C_3(T = 5)	Camera	3	5 (0.5 ± 0.005)	0.5 ± 0.005
C_3(T = 10)	Camera	3	10 (1.0 ± 0.010)	1.0 ± 0.010
C_5(T = 3)	Camera	5	3 (0.3 ± 0.003)	0.3 ± 0.003
C_5(T = 5)	Camera	5	5 (0.5 ± 0.005)	0.5 ± 0.005
C_5(T = 10)	Camera	5	10 (1.0 ± 0.010)	1.0 ± 0.010
P_3(T = 3)	Person	3	3 (0.3 ± 0.003)	0.3 ± 0.003
P_3(T = 5)	Person	3	5 (0.5 ± 0.005)	0.5 ± 0.005
P_3(T = 10)	Person	3	10 (1.0 ± 0.010)	1.0 ± 0.010
P_5(T = 3)	Person	5	3 (0.3 ± 0.003)	0.3 ± 0.003
P_5(T = 5)	Person	5	5 (0.5 ± 0.005)	0.5 ± 0.005
P_5(T = 10)	Person	5	10 (1.0 ± 0.010)	1.0 ± 0.010

successively. Then, future 3D pose is predicted at each frame successively. The sequence of predicted 3D pose shows the motion of the action. In Fig. 4 and Fig. 5, 6 skeletons are selected to roughly show the motion of the action. From the figures, we can see the future predicted 2D pose almost match to the measured 2D pose. We also find that the predicted error becomes large in the middle of the frames especially in the action "sitting on a chair".

In discussion about personal motion habit and features in an action, we found the prediction error becomes large in the case the person has similar motions in different daily actions. This may be caused by the personal motion habit. For example, in the middle of Fig. 5, the top of the body in prediction, green colored skeleton, is shifted to left lower direction. It means the top of the body directs to the right lower in the person-centered coordinate system. In the action "picking up a box" of the action data set, the person picks up the box in front of the right foot. In this action, the top of the body of the person directs to right lower in the person-centered coordinate system. On the other hand, we examined the action data set, and found the person has a personal motion habit that the top of the body directs to right lower at the beginning of the action "sitting on a chair". Then we found the motions are similar to each other at the beginning of the action "picking up a box" and "sitting on a chair". This may be the reason that the prediction error becomes large in the middle of the sitting action of Fig. 5. This kind of error may happen more in prediction in the person-centered coordinate system. This kind of error will decrease if more action data are used for training. This will also show the possibility that the motion predictor can predict the person's motion including personal motion habit by learning the many motion data in many actions. Furthermore, we can see that, even when the predicted pose has large error at the beginning of the action like this, the predicted pose gradually converged to the sitting pose by successive input of the measured data frames of sitting action.

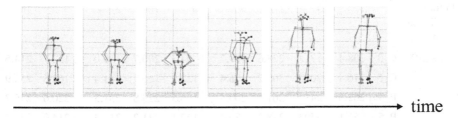

→ time

Fig. 4. Example of predicted pose by P_3(T = 5) for standing (Color figure online)

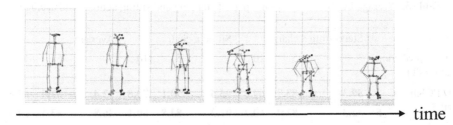

→ time

Fig. 5. Example of predicted pose by P_3(T = 5) for sitting (Color figure online)

As evaluation indices, we use MPJPE and PCKh@0.5. MPJPE is a mean distance between the predicted 3D keypoint and the measured corresponding 3D keypoint for the whole 25 keypoints of the body skeleton. PCKh@0.5 is a ratio that predicted keypoint is within the radius of the sphere which approximates the head of the person. In the experiment the radius is 0.110 m. We compare these indices for evaluation.

We show the evaluation of prediction error for 7 pair of actions "standing from a chair" and "sitting on a chair". When we evaluate the prediction error for a pair of "standing from a chair" and "sitting on a chair", the pair of action data are removed from the training action data set. Then the remaining 39 action data are used for training, and the pair of "standing from a chair" and "sitting on a chair" is used for test.

For evaluation, first we calculate the moving average of MPJPE for the predicted actions "standing from a chair" and "sitting on a chair" in the time length of the action (the number of data frames NF). Next we calculate the average MPJPE of the moving average for 7 actions of "standing from a chair" and "sitting on a chair" each. We show the average MPJPE of 7 actions of in Table 2. In Table 2, if the average MPJPE is under 110 mm, it is represented in bold. We can find a tendency the prediction error becomes large in the far future prediction from Table 2. Table 3 shows the average PCKh@0.5 in 2D and 3D. The average PCKh@0.5 is the average of 7 moving average of PCKh@0.5. In Table 3, if the average PCKh@0.5 is over 50%, it is represented in bold. In 2D image space, all the cases are over 50%. The average PCKh@0.5 in 3D is relatively lower than that in 2D. This shows the prediction error is large in depth. We can find a tendency the prediction error becomes large roughly in the far future prediction also from Table 3.

Table 2. Average MPJPE [mm] for 7 pair of standing and sitting actions, and average

Action		Stand from a chair			Sit on a chair			Average		
Future prediction frames (T)		3	5	10	3	5	10	3	5	10
Ours (3D)	C_3	208.4	171.0	228.6	**109.7**	**94.3**	200.7	159.0	132.7	214.6
	C_5	176.4	202.3	303.2	**106.4**	122.9	166.7	141.4	162.6	234.9
	P_3	262.6	243.5	261.6	200.3	196.5	229.8	231.5	220.0	245.7
	P_5	256.1	250.9	248.2	176.4	177.1	241.2	216.3	214.0	244.7

Table 3. Average PCKh@0.5 [%] for 7 pair of standing and sitting actions, and average

Action		Stand from a chair			Sit on a chair			Average		
Future prediction frames (T)		3	5	10	3	5	10	3	5	10
2D PCKh @0.5	C_3	**89.3**	**83.9**	**64.8**	**95.4**	**90.1**	**70.8**	**92.4**	**87.0**	**67.8**
	C_5	**82.3**	**82.9**	**60.2**	**91.3**	**83.8**	**69.1**	**86.8**	**83.4**	**64.7**
	P_3	**92.0**	**93.7**	**70.5**	**90.2**	**89.6**	**64.7**	**91.1**	**91.6**	**67.6**
	P_5	**97.9**	**92.4**	**78.5**	**91.3**	**87.4**	**65.5**	**94.6**	**89.9**	**72.0**
3D PCKh @0.5	C_3	**72.9**	**54.7**	**42.2**	**73.9**	**56.8**	**42.2**	**73.4**	**55.7**	**42.2**
	C_5	**69.9**	**52.1**	**43.0**	**63.6**	**53.8**	**42.9**	**66.8**	**52.9**	**42.9**
	P_3	**63.4**	**66.1**	**42.6**	48.4	**63.4**	**50.0**	**55.9**	**64.8**	46.3
	P_5	**68.1**	**59.0**	**51.0**	**67.4**	**54.0**	37.2	**67.7**	**56.5**	44.1

4 Conclusion

We proposed a method for prediction of 3D human pose using an RGB-D camera in real-time for personal use. We measured a 3D pose of a person using an RGB-D camera and trained RNN to predict future 3D pose as an output from a sequence of input data frames in real-time. The trained RNN outputs a future predicted 3D pose in real-time from a successively input sequence of 3D pose measured RGB-D camera.

We evaluated the prediction error for two motions "sitting on a chair" and "standing from a chair" as representative daily actions. In the experiment, we used a small action data set for training RNN. The action data set includes only 41 actions. Even from the small action data set, the predicted future pose approximately matches to the measured pose, especially in 2D pose. We can expect prediction of future pose will be more precise using the bigger action data. Then the prediction of motion will be more personalized.

Acknowledgments. This work was supported by JSPS KAKENHI Grant Number JP17H01801. This work was supported in part by the Kansai University Educational Research Enhancement Fund in 2020 under the title of "Real-World Service Innovation through the AI Robot Challenge."

References

1. Pishchulin, L., et al.: DeepCut: joint subset partition and labeling for multi person pose estimation. In: CVPR2016, pp. 4929–4937 (2016)
2. Cao, Z., Simon, T., Wei, S., Sheikh, Y.: Realtime multi-person 2D pose estimation using part affinity fields. In: CVPR2017, pp. 1302–1310 (2017)
3. Papandreou, G., Zhu, T., Chen, L., Gidaris, S., Tompson, J., Murphy, K.: PersonLab: person pose estimation and instance segmentation with a bottom-up, part-based, geometric embedding model. In: Ferrari, V., Hebert, M., Sminchisescu, C., Weiss, Y. (eds.) ECCV 2018. LNCS, vol. 11218, pp. 282–299. Springer, Cham (2018). https://doi.org/10.1007/978-3-030-01264-9_17
4. Chen, C.-H., Ramanan, D.: 3D human pose estimation = 2D pose estimation + matching. In: CVPR2017, pp. 5759–5767 (2017)
5. Pavlakos, G., Zhou, X., Derpanis, K.G., Daniilidis, K.: Coarse-to-fine volumetric prediction for single-image 3D human pose. In: CVPR2017, pp. 1263–1272 (2017)
6. Zhou, X., Huang, Q., Sun, X., Xue, X., Wei, Y.: Towards 3D human pose estimation in the wild: a weakly-supervised approach. In: ICCV2017, pp. 398–407 (2017)
7. Novotny, D., Ravi, N., Graham, B., Neverova, N., Vedaldi, A.: C3DPO: canonical 3D pose networks for non-rigid structure from motion. In: ICCV2019, pp. 7687–7696 (2019)
8. Cheng, Y., Yang, B., Wang, B., Yan, W., Tan, R.T.: Occlusion-aware networks for 3D human pose estimation in video. In: ICCV2019, pp. 723–732 (2019)
9. Moon, G., Chang, J.Y., Lee, K.M.: Camera distance-aware top-down approach for 3D multi-person pose estimation from a single RGB image. In: ICCV2019, pp. 10132–10141 (2019)
10. Martinez, J., Hossain, R., Romero, J., Little, J.J.: A simple yet effective baseline for 3D human pose estimation. In: ICCV2017, pp. 2659–2668 (2017)
11. Chiu, H., Adeli, E., Wang, B., Huang, D., Niebles, J.C.: Action-agnostic human pose forecasting. In: 2019 IEEE Winter Conference on Applications of Computer Vision, pp. 1423–1432 (2019)
12. Wu, J., et al.: Single image 3D interpreter network. In: Leibe, B., Matas, J., Sebe, N., Welling, M. (eds.) ECCV 2016. LNCS, vol. 9910, pp. 365–382. Springer, Cham (2016). https://doi.org/10.1007/978-3-319-46466-4_22
13. Newell, A., Yang, K., Deng, J.: Stacked hourglass networks for human pose estimation. In: Leibe, B., Matas, J., Sebe, N., Welling, M. (eds.) ECCV 2016. LNCS, vol. 9912, pp. 483–499. Springer, Cham (2016). https://doi.org/10.1007/978-3-319-46484-8_29
14. Chao, Y.-W., Yang, J., Price, B., Cohen, S., Deng, J.: Forecasting human dynamics from static images. In: CVPR2017, pp. 3643–3651 (2017)
15. Horiuchi, Y., Makino, Y., Shinoda, H.: Computational foresight: forecasting human body motion in real-time for reducing delays in interactive system. In: Proceedings of 2017 ACM International Conference on Interactive Surfaces and Spaces, pp. 312–317 (2017)

ROS2-Based Distributed System Implementation for Logging Indoor Human Activities

Kyohei Yoshida, Tadashi Matsuo, and Nobutaka Shimada[✉]

Ritsumeikan University, 1-1-1 Noji-Higashi, Kusatsu, Shiga 525-8577, Japan
nshimada@is.ritsumei.ac.jp

Abstract. This research implements a system that detects and records various human activities in indoor scenes. For example, it detects who brings in or takes out an object and the handled object's image with the incident timestamp. It's constructed over ROS2, a widely used distributed communication framework for robotic implementation based on micro-services architecture, so that it can separate each subprocess of detection and improve the maintainability of each module. This paper reports the constructed system with visual human and pose detection, object detection, and recognition of object handling activities. Since the system was able to separate hardware not only service process, it was able to employ computationally heavy machine learning models simultaneously on multiple PCs with GPU.

Keywords: Interactive human-space design and intelligence · Human-robot interaction · ROS2

1 Introduction

1.1 Research Background and Objectives

In recent years, cameras have become smaller and higher quality, and camera devices such as surveillance cameras and drive recorders have been installed everywhere. Devices that can simultaneously acquire both color images and depth information, such as the Intel RealSense [2] and some recent models of smart phones like iPhone12 and Galaxy have been introduced. There is a limit to how much of this information can be used by humans to monitor and accurately detect important events. In addition, recent developments in technology and SDK libraries have made it possible to recognize transition in indoor or outdoor scene, people's activities, and various objects in real time from color and depth information.

In a previous study, Maki et al. [8] proposed an integrated system to automatically detect and store indoor scene changes. This system automatically detects and stores changes in indoor events such as objects being brought in and out of a room. However, the implementation of most systems is tightly coupled, which makes it difficult to maintain. In addition, they are not scalable. Even if a new algorithm is proposed to detect an event,

D.-S. Huang et al. (Eds.): ICIC 2021, LNCS 12836, pp. 862–873, 2021.
https://doi.org/10.1007/978-3-030-84522-3_70

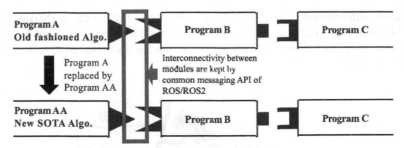

Fig. 1. Loose-coupling nodes in distributed architectures

it is difficult to replace the implementation or to add a new function to detect a new event.

In order to solve this problem, we attempt to design and implement a system that automatically detects and stores changes in indoor events using a micro-service architecture (hereinafter referred to as the "indoor event detection system"). This system is expected to make all functions in the system loosely coupled. ROS [9] and its next generation, ROS2 [5], are the de facto standard communication framework for such purposes in robotics field.

Loosely-coupled programs have an advantage of requiring minimal changes when their modules inside are upgraded as shown in Fig. 1. If we can sophisticatedly design common "API" between any modules, we can replace any module to another as long as the module can be accessed via the common API. If the implementation of program A is upgraded, program B needs only to be modified for switching its information source, and program C is not affected at all. This architecture is similar to networking protocol (TCP/IP). In addition, since the hardware can be divided in the case of distributed cooperation through process communication, it is possible to scale computational loads that could not be handled by a single piece of hardware like GPU. The implementation of loosely coupled programs is less complicated than that of tightly coupled systems. In this paper, a loosely coupled program design is adopted because the system is designed for under-development and growing systems for long-term operation. We aim to create a system that is easy to maintain and expand, and easy to replace functions, which has been a problem in previous studies. For such a purpose we adopt ROS2 as the core platform of the whole system.

2 System Overview and System Configuration

2.1 System Overview

Figure 2 shows an overview of the system we have been building in this research. The system is roughly divided into an input section, an event detection and interpretation section, a plug-in section, and an output section.

The input section acquires information about the indoor situation and provides it to the processing server. This system requires color image information, depth information, and camera information (e.g., intrinsics and extrinsics of each camera). Any type of device fixed to any fixed place in the space or even on moving robots can be used. In this research, we use an Intel® RealSensc™ [2] (hereinafter referred to as RealSense) fixed to the ceiling.

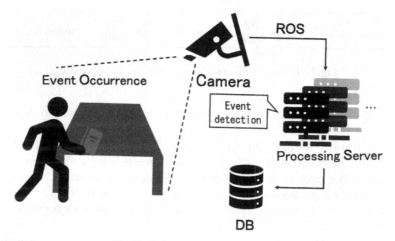

Fig. 2. Indoor scene logging system

The event detection/interpretation section detects events occurring in the video for human actions and objects based on the color image and depth information obtained from the input section. Here, targeted events are defined as "when, Where, Who, What, or What happened". The event detection and interpretation section obtains information in real time, processes it sequentially, and notifies the output section when an event is detected. The internal processing is described later.

The plug-in part provides functions to support the event detection and interpretation part. This section includes functions that use external libraries and machine-learning-based estimation, which are complicated to install. We have made it easier to separate the hardware by configuring them as plug-ins instead of including them in the event detection part. Plug-ins require only the construction of an environment for the plug-in, which reduces the cost of system installation. For example, the use of estimation for machine learning generally requires a machine equipped with a GPU, but the machine running the event detection and interpretation part does not need a GPU.

The output section displays and stores the detection results output from the event detection and interpretation section. It also has functions to search and retrieve past events from outside. In this study, we implement a function to save event information in a DB, a function to save color images and depth information before and after an event, and a function to retrieve past events via a Web API.

2.2 System Configuration

In this study, the hardware and software to run the system were configured as follows.

Server for Input Section

OS: Ubuntu 18.04 LTS
ROS distribution: ROS2 Dashing Diademata
Input device: Intel® RealSense2.0 [1]
Python version: 3.6.9
Image processing: OpenCV-Python 4.4.0.46 [3]

Server for Event Detection and Interpretation Section

OS: Ubuntu 18.04 LTS
ROS distribution: ROS2 Dashing Diademata
Python version: 3.6.9
Image Processing: OpenCV-Python 4.4.0.4

Server for Plugin Section (Human Skeleton Estimation, Human Region Extraction)

OS: Ubuntu 18.04 LTS
ROS distribution: ROS2 Dashing Diademata
Python version: 3.6.9
Estimation Model: OpenPose v1.7.0 [7] (human skeleton estimation), YOLACT [6] (human region estimation)
GPU: GeForce RTX 2070 SUPER

3 System Design

This section describes the design of the system implementation. In this section, we describe three aspects of the system: the ROS package configuration, the node configuration, and the internal configuration of the event detection and interpretation unit.

3.1 ROS Package Configuration

The ROS package configuration of this system is shown in Fig. 3. ROS packages are created for the input section, the event detection and interpretation section, and the plug-in section. However, the output part is included in the event detection and interpretation part package. For the plug-in part, one package is created for each plug-in. The green package corresponds to the input part, the orange package to the event detection and interpretation part and the output part, and the blue package to the plug-in part. In particular, in this study, the package of the event detection and interpretation part including the output part is called the core system package, and the group of these functions is called the core system "Shigure".

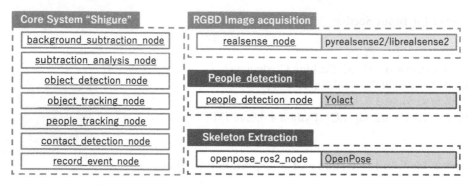

Fig. 3. ROS packages configurations (Color figure online)

3.2   Node Configuration

In this system, multiple ROS nodes send and receive messages from each other and operate cooperatively to form a single large system, and the "service" described by Thönnes [10] is realized in each node in this system. In other words, one node per function is implemented as a single node. This makes the services loosely coupled to each other, which facilitates functional expansion and maintenance. For example, when a new method is proposed for an object detection algorithm, only that node needs to be rewritten, thus allowing for low-cost functional expansion.

Figure 4 shows the node configuration of our system. Green is the input node, orange is the object detection node, blue is the person detection node, and the black area surrounded by orange is the event detection and storage node.

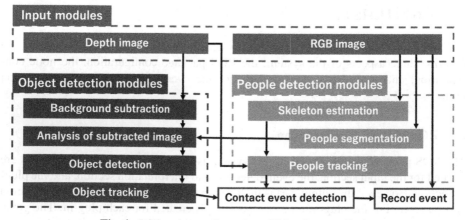

Fig. 4. ROS nodes configurations (Color figure online)

3.3 Internal Structure of Event Detection and Interpretation Section

The event detection/interpretation section is further divided into subsections due to the large number of processing modules. In this study, the nodes were designed based on hierarchical event detection. Figure 5 illustrates the hierarchical event detection architecture. In this architecture, information obtained from input devices is defined as lower-level information, and more specific information is defined at higher levels. In this study, we implement LV4 as the responsibility of the core system. The process flow of the core system is shown in Fig. 6.

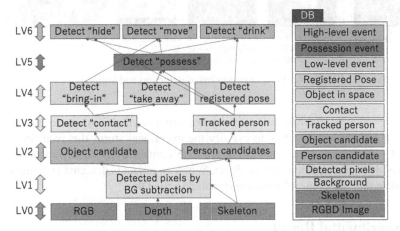

Fig. 5. Hierarchical event detection

This section describes the flow of event detection. Event detection can be roughly divided into object detection, person detection, and event judgment and storage. The parts other than the input part of Fig. 4 fall under this category.

The object detection process is to detect objects that have been brought in or brought out. The object detection process uses the depth information. The background difference extraction node estimates the background from the depth information, compares it with the current frame, and extracts the difference. The background difference analysis node receives the result, and analyzes the part of the frame that has a temporal difference and is not the human area as an object candidate. The object estimation node then estimates the object region by labeling the background subtraction analysis image. Finally, the object tracking node assigns an object ID for each detected object, and the unique ID is associated to the objects that are considered to be identical by tracking.

The person detection process involves estimating the skeletal structure and area of a person, as well as the identity of the person. The person detection process uses a machine learning model with color images as input. The person skeleton estimation node uses OpenPose [7] to estimate the skeleton of a person from a color image. The person region estimation node uses YOLACT [6] to estimate the region of the person from the color image. The person tracking node estimates the same person from the depth image and the person skeleton estimation information, and assigns a unique person ID.

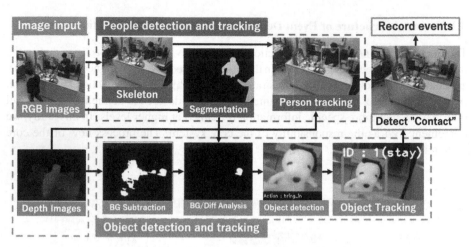

Fig. 6. Flow of image based detection and logging

The event type classification and storage process combines the results of object detection and person detection to classify the event, and stores the event information when it is judged that an event occurred. The contact event detection node detects "contact" between a person and an object. When an object is detected as being brought in or taken out, the event information is stored in the DB.

4   Experimental Result

4.1   Object Detection Process

The object detection process cascades modules and treats the depth information obtained from the input section in the following order: background difference extraction node, background difference analysis node, object estimation node, and object tracking node. The object detection process treats the depth information obtained from the input section in the following order: background subtraction extraction node, background subtraction analysis node, object estimation node, object tracking node. The process finally converts it into information about the object, such as the area where the object exists and the state of the object (brought in, brought out, existing in the scene). In this section, we describe the processing results of each node.

4.2   Background Subtraction Extraction Node

The pixel region with temporal changes is extracted by background subtraction as shown in Fig. 7. The background image is continuously changed due to human activity or temporal progress. For robust background subtraction we maintain the background image by averaging recent frames of input. The threshold of subtraction is dynamically determined by considering standard deviation for each pixel. In temporal averaging pixels, we remove pixels on which has no depth available.

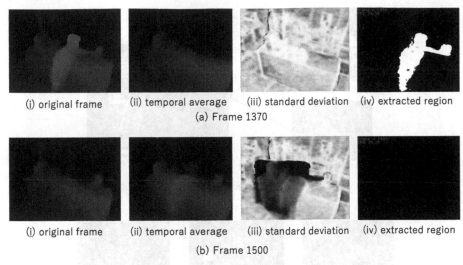

(i) original frame   (ii) temporal average   (iii) standard deviation   (iv) extracted region

(a) Frame 1370

(i) original frame   (ii) temporal average   (iii) standard deviation   (iv) extracted region

(b) Frame 1500

Fig. 7. Robust background subtraction

Frame 1370 is a scene where a person enters the space and places an object on a desk. Since the background information does not contain depth information between the person and the object, it is extracted as a difference. In frame 1500, the background information is updated to show that the object is now blended into the background.

4.3 Background Subtraction Analysis Node

In the background subtraction analysis node, only the differences that changed over a certain period of time were obtained from the background subtraction image from which the human region was deleted. In addition, the information for estimating the human region was obtained from the human region estimation node. All processing is implemented by OpenCV standard API [3]. The processing results are shown in Fig. 8.

Object Detection Node. In the object Detection node, the background subtraction image after the subtraction analysis process was labeled and divided into object regions. At this time, events that occurred to the object, such as being brought in or taken out, were also determined: if the depth value of the region decreases (namely the region gets nearer to camera) the action should be "bring-in", otherwise "take-away") Since the same object is detected in consecutive frames of the background subtraction, the node verifies whether the object is the same as the one detected in the previous frame by detecting the start and end frame of the subtracted region is existing. The results are shown in Fig. 9. It can be seen that the object is clipped and its detected action type is shown.

Object Tracking Node. The object tracking node issues an object ID and links the events of bringing in and taking out with the same object. The tracking results are shown in Fig. 10, which shows that an object brought in at frame 136 is detected and put into the waiting state, and is tracked as the same object until the object is taken out at frame 941.

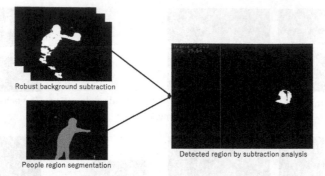

Fig. 8. Object and person extraction by background subtraction

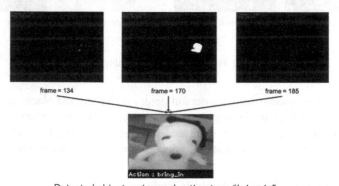

Detected object region and action type "bring-in"

Fig. 9. Non-registered object detection

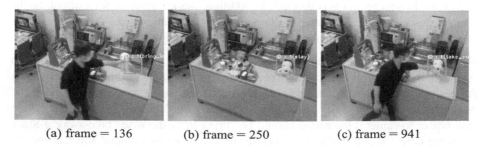

(a) frame = 136   (b) frame = 250   (c) frame = 941

Fig. 10. Feature-based object tracking

4.4 Person Detection Process

In the person detection process, the color image obtained from the input section is sent to the person skeleton estimation node and the person region estimation node. The results of the processing of the human skeletal estimation node are sent to the person tracking node. The final result is converted into information such as the range of the person and the person ID. In this section, we describe the processing results of each node.

People Region Segmentation Node. We employed YOLACT [6] to segment people regions from color images. The estimation result is shown in Fig. 11(a) and (b).

Personal Skeleton Estimation Node. We estimated the skeleton of a person from a color image using OpenPose [7]. The implementation of OpenPose functions for ROS platform is provided from [4] and here it was imported to the system. The estimation result is shown in Fig. 11(a) and (c).

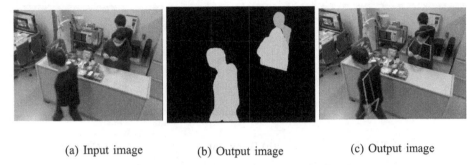

(a) Input image   (b) Output image   (c) Output image

Fig. 11. Person segmentation skeleton detection by deep neural model

Person Tracking Node. This node is used to track a person by combining the person information of consecutive frames using the person skeleton estimation information and depth information. The person ID is issued at this node so that the user can make queries for that person later using DB. The tracking results are shown in Fig. 12. It can be seen that the same person is tracked as the same person across frames.

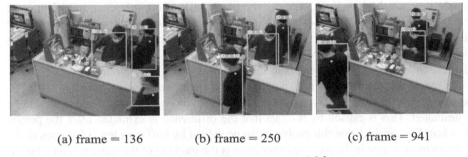

(a) frame = 136   (b) frame = 250   (c) frame = 941

Fig. 12. People tracking over sequential frames

4.5 Event Detection and Storing to DB

In the event judgment and storage process, the contact event detection node detects the contact between an object and a person using the information of the object tracking node and the person tracking node. It also stores the detection results such as "who, what, and how" in the DB.

In this study, the detection events are limited to taking out, bringing in, and touching of objects, however, it can be easily expanded to other activities using machine learning technique. According to ROS-based distributed architecture, hardware constraints (ex. the number of GPUs) are comparatively relaxed. The contact detection result is shown in Fig. 13. It can be seen that bringing in, taking out, and contact are detected.

(a) Take in   (b) Bring out   (c) Contact

Fig. 13. Detection of human-object interactions

5 Conclusion

In this study, we implemented a system that can detect and record various events involving objects and people that occur in an indoor scene, and search for event scenes about that the user make queries, according to a ROS-based micro-service architecture. By constructing the total system from the micro-service, we were able to implement a complicated system on the distributed hardware resources. As the side effects of the partitioning of the hardware resources, the visibility of the implementation code was also improved, which made it relatively easy to incorporate the machine learning inference model into the image understanding processes. In addition, the frame rate was improved compared to the previous monolithic system because the ROS-based architecture enables to process heavy computations of GPU-driven DL methods in the distributed environment. We believe that this will make it easier to scale up the system in the future.

One of the future tasks is to upgrade each image processing modules to SOTA methods. The false positives that occur when a person remains in the space should be eliminated. This is caused by the fact that the difference is extracted after the person has left. We consider that this problem can be solved by buffering tens of frames of the person mask image in the past. Another issue is the tracking of the same person who has left the screen once. In the current implementation, a new ID is assigned to the person. To avoid this problem, we need to consider using color information or combining it with face recognition for example.

References

1. IntelRealSense librealsense2: Intel® RealSense™ SDK. https://github.com/IntelRealSense/librealsense. Accessed 01 Mar 2021

2. Intel® RealSense™ technology. https://www.intel.co.jp/content/www/jp/ja/architecture-and-technology/realsense-overview.html. Accessed 01 Mar 2021
3. OpenCV-Python PyPI. https://pypi.org/project/opencv-python/. Accessed 01 Mar 2021
4. OpenPose ROS2. https://github.com/Rits-Interaction-Laboratory/openpose_ros2. Accessed 01 Mar 2021
5. ROS 2 overview. https://index.ros.org/doc/ros2/. Accessed 01 Mar 2021
6. Bolya, D., Zhou, C., Xiao, F., Lee, Y.J.: YOLACT: real-time instance segmentation. In: ICCV International Conference on Computer Vision, pp. 9157–9166 (2019)
7. Cao, Z., Hidalgo Martinez, G., Simon, T., Wei, S., Sheikh, Y.A.: OpenPose: real-time multi-person 2D pose estimation using part affinity fields. IEEE Trans. Pattern Anal. Mach. Intell. **43**(1), 172–186 (2019)
8. Maki, K., Katayama, N., Shimada, N., Shirai, Y.: Image-based automatic detection of indoor scene events and interactive inquiry. In: Proceedings of 19th International Conference on Pattern Recognition ICPR 2008, pp. 1–4 (2008)
9. Quigley, M., et al.: ROS: an open-source robot operating system. In: ICRA Workshop on Open Source Software, Kobe, Japan, vol. 3, p. 5 (2009)
10. Thönes, J.: Microservices. IEEE Softw. **32**(1), 116 (2015)

The Virtual Camera Path in 3D Animation

Jingjing Tang, Liang Song$^{(\boxtimes)}$, Jiabao Zeng, and Juncong Lin

School of Informatics, Xiamen University, Xiamen 361005, China
songliang@xmu.edu.cn

Abstract. Unlike real-world filmmaking, the camera track in 3D animated films is more complicated. The location and shooting angle of the virtual camera need to be more accurately computed to achieve the correct delivery of information to the audience. Although there are many technologies that study camera movement in a virtual scene, they have not considered how to automatically generate camera path in complex situations. Based on the above consideration, this paper proposes a new method to automatically compute the camera path in a 3D animation scene. Our method first samples the initial camera path that reflects the complex motion of the character in the scene, then processes the complex original path by using the segmented and differentiable Bezier curve algorithm and finally uses nonlinear programming to further optimize the motion. We verify and advantage of the method through experiments, which demonstrate its effectiveness in the production of 3D animated moves.

Keywords: Camera path · Sampling method · Segmented and differentiable Bezier curve · Nonlinear programming

1 Introduction

Nowadays, there are many modeling engines that make it easier to create 3D animation movies [1, 2], such as the movie Baymax Dreams by Unity's Cinemachine [3] and cutscenes of the game Diablo made in Unreal Engine [4]. However, most filmmaking techniques are to edit multiple simple scenes or relying on script-based animation shooting, which produce a relatively blunt video effect. The scene processing in [5] is based on the standard cinematic idioms to represent the movement of a character, rather than controlling the camera in the scene to achieve the desired shooting effect. The rule-based engine in [6] is used to interpret the text script into the relevant movie effect, but the movement connection of the character is rigid and the shooting angle cannot be adjusted flexibly. Moreover, when faced with more rich scenes, such as passing through obstacles or dramatic movement of characters, it is more difficult to be dealt with and the results are often relatively simple and crude, because of the complex computation of the virtual camera paths. Li and Jin [7] model the camera path to achieve video stabilization such as linear and parabolic, but SFM they proposed is restricted to partially static scene

This work was supported in part by China Scholarship Council, and in part by the Fundamental Research Funds for the Central Universities of China under Grant No. 20720190028.

© Springer Nature Switzerland AG 2021
D.-S. Huang et al. (Eds.): ICIC 2021, LNCS 12836, pp. 874–882, 2021.
https://doi.org/10.1007/978-3-030-84522-3_71

videos, otherwise, camera motion would be blurred. Based on a binary tree structure and a greedy traversal strategy the method in [8], they obtain high-quality viewpoints to approximately compute a smooth camera path, and presents the motion features of the characters.

The virtual camera is required to capture the movement process of the 3D character, and the camera itself needs to move continuously and steadily. Therefore, the generation of the virtual camera path first computes a rough camera path according to the requirements of animation shooting (for example, the positions and angles of virtual cameras), and then generates a smooth camera moving track based on this path, and finally optimizes the camera motion details between frames to ensure that the visual result is reasonable and smooth. In [9], only a Bezier curve was used to fit the original camera path with multiple turns, resulting in the camera path ignoring these turning details. The method [8] based on the binary tree structure and the greedy traversal strategy requires a larger amount of computation to get the camera position of each frame and does not deal with the problem of audience perception.

Our method provides the ability to extract virtual camera motion attributes from long-time scenes and to automatically compute the camera motion path as natural as possible. Of course, it can also connect multiple long shots into a complete scene that continuously cover the whole movement of the character.

Contributions of this paper are as follows:

- Sampling method on the raw paths of the virtual camera: conditional selection considering the movement changes of a character to automatically compute the raw path, which reduces the computing burden for generating Bezier curves;
- Employing segmented Bezier and differentiable curves: for complex motion, it is divided into multiple motion tracks and smoothly connected.

2 Problem Formulation

Our goal is to provide a complete computation method from input X to output Y, in which character positions are $\{x1_i\}_{i=0}^{N}$, body orientations are $\{x2_i\}_{i=0}^{N}$, and camera positions are $\{y_i\}_{i=0}^{N}$ where $\{x1_i, x2_i\} \in X$ and $y_i \in Y$. Our method includes three transformations $F_{Sample}: X \rightarrow R$, $F_{Bezier}: R \rightarrow B$, and $F_{nonLinear}: B \rightarrow Y$, where R represents the original camera path by sampling strategy, B represents the camera path obtained by using the segmented and differentiable Bezier curve algorithm.

2.1 Raw Path

We apply the filtering conditions to generate a reasonable part of positions from all the inputs, which can represent the complete movement as much as possible. Then the camera's raw path can be computed based on sampled information. For the filter function S_1 and computation function of raw path F_{Sample}, we expressed the objective as:

$$R = F_{Sample}(S_1(X, Cond), d) \tag{1}$$

where $Cond$ includes conditions, and d is the fixed distance between the camera and the character.

2.2 Segmented and Differentiable Bezier Curves

The raw path obtained by sampling is discrete and is not suitable as directly camera path. We need a curve fitting process to get a smooth curve. Considering the large turning points in the raw path, a smoothing is used by controlling the derivative at the turning points. Therefore, the raw path is divided into s segments, and each segment is computed separately for Bezier curve. The formula of cubic Bezier curve [10] is as follows:

$$F_{Bezier}(u) = U * M_{bezier} * P,$$

$$M_{bezier} = \begin{bmatrix} -1 & 3 & -3 & 1 \\ 3 & -6 & 3 & 0 \\ -3 & 3 & 0 & 0 \\ 1 & 0 & 0 & 0 \end{bmatrix} \quad (2)$$

where $U = [u^3, u^2, u, 1]$ is the parameter matrix containing the position of points on the curve, M_{bezier} is the cubic Bezier curve coefficient matrix, and P is the control point matrix $(P = [P_0, P_1, P_2, P_3]^T)$. In order to obtain as close to the raw path as possible while keeping the curves smooth at the connection points, the error function ϵ [11] and the derivative at the junction of the curve are introduced:

$$\min\epsilon = \sum_{i=0}^{N}(B(u_i) - R_i)^2 \quad (3)$$

Constrains

$$\begin{aligned} \overrightarrow{P01} \times \overrightarrow{N_i} &= 0, \\ \overrightarrow{P32} \times \overrightarrow{N_{i+1}} &= 0 \end{aligned} \quad (4)$$

Where

$$\begin{aligned} \overrightarrow{N_i} &= N_{i+1} - N_i, \\ \overrightarrow{N_{i+1}} &= N_{i+2} - N_{i+3} \end{aligned} \quad (5)$$

Decision variables

$$\begin{aligned} \overrightarrow{P01} &= P_1 - P_0, \\ \overrightarrow{P32} &= P_2 - P_3 \end{aligned} \quad (6)$$

As shown above, the Eq. 3 is our objective function, in which $B(u_i)$ is the cubic Bezier curve computed by two known control points (P_0, P_3) and two unknowns (P_1, P_2), and R_i is the position on the raw path, and our goal is to select the appropriate P_1 and P_2 to make the Bezier curve fit the original path as much as possible. The cross in Eq. 4 between vectors is designed to remain the same direction from the end control point to its neighboring control point as the direction of the derivation at this connection point. $\overrightarrow{N_i}$ and $\overrightarrow{N_{i+1}}$ in Eq. 5 represent the directions of derivation and compute as the vector of the angle bisector at the start and end of the curve at segment i using analytic geometric method (see Algorithm 2). The decision variables of this error function are $\overrightarrow{P01}$ and $\overrightarrow{P32}$ which represent the directions from the end control points P_0 and P_3 to their neighboring control points P_1 and P_2 separately.

2.3 Camera Path

In order to have a smoother visual effect, the constraints of velocity and acceleration are added to the nonlinear programming to improve the process of camera movement on the computation curve. The nonlinear programming function [11] is expressed as follows:

$$Y = \text{F}_{\text{nonLinear}}(B, v_{max}, a_{max}) \qquad (7)$$

where, v_{max} represents velocity constraint and a_{max} represents acceleration constraint.

3 Path Computation of Virtual Cameras

In this section, we introduce three important algorithms in the process of computing the camera path based on the moving character in the virtual scene.

3.1 The Sampling Strategy

In the sampling process, we consider two aspects: (1) retain as little position data as possible, that is, to construct a more complete target path with as few points as possible; (2) keep orientation information as much as possible in order to describe more complete orientation changes of the character. For the above requirements, we use two parameters distance and perspective rotation. Samples are taken when the cumulative distance d_i and straight-line distance r_i are greater than or equal to the threshold δd and δr, and when the angle of view change α_i is greater than or equal to the threshold $\delta \alpha$. The sampling algorithm is detailed in Algorithm 1.

3.2 Segmented and Differentiable Bezier Curve Algorithm

We set the segmenting conditions to automatically identify the locations, where original paths were divided into s segments, each of which will be used for the Bezier curve algorithm. Cubic Bezier curve needs four control points (N_0, N_1, N_2, N_3) as shown in Eq. 2. The algorithm [12] inserts two control points (N_1, N_2) between N_0 and N_3 and ensures that the connection point and the newly inserted control points on both sides remain collinear. Also, there is a derivative at the point of connection as shown in Fig. 1. The construction algorithm of control points uses the principle of angular bisector [12].

Algorithm 1. Sampling Strategy.

Input:	The set of character data including orientation and position, X;
	$\delta d; \delta r; \delta \alpha$;
Output:	The set of sampling points, X';

1. initialize the number of sampling points k;
2. for each $i \in [0, N]$ do
3.   if $r_i \geq \delta r$ then
4. $X'(k) = x_i$;
5.   else if $d_i \geq \delta d$ then
6. $X'(k) = x_i$;
7.   end if
8.   if $\alpha_i \geq \delta \alpha$ then
9. $X'(k) = x_i$;
10.   end if
11. end for

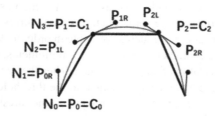

Fig. 1. The segmentation points and inserted points of segmented Bezier curve [12]

3.3 Control Points and Nonlinear Programming

Four control points (P_0, P_1, P_2, P_3) are needed to fit the raw path with Cubic Bezier curve, where P_0, P_3 are the end control points of the curve and represent the endpoints of the raw path, and P_1, P_2 represent the points on the tangent of the two endpoints. The matrix formula is the cubic Bezier curve is shown in Eq. 2.

By controlling the two control points P_1, P_2, the fitting curve will have different curvature and shape. In order to fit the raw path as much as possible, an error function ϵ is defined as shown in Eq. 3. This equation represents the sum of the squares of distances between the position $B(u_i)$ on the Bezier curve and the position R_i on the raw path for N frames. The better the fit between the Bezier curve and the raw path, the smaller the error function value. Since P_1 and P_2 are unknown, these two control points and Bezier curves can be obtained by combining Eq. 3 to Eq. 6.

In the process of optimizing the camera position for each frame on the curve, we took nonlinear programming from [11], adding acceleration and velocity constraints. Assuming the input duration d including N frames, the camera position is represented by u_i, the optimized camera position is u'_i, and the main part of nonlinear programming is as follows:

$$\min \sum_{i=0}^{N} |u'_i - u_i| \qquad (8)$$

Constraints

$$\begin{aligned}
\left| u'_i - u'_{i-1} \right| &\leq \tfrac{v_{max} \cdot dt}{L} \\
\left| 2u'_{i-1} - u'_i - u'_{i-2} \right| &\leq \tfrac{a_{max} \cdot dt^2}{L} \\
u'_1 - u'_0 &\leq \tfrac{a_{max} \cdot dt^2}{L} \\
u'_N - u'_{N-1} &\leq \tfrac{a_{max} \cdot dt^2}{L}
\end{aligned} \qquad (9)$$

The velocity v_{max} and acceleration a_{max} define implicit constraints. When the minimum acceleration and velocity are satisfied, the camera accelerates before reaching half and then decelerates after reaching halfway of the distance with the beginning and end velocity of zero.

4   Experimental Results

In this section, we apply our method on a dataset introduced by [9] and conducted ablation experiments. The computation program of the virtual camera path is written by MATLAB R2020b, the scene rendering engine is the 2018.4.31f1 version of the Unity, and the operating system is Windows 10.

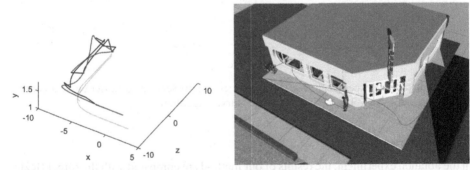

Fig. 2. The results of our method: the movement of the character is green; the camera raw path is blue; the computed camera path is red. (Color figure online)

4.1   Performance

We implement a camera 3D path planning for a single scene on the data set. It is a complex scene in which the character rides off on a bicycle for 25 s and consists of two dramatic turns, the first being a 180° turn and the second a 90° turn. Using a conditional sampling strategy, the camera's raw path adds loop lens rotation to the cycling action, which is similar to a close-up movie effect. Figure 2 shows the movement of the character, the camera raw path, and the computed camera path.

880   J. Tang et al.

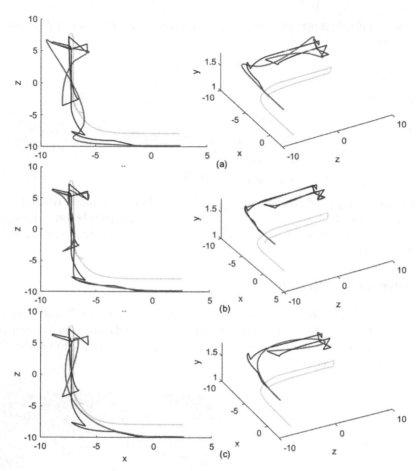

Fig. 3. Comparative experiment results: (a) Our method; (b) Segmented Bezier curve algorithm without derivative control; (c) The direct Bezier curves algorithm.

4.2 Comparison

In the ablation experiment, the results of our method are compared with the direct Bezier curves algorithm [11] and piecewise Bezier curve algorithm uncontrolled curvature at the piecewise points. The results are shown in Fig. 3. It is obvious from the results of the unsegmented Bezier curve that the camera path is poorly fitted and lacks a cinematic close-up effect. In the case of only segmenting but not controlling the derivative at the segmenting points, the situation of over-fitting occurs, and a relatively sharp turning occurs at the dividing points, which affects the visual effect. Therefore, the proposed method has good fitting performance and better visual effect.

4.3 The Segmentation Results

In segmenting the raw path, it takes many experiments to adjust the parameters to get the best segmentation results in order to better use the segmented and differentiable Bezier

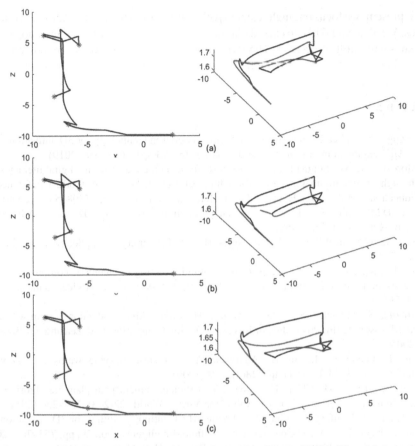

Fig. 4. The segmentation results: (a): $\delta d = 18$, $\delta r = 15$ and $\delta \alpha = 60$; (b): $\delta d = 18$, $\delta r = 15$ and $\delta \alpha = 30$; (c): $\delta d = 20$, $\delta r = 20$ and $\delta \alpha = 60$. The segmentation results are the left; the fitting results with derivative control at the junction of curve are the right.

curve and partial results are shown in Fig. 4. When each segment after segmentation roughly presents a trapezoid as shown in Fig. 4(a), the fitting effect is the best, because it is more conducive to obtain the four control points in the cubic Bezier curve.

5 Conclusion

Camera path planning considering character movement in 3D animation is a significant challenge. In this paper, we have proposed a new way which automatically computes and smooths camera path with sampling movement features of the character. We rely on the sampling strategy to obtain the original path which can simplify complex scenes. The camera paths are expressed by segmented and differentiable Bezier curve. The camera motion details are then optimized by nonlinear programming with velocity and acceleration constraints.

At present, we focus on single-camera path computation for single-character complex scenes. We also would like to extend our method to the case of multi-camera cooperation or scenes with multi-objectives. The switching of perspectives will be smoother and the connection of multi-scenes will be more natural by multi-camera coordination.

References

1. Zhang, L.: Application research of automatic generation technology for 3D animation based on UE4 engine in marine animation. J. Coastal Res. **93**(sp1), 652–658 (2019)
2. Morozoc, A.: Machinima learning: prospects for teaching and learning digital literacy skills through virtual filmmaking. In: Proceedings of ED-MEDIA 2008-World Conference on Educational Multimedia, Hypermedia and Telecommunications, pp. 5898–5907 (2008)
3. Lee, D.M., Shin, S.J.: Real-time virtual production using unity 3D. Int. J. Adv. Smart Convergence **7**(4), 138–146 (2018)
4. Keo, M.: Graphical Style in Video Games. HAMK University of Applied Sciences, Finland (2017)
5. Kardan, K., Casanova, H.: Heuristics for continuity editing of cinematic computer graphics scenes. In: Proceedings of the 2009 ACM SIGGRAPH Symposium on Video Games, pp. 63–69 (2009)
6. Jinhong, S., Miyazaki, S., Aoki, T., Yasuda, H.: Filmmaking production system with rule-based reasoning. In: Proceeding of Image and Vision Computing New Zealand, pp. 366–371 (2003)
7. Liu, F., Gleicher, M., Jin, H., Agarwala, A.: Content-preserving warps for 3D video stabilization. ACM Trans. Graph. **28**(3), 1–9 (2009)
8. Yeh, I.C., Lin, C.H., Chien, H.J., Lee, T.Y.: Efficient camera path planning algorithm for human motion overview. Comput. Animation Virtual Worlds **22**(2–3), 239–250 (2011)
9. Galvane, Q., Ronfard, R., Lino, C., Christie, M.: Continuity editing for 3D animation. In: Proceedings of the AAAI Conference on Artificial Intelligence. vol. 29, pp. 753–761 (2015)
10. Hearn, D., Baker, M.P.: Computer Graphics. Prentice Hall, New Jersey (1994)
11. Galvane, Q., Christie, M., Lino, C., Ronfard, R.: Camera-on-rails: automated computation of constrained camera paths. In: Proceedings of the 8th ACM SIGGRAPH Conference on Motion in Games, pp. 151–157 (2015)
12. Wang, J.R., Zhao, N.S., Hua, W.Y., Wang, Y.M.: Construction of cubic Bezier continuous subsection curve's control points. Comput. Eng. Appl. **46**(22), 190–193 (2010)

Author Index

Printed in the United States
by Baker & Taylor Publisher Services

Printed in the United States
by Baker & Taylor Publisher Services